Coastal and Marine Pollution

Coastal and Marine Pollution

Source to Sink, Mitigation and Management

Edited by

Meththika Vithanage
University of Sri Jayewardenepura
Nugegoda
WP, Sri Lanka

Sameera M. Samarasekara
University of Sri Jayewardenepura
Nugegoda
WP, Sri Lanka

Bryan D. James
Woods Hole Oceanographic Institution
Woods Hole
MA, US

Christopher M. Reddy
Woods Hole Oceanographic Institution
Woods Hole
MA, US

WILEY

This edition first published 2025

Registered Offices
John Wiley & Sons, Inc., 111 River Street, Hoboken, NJ 07030, USA
John Wiley & Sons Ltd, New Era House, 8 Oldlands Way, Bognor Regis, West Sussex, PO22 9NQ, UK

For details of our global editorial offices, customer services, and more information about Wiley products visit us at www.wiley.com.

The manufacturer's authorized representative according to the EU General Product Safety Regulation is Wiley-VCH GmbH, Boschstr. 12, 69469 Weinheim, Germany, e-mail: Product_Safety@wiley.com.

Wiley also publishes its books in a variety of electronic formats and by print-on-demand. Some content that appears in standard print versions of this book may not be available in other formats.

Library of Congress Cataloging-in-Publication Data

Names: Vithanage, Meththika, editor. | Samarasekara, Sameera M, editor. | James, Bryan D., editor. | Reddy, Christopher M., editor.
Title: Coastal and marine pollution : source to sink, mitigation and management / edited by Meththika Vithanage, Sameera M. Samarasekara, Bryan D. James, Christopher M. Reddy.
Description: Hoboken, NJ : Wiley, 2025. | Includes bibliographical references.
Identifiers: LCCN 2024039256 (print) | LCCN 2024039257 (ebook) | ISBN 9781394236992 (hardback) | ISBN 9781394237012 (adobe pdf) | ISBN 9781394237005 (epub)
Subjects: LCSH: Marine pollution. | Coastal ecosystem health.
Classification: LCC GC1085 .C583 2025 (print) | LCC GC1085 (ebook) | DDC 363.739/409146–dc23/eng/20241025
LC record available at https://lccn.loc.gov/2024039256
LC ebook record available at https://lccn.loc.gov/2024039257

Cover Design: Wiley
Cover Image: @Fotos593/Shutterstock

Set in 9.5/12.5pt STIXTwoText by Straive, Pondicherry, India
Printed and bound by CPI Group (UK) Ltd, Croydon, CR0 4YY

C9781394236992_150325

Contents

Notes on Editors

Meththika Vithanage
20/10/1976
Meththika Vithanage is a Professor of Natural Resources and the founding director of the Ecosphere Resilience Research Centre at the University of Sri Jayewardenepura, Sri Lanka. She also serves as an Adjunct Professor at the Institute of Agriculture, University of Western Australia, and the National Institute of Fundamental Studies, Sri Lanka. Recognized as a Highly Cited Researcher in 2021 by Clarivate, she has received numerous prestigious awards. From 2017 to 2023, she has been listed among the top 2% of the most cited scientists globally across various disciplines.

Ratnayakage Sameera Maduranga Samarasekara
03/10/1987
R.S.M. Samarasekara, PhD, is a Senior Lecturer at the Faculty of Engineering, University of Sri Jayewardenepura, and a Chartered Engineer at the Institution of Engineers in Sri Lanka. He is also a visiting lecturer at the Ocean University of Sri Lanka. He has received numerous awards and research grants. His interests focus on ocean water quality, coastal erosion management, coastal infrastructure, ocean energy, and coastal engineering.

Bryan Daniel James
03/06/1994
Bryan D. James, PhD, is a Postdoctoral Scholar at the Woods Hole Oceanographic Institution (USA) in the Department of Marine Chemistry and Geochemistry and the Department of Biology, and an incoming Assistant Professor in the Department of Chemical Engineering at Northeastern University (USA). He has received numerous awards and accolades including being named a CAS Future Leader '23. His interests focus on the intersection of environmental, material, and biomedical sciences to design safe and sustainable chemicals and materials.

Christopher Michael Reddy
08/07/1969
Christopher M. Reddy, PhD, is a Senior Scientist at the Woods Hole Oceanographic Institution (USA). He studies marine pollution and applies that knowledge to develop more sustainable materials. The nature of his work has connected him with the media, responders, officials, and the public. In his recent book, "Science Communication in a Crisis: An Insider's Guide," Reddy presents a clear pathway to effective and collaborative communication among various stakeholders.

List of Contributors

Bandara Abeysinghe
Department of Earth Resources Engineering
University of Moratuwa
Moratuwa, Sri Lanka

Kaushani Amarasena
School of Science
University of Wales Trinity Saint David
London, United Kingdom

Gayan Amarasooriya
Department of Applied Earth Sciences
Faculty of Applied Sciences
Uva Wellassa University
Badulla, Sri Lanka

Nadeera Batapola
Department of Earth Resources Engineering
University of Moratuwa
Moratuwa, Sri Lanka

Bellanthudawage Kushan Aravinda Bellanthudawa
Department of Agricultural Engineering
and Environmental Technology
Faculty of Agriculture
University of Ruhuna, Matara
Southern Province, Sri Lanka
University of Chinese Academy of Sciences
Beijing, China

Research Center for Eco-Environmental
Sciences
Chinese Academy of Sciences
Beijing, China

Shermila M. Botheju
Department of Indigenous Medical Resources
Faculty of Indigenous Health Sciences
and Technology
Gampaha Wickramarachchi University
of Indigenous Medicine
Yakkala, Sri Lanka

Chandrasekara SSK
Department of Agricultural Engineering
Faculty of Agriculture
University of Peradeniya
Peradeniya, Sri Lanka

Titus Cooray
Department of Applied Earth Sciences
Faculty of Applied Sciences
Uva Wellassa University
Badulla, Sri Lanka

Dileepa de Croos
Department of Aquaculture and Fisheries
Faculty of Livestock, Fisheries and Nutrition
Wayamba University of Sri Lanka
Gonawila (NWP), Sri Lanka

Sandun Dassanayake
Department of Decision Sciences
Faculty of Business
University of Moratuwa
Katubedda, Sri Lanka

S.M. Dassanayake
Department of Decision Sciences
University of Moratuwa
Moratuwa, Sri Lanka

Nalika R. Dayananda
Department of Indigenous Medical Resources
Faculty of Indigenous Health Sciences and Technology
Gampaha Wickramarachchi University of Indigenous Medicine
Yakkala, Sri Lanka

Pubudi Dilshara
Department of Earth Resources Engineering
University of Moratuwa
Moratuwa, Sri Lanka

Nimila Dushyantha
Department of Applied Earth Sciences
Faculty of Applied Sciences
Uva Wellassa University
Badulla, Sri Lanka

Suchima Gonapinuwala
Department of Aquaculture and Fisheries
Faculty of Livestock, Fisheries and Nutrition
Wayamba University of Sri Lanka
Gonawila (NWP), Sri Lanka

Jonas Gunnarsson
Department of Ecology
Environment and Plant Sciences (DEEP)
Stockholm University
Stockholm, Sweden

Mihiri Indunil Gunasekara
School of Civil and Environmental Engineering
Faculty of Engineering
Queensland University of Technology
Brisbane, QLD, Australia

Christopher W. Haley
ECOPOL
Mount Kembla
New South Wales, Australia

Bryan D. James
Department of Marine Chemistry and Geochemistry
Woods Hole Oceanographic Institution
Woods Hole, MA, USA

Gohagodage Nirmal Tharanga Jayalath
Department of Applied Earth Sciences
Faculty of Applied Sciences
Uva Wellassa University
Badulla, Sri Lanka

Harshi Jayasingha
Postgraduate Institute of Agriculture
University of Peradeniya
Peradeniya, Sri Lanka

C.L. Jayawardena
Department of Earth Resources Engineering
University of Moratuwa
Moratuwa, Sri Lanka

Abdullah İzzeddin Karabulut
Graduate School of Natural and Applied Sciences
Remote Sensing and Geographic Information Systems
100/2000 CoHE, Harran University
Sanliurfa, Turkey

Karunathilake, N.M.O.A
Department of Zoology and Aquatic Sciences
Centre for Marine Science and Technology
University of Sri Jayewardenepura
Nugegoda, Sri Lanka

Gayithri Niluka Kuruppu
Department of Industrial Management
Faculty of Business
University of Moratuwa
Moratuwa, Sri Lanka

Savinda Arambawatta Lekamge
School of Earth and Atmospheric Sciences
Faculty of Science
Queensland University of Technology
Brisbane, QLD, Australia

J.A. Liyanage
Department of Chemistry
University of Kelaniya
Kelaniya, Sri Lanka

Dilani Madhubhashini
School of Earth and Atmospheric Sciences
Faculty of Science
Queensland University of Technology
Brisbane, QLD, Australia

Siddihalu Lakshitha Madunil
Department of Indigenous Medical Resources
Faculty of Indigenous Health Sciences and Technology
Gampaha Wickramarachchi University of Indigenous Medicine
Yakkala, Sri Lanka

I. Mahakalanda
Department of Decision Sciences
University of Moratuwa
Moratuwa, Sri Lanka

Ishani Mahawaththa
Department of Biological Sciences
College of Science
University of North Texas
Denton, Texas, USA

R. John Morrison
School of Earth
Atmospheric and Life Sciences
University of Wollongong
Northfields Ave Wollongong
New South Wales, Australia

Chamila Jinendra Nanayakkara
Department of Earth Resources Engineering
University of Moratuwa
Moratuwa, Sri Lanka

Minh-Ky Nguyen
Faculty of Environment and Natural Resources
Nong Lam University
Ho Chi Minh City, Viet Nam

Divya Pal
Department of Ecology
Environment and Plant Sciences (DEEP)
Stockholm University
Stockholm, Sweden

G.K.M.R. Peter
Department of Earth Resources Engineering
University of Moratuwa
Moratuwa, Sri Lanka

Perera, S.P.K.K.A
Department of Zoology and Aquatic Sciences
Centre for Marine Science and Technology
University of Sri Jayewardenepura
Nugegoda, Sri Lanka

W.P.R.T. Perera
Department of Indigenous Medical Resources
Gampaha Wickramarachchi University of Indigenous Medicine
Yakkala, Sri Lanka

Iddamalgoda Jayawardenage Judi Udari Nisansala Perera
Department of Agricultural Engineering and Environmental Technology
Faculty of Agriculture
University of Ruhuna
Matara, Southern Province, Sri Lanka
South China Sea Institute of Oceanology
Chinese Academy of Sciences
Guangzhou, China
University of Chinese Academy of Sciences
Beijing, China

Udari Perera
Department of Chemical and Process Engineering
Faculty of Engineering
University of Moratuwa
Katubedda, Sri Lanka

Uswatta Liyanage Harsha Prabath Perera
Department of Applied Earth Sciences
Faculty of Applied Sciences
Uva Wellassa University
Badulla, Sri Lanka

Ranjith Premasiri
Department of Earth Resources Engineering
University of Moratuwa
Moratuwa, Sri Lanka

Dulanjalee Rajapaksha
School of Natural Sciences
University of Tasmania
Hobart, Australia

Ranatunga, R.R.M.K.P
Department of Zoology and Aquatic Sciences
Centre for Marine Science and Technology
University of Sri Jayewardenepura
Nugegoda, Sri Lanka

Mahinsasa Rathnayake
Department of Chemical and Process Engineering
Faculty of Engineering
University of Moratuwa
Katubedda, Sri Lanka

R.M.N. Priyanga Rathnayake
Radiation Protection and Technical Services Division
Sri Lanka Atomic Energy Board
Wellampitiya, Sri Lanka

Amila Sandaruwan Ratnayake
Department of Applied Earth Sciences
Faculty of Applied Sciences
Uva Wellassa University
Badulla, Sri Lanka

Nalin Ratnayake
Department of Earth Resources Engineering
University of Moratuwa
Moratuwa, Sri Lanka

Christopher M. Reddy
Department of Marine Chemistry and Geochemistry
Woods Hole Oceanographic Institution
Woods Hole, MA, USA

Leneka Terika Rhoden
Department of Life Sciences
Faculty of Science and Technology
The University of the West Indies
Mona, Jamaica

Ranepura Dewage Charuka Sandaruwan
Sri Lanka Wildlife Conservation Society
Pussallayaya, Handungamuwa
Matale, Central Province, Sri Lanka
Faculty of Graduate Studies
University of Kelaniya
Kelaniya, Sri Lanka

Lahiru Udayanga
Department of Biosystems Technology
Faculty of Agriculture and Plantation Management
Wayamba University of Sri Lanka
Gonawila (NWP), Sri Lanka

Menuka Udugama
Department of Agribusiness Management
Faculty of Agriculture and Plantation Management
Wayamba University of Sri Lanka
Gonawila (NWP), Sri Lanka

Meththika Vithanage
Ecosphere Resilience Research Center
Faculty of Applied Sciences
University of Sri Jayewardenepura
Nugegoda, Sri Lanka

Panchala Weerakoon
Department of Applied Earth Sciences
Faculty of Applied Sciences
Uva Wellassa University
Badulla, Sri Lanka

S.D.N.A.M.A.M. Weerasinghe
School of Biomedical Engineering
Department of Occupational Health and Safety Engineering/Disaster Management
Inje University
South Korea

Wijetunge, D.S.
Department of Zoology and Aquatic Sciences
Centre for Marine Science and Technology
University of Sri Jayewardenepura
Nugegoda, Sri Lanka

Hasintha Wijesekara
Department of Natural Resources
Sabaragamuwa University
Belihuloya, Sri Lanka

Madhuni Madhushika Wijesooriya
Department of Biological and Chemical Engineering
Aarhus University
Aarhus C, Denmark

Jithya Wijesinghe
Department of Indigenous Medical Resources
Faculty of Indigenous Health Sciences and Technology
Gampaha Wickramarachchi University of Indigenous Medicine
Yakkala, Sri Lanka

Pelin Soyertaş Yapıcıoğlu
Environmental Engineering Department
Harran University
Sanliurfa, Turkey

Mehmet İrfan Yeşilnacar
Environmental Engineering Department
Harran University
Sanliurfa, Turkey

Preface

The urgency of addressing coastal and marine pollution has never been greater. With escalating anthropogenic pressures and the looming effects of climate change, the challenges faced by marine ecosystems are complex and multifaceted. This book arises from a collective endeavour to provide a comprehensive and enduring resource on the sources, impacts, and innovative solutions for managing pollution in marine and coastal environments.

Recognizing the need to move beyond a fragmented understanding, this volume brings together leading experts from academia, industry, and research institutions worldwide. The contributions span a broad range of topics—from land-based and marine pollution sources to cutting-edge monitoring technologies, management strategies, and policy interventions. Together, they form a holistic framework designed to inform and inspire future research, policy-making, and practical applications.

One distinctive feature of this book is its interdisciplinary nature. Chapters explore pollution from industrial runoff, shipping, and agricultural activities, delve into underappreciated issues such as personal care products and radioactive waste, and analyse the impacts of phenomena like ocean acidification. The volume also highlights innovative approaches, including the use of artificial intelligence, biochar applications, and GIS-based assessments, showcasing the potential for technological and methodological advancements in tackling pollution.

Equally notable is the geographic diversity of the contributors, representing institutions in Sri Lanka, Australia, Vietnam, Turkey, and beyond. This global perspective ensures that the book addresses both universal and region-specific challenges, making it a valuable resource for researchers, practitioners, and policymakers.

In addition to providing a thorough exploration of current issues, this book emphasizes the future, presenting case studies of disasters such as the X-Press Pearl incident and offering forward-looking discussions on microplastics, blue economy impacts, and transcending pollution thresholds. By fostering a deeper understanding of these pressing issues, we hope to inspire actionable solutions and spark innovations that will safeguard our marine ecosystems for generations to come.

Whether you are a scientist, a policymaker, or a student seeking to contribute to this critical field, we hope this book serves as a catalyst for meaningful change.

Acknowledgment

The ever-evolving challenges of coastal and marine pollution demand a resource that goes beyond documenting the present to laying a foundation for future advancements. With this goal in mind, we aimed to create a book that not only highlights the current state of knowledge but also provides a comprehensive framework for addressing emerging issues in this critical field. Achieving this required the generous contributions of leading experts, who took time from their demanding schedules to share their insights and expertise. We are deeply grateful to all the contributors for the depth and quality of their chapters, which have enriched this volume immensely.

A notable feature of this book is its multidisciplinary and international collaboration. The authors represent a wide range of expertise, from environmental engineering and marine chemistry to decision sciences and policy development. This diversity reflects the complexity of coastal and marine pollution, which requires integrated approaches involving advanced technologies, innovative management strategies, and effective governance. The volume also benefits from a balance between academic perspectives and industry insights, demonstrating that impactful solutions often arise at the intersection of research and application.

The importance of this book lies not just in its content but in its potential to influence the future. Coastal and marine pollution is a global issue, and its management requires collaborative efforts across sectors and disciplines. By presenting case studies, innovative tools, and future perspectives, we hope to inspire new research directions and practical solutions that will help mitigate the devastating impacts of pollution on marine ecosystems and the communities that depend on them.

This book is a collective effort, and its success is a testament to the dedication and collaboration of the contributing authors. If it sparks new ideas, improves existing methods, or simply broadens the understanding of the challenges and solutions associated with coastal and marine pollution, it will have achieved its purpose.

1

Overview of Coastal and Marine Pollution: Sources, Impacts, and Challenges

Gohagodage Nirmal Tharanga Jayalath and Amila Sandaruwan Ratnayake

Department of Applied Earth Sciences, Faculty of Applied Sciences, Uva Wellassa University, Badulla, Sri Lanka

1.1 Introduction

Marine environments are generally understood as aquatic ecosystems with high contents of dissolved salts, including the open ocean and coastal ecosystems such as estuaries. According to the United Nations Convention on the Law of the Sea, the "marine environment" is defined as "the physical, chemical, geological and biological components, conditions and factors which interact and determine the productivity, state, condition, and quality of the marine ecosystem, the waters of the seas and the oceans and the airspace immediately above those waters, as well as the seabed and ocean floor and subsoil thereof" (Valencia and Akimoto 2006). Coastal environments are described as transition zones where the terrestrial watersheds make close interactions with the open ocean and the atmosphere (Lavalle et al. 2011; Werner and Blanton 2019). Accordingly, coastal environments include some of the most complex and dynamic ecosystems on earth such as estuaries, bays, coral and other biogenic reefs, shallow near-shore waters, tidal wetlands, mudflats, mangrove swamps, and saltmarshes (Cowie and Woulds 2011; Cabral et al. 2019; Nunes and Leston 2020).

Throughout history, the coastal and marine environments have been an integral part of human civilization. Occupying over 70% of the earth's surface, coastal and marine environments play a crucial role in regulating global environmental conditions and processes. These include various physical, chemical, and biological processes such as the hydrological cycle, solar energy balance, global nutrient cycle, biological food chain, and global and regional climate patterns (Früh-Green et al. 2018; Ratnayake 2021a,b). In addition, many human activities rely on coastal and marine environments, such as shipping, fishing and food supply, recreational activities, tourism, and exploring for minerals and oil (Amalan et al. 2018; Ratnayake et al. 2018; Reichelt-Brushett 2023). Until recently, the oceans were thought to be an inexhaustible resource that was so vast and almost limitless in its capacity, and it could not possibly be impacted by human activities, and they were able to withstand any amount of waste directed toward them (Potters 2013; Weis 2015). Even when the limited nature of the terrestrial resources and their pollution levels were being acknowledged, many believed that the oceans have an unlimited capacity as the world's ultimate repositories. For instance, Rachel Carson in 1951, in her book *The Sea Around Us*, expressed that humans could

Coastal and Marine Pollution: Source to Sink, Mitigation and Management, First Edition.
Edited by Meththika Vithanage, Sameera M. Samarasekara, Bryan D. James, and Christopher M. Reddy.

not possibly alter the oceans in the same manner they had exploited the continents, but she changed her viewpoint later. The introductory chapter of this book will thus provide an overview of sources, impacts, and challenges in coastal and marine pollution.

1.2 Pollution in Coastal and Marine Environments

Pollution is defined, in a broader context, as the undesired presence of various toxic substances or any form of contamination, due to human activities with harmful effects on living beings and the environment (UNEP 1982; Willis et al. 2021). It is important to distinguish between the terms "contaminant" and "pollutant" since both words are used somewhat interchangeably by both scientific and mainstream media. A contaminant is a substance (biological, chemical, or physical) or energy that is typically not found or is uncommon in the environment. However, it can harm living organisms in high enough concentrations (Jones and Gomes 2013; Weis 2015). Simply, contaminant refers to the presence of a substance in a sample without any indication of harm, but with the potential to be harmful. A pollutant is defined as any form of a contaminant within an ecosystem that negatively affects or can affect individual organisms or communities by altering their growth rates and reproduction, or by disruptions to human well-being, health, comfort, or property values (Potters 2013). Therefore, all pollutants are contaminants, but not all contaminants are defined as pollutants.

Coastal and marine pollution has been defined in numerous contexts based on its use in different conventions and international treaties or on its relevance to oceanic areas of international significance (Tomczak 1984). In marine science, arguably, the most successful and commonly used definition for marine pollution is introduced by the Joint Group of Experts on Scientific Aspects of Marine Pollution (GESAMP), a United Nations advisory board set up in 1969. As part of the fundamental framework of the 1982 UN Convention on the Law of the Sea (UNCLOS), GESAMP defined marine pollution as, "the introduction by man, directly or indirectly, of substances or energy into the marine environment (including estuaries) resulting in such deleterious effects as harm to living resources, hazards to human health, hindrance to marine activities including fishing, impairment of quality for use of sea water, and reduction of amenities" (UNEP 1982; Tomczak 1984; Islam and Tanaka 2004). Since then, this definition has found its way to the protocols and constitutions of many international agreements and conventions. While the initially proposed definition focused only on the introduction of various substances into the marine environment, the later revisions incorporated the concept of energy, indicating that heat, radioactivity, and noise can also be considered pollutants (Wilhelmsson et al. 2013; Beiras 2018).

Coastal and marine environments are continuously being polluted (Riechers et al. 2021; Pilapitiya and Ratnayake 2024). Coastal and marine pollution alters the physical, chemical, and biological features of oceans and coastal ecosystems, posing potential threats to marine organisms, ecosystems, and biodiversity. Consequently, it impacts the quality and productivity of marine ecosystems (Wilhelmsson et al. 2013). For instance, oceans face significant threats from various impacts such as climate change, environmental pollution, over-exploitation of resources, increased maritime transportation, reduced river discharges, and alterations in river sediment dynamics (Halpern et al. 2008; Hossain 2019). This will result in numerous negative effects on human health, livelihood, food security, marine navigation, and biodiversity (Riechers et al. 2021; Willis et al. 2021). Additionally, it is also important to remember that oceans have more resources to offer than we are currently aware of, which requires sustenance of the ocean ecosystems, to explore them in the future. Hence, the conservation and sustenance of the oceans and the resources are crucial in achieving the ambitious aim of transforming the global economy into a blue economy based on the water and the oceans (Hossain 2019). These underscore the concerns and the necessity for immediate actions to minimize and rectify coastal and marine pollution problems.

1.3 Sources of Coastal and Marine Pollution

1.3.1 Sources and Distribution Pathways of Pollutants

Pollutants can originate from an array of pollution sources. Understanding these diverse origins is essential for effective management and mitigation efforts. All pollutants, whether they are physical, chemical, or biological, can originate from the ocean itself or land sources, reaching marine ecosystems through different pathways (Figure 1.1). However, the majority of the marine debris (about 80%) originates from land-based sources which include pollutants such as nutrients, sediments, pathogenic organisms, harmful chemicals like heavy metals, pesticides, industrial effluents, and pharmaceutical chemicals (Jambeck et al. 2015; Weis 2015; Gough 2017; Macko 2018). Since most contaminants originate from terrestrial sources through human activities, their origin and transportation into the sea are broadly described via three major pathways of sources: runoff via riverine inputs, atmospheric deposition, and anthropogenic pathways of direct inputs (Figure 1.2). The extent to which each of these pathways contributes to marine pollution varies significantly depending on the substance and specific circumstances. Accurate quantification of these processes is quite challenging due to the lack of data and the intricate nature of natural processes, particularly at the interfaces between land and sea and between sea and atmosphere (Wowk 2013; Cabral et al. 2019).

Sources of coastal and marine pollutants can also be broadly classified as point sources and nonpoint sources based on their degree of dispersion (Figure 1.1). Point sources are more localized sources where the origin can be traced back. These could be a waste discharge pipe from an industry or an oil spill. Nonpoint source pollution originates from a rather dissipated source which can be quite difficult to identify or trace back in time and space, for example, agricultural or urban runoffs (Jones and Gomes 2013; Perera et al. 2022). Unlike nonpoint sources, point sources thus can be easily monitored and legislated and can be controlled using novel technological applications and wastewater treatment methods (Weis 2015; Beiras 2018; Reichelt-Brushett 2023). This section further examines various sources of coastal and marine pollution, ranging from industrial activities to natural events, highlighting their impacts and implications for marine ecosystems.

1.3.2 Industrial Activities

Industrial processes are one of the major contributing sources of coastal and marine pollution (Kennish 2019). Industrialization is quantitatively related to pollution in less regulated parts of the world (Wilhelmsson et al. 2013; Beiras 2018). The more industrialized regions can be major sources of chemical pollution such as chemical industry, sewage treatments, and landfills compared to rural areas except for regions with heavy regulations and the industrial emissions being well controlled and monitored. For instance, the drastic increment in metal concentrations, especially in the coastal regions, is mostly caused by industrial and mining activities (Cabral et al. 2019). Numerous industrial products are potential marine waste if they are subjected to incorrect disposal on land or lost during transportation or loading/unloading at harbors (Pawar et al. 2016). Plastic pellets used in the manufacture of plastic products, products from offshore industrial platforms,

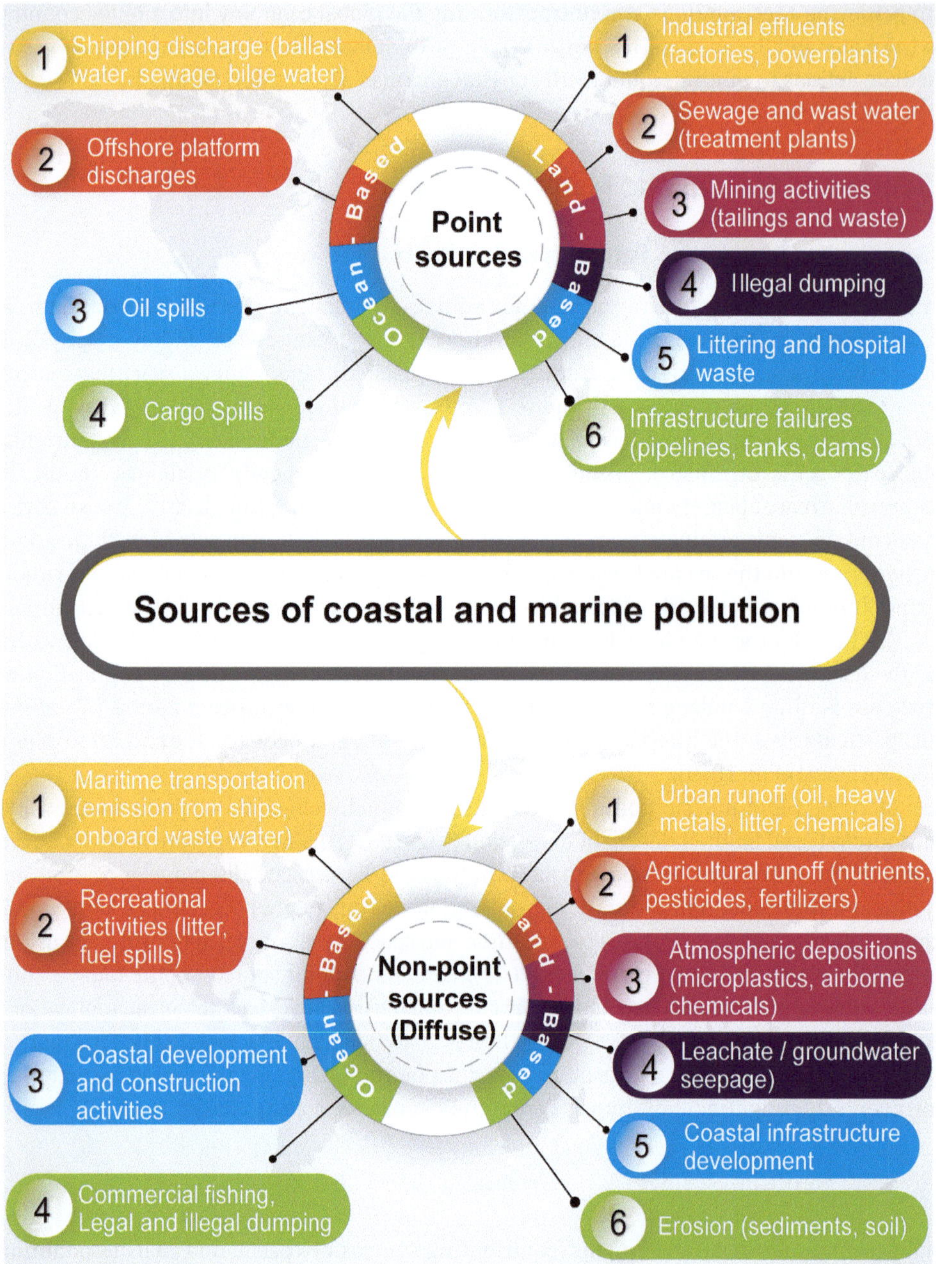

Figure 1.1 Summary of point and nonpoint sources of coastal and marine pollution based on their origin.

dyes, flame-retardants, waste from the tourism and recreational industry, and wastes from the pharmaceutical industry and fisheries industry are some examples of pollutants in industrial products (Riechers et al. 2021). Furthermore, almost all industrial processes generate wastewater including an array of pollutants such as hazardous chemicals, heavy metals, soluble organic substances, suspended solid particles, nitrogen and phosphorous, refractory substances, volatile substances, oil and floating material, and other aquatic toxins (Akankali and Elenwo 2015).

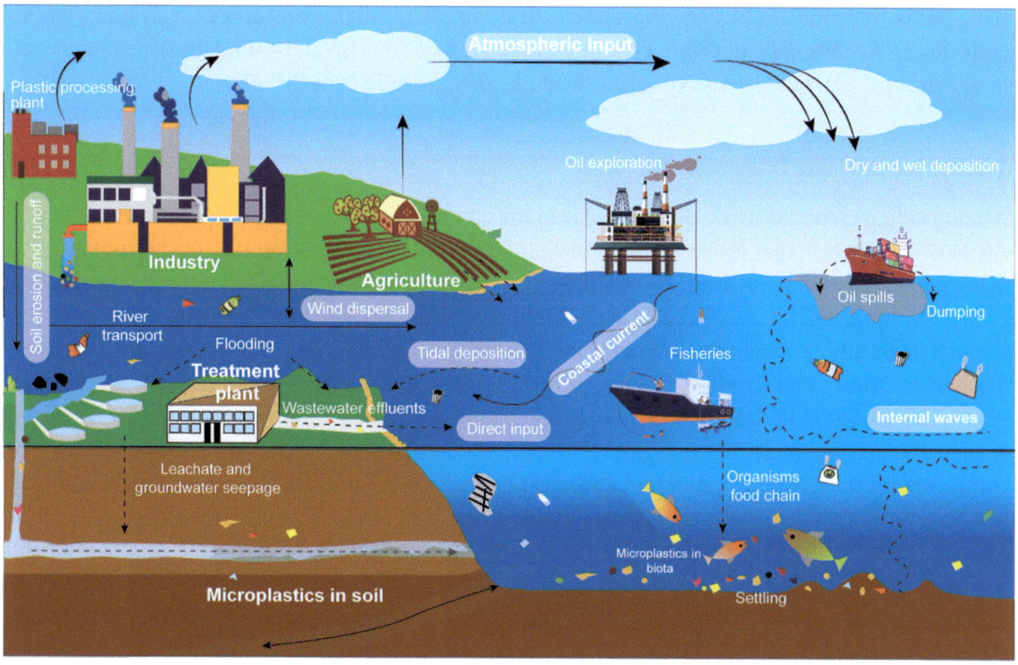

Figure 1.2 Different sources and major transportation pathways of coastal and marine pollutants.

The industrial effluents containing most of these pollutants are usually discharged from factories and manufacturing plants into rivers or coastal areas either without treatments or with improper treatments (Pawar et al. 2016). While pollution is largely concerned with the discharge of industrial wastewater as a pollution source due to its profound effects on the marine environment, thermal heat, radiations and energy forms like noise can also have adverse impacts on the marine environment (Akankali and Elenwo 2015).

1.3.3 Agricultural Runoff

Agricultural pollutants are another major contributor to coastal and marine pollution originating from land-based sources (Duraisamy and Latha 2011). Agricultural practices such as the use of fertilizers, pesticides, and herbicides contribute to pollution via land runoff into rivers that eventually reach the oceans (Akankali and Elenwo 2015; Nunes and Leston 2020). Since agricultural runoff arises from diffused sources, it poses greater risks to oceans than most point sources due to the difficulty of monitoring and regulating. Over the past 50–60 years, changes in agricultural practices have caused nutrients, pesticides, and soil washout into water bodies (Jones and Gomes 2013). Agricultural pollutants occur from two main categories, (i) agrochemicals (i.e., chemical fertilizers, herbicides, pesticides, and hormones), and (ii) agro-organic pollutants (i.e., organohalogen pesticides, organic fertilizers, or green manure derived from organic domestic wastes and composting methods) (Windom 1992; Akankali and Elenwo 2015). Agrochemicals can also discharge chemical pollutants like heavy metals into the soil and eventually into the marine environment (Nunes and Leston 2020). Nitrogen and phosphorous lead to eutrophication and hypoxia by blooming harmful algae in marine ecosystems (Lillebø et al. 2005; Cabral et al. 2019). While there can be several processes leading to enhanced nitrogen influx into coastal and marine ecosystems, riverine input carrying agricultural runoff is often considered to be the primary contributor. Accordingly, it is important to improve regulations in controlling the use of chemicals in agricultural practices (Weis 2015; Cabral et al. 2019; Micella et al. 2024).

1.3.4 Sewage and Wastewater

Among various types of waste discharged into the coastal and marine ecosystems, sewage is by far the most significant organic waste type in terms of quantity (Islam and Tanaka 2004; Kennish 2019). Sewage effluents comprise a variety of waste materials, including water and waste from household baths, washing machines, kitchenware, culinary wastes, animal remains, and slaughterhouse wastes. Generally, sewage effluents and wastewater, whether treated or untreated, are ultimately discharged into the oceans (Critchell et al. 2019). The enrichment of nutrients facilitated by the release of sewage and wastewater into the marine ecosystems regularly leads to oxygen level reduction, retarded plant life and decay, and heavy decline in the seawater quality resulting in complications such as eutrophication and algal blooms (Vikas and Dwarakish 2015). In addition, sewage and wastewater pose numerous threats to human health due to the release of harmful pathogens (Islam and Tanaka 2004; Kennish 2019). Sewage-derived pathogenic microorganisms such as bacteria, viruses, protozoans, and helminths found in sewage are known to cause different infectious diseases such as cholera, diarrhea, dysentery, hepatitis, and typhoid (Sheavly and Register 2007; Potters 2013; Weis 2015; Kennish 2019). Urban stormwater runoff and treated municipal wastewater have been found to contain about 100 human gastrointestinal pathogens, bacteria, viruses, and parasites such as *Salmonella* and *Shigella* which are responsible for typhoid and dysentery, respectively (Sindermann 1995). The amount of anthropogenic carbon added to contaminated water through raw sewage and sewage sludge is also known to raise the biological oxygen demand (BOD) considerably. For example, the highest quantities of dissolved and particulate organic carbon are seen in estuaries and nearshore areas. Dissolved organic carbon levels can even reach up to 100 mg/L in sewage waste-contaminated estuaries and nearshore areas, whereas the values are around 1–5 mg/L in uncontaminated areas (Kennish 2019).

1.3.5 Oil Spills

Liquid petroleum hydrocarbons that are released into the ocean or coastal regions because of human activity are commonly referred to as marine oil spills. These include crude oil spillage from tankers, offshore platforms, drilling rigs, and oil wells, including spills of refined petroleum products (such as diesel and petrol) and their byproducts; heavier fuels used by large ships (such as bunker fuel); and spills of any oily white waste (Zhang et al. 2019). In marine environments, oil spills occur frequently, but many minor spills go unreported, mostly in areas lacking environmental regulations and jurisdiction (Yoshioka et al. 1985). Small oil spills from ships, which involve less than 700 tons, typically happen due to mistakes made by people during routine tasks, such as loading and unloading. Larger spills (more than 700 tons) are most frequently caused by accidents like collisions and groundings (Wilhelmsson et al. 2013). Throughout the petroleum exploration, extraction, and shipping processes, spills can happen for several reasons, including overpressurization, mechanical failure, pipeline corrosion, and ship collisions (Zhang et al. 2019). The Exxon Valdez oil spill in Alaska in 1989, the Deepwater Horizon oil spill in 2010, and MT New Diamond in the Indian Ocean in 2020 are some examples that captured global attention (Riechers et al. 2021; Ratnayake and Perera 2022). However, despite the publicity and attention, events of such magnitude are quite uncommon and are almost indiscernible to the public in a short time of two to five years (Windom 1992). This is because most spills to date have been in tropical or temperate waters, which show greater rates of biological and physical weathering (Preston 2013). According to Preston (2013), the only major exception to this generalization is the Exxon Valdez spill which took place in the Alaskan region having low temperatures and therefore had low degradation rates. However, the oil spills at the sea are not the primary source through which oil enters the sea. More than half of the oil entering the sea is found to originate from land-based sources, while only a quarter results from maritime transports, and the rest are through atmospheric sources (Beiras 2018; Macko 2018; Kennish 2019).

1.3.6 Maritime Transport

Maritime transport is undoubtedly one of the major cornerstones of global trade and economy connecting nations across the world. It is also, however, a strong anthropogenic contributor to coastal and marine pollution forms which has significantly influenced marine ecosystems. Tankers, cargos, container ships, bulk carriers, ferries, cruise ships, recreational boats, and fishing vessels are some of the main categories of ships involved in maritime transport (Figure 1.3) (Andersson et al. 2016; Perera et al. 2022). While oil spillages are one of the primary concerns associated with the use of these ships, they also generate many other pollutant forms in the ocean such as solid waste, cargo hold cleaning waters, ballast water, bilge water, sewage, noise, biocidal antifouling paints, exhaust emissions from engine, and refrigerants (Figure 1.3). Ballast water discharged into the seas often may contain many exotic species of invasive and pathogenic microbes (Drake et al. 2007; Lindgren et al. 2016). Conversely, bilge water may include solid

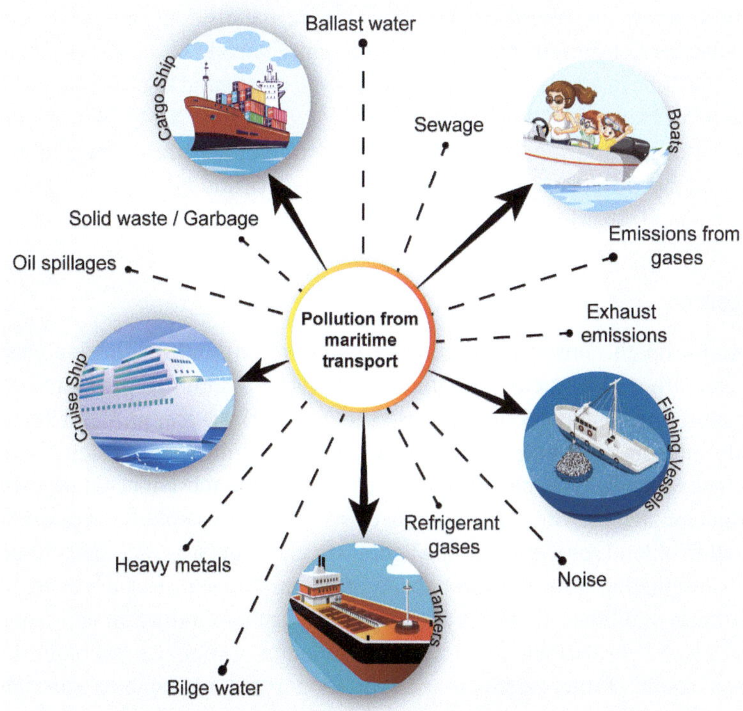

Figure 1.3 Pollutant forms originating from maritime transportation.

waste, oil, petrol, chemicals, and other pollutants that could end up in the ocean. Gases such as carbon dioxide, carbon monoxide, nitrogen oxide, sulfur dioxide, and hydrocarbons emitted from diesel engines of most ships can lead to pollutants via atmospheric depositions (Aakko-Saksa et al. 2023). Apart from the above pollutant forms and their adverse impacts shipping activities can also result in wildlife and habitat destruction from pollutant forms such as biocidal antifouling paints and noise (Dotinga and Elferink 2000; Weis 2015; Kennish 2019).

1.3.7 Plastic Pollution

Plastics are perhaps by far the most significant and dangerous pollutant form of marine debris that harms coastal and marine ecosystems (Islam and Tanaka 2004; Lithner et al. 2011; Pilapitiya and Ratnayake 2024). They pose serious harm to marine wildlife such as seabirds, mammals, turtles, and other aquatic organisms when entangled in plastic pollutants like fishing nets and lines, packaging, and wrapping materials. Plastics also lead to habitat fouling through accumulation, and their slow degradation rates further intensify their interferences with ecosystem functions (Derraik 2002; Vegter et al. 2014). Additionally, plastics often contain harmful chemicals or toxic chemicals attached to their surfaces, and when consumed by marine organisms can block their air and digestive tracts causing complications such as suffocation, starvation, and ultimately death (Crain et al. 2009). Of all the marine litter, plastics account for approximately 70% of litter found on shorelines, 68–99% of litter found in the water column, and 23–89% of marine litter found on the seafloor of the continental shelf (Kennish 2019; Galgani et al. 2022). Recently, microplastics have become the target of scientific research as an emerging contaminant due to their widespread nature in the estuarine and marine environment in quantities and pathways which have not yet been fully understood (Da Costa et al. 2023). Wastewater containing health and medical products, textiles such as nylon and polyester fibers, plastic products, cosmetics, and other hygiene products are some sources of microplastics entering the marine environment (Micella et al. 2024). Their smaller size and high concentrations in coastal and marine ecosystems are found to pose threats of accumulation in marine food webs passing down to humans (Micella et al. 2024). Hence, the threats to the biodiversity of coastal and marine ecosystems and numerous health impacts on humans are expected to rise further with the continued growth of plastics and microplastic pollution (Wilcox et al. 2019; Verma et al. 2020).

1.3.8 Atmospheric Deposition

Atmospheric deposition occurs on the entire ocean surface without confining to the land/sea interface where both dry deposition of particulate material and wet deposition of a mixture of dissolved and particulate material are conveyed through wind (Figure 1.2). Hence, unlike riverine inputs, atmospheric inputs do not require a passage through the estuarine environment (Preston 2013). However, the coastal areas nearest to the continental source material are the most affected, as they exhibit the strongest atmospheric signals. The contaminants delivered via atmospheric deposition are derived from the evaporation of land sources including light hydrocarbons, gases, and particulate matter carrying trace metals originating from the combustion of organic substances such as fossil fuels, pesticides from agriculture, and other aerosol products originating from industrial emissions and urban areas (Potters 2013; Reichelt-Brushett 2023). The accumulation of chemical pollutants in remote oceanic regions is a result of the extensive transportation pathways enabled through atmospheric dynamics (Weis 2015; Beiras 2018). Once in the marine environment, these can impact marine life, water quality, and the health of coastal and marine ecosystems.

1.3.9 Coastal Developments

Coastal development projects such as the construction of ports, hotels, resorts, and other coastal infrastructure as well as coastal erosion control measures can severely impact coastal ecosystems. These development projects can lead to increased runoff of pollutants into the oceans carrying sediments, nutrients, heavy metals, and other wastes, thus resulting in reduced water quality, destruction of the natural beauty of the coasts, and contributing toward eutrophication and algal blooms (Diaz et al. 2013; Saengsupavanich et al. 2023). Development activities involving dredging, filling, and changing coastal habitats such as wetlands, mangroves, marshes, and coral reefs can lead to habitat destruction and fragmentation which can result in decreased biodiversity and compromise the aesthetic beauty (Crain et al. 2009; Cheung and Sumaila 2013). In addition, coastal development projects can negatively impact important ecosystem functions such as water filtration properties and erosion regulation due to the hardening of estuarine and coastal shorelines (Crain et al. 2009; Kennish 2019; Saengsupavanich et al. 2023).

1.3.10 Mining Activities

Coastal and marine mining operations are carried out to extract many marine resources including oil and gas, sand, gravel, and several other seabed minerals such as heavy minerals, cobalt-rich crusts, and phosphorite deposits extending from shorelines into the deep sea (Kennish 2019). These coastal and offshore mining activities can have serious impacts on the coastal and marine biodiversity such as alteration and habitat loss of aquatic organisms such as planktons, decrease in water quality due to turbidity, and release of toxic chemicals affecting life stages of such organisms (Macko 2018). Mining operations on land also give rise to tailings and other waste accumulations in huge quantities and facilitate the release of sediments which can damage the coastal and marine ecosystems and affect their biodiversity (Wilhelmsson et al. 2013). In addition, most mining operations on land also result in problems like acid mine drainage through which acidic leachates concentrated in heavy metals and metalloids such as arsenic, cadmium, copper, and lead can leach into the water columns which can ultimately find their way to oceans and impact the water quality and oxygen level imbalances affecting the aquatic life (Duraisamy and Latha 2011; Weis 2015).

1.3.11 Invasive Species

Activities such as shipping, aquaculture, fisheries, and ornamental fish trading have caused biological pollution resulting from invasive and non-native species (Crain et al. 2009). These species are described as sources of pollutants based on their potential to degrade existing biological systems ranging from cellular to ecosystem levels (Islam and Tanaka 2004). The introduction of these species may occur in several ways such as through the discharge of ballast waters in ships, through urban and agricultural runoff, mine leachates and effluents, and aquaculture (Drake et al. 2007). Once in their new environment, these species often rapidly outcompete the native species for nutrition, growth, and reproduction (Mayes 2013). This leads to significant reductions in biodiversity and habitat alterations leading to the elimination of native species in extreme scenarios. The proliferation of invasive species ultimately results in total loss of biodiversity, and fundamental changes in trophic levels in food webs through predation and competition leading to the collapse of the impacted coastal and marine ecosystems (Crain et al. 2009; Weis 2015; Kennish 2019).

1.3.12 Nuclear Fallout

Similar to most other pollutants, radioactive substances are also introduced into marine environments via runoff, atmospheric deposition (explosions from nuclear disasters such as Chernobyl and Fukushima Daiichi, and nuclear testing), liquid effluents (heated water containing radioactive waste), and solid radioactive wastes (nuclear submarines and dumping) (Ansari and Matondkar 2014; Nunes and Leston 2020). In addition, until the 1970s, the oceanic floors were often the final destination for radioactive waste generated from nuclear testing which introduced anthropogenic radioactive substances into the environment in unparalleled quantities (Wilhelmsson et al. 2013; Beiras 2018). However, unlike other pollutants when radioactive isotopes accumulate in the food chains, they pose significant threats (Mayes 2013). Radioactive contamination in marine environments can harm and accumulate in marine life. Furthermore, by concentration in food chains and spumes, radioactive substances can be transferred into the terrestrial environments and finally in humans which is of great concern since it can lead to serious health impacts such as cancer and genetic disorders (Mayes 2013; Kennish 2019).

1.3.13 Natural Events

Natural events such as storms, earthquakes, tsunamis, volcanic eruptions, and algal blooms can act as sources of coastal and marine pollution, although they are not as frequent as anthropogenic sources (Cooke et al. 2022; Ratnayake et al. 2023). Natural events can release a variety of potential pollutants such as sediments, organic debris, gases, ashes, and harmful toxins into the marine and terrestrial/coastal environments (Tapia et al. 2019). For instance, storms and hurricanes can result in flooding, erosion, and infrastructure damage and wash off debris from the land to the oceans (Knap and Rusyn 2016). Earthquakes and tsunamis can lead to disasters in ocean beds, volcanic eruptions, and collect debris in the marine and coastal environments. The release of ash, toxic gases, debris avalanches, and lava mudflows into oceanic environments due to volcanic eruptions can significantly alter the water chemistry and pH levels affecting the local biodiversity (Cooke et al. 2022). Algal blooms originating naturally can lead to eutrophication and hypoxia in localized oceanic regions (Diaz et al. 2013; Knap and Rusyn 2016).

1.3.14 Climate Change

The climate changes of the world brought about by humans are another source leading to major coastal and marine pollution effects. Effects of changes in climatic conditions are observed both regionally and globally such as the rise of ocean temperatures, consequent biochemical changes accompanied by sea temperature rise, and ocean acidification resulting from increased inputs of carbon dioxide (Kennish 2019). Climate change will add more stress on coastal and marine ecosystems, leading to loss of biodiversity, and ecological imbalances, making them more vulnerable to pollution. For example, with increasing global precipitation levels, the nutrient and sediment discharges into the ocean ecosystems are predicted to rise further (Jennerjahn 2012). This will in turn lead to further intensification of eutrophication levels, the occurrence of algal blooms, increased turbidity levels, and reduced water quality in oceans (Mayes 2013; Shine 2013). Potential loss of coastal habitats such as wetlands, marshes, lagoons, and mangroves is another adverse impact that may result from coastal inundation from rising sea levels (Cabral et al. 2019). Furthermore, climate change will lead to huge alterations in estuarine and oceanic circulation patterns, tidal patterns, freshwater inputs, temperature and salinity patterns, and oceanic currents. These changes can ultimately bring about major alterations in biodiversity, migration patterns of marine life, and coastal and marine ecosystem dynamics (Wilhelmsson et al. 2013; Riechers et al. 2021).

1.4 Impacts of Coastal and Marine Pollution

1.4.1 Harm to Marine Life

1.4.1.1 Bioaccumulation and Biomagnification

Bioaccumulation is the concentration of pollutants or toxins (e.g., heavy metals like mercury and lead, persistent organic pollutants [POPs] such as organohalogens) in the tissues of marine organisms (Nunes and Leston 2020; Verma et al. 2020). Toxic pollutants are built up within organisms such as fish and shellfish in higher concentrations. As the smaller organisms are consumed by predators in high levels of biological organization in the food chains, the pollutants are subjected to biomagnification leading to progressively higher concentrations (Shine 2013). Bioaccumulation of such toxins as polychlorinated biphenyls (PCBs) can affect the physiology of the organisms, impair the reproduction process, weaken immunity, lead to abnormal developments, and hence affect the biodiversity of the impacted area (Cabral et al. 2019). However, the most serious issues arise when such organisms are used for human consumption. Once consumed by humans, bioaccumulation and biomagnification can lead to major health issues such as neurological damage, severe complications in unborn children, bone weakening, carcinogenesis, endocrine disruptions, and chronic renal diseases (Nunes and Leston 2020). For example, the mercury poisoning of tens of thousands of people in Minamata Bay of southwest Japan during the mid-twentieth century resulted from the consumption of methyl-mercury accumulated fish and shellfish from the release of mercury from chemical corporations (Shine 2013). Efforts to address these impacts can involve reduced discharges of potential pollutants, hazardous material regulations, monitoring pollutant levels, and the use of sustainable fisheries management techniques (Hossain 2019).

1.4.1.2 Habitat Degradation and Fisheries Decline

Crucial habitats found in coastal and marine ecosystems, like coral reefs, seagrass beds, estuaries, marshes, and mangroves, that act as nurseries and breeding grounds for many species including fish, birds, plankton, and other organisms can be damaged due to pollution and human stressors (Weis 2015). Pollutants arising from runoff, sewage and wastewater, oil spills, and heated and acidic water can lead to lowered water quality due to high turbidity, low pH levels, and rise in temperatures which can disrupt biochemical balances, and ecological functions leading to habitat degradation (Potters 2013; Weis 2015). Sediment accumulation can cover coral reefs and seagrass beds whereas nutrients in runoff can cause eutrophication and hypoxia leading to oxygen depletion and penetration of light into deeper seabeds (Diaz et al. 2013). Habitat loss can also occur due to coastal infrastructure development projects, mining, and dredging activities. Habitat degradation results in extreme ecological consequences including biodiversity loss, reduced habitat complexities and quantities, competition for habitats, and changes in species distribution. To ensure healthy ecosystems, mitigation efforts like habitat restoration methods, marine and coast protection zones, and coastal zone management methods are highly recommended.

Fisheries decline in marine environments is due to habitat destruction, polluted waters, climate change impacts, invasive species, and overfishing. Coastal and marine pollution such as oil spillages affect habitat degradation leading to the loss of nursery habitats and spawning grounds of fish (Preston 2013). This leads to a loss of fish diversity and abundance. Climate change effects such as ocean warming, and ocean acidification can also alter fish reproduction and distribution greatly (Shine 2013). Reduced fish stocks can affect the livelihood of most coastal communities worldwide which can impact the global seafood supply as well (Willis et al. 2021). Habitat restorations, sustainable fisheries management, and fish conservation strategies are hence required to ensure the sustainability of fish resources.

1.4.2 Biodiversity Loss

The loss of biodiversity in coastal and marine environments is contributed by several sources. Habitat destruction, introduction of invasive species, and pollution from oil spills and other chemicals and debris can decline in many marine species including endangered and vulnerable ones ultimately causing their extinction (Farrington 2013; Preston 2013). Depletion of fish stock due to overfishing can disrupt food webs in marine ecosystems threatening their sensitive ecological balances (Mostofa et al. 2013). Climate change impacts such as sea level rise, ocean acidification, and warming can further alter marine ecosystems where they can bring about major shifts in species compositions, adding up many environmental stressors and leading to biodiversity loss (Ansari and Matondkar 2014). Conservation efforts, such as pollution prevention, habitat restoration, and species protection measures, are important measures for addressing biodiversity loss and preserving the health and integrity of coastal and marine environments.

1.4.3 Economic Consequences

Coastal and marine ecosystems are constantly being polluted with some significant economic consequences, with tourism and fisheries being affected the most. Coastal pollution impacts such as loss of natural aesthetic values of beaches, loss of biodiversity such as coral reefs, and polluted waters can deter tourists who often seek out natural, exotic, and clean environments for recreational activities such as swimming and snorkeling (Sheavly and Register 2007; Gelcich et al. 2014). Such polluted coastal environments may experience dwindled tourist numbers which can impact the economies of the coastal communities that rely on tourism, leading to revenue losses for local businesses, hotels, and tour guides (Pawar et al. 2016). The negative publicity can impact the reputation of such areas as tourist attractions in the long run.

Other than tourism, coastal and marine pollution can heavily affect the fisheries and aquacultural economy as well. Polluted waters, loss of habitats, eutrophication, and invasive species can reduce fishery yields, greatly affecting the causing economic losses for the local fishing communities, and other associated businesses (Pawar et al. 2016; Willis et al. 2021). Aquaculture farms that rely on healthy aquatic environments can also experience severe financial blows due to pollution impacts such as water contamination, and outbreaks of diseases (Diaz et al. 2013). Such economic impacts can heavily affect coastal communities and nations that depend on fisheries and aquaculture as a primary income source and their food security (Wowk 2013). Mitigation of such economic consequences requires the application of effective coastal and pollution management and conservation methods coupled with long-term economic plans.

1.4.4 Human Health Risks

Pollution of coastal and marine environments can significantly impact human health primarily through the consumption of contaminated seafood. Consumption of seafood such as fish and shellfish contaminated from pollutants including heavy metals (e.g., arsenic, mercury, cadmium), organic substances, and pathogenic organisms such as bacteria, viruses, and other parasites can lead to many health hazards (Preston 2013). Human health risks may range from acute disorders to even death and include food poisoning, gastrointestinal disorders, infectious diseases such as cholera, diarrhea, typhoid, dysentery, and long-term health effects affecting neurology and development (Sheavly and Register 2007). Some persistent contaminants such as POPs, polycyclic aromatic hydrocarbons (PAHs), PCBs, and heavy metals can accumulate in food chains leading to severe health issues including neurological damage, carcinogenesis, genetic mutations, and may also affect vulnerable communities such as pregnant women and children (Nunes and Leston 2020).

Polluted coastal water can also pose several other health risks to people such as swimmers, beachgoers, fishermen, and persons involved in recreational activities (Pawar et al. 2016). Polluted coastal water may cause risks of waterborne diseases, skin infections, and respiratory disorders (Sindermann 1995). An increase in invasive and some other marine species such as jellyfish and algal toxins produced by algal blooms in some coastal waters can lead to several health complications including skin irritations, and neurological and respiratory complications (Mostofa et al. 2013). Regulations to protect water quality and prevent pollution methods are necessary for safeguarding human health and promoting safe recreational and other economic activities in coastal and marine environments.

1.4.5 Eutrophication

Eutrophication results from an increasing rate of primary production and accumulation of organic carbon over normal processing levels in marine ecosystems. As a visible ecosystem response to excess nutrient enrichment, particularly nitrogen and phosphorous, excessive growth in phytoplankton, macroalgae, and other aquatic vegetation occurs in the affected areas (Mayes 2013; Mostofa et al. 2013; Potters 2013). Eutrophication can add multiple stressors on the impacted ecosystem such as blockage of sunlight penetration into deeper seas, affecting the growth of seagrass beds and other light-dependent organisms. Therefore, eutrophication affects photosynthetic production, habitat degradation of many organisms, and collapses in food chains (Mayes 2013). The most critical threat of eutrophication is perhaps the decline of oxygen levels in deeper water due to increased oxygen consumption by the bacteria for the decomposition of enhanced influxes of organic matter in the seabed (Duraisamy and Latha 2011; Cabral et al. 2019). This leads to hypoxic or anoxic conditions in the impacted areas resulting in oxygen-depleted "dead zones" which can add further stresses to aquatic ecosystems by changing ecosystem dynamics, disrupting of food webs, and leading to loss of biodiversity (Diaz et al. 2013). The Gulf of Mexico, the Chesapeake Bay, and the Baltic Sea are some examples of enormous dead zones caused by eutrophication (Macko 2018). Careful management of nutrient influxes, sewage discharges, and remedial measures are hence necessary for minimizing eutrophication and its associated stressors in coastal and marine environments.

1.4.6 Coastal and Infrastructure Damage

Damage to coastal infrastructure such as corrosion of piers, bridges, buildings, and pipelines may arise due to factors like sea level rise or pollutants in seawater undermining their structural integrity (De Almeida and Mostafavi 2016). Infrastructure damage may pose risks of accidents to public safety and hence requires frequent monitoring, maintenance, and additional protective measures to ensure their integrity and functions. Long-term pollution prevention strategies and regulations to enable sustainable coastal resource utilization will be essential in preventing and minimizing coastal and infrastructure damages.

1.4.7 Climate Change Amplification

This mainly refers to the associated impacts due to greenhouse gas emissions such as global warming, ocean acidification, and sea level rise which intensify the existing environmental stressors and their impacts on the coastal and marine ecosystems (Cabral et al. 2019). For example, the increase in global temperatures leads to frequent and severe heatwaves. Pollutants such as black carbon emitted from incomplete fossil fuel combustions can be deposited on ice and snow reducing their reflectivity and accelerating melting. Black carbon when deposited in coastal areas can thus lead to the retreat of glaciers and melting, thereby amplifying sea level rise (Ramanathan and Carmichael 2008). Similarly, methane, a greenhouse gas emitted through human activities such as livestock and agriculture, when in coastal and marine environments, is known to accelerate permafrost thawing which in turn releases more greenhouse gases including methane, thereby further enhancing global warming effects (Xie et al. 2023). Therefore, mitigation of climate change amplification requires a greater understanding of the complex interactions between coastal and marine ecosystems and human activities.

1.4.8 Loss of Aesthetic and Cultural Value

Discharge of pollutants such as litter, sewage, and wastewater, including oil spills in the coastal and marine environments can be esthetically unpleasant, especially in tourist attraction areas like beaches (Lindgren et al. 2016). As a result, marine pollution may lose aesthetic value which can discourage people from engaging in recreational activities (Sheavly and Register 2007). In addition, to some communities in the world whose livelihood depends on the coastal ecosystems, they are of cultural significance involving traditional ceremonies and cultural practices. Pollution of such environments may strip them of their cultural and traditional significance, undermining their identity and connection to those environments. Furthermore, the economy of most coastal communities depends on cultural and aesthetic values, the degradation of which can be a severe blow to their livelihoods.

1.4.9 Social and Environmental Justice Issues

Coastal and marine pollution result in environmental injustices across the world where vulnerable and marginalized communities, especially nearby coastal areas, bear an unfair share of these economic and health consequences (Donohoe 2003). Environmental injustices like hazards and harm due to coastal and marine pollution often impact cumulatively and differentially on several aspects of the human well-being of coastal communities across the world (Bennett et al. 2023). This leads to many social injustices in aspects such as health, livelihood, economics, culture, community infrastructure, food security, and other human rights (Bennett et al. 2023). To address social and environmental injustices, it is necessary to create more globally integrated discussions and actions on equitable access to resources, inclusive decision-making, and community engagement, prioritizing the necessities of such communities.

1.4.10 Marine Debris and Plastic Pollution

Marine debris or litter and plastic pollution is a persistent problem that worsens with time (Jambeck et al. 2015). Plastics are probably the most dangerous marine pollutant since they cause significant habitat fouling, disrupt ecosystem dynamics, and persist in environments for prolonged times due to very slow decay (Akankali and Elenwo 2015). Marine plastic debris creates serious threats to marine and coastal organisms like seabirds, turtles, mammals, and other creatures that consume them or get caught in plastic items like fishing gear, nets, packaging, and other synthetic waste (Islam and Tanaka 2004). Additionally, marine debris often contains or is coated with toxic chemicals such as PAHs, PCBs, and others (Crain et al. 2009). Ingestion of plastics by marine organisms can cause blockages in digestive tracts and air passage obstructions, resulting in suffocation, malnutrition, and eventual death. Furthermore, the breakdown of plastics into microplastics and their accumulation in marine organisms leads to mobilization to humans through food webs (Micella et al. 2024). Despite their smaller sizes, microplastics can persist in marine environments for longer times and carry unknown ecological and health risks in both humans and marine organisms alike (Da Costa et al. 2023). Addressing plastic pollution demands coordinated global efforts, including minimizing the use of disposable plastics, establishing efficient waste management systems, and advocating for environmentally friendly practices to safeguard marine ecosystems and biodiversity.

1.5 Challenges in Coastal and Marine Pollution

1.5.1 Lack of Global Regulation and Enforcement

Coastal and marine environments are transboundary, and pollution often originates in one country but affects others downstream (Nunes and Leston 2020; Pilapitiya and Ratnayake 2024). The absence of effective global regulations and enforcement mechanisms makes it difficult to hold polluters accountable. Despite the presence of basic regulatory and management frameworks, coastal and marine pollution continues to increase worldwide where the coastlines, estuaries and mangrove systems, ocean gyres, and seabed are being degraded at increasing rates (Da Costa et al. 2020). The existing frameworks are not properly implemented owing to the insufficient participation of many countries and states both internationally and regionally without strong cooperation (Wowk 2013). Although many international agreements and initiatives such as the United Nations Convention on the Law of the Sea (UNCLOS) and the International Maritime Organization (IMO) conventions already address specific aspects of marine pollution, integrated action on coastal and marine pollution appears to be lagging and leaving space for loopholes (Agamuthu et al. 2019; Da Costa et al. 2020). For example, there are no legal instruments dedicated to the management of marine litter yet in the Black Sea, even though the Bucharest Convention contains several articles about marine debris (Chen 2015). Furthermore, disparities in environmental regulations and enforcement capacities among developed and developing countries contribute to inconsistencies in pollution control measures and accountability (Howes et al. 2017). Consequently, enhanced international cooperation and coordination, strengthening existing international agreements, promoting information sharing and technology transfer, and supporting capacity-building initiatives are essential to facilitate collective actions toward coastal and marine pollution prevention.

1.5.2 Data and Monitoring Gaps

Challenges in data and monitoring gaps of pollutants and their impacts are another significant obstacle in addressing coastal and marine pollution. Understanding their extent and impacts requires adequate monitoring, yet significant gaps exist in data collection and long-term observations, particularly in remote or poorly regulated areas (Wowk 2013). These gaps hinder the ability to assess the full scope of marine pollution, prioritize resources, and implement targeted interventions. Insufficient data may lead to underestimations of pollution levels and delay urgent action. Addressing this challenge requires investment in monitoring infrastructure, capacity building, and innovative technologies for data collection and analysis such as projection models (Alava et al. 2017; Pilapitiya and Ratnayake 2024). Collaborative efforts among governments, research institutions, nongovernmental organizations, and industry stakeholders are crucial to fill data gaps, improve monitoring coverage, and gain a more holistic and comprehensive understanding of coastal and marine pollution dynamics and interactions (Cabral et al. 2019; Saengsupavanich et al. 2023).

1.5.3 Complexity of Pollution Sources

Coastal and marine pollutants originate from a diverse range of sources originating from both terrestrial and marine environments (Figures 1.1 and 1.2). Industrial effluents, agricultural and urban runoff, maritime transportation, atmospheric depositions, and marine debris are a few such sources contributing to coastal and marine pollution (Figures 1.1 and 1.2). The complex nature of pollution sources, pollutants, their diffuse nature, dynamic interactions with the coastal and marine ecosystems, and changes in response to factors such as climate change make it challenging to correctly identify and prioritize mitigation strategies effectively (Cabral et al. 2019; Kennish 2019). Therefore, regulation and control efforts require coordinated efforts through interdisciplinary collaborations and integrated regulatory approaches. Further, the transboundary nature of coastal pollution makes it necessary for improved international coordination and understanding among nations to address the issues arising from the complexities of coastal and marine pollution (Nunes and Leston 2020).

1.5.4 Technological and Financial Constraints

Most efforts and incentives to address coastal and marine pollution fall short of technological and financial constraints, particularly in developing nations in regions like South Asia, Southeast Asia, and Africa. Tackling pollution problems effectively requires advanced pollution control methods and novel technological innovations which often require substantial financial support. Many developing countries struggle to secure necessary financial resources and lack the technical expertise required for the implementation of advanced technologies in combating coastal and marine pollution (Islam and Tanaka 2004). For instance, the lack of a strong financial structure is a common problem from which most regional environmental programs in Southeast Asia suffer (Tiquio et al. 2017). Such financial challenges result in weak implementation of environmental policies and regulations, and difficulties in applying environmental management techniques and sustainable technologies. In addition, the lack of trained staff and equipment for conducting necessary baseline studies and long-term pollution monitoring leads to fragmented regional projects, and programs having little to no impact or even overlapping activities (Elliott 2011; Gunawardhana et al. 2024). Therefore, international cooperation and support structures are necessary to bridge the gaps in technological and financial resources to enable technology transfer, capacity building, and financial assistance for integrated coastal and marine pollution management in developing nations.

1.5.5 Plastic Pollution

Plastic pollution is one of the serious challenges affecting the coastal and marine environments (Jambeck et al. 2015; Vikas and Dwarakish 2015). Due to their widespread use in everyday products, plastic pollution reduction has become quite difficult, hence raising awareness, and promoting major sustainable consumption and behavioral practices among the public (Willis et al. 2021). In addition, improved waste management infrastructure, implementation of recycling, and reusable programs are vital for the mitigation of existing and future issues related to plastic pollution in marine environments (Wowk 2013). Furthermore, investments in research and innovation are essential for tackling emerging issues such as microplastic impacts, and the development of sustainable alternatives such as biodegradable plastics (Weis 2015; Pawar et al. 2016). Therefore, mitigation and combat of plastic pollution and associated issues demand cooperative efforts from governments, industries, and the public to implement solutions, enact policies, and enable the transition to sustainable lifestyles.

1.5.6 Long-Term and Cumulative Effects

The long-lasting nature of most pollutants in the coastal and marine ecosystems is a critical challenge in minimizing and regulating coastal and marine pollution (Hall et al. 2013). In most of the coastal regions, many anthropogenic stressors have cumulative, antagonistic, or synergistic effects. This creates unknown complexities such as overlapping threats (Alava et al. 2017). This long-lasting nature and the cumulative effects of most pollutants render their impacts and damages on marine ecosystems almost irreversible. Pollutants like POPs, and plastics which can break into microplastics and even nanoplastics can exist in the environment for extended periods and slowly accumulate in marine organisms (Crain et al. 2009). To address the negative impacts of these marine pollution, prevention and early intervention strategies are more effective than remediation (Pettipas et al. 2016). Implementation of strict pollution control methods, sustainable lifestyle and consumption practices, and efforts for ecosystem restorations are essential in mitigating such long-term and cumulative effects of coastal and marine pollution.

1.5.7 Climate Change Interactions

The interaction of coastal and marine pollution with climate change can lead to many challenges, either climate change dominant or contaminant dominant (Alava et al. 2017; Cabral et al. 2019). These interactions can result in additive or synergistic effects leading to the amplification of climate change effects adding further stressors to the ecosystems (Alava et al. 2017). For instance, ocean warming can accelerate the release of certain pollutant chemicals such as heavy metals which may have been stored in sediments over centuries (Shine 2013). Climate change impacts like ocean acidification from the dissolution of atmospheric carbon dioxide by seawater can lead to weakened shells in calcifying organisms due to reduced calcification rates making them more vulnerable to impacts of pollution (Doney et al. 2009).

1.5.8 Political and Economic Interests

Economic interests such as maritime transport activities and offshore resource extraction can interfere with environmental protection aspects. The economic exploitation of coastal and marine environments does not always align with coastal and marine conservation efforts leading to environmental pollution and degradation (Bigg 2013). Thus, policymaking and regulation enforcement can be heavily influenced by economic and political factors and pressures from industry stakeholders. Economic prosperity and environmental sustainability require a balance between the two aspects through transparent decision-making, stakeholder engagement, and sustainable practices prioritizing environmental health (Wowk 2013; Gough 2017).

1.5.9 Inadequate Waste Management

In many regions, particularly in developing countries, coastal and marine pollution is amplified by insufficient waste management systems (Islam and Tanaka 2004; Micella et al. 2024). Inadequate waste collection methods, treatments, and disposal facilities are some major reasons for the improper disposal of sewage and debris which releases many pollutants into rivers and oceans through runoff (Sheavly and Register 2007). Lack of proper waste management can further intensify the accumulation of pollutants compromising water quality, biodiversity, and human health (Bennett et al. 2023). Increased public awareness, proper infrastructure development for waste management, policymaking, and implementation are necessary to promote responsible waste disposal in marine environments.

1.5.10 Community Engagement

Since most of the coastal communities rely on industries that contribute heavily to coastal and marine pollution, convincing and engaging them in pollution control efforts can be quite a challenge. Lack of awareness of the impacts of pollution on coastal environments is another reason for inadequate community engagement (Nugraha 2023). Thus, pollution control efforts should target empowering communities through capacity building, improved awareness, and shared partnerships. Incorporating traditional knowledge, promoting dialogue, and inclusion can be productive in enhancing community engagement to actively participate in coastal and marine pollution control (Crain et al. 2009).

1.5.11 Enforcement and Legal Frameworks

Law enforcement in relation to coastal and marine pollution often meets several challenges that hinder and compromise effective pollution control implementation. Limited financial and technological resources, lack of political interest, and competing interests make the enforcement of the existing legal framework inadequate (Islam and Tanaka 2004; Sarwar et al. 2021). In addition, existing legal loopholes, corruption, and issues in jurisdiction can compromise the enforcement of many laws, especially in developing nations (Tiquio et al. 2017). Strong enforcement mechanisms, transparent governance, and active stakeholder participation are necessary to achieve effective legal frameworks along with strengthened capacities, enhanced coordination amongst regulatory institutions, and public accountability (Crain et al. 2009; Wowk 2013).

1.5.12 Emerging Pollutants

Emerging pollutants refer to a diverse group of persistent or pseudo-persistent organic and inorganic chemicals, typically occurring in very low concentrations with potential impacts on the health of organisms including humans, fish, and wildlife, along with other adverse effects on the ecosystems (Jones and Gomes 2013; Mostofa et al. 2013). These include a range of pollutants such as microplastics, pharmaceuticals such as painkillers, personal care products (PCPs) such as cosmetics, endocrine-disrupting compounds (EDCs), steroids and hormones, sunscreens/UV filters, surfactants, antiseptics, among others (Mostofa et al. 2013; Reichelt-Brushett 2023). These enter marine environments via different pathways including urban and agricultural runoff, atmospheric deposition, and natural productions like algal blooms, and have impacts on both ecosystems and organisms, which are not yet fully understood (Jones and Gomes 2013). These compounds can get bioaccumulated in marine organisms and have the potential to cause hormonal imbalances, affecting their health and reproduction (Cabral et al. 2019). Emerging pollutants like microplastics may pose numerous impacts including physical harm and leach of chemical additives in organisms and humans which are yet to be understood fully. Small size and the ubiquitous nature of microplastics along with their chemical properties allow them to sorb and accumulate other pollutants which can have synergistic adverse effects on the ecosystems as well as on the organisms (Vikas and Dwarakish 2015; Cabral et al. 2019). Furthermore, the presence of emerging pollutants in soil and groundwater may have potential new forms of land contamination which can ultimately affect the marine environment via runoff (Jones and Gomes 2013). Addressing the issues of emerging pollutants demands comprehensive research to facilitate a better understanding of their sources, transport pathways, and impacts on marine and coastal ecosystems (Micella et al. 2024).

1.6 Summary

This chapter has provided an overview of coastal and marine pollution along with their sources, impacts, and challenges. It highlights the definitions and concerns of coastal and marine pollution, many sources of coastal and marine pollution and their pathways, impacts on coastal and marine ecosystems as well as human beings and mitigation methods, and finally the challenges surrounding them. Coastal and marine pollution is mainly an anthropogenic process, with most pollutants originating from land-based sources. Numerous sources of coastal and marine pollution include industrial processes, agricultural and urban runoff, sewage and wastewater, oil spills and maritime transport, and plastic pollution. These result in a variety of impacts of varying degrees to both ecosystems and humans, such as harm to marine life, economic losses, health risks to humans, coastal and infrastructure damages, and climate change amplification. Prevention and mitigation of these impacts face numerous challenges like lack of global regulations, gaps in knowledge, data and monitoring, complex and diffuse nature of many pollutant sources, lack of financial and technological resources, climate change interactions, and emerging contaminants. The prevention and control of coastal and marine pollution therefore requires establishing more holistic and comprehensive global frameworks, strict implementation of new and stronger policies, raising awareness, research and innovation, sustainable and greener economies and lifestyles, and integrated management systems, along with global and regional coordination.

References

Aakko-Saksa, P.T., Lehtoranta, K., Kuittinen, N. et al. (2023). Reduction in greenhouse gas and other emissions from ship engines: current trends and future options. *Progress in Energy and Combustion Science* 94: 101055. https://doi.org/10.1016/j.pecs.2022.101055.

Agamuthu, P., Mehran, S.B., Norkhairah, A., and Norkhairiyah, A. (2019). Marine debris: a review of impacts and global initiatives. *Waste Management and Research* 37: 987–1002. https://doi.org/10.1177/0734242X19845041.

Akankali, J.A. and Elenwo, E.I. (2015). Sources of marine pollution on Nigerian coastal resources: an overview. *Open Journal of Marine Science* 5: 226–236. https://doi.org/10.4236/ojms.2015.52018.

Alava, J.J., Cheung, W.W., Ross, P.S., and Sumaila, U.R. (2017). Climate change–contaminant interactions in marine food webs: toward a conceptual framework. *Global Change Biology* 23: 3984–4001. https://doi.org/10.1111/gcb.13667.

Amalan, K., Ratnayake, A.S., Ratnayake, N.P. et al. (2018). Influence of nearshore sediment dynamics on the distribution of heavy mineral placer deposits in Sri Lanka. *Environmental Earth Sciences* 77: 737. https://doi.org/10.1007/s12665-018-7914-4.

Andersson, K., Baldi, F., Brynolf, S. et al. (2016). Shipping and the environment. In: *Shipping and the Environment* (ed. K. Andersson, S. Brynolf, J.F. Lindgren, and M. Wilewski-Bien), 3–27. Berlin: Springer-Verlag https://doi.org/10.1007/978-3-662-49045-7_1.

Ansari, Z.A. and Matondkar, S.G.P. (2014). Anthropogenic activities including pollution and contamination of coastal marine environment. *Journal of Ecophysiology and Occupational Health* 14: 71–78. https://doi.org/10.15512/joeoh/2014/v14i1-2/50743.

Beiras, R. (2018). *Marine Pollution: Sources, Fate and Effects of Pollutants in Coastal Ecosystems*, 1e. Amsterdam: Elsevier https://doi.org/10.1016/C2017-0-00260-4.

Bennett, N.J., Alava, J.J., Ferguson, C.E. et al. (2023). Environmental (in) justice in the Anthropocene Ocean. *Marine Policy* 147: 105383. https://doi.org/10.1016/j.marpol.2022.105383.

Bigg, M.G. (2013). The regulation of industrial pollution. In: *Pollution: Causes, Effects and Control*, 5e (ed. R.M. Harrison), 522–550. The Royal Society of Chemistry https://doi.org/10.1039/BK9781849736480-00522.

Cabral, H., Fonseca, V., Sousa, T., and Costa Leal, M. (2019). Synergistic effects of climate change and marine pollution: an overlooked interaction in coastal and estuarine areas. *International Journal of Environmental Research and Public Health* 16: 2737. https://doi.org/10.3390/ijerph16152737.

Chen, C.L. (2015). Regulation and management of marine litter. In: *Marine Anthropogenic Litter* (ed. M. Bergmann, L. Gutow, and M. Klages). Cham: Springer https://doi.org/10.1007/978-3-319-16510-3_15.

Cheung, W.W.L. and Sumaila, U.R. (2013). Paths to sustainable ocean resources. In: *Managing Ocean Environments in a Changing Climate: Sustainability and Economic Perspectives*, 1e (ed. K.J. Noone, U.R. Sumaila, and R.J. Diaz), 277–299. Burlington: Elsevier https://doi.org/10.1016/B978-0-12-407668-6.00011-2.

Cooke, S.J., Galassi, D.M., Gillanders, B.M. et al. (2022). Consequences of "natural" disasters on aquatic life and habitats. *Environmental Reviews* 31: 122–140. https://doi.org/10.1139/er-2022-0050.

Cowie, G. and Woulds, C. (2011). Tracer studies of benthic communities and biogeochemical processes in coastal and estuarine marine environments. In: *Treatise on Estuarine and Coastal Science*, 1e (ed. D. Mclusky and E. Wolanski), 39–70. San Diego: Elsevier Academic Press https://doi.org/10.1016/B978-0-12-374711-2.00403-4.

Crain, C.M., Halpern, B.S., Beck, M.W., and Kappel, C.V. (2009). Understanding and managing human threats to the coastal marine environment. *Annals of the New York Academy of Sciences* 1162: 39–62. https://doi.org/10.1111/j.1749-6632.2009.04496.x.

Critchell, K., Bauer-Civiello, A., Benham, C. et al. (2019). Plastic pollution in the coastal environment: current challenges and future solutions. In: *Coasts and Estuaries*, 1e (ed. E. Wolanski, J.W. Day, M. Elliott, and R. Ramachandran), 595–609. Elsevier https://doi.org/10.1016/B978-0-12-814003-1.00034-4.

Da Costa, J.P., Mouneyrac, C., Costa, M. et al. (2020). The role of legislation, regulatory initiatives and guidelines on the control of plastic pollution. *Frontiers in Environmental Science* 8: 104. https://doi.org/10.3389/fenvs.2020.00104.

Da Costa, J.P., Avellan, A., Mouneyrac, C. et al. (2023). Plastic additives and microplastics as emerging contaminants: mechanisms and analytical assessment. *TrAC Trends in Analytical Chemistry* 158: 116898. https://doi.org/10.1016/j.trac.2022.116898.

De Almeida, B.A. and Mostafavi, A. (2016). Resilience of infrastructure systems to sea-level rise in coastal areas: impacts, adaptation measures, and implementation challenges. *Sustainability* 8: 1115. https://doi.org/10.3390/su8111115.

Derraik, J.G. (2002). The pollution of the marine environment by plastic debris: a review. *Marine Pollution Bulletin* 44: 842–852. https://doi.org/10.1016/S0025-326X(02)00220-5.

Diaz, R.J., Eriksson-Hägg, H., and Rosenberg, R. (2013). Hypoxia. In: *Managing Ocean Environments in a Changing Climate: Sustainability and Economic Perspectives*, vol. 1 (ed. K.J. Noone, U.R. Sumaila, and R.J. Diaz), 67–96. Burlington: Elsevier https://doi.org/10.1016/B978-0-12-407668-6.00004-5.

Doney, S.C., Fabry, V.J., Feely, R.A., and Kleypas, J.A. (2009). Ocean acidification: the other CO_2 problem. *Annual Review of Marine Science* 1: 169–192. https://doi.org/10.1146/annurev.marine.010908.163834.

Donohoe, M. (2003). Causes and health consequences of environmental degradation and social injustice. *Social Science and Medicine* 56: 573–587. https://doi.org/10.1016/S0277-9536(02)00055-2.

Dotinga, H.M. and Elferink, A.G.O. (2000). Acoustic pollution in the oceans: the search for legal standards. *Ocean Development and International Law* 31: 151–182. https://doi.org/10.1080/009083200276102.

Drake, L.A., Doblin, M.A., and Dobbs, F.C. (2007). Potential microbial bioinvasions via ships' ballast water, sediment, and biofilm. *Marine Pollution Bulletin* 55: 333–341. https://doi.org/10.1016/j.marpolbul.2006.11.007.

Duraisamy, A. and Latha, S. (2011). Impact of pollution on marine environment – a case study of coastal Chennai. *Indian Journal of Science and Technology* 4: 259–262. https://doi.org/10.17485/ijst/2011/v4i3/29978.

Elliott, L. (2011). ASEAN and environmental governance: rethinking networked regionalism in Southeast Asia. *Procedia – Social and Behavioral Sciences* 14: 61–64. https://doi.org/10.1016/j.sbspro.2011.03.023.

Farrington, J.W. (2013). Oil pollution in the marine environment I: inputs, big spills, small spills, and dribbles. *Environment: Science and Policy for Sustainable Development* 55: 3–13. https://doi.org/10.1080/00139157.2013.843980.

Früh-Green, G.L., Orcutt, B.N., Rouméjon, S. et al. (2018). Magmatism, serpentinization and life: insights through drilling the Atlantis Massif (IODP Expedition 357). *Lithos* 323: 137–155. https://doi.org/10.1016/j.lithos.2018.09.012.

Galgani, F., Michela, A., Gérigny, O. et al. (2022). Marine litter, plastic, and microplastics on the seafloor. In: *Plastics and the Ocean: Origin, Characterization, Fate, and Impacts*, 1e (ed. A.L. Andrady), 151–197. Hoboken: Wiley https://doi.org/10.1002/9781119768432.ch6.

Gelcich, S., Buckley, P., Pinnegar, J.K. et al. (2014). Public awareness, concerns, and priorities about anthropogenic impacts on marine environments. *Proceedings of the National Academy of Sciences of the United States of America* 111: 15042–15047. https://doi.org/10.1073/pnas.1417344111.

Gough, A. (2017). Educating for the marine environment: challenges for schools and scientists. *Marine Pollution Bulletin* 124: 633–638. https://doi.org/10.1016/j.marpolbul.2017.06.069.

Gunawardhana, G.M.S.S., Perera, U.L.H.P., Ratnayake, A.S. et al. (2024). Formation of secondary microplastics during degradation of plastics originating from the MV X-Press Pearl maritime disaster. *Discover Environment* 2: 22. https://doi.org/10.1007/s44274-024-00044-2.

Hall, J., Diaz, R.J., Gruber, N., and Wilhelmsson, D. (2013). Impacts of multiple stressors. In: *Managing Ocean Environments in a Changing Climate: Sustainability and Economic Perspectives*,

1e (ed. K.J. Noone, U.R. Sumaila, and R.J. Diaz), 193–221. Burlington: Elsevier https://doi.org/10.1016/B978-0-12-407668-6.00008-2.

Halpern, B.S., Walbridge, S., Selkoe, K.A. et al. (2008). A global map of human impact on marine ecosystems. *Science* 319: 948. https://doi.org/10.1126/science.1149345.

Hossain, M.M. (2019). Future importance of healthy oceans: ecosystem functions and biodiversity, marine pollution, carbon sequestration, ecosystem goods and services. *Journal of Ocean and Coastal Economics* 6: 1–33. https://doi.org/10.15351/2373-8456.1103.

Howes, M., Wortley, L., Potts, R. et al. (2017). Environmental sustainability: a case of policy implementation failure? *Sustainability* 9: 165. https://doi.org/10.3390/su9020165.

Islam, M.S. and Tanaka, M. (2004). Impacts of pollution on coastal and marine ecosystems including coastal and marine fisheries and approach for management: a review and synthesis. *Marine Pollution Bulletin* 48: 624–649. https://doi.org/10.1016/j.marpolbul.2003.12.004.

Jambeck, J.R., Geyer, R., Wilcox, C. et al. (2015). Plastic waste inputs from land into the ocean. *Science* 347: 768–771. https://doi.org/10.1126/science.1260352.

Jennerjahn, T.C. (2012). Biogeochemical response of tropical coastal systems to present and past environmental change. *Earth-Science Reviews* 114: 19–41. https://doi.org/10.1016/j.earscirev.2012.04.005.

Jones, O.A.H. and Gomes, R.L. (2013). Chemical pollution of the aquatic environment by priority pollutants and its control. In: *Pollution: Causes, Effects and Control*, 5e (ed. R.M. Harrison), 1–28. The Royal Society of Chemistry https://doi.org/10.1039/BK9781849736480-00001.

Kennish, M.J. (2019). *Practical Handbook of Marine Science*, 4e, 335–457. Boca Raton: CRC Press https://doi.org/10.1201/b22246.

Knap, A.H. and Rusyn, I. (2016). Environmental exposures due to natural disasters. *Reviews on Environmental Health* 31: 89–92. https://doi.org/10.1515/reveh-2016-0010.

Lavalle, C., Rocha Gomes, C., Baranzelli, C., and Batista Silva, F. (2011). *Coastal Zones: Policy Alternatives Impacts on European Coastal Zones 2000–2050*. Luxembourg: European Commission, Joint Research Centre. Institute for Environment and Sustainability https://doi.org/https://doi.org/10.2788/93357.

Lillebø, A.I., Neto, J.M., Martins, I. et al. (2005). Management of a shallow temperate estuary to control eutrophication: the effect of hydrodynamics on the system's nutrient loading. *Estuarine, Coastal and Shelf Science* 65: 697–707. https://doi.org/10.1016/j.ecss.2005.07.009.

Lindgren, J.F., Wilewska-Bien, M., Granhag, L. et al. (2016). Discharges to the sea. In: *Shipping and the Environment* (ed. K. Andersson, S. Brynolf, J. Lindgren, and M. Wilewska-Bien), 125–168. Berlin: Springer-Verlag https://doi.org/10.1007/978-3-662-49045-7_4.

Lithner, D., Larsson, Å., and Dave, G. (2011). Environmental and health hazard ranking and assessment of plastic polymers based on chemical composition. *Science of the Total Environment* 409: 3309–3324. https://doi.org/10.1016/j.scitotenv.2011.04.038.

Macko, S.A. (2018). A perspective on marine pollution. In: *The Marine Environment and United Nations Sustainable Development Goal 14*, 1e (ed. M.H. Nordquist, J.N. Moore, and R. Long), 291–308. Leiden: Brill Nijhoff https://doi.org/10.1163/9789004366619_020.

Mayes, W.M. (2013). Water pollution biology. In: *Pollution: Causes, Effects and Control*, 5e (ed. R.M. Harrison), 80–114. The Royal Society of Chemistry https://doi.org/10.1039/BK9781849736480-00080.

Micella, I., Kroeze, C., Bak, M.P., and Strokal, M. (2024). Causes of coastal waters pollution with nutrients, chemicals and plastics worldwide. *Marine Pollution Bulletin* 198: 115902. https://doi.org/10.1016/j.marpolbul.2023.115902.

Mostofa, K.M., Liu, C.Q., Vione, D. et al. (2013). Sources, factors, mechanisms and possible solutions to pollutants in marine ecosystems. *Environmental Pollution* 182: 461–478. https://doi.org/10.1016/j.envpol.2013.08.005.

Nugraha, A. (2023). Integrated coastal management in the current regional autonomy law regime in Indonesia: context of community engagement. *Australian Journal of Maritime and Ocean Affairs* 1–20. https://doi.org/10.1080/18366503.2023.2283355.

Nunes, M. and Leston, S. (2020). Coastal pollution: an overview. In: *Life Below Water*, 1e (ed. W.L. Filho, A.M. Azul, L. Brandli, et al.), 1–11. Springer Cham https://doi.org/10.1007/978-3-319-71064-8_9-1.

Pawar, P.R., Shirgaonkar, S.S., and Patil, R.B. (2016). Plastic marine debris: sources, distribution and impacts on coastal and ocean biodiversity. *PENCIL Publication of Biological Sciences* 3: 40–54. http://www.pencilacademicpress.org/ppbs.

Perera, U.L.H.P., Subasinghe, H.C.S., Ratnayake, A.S. et al. (2022). Maritime pollution in the Indian Ocean after the MV X-Press Pearl accident. *Marine Pollution Bulletin* 182: 114301. https://doi.org/10.1016/j.marpolbul.2022.114301.

Pettipas, S., Bernier, M., and Walker, T.R. (2016). A Canadian policy framework to mitigate plastic marine pollution. *Marine Policy* 68: 117–122. https://doi.org/10.1016/j.marpol.2016.02.025.

Pilapitiya, P.G.C.N.T. and Ratnayake, A.S. (2024). The world of plastic waste: a review. *Cleaner Materials* 11: 100220. https://doi.org/10.1016/j.clema.2024.100220.

Potters, G. (2013). *Marine Pollution*, 1e. London: Bookboon.

Preston, M.R. (2013). Chemistry and pollution of the marine environment. In: *Pollution: Causes, Effects and Control*, 5e (ed. R.M. Harrison), 29–59. The Royal Society of Chemistry https://doi.org/10.1039/BK9781849736480-00029.

Ramanathan, V. and Carmichael, G. (2008). Global and regional climate changes due to black carbon. *Nature Geoscience* 1: 221–227. https://doi.org/10.1038/ngeo156.

Ratnayake, A.S. (2021a). Late Cretaceous to Miocene paleoclimatic changes in the Indian Ocean: insights from the deepwater Mannar Basin, Sri Lanka. *Geo-Marine Letters* 41: 37. https://doi.org/10.1007/s00367-021-00710-x.

Ratnayake, A.S. (2021b). Paleoenvironments and source rock potential of Dorado North well in the Mannar Basin (Indian Ocean). *Arabian Journal of Geosciences* 14: 792. https://doi.org/10.1007/s12517-021-07127-x.

Ratnayake, A.S. and Perera, U.L.H.P. (2022). Coastal zone management in Sri Lanka: a lesson after recent naval accidents. *Marine Pollution Bulletin* 182: 113994. https://doi.org/10.1016/j.marpolbul.2022.113994.

Ratnayake, A.S., Kularathne, C.W., and Sampei, Y. (2018). Assessment of hydrocarbon generation potential and thermal maturity of the offshore Mannar Basin, Sri Lanka. *The Journal of Petroleum Exploration and Production Technology* 8: 641–654. https://doi.org/10.1007/s13202-017-0408-1.

Ratnayake, A.S., Wijewardhana, T.D.U., Haraguchi, T. et al. (2023). Sedimentological observations and geochemical characteristics of paleo-tsunami deposits along the east coast of Sri Lanka in the Indian Ocean. *Quaternary International* 661: 49–59. https://doi.org/10.1016/j.quaint.2023.02.015.

Reichelt-Brushett, A. (2023). Marine pollution in context. In: *Marine Pollution–Monitoring, Management and Mitigation*, 1e (ed. A. Reichelt-Brushett), 1–22. Springer Cham https://doi.org/10.1007/978-3-031-10127-4_1.

Riechers, M., Brunner, B.P., Dajka, J.C. et al. (2021). Leverage points for addressing marine and coastal pollution: a review. *Marine Pollution Bulletin* 167: 112263. https://doi.org/10.1016/j.marpolbul.2021.112263.

Saengsupavanich, C., Ratnayake, A.S., Yun, L.S., and Ariffin, E.H. (2023). Current challenges in coastal erosion management for southern Asian regions: examples from Thailand, Malaysia, and Sri Lanka. *Anthropocene Coasts* 6: 15. https://doi.org/10.1007/s44218-023-00030-w.

Sarwar, F., Ali, S., Bhatti, S.H., and ur Rehman, S. (2021). Legal approaches to reduce plastic marine pollution: challenges and global governance. *Annals of Social Sciences and Perspective* 2: 15–20. https://doi.org/10.52700/assap.v2i1.32.

Sheavly, S.B. and Register, K.M. (2007). Marine debris and plastics: environmental concerns, sources, impacts and solutions. *Journal of Polymers and the Environment* 15: 301–305. https://doi.org/10.1007/s10924-007-0074-3.

Shine, K.P. (2013). Climate change. In: *Pollution: Causes, Effects and Control*, 5e (ed. R.M. Harrison), 326–339. The Royal Society of Chemistry https://doi.org/10.1039/BK9781849736480-00326.

Sindermann, C.J. (1995). *Ocean Pollution: Effects on Living Resources and Humans*, 1e, 1–304. Boca Raton: CRC Press https://doi.org/10.1201/b15709.

Tapia, J., Villagra, F., Bertrán, C. et al. (2019). Effect of the earthquake-tsunami (Chile, 2010) on toxic metal content in the Chilean abalone mollusc *Concholepas*. *Ecotoxicology and Environmental Safety* 169: 418–424. https://doi.org/10.1016/j.ecoenv.2018.11.040.

Tiquio, M.G.J.P., Marmier, N., and Francour, P. (2017). Management frameworks for coastal and marine pollution in the European and South East Asian regions. *Ocean and Coastal Management* 135: 65–78. https://doi.org/10.1016/j.ocecoaman.2016.11.003.

Tomczak, M. (1984). Defining marine pollution. *Marine Policy* 8: 311–322. https://doi.org/10.1016/0308-597X(84)90023-X.

United Nations Environment Programme (1982). Marine pollution. *UNEP Regional Seas Reports and Studies No. 25*. https://wedocs.unep.org/20.500.11822/25355

Valencia, M.J. and Akimoto, K. (2006). Guidelines for navigation and overflight in the exclusive economic zone. *Marine Policy* 30: 704–711. https://doi.org/10.1016/j.marpol.2005.11.002.

Vegter, A.C., Barletta, M., Beck, C. et al. (2014). Global research priorities to mitigate plastic pollution impacts on marine wildlife. *Endangered Species Research* 25: 225–247. https://doi.org/10.3354/esr00623.

Verma, J., Pant, H., Sing, S., and Tiwari, A. (2020). Three major dimensions of life: environment, agriculture and health. In: *Marine Pollution, Sources, Effect and Management* (ed. H. Pant), 270–276. Society of Biological Sciences and Rural Development.

Vikas, M. and Dwarakish, G.S. (2015). Coastal pollution: a review. *Aquatic Procedia* 4: 381–388. https://doi.org/10.1016/j.aqpro.2015.02.051.

Weis, J.S. (2015). *Marine Pollution: What Everyone Needs to Know*, 1e. New York: Oxford University Press https://doi.org/10.1093/wentk/9780199996698.001.0001.

Werner, F. and Blanton, J.O. (2019). Coastal circulation models. In: *Encyclopedia of Ocean Sciences*, 3e (ed. J.K. Cochran, H.J. Bokuniewicz, and P.L. Yager), 467–474. Elsevier Academic Press https://doi.org/10.1016/B978-0-12-409548-9.11412-5.

Wilcox, C., Hardesty, B.D., and Law, K.L. (2019). Abundance of floating plastic particles is increasing in the Western North Atlantic Ocean. *Environmental Science and Technology* 54: 790–796. https://doi.org/10.1021/acs.est.9b04812.

Wilhelmsson, D., Thompson, R.C., Holmström, K. et al. (2013). Marine pollution. In: *Managing Ocean Environments in a Changing Climate: Sustainability and Economic Perspectives*, 1e (ed. K.J. Noone, U.R. Sumaila, and R.J. Diaz), 127–169. Burlington: Elsevier https://doi.org/10.1016/B978-0-12-407668-6.00006-9.

Willis, K.A., Serra-Gonçalves, C., Richardson, K. et al. (2021). Cleaner seas: reducing marine pollution. *Reviews in Fish Biology and Fisheries* 32: 145–160. https://doi.org/10.1007/s11160-021-09674-8.

Windom, H.L. (1992). Contamination of the marine environment from land-based sources. *Marine Pollution Bulletin* 25: 32–36. https://doi.org/10.1016/0025-326X(92)90180-E.

Wowk, K.M. (2013). Paths to sustainable ocean resources. In: *Managing Ocean Environments in a Changing Climate: Sustainability and Economic Perspectives*, 1e (ed. K.J. Noone, U.R. Sumaila, and R.J. Diaz), 301–348. Burlington: Elsevier https://doi.org/10.1016/B978-0-12-407668-6.00012-4.

Xie, G.Z., Zhang, L.P., Li, C.Y., and Sun, W.D. (2023). Accelerated methane emission from permafrost regions since the 20th century. *Deep Sea Research Part I: Oceanographic Research Papers* 195, 103981. https://doi.org/10.1016/j.dsr.2023.103981

Yoshioka, G.A., Franzoni, A.J., Kooyoomjian, K.J. et al. (1985). Patterns and trends in reported small oil spills. *International Oil Spill Conference Proceedings*, 141–148. https://doi.org/10.7901/2169-3358-1985-1-141

Zhang, B., Matchinski, E.J., Chen, B. et al. (2019). Marine oil spills-oil pollution, sources and effects. In: *World Seas: An Environmental Evaluation*, 2e (ed. C. Sheppard), 391–406. London: Elsevier Academic Press https://doi.org/10.1016/B978-0-12-805052-1.00024-3.

2

Pollution from Land-Based Sources: Industrial and Urban Runoff

Mihiri Indunil Gunasekara[1], *Ishani Mahawaththa*[2], *Dilani Madhubhashini*[3], *and Kaushani Amarasena*[4]

[1] *School of Civil and Environmental Engineering, Faculty of Engineering, Queensland University of Technology, Brisbane, QLD, Australia*
[2] *Department of Biological Sciences, University of North Texas, Denton, TX, 76201, USA*
[3] *School of Earth and Atmospheric Sciences, Faculty of Science, Queensland University of Technology, Brisbane, QLD, Australia*
[4] *School of Science, University of Wales Trinity Saint David, London, United Kingdom*

The world population is estimated to reach 9.7 billion by 2050. The growing population result in increased consumption of resources, waste generation, and deteriorating environmental conditions. The majority of megacities those with populations of over 2.5 million are found along coastlines. Approximately 20 billion tons of waste is dumped into the ocean annually without any kind of previous processing (Immig 2018). Many different types of contaminants and pollutants are capable of causing pollution in the marine and coastal environments. There are several methods to classify those pollutants. Three key approaches to categorizing coastal and marine pollutants were highlighted (Potters, 2013) according to their physicochemical constitution, physical state, and persistence in the environment. Based on the physicochemical constitution of the pollutants, they can be classified as organic, inorganic, and not in chemical origin. There are three types of pollutants according to their physical states: solid, liquid, and gaseous. A third group of pollutants is conservative or persistent.

2.1 Organic Marine and Coastal Pollutants

Organic pollutants are defined as carbon-based chemicals and contaminants. They originate mainly from sewage, urban runoff, industrial effluents, and agricultural wastewater. When organic pollutants decompose in a water body, the dissolved oxygen in the water may be consumed at a greater rate than it can be replenished, leading to oxygen depletion and severe effects on the aquatic ecosystem (Nageeb 2013). Pesticides, fertilizers, hydrocarbons, phenols, biphenyls, plasticizers, detergents, oils, greasers, pharmaceuticals, proteins, and carbohydrates are some examples of organic pollutants. Polyaromatic hydrocarbons, organic pesticides, polycyclic biphenyls (PCBPs), and dioxins are among the most significant organic pollutants in the marine environment (Potters 2013). Many of these substances have the properties of being lipophilic, toxic, and extremely persistent, which makes them suitable for bioaccumulation.

Coastal and Marine Pollution: Source to Sink, Mitigation and Management, First Edition.
Edited by Meththika Vithanage, Sameera M. Samarasekara, Bryan D. James, and Christopher M. Reddy.

Polycyclic aromatic hydrocarbons (PAHs) are substances that have two or more fused aromatic rings arranged in a clustered, angular, or linear structure (Juhasz and Naidu 2000). PAHs are widely distributed across the marine ecosystem, with the majority of urbanized coastal areas exhibiting the highest concentrations of these compounds due to industrial operations or human activities (Perugini et al. 2007). These pollutants are mostly introduced into the marine and coastal environments by oil spills, incomplete combustion of organic materials, and the usage of coal tar creosote. When PAHs are discharged into the ocean, they usually adsorb rapidly on suspended particles and sediments. Fish and other marine creatures can obtain these PAHs through the food chain, along with contaminated sediments and waterborne substances (Perugini et al. 2007).

The number of PAHs in marine ecosystems has increased significantly due to human activity, causing serious issues to human health and the environment. Several PAHs and their associated byproducts are regarded as highly toxic carcinogens and mutagens. Pyrogenic PAHs are a type of chronic pollutant that mainly reaches marine ecosystems by atmospheric deposition and run-off from rivers (Liu et al. 2012). Oil spills cause the release of petrogenic PAHs, which are acute hydrocarbon pollutants with aromatic components (Duran and Cravo-Laureau 2016).

Halogenated hydrocarbons are another type of organic marine pollutant consisting of hydrocarbon chains bearing one or more halogen atoms. They are highly persistent in the environment and, following biomagnification in the food chain, they will ultimately find their way into the live tissue of marine animals (Potters 2013). The most abundant halogenated hydrocarbons in the marine and coastal areas include aldrin, chlordane, dichlorodiphenyltrichloroethane (DDT), endrin, dieldrin, heptachlor, hexachlorobenzene, polychlorinated biphenyls (PCBs), mirex, toxaphene, and polychlorinated dioxins/furans. These hydrocarbons are included in the group of the Dirty Dozen, a collection of 12 substances that received priority attention in the Stockholm Convention in 2004 (Potters 2013).

The term "Persistent Organic Pollutants" (POPs) refers to harmful carbon-based substances that have damaged the environment including marine and coastal ecosystems. They can bioaccumulate in significant amounts in the tissues of marine species due to their high lipophilicity. The Southern Ocean in the Indian Pacific region acts as an environmental sink for POPs under climate conditions (Immig 2018).

2.2 Inorganic Marine and Coastal Pollutants

Inorganic pollutants are nonbiodegradable compounds, frequently sourced from residential, commercial, and industrial settings. These substances range in composition from salts like phosphates, sulfates, and nitrates to heavy metals such as arsenic, lead, and mercury. Anions and cations, radioactive materials, halides, and heavy metals are a few typical examples of inorganic substances. Since these inorganic pollutants have always been a component of marine cycles and ecosystems, they have a special role than other pollutants. Inorganic contaminants can persist longer in aquatic environments and worsen the state of the water due to their nonbiodegradability.

One of the main organic pollutants which adversely impacts the ocean biota is heavy metal contamination. Over 300 million tons of heavy metals, such as Cr, Cu, Zn, As, Cd, Pb, and Sn, from consumer and industrial items end up in the ocean annually (Bandara and Manage 2022). Heavy metals are a serious problem in marine ecosystems because of their toxicity, nonbiodegradability, biological accumulation, and carcinogenic nature. Human activities along the shore will be directly affected by the increase in the concentration of metals in coastal waters. The widely publicized instance of heavy metal poisoning included the disposal of industry effluent in Japan's Minimata

Bay between 1932 and 1968. Exposure to mercury poisoning resulted in almost 1700 human deaths, and the consequences persist even today (Ekino et al. 2007).

Inorganic nutrients, including phosphate and nitrate, that are introduced from the land have an impact on the growth of marine algae, which ultimately contributes to marine eutrophication. Furthermore, this increased algal bloom may result in toxic chemicals that degrade the health of marine life. One such instance is the Florida coast algal bloom in 2004, which produced the neurotoxic chemical brevotoxins, which harmed many species of fish (Bricelj et al. 2012).

2.3 Newer Forms of Marine and Coastal Pollutants

2.3.1 Plastics

Most types of plastics are made up of complex combinations of chemical substances, residual monomers, and polymers. Due to the unique characteristics of plastics, plastic pollutants continue to persist in ocean basins. The improper disposal of waste has resulted in five trillion pieces of plastic debris, totaling approximately 260 000 tons, drifting over the world's ocean surface (Eriksen et al. 2014). Plastics include toxic compounds that can accumulate in marine species, including metal ions, oligomers, flame retardants, bisphenol A (BPA), and monomers. Since plastics are buoyant, they are easily carried by ocean currents and transferred throughout ocean basins; as a result, their contamination can be found on even the most remote islands, from the poles to the equator, and from coastlines to the deepest sections of the sea (Wabnitz and Nichols 2010). Although most plastic waste ends up floating on the ocean's surface, there have been several reports of plastics sinking to a range of depths in the marine environment. The most surprising estimate is that "Unless the entire world takes immediate action, plastic waste would outweigh fish in the global oceans by 2050" (Immig 2018).

The deposited plastics in the ocean basins can be generally classified into four types depending on their sizes: megaplastics, macroplastics, mesoplastics, and microplastics (Thushari and Senevirathna 2020). Microplastics are either created by breaking down larger plastic items into smaller pieces or produced as tiny plastic particles to make abrasives for blasting, scrubbers for cosmetics, or resin pellets. Since they are smaller, these particles can be found extensively and have a higher chance of being consumed by marine organisms (Lee et al. 2013). In comparison to ambient seawater, POPs and other ocean pollutants can concentrate many times more in and on microplastic particles. Moreover, microplastics in aquatic environments can cause biofouling, or the growth of organisms on the surface.

Oceanic regions that accumulate plastic waste and other marine debris are known as garbage patches. The world's greatest concentration of plastic debris in the ocean is called the Great Pacific Garbage Patch (GPGP), and it is located between Hawaii and California. According to recent studies, there is evidence that the amount of plastic contamination in the GPGP, which is thought to extend over 1.6 million square kilometers, is increasing rapidly compared to the surrounding areas (Immig 2018). Other recognized patches include the North Atlantic garbage patch, the South Atlantic garbage patch, the South Pacific garbage patch, and the Indian Ocean garbage patch.

2.3.2 Noise as a Marine Pollutant

Even though noise pollution is one of the less well-known types, it nevertheless seriously affects marine life. Natural and anthropogenic activities, such as shipping, vessel traffic, high-powered sonar technologies, explosive ordnance, offshore hydrocarbon seismic investigations, drilling processes, and pile driving for offshore constructions, are the main causes of the rise in oceanic noise pollution (Chahouri et al. 2022). Over the past 50–60 years, these anthropogenic activities have caused the average noise level underwater to double every decade (Potters 2013). Under the correct circumstances, high source levels of noise pollution can spread over hundreds of kilometers. Noise pollution is one of the chronic sources of pollutants, which is difficult to control or regulate and needs to be addressed immediately to safeguard marine environmental health. The majority of marine creatures rely on sound to perform essential tasks. However, the volume of noise in the water is increasing, which makes it more and more difficult for many marine organisms to survive. From the smallest plankton to the largest whale, almost all marine species are impacted by this.

2.3.3 Invasive Species as Marine Pollutants

On a global scale, invasive species are considered one of the biggest threats to the diversity of marine and coastal ecosystems. Any species that has been unintentionally or purposefully introduced to regions outside of its native habitat is considered invasive. These species have a high rate of reproduction, which allows them to suffocate native species in their new habitat. According to IUCN (2024), there are 63 invasive species in the marine environment. Marine biological invasions can occur via several routes, including transport of species via inland waterways via navigational canals, intentional and unintentional release of exotic species in the aquarium trade, shipping through ballast water and contamination of ship hulls, the aquaculture overspill of alien species introduced for farming, and the plastic pollution transfer of invasive species attached to plastic debris. The green crab (*Carcinus maenas*), the rosy wolf snail (*Euglandina rosea*), the zebra mussel (*Dreissena polymorpha*), the killer algae (*Caulerpa taxifolia*), and the sea walnut (*Mnemiopsis leidyi*) are a few well-known examples of invasive species (Potters 2013). Marine invasive species are extremely difficult and costly to eradicate, and they can have a catastrophic effect on human health, fishing, tourism, biodiversity, and coastal development.

2.4 Climate Change and Marine and Coastal Pollution

Numerous physical and chemical aspects of ocean parameters, such as temperature, pH, salinity, oxygen content, ice cover, and currents, are altered by climate change. These alterations are additionally revising the behavior of pollutants in the marine environment, which has an impact on their potential for toxicity and distribution. Accordingly, fluctuations in the properties of seawater have been associated with changes in the fate and global circulation of chemical pollutants, their bioaccumulation, their bioavailability, and their toxicity to marine life (Cabral et al. 2019). There has been a projected 50% rise in marine heat waves during the last 30 years. The Intergovernmental Panel on Climate Change estimates that the oceans have absorbed 90% of the additional heat caused by global warming. Because of this, as the average sea temperature rises globally, any naturally occurring marine heat waves will now be more intense.

The marine ecosystem is being impacted by these changes. The loss of marine species and habitats can result from rapid increases in temperature and acidification. Sea temperature can rise quickly during marine heat waves, frequently brought on by modifications to the local wind patterns. These oceanic heat waves can seriously affect fisheries, fish stock health and abundance, and ultimately the number of fish caught.

2.5 Pollutant Transport from Land-Based Sources

The three main pathways of coastal and marine pollution are the direct discharge of garbage into the ocean, surface runoff, and pollutants released from the atmosphere. Most of the pollutants (80%) originate on land and are caused by humans. Rivers are the main source for these land-based sources. The remaining portion (20%) comes from marine-related activities such as illegal dumping and shipping (Kansuah 2020; Mihai et al. 2022).

Coastal and marine pollution happens in two ways. They are chemical pollution and trash pollution (Akankali and Elenwo 2015). Chemical pollution, also known as nutrient pollution, mostly occurs because of anthropogenic activities which lead to runoff. Marine trash contains all synthetic products such as plastic which leads to trash pollution (Katsanevakis 2008).

Mainly industrial activities, urban runoff, agricultural runoff, oil spills, improperly treated or untreated sewage and wastewater discharge, plastic pollution, airborne pollutants, shipping activities, offshore activities, improper disposal of solid waste into coastal areas, and illegal dumping of waste to the coastal areas and ocean are the main sources for marine and plastic pollution (Zutshi and Prasad 2008).

Through the direct release of several contaminants into water bodies, industrial activities play a considerable role in the contamination of the coast and marine environments. A mixture of pollutants is released into rivers and estuaries by manufacturing activities that release untreated or insufficiently treated effluent, eventually reaching coastal zones. The release of heavy metals from industrial sources is one instance. These releases, which are primarily composed of heavy metals and chemicals, have a negative impact on marine ecosystems and endanger aquatic life as well as the general health of the environment (Bashir et al. 2020). Toxic substances including mercury, lead, and cadmium that can poison coastal waters and harm marine life are frequently found in industrial effluents. Due to their persistent nature, heavy metals can build up in the tissues and sediments of marine life, which can result in bioaccumulation and biomagnification (Baby et al. 2010).

2.5.1 Industrial Runoff

Industrial runoff, a critical side of the industrial environment, embodies the discharge of water laden with diverse contaminants from industrial facilities into surrounding water bodies or drainage systems (Fredston-Hermann et al. 2016). Urban runoff, a significant component of the urban hydrological cycle, refers to the process by which precipitation interacts with impervious surfaces within built environments, leading to the rapid conveyance of surface water to drainage systems (Willis et al. 2017). Both of these encompass a spectrum of pollutants, including heavy metals, organic chemicals, nutrients, and suspended solids, derived from industrial processes such as manufacturing, mining, and chemical production, originating from anthropogenic activities and materials present in urban landscapes (Cornelissen et al. 2008).

In the context of marine pollution, both industrial and urban runoff represents a critical environmental challenge with far-reaching consequences for coastal and marine ecosystems (Edwards et al. 2001; Willis et al. 2017). Mainly these two factors play a critical role in transporting marine pollutants from land-based sources. The interaction between human activities and aquatic environments engenders the discharge of diverse pollutants into marine ecosystems, stemming primarily from industrial processes and urbanization. Industrial runoff denotes the release of contaminants from industrial facilities into nearby water bodies, while urban runoff refers to the mobilization of pollutants from urbanized landscapes into aquatic environments during precipitation events. Both forms of runoff act as conduits for a plethora of pollutants, including heavy metals, organic chemicals, nutrients, microplastics, and pathogens, which can accumulate in marine sediments, water columns, and biota, posing significant ecological risks (Fredston-Hermann et al. 2016). This comprehensive review explains the sources, pathways, and impacts of industrial and urban runoff on marine pollution, exploring the mechanisms underlying pollutant transport and transformation, as well as the ecological consequences for marine ecosystems. Furthermore, it evaluates current mitigation strategies and identifies future research needs to address the complex challenges posed by industrial and urban runoff and safeguard the health and integrity of marine environments.

Industrial runoff forms a major source of marine pollution, stemming from various industrial activities such as manufacturing, mining, chemical production, and wastewater or sewage treatment (Müller et al. 2020). These processes release numerous pollutants into marine environments, including heavy metals (e.g., mercury, lead, and cadmium), organic chemicals (e.g., polycyclic aromatic hydrocarbons, pesticides, and industrial solvents), nutrients (e.g., nitrogen, and phosphorus), and suspended solids. The discharge of untreated or inadequately treated industrial effluents directly into rivers, estuaries, and coastal waters introduces contaminants that can persist in the marine environment for extended periods, posing risks to marine organisms and ecosystems (Boesch et al. 2001). Additionally, accidental spills, leaks, and discharges from industrial facilities can exacerbate pollution levels, leading to acute and chronic ecological impacts on marine habitats, biodiversity, and ecosystem functioning. Understanding the sources and composition of industrial runoff is essential for devising effective pollution control measures and mitigating its adverse effects on marine environments (Fabricius 2005).

2.5.2 Urban Runoff

Equally, urban runoff plays a significant role in marine pollution by mobilizing pollutants from urbanized landscapes into adjacent aquatic ecosystems. As rainwater and snowmelt traverse impervious surfaces such as roads, rooftops, parking lots, and sidewalks, they accumulate various contaminants, including oil, grease, heavy metals, pesticides, fertilizers, road salts, and litter (Verma et al. 2020). These pollutants are then conveyed through stormwater drainage systems, and ultimately discharged into rivers, estuaries, and coastal waters. The widespread urbanization and expansion of impermeable surfaces exacerbate the volume and velocity of urban runoff, intensifying its impact on marine environments (Boesch et al. 2001). For instance, Willis et al. (2017) reported marine pollution directly associated with the runoff, population density of the surrounding catchment, and drainage systems connected to the sea. Furthermore, urban runoff can exacerbate flooding, erosion, and sedimentation in coastal areas, exacerbating habitat degradation and compromising water quality. Addressing the sources and pathways of urban runoff is crucial for implementing sustainable stormwater management practices and reducing the influx of pollutants into marine ecosystems (Müller et al. 2020).

The usage of fertilizers in cities increases the amount of nutrients that runoff contains. The overabundance of nutrients, especially nitrogen and phosphorus, can cause nutrient pollution in coastal waters, which can cause aquatic ecosystems to experience oxygen depletion and algal blooms (Ayala 2021). The sediments from construction sites and degraded surfaces are frequently carried by urban runoff. By decreasing light penetration and changing the nature of the sediment, excessive sedimentation can destroy marine life, suffocate benthic ecosystems, and deteriorate the quality of the water (Roberts 2012).

2.5.3 Agricultural Runoff

Agriculture runoff is one of the major causes of ocean and marine pollution. Agricultural activities are essential for food production but can also be a significant source of marine pollution. The use of fertilizers, pesticides, and herbicides in farming contributes to runoff that carries these chemicals into rivers and estuaries (Micella et al. 2024). Excessive nutrient runoff, particularly nitrogen and phosphorus, can lead to harmful algal blooms, oxygen depletion, and "dead zones" in coastal areas (Ito et al. 2024). Erosion from agricultural fields contributes to the transport of sediment into water bodies through runoff. Sedimentation can degrade water quality, clouding the water and reducing light penetration. This runoff also can carry microbial contaminants such as bacteria and pathogens from animal waste posing a risk to marine organisms. Animal feces from factory farms are sometimes kept in enormous, open-air lagoons that are prone to spills and leaks. These lagoons are frequently as large as multiple football fields. For example, over 110 000 fish were killed when 200 000 gal of manure from an Illinois hog farm flowed into a waterway in 2011. Farmers frequently choose to apply manure to nearby regions when lagoons are full rather than paying to have the trash removed off-site. The United States Department of Agriculture (USDA) claims that when animal feces are over-applied to land, it can taint water sources and release toxic chemicals into the environment (FAIRR 2019).

The literature underscores the pivotal role of runoff in facilitating the transportation of diverse pollutants from terrestrial sources to marine ecosystems (Willis et al. 2017). Understanding the dynamics of this process necessitates the identification and examination of influential factors associated with industrial, urban, and agricultural runoff (Brodie et al. 2012). Anthropogenic activities, catchment characteristics, and climatic variables emerge as key determinants influencing the fate and transport of pollutants from land-based sources to the sea. Figure 2.1 illustrates the factors affecting the transport of pollutants from land-based sources.

Anthropogenic activities, encompassing industrial processes, urban development, and agricultural practices, significantly contribute to the introduction of pollutants into terrestrial environments. Industrial operations generate various contaminants, including heavy metals, organic compounds, and industrial chemicals, which can be washed into water bodies during rainfall events (Reuer and Weiss 2002). Urban areas act as concentrated sources of pollutants such as sediments, nutrients, heavy metals, and hydrocarbons, primarily through surface runoff from impermeable surfaces and sewage systems. Likewise, agricultural activities introduce pollutants such as fertilizers, pesticides, and sediments into water bodies through runoff from cultivated fields and livestock operations. The types of chemicals that will be discharged from a certain point are determined by the land use patterns of the surroundings (Brodie et al. 2012). The types of pollutants in industrial and commercial sources differ from residential sources. For instance, Zhang et al. (2022) identified the highest number of microplastics recorded in agricultural spoils than other land use patterns. He emphasized the abundance of microplastics determined by the different land use patterns.

Catchment characteristics, comprising topography, soil properties, land use/land cover, and drainage patterns, exert considerable influence on the transport pathways and retention of pollutants within terrestrial environments. Topographical features, such as slope steepness and aspect, influence the velocity and volume of runoff, thereby affecting the transport capacity of pollutants. Soil properties, including texture, permeability, and organic matter content, govern infiltration rates and runoff generation, influencing the potential for pollutant mobilization and retention (Müller et al. 2020). Keller et al. (2010) investigated the potential correlation between geological factors, specifically elevation, and the accumulation of marine litter along the US

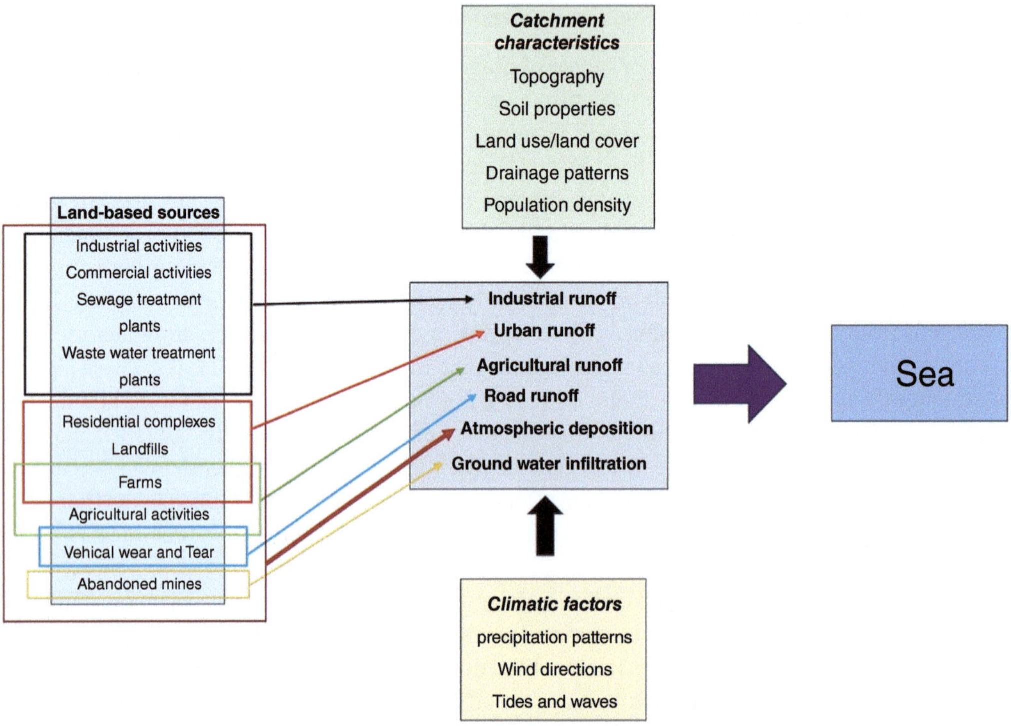

Figure 2.1 Factors exerting influence on the transport of pollutants from terrestrial origins to the oceanic environment.

West Coast. Utilizing geospatial analysis and statistical methods, data pertaining to elevation and marine litter accumulation were collected and analyzed. Results suggest a significant correlation between elevation and the presence of excessive marine litter, highlighting the influence of geological features on coastal pollution dynamics. Findings from this research contribute to a better understanding of the complex interactions between natural geomorphological characteristics and anthropogenic pollution in coastal environments. Land use/land cover types determine the extent and intensity of pollutant sources, with urban and agricultural areas typically exhibiting higher pollutant loading compared to natural or forested landscapes. Drainage patterns, including stream networks and wetland areas, serve as conduits or sinks for pollutant transport, modulating the dispersion and deposition of contaminants within the landscape.

Climatic factors, encompassing precipitation patterns, intensity, duration, and frequency, play a crucial role in regulating the timing and magnitude of runoff events, thereby influencing the mobilization and transport of pollutants from land-based sources to marine environments (Hou and Zhan 2023). Heavy precipitation events can lead to increased surface runoff and erosion, resulting in elevated pollutant loads entering water bodies. Similarly, extended periods of dry weather can enhance the accumulation of pollutants on impermeable surfaces, which are subsequently washed into aquatic systems during subsequent rainfall events. Therefore, rainfall is considered the main influential factor among the meteorological factors. Axelsson and van Sebille (2017) identified that intensive rainfall events can increase the pollution loads transported to oceans (Keller et al. 2010). Moreover, changes in climate patterns, such as alterations in precipitation regimes or increased frequency of extreme weather events, may exacerbate the risks associated with pollutant transport and contamination of marine ecosystems.

Population density of the catchments plays a crucial role in the abundance of the pollutants and that may contribute to the accumulation of the myriad of pollutants in the ocean. The area, slope, elevation, and geology of the catchments also influence the runoff (Müller et al. 2020). According to the survey, studies carried out associated with the coastal areas of Australia highlight the positive correlation between land-based sources on marine pollution (Willis et al. 2017).

The transport and fate of pollutants in industrial and urban runoff are governed by complex physical, chemical, and biological processes that influence their distribution, transformation, and persistence in marine environments. Once discharged into water bodies, pollutants can undergo various fate processes, including dilution, dispersion, adsorption, absorption, volatilization, photolysis, biodegradation, and bioaccumulation, biomagnification which determine their environmental fate and toxicity (Hale et al. 2024). For instance, reabsorbs can absorb sediment particles, accumulate in benthic habitats, and bioaccumulate in marine organisms, leading to chronic exposure and adverse effects on aquatic biota (Latimer and Zheng 2003). Similarly, organic chemicals such as pesticides and PAHs can undergo degradation and transformation through microbial metabolism, photolysis, and chemical reactions, altering their toxicity and bioavailability in marine ecosystems. Most ubiquity pollutant microplastics can act as a passage agent to transport diverse chemicals by absorbing or adsorbing them (Müller et al. 2020). Additionally, transport and fate of pollutants from inland waterways to marine environments are influenced by a myriad of interconnected factors (Alimi et al. 2018).

Hydrological characteristics, such as river discharge and flow velocity, dictate the movement of pollutants downstream, with high-flow events mobilizing sediment-bound contaminants and low-flow periods concentrating pollutants in water bodies. Topographic features and geomorphological characteristics, including slope and riverbed composition, determine the pathways and rates of pollutant transport, shaping erosional and depositional processes (Sheppard 1993). Anthropogenic activities, such as agriculture, urbanization, and industrial development, introduce various contaminants into inland waterways through point and nonpoint sources, affecting their transport to marine environments (Beyer and Biziuk 2009). Climate and weather patterns, including precipitation and storm events, play a significant role in pollutant transport dynamics, with extreme weather events leading to sudden influxes of pollutants into waterways (Alves et al. 2021). Hydrodynamic processes, such as currents and tides, influence the movement and distribution of pollutants within water bodies, affecting their dilution, settling, or suspension (Zheng et al. 2011). Biological interactions, including bioaccumulation and biodegradation, can modify the fate of pollutants in inland waterways and marine environments, altering their concentrations and toxicity levels (Cai et al. 2023). Understanding the behavior and fate of pollutants in marine environments is essential for assessing their ecological risks and developing effective management strategies to mitigate their impacts on marine organisms and ecosystems.

2.6 Sources of Marine Pollution

The major environmentally devastating source of marine pollution is oil spills. In recent times, these can result from accidents during offshore drilling, tanker discharges, or deliberate illegal dumping of oil (Abou Samra and Ali 2024). Oil spills create a physical coating on the surface of the water, forming an oil slick that can extend over large areas. Liquid petroleum hydrocarbon contaminates the water in an oil spill, and the effects of spills extend far beyond their original site. They have immediate and long-term effects on marine ecosystems, causing harm to marine life, disrupting food chains, and damaging habitats (Tedesco et al. 2024). Crude oil and its components contain toxic substances such as PAHs that can have immediate toxic effects on marine organisms (Pal and Sen 2024). The 1989

Exxon Valdez oil disaster in Alaska's Prince William Sound covered almost a thousand miles of pristine shoreline and killed hundreds of millions of fish, birds, and other creatures. Despite massive clean-up efforts, crude oil is still present under the sand and rocks on the beaches in southwest Alaska many years after the catastrophe, and the damage the spills caused to numerous local species is still evident. Billions of dollars have also been spent on cleanup efforts (SLYCAN 2018).

Improperly treated or untreated sewage and wastewater discharge pose a significant threat to marine ecosystems. This is a serious concern as the widespread practice of dumping sewage without primary treatment, can carry significant amounts of human organic waste products such as feces, which seriously contaminate the coastal environment and beach sand (Victoria et al. 2022). They also contain high concentrations of suspended solids and nutrients. Municipalities and industries contribute to this issue by releasing effluents containing pathogens, nutrients, and harmful chemicals such as pharmaceuticals, heavy metals, and industrial chemicals into rivers and coastal waters. The introduction of these pollutants can lead to waterborne diseases, algal blooms, and overall degradation of water quality (Ben-Haddad et al. 2024).

The prevalent issue of plastic pollution has garnered global attention due to its widespread impact on marine ecosystems (Jambeck et al. 2015). Improper disposal and mismanagement of plastic waste result in plastic debris entering oceans and coastal areas. Marine life is directly threatened by plastic trash, which takes the form of bottles, bags, packaging, and microplastic particles (Thacharodi et al. 2024). This can have severe consequences for marine life, as animals may consume or become entangled in plastic, leading to injury or death. Also, plastics have the ability to release dangerous chemicals into the water around them, polluting the environment and potentially harming marine ecosystems. The health of marine species may be impacted by water contamination caused by plastic additives and POPs (Jambeck et al. 2015).

Airborne pollutants, including heavy metals and POPs, can be deposited into oceans and coastal regions through atmospheric processes (Hellar-Kihampa 2024). This form of pollution, known as atmospheric deposition, can have far-reaching effects on marine ecosystems, affecting both water and sediment qualities. Some chemical molecules can travel great distances via the air, including pesticides and PCBs. Rainfall or atmospheric processes can then dump these pollutants into marine and coastal ecosystems. POPs have the ability to pollute aquatic environments once they are in the water, harming marine life and possibly making their way up the food chain (Sardenne et al. 2024). In addition to that, through atmospheric deposition, nitrogen oxides and other nitrogen-containing substances in the air can end up in coastal seas (Valigura et al. 1996). This may have an impact on the marine environment's nutrient loading and subsequent eutrophication. The overabundance of nutrients can cause algal blooms, lower oxygen levels, and harm coastal ecosystems (Qi et al. 2013). Acid rain is another method of harming the coastal and marine environment. Acid rain can be caused by airborne pollutants such as nitrogen oxides and sulfur dioxide, which can cause acid deposition. The pH of coastal waters can be lowered by acid rain, which can have an impact on marine life including corals, shellfish, and other calcifying species. Marine organisms, especially those with skeletons or shells made of calcium carbonate, can have growth and development disruptions as a result of acidification (Vong et al. 1988). The wind has the ability to carry fine particulate matter from natural sources, transportation, and industrial processes and deposit it in maritime habitats. This may lead to increased sedimentation and turbidity, which may impair light penetration and have an adverse effect on coral reef health and benthic creature health (Vong et al. 1988).

Shipping activities are another major concern that contribute to ocean and marine pollution. Shipping operations contribute to marine pollution through various mechanisms (Bello 2022). Accidental spills of oil and other hazardous materials during transportation, as well as the intentional discharge of ballast water containing invasive species, are significant concerns. These activities can introduce pollutants and disrupt local ecosystems. Indirectly, pollutants including sulfur oxides (SOx), nitrogen oxides (NOx), and particulate matter are released by ships' exhaust, which leads to air pollution and atmospheric deposition onto coastal seas. And through vessel traffic, shipping operations produce underwater noise that may interfere with fish and marine mammals' ability to communicate and navigate (Erbe et al. 2019). Improper disposal of solid waste, including hazardous materials, from ships can lead to pollution in coastal waters and damage to marine habitats.

Offshore activities, including drilling, exploration, and extraction of resources, can lead to the release of pollutants into marine environments. The discharge of chemicals used in drilling fluids, as well as accidental oil spills from offshore platforms, poses significant threats to marine ecosystems. Large volumes of crude oil were accidentally spilled into marine habitats during offshore drilling operations, causing both short-term and long-term ecological damages. In addition to that, underwater noise produced by seismic surveying and drilling equipment can disrupt fish, marine animals, and other aquatic life (Weilgart 2007). Chemicals used in production and drilling operations, including biocides, corrosion inhibitors, and lubricants, can leak into water and sediments causing severe pollution in marine and coastal environments. Physical disturbance of the seafloor and related habitats, such as coral reefs and benthic ecosystems, may result from offshore drilling operations and can lead to severe pollution.

Improper disposal of solid waste in coastal areas is a direct and visible form of pollution. This can include the dumping of industrial waste, construction debris, and household garbage directly into rivers and coastal waters, leading to habitat degradation and water contamination. Solid trash can physically harm coastal environments, such as mangroves and sea grasses, reducing the biodiversity and functionality of these ecosystems. Large garbage items, such as tires or abandoned vessels, are especially dangerous in this regard (Kelleway et al. 2017). This also led to microplastic pollution and hazardous elements found in certain solid waste products have the potential to contaminate coastal waters with chemicals. Also, improperly disposed waste can contaminate coastal areas with pathogens and pollutants. Historically, garbage from industries and chemicals, radioactive materials, rubbish, weapons, sewage sludge, and contaminated dredging material have all been disposed of in the ocean by communities worldwide. The detrimental effects of garbage disposal on the maritime ecosystem received little attention. The possibility of recycling or reusing such materials received even less attention (Kumartasli and Avinc 2020). Because it was believed that marine waters could mix and disseminate contaminants indefinitely, wastes were often deposited in coastal and ocean waterways. Some oceanic regions were contaminated with large amounts of dangerous contaminants, such as heavy metals, inorganic fertilizers, and chlorinated petrochemicals, after decades of unchecked dumping. Certain ocean waters had significant oxygen depletion as a result of unregulated ocean dumping (Kumar et al. 2021).

Illegal dumping of waste, including chemicals and other pollutants, is an undercover source of marine pollution. Garbage, including sewage disposal in the ocean, has long been seen to be the least expensive and most convenient method of getting rid of trash. Individuals or entities engaging in illegal activities may discharge hazardous materials directly into rivers or coastal waters, evading proper waste disposal procedures (Kindt 1984; Weldemichael 2012; Ryan 2023).

2.7 Effects Related to Marine Pollution

According to the literature, it has been specified that all above-discussed land-based sources have become a burden to the marine environment and have caused huge damage. It is necessary to identify the impact of these land-based pollutants to mitigate the risk regarding. For instance, the issue of coral reef die-off resulting from pollution originating from land has been overshadowed since the 1998 ocean warming events that were associated with global warming (DeGeorges et al. 2010). In Southeast Asia, land-based pollution has had a catastrophic effect on the nearshore fringing reef. Low species diversity, low live coral cover, high coral mortality, and restricted coral morphological composition are the effects that are seen in the transect data. Net reef erosion is putting the polluted reefs in grave danger from a carbonate budget perspective (Edinger and Risk 2013). The ecosystem experiences trophic changes due to nutrient enrichment, increased turbidity, and sedimentation, particularly of organic-rich suspended matter. Additionally, coral recruitment and diversity are decreased, and corals are replaced by organisms that compete with corals for space and resources, such as macroalgae, filter feeders, and coral-eating crown-of-thorns starfish outbreaks occur more frequently (Brodie et al. 2012). Coral reef fish survival and settlement success may be negatively impacted by rising suspended sediment levels in coral reef habitats. The disturbance of habitat choice brought on by sedimentation could exacerbate the impacts of habitat loss on coral reefs (Wenger et al. 2011). Lateritic soil sediments and the surrounding water can combine to create chemically changed conditions that are unfavorable for fish survival (O'Connor et al. 2016). Both directly by sediment ingesting and indirectly through the food chain, toxic substances bioaccumulate in fish and other apex predators (Islam and Tanaka 2004). Organic matter found in suspended sediment may act as an energy source for coral growth, with a maximum continuous concentration of 10–20 $ml{-}^{1}$ or a depositional rate of 200–1000 $mg/cm^2/d$ possible. Coral development rates are slowed down by smothering and lower light levels when chronic concentrations or depositional rates exceed these thresholds (Larsen and Webb 2009). In individual corals, reduced growth rates, reduced calcification, and increased mortality; in communities, reductions in species richness and coral cover; and in ecosystems, net productivity and accretion rates can be found as the specific stress responses associated with sedimentation (Wear and Thurber 2015). High siliciclastic content in reef sediment, high insoluble residue content in coral skeletons, particularly open-lattice species such as massive *Porites*, high Ba/Ca ratio in coral skeletons, low recruitment rates of corals, often resulting in size distributions skewed to large colonies, low species diversity of corals, relative to reference sites in same region, low percentage live coral cover, combined with high cover of dead coral (high mortality index), predominance of stress-tolerant coral species and growth forms, especially multilobate massive colonies and high-extension, low-density skeletal growth are the features that take place in the sedimentation-prone reefs with or without stress factors (Risk and Edinger 2011).

The impacts of short-term (up to 96 hours) exposure to low concentrations of insecticides, including endosulfan, chlorpyrifos, and profenofos, on different life history stages of several branching coral species, like *Acropora millepora* impacted its photosynthetic efficiency and/or zooxanthellae density within mature branches (van Dam et al. 2011). Pesticide exposure can have serious physiological effects, which are particularly alarming for species that are site attached and may not be able to escape contaminated areas. Although little research has concentrated on coral reef species, there is growing confirmation that pesticides can interfere with fish reproduction (Wenger et al. 2015). The majority of marine species go through a microscopic, planktonic stage in their growth, and this is when populations may be most vulnerable to the harmful effects of nitrogen and phosphorus (Baker et al. 2014). Due to nitrogen and phosphorus from the industrial discharge, phytoplankton blooms leading to hypoxia, jellyfish blooms and macroalgae blooms that can outcompete seagrasses and corals, can occur (Todd et al. 2010).

The amount of evidence linking plastic to toxicity has increased recently. Although the physical effects of consuming plastic can lead to direct toxicity, the release of hazardous substances from plastics can cause indirect toxicity. The majority of plastic additives can be discharged into the environment and made available to organisms because they are physically rather than chemically attached to the plastic (Hermabessiere et al. 2017). Microplastics are highly bioavailable and easily eaten by aquatic species, including green sea turtles, seabirds, mollusks, echinoderms, corals, and larval stages of vertebrates, and this has a negative impact on their health because of their small size (Häder et al. 2020). Moreover, microplastics may not only affect species at the organism level; it also has the capacity to modify population structure; for instance, it is an important oviposition resource for pelagic insects (Wright et al. 2013). Microplastics (polyethylene spheres, 1–5 μm in diameter) have the ability to reduce acetylcholinesterase activity in juveniles (0+ group, 1.0–1.2 cm long) of the common goby *Pomatoschistus microps* together with oxidative damage. The mixture also decreased isocitrate dehydrogenase activity, showing a toxicological interaction between microplastics and pyrene, while increasing the concentration of bile pyrene metabolites, suggesting that microplastics could modulate both the bioavailability and biotransformation of pyrene (Guzzetti et al. 2018).

The distribution of metals was strongly associated with the amounts of nutrients, mud contents, organic matter contents, and low pH levels (Vezzone et al. 2019). Ocean acidification-related changes in seawater chemistry can have an impact on the solubility, speciation, and distribution of metals in water and sediments, which may have an impact on the toxicity of those metals to marine life (Ivanina and Sokolova 2015). Toxic elements frequently have unwanted consequences in organisms at various trophic levels and interfere with the roles that vital metals play in the body. Certain metals, such as Zn and Cu, are essential for the correct operation of metabolic processes since they are components of enzymes and hemocyanin. Certain metals, such as Pb, Cd, Ni, and Hg, can be fatal at high concentrations and have negative effects such as endocrine and behavioral disorders; for instance, internal damages from the toxicity of Hg involve neurological damage (brain damage in fish), fetal deformity of the fingerlings, permanent disabilities, fatty liver and degeneration, and immune system damage. External damages include loss of equilibrium, increased opercular movement, irregular vertical movements, and, in the worst cases, death (Zeitoun and Mehana 2014).

In summary, the adverse effects of pollution originating from land-based sources on both ecosystems and human health are unequivocal. Urgent intervention, characterized by stringent regulatory measures, adoption of sustainable methodologies, and deployment of advanced technologies, is imperative to counteract these challenges. Of particular significance in facilitating the transport of land-based pollutants to marine environments is the phenomenon of runoff, with urban and industrial runoff playing pivotal roles. To effectively mitigate these threats, it is essential to comprehensively assess and address the influential factors associated with runoff. Future research studies should be focused on assessing these influential factors to have a better understanding of the behavior and dispersion of the pollutants. Thus, concerted efforts are indispensable for the preservation of biodiversity, the protection of public health, and the sustenance of ecosystem integrity, ensuring the welfare of present and future generations.

References

Abou Samra, R.M. and Ali, R. (2024). Tracking the behavior of an accidental oil spill and its impacts on the marine environment in the Eastern Mediterranean. *Marine Pollution Bulletin* 198: 115887.

Akankali, J. and Elenwo, E. (2015). Sources of marine pollution on Nigerian coastal resources: an overview. *Open Journal of Marine Science* 5 (02): 226.

Alimi, O.S., Farner Budarz, J., Hernandez, L.M., and Tufenkji, N. (2018). Microplastics and nanoplastics in aquatic environments: aggregation, deposition, and enhanced contaminant transport. *Environmental Science and Technology* 52 (4): 1704–1724.

Alves, T.M., Kokinou, E., Ekström, M. et al. (2021). Scientific, societal and pedagogical approaches to tackle the impact of climate change on marine pollution. *Scientific Reports* 11 (1): 2927.

Axelsson, C. and van Sebille, E. (2017). Prevention through policy: urban macroplastic leakages to the marine environment during extreme rainfall events. *Marine Pollution Bulletin* 124 (1): 211–227.

Ayala, T. (2021). *Sources and Impacts of Nutrient Pollution on Coastal Ecosystems*. University of Florida.

Baby, J., Raj, J.S., Biby, E.T. et al. (2010). Toxic effect of heavy metals on aquatic environment. *International Journal of Biological and Chemical Sciences* 4 (4): 939–952.

Baker, T.J., Tyler, C.R., and Galloway, T.S. (2014). Impacts of metal and metal oxide nanoparticles on marine organisms. *Environmental Pollution* 186: 257–271.

Bandara, K.R. and Manage, P.M. (2023). Heavy metal contamination in the coastal environment and trace level identification. *IntechOpen*. https://doi.org/10.5772/intechopen.106653.

Bashir, I., Lone, F.A., Bhat, R.A. et al. (2020). Concerns and threats of contamination on aquatic ecosystems. *Bioremediation and Biotechnology* 1–26. https://doi.org/10.1007/978-3-030-35691-0_1.

Bello, F.A. (2022). Critical analysis of policies on single-use plastics disposal from ships as a source of pollution to the marine environment. A dissertation submitted to the World Maritime University in partial fulfilment of the requirements for the award of the degree of FATAI ABIOLA BELLO Nigeria MASTER OF SCIENCE in MARITIME AFFARS. https://commons.wmu.se/cgi/viewcontent.cgi?article=3141&context=all_dissertations

Ben-Haddad, M., Abelouah, M.R., Hajji, S. et al. (2024). Macroalgal blooms on Moroccan coasts: plastic trapping and tourism challenges. *Regional Studies in Marine Science* 70: 103390.

Beyer, A. and Biziuk, M. (2009). Environmental fate and global distribution of polychlorinated biphenyls. *Reviews of Environmental Contamination and Toxicology* 201: 137–158.

Boesch, D.F., Burroughs, R.H., Baker, J.E. et al. (2001). *Marine Pollution in the United States*. Arlington, VA: Pew Oceans Commission.

Bricelj, V., Haubois, A.-G., Sengco, M. et al. (2012). Trophic transfer of brevetoxins to the benthic macrofaunal community during a bloom of the harmful dinoflagellate *Karenia brevis* in Sarasota Bay, Florida. *Harmful Algae* 16: 27–34.

Brodie, J.E., Kroon, F., Schaffelke, B. et al. (2012). Terrestrial pollutant runoff to the Great Barrier Reef: an update of issues, priorities and management responses. *Marine Pollution Bulletin* 65 (4–9): 81–100.

Cabral, H., Fonseca, V., Sousa, T., and Costa Leal, M. (2019). Synergistic effects of climate change and marine pollution: an overlooked interaction in coastal and estuarine areas. *International Journal of Environmental Research and Public Health* 16 (15): 2737.

Cai, Z., Li, M., Zhu, Z. et al. (2023). Biological degradation of plastics and microplastics: a recent perspective on associated mechanisms and influencing factors. *Microorganisms* 11 (7): 1661.

Chahouri, A., Elouahmani, N., and Ouchene, H. (2022). Recent progress in marine noise pollution: a thorough review. *Chemosphere* 291: 132983.

Cornelissen, G., Pettersen, A., Nesse, E. et al. (2008). The contribution of urban runoff to organic contaminant levels in harbour sediments near two Norwegian cities. *Marine Pollution Bulletin* 56 (3): 565–573.

van Dam, J.W., Negri, A.P., Uthicke, S., and Mueller, J.F. (2011). Chemical pollution on coral reefs: exposure and ecological effects. In: *Ecological Impacts of Toxic Chemicals (187-211)* (ed. F. Sánchez-Bayo, P.J. van den Brink, and R.M. Mann). Bentham Science Publishers https://library.oapen.org/bitstream/handle/20.500.12657/75863/1/9781608051212.pdf#page=193.

DeGeorges, A., Goreau, T.J., and Reilly, B. (2010). Land-sourced pollution with an emphasis on domestic sewage: lessons from the Caribbean and implications for coastal development on Indian Ocean and Pacific coral reefs. *Sustainability* 2 (9): 2919–2949.

Duran, R. and Cravo-Laureau, C. (2016). Role of environmental factors and microorganisms in determining the fate of polycyclic aromatic hydrocarbons in the marine environment. *FEMS Microbiology Reviews* 40 (6): 814–830.

Edinger, E.N. and Risk, M.J. (2013). Effect of land-based pollution on central Java coral reefs. *Journal of Coastal Development* 3 (2): 593–613.

Edwards, J., Edyvane, K., Boxall, V. et al. (2001). Metal levels in seston and marine fish flesh near industrial and metropolitan centres in South Australia. *Marine Pollution Bulletin* 42 (5): 389–396.

Ekino, S., Susa, M., Ninomiya, T. et al. (2007). Minamata disease revisited: an update on the acute and chronic manifestations of methyl mercury poisoning. *Journal of the Neurological Sciences* 262 (1–2): 131–144.

Erbe, C., Marley, S.A., Schoeman, R.P. et al. (2019). The effects of ship noise on marine mammals – a review. *Frontiers in Marine Science* 6: 476898.

Eriksen, M., Lebreton, L.C., Carson, H.S. et al. (2014). Plastic pollution in the world's oceans: more than 5 trillion plastic pieces weighing over 250,000 tons afloat at sea. *PloS One* 9 (12): e111913.

Fabricius, K.E. (2005). Effects of terrestrial runoff on the ecology of corals and coral reefs: review and synthesis. *Marine Pollution Bulletin* 50 (2): 125–146.

FAIRR. (2019). Marine pollution as a result of intensive farming. FAIRR. https://www.fairr.org/resources/knowledge-hub/intensive-farming-pollution/marine-pollution-as-a-result-of-intensive-farming (accessed 20 February 2024).

Fredston-Hermann, A., Brown, C.J., Albert, S. et al. (2016). Where does river runoff matter for coastal marine conservation? *Frontiers in Marine Science* 3: 273.

Guzzetti, E., Sureda, A., Tejada, S., and Faggio, C. (2018). Microplastic in marine organism: environmental and toxicological effects. *Environmental Toxicology and Pharmacology* 64: 164–171.

Häder, D.-P., Banaszak, A.T., Villafañe, V.E. et al. (2020). Anthropogenic pollution of aquatic ecosystems: emerging problems with global implications. *Science of the Total Environment* 713: 136586.

Hale, R.C., King, A.E., and Ramirez, J.M. (2024). Plastic debris: an overview of composition, sources, environmental occurrence, transport, and fate. *Microplastic Contamination in Aquatic Environments* 1–31. https://doi.org/10.1016/B978-0-443-15332-7.00004-1.

Hellar-Kihampa, H. (2024). Reviewing the interdecadal dynamics of micropollutants in the Tanzanian coastal zone from 2002 to 2022. *Environmental Science and Pollution Research* 1–19.

Hermabessiere, L., Dehaut, A., Paul-Pont, I. et al. (2017). Occurrence and effects of plastic additives on marine environments and organisms: a review. *Chemosphere* 182: 781–793.

Hou, G. and Zhan, B. (2023). Relationship among marine pollution, outdoor activities, and climate change: fresh evidence from panel threshold regression analysis from coastal regions of China. *Environmental Research* 236: 116668.

Immig, J. (2018). *Ocean Pollutants Guide: Toxic Threats to Human Health and Marine Life*. Ipen.

Islam, M.S. and Tanaka, M. (2004). Impacts of pollution on coastal and marine ecosystems including coastal and marine fisheries and approach for management: a review and synthesis. *Marine Pollution Bulletin* 48 (7–8): 624–649.

Ito, K., Sonoki, S., Minami, K. et al. (2024). Spatial and economic quantification of provisioning service by eelgrass beds in Lake Notoro, Hokkaido, Japan. *Scientific Reports* 14 (1): 3742.

IUCN. (2024). Global Invasive Species Database. https://www.iucngisd.org/gisd/index.php (accessed 3 February 2024).

Ivanina, A.V. and Sokolova, I.M. (2015). Interactive effects of metal pollution and ocean acidification on physiology of marine organisms. *Current Zoology* 61 (4): 653–668.

Jambeck, J.R., Geyer, R., Wilcox, C. et al. (2015). Plastic waste inputs from land into the ocean. *Science* 347 (6223): 768–771.

Juhasz, A.L. and Naidu, R. (2000). Bioremediation of high molecular weight polycyclic aromatic hydrocarbons: a review of the microbial degradation of benzo [a] pyrene. *International Biodeterioration and Biodegradation* 45 (1–2): 57–88.

Kansuah, K. B. (2020). A national approach to address the impact of land-based pollution on marine environment in Liberia. The Maritime Commons: Digital Repository of the World Maritime University. https://commons.wmu.se/cgi/viewcontent.cgi?article=2366&context=all_dissertations

Katsanevakis, S. (2008). Marine debris, a growing problem: sources, distribution, composition, and impacts. In: *Marine Pollution: New Research* (ed. T.N. Hofer), 53–100. New York: Nova Science Publishers.

Keller, A.A., Fruh, E.L., Johnson, M.M. et al. (2010). Distribution and abundance of anthropogenic marine debris along the shelf and slope of the US West Coast. *Marine Pollution Bulletin* 60 (5): 692–700.

Kelleway, J.J., Cavanaugh, K., Rogers, K. et al. (2017). Review of the ecosystem service implications of mangrove encroachment into salt marshes. *Global Change Biology* 23 (10): 3967–3983.

Kindt, J.W. (1984). Solid wastes and marine pollution. *Catholic University Law Review* 34: 37.

Kumar, R., Verma, A., Shome, A. et al. (2021). Impacts of plastic pollution on ecosystem services, sustainable development goals, and need to focus on circular economy and policy interventions. *Sustainability* 13 (17): 9963.

Kumartasli, S. and Avinc, O. (2020). Recycling of marine litter and ocean plastics: a vital sustainable solution for increasing ecology and health problem. In: *Sustainability in the Textile and Apparel Industries: Sourcing Synthetic and Novel Alternative Raw Materials* (ed. S.S. Muthu and M.A. Gardetti), 117–137. Springer https://www.researchgate.net/profile/Munir-Tasdemir/publication/340302890_Sustainable_Green-Based_Composites_from_Renewable_Resources_in_Textile_Industrial_Cotton_Wastes/links/670f64217797480e894bf832/Sustainable-Green-Based-Composites-from-Renewable-Resources-in-Textile-Industrial-Cotton-Wastes.pdf#page=121.

Larsen, M.C. and Webb, R.M. (2009). Potential effects of runoff, fluvial sediment, and nutrient discharges on the coral reefs of Puerto Rico. *Journal of Coastal Research* 25 (1): 189–208.

Latimer, J.S. and Zheng, J. (2003). The sources, transport, and fate of PAHs in the marine environment. In: *PAHs: An Ecotoxicological Perspective* (ed. P.E.T. Douben), 7–33. Wiley.

Lee, J., Hong, S., Song, Y.K. et al. (2013). Relationships among the abundances of plastic debris in different size classes on beaches in South Korea. *Marine Pollution Bulletin* 77 (1–2): 349–354.

Liu, L.-Y., Wang, J.-Z., Wei, G.-L. et al. (2012). Polycyclic aromatic hydrocarbons (PAHs) in continental shelf sediment of China: implications for anthropogenic influences on coastal marine environment. *Environmental Pollution* 167: 155–162.

Micella, I., Kroeze, C., Bak, M.P., and Strokal, M. (2024). Causes of coastal waters pollution with nutrients, chemicals and plastics worldwide. *Marine Pollution Bulletin* 198: 115902.

Mihai, F.-C., Gündoğdu, S., Khan, F.R. et al. (2022). Plastic pollution in marine and freshwater environments: abundance, sources, and mitigation. In: *Emerging Contaminants in the Environment*, 241–274. Elsevier https://doi.org/10.1016/B978-0-323-85160-2.00016-0.

Müller, A., Österlund, H., Marsalek, J., and Viklander, M. (2020). The pollution conveyed by urban runoff: a review of sources. *Science of the Total Environment* 709: 136125.

Nageeb, M. (2013). *Adsorption Technique for the Removal of Organic Pollutants from Water and Wastewater*. InTech: https://doi.org/10.5772/54048.

O'Connor, J.J., Lecchini, D., Beck, H.J. et al. (2016). Sediment pollution impacts sensory ability and performance of settling coral-reef fish. *Oecologia* 180: 11–21.

Pal, D. and Sen, S. (2023). Emerging petroleum pollutants and their adverse effects on the environment. In: *Impact of Petroleum Waste on Environmental Pollution and its Sustainable*

Management Through Circular Economy, Environmental Science and Engineering (ed. I.D. Behera and A.P. Das). Cham: Springer https://doi.org/10.1007/978-3-031-48220-5_5.

Perugini, M., Visciano, P., Giammarino, A. et al. (2007). Polycyclic aromatic hydrocarbons in marine organisms from the Adriatic Sea, Italy. *Chemosphere* 66 (10): 1904–1910.

Potters, G. (2013). *Marine Pollution*. Bookboon.

Qi, J., Shi, J., Gao, H., and Sun, Z. (2013). Atmospheric dry and wet deposition of nitrogen species and its implication for primary productivity in coastal region of the Yellow Sea, China. *Atmospheric Environment* 81: 600–608.

Reuer, M.K. and Weiss, D.J. (2002). Anthropogenic lead dynamics in the terrestrial and marine environment. *Philosophical Transactions of the Royal Society of London, Series A: Mathematical, Physical and Engineering Sciences* 360 (1801): 2889–2904.

Risk, M.J. and Edinger, E. (2011). Impacts of sediment on coral reefs. In: *Encyclopedia of Modern Coral Reefs*, 575–586. Netherlands: Springer https://www.researchgate.net/profile/Michael-Risk/publication/278716776_Impacts_of_Sediment_on_Coral_Reefs/links/5b3134520f7e9b0df5c85e5d/Impacts-of-Sediment-on-Coral-Reefs.pdf.

Roberts, D.A. (2012). Causes and ecological effects of resuspended contaminated sediments (RCS) in marine environments. *Environment International* 40: 230–243.

Ryan, P.G. (2023). Illegal dumping from ships is responsible for most drink bottle litter even far from shipping lanes. *Marine Pollution Bulletin* 197: 115751.

Sardenne, F., Le Loc'h, F., Bodin, N. et al. (2024). Persistent organic pollutants and trace metals in selected marine organisms from the Akanda National Park, Gabon (Central Africa). *Marine Pollution Bulletin* 199: 116009.

Sheppard, C.R. (1993). Physical environment of the Gulf relevant to marine pollution: an overview. *Marine Pollution Bulletin* 27: 3–8.

SLYCAN. (2018). Oil spills and their devastating effects to our oceans. SLYCAN Trust. https://www.slycantrust.org/post/oil-spills-and-their-devastating-effects-to-our-oceans

Tedesco, P., Balzano, S., Coppola, D. et al. (2024). Bioremediation for the recovery of oil polluted marine environment, opportunities and challenges approaching the Blue Growth. *Marine Pollution Bulletin* 200: 116157.

Thacharodi, A., Meenatchi, R., Hassan, S. et al. (2024). Microplastics in the environment: a critical overview on its fate, toxicity, implications, management, and bioremediation strategies. *Journal of Environmental Management* 349: 119433.

Thushari, G.G.N. and Senevirathna, J.D.M. (2020). Plastic pollution in the marine environment. *Heliyon* 6 (8).

Todd, P.A., Ong, X., and Chou, L.M. (2010). Impacts of pollution on marine life in Southeast Asia. *Biodiversity and Conservation* 19: 1063–1082.

Valigura, R. A., Luke, W. T., Artz, R. S., and Hicks, B. B. (1996). Atmospheric nutrient input to coastal areas: reducing the uncertainties. https://repository.library.noaa.gov/view/noaa/1782/noaa_1782_DS1.pdf

Verma, J., Pant, H., Sign, S., and Tiwari, A. (2020). Marine pollution, sources, effect and management. In: *Three Major Dimensions of Life: Environment*, 270–276. Prayagraj, India: Agriculture and Health; Society of Biological Sciences and Rural Development https://www.researchgate.net/profile/Hemlata-Pant-3/publication/345674343_MARINE_POLLUTION_SOURCES_EFFECT_AND_MANAGEMENT/links/5faa868192851cc286a504df/MARINE-POLLUTION-SOURCES-EFFECT-AND-MANAGEMENT.pdf.

Vezzone, M., Cesar, R., de Souza Abessa, D.M. et al. (2019). Metal pollution in surface sediments from Rodrigo de Freitas Lagoon (Rio de Janeiro, Brazil): toxic effects on marine organisms. *Environmental Pollution* 252: 270–280.

Victoria, N.S., Kumari, T.S.D., and Lazarus, B. (2022). Assessment on impact of sewage in coastal pollution and distribution of fecal pathogenic bacteria with reference to antibiotic resistance in the coastal area of Cape Comorin, India. *Marine Pollution Bulletin* 175: 113123.

Vong, R., Hansson, H.C., Covert, D., and Charlson, R. (1988). Acid rain: simultaneous observations of a natural marine background and its acidic sulfate aerosol precursor. *Geophysical Research Letters* 15 (4): 338–341.

Wabnitz, C. and Nichols, W.J. (2010). Plastic pollution: an ocean emergency. *Marine Turtle Newsletter* 129: 1.

Wear, S.L. and Thurber, R.V. (2015). Sewage pollution: mitigation is key for coral reef stewardship. *Annals of the New York Academy of Sciences* 1355 (1): 15–30.

Weilgart, L.S. (2007). The impacts of anthropogenic ocean noise on cetaceans and implications for management. *Canadian Journal of Zoology* 85 (11): 1091–1116.

Weldemichael, A.T. (2012). Maritime corporate terrorism and its consequences in the western Indian Ocean: illegal fishing, waste dumping and piracy in twenty-first-century Somalia. *Journal of the Indian Ocean Region* 8 (2): 110–126. https://doi.org/10.1080/19480881.2012.730747.

Wenger, A.S., Johansen, J., and Jones, G. (2011). Suspended sediment impairs habitat choice and chemosensory discrimination in two coral reef fishes. *Coral Reefs* 30: 879–887.

Wenger, A.S., Fabricius, K.E., Jones, G.P., and JE, B. (2015). Effects of sedimentation, eutrophication, and chemical pollution on coral reef fishes. In: *Ecology of Fishes on Coral Reefs*, vol. 145. https://diving-rov-specialists.com/index_htm_files/env_55-effects-sedimentation-eutrophication-and-chemical-pollution-coral-reef-fishes.pdf.

Willis, K., Denise Hardesty, B., Kriwoken, L., and Wilcox, C. (2017). Differentiating littering, urban runoff and marine transport as sources of marine debris in coastal and estuarine environments. *Scientific Reports* 7 (1): 44479.

Wright, S.L., Thompson, R.C., and Galloway, T.S. (2013). The physical impacts of microplastics on marine organisms: a review. *Environmental Pollution* 178: 483–492.

Zeitoun, M.M. and Mehana, E.E. (2014). Impact of water pollution with heavy metals on fish health: overview and updates. *Global Veterinaria* 12 (2): 219–231.

Zhang, Y., Wang, K., Chen, W. et al. (2022). Effects of land use and landscape on the occurrence and distribution of microplastics in soil, China. *Science of the Total Environment* 847: 157598.

Zheng, B., Zhao, X., Liu, L. et al. (2011). Effects of hydrodynamics on the distribution of trace persistent organic pollutants and macrobenthic communities in Bohai Bay. *Chemosphere* 84 (3): 336–341.

Zutshi, B. and Prasad, S.R. (2008). Impact of pollution on fresh-and marine water resources – a review. *Pollution Research* 27 (3): 461–466.

3

Marine Pollution Issues Relating to Shipping, Ports, and Use of Marine Coatings

Christopher W. Haley[1] *and R. John Morrison*[2]

[1] *ECOPOL, Mount Kembla, New South Wales, Australia*
[2] *School of Earth, Atmospheric and Life Sciences, University of Wollongong, Northfields Ave Wollongong, New South Wales, 2515, Australia*

3.1 Introduction

3.1.1 Ships and Shipping Impacts

A wide range of boats and ships move into and through the global oceans with several authors reporting that about 90% of global freight transport is by ships (e.g., Cirimanna et al. 2015; Li and Ning 2019; Schnurr and Walker 2019). The large number of ports used by these vessels leads to an extremely diverse set of impacts. The vessels can vary from 5 m local fishing and dive support boats to ocean-going cruise ships, container ships, bulk carriers to oil supertankers that are now up to 400 m in length and carry a load exceeding 400 000 tonnes. For this chapter, we consider ships as vessels of length greater than 20 m. United Nations (UN) data indicates that the global merchant fleet is about 100 000 vessels (see Table 3.1, International Underwriters Association 2022). In addition to the impacts of the vessels directly on the marine environment, shipping also leads to impacts from the development and maintenance of ship handling facilities onshore. These onshore/nearshore construction and related activities are discussed below, and this section will consider the impacts of the ships themselves.

Although ships are the most energy-efficient method of moving a given mass of cargo a given distance, the increasing size of the industry means that it has a significant effect on the environment. The annual increase in the amount of shipping overwhelms gains in efficiency, such as, from slow steaming or other technical developments. The growth in tonne-kilometers of sea shipment has averaged 4% yearly since the 1990s, and it has grown by a factor of three since the 1970s. Shipping is a major global enterprise involving trade and commerce in many sectors of the world economy, with many countries owning vessels privately and in government, and crews coming from numerous countries.

As a result, many agreements must be prepared to facilitate the complex arrangements for international trade to operate efficiently. In order to achieve this, the United Nations established the International Maritime Organization (IMO) to develop international treaties on maritime safety, to avoid unfair practices in shipping, and to reduce maritime pollution. The IMO was created in 1948 and has grown to more than 170 member countries, managed through

Coastal and Marine Pollution: Source to Sink, Mitigation and Management, First Edition.
Edited by Meththika Vithanage, Sameera M. Samarasekara, Bryan D. James, and Christopher M. Reddy.

Table 3.1 Numbers of vessels in the global merchant fleet (International Underwriting Association 2022).

Vessel type	Number 2022
Oil tankers	2206
Tanker ships	9809
Chemical carriers	3991
Bulk carriers	12766
LPG carriers	1532
LNG carriers	685
Container ships	5606
Multipurpose freighters	3188
General cargo ships	16112
RoRo ships	1586
Total cargo ships	*57481*
Offshore vessels	8926
Services	2112
Tugs	21288
Passenger ships	8938
Other non-cargo ships	4529
Total	*45523*
Grand total	103044

56 conventions/agreements. Many of the major environmental issues relating to global shipping have been addressed and resolved through the Marine Environment Protection Committee (MEPC) of the IMO, which has become, in effect, a body for resolving marine environment and safety issues. Pollution is a multi-sectorial, multicomponent topic which is difficult to define precisely; one widely accepted version is that used in the UN Convention of the Law of the Sea: "Pollution is the introduction by humans, directly or indirectly, of substances or energy into the marine environment which results in deleterious effects such as harm to living organisms and marine life, hazards to human health, hindrance to marine activities, including fishing and other legitimate uses of the sea, impairment of the quality of seawater and reduction of amenities" (UNCLOS 1982).

In this chapter, we are primarily concerned with the impacts of human activities on the marine environment, including the open oceans, coastal areas, ports, harbors, and estuaries. Some of these impacts and their causes are clearly obvious (Figure 3.1), but many are not. Some pollution is deliberate, some is accidental, but a great deal is the result of people being unaware they are creating a problem. This includes seafarers, and hence an important contribution to marine pollution control is to upgrade their knowledge and assist them to reduce the impacts of their activities on the marine environment.

Sustainable governance of the marine environment requires a multidisciplinary, scientific understanding of marine ecosystems and human activities that affect them. This is an ongoing concern for the United Nations organization as a whole and there was an identified need for an independent high-level source of technical advice. The Joint Group of Experts on Scientific Aspects

Figure 3.1 Much pollution from shipping is a result of vessels running aground on rocky outcrops, coral reefs, and unmapped sandbars. Recovery of cargoes, particularly from isolated locations, may often be uneconomic. *Source:* Photo by R.J. Morrison.

of Marine Pollution (GESAMP) was created to carry out this role. GESAMP was to ensure that its sponsoring organizations:

- Identify areas of common scientific interest
- Reduce costs by cosponsoring
- Receive consistency in the advice provided
- Place their specialized activities in a broader perspective

3.2 Marine Pollution from Shipping

Marine pollution from shipping covers the deliberate discharge, the accidental spillage, or release of one or more of numerous substances into the marine environment under conditions where the additives are likely to harm the environment. The most common varieties of marine pollutants are oils, hazardous chemicals, plastics, biological organisms, bulk ores (coal, iron ore, bauxite, and phosphate rock), and grains. Shipping can harm the environment by adding heat, light, and noise.

Heat and light pollution normally are the most important in ports; heat pollution is produced by discharge of hot water into a water body which has limited mixing with open oceans or bays. Increasing the water temperature causes a decrease in dissolved oxygen and will disturb the thermal equilibrium of some fish species. Light pollution is the result of an area of water receiving continuous illumination from anthropogenic sources resulting from port infrastructure safety and security lighting plus light from ships in the port. Under these conditions, the life cycle of fish species that require a dark phase for completion may struggle to survive.

Noise is an ever-present occurrence in water bodies due to waves, marine animal communication, resource exploitation operations, and military activities. In the oceans, sound can be heard for hundreds of kilometers. Many marine species use sound in all aspects of their life cycles including reproduction, feeding, communication, and navigation. Marine transport is a major anthropogenic source of low-frequency background noise, a result of cavitation, which is the formation of water vapor cavities as water passes over propeller blades. This effect can be reduced by selecting appropriately designed propellers, ensuring regular cleaning and maintenance, and adjusting vessel speeds to an optimum with regard to noise propagation (IMO 2014).

3.2.1 Air Pollution

Ships are a major source of air pollution. They utilize fuels like bunker and heavy fuel oil in increasing quantities discharging, on combustion, greenhouse gases (GHGs) notably CO_2, N_2O, and methane. Given the large number of vessels traversing the oceans and the distances traveled, the quantities of GHG generated by shipping are substantial. The annual GHG emissions from shipping increased from 977 million tonnes in 2012 to 1076 million tonnes in 2018 (Fourth IMO GHG Study 2020) representing a 9.6% increase. The share of shipping in the global anthropogenic emissions of CO_2 increased from 2.76% in 2012 to 2.89% in 2018. If shipping was included in the global list of country emissions of GHG, it would be in sixth place (behind the United States, China, Russia, India, and Japan) indicating the importance of shipping as a pollutant source (Shi 2016). Shipping also contributes 10–15% of global anthropogenic emissions of SOx and NOx (Lindstad and Eskeland 2016).

If these GHGs remain in the atmosphere, it could be argued they are not marine pollutants (Shi 2016). In the oceanic environment, however, there are frequent rain events, which lead to interactions between rainwater and certain gases, e.g.,

$$CO_2(g) + H_2O(l) <=> H_2CO_3(l)$$

$$NOx(g) + H_2O(l) <=> HNO_3(l)$$

$$SOx(g) + H_2O(l) <=> H_2SO_3(l) + H_2SO_4(l)$$

These "washout" products enter the surface ocean, leading to increased acidity there.

A further consideration should be noted specifically for CO_2. The massive increase in CO_2 concentration in the atmosphere over the last 200 years has been mitigated to some extent by the fact that CO_2 has an appreciable solubility in seawater due to:

a) Reaction with CO_3^{2-}

$$CO_3^{2-}(l) + H_2O(l) + CO_2(g) <=> 2HCO_3^-(l)$$

with the equilibrium constant for this reaction being so large that a substantial proportion of the CO_2 is converted to bicarbonate; and,

b) the sinking of particulate organic matter (POM) and $CaCO_3$ into the deep ocean removes the carbon from contact with the atmosphere preventing carbon emissions back to the atmosphere until some upwelling occurs.

In the surface waters, a further reaction sequence occurs

$$CO_2(g) + H_2O(l) <=> H^+(l) + HCO_3^-(l) <=> 2H^+(l) + CO_3^{2-}(l)$$

Thus, H^+ ions are produced, and this leads to an overall increase in H^+ concentration and a decrease in pH, that is, acidification of the ocean waters. As $[H^+]$ and $[CO_3^{2}-]$ are critical for the synthesis and stability of carbonate-based marine structures, acidification of the oceans is a long-term issue for marine scientists (Doney et al. 2016).

3.2.1.1 Management of Air Pollution from Ships

A number of different strategies have been adopted. These include technology change, organic matter (POM) discharges, upgrading engines, use of better quality fuels with low pollutant contents (such as low sulfur products), seawater scrubbing, incorporation of selective catalytic reduction (SCR) systems that dramatically reduce exhaust gas concentrations of NOx and SOx. Other mechanisms include operational changes, greater use of shore-side power, reduction of vessel speeds, and market-based strategies, for example, variable port fees to reward low emitters. Rising sea levels are now a significant concern for millions of people who live in low-lying coastal areas and on small islands. Estimates of the likely changes in sea level are 25–30 cm by 2050 and by up to 60 cm by 2100 (IPCC 2001). Many of the relevant administrations responsible for planning and development in these zones are having to divert limited available resources to put in place actions that will enable humans to adapt to the changing situation. Although shipping is a small contributor to the atmospheric GHG increases, every effort should be made by the maritime community to reduce their contributions as an incentive to other sectors to adopt a similar approach.

3.2.2 Marine Pollutants

There is a wide range of substances that, when introduced into the marine environment, cause harm. These include petroleum products of all kinds (crude oils, hydrocarbon products, lubricating oil, and oily wastes), chemicals (pesticides, fertilizers, and industrial chemicals), metals and metal industry wastes, sewage, plastics, and solid wastes of all kinds from garbage. In the sections below, we will consider some case studies to illustrate the sources, amounts, impacts, and management of pollutants.

3.2.2.1 Oil Pollution

The best-known and understood marine pollution issue is that related to oil pollution; in this context, the definition of oil has been broadened to mean petroleum in any form, including crude oil, fuel oil, sludge, oil refuse, and refined products (other than petrochemicals) (Gold 1997). The primacy of oil is not surprising given the quantities transported and the publicity associated with incidents like the *Torrey Canyon* (English Channel, 1965), *Amoco Cadiz* (Portsall Rocks, Brittany, 1978), *Exxon Valdez* (Prince William Sound, Alaska, 1989), and *Braer* (Shetland 1993). These incidents led to extensive media publicity on the spills and the impacts on the physical and biological environments of the surrounding areas. Many smaller incidents, occurring during terminal operations when oil is being loaded or discharged, lead to spillages of substantial quantities of oil in ecologically sensitive areas. The serious nature of the incidents led to IMO coordinating the first global agreement to minimize the problems associated with oil pollution from ships (MARPOL 73/78) through an overarching agreement and a series of Protocols and Annexes (IMO 2023). The significant improvement of oil spill management via procedures developed through the IMO enhanced the credibility of IMO as the major global agency for the management of international agreements on marine environmental management through the activities of the MEPC.

Oil transport by sea had been increasing in the years following World War II, such that by the 1970s the annual volumes reached 1450 million tonnes (~1600 million m^3, or 10 150 million barrels) (ITOPF 2022). With the ever-increasing volumes being transported, there was also a dramatic increase in spillages. By 2010, the quantities transported globally had reached 2900 million tonnes per year, and spillages averaged 15 000 tonnes per year (ITOPF 2022). The main reasons for the tanker spills worldwide between 1970 and 2022 were ship grounding (32%), allision/collision (30%), and hull failure (75%).

3.2.2.2 Behavior of Oil in Water

Spilled oil is a mixture of components with numerous individual compounds having a wide range of chemical and physical properties. The different compounds behave differently—if the spill is of "light" crude, a proportion of the spilled material will spread over the surface as a very thin layer. If the spill is of "heavy" crude, a greater proportion of the spill may sink into the water column, and the remaining material forms "pancakes" several centimeters thick. The behavior of spills is also dependent on the geomorphology of the surrounding environment—open sea, enclosed bay or port, flat coastline with mangroves or mudflats, and cliffs (Gold 1997). The response of the spilled oil will be determined by a series of biological, chemical, and physical processes as shown in Figure 3.2.

A proportion of the oil on the surface will evaporate depending on the oil composition and the atmospheric temperature. Some oils will dissolve in water, some may get incorporated into sediments. The dissolved oil will be microbially degraded. Some of the oil will form an emulsion (mousse) with water, which will degrade at a rate depending on the temperature, surface roughness,

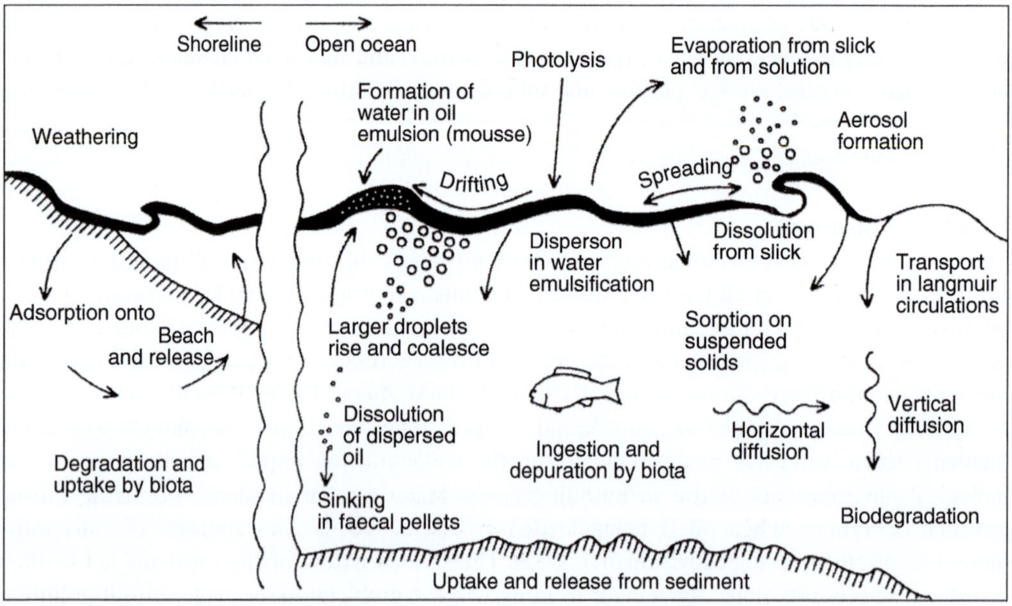

Figure 3.2 Interactions of oil with the marine environment after a spill. *Source:* Adapted from GESAMP (1993).

and oil spill composition. Some oil may wash ashore where its persistence will depend on the type of oil, climate, and geomorphology of the coast. Further details on the processes occurring are given by Gold (1997, pp. 254–286).

3.2.3 Hazardous and Noxious Substances (HNS)

Hundreds of HNS are transported by sea. A wide range of substances is considered under this umbrella, including biological organisms, explosives, flammable materials, corrosives, oxidizing, and radioactive materials. Shipping is the only long-distance transportation method for hazardous materials that cannot be sent by air. These materials are required to be packaged in containers that can withstand impacts if the ship founders and many packages are required to be stored on deck. The quantities involved have been increasing, causing concern about the impacts of HNS, especially if the transporting vessels are involved in collisions or have fires aboard. Ships regularly transporting such materials will have appropriate modifications to protect the crew. The International Maritime Dangerous Goods Code (IMDG) developed by IMO provides guidance on the minimum standards for transport of dangerous goods. Ships should have a HNS management strategy for the substances they carry which gives guidance about handling during loading and unloading and spill management.

3.2.4 Ballast Water Issues (Aquatic Invasive Species, AIS)

Harmful species transfers undoubtedly present one of the most serious anthropogenic threats to the oceanic world. One of the chief mechanisms through which such transfers are facilitated is via ballast water operations. When a ship's cargo hold is empty, the maintenance of balance is difficult while the vessel is at sea because it is top-heavy. When wooden-hulled ships were used, rocks and boulders were manually added to the hold to improve ship stability. When the next cargo was ready for loading, the ballast was manually removed, and the cargo was loaded. Since the introduction of steel-hulled ships, water has been used to balance and stabilize ships at sea. This ballast water is discharged when the ship is in position to load the next cargo. As water is pumped into the ballast tank, any small organisms and suspended sediments in the water will be pumped into the ballast tank. When the ballast water is pumped out again, many of the organisms are still alive and sometimes they are very unwelcome in the new surroundings. These species can include bacteria, microbes, small invertebrates, eggs, cysts, and larvae of several species. The transferred organisms may survive and establish a reproductive population in the host environment, and in the absence of suitable predators, become invasive, out-competing native species and multiplying into pest proportions.

Scientists first recognized the signs of an alien species transfer following an outbreak of Asian phytoplankton algae in the North Sea in 1903, but it was not until the 1970s that the scientific community began serious consideration of the issue. In the 1980s, the concerns were brought to the attention of the MEPC at IMO (see Table 3.2). In 2004, the International Convention for the Control and Management of Ships' Ballast Water and Sediments (BWM Convention) was adopted to provide globally unified prevention measures. Even when the BWM Convention standards are met, a certain number of organisms is acceptable in the discharge. Organisms under 10 μm remain unregulated. Prevention of organism transfer is far from easy, and so the implementation of the BWM Convention remains a challenge. Several techniques of sterilization have been tried with some success, but the most common method involves the exchange of water in deep zones outside

Table 3.2 Examples of marine species invasions **via** ballast water transfers.

Species	Origin	Introduced to	Comment
Sea walnut *Ctenophora*	North America	Black Sea	Deprived local species of zooplankton, continued to spread to the Caspian Sea
Golden muscle *Limnoperna*	South East Asia	South America	Spreading through rivers and affecting hydroelectric facilities
Hemimysis anomala	Eastern Europe	North Europe Great Lakes North America	This shrimp-like crustacean native to the Caspian Sea region has been spreading across Europe since the 1950s and reached North America in 2006
Pontederia crassipes-water hyacinth	South America	Europe, New Zealand, North America, Australia	Free-floating perennial aquatic native to tropical and subtropical South America
Zebra Mussel *Dreissenaen spreading polymorpha*	Eastern Europe (Black Sea)	Western Europe North America	Fouls hard surfaces in large numbers Displaces native species Blocks water pipes for supply and drainage

the EEZ (Mid-Ocean Exchange, MOE). It is considered that any organisms in the port water used as ballast will not survive in the very different open ocean waters. The BWM Convention entered into force in 2017 and requires all ships operating internationally to manage their ballast water, including maintaining a ballast water record book and an International Ballast Water Management Certificate.

3.3 Pollution from Ship Wastes

3.3.1 Marine Litter

Marine litter is defined as "any persistent, manufactured or processed solid material discarded, disposed of or abandoned in the marine and coastal environment as a result of human activity" and is commonly referred to as "marine debris" (Galgani et al. 2018). The United Nations 2030 Agenda for Sustainable Development (UNSDG 2030) includes Sustainable Development Goal 14.1—to significantly reduce marine pollution of all kinds including marine debris by 2025.

3.3.2 Plastic Pollution

Of particular concern is plastics, given their inherent strength and durability that allows them to persist in the oceanic environment for indefinite periods of time, compounded by the sheer quantity of plastics that have been manufactured, used, and discarded globally since their advent in the 1950s. Plastics, in many ways, contribute positively to society, but there is an unwelcome side effect to the way we produce, use, and dispose of plastics; they are polluting ecosystems, creating risks for human and animal health, and destabilizing the climate. Plastics are one of the most extensive types of marine debris. They are commonly used in many items, and as society has developed new uses for them, the variety and quantity of plastic items found in the marine environment has increased dramatically. Plastics are now known to break down into smaller components, called microplastics (see below). Microplastics (having diameters <0.5 mm) and their associated toxic chemical components contribute to human and wildlife health risks as the toxic microplastics are ingested and move through the marine food web.

Global plastics production and use have grown exponentially between 1950 and 2022, the annual global production of materials expanded from 1.5 million tonnes to about 400 million tonnes (Statistica 2023). This current figure is driven by both the high levels of material consumption in rich countries and the rapidly growing needs of emerging economies since the 1950s. Around nine million people are employed globally in the polymer production and plastic processing industries. Light, strong, and seemingly inexpensive, plastics have permeated our lives, societies, and economies but at a pace that has escalated into significant costs to the environment, human health, and the economy. Currently, the world produces 430 million tonnes of plastics each year of which more than two-thirds are short-lived products which soon become waste, and a growing amount (139 million tonnes in 2021) after one single use. For these reasons, the international community is negotiating an agreement to end plastic pollution, which is due by 2024. In a historic decision at the United Nations General Assembly in March 2022, all 193 UN Member States decided to end plastic pollution. With negotiations on a binding legal agreement by 2024 now underway, the question is how speedily will that goal be realized.

Thousands of scientific papers documenting the presence of plastics in the ocean and their distribution, composition, physical and biological chemistries, and toxicology, as well as impacts on marine biota and habitats, have been published. It is generally assumed that most of the plastics entering the world oceans come from land-based sources—of an estimated 325 million tonnes generated by the 193 coastal nations in 2010 alone, approximately 4.8–12.7 million tonnes entered the ocean as litter particles in that year (Worm et al. 2017).

Research completed in European seas suggests that land-based litter accounts for more litter than sea-based material (Interwies et al. 2013) with sea-based sources of marine litter comprising an average 32%–50% of the total litter in some European basins, but estimates vary by region (Eunomia 2016). In the Adriatic Sea, for example, on an aggregated basis at the regional level, litter items derived from sea-based activities accounted for 6.3%, compared to 34.7% of total marine material being attributed to land-based sources (Vlachogianni et al. 2018). Using data on 17 974 litter items collected between 2011 and 2017, Schafer et al. (2019) estimated that 60% of the litter was derived from local and regional sea-based sources.

The plastics in the ocean occur in a wide range of sizes from large pieces of gear, for example, from discarded fishing nets up to the size of a football field to microplastics. One of the major ocean plastic pollution problems includes the assessment of abandoned, lost, or otherwise discarded fishing gear (ALDFG). Richardson et al. (2019) reviewed the scientific literature to estimate the proportion of commercially deployed fishing gear worldwide that becomes abandoned, lost, and discarded in the ocean. While this review was able to identify estimated losses for categories of gear, extrapolating to a quantitative assessment of ALDFG entering the world ocean annually was beyond the scope of the study. In the absence of definitive data, the "80/20" (ratio of land-sourced versus sea-sourced plastics) and "640 000 tonnes" (total amount of plastics entering the marine environment per year) figures are cited in numerous papers and reports.

Derelict fishing gear (*ghost gear*) continues to trap fish and animals, entangle and potentially kill marine life, smother habitat, and act as a hazard to navigation particularly for small craft where the netting fouls the propellers. Vessels involved in fishing are reputed to create fuel spillage and waste disposal problems, but they often leave behind pieces of fishing tackle and broken nets that create major problems for fish, turtles, and seabirds. Fishers may also use dynamite leading to severe mechanical impact on the surrounding ecosystem (Riegl 2001), and cyanide and natural toxins, for example, *Barringtonia acutangula* (L.) Gaertn, in both the catch and live fish trades which caused the death of many nontarget species and damage to coastal ecosystems (Johannes and Riepen 1995). Among the sea-based contributors to the problem of marine litter, the fishing sector features quite markedly, for example, Arcadis (2012) estimated a 65% share for the fishing sector alone, with the recreational sector also comprising a significant share, and the remaining sea-based marine litter coming from the merchant shipping sector (Eunomia 2016).

About 640 000 tonnes of abandoned, lost or otherwise discarded fishing gear lost in the ocean every year is similarly oft-quoted, but is poorly substantiated (Macfadyen et al. 2009; NAS 1975). Unfortunately, since those studies were completed, there have been few attempts, other than Jambeck et al. (2015), to estimate global inputs of plastic litter into the ocean from land, and there have been no efforts to determine inputs of marine litter from sea-based sources on a global scale.

There is a problem with the enforcement of the regulations regarding controls on derelict fishing gear. It is difficult to monitor and enforce the prohibition on plastic pollution for vessels on the high seas. Flag states often lack any incentive to do so. Practical measures such as the marking of gear and particular stowage technologies to reduce waste are often contained in nonbinding guidelines rather than mandatory rules. Also, there are insufficient incentives to persuade vessels to retrieve abandoned gear they come across while fishing. The 2018 FAO Committee on Fisheries meeting received a report that approximately 46% of the 79 000 tonnes of ocean plastic in the Great Pacific Garbage Patch is made up of fishing nets, some as large as football fields (Giskes et al. 2018).

Ocean plastic pollution was a focus at the 2022 UN Oceans Conference, which issued a declaration in support of an earlier decision by the UN General Assembly to start negotiations for a global plastics treaty. This initiative has been welcomed almost universally, but it must not distract from the fact we already have good international laws regulating ships where plastics are thrown overboard—these are not being enforced properly! An estimated half of ocean plastic pollution comes from some 4.5 million small fishing vessels operating in national and international waters. Recent research suggests more than 50 000 kg of plastic enters the oceans annually from industrial fishing gear alone. These numbers give an indication as to why plastics have been found all over the globe from the deepest part of the ocean in the Mariana Trench, in the islands of Alaska, and in remote regions such as the Pitcairn Islands in the South Pacific.

Better implementation and enforcement of existing laws would be a much faster way of addressing ship-sourced plastic pollution than waiting for a new treaty to be adopted. Two principal regimes have been developed under the IMO. One is the London (dumping convention) regime, which regulates the deliberate dumping of plastic waste at sea from vessels and platforms. The other is MARPOL, which regulates both deliberate and accidental discharge of plastics from vessels. Plastic pollution has also become an issue for regional fisheries management organizations (RFMOs). They collaborate with the UN Food and Agricultural Organization (FAO) on various initiatives to minimize the loss of fishing gear and the effects of ghost fishing. For example, the Commission for the Conservation of Antarctic Marine Living Resources (CCAMLR), which manages fisheries in the Southern Ocean, bans the use of plastic packaging bands on most vessels.

There are approximately 50–75 trillion pieces of plastic and microplastics currently in the ocean. Approximately 270 000 tonnes of plastic pollution float on the ocean's surface; about 1% of all plastic pollution in the oceans is on the surface. The other 99% stays below the surface. As of 2021, there are at least 165 million tonnes of plastic pollution in the world's oceans (GreenPeace 2019).

Plastics are one of the most extensive types of marine debris. They are commonly used in many items, and as society has developed new uses for them, the variety and quantity of plastic items found in the marine environment has increased dramatically. Plastics are now known to break down into smaller components, called microplastics (see below) having diameters <0.5 mm, and their associated toxic chemical components contribute to human and wildlife health risks as the toxic microplastics are ingested and move through the marine food web. Globally, approximately eight million pieces of plastic enter the ocean every single day, some of it wind-blown but most of it dumped deliberately. Approximately 60–95% of the waste currently in the oceans is plastic waste (Ritchie et al. 2023). Approximately 20% of the plastic pollution in the oceans is from marine activities, like commercial fishing.

Seabirds are especially vulnerable to plastic pollution; a recent study found plastic in 90% of seabirds (Clark et al. 2023). In addition to plastic ingestion, other debris such as packing bands, balloon strings, rubber bands, six-pack rings, and mesh bags can lead to entanglement. Entanglement affects many different types of animals: many sea bird species, 9 cetacean species, 11 pinniped species, 31 invertebrate species/taxa, and 6 sea turtle species reported as entangled in marine debris in the United States. Recovery plans for endangered species, including loggerhead, leatherback, green, and sea turtles, all identify marine debris as one of the highest priority threats facing these protected species.

3.3.3 Data Considerations

Despite the absence of global estimations of sea-based marine litter, the number, geographic spread, quality, and consistency of research studies documenting the distribution of litter and microplastics have increased in recent years. In most litter studies, results are reported as numeric counts of the abundance or density of plastic items or particles, sometimes sub-divided by size class or by type (e.g., fibers, fragments, and films) or sometimes only as aggregate values for the entire size range included in the counts (GESAMP 2019). While these reporting methods allow for comparison of plastic contaminant loadings within and among studies, they do not provide a precise estimate of pollution from this source.

Data and knowledge gaps in our understanding of ocean use other than fishing, aquaculture, shipping and boating, and ocean dumping as sea-based sources of litter include the following:

- Contributions of plastics from offshore oil and gas exploration and extraction.
- Quantifying levels and frequency of marine litter abandoned from the offshore oil and industry require further study to accurately assess the types for discharge into the marine environment.

The most comprehensive data available is for North Sea operations (this is the highest concentration of offshore oil and gas activity exploration); however, different materials and chemicals may be used in different regions and thus further global studies on marine debris originating from the ocean industrial activities are required.

- Shark and stinger net loss: While the use of shark and stinger nets is regionally described, the quantity or frequency of loss, either partial or complete, is not reported and there is little to no obligation to report loss to national authorities.
- Quantity of weather balloons lost at sea: Most weather balloons released at sea will not be retrieved and the majority become marine debris.
- Global estimates of fireworks as sources of marine litter: Understanding contributions to marine debris from at-sea fireworks displays is challenging.
- Quantity of abandoned scientific research equipment: while not a significant contributor of global marine litter in compared to other sources (e.g., oil and gas exploration and drilling).

3.3.4 Microplastics

Most plastics in the ocean break up into very small particles. Microplastics come from a variety of sources, including larger plastic debris that degrades into smaller and smaller pieces under the effects of UV light. These small plastic bits are called "microplastics." Other plastics are intentionally designed to be small. They pass unchanged through waterways into the ocean. Different groups of primary microplastics exist: (i) those that are specifically created with a size less than 5 mm, for particular purposes; (ii) secondary microplastics are those that originate from the disintegration or fragmentation of macro and microplastics due to the action of physical agents and UV rays (fibers, fragments, films, and foams); and (iii) tertiary microplastics are those used for the preproduction of plastics and reach the environment in the same state in which they were produced (pellets) (Anderson et al. 2016). It is *recommended* that a common methodology be developed for the study of microplastics in marine biota, including quality control studies that prevent underestimates of the impact of contamination due to the methodology. In this sense, we suggest that all studies of microplastics are conducted using modern analytical methods.

3.3.4.1 Plastic Pollution Solutions

Flag States should use their legal powers under the UNCLOS to take action against vessels entering their ports if there is evidence they have abandoned or negligently lost fishing gear at sea. They should require their own vessels to mark their equipment and create financial incentives so that floating fishing gear can be retrieved and safely disposed of.

The London regime has a robust compliance process that could be more regularly used to address the dumping of fishing gear and highlight this issue at an international level. While the new plastics treaty may ultimately play an important role in addressing ocean plastics, we do not have to wait until then to better address plastics pollution from ships, better enforcement of current laws would be a great advancement on what we already have.

At the UNEP Conference in February 2023, nations delivered a win for the planet by agreeing to negotiate a global agreement to end plastic pollution, including in the marine environment. The Intergovernmental Negotiating Committee developing a legally binding instrument on plastic pollution held its first meeting in Uruguay in late November 2023. Many governments confirmed their commitment to developing an instrument that addresses the full life cycle of plastics, protects human health and the environment, and pays special attention to the circumstances of countries most in need.

3.4 Sewage

The discharge of raw sewage into the sea can create a serious health hazard. Sewage can also lead to oxygen depletion and can be an obvious visual pollution in coastal areas—a major problem for countries dependent on tourist industries. The main sources of human-produced sewage are land based—such as municipal sewers or treatment plants, but the discharge of sewage into the sea from ships also contributes to marine pollution.

MARPOL (Annex IV) contains a set of regulations regarding the discharge of sewage into the sea from ships, including regulations regarding the ships' equipment and systems for the control of sewage discharge, the provision of port reception facilities for sewage, and requirements for survey and certification. It is generally considered that on the high seas, the oceans are capable of assimilating and dealing with raw sewage through natural bacterial action. Therefore, the regulations in

MARPOL Annex IV prohibit the discharge of sewage into the sea within a specified distance from the nearest land, unless otherwise provided.

Governments are required to ensure the provision of adequate facilities at ports and terminals for the reception of sewage, without causing delay to ships. Annex IV entered into force on 27 September 2003. A revised Annex IV was adopted on 1 April 2004 and entered into force on 1 August 2005. The revised Annex applies to ships, engaged in international voyages, of 400 gross tonnage and above, or which are certified to carry more than 15 persons. Annex IV requires ships to be equipped with either an approved sewage treatment plant or an approved sewage and comminuted disinfecting system or a sewage holding tank. Sewage that is not comminuted or disinfected may be discharged at a distance of more than 12 nautical miles from the nearest land when the ship is then subject to various regulations. In 2011, MEPC 62 adopted (by resolution MEPC.200(62)) an amendment to MARPOL Annex IV, which entered into force on 1 January 2013. The amendment introduced a definition for a "*Special Area*" as well as relevant requirements for the discharge of sewage from passenger ships in special areas and for port reception facilities.

The discharge of sewage from passenger ships within a special area is generally prohibited under the new regulations, except when the ship has in operation an approved sewage treatment plant that has been certified by the Administration (see resolution MEPC.227(64)). The sewage treatment plant installed on a passenger ship intending to discharge sewage effluent in special areas should additionally meet the nitrogen and phosphorus removal standard when tested for its Certificate of Type Approval by the Administration (resolution MEPC.227(64), Section 3.2) (IMO 2023).

3.5 Ship Collisions and Pollution

Ship collisions, sunken ships, and groundings on coastal landforms (see Figure 3.1) can lead to a range of impacts including:

- collision impact, physical damage, direct death of organisms due to pressure impact;
- suspended sediment plumes released following collisions, smothering nearby organisms, and impacting larval recruitment and development;
- fuel oil leakage with direct toxicity to some organisms and possible smearing of surficial organisms;
- cargo spillage (oils, solids such as iron ore, coal—impacts of fine materials), toxic chemicals (cyanide, arsenic, acids, etc.), impacts of containers and drums (more physical impact outcomes) (Jones 2007);
- release of biofouling anti-barnacle chemicals (e.g., copper-based) (Negri et al. 2002);
- release of ballast water with exotic organisms (Coles and Eldredge 2002);
- long-term impacts if grounded vessel is not removed—slow release of toxic material.

3.5.1 Ship Collisions with Megafauna

Increasing global trade has led to increasing numbers and sizes of merchant vessels traveling the world's oceans. At the same time, improved global strategies for the protection of megafauna are leading to higher populations and increased frequency of collisions between ships and fauna. Whales are the most impacted organisms, but manatees, basking sharks, dugongs, dolphins, and turtles are also at risk. Ship strikes have been reported to kill more than 20 000 whales each year

(MSC 2021). The problem is worldwide, but a geographic analysis of the strike data (Tournadre 2014) indicated that there are key spots where strikes are extremely common. As might be expected these zones are where major shipping routes overlap with whale prime feeding areas (Halpern et al. 2015). Further modeling by Pirotta et al. (2018) indicated that the application of a "road ecology" model could provide ships with better predictions on the detrimental effects of future expansion of shipping routes on marine megafauna. Other actions that have been considered potentially helpful include: reducing the speed at key locations; altering shipping routes at key times of whale migration; and improving communication systems between ship crew and coastal observers to warn of potential problems.

3.5.2 Cruise Ships

One class of ships that is rapidly increasing in numbers and potential for the creation of pollution problems is cruise liners. In the past 20 years, cruise liner numbers have doubled to 450 and the size and capacity have also greatly expanded. These ships are managed by over 50 companies. The ships have the capacity to cater for about 673 000 passengers, with the largest ships now about 240 000 tonnes, 350 m in length, and carrying more than 5000 passengers. Two components of this activity are the impacts of the vessels themselves and the impacts of passengers when involved in off-ship activities. The direct impacts of the cruise ships include the (mostly illegal) spills of oil, sewage, and garbage. Ships are supposed to offload sewage and solid wastes into designated facilities in ports, but there is substantial anecdotal evidence of at-sea disposal (C. Haley and J. Morrison, personal observations). Given that a modern cruise ship with 3000 passengers will generate 55 000–75 000 l of wastewater per day, the cost of downloading and treating such volumes will be a significant impost.

One particular group of cruise vessels requires special mention—those linked to reef tourism (Smith 1988). These create many of the common ship impacts (anchoring and chain damage, fuel spillage, and waste discharge), but in addition, they bring tourists in close proximity to reefs through SCUBA diving and reef walking activities. SCUBA diving was, for a long time, thought to have minimal impact on reefs, but evidence of broken coral fragments, re-attached corals, abraded corals, and dead organisms has been commonly found at heavily used sites globally (Barker and Roberts 2004; Broad et al. 2023), with the extent of damage linked to the types of corals present. Although many divers touch the reef during dives, a minority cause most of the damage with inexperienced divers thought to create many problems. Most damage is caused by flippers (direct damage to substrate and mobilization of sediment), followed by hands, knees, and equipment collisions, and even camera users have been observed to create reef damage. Reef walking also causes significant mechanical damage, along with the removal of materials (Woodland and Hooper 1977, Williamson et al. 2017). The extent of damage depends on the amount of activity and also the coral species impacted. Recovery rates also depend on species impacted and length of time between walking groups.

3.6 Ports

Ports are a vital constituent of global shipping infrastructure. They are the locations where cargoes are loaded and unloaded from ships, where connections between offshore and inland freight are made, if sufficient appropriate land is available, port areas can be used for manufacturing, shipbuilding, and repair yards. Ports have important characteristics; they are located in sheltered areas, protected from storms so that ship operations can be carried out even when weather conditions are unfavorable. They have zones of deep water to facilitate anchorage at all phases of local tides. They should be capable of handling and appropriate disposal of sewage and solid waste (including plastics) from docked ships. Shore-based electrical power supplies are also useful as they can connect to docked vessels, enabling the ship engines to be turned off, thus reducing air pollution in the port.

The protective nature of ports whereby some natural or man-made barriers restrict wave and tidal movements of water into the body of the port may have less welcome effects in the port whereby pollutants that have entered the port are restricted from escaping and they will remain in the port water until eventually they get adsorbed into the sediments where they will remain for a long time, contaminating port organisms and interfering in their life cycles.

While efficient ports are vital to the economic development of their surrounding areas, the related ship traffic, the handling of the goods in the ports, and the hinterland distribution can cause several negative environmental impacts. Shipping has an environmental impact both in ports, as well as in the immediate vicinity of the ports. Examples of these impacts are noise from ship engines and machinery used for cargo handling, GHG exhausts from the ship's main and auxiliary engines, and dust from the handling of substances such as iron ore, sand, and coal. Poor management structures for ports can lead to contamination of the port area and unsightly views (Figure 3.3). Many ports have developed land in areas adjacent to the port for industries that can integrate their operations with cargo coming to the port. Ships bring in raw materials and industry turns them into exportable products and further dispatches them via ships (see Figure 3.4).

Road and rail traffic to and from the port area cause additional environmental problems. Public authorities at various administrative levels have put in place a wide range of policy instruments to

Figure 3.3 Poor management structures for ports in developing countries have often led to unsightly and toxic nearshore zones such as this in Walu Bay, Fiji. *Source:* Photo by R.J. Morrison.

Figure 3.4 Dalian, China. Many ports are surrounded by industrial zones where imports are used as starting materials and manufactured products are exported. This is an example of a port with large petroleum-based manufacturing operations contributing to pollution by hydrocarbons. *Source:* Photo by R.J. Morrison.

limit negative environmental impacts from ports in relation to: near-port shipping activities as such (e.g., the fuels that may be used, requirements regarding the treatment of ballast water); the handling of the goods in the ports (e.g., emission standards for the handling equipment and limits on permitted noise levels); and the transport of the goods to the hinterland (e.g., emission standards for vehicles used in the transport and improvements in better road and rail infrastructure).

3.6.1 The Effect of Port Changes for Shipping on Biodiversity

Habitat loss was recognized by the International Union for the Conservation of Nature (IUCN) as the first cause of biodiversity loss. Habitats were never really considered in these port biodiversity studies, apart from the studies on artificial structures in harbors to encourage the growth of native species populations or maintain species assemblages and interactions. One of the most fundamental relationships in ecology is the relationship between an increase in available areas for habitat and an increase in species richness. There seem to be few studies on the effect of port size on biodiversity—do larger ports host greater biodiversity? In fact, there is a significant deficiency in the information regarding the biodiversity of ports; Madon et al. (2023) found that prior to 2016, there were very few papers (1 or 2 per year), and from 2015 to 2021, there were on average 18 papers per year. The limited number of pre-2016 papers may have arisen from the reluctance of ecologists to study areas with prior high human contact. Such studies are affected by the influence of port activities on diversity, for example, types of cargoes, size of ships, type, and extent of dredging.

Recent studies have started to highlight that the surrogate artificial habitats in ports can provide key heterogeneous microhabitats and structural refugia for many species. No study assessing the effect of port populations on adjacent populations, habitats, and food webs was found. It is very unlikely that port "resident" populations have no influence on the survival or fitness of "newly arrived" populations. Ports may allow the establishment and coexistence of more resistant species but there was little on source-sink dynamics, on ports as potential ecological traps, disease pathways, or prey–predator dynamics.

Ports cannot be considered isolated and further better knowledge of these dynamics through the integration of ports in the ecological dynamics of coastal ecosystems is a necessary step toward the development of relevant measures for coastal biodiversity protection. The surfaces of various marine vehicles and almost all marine infrastructure in contact with seawater are armored by protective coatings against corrosion and fouling. Antifouling agents (e.g., copper, zinc, tin, and synthetic organic compounds) are designed to be toxic to a range of marine organisms, inhibiting photosynthesis and growth of bacteria, algae, barnacles, etc.; they are designed to be released slowly (1–3 $\mu g/cm^2$/day, Takahashi 2009) into the ambient environment and persist in seawater. Ports, harbors, marinas, and tourist spots are locations of high densities of marine vessels, and adjacent coral reefs and mangroves may become vulnerable to these toxic chemicals when released from the original treatment site.

3.7 Marine Coatings

As noted earlier, shipping accounts for about 90% of global freight transport. This activity is carried out despite the marine environment being very corrosive for many of the materials used in ocean industries. Seawater is an electrolyte (dissolved salt concentrations averaging 34 000 mg/kg) and carries substantial quantities of marine organisms and their metabolites that enhance the erosivity of the marine environment. Once unprotected hard surfaces are immersed in seawater, two complementary mechanisms occur, contributing to the problem for shipping:

a) microbiologically influenced corrosion (MIC) is the corrosion of metal surfaces that is accelerated directly by the life activities of microorganisms or indirectly by their metabolites (Videla and Herrara 2005);
b) marine biofouling (MBF) is the undesirable settlement and accumulation of marine microorganisms, plants, and animals on submerged surfaces. This increases the frictional resistance of the surface, causing increased fuel consumption and accelerating corrosion of the solid surfaces (Lehars et al. 2012).

Metal surfaces in contact with seawater will rapidly adsorb fouling organisms (organisms that roughen surfaces, increasing friction between water and solid surfaces), and a marine biofouling process is induced, whereby a community of organisms settles and grows on external submerged surfaces.

Barnacles are a good indicator species to illustrate the issues arising from the use of coatings in the marine environment. They have been linked to damage to ship hulls since ancient times. Over 1000 different species have been identified living in ship communities. Barnacles start life as much more mobile organisms as they are released about tens of thousands of tiny larvae. After floating in the sea and developing into adults, they adsorb onto a solid surface. This is a step in the biofouling process that starts when the surface is placed into seawater. Very quickly organic compounds in the water start adsorbing on the surface and the evolution of a surface ecosystem involving algae, barnacles, and oysters occurs. This coating of marine micro- and meso-biota very quickly becomes thick enough to significantly increase drag. The search for some mechanism to reduce drag has been a target of marine ecological/mechanical research for a very long time.

Biofouling occurs throughout the marine environment and is a major problem for shipping as the growth of organisms on vessel hulls increases frictional drag which in turn reduces speed or involves increased fuel consumption to maintain speed—as fuel accounts for 60% of ship operating costs, biofouling is a significant issue for ship owners/operators. A range of technologies has been used to address the problem, starting with the Phoenicians (1500–500 BCE) who used lead and copper sheets to minimize biofouling in their wooden vessels. By the mid-1800s, coatings containing arsenic, copper, lead, and mercury were increasingly common, but while these were effective biocides, they were short-lived requiring frequent hull cleaning and repainting (Daffron et al. 2011; WHOI 1952).

A revolutionary change took place with the discovery of an effective antifoulant based on trialkyltins, especially tributyl tin (TBT) in the 1950s. Starting from the 1950s, antifouling (AF) paints incorporating tributyltin (TBT) and formulated in copolymer paints with cuprous oxide (Cu_2O) were routinely used; these were very effective as the TBT copolymer provides both the biocide and the paint matrix, and this hydrolyses in seawater to release the TBT leaving an unstable surface layer that gradually erodes to expose a fresh layer of active paint (Lewis 1998).

Unfortunately, TBT proved to be too effective in certain situations. By the 1980s, several oyster farmers in France noticed major declines in numbers due to increased spatfall anomalies in larval development and shell malformations affecting 80–100% of individual oysters. TBT was subsequently prohibited on vessels <25 m in France (Alzieu et al. 1986), thus mainly targeting recreational vessels that spend long periods in enclosed marinas allow passive leaching of TBT (Okori et al. 2011) to reach concentrations >1 μg/l.

At about the same time, Imposex (the development of male characteristics on female gastropods) was identified and linked to TBT contamination (Gibbs and Bryan 1986) and this led to further TBT restrictions on small vessels in the United Kingdom, the United States, Canada, Australia, and the EU (Champ 2001). The IMO intervened and facilitated the development and adoption of the International Convention on the Control of Harmful Anti-fouling Systems (AFS Convention 2001), which led to a ban on the application of TBT to all vessels.

The surfaces of various marine vehicles and almost all marine infrastructure under 25 m must be protected. Organic coatings are generally effective, relatively easy to apply, and moderately inexpensive, so they are often a first choice. They must be extremely robust as they have to operate in conditions of high salinity, high UV radiation, hot–cold temperature variations, wet-dry cycling, and for marine applications antifouling. Generally, coatings play two distinct roles: (i) isolating the substrate from the environment; (ii) changing the substrate surface to the desired properties. The barrier effect in the first role is responsible for the corrosion resistance while the antifungal properties are dependent on the surface functionality in the second role (Song and Feng 2020).

The final properties of the coating can be affected by the conditions during application. For example, salt crystals may be precipitated in pores or defects during the drying segment of a wet-dry cycle. This may lead to salt crystal-assisted degradation of the organic coating. High UV during application can cause changes in the polymer structure leading to changes in the final product. Coating developers are continually seeking new environmentally friendly additives such as antibiotics, quaternary ammonium ions, or biocides. The organic coating industry has been searching for green products for decades; major efforts have been made, but unfortunately, the antifouling requirements severely limit the options of coating materials and manufacturing procedures.

3.8 Conclusions

Marine transport is an essential part of our everyday lives. Massive quantities of materials are moved around the world's oceans as part of the commercial activities providing food, raw materials, finished products, employment, and recreation that people often take for granted. This transport involves a large number of vessels of varying sizes, shapes, and capacities. Despite the best intentions of many operators and crews, pollution incidents occur regularly including discharges following ship collisions, accidental discharges as part of ship operations and bunkering, inappropriate disposal of ship wastes (waste oil, sewage, and garbage), and transfer of petroleum products from ship to shore. The explanations for these failures are complex and multifaceted but include ship breakdown due to poor servicing/maintenance, failure of crew to follow operational guidelines, and insufficient enforcement of regulations designed to minimize pollution problems.

Inadvertent transfer of marine pests through ballast water on ships is an ever-present serious problem and quarantine agencies, especially in ports, must maintain surveillance to minimize further transfers. The impacts of ships include those caused by maritime transport in general, for example, poor waste disposal (plastics, chemicals, and organic matter), fuel spillages and leakages, anchor impairment and damage created by collisions with other ships, coastal infrastructure, and landforms. The problems with ships are also found in ports with the additional interaction with land-based activities in port zones. This can lead to serious issues for port environments if the area is not managed carefully. Ships are prone to biofouling if the hulls are not cleaned and covered with protective coatings. Getting suitable coating materials is difficult as the top-quality items are costly. A balance between cost and effectiveness is an owner's dilemma.

The global cost of marine pollution from shipping is impossible to assess. Direct costs, including damage to ships, marine infrastructure, and equipment are in the hundreds of millions of dollars per year. Indirect costs, such as cleaning up oil spills, etc., vary, but are probably in the tens of millions of dollars per year, while the costs of impacts on biological systems are inestimable (UN Industrial Development Organization, UNIDO at the following website. https://www.unido.org/stories/vision-we-need-broaden-discussion-circular-economy).

The capacity of regulatory authorities to monitor ships when out at sea to determine if they are complying with local regulations is limited. There may be financial constraints and weather restrictions. The recent rapid development of drone technology may be utilized to improve monitoring activities. Another important problem is the very long time to get legislation drafted or updated for controlling marine pollution. Regulators should look for the application of current regulations to see if they cannot be implemented more effectively while updates are being prepared. Strong support should be extended to the current process to develop a global treaty to manage plastics for improved protection of the environment.

References

Alzieu, C.L., Sanjuan, J., Deltrelt, J.P., and Borel, M. (1986). Tin contamination in Arcachon Bay: effects on oyster shell anomalies. *Marine Pollution Bulletin* 17: 494–498.

Anderson, J.C., Park, B.J., and Palace, V.P. (2016). Microplastics in aquatic environments: implications for Canadian ecosystems. *Environmental. Pollution* 218: 269–280. https://doi.org/10.1016/j.envpol.2016.06.074.

Arcadis (2012). Final Report: Marine Litter study to support the establishment of an initial quantitative headline reduction target – SFRA0025 European Commission DG Environment Project number BE0113.000668, 315 p. https://ec.europa./eu/environment/marine/good-environmental-status/descriptor-10/pdf/final_report.pdf

Barker, N.H.L. and Roberts, C.M. (2004). SCUBA diver behavior and the management of diving impacts on coral reefs. *Biological Conservation* 120: 481–489.

Broad, A., Rees, M., and Davis, A.R. (2023). Anchor and chain scour as disturbance agents in benthic environments: trends in the literature and charting a course to more sustainable boating and shipping. *Marine Pollution Bulletin* 161: Article Number 111683.

Champp, M.A. (2001) The status of the treaty to ban TBT in marine antifouling paints and alternatives. *Proceedings of the 24th UJNR (US/Japan) Marine Facilities Panel Meeting*, Hawaii (7–8 November 2001).

Cirimanna, R., Bright, F.V., and Pagliaro, M. (2015). Ecofriendly antifouling marine coatings. *Sustainable Chemistry and Engineering* 3: 559–565.

Clark, B.L., Carnerio, A.P.E., Reiman, F.L. et al. (2023). Global assessment of marine plastic exposure risk for oceanic seabirds. *Nature Communications* 14: Article Number 3665.

Coles, S.L. and Eldredge, L.G. (2002). Nonindigenous species introductions on coral reefs. A need for information. *Pacific Science* 56: 19–209.

Daffron, K.A., Lewis, J.A., and Johnston, E.L. (2011). Antifouling strategies: history and regulation, ecological impacts and mitigation. *Marine Pollution Bulletin* 82: 453–465.

Doney, S.C., Fabry, V.J., Feely, R.A., and Keplas, J.A. (2016). Ocean acidification: the other CO_2 problem? *Washington Journal of Environmental Law and Policy* 6 (2): 212.

Eunomia (2016). European Commission (DG ENV) study "to support the development of measures to combat a range of marine litter resources". UNEP European Office, Geneva, 13 p..

Galgani, F., Pham, C.K., Claro, F., and Consoli, P. (2018). Marine animal forests as useful indicators of entanglement by marine litter. *Marine Pollution Bulletin* 135: 735–738.

GESAMP, (2019). Joint Group of Experts on Scientific Aspects of Marine Pollution. Guidelines for the Monitoring and Assessment of Plastic Litter in the Ocean. *GESAMP Reports and Studies No. 99*, UNEP, Nairobi, 123 p.

GESAMP (1993). GESAMP Joint Group of Experts on Scientific Aspects of Marine Pollution: Impact of Oil and Related Chemicals and Wastes on the Marie Environment, London, 180 p.

Gibbs, P.E. and Bryan, G.W. (1986). Reproductive failure in populations of the dogwhelk *Nucella lapillus* caused by Imposex induced by the tributyltin from antifouling paints. *Journal of the Marine Biological Association of the United Kingdom* 266: 767–777.

Giskes, I., Baziuk, J., and Pragnell-Raash, H. (2018). *Report on Good Practices to Prevent and Reduce Marine Plastic Litter from Fishing Activities*. Rome: FAO and IMO.

Gold, E. (1997). *Gard Handbook on Marine Pollution*, 2e. Arendal: AssuranceForeneigen Gard 479 p.

GreenPeace (2019). *Ghost Gear: The Abandoned Nets Haunting Our Oceans*. Amsterdam: GreenPeace.

Halpern, B.S., Frazer, M., Potapencho, J. et al. (2015). Spatial and temporal changes in cumulative human impacts on the world's ocean. *Nature Communications* 6: Article No. 7615.

International Underwriting Association (2022). Global Shipping Inventory, IUA Annual Review 2020/2021, Minster Court, Mincing Lane, London EC3R 7AA.

IMO (2014). Guidelines for the reduction of underwater noise from commercial shipping to address adverse impacts on marine life. *21st ASCORBANS Advisory Meeting MEPC.1/Circ.8.33.*

IMO (2023). https://www.imo.org/en/AboutConventions/Pages/International-onvention-for-the-Prevention-of-Pollution-from-Ships(MARPOL)aspx.

Interwies, E., Görlitz, S., Stöfen, A. et al. (2013). Issue Paper to the International Conference on Prevention and Management of Marine Litter in European Seas. Plastics in the Marine Environment, Safety4Sea, UK Publications Office of the European Commission, Bristol, 13 p.

IPCC (2001). *IPCC Third Assessment Report*. Cambridge/New York: Cambridge University Press, 598 p.

ITOPF (2022). *Oil Tanker Spill Statistics 2022*. London: International Tankers Owners Pollution Federation.

Jambeck, J.R., Geyer, R., Wilcox, C. et al. (2015). Plastic waste inputs from land into the ocean. *Science* 347 (6223): 768–771. LC (London Convention) 1972.

Johannes, R.E. and Riepen, M. (1995). Environmental, economic and social implications of the live reef fish trade in Asia and the Western Pacific. Report funded by The Nature Conservancy, the

South Pacific Forum Fisheries Agency and Pew Scholarship in Conservation and the Environment, 83 p.

Jones, R. (2007). Chemical contamination of a coral reef by the grounding of a cruise ship in Bermuda. *Marine Pollution Bulletin* 54: 905–911.

Lehars, M., Margaillan, A., and Brassy, C. (2012). Fouling release coatings: a non-toxic alternative to biocidal antifouling coatings. *Chemical Reviews* 112: 4347–4390.

Lewis, J.A. (1998). Marine fouling and its prevention on underwater surfaces. *Materials Forum* 22: 41–61.

Li, Y. and Ning, C. (2019). Latest research progress of marine microbiological corrosion and bio-fouling, and new approaches of marine anti-corrosion and anti-fouling. *Bioactive Materials* 4: 189–195.

Lindstad, H.E. and Eskeland, G.S. (2016). Environmental regulations in shipping: priorities leaning towards globalization of scrubbers deserve scrutiny. *Transportation Research* D47: 67–76.

Macfadyen, G., Huntington, T., and Cappell, R. (2009). Abandoned, lost or otherwise discarded fishing gear. *UNEP Regional Seas Reports and Studies N. 185; FAO Fisheries and Aquaculture Technical Paper No. 523*, Rome, UNEP/FAO, 115 p.

Madon, B., Romain, D., Torrabla, A. et al. (2023). A review of biodiversity research in ports: lets not overlook everyday nature. *Ocean and Coastal Management* 242: 106623.

MSC (2021). Marine Stewardship Council. https://friendofthesea.org/marine-conservation-projects-and-awareness/save-the-whales-2/site (accessed 9 August 2023).

NAS (National Academy of Sciences) (1975). *Assessing Ocean Pollutants: A Report of the Study Panel on Assessing Ocean Pollutants to the Ocean Affairs Board, Commission on Natural Resources, National Research Council*. Washington: NAS, 10.

Negri, A.P., Smith, L.D., Webster, N.S., and Heyward, A.J. (2002). Understanding ship-grounding impacts on a coral reef: potential effects of anti-fouling pant contamination on coral recruitment. *Marine Pollution Bulletin* 44: 111–117.

Okori, H.K., Fatoki, O.S., and Adekola, F.A. (2011). Sources, environmental levels and toxicity of organotins in the marine environment – a review. *Asian Journal of Chemistry* 23: 473–482.

Pirotta, V., Grech, A., Jonsen, I.D. et al. (2018). Consequences of global shipping traffic for marine giants. *Frontiers in Ecology and the Environment* 17: 39–47.

Richardson, K., Hardesty, B.D., and Wilcox, C. (2019). Estimates of fishing gear loss rates at a global scale: a literature review and meta-analysis. *Fish and Fisheries* 20: 1218–1231.

Riegl, B. (2001). Degradation of reef structure, coral and fish communities in the Red Sea by ship groundings and dynamite fishing. *Bulletin of Marine Science* 69: 595–611.

Ritchie, H., Samborska, V., and Roser, M. (2023). Plastic pollution. www.ourworldindata.org (accessed 12 December 2023)

Schafer, E., Scheele, U, and Papenjohann, M. (2019) Identifying Sources of Marine Litter: Application of the Matrix Scoring Technique to the German North Sea Region. *OSPAR Technical Report*. https://doi.org/10.13140/RG.2.2.23106.8544

Schnurr, R.E. and Walker, T.R. (2019). Marine transportation and energy use. *Earth Systems and Environmental Sciences* https://doi.org/10.1016/B978-0-12-409548-9.-092780-8.

Shi, Y. (2016). Reducing greenhouse gas emissions from international shipping: is it time to consider market forces measures? *Marine Policy* 64: 123–134.

Smith, S.H. (1988). Cruise ships: a serious threat to coral reefs and associated organisms. *Ocean and Coastal Management* 11: 231–248.

Song, G.-L. and Feng, Z. (2020). Modification, degradation and evaluation of a few organic coatings for some marine applications. *Corrosion and Materials Deterioration* 1: 408–442.

Statistica (2023). Annual production of plastics worldwide 1950–2022. https://www.statis/10.com/statistics/282732/global-production-of-plastics-since-1950 (accessed 28 November 2023)

Takahashi, K. (2009). Release rate of biocides from antifouling paints. In: *Ecotoxicology of Antifouling Paints* (ed. T. Arai, H. Harino, M. Ohji, and W.J. Langston), 18X. Tokyo: Spring.

Tournadre, J. (2014). Anthropogenic pressure on the open ocean: the growth of ship traffic revealed by altimeter data analysis. *Geophysical Research Letters* 41: 7924–7932.

UNCLOS (1982). *United Nations Convention on the Law of the Sea*. New York: United Nations Division for Ocean Affairs and the Law of the Sea, 194 p.

Videla, H.A. and Herrara, L.K. (2005). Microbiologically influenced corrosion: looking to the future. *International Journal of Microbiology* 8: 169–190.

Vlachogianni, T., Fortibuoni, T., Ronchi, F. et al. (2018). Marine litter on the beaches of the Adriatic and Ionian seas: an assessment of their abundance, composition and sources. *Marine Pollution Bulletin* 131: 745–756. https://doi.org/10.1016/j.marpolbul.2018.05.006.

WHOI (1952). *Marine Fouling and Its Prevention*. Falmouth, MA: Woods Hole Oceanographic Institute.

Williamson, J.L., Brynes, E.E., Clark, J.A. et al. (2017). Ecological impacts and management implications of reef walking on a tropical reef flat community. *Marine Pollution Bulletin* 114: 742–750.

Woodland, D.J. and Hooper, J.N.A. (1977). The effects of human trampling on coral reefs. *Biological Conservation* 11: 1–4.

Worm, B., Lotze, H.K., Jubinville, J. et al. (2017). Plastic as a persistent marine pollutant. *Annual Review of Environmental Resources* 42: 1–26.

4

Offshore Oil and Gas Operations: Environmental Impacts and Mitigation Methods

Savinda Arambawatta Lekamge[1], *Ishani Mahawaththa*[2], *and S.D.N.A.M.A.M. Weerasinghe*[3]

[1] *School of Earth and Atmospheric Sciences, Faculty of Science, Queensland University of Technology, Brisbane, Queensland, 4000, Australia*
[2] *Department of Biological Sciences, University of North Texas, Denton, TX, 76201, USA*
[3] *School of Biomedical Engineering, Department of Occupational Health and Safety Engineering/Disaster Management, Inje University, 50834, South Korea*

4.1 Introduction

Offshore oil and gas exploration plays a major role in supplying the world's expanding energy needs. According to the U.S. Energy Information Administration, in 2024, the total world production of petroleum and other liquids is 102.76 million barrels per day, and the world's total consumption is 102.84 million barrels per day. Among all the main regions in the world, the United States alone holds the highest consumption of oil per day, which is 20.37 million barrels per day. However, numerous environmental challenges stemming from the oil and gas sectors and associated services have led to the deterioration of natural habitats, biodiversity loss, diminished water and air quality, and exacerbated climate change. Despite these detrimental effects, this industry has also presented opportunities for enhancing the economic well-being of producing nations and bolstering the global industrial landscape. Consequently, a balanced approach is imperative, that upholds environmental standards, addresses climate change, and sustains hydrocarbon production, or else other sources of energy must be sought to replace oil and gas.

The rapid expansion of oil and gas activities, spanning exploration, extraction, and facility operations, has intensified concerns regarding environmental repercussions. The mounting pressure from both local and international communities necessitates adherence to stringent environmental regulations and a concerted effort to mitigate the adverse impacts across all stages of operations.

Future expansion and advancements within the oil and gas sector hinge on investors' capacity to innovate towards sustainability, deploying cleaner technologies and processes that minimize emissions, liquid discharge, and waste generation. This entails alignment with environmental standards and corporate social responsibility principles. Effective management of emissions, discharges, and waste disposal are indispensable in preventing or mitigating adverse effects and their associated repercussions. Additionally, there is a critical need to establish comprehensive emergency response plans, develop containment facilities, and prioritize the reuse of treated industrial water and waste to minimize environmental release.

Coastal and Marine Pollution: Source to Sink, Mitigation and Management, First Edition.
Edited by Meththika Vithanage, Sameera M. Samarasekara, Bryan D. James, and Christopher M. Reddy.

4.2 Global Status of Offshore Oil and Gas Explorations

Offshore oil and gas operations play a crucial role in satisfying the world's ever-growing energy needs (Figure 4.1). These operations involve extracting oil and gas reserves located beneath the seabed, typically in ocean waters. The significance of offshore operations stems from several factors including abundant resources, technological advancements, diversification of supply, economic impact, environmental and regulatory challenges, and exploration and development (Albeldawi, 2023). These operations encompass exploration, drilling, extraction, and production of oil and natural gas reserves situated beneath the seabed in oceans and seas across the globe. Over the decades, the development of offshore fields has experienced remarkable growth, propelled by advancements in technology, escalating energy requirements, and the depletion of onshore reserves (Dou et al., 2022).

The distribution of wildcats (oil wells drilled without prior geological information), both onshore and offshore, varies greatly worldwide. In 2021, there was a slight uptick in the number of onshore wildcats drilled compared to 2020, totalling 423. However, the success rate of these exploratory wells dipped slightly to 44.9%. Notably, the Gulf of Mexico and the Guyana basin saw the most activity, with 21 and 11 wildcats completed, respectively, surpassing other regions. While deep water and ultra-deep-water drilling represent a smaller proportion of overall wildcats, the reserves discovered in these areas tend to be larger than those found in onshore or shallow water regions. Consequently, they contribute significantly to global reserves, with many new discoveries emerging in 2021 (Dou et al., 2022).

The global status of offshore oil and gas operations reflects a multifaceted interplay of factors including geological characteristics, economic variables, regulatory frameworks, environmental considerations, and technological prowess. Regions endowed with extensive offshore resources, such as the North Sea, the Gulf of Mexico, and offshore areas of Brazil, West Africa, and Southeast Asia, have witnessed substantial investments and activity in offshore exploration and production (Harris et al., 2016).

Offshore operations demand intricate technologies and engineering solutions to surmount the challenges posed by harsh marine environments, encompassing deep waters, high pressures, and corrosive conditions. Advanced drilling techniques, subsea equipment, floating production systems, and pipelines are deployed to access and extract hydrocarbons from beneath the seabed (Narula & Narula, 2019). The global status of offshore oil and gas operations is shaped by market dynamics, geopolitical considerations, and environmental sustainability concerns. Fluctuations in oil and gas prices, regulatory policies, and environmental regulations exert influence on investment decisions and operational strategies in the offshore sector. Additionally, apprehensions regarding the environmental impact of offshore activities, such as potential oil spills, habitat disturbance, and greenhouse gas emissions, have prompted heightened scrutiny and calls for stricter regulations and the adoption of cleaner technologies (UNEP, 2022).

Despite challenges, offshore oil and gas operations remain integral to the global energy mix, contributing significantly to the world's oil and natural gas production. Nevertheless, the industry confronts persistent pressures to balance energy security, economic viability, and environmental sustainability in the pursuit of offshore resources. As the world transitions towards cleaner energy sources and decarburization initiatives gain momentum, the future trajectory of offshore oil and gas operations is contingent upon evolving market dynamics, technological breakthroughs, and regulatory frameworks (IRENA, 2022).

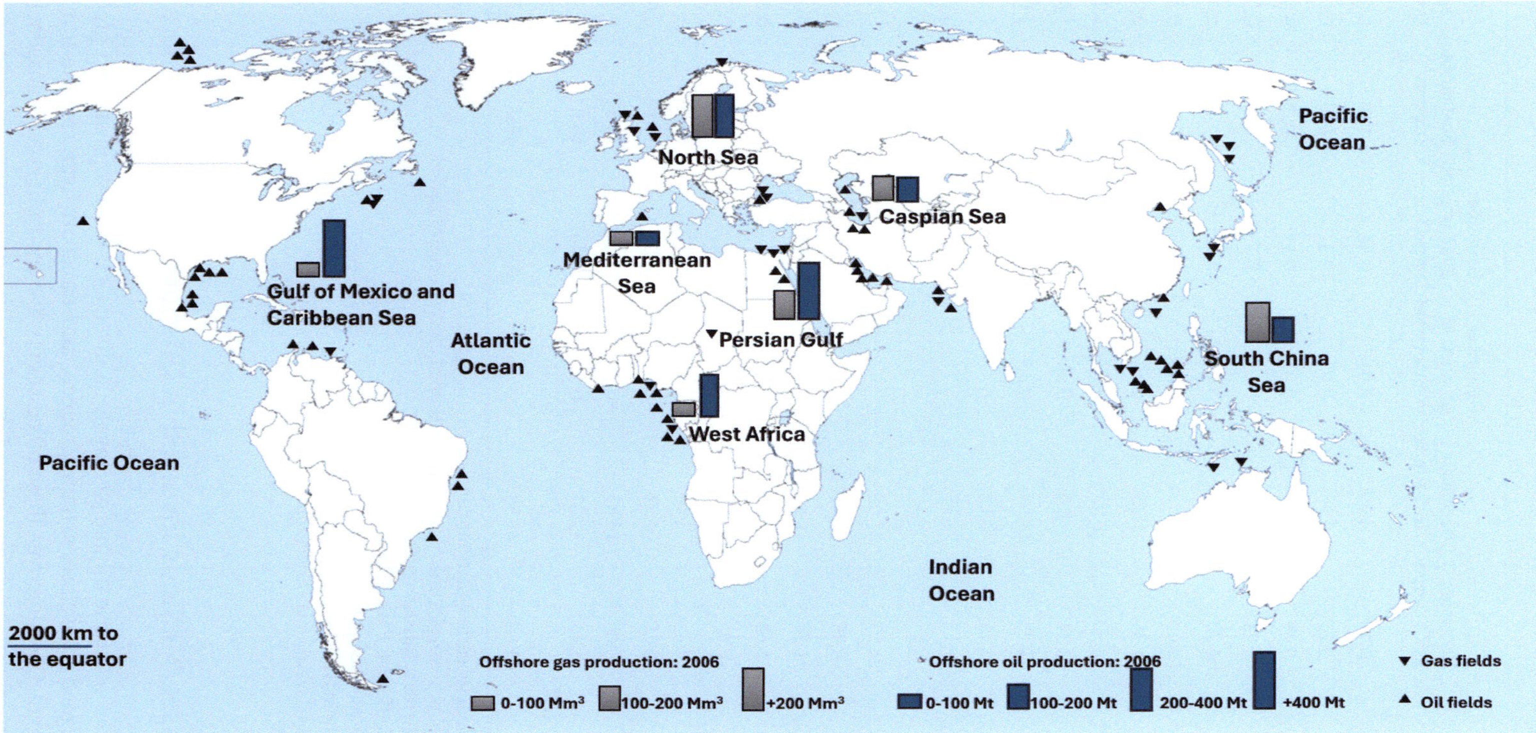

Figure 4.1 World map indicating major offshore fields along with their oil and gas potential.

Table 4.1 OPEC+ crude oil production million barrels per day.

Country	November 2023 supply	December 2023 supply	December production versus target	December 2023 implied target[a]	Sustainable capacity[b]	Effective spare capacity versus December[c]
Algeria	0.96	0.95	−0.01	0.96	1.0	0.05
Congo	0.25	0.26	−0.05	0.31	0.27	0.01
Equatorial Guinea	0.05	0.05	−0.07	0.12	0.06	0.02
Gabon	0.23	0.22	−0.05	0.17	0.23	0.01
Iraq	4.29	4.33	0.11	4.22	4.78	0.45
Kuwait	2.6	2.55	0.0	2.55	2.83	0.28
Nigeria	1.25	1.35	−0.39	1.74	1.38	0.03
Saudi Arabia	8.92	8.95	−0.03	8.98	12.17	3.22
UAE	3.24	3.23	0.36	2.88	4.23	1.0
Total OPEC-9[d]	21.79	21.89	−0.03	21.92	26.96	5.07
Iran[e]	3.21	3.15			3.8	
Libya[e]	1.17	1.18			1.23	0.05
Venezuela[e]	0.8	0.8			0.82	0.02
Total OPEC	26.97	27.02			32.81	5.14
Azerbaijan	0.49	0.48	−0.2	0.68	0.54	0.06
Kazakhstan	1.57	1.6	0.05	1.55	1.67	0.07
Mexico[f]	1.64	1.65			1.68	0.03
Oman	0.8	0.8	0	0.8	0.85	0.05
Russia	9.5	9.48	0.03	9.45	9.98	
Others[g]	0.85	0.82	−0.24	1.06	0.87	0.05
Total Non-OPEC	14.86	14.83	−0.36	13.54	15.58	0.25
OPEC+18 in November 2022 deal[e]	35.01	35.07	−0.39	35.46	40.87	5.3
Total OPEC+	41.83	41.85			48.4	5.39

[a] Includes extra voluntary curbs where announced.
[b] Capacity levels can be reached within 90 days and sustained for an extended period.
[c] Excludes shut in Iranian, Russian crude.
[d] Angola left OPEC effective 1 Jan 2024.
[e] Iran, Libya, Venezuela exempt from cuts.
[f] Mexico excluded from OPEC+ compliance.
[g] Bahrain, Brunei, Malaysia, Sudan and South Sudan.
Source: IEA (2023).

4.3 Processes/Methods Involved in Offshore Oil and Gas Operations

Offshore activities related to oil and gas involve a diverse range of sophisticated procedures and techniques designed to explore, extract, produce, and transport hydrocarbons from below the ocean floor (Figure 4.1). Moreover, these processes involve environmental management activities that ensure the safe and efficient extraction of hydrocarbons from offshore reserves while

Table 4.2 Global oil and gas exploration investment in recent years.

	Annual investment/10^8 USD				
Regions	**2017**	**2018**	**2019**	**2020**	**2021**
Africa	25.9	20.6	20.9	16.2	12.8
Americas	240.7	250.7	212.2	142.7	141.0
Europe	50.2	45.3	49.9	20.3	21.4
Asia Pacific	102.2	107.1	130.9	107.0	131.1
Middle East	1.6	0.6	1.9	1.0	0.3
Central Asia Russia	17.5	9.4	22.3	25.2	6.2
Total	438.1	433.7	438.2	312.5	312.8

Source: IEA (2023).

Table 4.3 Exploration of workload of each regions.

Regions	2D seismic/10^4 km	3D seismic/ 10^4 km^2	Wildcats	Outpost	Major basins	Wildcats of major basins
Africa	1.04	3.35	88	35	Abu Gharadiq	11
					Eastern Niger	11
					Cyrenaica	10
					Northern Egypt	10
Americas	0.98	4.63	139	58	Sureste	26
					Deepwater Guil of Mexico	21
					Guyana	11
Europe	0.92	3.86	96	134	North Sea	30
					Dnieper- Donets	13
					Pannonian	9
Asia Pacific	1.94	3.91	158	158	Indus	22
					Bohai Guil	18
					Eromanga	17
Middle East	0.62	0.54	87	47	Oman	60
					Zagros	12
					Central Arabian	5
					Rub' Al Khali	5
Central Asia-Russia	0.13	1.24	76	159	West Suberian	33
					Amu-Darya	16
					Volga-Urais	12
Total	5.64	17.54	644	591		352

Source: IEA (2023).

Table 4.4 Core fields and exploration characteristics of oil and gas upstream business of international oil companies.

Company name	Upstream core field types	Regions/Countries/Basins	Notes
Shell	Deep water, Integrated natural gas	USA, Brazil, Nigeria, East Africa	Exploration investment is focused on 8 core countries and regions, and 80% of exploration expenditure is used for deep water exploration on both sides of the Atlantic
ExxonMobil	Deep water, Gas, Tight oil	Guyana, Brazil, Mozambique, Papua New Guinea, Permian Basin	Focusing on deepwater venture exploration
bp	Deep water, Gas, Unconventional	Russia, USA Mexico Gulf, Angola, Australia, U.K, Norway, Argentina	Exploration core is focused on rolling exploration in current production areas
Chevron	Unconventional, Gas	USA Mexico Gulf, Brazil, Eastern Mediterranean, West Africa, Australia	Focusing on unconventional oil and gas exploration in North America
Total energies	Conventional, Gas	The United Arab Emirates, West Africa, Brazil, Norway, USA, Mozambique	Actively implement exploration in Africa
Eni	Conventional, Gas	Nigeria, Niger, Angola, Congo and USA Mexico Gulf	Independent exploration, "dual exploration model", focusing on low-risk and high potential areas
Equinor	Deep water, Unconventional	USA, Mexico gulf, Brazil, Canada and Angola	The focus of exploration will decrease from 30 countries in 2017 to 10 countries in 2021

Source: IEA (2023).

minimizing the environmental impacts. These operations are vital in fulfilling worldwide energy needs, yet they present notable technical, logistical, and environmental challenges. Detailed descriptions of each process are given below.

4.3.1 Exploration

The objective of exploration activities is to identify formations containing crude oil and natural gas and gain insights into the reservoir's composition before drilling. Through the application of exploration techniques, geological data is gathered to assess subsurface conditions and determine the potential of a formation to contain hydrocarbons. This information enables exploration teams to evaluate whether further exploration activities are warranted to confirm initial assessments of a significant reservoir of hydrocarbons. Consequently, a pre-drilling image of the reservoir can be acquired, facilitating informed decisions regarding the potential and magnitude of crude oil or natural gas reserves, as well as the initiation of more intensive exploration efforts in promising areas.

In a standard seismic exploration procedure, a series of data receivers (geophones for land operations, and hydrophones for marine operations) are deployed. Explosives or mechanical vibrators are commonly utilized on land, while air guns are employed in marine or offshore settings. Seismic surveys are employed by the offshore oil and gas industry to pinpoint the locations of oil and gas reservoirs beneath the seabed.

Like onshore exploration, the choice of exploration methods utilized for offshore exploration to detect crude oil and natural gas presence is decided based on the characteristics of the site. Indeed, the techniques deployed at a particular site are determined by its specific attributes, making them site-specific. Therefore, there is no single technique capable of providing all the required information to precisely delineate formations and facilitate immediate and successful drilling. For instance, in regions where limited knowledge exists about subsurface formations, preliminary reconnaissance methods are essential to pinpoint potential reservoir systems, subsequently meriting more comprehensive (albeit costly) geophysical investigations.

After identifying an area for further examination, more intricate techniques, like the seismic reflection method, are utilized. Drilling represents the ultimate phase of the exploration program and is indeed the sole method through which a petroleum reservoir deemed viable for development can be confirmed.

The primary approaches utilized for defining and characterizing a site include magnetism (measured by a magnetometer), gravity (assessed by a gravimeter), and sound waves (recorded by a seismograph). These methodologies rely on the physical attributes of subsurface sediments and structures that are amenable to measurement and estimation of the formation's occurrence or production potential. They encompass techniques responsive to applied geophysics principles and can be categorized based on gravitational, magnetic, seismic, electrical, electromagnetic, and radioactive properties. These geophysical methods can be further classified into two categories: those lacking depth control and those providing depth control.

4.3.2 Extraction

The initial step in extracting crude oil from an underground reservoir involves drilling a well into the reservoir (Speight, 2015). Frequently, multiple wells, known as multilateral wells, are drilled into the same reservoir to ensure that the extraction rate remains economically feasible. Additionally, certain wells, referred to as secondary wells, may be utilized to inject water, steam, acids, or diverse gas mixtures into the reservoir. This injection helps elevate or sustain reservoir pressure, thereby ensuring an economically viable extraction rate.

Aside from developing drilling rigs capable of operating at significant water depths, new drilling methods have emerged, aiming to enhance productivity and reduce unit costs. The advancement of directional and horizontal drilling techniques, allowing for the penetration of multiple diverse pay targets, exemplifies technological progress in offshore drilling. The industry now has the capability to minimize costs by utilizing fewer wells to access producing reservoirs at their optimal locations. Horizontal completions within formations further extend the reach of each well through crude oil-bearing (and/or natural gas-bearing) formations, resulting in increased flow rates compared to those achieved through simple vertical completions. These advancements stem from various developments, including the evolution of retrievable whipstocks, enabling drillers to exit cased wells without forfeiting potential production from existing wellbores. Moreover, top drive systems facilitate keeping the bit in the sidetracked hole, while enhancements in mud motor technology permit drilling up to 60° per 100-foot-radius holes without articulated systems. Additionally, pay zone steering systems demonstrate the ability to remain within pay zone boundaries.

Recent advancements in drilling technology encompass the implementation of multilateral and multibranch wells. A multilateral well comprises multiple horizontal (or nearly horizontal) laterals drilled from a single location and linked to a solitary wellbore. Conversely, a multibranch well involves multiple branches drilled from a single site and connected to a single wellbore. While not as prevalent in offshore environments as in onshore settings due to the requirement for pressure-sealed systems, multilateral and multibranch wells play increasingly significant roles in present and prospective offshore development initiatives.

Lastly, prior to undertaking drilling or any offshore activity, comprehensive planning should encompass environmental considerations. In fact, as with any other project with the potential to impact the environment, it is imperative to develop plans and obtain permits before mobilizing equipment to the work site. Following the acquisition of plans and permits, the pre-spud meeting, which typically addresses aspects like well depth, casing points, and rig selection, should also encompass discussions pertinent to the environmental management of the drilling and completion operations. Each site presents unique factors distinct from those encountered elsewhere (site specificity), necessitating strict compliance with site-specific regulations. Moreover, owing to public interest in projects with potential environmental implications, engagement in public forums and early disclosure of project details, even during the pending stage, is advisable. Well-crafted public disclosure statements disseminated to the potentially affected community may not assuage all concerns but can provide clarity regarding project objectives. In such instances, the public may offer viable alternatives that benefit the project developer. Eliminating the element of public surprise, often a catalyst for objections, is essential.

4.3.3 Production

Crude oil and natural gas gradually accumulate over geological times within porous underground rock formations known as reservoirs, located at diverse depths within the Earth's crust. Often, sophisticated equipment is necessary to extract (produce) these hydrocarbons (Gary et al., 2007; Speight, 2015). Typically, oil is found trapped within a stratum of porous sandstone, situated just beneath a dome-shaped anticline (see Figure 4.2) or folded layer of impermeable rock like limestone. Alternatively, in certain formations, oil is trapped along a fault, or fracture in the crust's layers. Whether offshore or onshore, the composition of a reservoir is intricate and distinctly tailored to each site (reservoir-specific), manifesting variances at both microscopic and

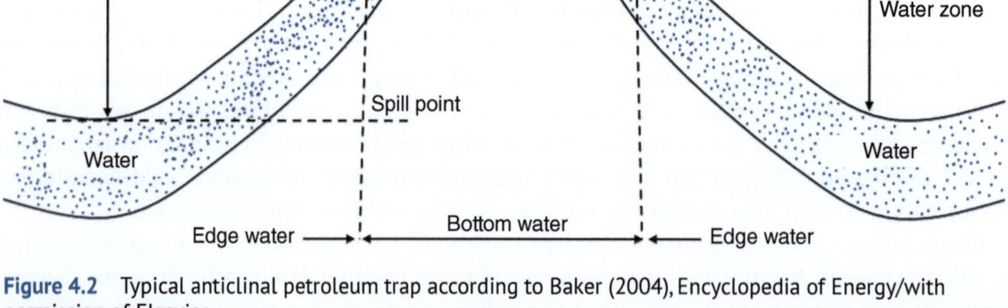

Figure 4.2 Typical anticlinal petroleum trap according to Baker (2004), Encyclopedia of Energy/with permission of Elsevier.

macroscopic levels. Due to the diverse types of accumulations and the wide spectrum of rock and fluid properties, reservoirs exhibit unique responses and necessitate individualized treatment.

Contemporary offshore extraction of crude oil and natural gas occurs within the territorial waters of coastal nations worldwide, encompassing activities both in oceans and lakes. Unlike the past, offshore operators are no longer restricted to traditional wharves or jetties extending into the ocean. Instead, wells are drilled from modern steel or concrete structures, many of which are movable and capable of floating while being transported. These structures facilitate drilling operations in waters exceeding depths of 7500 feet and are often situated more than 200 miles away from the shore. Offshore development strategies encompass a range of concepts, including fixed platforms, floating platforms, subsea wells, and various other specialized units.

Subsea production systems, also known as subsea wells, are generally positioned directly on the seafloor. However, in the shallower Arctic waters, there is a risk of ice keels or icebergs colliding with the equipment. To mitigate this risk, the subsea equipment may be placed in a depression dredged into the seafloor, allowing the ice to pass over it without causing damage. In deeper water depths, pipeline and subsea equipment can be safely positioned directly on the seafloor, below the potential threat of ice keels.

Extracting oil and gas offshore poses greater challenges compared to land-based installations, primarily because of the remote and harsher environmental conditions. Consequently, a significant portion of innovation in the offshore petroleum and gas exploration and production sector is focused on addressing these challenges, which includes the requirement to establish extensive production facilities.

4.3.4 Transport

Most of the crude oil and natural gas fields are situated far away from refineries, where crude oil undergoes conversion into usable products. Consequently, oil must be transported via pipelines and tankers to reach these refineries. However, before crude oil can be transported over long distances, it typically requires some form of treatment near the reservoir. Additionally, railroad cars and motor vehicles are extensively utilized for the transportation of petroleum products.

Fluids extracted from a well typically comprise more than just crude oil, often containing varying amounts of gas, saltwater, or even sand. Processing facilities located on offshore platforms are responsible for separating these undesirable components from the crude oil before it is stored and transported. Thus, separation is a crucial step that must be completed before transportation. This separation and purification process commonly occurs at a central facility that gathers oil from multiple wells.

Gas can be conveniently separated at the wellhead. When the pressure of the gas within the crude oil at the surface is manageable, a simple flow tank equipped with baffles can be employed to separate the gas from the oil under atmospheric pressure. However, if a significant volume of gas is present, particularly in cases where the crude oil is under substantial pressure, a series of flow tanks becomes necessary. The natural gas itself may contain various non hydrocarbon substances as impurities, with hydrogen sulfide being the most prevalent. This compound imparts a noticeable odor to the gas. While a small amount of hydrogen sulfide is considered beneficial as it serves as an indicator of leaks and their location, excessive amounts can be problematic.

Another essential step in preparing crude oil for transportation involves the removal of excess water. Typically, crude oil extracted at the wellhead contains emulsified water, sometimes in proportions as high as 80–90%. However, it is typically required that crude oil transported via pipeline contains significantly less water than what may be present at the wellhead. Specifications often stipulate that the maximum tolerable water content in crude oil for pipeline transportation ranges from 0.5% to 2.0%. Hence, it becomes imperative to eliminate the surplus water from the crude oil before it can be transported.

In an emulsion, the globules of one phase are typically enveloped by a thin layer of an emulsifying agent, preventing them from coalescing into larger droplets. In the case of an oil-water emulsion, the emulsifying agent may consist of the heavier (asphaltic) and more polar constituents. This layer can be disrupted through mechanical, electrical, or chemical means, including the use of demulsifying agents, thereby reducing the proportion of water in the oil to the specified levels, rendering the crude oil suitable for transportation.

The process of transporting crude oil can be further streamlined by blending crude oils from multiple wells, thereby achieving a more uniform composition for delivery to the refinery. Typically, crude oils with similar characteristics are blended, although fluctuations in the properties of individual crude oils may lead to notable variations in the blend's properties over time. However, the practice of blending multiple crude oils before transportation, or even after transportation but before refining, can alleviate the frequent need to adjust processing conditions, which would otherwise be necessary for each crude oil individually.

4.4 Environmental Impacts

Offshore oil and gas operations have a range of environmental impacts, both stemming from the various methods used to explore the site prior to drilling and from the drilling activity itself. These different impacts will be discussed here, and a summary of the impacts is given in Figures 4.3 and 4.4.

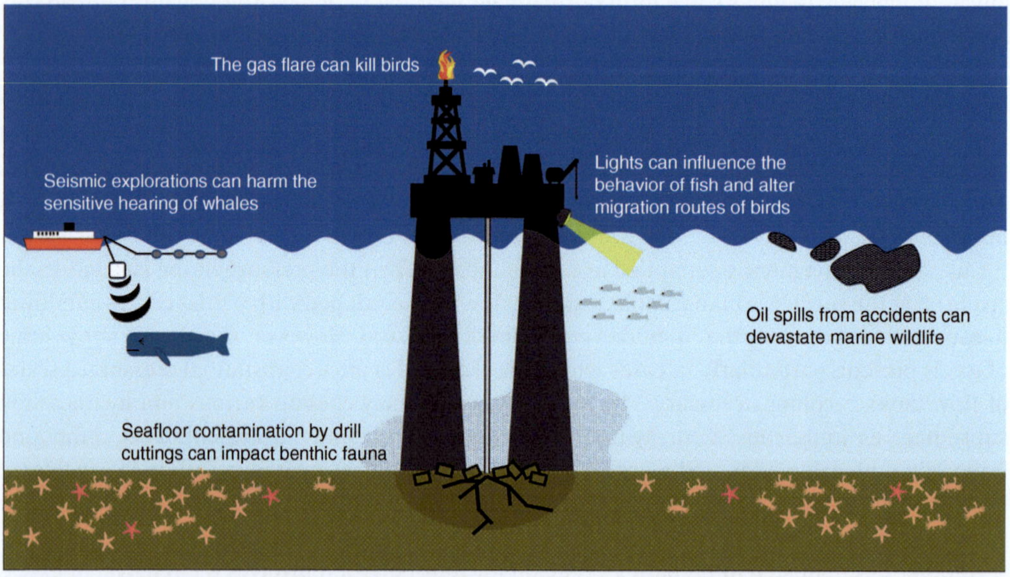

Figure 4.3 Summary of environmental impacts from offshore oil and gas operations.

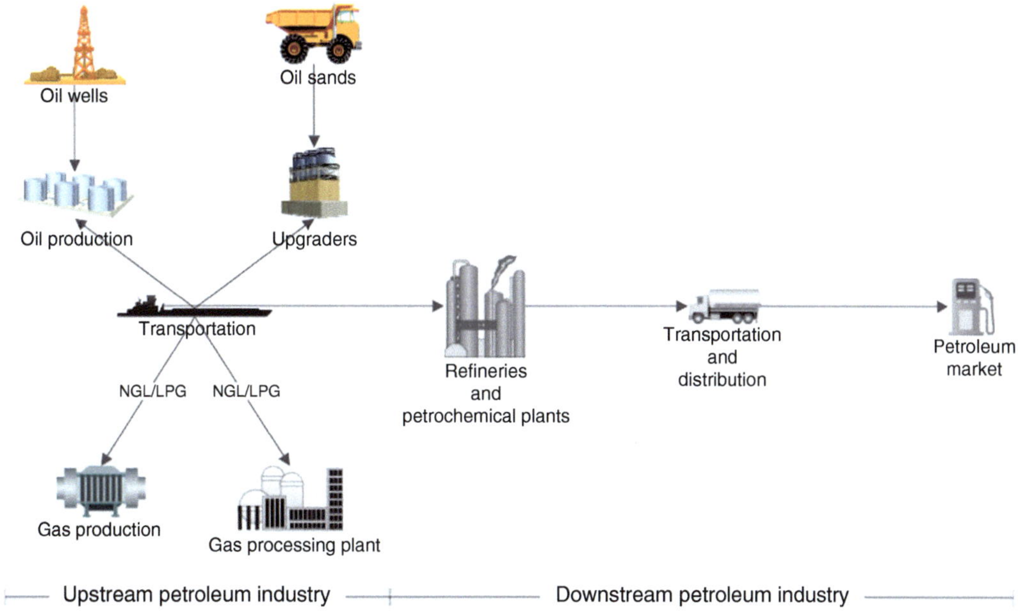

Figure 4.4 Flow diagram of the offshore oil and gas production.

4.4.1 Noise

Two broad categories can be used to classify the primary sources of noise resulting from offshore oil and gas operational activities: (i) construction and preparation (such as road construction, site and well pad preparation, and truck traffic), and (ii) production and completion (such as flaring operations, drilling, hydraulic fracturing, compressor stations). The noise levels associated with different operations vary. Seismic studies, used to identify possible reservoirs and uncover the underlying geology are one of the first ways deep-water oil and gas production activities impact the environment, because of increased vessel activity and emissions of light and sound underwater (Cordes et al., 2016). Broadband, loud noises generated by seismic airgun pulses have the potential to negatively impact baleen whales directly through physical effects like temporary or permanent shifts in hearing thresholds (deafness) or indirectly through behavioral changes in their prey (startle response, moving away from the source, altering vocalizations, or masking vocalizations) (Nieukirk et al., 2012). Moreover, seismic surveys tend to change the behavior of other marine species, including fish and cephalopods; leaving the area of the noise through changes in depth distribution, schooling behavior, and startle responses to short-range start-up or high-level sounds; sirenians may be displaced from key feeding habitats; and pinnipeds may abandon colonies (Prideaux and Prideaux 2015). Fish and marine mammals exposed to noise pollution suffer from pathogenic effects, behavioral abnormalities, and disturbances in feeding, mating, breeding, and nursing. Other potential effects include tissue damage, stress, trauma, temporary and permanent hearing loss, and disturbances to the food chain (Sultana & Zhi, 2007). There are potential public health impacts associated with the noise from oil and gas operations. At lower noise thresholds, there is a great risk of annoyance and sleep disturbance. With higher dBAs, hypertension and cardiovascular disease can occur. Numerous operational noise sources have generated noise at levels that are known to be linked to these results (Hays et al., 2017).

4.4.2 Light

High-pressure sodium or mercury vapor lights are used to illuminate lit platforms, and they run from sunset to sunrise. The effects of light are not well understood, particularly regarding fish and invertebrates in the marine environment (Hawkins et al. 2014). However, the vertical relief offered by the structure and the positive phototaxis behavior of many fish species, mean that oil and gas platforms serve as fish attraction sites, particularly at night. Fish behavior around illuminated and unlit oil and gas platforms exhibits several different tendencies that are explained by the influence of light (Barker & Cowan, 2018). Another form of illumination comes from the flares, created when burning excess gas. Bird deaths from collisions with infrastructure and overheating from flare incineration are the most evident and immediate effects of the flares. Although specialized "flare watches" at other platforms saw no direct fatalities, early reports on migrating landbirds emphasized unusual occurrences where hundreds or thousands of birds were burned by flares (Ronconi et al., 2015). For instance, Lucas et al. (2012) reported that the results of the 1993–2009 surveys were conducted on Sable Island showed that there was a declining trend in the oiling rate of beached birds. Migratory birds are attracted to the light at night during their migratory journey, and they become deflected off their intended course and are either trapped by the light, which causes them to circle around the installations, or they find refuge inside, which ultimately prevents them from completing their migration route. There is evidence indicating that their reproduction success and winter survival have decreased (Pragasam, 2021).

4.4.3 Oil Spills

When oil spills occur in the sea, the oil undergoes several physical and chemical processes, including spreading, evaporation, dispersion, emulsification, dissolution, photo-oxidation, sedimentation, sinking, biodegradation, shoreline interaction, and ingestion (Passow & Overton, 2021). Spilled oil on marine organisms can coat marine mammals and birds, which can impair the insulating capacity of the fur on animals such as otters and seals, and reduce the water repulsion of seabird feathers, causing these animals to die from hypothermia. Oil can be ingested by marine animals poisoning them, and dissolved oil can be incorporated into the body across membranes (e.g. gill surfaces). An economic impact stems from the adverse effects on the marketing of commercially valuable species (Farrington, 2014). Because of their lipophilic properties, plankton can accumulate PAHs and give narcotic effects to copepods (*Oithona davisae*) and other planktons. Oil spills result in the large-scale death of marine organisms, and the organisms that survive frequently exhibit aberrant reproductive and developmental patterns (Saadoun, 2015). Because hydrocarbons are volatile, breathing them in causes respiratory tract irritation and narcosis in birds and mammals. All species experience toxicity after consuming oil or petroleum-based compounds, which is caused by the action of the liver to break down the oil-based PAHs. Liver enzymes often do not detoxify oil-derived PAHs but instead convert them into more reactive and hazardous compounds, many of which are carcinogenic (Chen & Denison, 2011). Various methods are used to disperse oil spills, such as adding chemicals that help to break the oil up into smaller droplets and mechanical dispersion using methods such as skimmers. When the exposure doses are not significantly different, crustaceans are significantly more susceptible to mechanically dispersed oil than to chemically dispersed oil. However, mollusks appeared to be more susceptible to chemically distributed oil than mechanically dispersed oil, although the data suggesting this was limited. Compared to earlier life stages, adults had a greater susceptibility to oil exposure (Kalter & Passow, 2023). The most noticeable long-term consequences on killer whales, sea otters, and certain marine birds are on populations involving delayed population recovery, especially in the nearshore ecosystem) and individuals such as decreased health and reproduction (Helm et al., 2014). Health risks associated with oil spills can affect both those involved in the cleanup efforts and the residents of the affected coastal areas. These risks can include effects on mental health, such as the prevalence of anxiety disorders, depression, and posttraumatic stress disorder in the populations exposed to oil; physical and physiological effects, such as respiratory issues, skin, eye, and throat irritations; neurological effects, such as headaches, nausea, vomiting, and dizziness; trauma-related symptoms, such as backpain; and genotoxic, immunotoxin, and endocrine toxicity, structural chromosomal alterations, chromosomal imbalances, hematological cancer, and increased risk of leukemia and lymphomas (Laffón et al., 2016).

The devastation caused by oil spills was particularly apparent during the Deepwater Horizon oil spill, the largest oil spill to have occurred at the time of writing. The spill occurred when an offshore oil rig exploded, releasing 210 million gallons of oil into the ocean over the course of 87 days, which contaminated over 2113 km of Gulk of Mexico shoreline. The damage was apparent in the loss of a wide range of wildlife from the shoreline, with the loss of over 99% of the nearshore oyster cover (Powers et al., 2017) being just one indication of the severe damage to the shoreline habitat. Whilst a large proportion of spilled oil sits on the surface of the water, some contamination reaches the seafloor impacting benthic fauna, and the damage to benthic fauna caused by the Deepwater Horizon oil spill took years to fully heal (Schwing et al., 2020). It is also estimated that around 200 000 seabirds died during the disaster (Haney et al., 2014), although this estimate only includes birds that died during the initial oil release. Oil exposure may have caused long-term injury and illness to many seabirds, resulting in increased mortality for several years after the disaster.

4.4.4 Environmental Contamination by the Platform

PW contains both organic and inorganic substances that, even in small amounts, can affect the water, soils, sediments, and biota in the surrounding ecosystem (Atoufi & Lampert, 2020). Certain metals and higher molecular weight aromatic and saturated hydrocarbons may build up in the sediments close to the produced water outflow if the water is dumped into shallow estuaries and marine seas. This could have an adverse effect on the benthic fauna (biological communities that live on the ocean floor) (Neff et al., 2011). Biochemical reactions such as lysosomal membrane stability (LMS, neutral red retention time assays), DNA damage (micronuclei frequency), and digestive gland histochemistry have been integrated with the information on PAH body burden in mussels (Bakke et al., 2013). Despite the low concentration of phenols in PW, the APs exhibit a strong affinity for fat, which leads to their tendency to bioaccumulate in animal tissue and to degrade slowly (Zheng et al., 2016).

The impact of hydrocarbon contamination and seafloor disruption on the surrounding benthic fauna has been investigated in numerous studies over the decades. These studies often use videos or photographs of the sea floor, taken at locations close to the drilling site, to allow the researchers to quantify the diversity and richness of the benthic fauna as it relates to distance from the drilling site. Some studies do not show clear impacts, with the dominant fauna not seeming to be affected by the drilling activity (Neff et al., 1989), but the majority show significant disruption. Some studies show as much as an 88% decline in the abundance of dominant species around the drilling site (Currie & Isaacs, 2005), although typically the effect is only seen within 100 m around the drilling site. The decreases in benthic fauna often correspond to increases in the contamination of the site by non-aqueous fluids (NAFs) (Santos & Andrade, 2009) and barium. The decrease in benthic fauna abundance as it relates to distance from the drilling site is demonstrated in Figure 4.5, but it

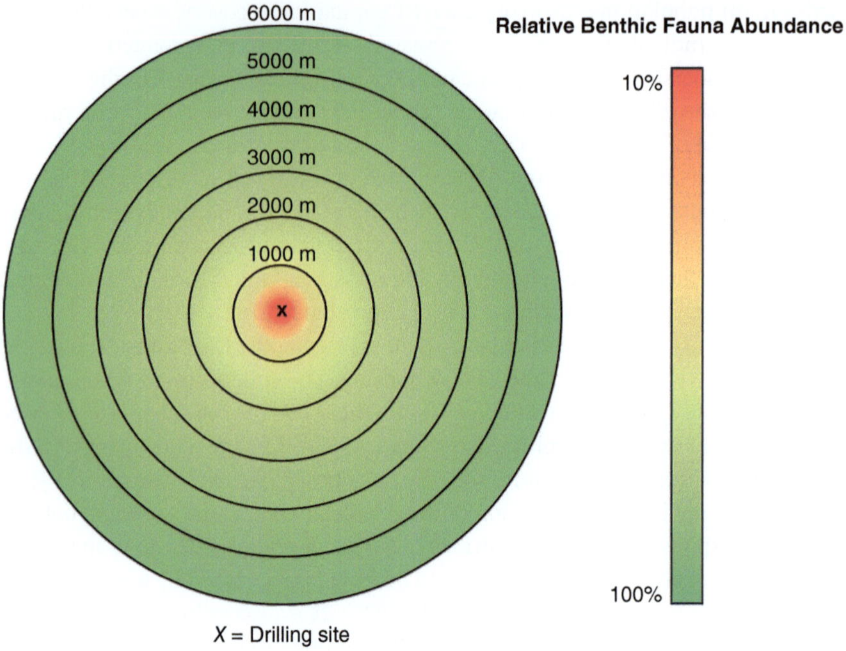

Figure 4.5 Relative benthic fauna abundance with the distance from the drilling site.

should be noted that the studies used to generate this figure vary somewhat in their design (Kingston, 1987). Recovery of the benthic fauna does occur after drilling operations stop, but some studies show that the populations do not fully recover in the immediate vicinity of the drilling site, even one year later (Santos & Andrade, 2009), especially in areas where visible drilling cuttings are still observable (Gates & Jones, 2012). Whilst these more acute impacts of drilling on benthic fauna have become well characterized, more research is needed to fully understand the longer term impacts (Ellis et al., 2012).

The amount of gas flared and its emissions vary, whether from complete or incomplete combustion, and sweet or sulfur-containing hydrocarbons. This presents an ongoing challenge in figuring out the true effects of gas flaring and its emissions on human health as well as its part in local and global environmental degradation (Ismail & Umukoro, 2012). Corrosion and soil acidification can be aided by high relative humidity, industrial SO_2 emissions, and chlorides from sea spray in particular areas (Villaseñor et al., 2003). The formation of ozone and air toxicity are two direct effects of emissions linked to oil and gas supply chains on regional air quality, although these effects differ by region (Allen, 2016). Ground-level ozone production, acidification, and eutrophication are all possible outcomes of NO_X release. While CO_2 and CH_4 can contribute to global warming. Determining the air quality in coastal environments is extremely crucial, since the atmosphere produces several secondary pollutants, including nitric and sulfuric acids, in addition to NO_x and SO_2 emissions. The region's flora and animals are impacted by these pollutants, which dissolve in water and are transported by rainfall into the soil and the ocean (Schifter et al., 2005).

Removing the infrastructure and machinery utilized in the exploration and extraction of oil and gas from the marine environment is known as decommissioning offshore oil and gas installations. Decommissioning of oil pipelines has significant environmental effects, such as noise pollution, disturbance of the seabed, biodiversity, waste disposal, water pollution, and air pollution (Shams et al., 2023).

4.4.5 Summary of Environmental Impacts

The various environmental impacts of offshore oil and gas exploration are summarized in Figure 4.5. The damage caused by these activities is far-reaching, having an impact on everything from the tiniest benthic crustacean, all the way up to larger animals such as whales. The complexity of marine ecosystems is such that the full extent of these impacts can likely never be fully understood, with the delicate balance of ocean life potentially being irreversibly harmed if measures are not taken to reduce the negative consequences of offshore oil and gas exploration.

4.5 Mitigation Methods

Finding ways of mitigating the environmental damage caused by oil and gas operations is critical to ensuring their sustainability. For each type of damage caused to the environment, there are different approaches to limiting this damage. Seismic survey mitigations, pile driving mitigations, considerations in acquiring seabed imagery, infrastructure planning and setbacks from areas with defined benthic conservation objectives, drill cuttings and fluids mitigations, anchors and chains mitigations, pipelines and flowlines mitigations, PW mitigations, drilling, and production mitigation technologies are the important mitigating strategies that should be taken into account to reduce the possible effects of production and exploration activities in areas with established benthic conservation (Oak, 2020).

4.5.1 Regulations

Well-crafted regulations can effectively reduce noise while preventing needless disruption and can enable the industry to solve the remaining technological difficulties by unleashing its creative potential through the prudent application of incentive-based and command-and-control methods

Table 4.5 The summary of each event and its impact.

Event	Impact
Noise	Temporary or permanent deafness (Nieukirk et al., 2012), behavioral abnormalities, disturbances in feeding, mating, breeding, and nursing, tissue damage, stress, trauma, and disturbances to the food chain (Sultana & Zhi, 2007).
Light	Bird deaths from collisions with infrastructure and overheating (Ronconi et al., 2015), unusual occurrences of birds around the lights, migratory birds are deflected off and are either trapped by the light (Pragasam, 2021).
Oil spills	Accumulation of toxins causes; death, aberrant reproductive and developmental patterns in plankton and invertebrates (Saadoun, 2015), delayed population recovery in marine vertebrates (Chen & Denison, 2011), the prevalence of anxiety disorders, depression, posttraumatic stress disorder, respiratory issues, skin, eye, and throat irritations, headaches, nausea, vomiting, and dizziness, back pain, genotoxic, immunotoxin, and endocrine toxicity, structural chromosomal alterations, chromosomal imbalances, hematological cancer, and increased risk of leukemia and lymphomas in humans (Laffon et al., 2016).
Produced water	Effect on the lysosomal membrane stability, DNA damage, and effect on the digestive gland in mussels (Bakke et al., 2013), the tendency to bioaccumulate in animal tissue and to degrade slowly (Zheng et al., 2016).
Gas	Acceleration of global warming, rapidly decreasing the air quality in coastal environments, and ocean acidification (Schifter et al., 2005).

(Merchant, 2019). Utilizing a smaller number of sources, such as a smaller number or size of airguns, to achieve the exploratory targets in the seismic survey would be an ideal solution to mitigate the effects of noise (Oak, 2020). According to the International Whaling Commission's Scientific Committee, there is a positive contribution of prohibiting seismic surveys in some coastal areas, especially Bahia and the Espı́rito Santo coast in Brazil, during the humpback whale breeding season (Weilgart, 2007). An example of the influence of regulations can be seen when comparing the implementation of oil dispersants by Norway which has a clearly documented and regularly tested procedure, and Russia which does not, with Russia's use of dispersants to combat oil spills often being delayed by the permitting process thereby reducing their effectiveness when compared to Norwegian use (Belkina et al., 2015).

4.5.2 Protecting Birds From Platform Lights

In order to reduce the mortality of sea birds caused by light and gain a better understanding of its impact, Rodríguez et al. (2015) advise; launching a rescue effort during the fledgling seasons, organizing public awareness campaigns aimed at educating people about the negative effects of light pollution on the economy, ecology, and health, and implementing legislation aimed at protecting the natural night sky with a particular focus on lowering light pollution levels in the vicinity of their colonies. Using satellite-derived light information, turning off the lights when they are not needed, shielding lights to reduce skyward illumination, and using lights of different spectra to reduce the attraction of migratory land birds are other suggestions to mitigate the effects of light impacts (Gjerdrum et al., 2021). The platform illumination with a blue-green light source, with 5% red in the spectrum, is secure to the human eye, improves safe working conditions, and reduces the accumulation of birds by as much as 2–10 times compared to conventionally lit platforms, as seen when the approach was implemented on oil platform L15 in the North Sea. (Marquenie et al., 2014).

4.5.3 Reducing Environmental Contamination

A novel process of treating environmental contamination is proposed by combining novel technologies: tubular separation, combined fiber coalescence, and membrane separation; these are incorporated in the three main treatment steps: pretreatment, standard-reaching treatment, and advanced purification treatment, respectively. This overcomes the drawbacks of conventional PW treatment processes, which include lengthy procedures, large space occupation, and limited treatment efficiency (Liu et al., 2021). When there is a significant volume of PW present in arid places, irrigation using that water is frequently an option if the PW has been sufficiently treated (Salem and Thiemann, 2022). Based on factors such as the chemistry of PW, cost-effectiveness, space availability, plans for reuse and discharge, long-term operation, and byproducts, the optimum technology for treating PW is chosen (Fakhru'l-Razi et al., 2009).

More research is required to fully understand the impacts of oil drilling on benthic fauna and therefore to understand the best approaches to limiting these impacts. However, it has been suggested that burying the drill cuttings (termed cuttings re-injection or CRI) rather than depositing them onto the sea floor will help to alleviate some of the damage caused by the benthic fauna (Gumarov et al., 2014). As the loss of benthic fauna is linked to environmental contamination, careful management of waste and minimizing contamination by drilling fluid will also help to preserve benthic fauna. CRI was successfully used in the Gulf of Mexico, North Sea and Alaska in the 1990s (Zhang et al., 2022) and was successfully implemented in the offshore oil field of Saudi Arabia in 2011 (Ezell et al., 2011).

4.5.4 Oil Spill Responses

An oil spill response method called "in situ burning" involves igniting and burning oil to remove it off the water surface. Large amounts of oil are quickly and efficiently removed with this procedure. Under some circumstances, burning is inexpensive, readily managed, and produces few environmental issues (Ko & Day, 2004). The gases released during an ISB operation typically pose less of a concern to human health than particulate matter does (Faksness et al., 2022). Oil analysis techniques are an essential component of the scientific, environmental, and technical aspects of oil spills. The most promising environmental forensic methods (such as oil fingerprinting and biomarking) at various response levels and habitat areas must receive priority funding when it comes to oil spill analysis (Li et al., 2016).

4.5.5 Platform Decommissioning

The effects of decommissioning must be identified, and appropriate mitigation measures must be implemented to reduce the adverse effects over the project's life cycle. Shams et al. (2023) state that numerous variables, such as the location of the platform, the type of platform, and environmental circumstances, will affect the optimal decommissioning settings depending on choosing the best method of decommissioning. According to their study, the least disruptive and environmentally friendly way to decommission offshore oil and gas platforms is by leave-in-situ decommissioning. Decommissioned oil well infrastructure has been demonstrated to serve as a favorable habitat for marine wildlife, providing shelter for smaller fish and surfaces for the adherence of plants and mollusks.

4.6 Conclusions

Offshore oil and gas operations have been demonstrated to have serious environmental impacts, stemming from different stages of the process. Some of these impacts only occur as a result of faults or accidents (i.e. oil spills) but many of them occur just as a result of the usual operation of oil wells. The effects described in this chapter are all also the more direct impacts of oil and gas operations. Not considered here are the indirect impacts caused by the global warming that has resulted from the burning of the fossil fuels that result from oil and gas operations. The literature covered in this chapter does suggest a wide range of approaches that can mitigate these direct environmental impacts, although more research is needed in some of these areas to fully understand the best approaches to this. However, none of these approaches will avert the environmental impacts of global warming, for which the only effective mitigation strategy is to completely cease all oil and gas operations. Whether this is a feasible accomplishment, and if so on what timescale this can be achieved, is beyond the scope of this chapter.

References

Albeldawi, M. (2023). Chapter 10 – Environmental impacts and mitigation measures of offshore oil and gas activities. In: *Developments in Petroleum Science*, vol. 78 (ed. I.A. Hussein and M. Mahmoud), 313–352. Elsevier https://doi.org/10.1016/B978-0-323-99285-5.00002-8.

Allen, D.T. (2016). Emissions from oil and gas operations in the United States and their air quality implications. *Journal of the Air & Waste Management Association* 66 (6): 549–575. https://doi.org/10.1080/10962247.2016.1171263.

Atoufi, H.D. and Lampert, D.J. (2020). Impacts of oil and gas production on contaminant levels in sediments. *Current Pollution Reports* 6 (2): 43–53. https://doi.org/10.1007/s40726-020-00137-5.

Baker, R. (2004). Oil and natural gas: offshore operations. In: *Encyclopedia of Energy* (ed. C.J. Cleveland), 581–594. Elsevier https://doi.org/10.1016/B0-12-176480-X/00258-8.

Bakke, T., Klungsøyr, J., and Sanni, S. (2013). Environmental impacts of produced water and drilling waste discharges from the Norwegian offshore petroleum industry. *Marine Environmental Research* 92: 154–169. https://doi.org/10.1016/j.marenvres.2013.09.012.

Barker, V.A. and Cowan, J.H. (2018). The effect of artificial light on the community structure of reef-associated fishes at oil and gas platforms in the northern Gulf of Mexico. *Environmental Biology of Fishes* 101: 153–166.

Belkina, N., Sarkova, O., and Jensen, S. (2015). Regulatory policies for using oil dispersants in the Barents Sea. *Polar Research* 34 (1): 24326.

Chen, J., & Denison, M. S. (2011). The Deepwater Horizon oil spill: environmental fate of the oil and the toxicological effects on marine organisms. *Horizon (indicated in red).*

Cordes, E.E., Jones, D.O., Schlacher, T.A. et al. (2016). Environmental impacts of the deep-water oil and gas industry: a review to guide management strategies. *Frontiers in Environmental Science* 4: 58.

Currie, D.R. and Isaacs, L.R. (2005). Impact of exploratory offshore drilling on benthic communities in the Minerva gas field, Port Campbell, Australia. *Marine Environmental Research* 59 (3): 217–233. https://doi.org/10.1016/j.marenvres.2004.05.001.

Dou, L., Wen, Z., Wang, J. et al. (2022). Analysis of the world oil and gas exploration situation in 2021. *Petroleum Exploration and Development* 49 (5): 1195–1209. https://doi.org/10.1016/S1876-3804(22)60343-4.

Ellis, J., Fraser, G., and Russell, J. (2012). Discharged drilling waste from oil and gas platforms and its effects on benthic communities. *Marine Ecology Progress Series* 456: 285–302. https://doi.org/10.3354/meps09622.

Ezell, R., Quinn, F., Chima, J., & Baim, A. (2011). First Successful Field Utilization of Cuttings Re-Injection (CRI) in the Offshore Field of Saudi Arabia as an Environmentally Friendly and Cost Effective Waste Disposal Method. SPE Annual Technical Conference and Exhibition,

Fakhru'l-Razi, A., Pendashteh, A., Abdullah, L.C. et al. (2009). Review of technologies for oil and gas produced water treatment. *Journal of Hazardous Materials* 170 (2–3): 530–551.

Faksness, L.-G., Leirvik, F., Taban, I.C. et al. (2022). Offshore field experiments with in-situ burning of oil: emissions and burn efficiency. *Environmental Research* 205: 112419.

Farrington, J.W. (2014). Oil pollution in the Marine Environment II: fates and effects of oil spills. *Environment: Science and Policy for Sustainable Development* 56 (4): 16–31. https://doi.org/10.1080/00139157.2014.922382.

Gary, J.H., Handwerk, J.H., Kaiser, M.J., and Geddes, D. (2007). *Petroleum Refining: Technology and Economics*. CRC press.

Gates, A.R. and Jones, D.O.B. (2012). Recovery of Benthic Megafauna from anthropogenic disturbance at a hydrocarbon drilling well (380 m depth in the Norwegian sea). *PLoS ONE* 7 (10): e44114. https://doi.org/10.1371/journal.pone.0044114.

Gjerdrum, C., Ronconi, R.A., Turner, K.S., and Hamer, T.E. (2021). Bird strandings and bright lights at coastal and offshore industrial sites in Atlantic Canada. *Avian Conservation and Ecology* 16 (1): https://doi.org/10.5751/ace-01860-160122.

Gumarov, S. M., Shokanov, T. A., Simmons, S., Anokhin, V. V., Benelkadi, S., & Ji, L. (2014). Drill Cuttings Re-Injection Well Design and Completion: Best Practices and Lessons Learned. IADC/SPE Drilling Conference and Exhibition,

Haney, J.C., Geiger, H.J., and Short, J.W. (2014). Bird mortality from the Deepwater Horizon oil spill. I. Exposure probability in the offshore Gulf of Mexico. *Marine Ecology Progress Series* 513: 225–237.

Harris, P., Alo, B., Bera, A. et al. (2016). *Offshore hydrocarbon industries*. United Nations, New York: *The First Global Integrated Marine Assessment*.

Hawkins, A.D., Pembroke, A.E., and Popper, A.N. (2014). Information gaps in understanding the effects of noise on fishes and invertebrates. *Reviews in Fish Biology and Fisheries* 25 (1): 39–64. https://doi.org/10.1007/s11160-014-9369-3.

Hays, J., McCawley, M., and Shonkoff, S.B. (2017). Public health implications of environmental noise associated with unconventional oil and gas development. *Science of the Total Environment* 580: 448–456. https://doi.org/10.1016/j.scitotenv.2016.11.118.

Helm, R.C., Costa, D.P., DeBruyn, T.D. et al. (2014). Overview of effects of oil spills on marine mammals. In: *Handbook of Oil Spill Science and Technology* (ed. M. Fingas), 455–475. Wiley https://doi.org/10.1002/9781118989982.ch18.

IEA. (2023). *World Energy Outlook*. https://www.iea.org/reports/world-energy-outlook-2023.

IRENA (2022, November 11). The alliance for industry decarbonization outlines core focus topics, and welcomes new partners at Maiden COP27 meeting. *Decarbonization* [Press release]. https://www.irena.org/News/pressreleases/2022/Nov/The-Alliance-for-Industry-Decarbonization-outlines-core-focus-topics-welcomes-new-partners-at-COP27

Ismail, O.S. and Umukoro, G.E. (2012). Global impact of gas flaring. *Energy and Power Engineering* 04 (04): 290–302. https://doi.org/10.4236/epe.2012.44039.

Kalter, V., & Passow, U. (2023). Quantitative review summarizing the effects of oil pollution on subarctic and arctic marine invertebrates. *Environmental Pollution*, 319, 120960. https://doi.org/10.1016/j.envpol.2022.120960

Kingston, P.F. (1987). Field effects of platform discharges on benthic macrofauna. *Philosophical Transactions of the Royal Society of London. B, Biological Sciences* 316 (1181): 545–565.

Ko, J. and Day, J.W. (2004). A review of ecological impacts of oil and gas development on coastal ecosystems in the Mississippi Delta. *Ocean & Coastal Management* 47 (11–12): 597–623. https://doi.org/10.1016/j.ocecoaman.2004.12.004.

Laffón, B., Pásaro, E., and Valdiglesias, V. (2016). Effects of exposure to oil spills on human health: Updated review. *Journal of Toxicology & Environmental Health Part B: Critical Reviews* 19 (3–4): 105–128. https://doi.org/10.1080/10937404.2016.1168730.

Li, P., Cai, Q., Lin, W. et al. (2016). Offshore oil spill response practices and emerging challenges. *Marine Pollution Bulletin* 110 (1): 6–27. https://doi.org/10.1016/j.marpolbul.2016.06.020.

Liu, Y., Lu, H., Li, Y., Xu, H., Pan, Z., Dai, P., Wang, H., & Yang, Q. (2021). A review of treatment technologies for produced water in offshore oil and gas fields. *Science of the Total Environment*, 775, 145485. https://doi.org/10.1016/j.scitotenv.2021.145485

Lucas, Z., Horn, A., and Freedman, B. (2012). Beached bird surveys on Sable Island, Nova Scotia, 1993 to 2009, show a decline in the incidence of oiling. *Proceedings of the Nova Scotian Institute of Science (NSIS)* 47 (1).

Marquenie, J.M., Wagner, J., Stephenson, M.T., & Lucas, L. (2014). Green lighting the way: managing impacts from offshore platform lighting on migratory birds. Paper presented at the SPE International Conference on Health, Safety, and Environment, Long Beach, California, USA. https: //doi.org/10.2118/168350-MS

Merchant, N.D. (2019). Underwater noise abatement: economic factors and policy options. *Environmental Science & Policy* 92: 116–123. https://doi.org/10.1016/j.envsci.2018.11.014.

Narula, K. and Narula, K. (2019). Oceans as a Source of Hydrocarbon Energy. *The Maritime Dimension of Sustainable Energy Security* 145–162.

Neff, J.M., Bothner, M.H., Maciolek, N.J., and Grassle, J.F. (1989). Impacts of exploratory drilling for oil and gas on the benthic environment of Georges Bank. *Marine Environmental Research* 27 (2): 77–114.

Neff, J.M., Lee, K., and DeBlois, E.M. (2011). Produced Water: Overview of composition, fates, and effects. In: *Produced water: Environmental risks and advances in mitigation technologies* (pp. 3–54). Springer. https://doi.org/10.1007/978-1-4614-0046-2_1.

Nieukirk, S.L., Mellinger, D.K., Moore, S.E. et al. (2012). Sounds from airguns and fin whales recorded in the mid-Atlantic Ocean, 1999–2009. *Journal of the Acoustical Society of America* 131 (2): 1102–1112. https://doi.org/10.1121/1.3672648.

Oak, T. G. (2020). *Oil and gas exploration and production activities in areas with defined benthic conservation objectives: A review of potential impacts and mitigation measures.* Canadian Science Advisory Secretariat (CSAS).

Passow, U., & Overton, E. B. (2021). The complexity of spills: the fate of the deepwater horizon oil. *Annual Review of Marine Science*, 13(1), 109–136. https://doi.org/10.1146/annurev-marine-032320-095153

Powers, S.P., Peterson, C.H., Cebrian, J., and Heck, K.L. Jr. (2017). Response of nearshore ecosystems to the Deepwater Horizon oil spill. *Marine Ecology Progress Series* 576: 107–110.

Pragasam, S. (2021). Eco friendly lighting options for offshore installations and related design challenges. *Day 2 Tue, November 16, 2021*. https://doi.org/10.2118/207994-ms

Prideaux, G. and Prideaux, M. (2015). Environmental impact assessment guidelines for offshore petroleum exploration seismic surveys. *Impact Assessment and Project Appraisal* 34 (1): 33–43. https://doi.org/10.1080/14615517.2015.1096038.

Rodríguez, A., García, D.P., Rodríguez, B. et al. (2015). Artificial lights and seabirds: is light pollution a threat for the threatened Balearic petrels? *Journal of Ornithology* 156 (4): 893–902. https://doi.org/10.1007/s10336-015-1232-3.

Ronconi, R.A., Allard, K.A., and Taylor, P.D. (2015). Bird interactions with offshore oil and gas platforms: review of impacts and monitoring techniques. *Journal of Environmental Management* 147: 34–45. https://doi.org/10.1016/j.jenvman.2014.07.031.

Saadoun, I. (2015). Impact of oil spills on marine life. *In InTech eBooks.* https://doi.org/10.5772/60455.

Salem, F. and Thiemann, T. (2022). Produced water from oil and gas exploration—problems, solutions and opportunities. *Journal of Water Resource and Protection* 14 (02): 142–185. https://doi.org/10.4236/jwarp.2022.142009.

Santos, C.F. and Andrade, F. (2009). Environmental sensitivity of the Portuguese coast in the scope of oil spill events—comparing different assessment approaches. *Journal of Coastal Research* 885–889.

Schifter, I., González-Macías, C., Miranda, A. C., & López-Salinas, E. (2005). Air emissions assessment from offshore oil activities in Sonda de Campeche, Mexico. *Environmental Monitoring and Assessment*, 109 (1–3), 135–145. https://doi.org/10.1007/s10661-005-5844-6

Schwing, P.T., Montagna, P.A., Joye, S.B. et al. (2020). A synthesis of deep benthic faunal impacts and resilience following the Deepwater Horizon oil spill. *Frontiers in Marine Science* 7: 560012.

Shams, S., Prasad, D. M. R., Imteaz, M. A., Khan, M. M. H., Ahsan, A., & Karim, M. R. (2023). An assessment of environmental impact on offshore decommissioning of oil and gas pipelines. *Environments*, 10(6), 104. https://doi.org/10.3390/environments10060104

Speight, J.G. (2015). *Handbook of Offshore Oil and Gas Operations*, 1e. Elsevier.

Sultana, S. and Zhi, C. (2007). A review of noise impacts from offshore oil-gas production activities on the marine biota. *Canadian Acoustics* 35 (3): 174–176. Retrieved from https://jcaa.caa-aca.ca/index.php/jcaa/article/view/1957.

UNEP (2022). Harmful marine extractives: offshore oil & gas: understanding the risks & impacts of financing non-renewable extractive industries. https://www.unepfi.org/publications/harmful-marine-extractives-offshore-oil-gas/ (accessed on 29 February 2024)

Villaseñor, R., Magdaleno, M., Quintanar, A.I. et al. (2003). An air quality emission inventory of offshore operations for the exploration and production of petroleum by the Mexican oil industry. *Atmospheric Environment* 37 (26): 3713–3729. https://doi.org/10.1016/s1352-2310(03)00445-x.

Weilgart, L. (2007). The impacts of anthropogenic ocean noise on cetaceans and implications for management. *Canadian Journal of Zoology* 85 (11): 1091–1116. https://doi.org/10.1139/z07-101.

Zhang, S., Feng, Y., Li, B. et al. (2022). Fracture development during disposal of hazardous drilling cuttings by deep underground injection: a review. *Journal of Rock Mechanics and Geotechnical Engineering* 14 (5): 1652–1670. https://doi.org/10.1016/j.jrmge.2022.05.001.

Zheng, J., Chen, B., Thanyamanta, W., Hawboldt, K., Zhang, H., & Liu, B. (2016). Offshore produced water management: a review of current practice and challenges in harsh/Arctic environments. *Marine Pollution Bulletin*, 104 (1–2), 7–19. https://doi.org/10.1016/j.marpolbul.2016.01.004

5

Coastal and Marine Pollution from Agricultural Activities: Fertilizers and Pesticides

Jithya Wijesinghe, Shermila M. Botheju, and Nalika R. Dayananda

Department of Indigenous Medical Resources, Faculty of Indigenous Health Sciences and Technology, Gampaha Wickramarachchi University of Indigenous Medicine, Yakkala, Sri Lanka

5.1 Introduction

Marine pollution encompasses a wide range of sources, including hazardous chemicals, sewage, fertilizers, plastics, abandoned fishing nets, and even the noise generated by shipping and drilling activities. More than 80% of it comes from activities that occur on land, such as accidental spills, intentional dumping, unprocessed waste, pollutants that fall from the atmosphere, or the runoff from drains and rivers.

There are three prominent types of oceanic and coastal pollution: first, nitrogen-phosphorous pollution caused by agricultural activities, sewage, urban and industrial run-off; second, chemical pollution which includes pesticides, petroleum, pharmaceuticals and personal care products, heavy metals, and industrial discharge; and third, plastic-debris pollution. In all these types, the contribution of agriculture inputs is high and its adverse effects on the coastal and marine ecosystems have become one of the most pressing global challenges today. Chemical substances have become prevalent in the marine environment, as evidenced by the widespread presence of chemical pollutants across various regions and diverse marine ecosystems, including estuaries and deep oceanic regions. Most of these pollutants originate from terrestrial sources, initially serving specific purposes and providing benefits, such as being used as agricultural inputs, including fertilizers and pesticides.

Many constituents found in contemporary pesticide formulations can induce toxicity in marine organisms. Pesticides are introduced into the marine ecosystem via various pathways, including sewage and stormwater systems, rivers and streams, as well as through direct runoff, vapor, and spray drift originating from agriculture, forestry, aquaculture, golf courses, park and garden, etc. Agricultural runoff, particularly from chemicals leaching from fertilizer, threatens coastal and marine ecosystems substantially. The utilization of nutrients such as nitrogen and phosphorus, as well as the presence of persistent toxins and their accumulation, can result in the depletion of oxygen levels and the formation of marine dead zones. The utilization of fertilizers and pesticides coated with plastic materials represents a significant and avoidable contributor to the escalating presence of microplastic pollution within agricultural soils and subsequently finds its way into coastal and marine ecosystems because of agricultural runoff. All these pollutants have been

Coastal and Marine Pollution: Source to Sink, Mitigation and Management, First Edition.
Edited by Meththika Vithanage, Sameera M. Samarasekara, Bryan D. James, and Christopher M. Reddy.

documented to be associated with various adverse health effects, alterations in population dynamics and potential imbalances within ecosystems. Sustainable agriculture practices could address many of these issues. The UN's Sustainable Development Goal target 14.1 states pledge to: "By 2025, prevent and significantly reduce marine pollution of all kinds, from land-based activities, including marine debris and nutrient pollution" (UN SDG 2019).

With these thoughts in mind, this chapter aims to assess the current status of use of fertilizer and pesticides, their contribution to the pollution of coastal and marine ecosystems, and the potential impact on marine fauna, flora, and ecosystem processes. The chapter will also explain the many and varied ways in which this pollutant (intentionally or unintentionally) can find its way into oceans and coastal habitats. To end, the chapter will address management, monitoring, and regulation options and highlight knowledge gaps.

5.2 Current Status and Trends in Use of Fertilizer and Pesticides in Agriculture

In 2019, the agricultural use of nitrogen (N) and phosphorus (P) in fertilizers amounted to 105 TgN/ year and 43 TgP/year, respectively, according to the FAO. These inputs are widely acknowledged as significant contributors to water pollution, with approximately 32–45% of nitrogen and 10–20% of phosphorus being transported to the oceans through rivers (Beusen et al. 2016). In contrast, there has been significantly less focus on the use of pesticides and how they impact the environment. This is likely due to the fact that the yearly worldwide input of pesticide active substances, approximately 3 Tg/year, is a relatively small portion of nitrogen (N) and phosphorus (P) inputs (FAO 2020), resulting in less noticeable direct impacts on water and soil pollution. Out of the total annual net inputs of 0.94 Tg of pesticides, 90% of the losses occurred in the root zone (0–100 cm). Within this, 82% of the pesticides were biodegraded, 7.2% leached below the soil root zone, less than 0.1% converged into river reaches, and a negligible amount was released as direct runoff into oceans outside of rivers. The total amount of pesticides exported to the ocean may exceed 0.71 Gg (Maggi et al. 2023).

5.3 Pesticide Distribution in Coastal and Marine Environment

Pesticides encompass a wide range of chemical compounds developed to control or eliminate pests. These substances can be classified into different groups based on their main targets and chemical compositions such as insecticides, herbicides, fungicides, and nematicides.

Most pesticides found in marine environments come from land-based sources, including DDT (an organochlorine insecticide), Atrazine, Diuron (a PSII herbicide), and Dieldrin (a cyclodiene insecticide), among others. However, there are some exceptions, such as antifoulants. Pesticides are conveyed from their origin through water and sediment and through the atmosphere. Pesticides have been detected in marine environments worldwide for several decades, specifically in water, sediment, and biota. For example, Ojemaye et al. (2020) reported atrazine, alachlor, simazine, metolachlor, and butachlor in Cape Town, South Africa. Atrazine and acetochlor were found in Jiaozhou Bay, China (Ouyang et al. 2019) while pesticides like imidacloprid, metsulfuron-methyl, tebuconazole, propiconazole, pendimethalin, chlorpyrifos, endosulfan, malathion, and many more were found in Great Barrier Reef Lagoon, Australia (Thai et al. 2020). Furthermore, DDTs, HCH, endrin, dieldrin, naphthalene, acenaphthylene, anthracene, heptachlor, pyrene

(Mwevura et al. 2020), and organophosphate pesticides (e.g. chlorpyrifos, diazinon, dimethoate, ethion, malathion, parathion, and terbufos). Ponce-Vélez and de la Lanza-Espino (2019) were recorded in marine sediments in the West Coast of Tanzania and the Gulf of Mexico, respectively. When it comes to marine organisms, organochlorines: HCHs, DDTs, and chlordane were found in zooplankton, oyster, crab, and goby in Korea (Kim 2019) while Sundhar et al. (2020) reported HCH, DDTs1, heptachlor, endosulfan, dieldrin in commercial marine fishes, and seaweeds in India. Moreover, several pesticides were recorded in coral, dolphins, fish, invertebrates, mollusks, seagrass, and turtles in Australia (Cagnazzi et al. 2020). Once organisms absorb pesticides, they serve as transportation both spatially and through food chains.

5.4 Pesticide Entering Pathways to Coastal and Marine Environment

Most pesticides are applied on land, especially in farming. For instance, according to IbisWorld (2016), more than 70% of pesticides used in Australia are used on agricultural land. Nonetheless, termites, insects, and weeds in and around homes, fleas, and parasitic worms in household pets, as well as weeds in parks and waterways, are all well controlled using pesticides in metropolitan settings. The main routes that these urban pesticides end up in waterways are through wastewater treatment facilities (Zhang et al. 2020), stormwater drains (Chen et al. 2019), and surface runoff.

Surface water runoff is the primary method of transporting substances away from a site, which then flows into estuaries and marine waters. Groundwater also carries pesticides away from their original location, but their impact is usually smaller compared to surface water. However, during periods of low rainfall, their influence can be substantial. Pesticides are authorized for application on specific crops or types of agriculture, with a prescribed maximum rate of use. Hence, the specific crops or agricultural activities and the proportion of the catchment area they occupy play a crucial role in determining the nature, quantity, and intensity of pesticides. These factors ultimately govern the extent and speed at which surface runoff and eroded soil find their way into water bodies. Moreover, recent research has shown that land use directly influences the quantity of pesticides found, the toxicity of pesticide mixtures, and the potential risk they pose. This is particularly evident at the location where rivers flow into marine waters (Warne et al. 2020a; Warne et al. 2020b) (Figure 5.1). The proximity between the pesticide application site and the ocean directly affects the quantity and composition of pesticides that ultimately reach marine waters. For example, as shown in Figure 5.1, 62 100 km of rivers had average cumulative pesticide concentrations below detection limits of 10^{-3} μg/l, 140 300 km had concentrations up to 1 μg/l, and more than 13 000 km above 1 μg/l. The amount and speed at which surface runoff and eroded soil enter waterways are determined by a combination of factors, including the volume and intensity of rainfall, as well as environmental properties such as soil moisture content, vegetative ground cover, ground slope, and the presence of riparian vegetation.

In addition, pesticides can be entered into the coastal and marine environment via the atmosphere. In regions with high temperatures, some pesticides used will undergo evaporation from plants, soil, and water. They are subsequently transported by extensive atmospheric circulation patterns toward the polar regions in the north or south. Once the air temperature reaches the point of condensation, pesticides will condense onto various surfaces such as biota, ice, snow, soil, and water. Consequently, pesticides with lower condensation points are carried toward the poles more extensively than pesticides with higher condensation points, resulting in a spatial separation. This phenomenon is referred to as global fractionation. Air temperatures fluctuate annually in

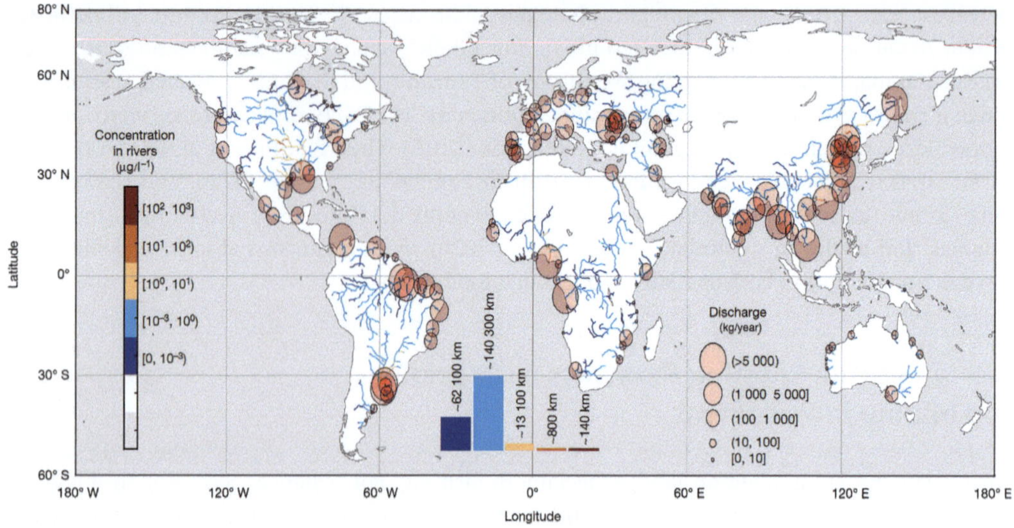

Figure 5.1 Average total pesticide active substances (PAS) concentration in river reaches and annual average PAS discharge to oceans. *Source:* Maggi et al. (2023)/Springer Nature. Vertical bars quantify the cumulative length of river reaches receiving PAS at specific concentrations.

accordance with the changing seasons. During the winter season, the temperature can drop to a point where a pesticide undergoes condensation. During the summer, the elevated atmospheric temperature may cause the pesticide to evaporate and disperse toward the poles until it reaches its condensation temperature. This annual evaporation, transport, and condensation cycle is called the grasshopper effect. The rate of degradation in the air also impacts the extent to which pesticides can be transported, with less persistent chemicals having a lower probability of being transported over long distances.

5.5 Fertilizer Use and Losses from Agricultural Land to Coastal and Marine Environment

The excessive nutrients resulting from the application of fertilizers, pollution discharge, and water regulation outflows from land to oceans have a significant negative impact on coastal ecosystems. The discharge of nutrient-polluted waters, containing primarily nitrogen (N) and phosphorus (P) compounds, from various sources such as animal manure and chemical fertilizers losses through rivers is causing severe and harmful impacts on coastal and marine ecosystems worldwide. The growing need for nitrogenous fertilizers, especially urea, in agriculture has led to a significant rise in the release of nitrogen into the marine environment (Lu and Tian 2017). Nitrogen and phosphorus fertilizer usage rates per unit of cropland area increased by approximately eight times and three times, respectively, since the year 1961 (Lu and Tian 2017). This increase in fertilizer nitrogen is compounded by the inefficient use of fertilizer on agricultural lands. More than half of the synthetic fertilizer applied to the world's fields has been applied in the past 30 years (Pearce 2018) but less than half of this fertilizer reaches the intended crops, with the remainder running off into rivers and eventually into the ocean.

Several studies have shown that crops have low nitrogen use efficiency, with only about half of the nitrogen applied to croplands being absorbed by plant biomass. The remaining nitrogen is lost through leaching (16%), soil erosion (15%), and gaseous emission (14%) (Liu et al. 2019). It is projected that the amount of nitrogen and phosphorus entering coastal and marine waters worldwide will at least double by 2050. This increase will be driven by factors such as the growing use of fertilizers, expansion of coastal aquaculture, population growth and the resulting increase in sewage and animal waste, as well as the additional nitrogen deposition from gaseous emissions produced by burning fossil fuels and other industrial activities (Johnson and Harrison 2015). The riverine nitrogen input to coastal seas has roughly doubled in the past few centuries (Beusen et al. 2016). The current input is primarily influenced by nitrate, which is a result of indirect land use inputs using fertilizers.

Furthermore, the substantial volume of manure generated by intensive animal production is typically utilized as fertilizer on land, stored in the feedlot, or stored in lagoons. Often, when there is an excessive amount of manure, it is used on crops more than required, which worsens the problem of nutrient runoff and leaching. Livestock manure is abundant in nutrients, especially nitrogen and phosphorus, and may also contain potentially harmful chemicals and heavy metals. In addition, various microorganisms, such as bacteria, viruses, and parasites found in livestock manure, can pose risks to both humans and marine ecosystems.

5.6 Impact on the Coastal and Marine Environment of Agrochemicals

5.6.1 Current Status and Trends

Coastal and marine pollution stemming from agricultural activities, particularly the use of fertilizers and pesticides, has become a significant ecological concern worldwide. These agrarian pollutants, while aiding agricultural productivity, often find their way into water bodies through various pathways such as agricultural runoff and leaching, leading to detrimental impacts on marine and coastal ecosystems. The current status and trends of this pollution express an alarming picture, indicating the urgent need for sustainable agricultural practices and stringent environmental regulations.

Fertilizers, primarily composed of nitrogen and phosphorus compounds, are essential for enhancing soil fertility and promoting plant growth. However, excess application or improper management of these chemicals results in runoff into nearby waterways during rainfall events. According to the United Nations Food and Agriculture Organization (FAO), approximately 60 million tons of nitrogen-based fertilizers are applied globally each year, with a significant portion ending up in coastal and marine environments (FAO 2009). Similarly, phosphorus runoff from agricultural lands contributes to eutrophication (Withers and Haygarth 2007), a process characterized by excessive nutrient enrichment in water bodies, leading to harmful algal blooms and oxygen depletion. The impact of fertilizer runoff on coastal and marine environments is profound. Elevated nutrient levels fuel the growth of algae, leading to the formation of harmful algal blooms (HABs). These blooms can release toxins, resulting in mass mortality events among marine life and posing risks to human health through seafood consumption. Moreover, when these algae die and decompose, the process consumes oxygen, creating hypoxic or dead zones where marine life cannot survive. The Gulf of Mexico's "Dead Zone," primarily fueled by agricultural runoff from the Mississippi River watershed (Brumfield Ii 2010), is one of the most infamous examples, spanning thousands of square kilometers.

Pesticides such as organochlorines, organophosphates, and glyphosate have been detected in marine ecosystems worldwide, posing threats to aquatic organisms' health and biodiversity.

Pesticide pollution in coastal and marine environments arises primarily from agricultural runoff, urban runoff, and aquaculture practices. According to the Environmental Protection Agency (EPA), approximately 347 million pounds of pesticides are used annually in the United States alone (Pimentel 2005), with significant amounts entering water bodies through runoff and leaching. The situation is even more alarming, with millions of tons of pesticides applied annually in agricultural fields worldwide (Stephenson et al. 2001). Despite regulatory efforts and best management practices, pesticide contamination in coastal and marine ecosystems remains widespread, driven by factors such as intensive agricultural practices, improper pesticide use, and inadequate enforcement of regulations.

Considering the recent statistics, about 3 Tg of pesticides are used annually for agricultural crop protection (FAO 2023). How much of these pesticides remain on land and reach the aquifer or the ocean is unknown, and the monitoring environmental fate of those synthetic chemicals is challenging because a detailed picture of their mobility in time and space is largely missing. But Maggi et al. (2023) developed a process-based model accounting for the hydrology and biogeochemistry (Figure 5.2) of the 92 most used agricultural pesticide active substances to assess their pathways through the principal catchments of the earth and interpreted a near-present picture of the global land and river budgets, including discharge to oceans. According to their findings, around 0.94 Tg net annual pesticide input in the year 2015, 82% is biologically degraded, 10% remains as residue in soil, and 7.2% leaches below the root zone.

According to those numbers, river systems receive 0.73 Gg of pesticides from their drainage or upper/lower catchment areas at a rate of 10 to more than 100 kg/year/km. By contrast to their fate

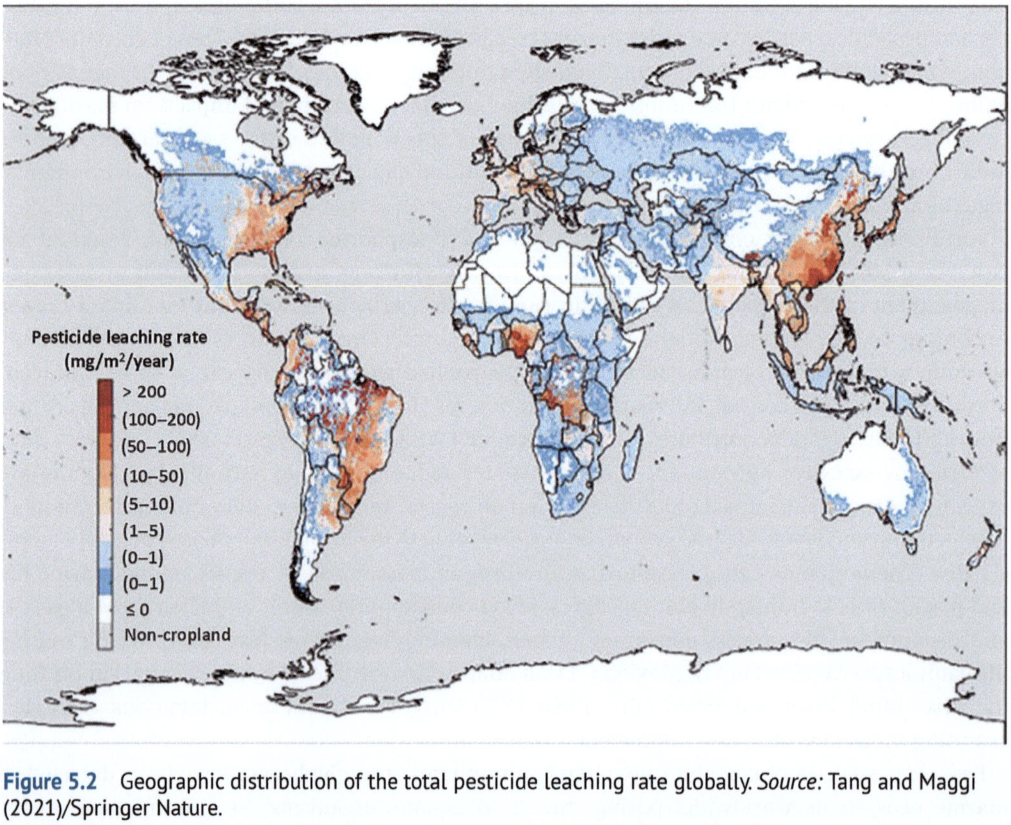

Figure 5.2 Geographic distribution of the total pesticide leaching rate globally. *Source:* Tang and Maggi (2021)/Springer Nature.

in soil, only 1.1% of pesticides entering rivers are degraded along streams, exceeding safety levels (concentrations >1 μg/l) in more than 13 000 km of river length, with 0.71 Gg of pesticide active ingredients released to oceans annually. Herbicides represent the prevalent pesticide residue on both land (72%) and river outlets (62%) (Maggi et al. 2023). Therefore, a certain number of pesticides can enter the ocean system through freshwater inputs and can directly harm aquatic organisms, including fish, invertebrates, and plankton, through acute toxicity or chronic exposure. Furthermore, pesticides can disrupt food webs and alter community dynamics, leading to shifts in species composition and biodiversity loss. For instance, studies have shown the adverse effects of pesticides on coral reefs, seagrass beds, and mangrove ecosystems, which are vital habitats for marine biodiversity and ecosystem services (Moberg and Rönnbäck 2003; Mitra et al. 2016; de los Santos et al. 2020). Furthermore, some pesticides, such as neonicotinoids, have been linked to the decline of pollinator populations critical for ecosystem health and agricultural productivity. Persistent pesticides, like DDT, can bioaccumulate in marine food webs, posing long-term risks to top predators, including seabirds and marine mammals (Corsolini and Sarà 2017).

Similarly, heavy metal contaminations can be added as additional matrices from agrochemicals such as fertilizers (especially from phosphate fertilizers) also pose significant threats to coastal and marine ecosystems, with profound implications for biodiversity and ecosystem resilience. Heavy metals, including mercury, lead, cadmium, and arsenic, can enter marine environments not only as a matrix but also with the pH changes of the agricultural soils, soil-bound metals can be released into the soil solutions, and they can be leached out. According to the World Health Organization (WHO) records, heavy metal pollution affects approximately 100 million people living in coastal areas worldwide, with serious health consequences such as neurological disorders, kidney damage, and cardiovascular diseases (Fernández-Luqueño et al. 2013). Heavy metals are nondegradable elements naturally occurring in coastal seas. They are not particularly toxic as the condensed free elements, but they are dangerous to living organisms in the form of cations with the capacity to bind with short carbon chains (Chowdhury 2013). In this form, they bioaccumulate in marine organisms and concentrate year after year.

The current status of coastal and marine pollution from agricultural activities reflects concerning trends. Despite efforts to mitigate nutrient and pesticide runoff through best management practices and regulatory measures, the scale of agricultural production continues to intensify globally. Rapid urbanization and expanding agricultural land further exacerbate the problem, increasing the pressure on coastal and marine ecosystems. Moreover, climate change-induced factors, such as altered precipitation patterns and sea level rise, may exacerbate nutrient runoff and pesticide transport, amplifying the challenges faced by marine environments.

5.7 Impacts of Agrarian-Based Synthetic Fertilizers and Pesticides on Marine and Coastal Life and Human Health

Coastal and marine pollution resulting from agricultural activities, particularly the use of fertilizers and pesticides, has profound effects on marine ecosystems, including mangroves, coral reefs, seagrasses, marine wildlife, and human health (Figure 5.3). Synthetic chemicals present in these agricultural inputs can react with the marine environment in various ways, exacerbating the impacts on marine life. As illustrated in Figure 5.3, once the fertilizers and pesticides enter the aquatic environment, bioaccumulation and magnification occur in marine wildlife. These residues enter into the human and animals through food webs causing detrimental health risks.

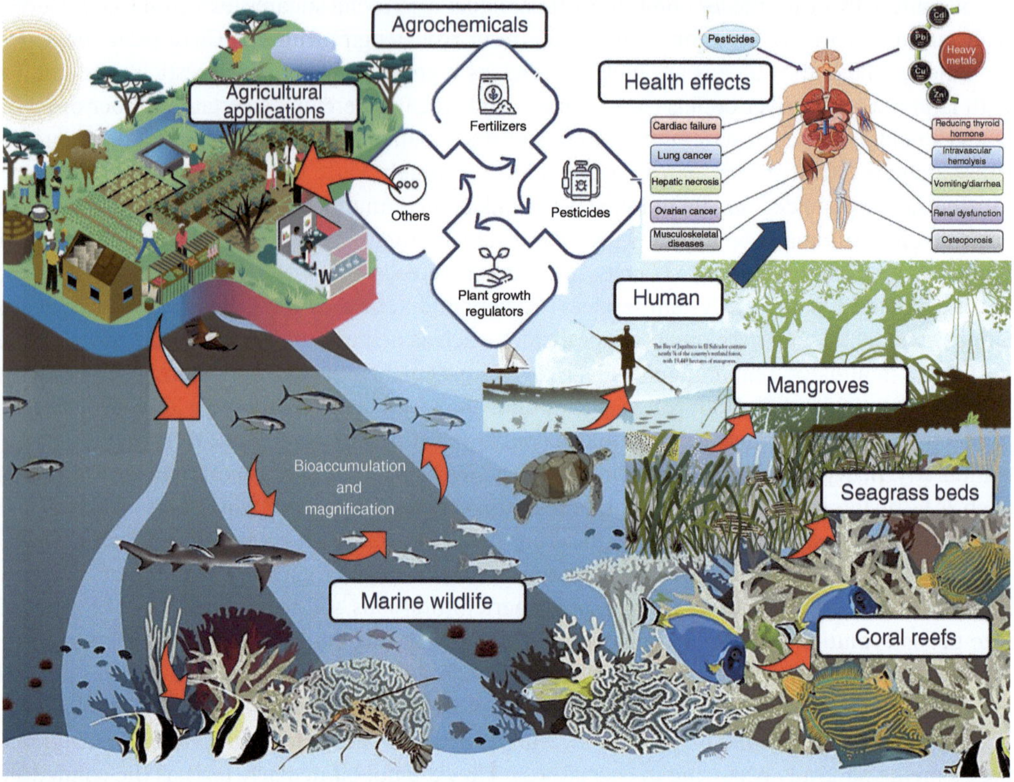

Figure 5.3 Impacts of agrarian-based synthetic fertilizers and pesticides on marine and estuarine ecosystems and human health.

Mangroves, critical coastal ecosystems, face significant threats from agricultural pollution. Synthetic fertilizers, containing nitrogen and phosphorus compounds, can trigger eutrophication when they enter water bodies. Excess nutrients fuel the growth of algae, leading to algal blooms that block sunlight and hinder mangrove growth. Additionally, pesticides, designed to control pests and weeds, can leach into waterways and directly harm mangrove trees, disrupting the intricate web of life they support. According to the available statistics, approximately 30% of mangrove loss worldwide is attributed to agricultural runoff, threatening the biodiversity and resilience of these vital ecosystems (Akram et al. 2023).

Coral reefs, often called the "rainforests of the sea," are sensitive to synthetic chemicals present in agricultural runoff (Mulhall 2008). Nitrogen and phosphorus compounds from fertilizers can stimulate algal growth, leading to algal blooms that smother coral colonies and interfere with their symbiotic relationships with algae. Pesticides, such as herbicides and insecticides, can directly harm corals by inhibiting their growth and reproduction. Some pesticides contain toxic chemicals such as organochlorines, which can accumulate in coral tissues and disrupt their physiological processes. The Global Coral Reef Monitoring Network (GCRMN) estimates that approximately 75% of coral reefs worldwide are threatened by various anthropogenic stressors, including agricultural pollution (Obura et al. 2019).

Seagrass meadows, essential coastal habitats, are also vulnerable to synthetic chemicals present in agricultural runoff. Nitrogen and phosphorus compounds from fertilizers can stimulate algal growth, shading seagrasses and inhibiting their photosynthesis (Burkholder et al. 2007). Pesticides, when carried by runoff or groundwater discharge, can accumulate in seagrass sediments, affecting seagrass health and the organisms that rely on these habitats for food and shelter. Therefore, agricultural runoff contributes to the decline of seagrass coverage by up to 50% in some regions of the world (Waycott et al. 2009), threatening the stability and productivity of these critical ecosystems.

When considering overall marine wildlife, synthetic chemicals present in agricultural runoff also impact marine wildlife populations, including fish, crustaceans, marine mammals, and seabirds. Nitrogen and phosphorus compounds can lead to harmful algal blooms that produce toxins harmful to fish and other marine organisms. Pesticides, particularly those containing toxic chemicals like organophosphates and pyrethroids, can directly harm marine life through acute toxicity or chronic exposure (Kadiru et al. 2022). Additionally, these synthetic persistent chemicals can disrupt food webs and reproductive cycles, leading to population declines and reduced biodiversity.

Potentially harmful chemical constituents in fertilizers and pesticides, which enter marine ecosystems primarily through runoff from agricultural lands, can contaminate seafood, water sources, and air, leading to a range of adverse health effects in humans. As an example, a chemical species commonly found in marine environments due to agricultural runoff is atrazine, a widely used herbicide. Atrazine can leach into water bodies and persist for extended periods, posing risks to human health through groundwater contamination and runoff into surface waters. Chronic exposure to atrazine has been linked to hormonal disruption, reproductive abnormalities, and developmental issues in humans (Pathak and Dikshit 2011). Another example is glyphosate, the active ingredient in many herbicides. Glyphosate can enter marine ecosystems through agricultural runoff and has been detected in coastal waters worldwide. Exposure to glyphosate has been associated with various health concerns, including cancer, kidney and liver damage, and disruptions to the endocrine system (Feltracco et al. 2022). Furthermore, pesticides containing organophosphates, such as chlorpyrifos and diazinon, are commonly used in agriculture and can contaminate marine environments through runoff and spray drift. These chemicals are neurotoxic and can cause acute poisoning in humans through ingestion, inhalation, or dermal contact (Kadiru et al. 2022). Chronic exposure to organophosphates has been linked to neurological disorders, respiratory problems, and developmental issues, particularly in children (Eskenazi et al. 1999). Furthermore, pyrethroid insecticides, such as permethrin and cypermethrin, are widely used in agriculture and can enter marine ecosystems through runoff and direct application. These chemicals are toxic to aquatic organisms and can bioaccumulate in seafood, posing risks to human health through contaminated seafood consumption. Exposure to pyrethroids has been associated with allergic reactions, respiratory irritation, and neurological symptoms in humans (Farag et al. 2021).

Moreover, heavy metals from agricultural outputs, especially lead and cadmium, which may be present in fertilizers and pesticides as impurities or additives, can contaminate marine environments and accumulate in oceans. The effects of metals on organisms are associated with interference in the metabolic processes involving sulfur-containing constituents (Kaczor-Kamińska et al. 2020) because most of the widely distributed heavy metals have high affinities for sulfur and tend to bind with sulfhydryl groups of proteins and enzymes in living beings. Heavy metal interference is reported to cause an increase in the permeability of the cell membrane in phytoplankton and other marine algae, leading to the loss of intracellular constituents and, therefore, cellular integrity. Once in the system, metals concentrate in protein-rich tissues such as the liver and muscle of organisms. High trace element burdens in marine mammals are associated with various responses. In addition, cadmium and lead are potential immuno-suppressants and marine mammals tend to accumulate in the liver at higher levels than other marine organisms, especially in marine mammals (Jakimska et al. 2011). Chronic exposure to heavy metals through contaminated seafood consumption can lead to a range of health problems, including neurological disorders, kidney damage, cardiovascular diseases, and developmental issues in children. Consumption of contaminated seafood can result in neurodevelopmental delays, cognitive impairments, and cardiovascular problems (Goyer and Clarkson 1996).

In summary, marine pollution from agrochemicals, including fertilizers and pesticides, can have profound effects on human health through exposure to chemical species such as atrazine, glyphosate, organophosphates, pyrethroids, and heavy metals. Addressing this issue requires concerted efforts to minimize pollution sources, promote sustainable agricultural practices, enhance monitoring and regulation of agrochemical use, and mitigate human health risks associated with marine pollution. Collaboration between government agencies, industry stakeholders, scientific institutions, and local communities is essential to protect human health and safeguard marine and coastal ecosystems from the impacts of agrochemical-related toxic pollutants.

5.8 Impact on the Blue Economy

The blue economy refers to a holistic approach to utilizing ocean resources sustainably for economic growth, job creation, and societal well-being while prioritizing the conservation and preservation of marine ecosystems. It encompasses various sectors and activities, including fisheries, aquaculture, tourism, shipping, renewable energy, marine biotechnology, coastal development, marine conservation, marine research, marine transportation, and ocean exploration (Patil et al. 2018).

Coastal and marine pollution resulting from agricultural activities, particularly the use of fertilizers and pesticides, poses significant challenges to the sectors of the blue economy. These pollutants, often transported by runoff from agricultural lands, find their way into rivers, estuaries, and ultimately coastal and marine environments, where they exert detrimental impacts on marine ecosystems and the industries dependent on them. This newest approach acknowledges the economic potential of oceans and coastal areas while emphasizing responsible management to ensure long-term environmental sustainability and social equity.

In the fisheries and aquaculture sector, agricultural pollution can have severe consequences for marine biodiversity and fish populations. Fertilizers containing nitrogen and phosphorus compounds promote the growth of algae, leading to harmful algal blooms that deplete oxygen levels in coastal waters, resulting in fish killings and disrupting the balance of marine ecosystems. According to scientific research, agricultural runoff is a primary source of marine pollution, contributing to eutrophication and habitat degradation. Pesticides from agricultural runoff can also accumulate in aquatic organisms, including commercially valuable fish and shellfish, posing risks to human health and diminishing the marketability of seafood products. The marine pollution from agricultural runoff costs the global economy billions of dollars annually in lost fisheries revenue and ecosystem services (Tegtmeier and Duffy 2004).

The tourism sector, which relies heavily on pristine coastal environments and marine biodiversity, is also vulnerable to the impacts of agricultural pollution. Pollution-driven phenomena such as algal blooms and coral reef degradation can deter tourists, tarnish destinations' reputations, and lead to economic losses for businesses reliant on tourism revenue. Marine pollution reduces the value of tourism assets and services, negatively impacting coastal communities and exacerbating poverty and inequality in developing countries. Studies have shown that marine pollution from agricultural runoff can significantly reduce the attractiveness and recreational value of coastal areas, resulting in decreased tourist arrivals and revenues (Bennett et al. 2021).

Agricultural pollution contributes to the formation of navigational hazards such as algal mats and dead zones, posing risks to vessel operations and maritime infrastructure in the shipping and maritime transportation sector. These contaminations can lead to harmful algal blooms, biofouling on ships and port infrastructure, increased sedimentation in shipping channels, degraded water quality, and ecological impacts, all of which hinder safe navigation, impact shipping routes, and disrupt port operations in marine ecosystems. Nutrient-rich runoff can promote the growth of biofouling organisms such as barnacles, algae, and mussels on ship hulls and port infrastructure. Biofouling increases drag on ships, leading to reduced fuel efficiency and increased operating costs. It can also clog the water intake systems of vessels and port facilities, impacting their functionality and requiring frequent maintenance and cleaning and disrupting maritime activities, such as marine spatial planning, offshore energy development, and marine biodiversity prospecting (Islam and Tanaka 2004).

Table 5.1 provides insights into the diverse sectors of the blue economy and how they are impacted by agriculture-driven marine pollution by highlighting the importance of addressing pollution to ensure the sustainability and resilience of ocean-based industries and activities.

Renewable energy development, especially receiving from wave currents in the blue economy, can also be impacted by coastal and marine pollution from agricultural activities. For example,

Table 5.1 Various sectors of the blue economy, their contributions to the global economy, and how they are affected by agriculture-driven marine pollution.

Sector of blue economy	Description	Contribution to the global economy	Impact of agriculture-driven marine pollution
Fisheries	Involves the harvesting, processing, and marketing of fish and seafood products.	Provides food security, employment, and income for millions of people worldwide. Global fish production reached 179 million tonnes in 2018, with a value of over $401 billion (Saleh et al. 2022).	Nutrient runoff from fertilizers fuels harmful algal blooms, leading to fish kills and habitat degradation. Pesticide contamination can affect fish health and marketability of seafood products.
Aquaculture	The farming of aquatic organisms such as fish, shellfish, and algae for commercial purposes.	Supplies a significant portion of the world's seafood and contributes to economic growth and food security. Global aquaculture production amounted to 82.1 million tonnes in 2018, valued at approximately $250 billion (Bartley 2022)	Nutrient runoff and pesticide contamination can affect water quality in aquaculture facilities, leading to decreased productivity and health issues in farmed organisms.
Tourism	Activities such as recreational boating, diving, snorkeling, and beach vacations in coastal and marine areas.	Generates revenue, employment, and economic growth in coastal regions. The global tourism industry contributed $8.9 trillion to the global GDP and supported 319 million jobs in 2019 (Bhatia et al. 2022)	Pollution-driven phenomena such as algal blooms and habitat degradation can deter tourists, leading to economic losses for tourism-dependent communities.
Shipping/ marine transportation	Transportation of goods and passengers through maritime vessels.	Facilitates global trade, contributing to economic development and connectivity. Shipping transports around 90% of global trade, contributing over $380 billion to the global economy annually (Wang et al. 2020).	Nutrient runoff and pesticide contamination can contribute to navigational hazards such as algal mats and dead zones, posing risks to vessel operations and maritime infrastructure.

(*Continued*)

Table 5.1 (Continued)

Sector of blue economy	Description	Contribution to the global economy	Impact of agriculture-driven marine pollution
Renewable energy	Generation of energy from marine resources such as wind, waves, and tides.	Provides clean and sustainable energy sources, reducing dependence on fossil fuels. The global renewable energy market was valued at $928.0 billion in 2017 and is projected to reach $1,512.3 billion by 2025 (Shouman 2020).	Pollution can disrupt marine ecosystems, affecting the efficiency and reliability of renewable energy infrastructure.
Marine biotechnology	Utilization of marine organisms and materials for medical, industrial, and research purposes.	Drives innovation and scientific advancements, contributing to economic growth and job creation. The global marine biotechnology market was valued at $4.1 billion in 2019 and is expected to reach $9.9 billion by 2027 (Mone and Trivedi 2023).	Pollution can degrade marine habitats and affect the health and viability of marine organisms used in biotechnological applications.
Coastal development	Infrastructure development, coastal engineering, and urban planning in coastal areas.	Supports coastal communities, tourism, and economic activities, enhancing resilience to natural hazards. Coastal tourism alone generates approximately $7.6 trillion in global revenue annually (Houston 2018).	Pollution can degrade coastal environments, affecting the quality of life, property values, and economic activities in coastal regions.
Marine conservation	Efforts to protect and preserve marine biodiversity, habitats, and ecosystems.	Sustains marine ecosystems, biodiversity, and ecosystem services, contributing to long-term environmental and economic benefits. The economic value of coral reefs is estimated at $9.9 trillion annually (Sing Wong 2023).	Pollution threatens the integrity of marine ecosystems, undermining conservation efforts and compromising the resilience of marine biodiversity.
Ocean exploration	Exploration and discovery of the ocean's depths, biodiversity, and geological features.	Expands scientific knowledge, inspires innovation, and fosters international collaboration, contributing to societal benefits and economic growth. The global deep-sea mining market is projected to reach $15.3 billion by 2030 (Jordan 2023).	Pollution can degrade marine environments and impact biodiversity, hindering exploration efforts and compromising scientific discoveries.

nutrient runoff from fertilizers can induce eutrophication, leading to algal blooms and oxygen depletion, which can foul equipment and impede the efficiency of marine renewable energy systems like offshore wind farms and tidal energy installations, requiring costly maintenance and reducing energy output. Similarly, pesticide contamination in coastal waters threatens marine organisms and ecosystems, disrupting food webs and reducing biodiversity. These impacts can compromise the viability of marine renewable energy projects by altering habitat suitability and ecosystem dynamics. Contaminated waters may deter investors, increase maintenance costs due to biofouling, and pose operational risks (Wenhai et al. 2019).

Marine biotechnology involves the utilization of marine organisms and materials for medical, industrial, and research purposes. Coastal and marine pollution from agricultural activities can impact marine biotechnology by contaminating marine organisms used in research and industrial processes. Pesticides and heavy metals can accumulate in these organisms, compromising their suitability for biotechnological applications and product development. Additionally, nutrient runoff leading to eutrophication can alter marine ecosystems, potentially affecting the availability and diversity of organisms used in biotechnology (Singh and Singh 2017). Such pollution-driven changes may disrupt research efforts, limit the development of biotechnological products, and compromise the sustainable utilization of marine resources, highlighting the need for pollution mitigation measures to safeguard marine biotechnology's effectiveness and sustainability.

In conclusion, coastal and marine pollution from agricultural activities, particularly the use of fertilizers and pesticides, poses significant challenges to the sectors of the blue economy. These pollutants can have detrimental impacts on marine ecosystems, biodiversity, and the industries and activities dependent on them. Addressing agricultural pollution requires concerted efforts to promote sustainable agricultural practices, enhance regulatory frameworks, and foster innovation and technology adoption. By safeguarding coastal and marine environments, we can ensure the long-term viability and prosperity of the blue economy, benefiting both current and future generations.

5.9 Innovative Solutions to Control and Prevent Coastal and Marine Pollution from Agricultural Activities

5.9.1 Regulatory Measures and Policies

Agricultural activities are considered the primary contributor to an increase in pollutant delivery to marine ecosystems. One of the largest sources of nitrogen (N) and phosphorus (P) pollution is industrialized agriculture in the form of animal manure, an inefficient nutrient application, bad irrigation practices, and soil erosion (Beman et al. 2005). It is important to determine the agricultural runoff hotspots and set policies on both local and regional levels due to the various vulnerabilities to agricultural runoff and the cost of prevention. The N and P input–output cycle may either remain confined within a single field or extend beyond boundaries due to the movement of air and water, alongside the global rise in N_2O concentrations. Consequently, addressing this issue demands solutions tailored to local and regional contexts while also prioritizing the enhancement of nutrient use efficiency on a global level (Krushelnytska 2018).

Based on the regulatory measures and policies that are aimed at addressing coastal and marine pollution due to agricultural activities, particularly related to fertilizers and pesticides, typically involve a combination of local, national, and international efforts. These measures are designed to reduce the adverse environmental and human health impacts associated with agricultural runoff (UN 1992). The establishment of water quality standards is one of the regulatory methods. Most countries establish water quality standards to regulate the permissible levels of pollutants in coastal and marine waters. These standards often include limits on nutrient concentrations such as nitrogen and phosphorus from fertilizers and pesticide residues (Howarth and Ramakrishna 2005). Compliance with these standards may be mandatory for agricultural operations, with enforcement mechanisms in place to ensure adherence. Further, determining and mitigating the amount of nutrients in animal feed and crop fertilizers increases the efficiency of their use and brings economic benefits. Based on the estimations of the Millennium Ecosystem Assessment, it is emphasized that the United States alone applies 20–30% more fertilizer to crops than is necessary (Krushelnytska 2018).

Environmental Impact Assessment (EIA) is another major factor that can be used as a regulatory measure. In areas where large-scale agricultural activities are conducted near coastal regions, the government may mandate EIAs for such projects. EIAs evaluate the potential environmental impacts of proposed activities, including pollution from fertilizers and pesticides, and propose mitigation measures to minimize harm to coastal and marine ecosystems (Gallardo 2024). Moreover, governments enact laws and regulations governing the sale, use, and disposal of pesticides and fertilizers. These regulations may include requirements for product labeling, usage instructions, and restrictions on certain chemicals deemed environmentally harmful. Also, governments may ban or phase out particularly hazardous pesticides or fertilizers. As an example, the voluntary and mandated controls in the EU for fertilizers which was established in 1991 are still relevant and can be used for nutrient management. As described by Van Grinsven et al. (2013), a cost–benefit analysis (CBA) of nitrogen use efficiency conducted for 29 EU countries has shown a large potential to enhance nitrogen efficiencies and decrease runoff from manure and fertilizer use with limited effects on agricultural production. Based on the CBA results, net social benefits would be created in both regions due to the translocation of agricultural production in the EU from northwest to east (Krushelnytska 2018).

Buffer zones along water bodies and riparian setbacks are regulatory tools used to protect coastal and marine environments from agricultural runoff. These measures require a certain distance between agricultural activities and water bodies to minimize the risk of pollution. They help to filter out pollutants before they reach coastal and marine ecosystems (Lincoln et al. 2023). The effectiveness of buffer zones and riparian setbacks depends on factors such as their width, vegetation composition, maintenance practices, and surrounding land use. Proper design, implementation, and management of these regulatory tools are essential to maximize their ecological and water quality benefits in protecting coastal and marine environments from agricultural pollution (Khanam et al. 2023; Majumdar and Avishek 2023). Further, many countries participate in various international agreements and treaties that aim to protect the marine environment from pollution. The United Nations Convention on the Law of the Sea (UNCLOS) and the International Convention for the Prevention of Pollution from Ships (MARPOL) are some examples of international agreements. Environmental laws and policies are important to strike a balance between economic development and the environment as well as promote the well-being of both current and future generations (Vanshika 2023).

In addition, selective sales taxes on polluting items may raise revenue and it may shift away from polluting products. Selective sales taxes on "polluting items" generate income and can influence consumer behavior away from products that contribute to pollution. Nutrient trading is an environmental market where participants voluntarily trade nutrient credits, which represent units of environmental improvement. The market operates based on regulatory requirements regarding nutrient levels permitted in water bodies, encouraging compliance with water quality regulations (Krushelnytska 2018). Subsidies and incentive programs are another type of method that may be important as a control measure. These programs may include subsidies for the adoption of precision agriculture techniques, organic farming methods, or the implementation of best management practices (BMPs) to minimize fertilizer and pesticide runoff (Tao and Liu 2024). Furthermore, the governments engage in public education and outreach efforts to raise awareness regarding the impacts of agricultural pollution on coastal and marine ecosystems, and these initiatives are such as educational campaigns, workshops, and training programs for farmers and other stakeholders on sustainable agricultural practices and pollution prevention measures (Orbih 2023).

Good agricultural practices that decrease the N and P runoff are another important control measure to mitigate ocean pollution due to agricultural waste. Planting perennial and winter crops, fertilizer application at the right time, conducting tillage, sedimentation reduction, rational irrigation, and pesticide management can be used as the basis for good agricultural practices (Krushelnytska 2018). Instead of the annual crops, planting perennial ones in highly sensitive areas may retain nitrogen in the soil and greatly reduce groundwater loss. Further, winter crop planting would decrease the nitrate leaching rate into the ground which occurs generally in the winter and spring due to heavier rainfall. It also provides an economic benefit due to the increased agricultural production and the added benefit of nitrogen fixation provided by the winter crops (Howarth and Ramakrishna 2005). Moreover, there are available technologies that can eliminate up to 95% of phosphorus and up to 90% of nitrogen from sewage, albeit at a considerable expense. However, constructing treatment plants specifically designed to address nitrogen and phosphorus from the outset can help mitigate costs. In addition, adjustments can be made to existing sewage plants or new ones can be constructed in densely populated areas, thereby distributing the expenses of implementing the new technology across a larger population base (Scheierling et al. 2010).

Techniques aimed at mitigating phosphorus runoff, which has a more localized impact compared to nitrogen and directly influences nearby ecosystems such as freshwater bodies and coastal lagoons, involve methods such as tillage, contour planting, and establishing buffer zones. Constructed wetlands, which effectively absorb various forms of nutrient runoff, offer an alternative nature-based approach to traditional wastewater treatment. These wetlands are estimated to diminish phosphorus levels by 11–49%, nitrogen by 26–78%, and various pharmaceutical pollutants by 5–90% (Krushelnytska 2018).

In conclusion, agricultural activities stand as an important contributor to the escalation of pollutant discharge into marine ecosystems. The primary sources of N and P pollution stem from industrialized agriculture, manifesting in forms such as animal manure, inefficient nutrient application, poor irrigation practices, and soil erosion (Smith and Siciliano 2015; Lasacco et al. 2021). To effectively address this scenario, it is imperative to determine agricultural runoff hotspots and implement policies at both local and regional levels, considering the diverse vulnerabilities to agricultural runoff as well as the associated prevention costs. The N and P input–output cycle, influenced by the movement of air and water, underscores the necessity for tailored solutions that cater to local and regional contexts while concurrently emphasizing the enhancement of nutrient use efficiency on a global scale (Javan et al. 2024).

Regulatory measures and policies play an important role in reducing the coastal and marine pollution attributable to agricultural activities, especially concerning fertilizers and pesticides. The establishment of water quality standards, EIAs, and regulations governing the sale and use of agricultural chemicals are crucial steps toward ensuring compliance and safeguarding marine environments (Monsurat et al. 2019). Additionally, international agreements and treaties contribute to global efforts aimed at protecting the marine environment from pollution. Innovative approaches including selective sales taxes on polluting items, nutrient trading markets, subsidies for sustainable agricultural practices, and public education initiatives further complement regulatory measures in mitigating agricultural pollution (Guyot 1994). Good agricultural practices, such as crop selection, fertilizer management, and erosion control, may serve as fundamental strategies to decrease nutrient runoff and preserve water quality.

Moreover, technological improvements in sewage treatment offer promising solutions to determine nutrient pollution, albeit at a considerable expense. Constructed wetlands are some of the technologies that provide nature-based alternatives to traditional wastewater treatment, effectively decreasing nutrient levels and pharmaceutical pollutants in water bodies. In summary, a multifaceted approach that integrates regulatory measures, innovative solutions, and sustainable practices is crucial to effectively reduce or mitigate agricultural pollution and safeguard coastal and marine ecosystems not only for the present but also for future generations. Collaboration among stakeholders, ongoing research, and continuous monitoring are imperative to ensure the long-term sustainability of marine environments in the face of agricultural pressures.

5.10 Innovative Solutions and Future Challenges

Innovative solutions as well as best management practices are important in reducing coastal and marine pollution from agricultural activities considering fertilizers and pesticides. Minimizing the use of harmful chemicals, reducing nutrient and pesticide runoff, and promoting sustainable agricultural practices are some of the approaches focused on best management practices. Precision agriculture, cover crops, crop rotation, integrated pest management, constructed wetlands, bioremediation, nutrient management plans, and soil health management can be considered some of the best innovative solutions and best management practices. Geographical Information Systems (GIS), Global Positioning Systems (GPS), and remote sensing are the technologies that can be employed for precision agriculture and these technologies can be used to optimize the application of fertilizers and pesticides. By precisely targeting inputs based on crop needs and environmental conditions, precision agriculture minimizes wastage and reduces the risk of nutrient and pesticide runoff into coastal and marine ecosystems.

Planting cover crops and crop rotation practices are best management practices that can be used to improve soil health, reduce erosion, and minimize the massive requirement for chemical fertilizers and pesticides. Legumes and grasses can be used as cover crops that can capture excess nutrients, prevent soil erosion, and enhance soil fertility. Importantly, crop rotation may help disrupt pests and disease cycles by reducing the reliance on pesticides (Schepers et al. 1985; Seta et al. 1993). Further, integrated pest management (IPM) is a holistic approach to the management of pests that can be used for the prevention, monitoring, and control using a combination of biological, cultural, physical, and chemical methods. IPM reduces the reliance on pesticides and minimizes their environmental impact by incorporating natural predators, crop rotation, trap cropping, and other nonchemical controls (Sharma et al. 2015; Stenberg 2017). The buffer stripes of vegetation along the water bodies and riparian zones may act as natural filters that can trap the sediments, excess nutrients, and pesticides before they reach the coastal and marine environments. In addition, planting the native vegetation in these areas may enhance biodiversity, stabilize stream banks, and reduce nutrient and pesticide runoff from agricultural fields (Barling and Moore 1994). Further, constructed wetlands are engineered ecosystems designed to mimic natural wetlands and treat agricultural runoff. These systems use plants, soil, and microorganisms to remove nutrients, pesticides, and sediments from the water before they enter coastal and marine habitats (Tang et al. 2021). Constructed wetlands also provide habitat for wildlife and contribute to biodiversity conservation.

Bioremediation involves using biological agents, including various microorganisms such as bacteria, fungi, and plants, to degrade or remove pollutants from soil and water. In agricultural settings, bioaugmentation with microbial inoculants or phytoremediation using pollutant-accumulating plants can help reduce the impacts of fertilizer and pesticide contamination on coastal and marine ecosystems (Abatenh et al. 2017). Implementing nutrient management plans helps farmers optimize fertilizer use, minimize nutrient runoff, and protect water quality. These plans typically involve soil testing, nutrient budgeting, and tailored fertilizer application based on crop requirements and environmental considerations (Phulpoto et al. 2016). Moreover, practices that enhance soil health, including conservation tillage, organic amendments, and agroforestry, can increase soil structure, water retention, and nutrient cycling. Healthy soils are more resilient to erosion, nutrient leaching, and pesticide runoff, thereby reducing their impact on coastal and marine environments. Agricultural stakeholders may lessen the environmental impact of farming operations and protect coastal and marine ecosystems from the harmful effects of pesticides and fertilizers by putting these creative solutions and best management practices into practice. In order to achieve long-term environmental sustainability and encourage the implementation of these sustainable methods, cooperation between farmers, academics, legislators, and conservation organizations is crucial.

Although there has been progress in tackling agricultural contamination of the coast and sea, problems still exist and will probably change in the future. Some of the predicted difficulties may be the climate change impacts, emerging contaminants, nonpoint source impacts, socioeconomic pressures, technological advances, policy and governance issues, cumulative impacts, globalization of agricultural trade, etc. A comprehensive, multidisciplinary strategy that incorporates community involvement, policy creation, scientific research, and stakeholder engagement is needed to address these upcoming difficulties. Stakeholders can create creative ideas and adaptable ways to protect coastal and marine ecosystems for future generations by cooperating and taking initiative. In conclusion, to adopt sustainable practices and save coastal and marine habitats for future generations, collaboration between farmers, researchers, lawmakers, and conservation organizations is crucial. To handle impending challenges and guarantee long-term environmental sustainability, innovative approaches and flexible tactics are required.

References

Abatenh, E., Gizaw, B., Tsegaye, Z., and Wassie, M. (2017). The role of microorganisms in bioremediation – a review. *Open Journal of Environmental Biology* 2: 038–046. https://doi.org/10.17352/ojeb.000007.

Akram, H., Hussain, S., Mazumdar, P. et al. (2023). Mangrove health: a review of functions, threats, and challenges associated with mangrove management practices. *Forests* 14: 1698. https://doi.org/10.3390/f14091698.

Barling, R.D. and Moore, I.D. (1994). Role of buffer strips in management of waterway pollution: a review. *Environmental Management* 18: 543–558. https://doi.org/10.1007/BF02400858.

Bartley, D.M. (2022). *World Aquaculture 2020 – A brief overview.* FAO Fisheries and Aquaculture Circular No. 1233. Rome: FAO https://doi.org/10.4060/cb7669en.

Beman, J.M., Arrigo, K.R., and Matson, P.A. (2005). Agricultural runoff fuels large phytoplankton blooms in vulnerable areas of the ocean. *Nature* 434: 211–214. https://doi.org/10.1038/nature03370.

Bennett, N.J., Blythe, J., White, C.S., and Campero, C. (2021). Blue growth and blue justice: ten risks and solutions for the ocean economy. *Marine Policy* 125: 104387. https://doi.org/10.1016/j.marpol.2020.104387.

Beusen, A.H.W., Bouwman, A.F., Van Beek, L.P.H. et al. (2016). Global riverine N and P transport to ocean increased during the 20th century despite increased retention along the aquatic continuum. *Biogeosciences* 13: 2441–2451. https://doi.org/10.5194/bg-13-2441-2016.

Bhatia, A., Roy, B., and Kumar, A. (2022). A review of tourism sustainability in the era of Covid-19. *Journal of Statistics & Management Systems* 25 (8): 1871–1888.

Brumfield Ii, J. (2010). *Dynamic Model of Ecological Factors Impacting the Gulf of Mexico Dead Zone (Doctor of Philosophy).* Louisiana State University and Agricultural and Mechanical College https://doi.org/10.31390/gradschool_dissertations.2963.

Burkholder, J.M., Tomasko, D.A., and Touchette, B.W. (2007). Seagrasses and eutrophication. *Journal of Experimental Marine Biology and Ecology* 350: 46–72. https://doi.org/10.1016/j.jembe.2007.06.024.

Cagnazzi, D., Harrison, P.L., Parra, G.J. et al. (2020). Geographic and temporal variation in persistent pollutants in Australian humpback and snubfin dolphins. *Ecological Indicators* 111: 105990. https://doi.org/10.1016/j.ecolind.2019.105990.

Chen, C., Guo, W., and Ngo, H.H. (2019). Pesticides in stormwater runoff—a mini review. *Frontiers of Environmental Science & Engineering* 13: 72. https://doi.org/10.1007/s11783-019-1150-3.

Chowdhury, Z.Z. (2013). *Preparation, Characterization and Adsorption Studies of Heavy Metals onto Activated Adsorbent Materials Derived from Agricultural Residues*. University of Malaya (Malaysia).

Corsolini, S. and Sarà, G. (2017). The trophic transfer of persistent pollutants (HCB, DDTs, PCBs) within polar marine food webs. *Chemosphere* 177: 189–199. https://doi.org/10.1016/j.chemosphere.2017.02.116.

Eskenazi, B., Bradman, A., and Castorina, R. (1999). Exposures of children to organophosphate pesticides and their potential adverse health effects. *Environmental Health Perspectives* 107: 409–419.

FAO (2009). *FAOSTAT Online Statistical Service; Food and Agriculture Organization of the United Nations (FAO)*. Rome, Italy: accessed 13 February 2024.

FAO 2020. Fertilizers by Nutrient (FAOSTAT, accessed 13 February 2024) https://www.fao.org/faostat/en/#data/RFN.

FAO 2023. Pesticides Use (FAOSTAT, accessed 13 February 2024). https://www.fao.org/faostat/en/#data/RP.

Farag, M.R., Alagawany, M., Bilal, R.M. et al. (2021). An overview on the potential hazards of pyrethroid insecticides in fish, with special emphasis on cypermethrin toxicity. *Animals* 11: 1880. https://doi.org/10.3390/ani11071880.

Feltracco, M., Barbaro, E., Morabito, E. et al. (2022). Assessing glyphosate in water, marine particulate matter, and sediments in the Lagoon of Venice. *Environmental Science and Pollution Research* 29: 16383–16391. https://doi.org/10.1007/s11356-021-16957-x.

Fernández-Luqueño, F., López-Valdez, F., Gamero-Melo, P. et al. (2013). Heavy metal pollution in drinking water - a global risk for human health: a review. *African Journal of Environmental Science and Technology* 7 (7): 567–584.

Gallardo, R.K. (2024). The environmental impacts of agriculture: a review. *International Review of Environmental and Resource Economics* 18: 165–235. https://doi.org/10.1561/101.00000166.

Goyer, R.A. and Clarkson, T.W. (1996). Toxic effects of metals. In: *Casarett and Doull's toxicology: The basic science of poisons*, 8e (ed. C.D. Klaassen), 691–736. McGraw-Hill Education.

Guyot, C. (1994). Strategies to minimize the pollution of water by pesticides; a critical review to plant protection measures in the past, present, and future. *Chemistry of Plant Protection* 9: 87–152.

Houston, J.R. (2018). The economic value of America's beaches—a 2018 update. *Shore & Beach* 862: 3–13.

Howarth, R. and Ramakrishna, K. (2005). Nutrient management. In: *Millennium Ecosystem Assessment, Ecosystems and Human Well-being: Policy Responses*. Washington, DC: Island Press.

IbisWorld (2016) Industry market research, reports, & statistics. https://www.ibisworld.com/au/market- size/pesticide-manufacturing/ (accessed 2 February 2024).

Islam, M.S. and Tanaka, M. (2004). Impacts of pollution on coastal and marine ecosystems including coastal and marine fisheries and approach for management: a review and synthesis. *Marine Pollution Bulletin* 48 (7–8): 624–649.

Jakimska, A., Konieczka, P., Skóra, K., and Namieśnik, J. (2011). Bioaccumulation of metals in tissues of marine animals, part I: the role and impact of heavy metals on organisms. *Polish Journal of Environmental Studies* 20 (5): 1127–1146.

Javan, K., Altaee, A., BaniHashemi, S. et al. (2024). A review of interconnected challenges in the water–energy–food nexus: urban pollution perspective towards sustainable development. *Science of the Total Environment* 912: 169319. https://doi.org/10.1016/j.scitotenv.2023.169319.

Johnson, A. and Harrison, M. (2015). The increasing problem of nutrient runoff on the coast: as development increases along coastlines worldwide, water quality—and everything that depends on it— degrades. *American Science* 103 (2): 98–102.

Jordan, S., 2023. Report on Building the Blue Innovation Pipeline: Lessons Learned from the United States and Sweden. Institute for Sustainable Development, Swedish-American Chambers of Commerce of the USA, Inc., and United States Embassy in Stockholm.

Kaczor-Kamińska, M., Sura, P., and Wróbel, M. (2020). Multidirectional changes in parameters related to sulfur metabolism in frog tissues exposed to heavy metal-related stress. *Biomolecules* 10: 574. https://doi.org/10.3390/biom10040574.

Kadiru, S., Patil, S., and D'Souza, R. (2022). Effect of pesticide toxicity in aquatic environments: a recent review. *International Journal of Fisheries and Aquatic Studies* 10: 113–118.

Khanam, Z., Sultana, F.M., and Mushtaq, F. (2023). Environmental pollution control measures and strategies: an overview of recent developments. In: *Geospatial Analytics for Environmental Pollution Modeling* (ed. F. Mushtaq, M. Farooq, A.B. Mukherjee, et al.). Cham: Springer https://doi.org/10.1007/978-3-031-45300-7_15.

Kim, S.-K. (2019). Trophic transfer of organochlorine pesticides through food-chain in coastal marine ecosystem. *Environmental Engineering Research* 25: 43–51. https://doi.org/10.4491/eer.2019.003.

Krushelnytska, O. (2018). Solving Marine Pollution: Successful models to reduce wastewater, agricultural runoff, and marine litter. *Pollution Management and Environmental Health* 1: 1–24.

Lasacco, D., Ancona, V., Paola, D.D. et al. (2021). Development of ecological strategies for the recovery of the main nitrogen agricultural pollutants: a review on environmental sustainability in agroecosystems. *Sustainability* 13 (13): 7163.

Lincoln, S., Chowdhury, P., Posen, P.E. et al. (2023). Interaction of climate change and marine pollution in Southern India: implications for coastal zone management practices and policies. *Science of the Total Environment* 902: 166061. https://doi.org/10.1016/j.scitotenv.2023.166061.

Liu, S., Xie, Z., Zeng, Y. et al. (2019). Effects of anthropogenic nitrogen discharge on dissolved inorganic nitrogen transport in global rivers. *Global Change Biology* 25 (4): 1493–1513.

de los Santos, C.B., Scott, A., Arias-Ortiz, A. et al. (2020). Seagrass ecosystem services: assessment and scale of benefits. In: *Out of the Blue: The Value of Seagrasses to the Environment and to People*, 19–21. United Nations Environment Programme.

Lu, C.C. and Tian, H. (2017). Global nitrogen and phosphorus fertilizer use for agriculture production in the past half century: shifted hot spots and nutrient imbalance. *Earth System Science Data* 9: 181.

Maggi, F., Tang, F.H.M., and Tubiello, F.N. (2023). Agricultural pesticide land budget and river discharge to oceans. *Nature* 620: 1013–1017. https://doi.org/10.1038/s41586-023-06296-x.

Majumdar, A. and Avishek, K. (2023). Riparian zone assessment and management: an integrated review using geospatial technology. *Water, Air, and Soil Pollution* 234: 319. https://doi.org/10.1007/s11270-023-06329-1.

Mitra, A., Zaman, S., Mitra, A., and Zaman, S. (2016). Threats to marine and estuarine ecosystems. *Basics of Marine and Estuarine Ecology* 365–417.

Moberg, F. and Rönnbäck, P. (2003). Ecosystem services of the tropical seascape: interactions, substitutions and restoration. *Ocean and Coastal Management* 46: 27–46. https://doi.org/10.1016/S0964-5691(02)00119-9.

Mone, D. and Trivedi, N. (2023). Marine extremophiles as a source of seaweed polysaccharide hydrolyzing enzymes. In: *Extremophiles*, 115–150. CRC Press.

Monsurat, S.R., Olalekan, R.M., and Olawale, S.H. (2019). A deep dive into the review of national environmental standards and regulations enforcement agency (NESREA) act. *International Research Journal of Applied Sciences* 1 (4): 108–125.

Mulhall, M. (2008). Saving the rainforest of the sea: an analysis of international efforts to conserve coral reefs. *Duke Environmental Law & Policy Forum* 19: 321–351.

Mwevura, H., Bouwman, H., Kylin, H. et al. (2020). Organochlorine pesticides and polycyclic aromatic hydrocarbons in marine sediments and polychaete worms from the west coast of Unguja island, Tanzania. *Regional Studies in Marine Science* 36: 101287. https://doi.org/10.1016/j.rsma.2020.101287.

Obura, D.O., Aeby, G., Amornthammarong, N. et al. (2019). Coral reef monitoring, reef assessment technologies, and ecosystem-based management. *Frontiers in Marine Science* 6: 580. https://doi.org/10.3389/fmars.2019.00580.

Ojemaye, C.Y., Onwordi, C.T., Pampanin, D.M. et al. (2020). Presence and risk assessment of herbicides in the marine environment of Camps Bay (Cape Town, South Africa). *Science of the Total Environment* 738: 140346.

Orbih, C. (2023). *Assessing the Drivers of Plastic Pollution on the Lagos Coastline: Strategies for Effective Policy Implementation and Mitigation*. World Maritime University dissertations.

Ouyang, W., Zhang, Y., Gu, X. et al. (2019). Occurrence, transportation, and distribution difference of typical herbicides from estuary to bay. *Environment International* 130: 104858. https://doi.org/10.1016/j.envint.2019.05.052.

Pathak, R.K. and Dikshit, A.K. (2011). Atrazine and human health. *International Journal of Ecosystem* 1: 14–23. https://doi.org/10.5923/j.ije.20110101.03.

Patil, P.G., Virdin, J., Colgan, C.S., Hussain, M.G., Failler, P., Vegh, T., 2018. Toward a blue economy.

Pearce, F. (2018). *When the Rivers Run Dry, Fully Revised and Updated Edition: Water-the Defining Crisis of the Twenty-First Century*, 328. Boston: Beacon Press.

Phulpoto, A.H., Qazi, M.A., Mangi, S. et al. (2016). Biodegradation of oil-based paint by *Bacillus* species monocultures isolated from the paint warehouses. *International journal of Environmental Science and Technology* 13: 125–134. https://doi.org/10.1007/s13762-015-0851-9.

Pimentel, D. (2005). Environmental and economic costs of the application of pesticides primarily in the United States. *Environment, Development and Sustainability* 7: 229–252. https://doi.org/10.1007/s10668-005-7314-2.

Ponce-Vélez, G. and de la Lanza-Espino, G. (2019). Organophosphate pesticides in coastal lagoon of the Gulf of Mexico. *Journal of Environmental Protection* 10 (02): 103–117.

Saleh, N.E., Wassef, E.A., and Abdel-Mohsen, H.H. (2022). Sustainable fish and seafood production and processing. In: *Sustainable Fish Production and Processing*, 259–291. Academic Press.

Scheierling, S.M., Bartone, C., Mara, D.D., and Drechsel, P. (2010). *Improving Wastewater Use in Agriculture: An Emerging Priority*, Policy Research Working Paper 5412. Washington, DC: World Bank.

Schepers, J.S., Francis, D.D., and Mielke, L.N. (1985). Water quality from erosion control structures in Nebraska. *Journal of Environmental Quality* 14: 186–190.

Seta, A.K., Blevins, R.L., Frye, W.W., and Barfield, B.J. (1993). Reducing soil erosion and agricultural chemical losses with conservation tillage. *Journal of Environmental Quality* 22: 661–665.

Sharma, R., Peshin, R., Shankar, U. et al. (2015). Impact evaluation indicators of an Integrated Pest Management program in vegetable crops in the subtropical region of Jammu and Kashmir, India. *Crop Protection* 67: 191–199. https://doi.org/10.1016/j.cropro.2014.10.014.

Shouman, E.R.M. (2020). Global prediction of wind energy market strategy for electricity generation. In: *Modeling*Simulation and Optimization of Wind Farms and Hybrid Systems (ed. K.Y. Maalawi), 1–30. Intech open.

Sing Wong, A. 2023. Temporal global trends of human population and dependency on coral reefs. Doctoral dissertation. University of Essex.

Singh, R.L. and Singh, P.K. (2017). Global environmental problems. In: *Principles and applications of environmental biotechnology for a sustainable future* (ed. R.L. Singh), 13–41. Singapore: Springer https://doi.org/10.1007/978-981-10-1866-4_2.

Smith, L.E.D. and Siciliano, G. (2015). A comprehensive review of constraints to improved management of fertilizers in China and mitigation of diffuse water pollution from agriculture. *Agriculture, Ecosystems and Environment* 209: 15–25.

Stenberg, J.A. (2017). A conceptual framework for integrated pest management. *Trends in Plant Science* 22: 759–769. https://doi.org/10.1016/j.tplants.2017.06.010.

Stephenson, G.R., Coats, J.R., and Yamamoto, H., 2001. Pesticide use and world food production: risks and benefits. In Expert Committee on Weeds Comité d'experts en malherbologie. Proceedings of the 2000 National Meeting 9-15.

Sundhar, S., Shakila, R.J., Jeyasekaran, G. et al. (2020). Risk assessment of organochlorine pesticides in seaweeds along the Gulf of Mannar, Southeast India. *Marine Pollution Bulletin* 161: 111709.

Tang, Z., Wood, J., Smith, D. et al. (2021). A review on constructed treatment wetlands for removal of pollutants in the agricultural runoff. *Sustainability* 13: 13578. https://doi.org/10.3390/su132413578.

Tao, Z. and Liu, G. (2024). Research progress and treatment status of agricultural non-point source pollution in China. *Academic Journal of Agriculture & Life Sciences* 5 (1): 1–6.

Tegtmeier, E.M. and Duffy, M.D. (2004). External costs of agricultural production in the United States. *International Journal of Agricultural Sustainability* 2: 1–20. https://doi.org/10.1080/14735903.2004.9684563.

Thai, P., Paxman, C., Prasad, P., et al. 2020. Marine monitoring program: annual report for inshore pesticide monitoring 2018–19, 69. Report to the Great Barrier Reef Marine Park Authority, Great Barrier Reef Marine Park Authority

UN Sustainable Development Goals 2019. Sustainable Development Goal 14. https://sustainabledevelopment.un.org/sdg14 (accessed 20 January 2024).

United Nations 1992. Report of the United Nations conference on environment and development. Conference on Environment and Development (accessed 20 January 2024).

Van Grinsven, H.J., Holland, M., Jacobsen, B.H. et al. (2013). Costs and benefits of nitrogen for Europe and implications for mitigation. *Environmental Science and Technology* 47 (8): 3571–3579. https://doi.org/10.1021/es303804g.

Vanshika, S. (2023). An overview of environmental laws and policies for sustainable human life. *Asia Pacific Law and Policy Review* 9: 33–50.

Wang, Z., Nong, D., Countryman, A.M. et al. (2020). Potential impacts of ballast water regulations on international trade, shipping patterns, and the global economy: an integrated transportation and economic modeling assessment. *Journal of Environmental Management* 275: 110892. https://doi.org/10.1016/j.jenvman.2020.110892.

Warne, M.S.J., Smith, R.A., and Turner, R.D.R. (2020a). Analysis of pesticide mixtures discharged to the lagoon of the Great Barrier Reef, Australia. *Environmental Pollution* 265: 114088. https://doi.org/10.1016/j.envpol.2020.114088.

Warne, MSt.J., Neelamraju, C., Strauss, J., et al. 2020b. Development of a method for estimating the toxicity of pesticide mixtures and a pesticide risk baseline for the reef 2050 water quality improvement plan. Report to the Queensland

Waycott, M., Duarte, C.M., Carruthers, T.J.B. et al. (2009). Accelerating loss of seagrasses across the globe threatens coastal ecosystems. *Proceedings of the National Academy of Sciences* 106: 12377–12381. https://doi.org/10.1073/pnas.0905620106.

Wenhai, L., Cusack, C., Baker, M. et al. (2019). Successful blue economy examples with an emphasis on international perspectives. *Frontiers in Marine Science* 6: 261. https://doi.org/10.3389/fmars.2019.00261.

Withers, P.J.A. and Haygarth, P.M. (2007). Agriculture, phosphorus and eutrophication: a European perspective. *Soil Use and Management* 23: 1–4. https://doi.org/10.1111/j.1475-2743.2007.00116.x.

Zhang, Y., Qin, P., Lu, S. et al. (2020). Occurrence and risk evaluation of organophosphorus pesticides in typical water bodies of Beijing China. *Environmental Science and Pollution Research International* 28 (2): 1454–1463.

6

Ocean Warming, Acidification, Plastic Pollution, and Water Quality Deterioration: A Multifaceted Crisis Unveiled

Nalika R. Dayananda[1], Jithya Wijesinghe[1], Shermila M. Botheju[1], W.P.R.T. Perera[1], and J.A. Liyanage[2]

[1] *Department of Indigenous Medical Resources, Gampaha Wickramarachchi University of Indigenous Medicine, Yakkala, Sri Lanka*
[2] *Department of Chemistry, University of Kelaniya, Kelaniya, Sri Lanka*

6.1 Ocean Warming: Impact on Global Ocean System and Cryosphere

6.1.1 The Oceans and Cryosphere: Intricately Linked Components of the Earth System

6.1.1.1 Global Oceans and Cryosphere

The interconnected systems of the global ocean and the cryosphere play fundamental roles in shaping the Earth's environment, sustaining life, and influencing various human activities. These vast expanses, encompassing the Arctic, Pacific, Atlantic, Indian, and Southern oceans alongside their peripheral regions, constitute a significant portion of the Earth's surface, covering approximately 71% (Maury 2023). Within this domain, the global ocean holds an astonishing 97% of the Earth's water, providing a critical foundation for life on our planet (Denny 2008). Moreover, it furnishes an overwhelming 99% of the biologically habitable space on Earth, fostering the proliferation of diverse ecosystems and supporting a myriad of species (Chaturvedi 2022). Additionally, the ocean contributes substantially to the primary production on Earth (Jiao et al. 2010), accounting for roughly half of this vital process, which is essential for sustaining life.

Complementing the vast expanse of the global ocean is the cryosphere, a collective term encompassing various frozen components of the Earth system found at or beneath the Earth's surface (Barry and Gan 2022). This includes snow, glaciers, ice sheets, ice shelves, icebergs, sea ice, lake ice, river ice, permafrost, and seasonally frozen ground. The cryosphere plays a crucial role in regulating the Earth's climate by reflecting solar radiation, maintaining global temperature equilibrium, and influencing weather patterns (Qin et al. 2018). Furthermore, it serves as a reservoir for freshwater, storing a significant portion of the world's freshwater resources in the form of ice and snow (Barry and Gan 2022), which is vital for sustaining terrestrial and aquatic ecosystems and supporting human societies.

6.1.1.2 Ocean and Cryosphere, and Their Linkages in the Earth System Through the Movement of Heat, Water, and Carbon

The oceans and cryosphere are intricately linked components of the Earth system, playing critical roles in regulating climate and supporting life on our planet. Their interactions involve complex

Coastal and Marine Pollution: Source to Sink, Mitigation and Management, First Edition.
Edited by Meththika Vithanage, Sameera M. Samarasekara, Bryan D. James, and Christopher M. Reddy.

processes of heat transfer, water circulation, and carbon cycling, which contribute to the overall stability and functioning of the Earth's environment (Benestad 2006). Heat exchange between the oceans and the cryosphere is a fundamental driver of climate patterns and oceanic circulation (Ussiri and Lal 2017). The cryosphere acts as a massive reservoir of cold temperatures, with ice sheets, glaciers, and sea ice reflecting incoming solar radiation back into space, thus helping to maintain global temperature balance (Qin et al. 2018). As ice melts, particularly in polar regions, it releases latent heat energy, contributing to regional and global warming (Benestad 2006). This process of ice-albedo feedback amplifies climate change, leading to accelerated melting and further heat absorption by the oceans (Winton 2008).

Water movement within the Earth system is facilitated by the interconnectedness of the oceans and cryosphere. The melting of glaciers and ice sheets contributes freshwater to the oceans, affecting sea level rise and altering ocean salinity gradients (Silvano et al. 2018). These changes can disrupt global thermohaline circulation patterns, which play a crucial role in redistributing heat around the planet (Rahmstorf 2006). Additionally, the melting of sea ice in polar regions influences oceanic currents and circulation, impacting climate variability on regional and even global scales (Figure 6.1).

Carbon cycling in the oceans and cryosphere is tightly linked to the movement of water and the exchange of gases between the atmosphere and the ocean surface (Parmentier et al. 2017). The oceans serve as a vast carbon sink, absorbing a significant portion of the carbon dioxide (CO_2) emitted into the atmosphere (Heinze et al. 2015) through human activities. Cold polar waters have a higher capacity to dissolve CO_2, making them important sites for carbon sequestration surfaces (Parmentier et al. 2017). However, as temperatures rise and ice melts, these regions may become less effective at absorbing CO_2, potentially exacerbating climate change (Ussiri and Lal 2017) (Figure 6.1).

The interactions between the oceans and cryosphere have far-reaching implications for the Earth system and human society. Changes in sea level, ocean circulation, and carbon cycling can

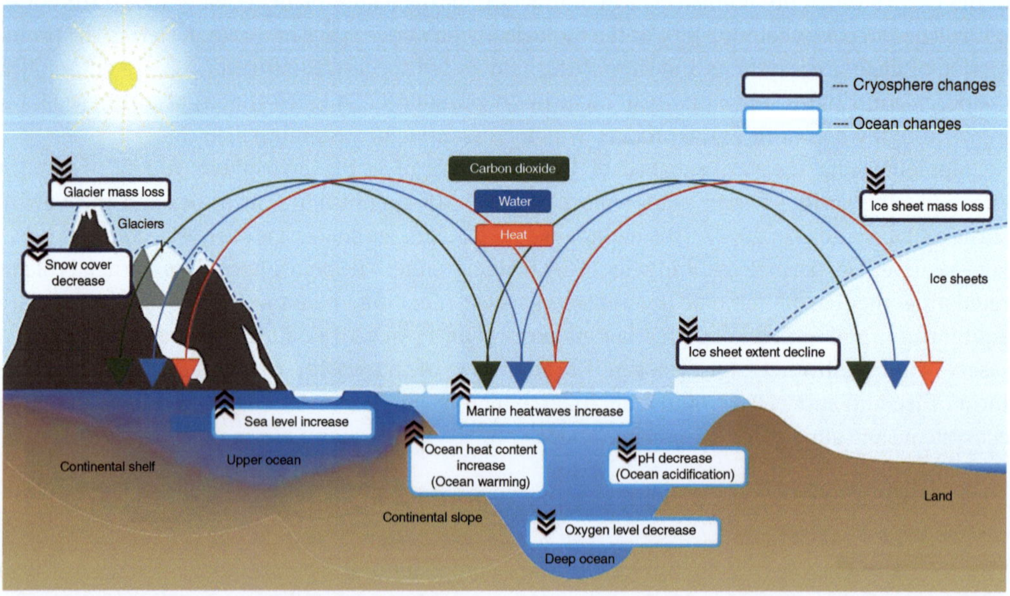

Figure 6.1 Schematic illustration of key components and changes of the ocean and cryosphere, and their linkages in the Earth system through the movement of heat, water, and carbon.

impact coastal communities through increased flooding, altered weather patterns, and shifts in marine ecosystems (Kennedy et al. 2002). Moreover, disruptions to the cryosphere can have cascading effects on global climate, biodiversity, and the availability of freshwater resources (Fountain et al. 2012). Furthermore, the cryosphere stores large quantities of organic carbon in the form of frozen soil and vegetation in permafrost regions (Ping et al. 2015). As permafrost thaws due to rising temperatures, microbial activity increases, leading to the decomposition of organic matter and the release of greenhouse gases such as methane and CO_2 into the atmosphere. This feedback loop, known as the permafrost carbon feedback, represents a significant potential source of additional carbon emissions, further amplifying climate change (Euskirchen et al. 2013).

6.1.2 Trends in Ocean Warming

Trends in ocean warming reveal a troubling reality of accelerating temperature increases and profound impacts on marine ecosystems. Over the past century, the world's oceans have absorbed approximately 90% of the excess heat trapped by greenhouse gases in the atmosphere, acting as a crucial buffer against the worst effects of climate change (Gruber 2011). However, this absorption comes at a cost, as it drives significant changes in ocean temperature profiles. Observational data collected over the past few decades indicate a clear trend of ocean warming. According to the Intergovernmental Panel on Climate Change (IPCC), the rate of ocean warming has accelerated since the 1970s, with the past few decades witnessing unprecedented increases in sea surface temperatures (IPCC 2013). Satellite observations and oceanographic measurements confirm this trend, revealing widespread warming of surface waters across the world's oceans.

When sunlight reaches the Earth's surface, the world's oceans absorb some of this energy and store it as heat. This heat is initially absorbed at the surface, but some of it eventually spreads to deeper waters. Currents also move this heat around the world. Water has a much higher heat capacity than air, meaning the oceans can absorb larger amounts of heat energy with only a slight increase in temperature (Johnson and Lyman 2020).

Increasing greenhouse gas concentrations are trapping more energy from the sun. Because changes in ocean systems occur over centuries, the oceans have not yet warmed as much as the atmosphere, even though they have absorbed more than 90% of the Earth's extra heat since 1955, and even as the rate of ocean heat uptake has doubled since 1993 (Figure 6.2). If not for the large heat-storage capacity provided by the oceans, the atmosphere would warm more rapidly (IPCC 2019). Increased heat absorption also changes ocean currents because many currents are driven by differences in temperature, which causes differences in density (Füssel 2009). These currents influence climate patterns and sustain ecosystems that depend on certain temperature ranges. Because water expands slightly as it gets warmer, an increase in ocean heat content will also increase the volume of water in the ocean, which is one of the major causes of the observed increases in sea level (Silvano et al. 2018). For all these reasons, ocean heat content is one of the most important indicators tracking the causes and responses of a changing climate.

The total amount of heat stored by the oceans is called "ocean heat content," and measurements of water temperature reflect the amount of heat in the water at a particular time and location. Ocean temperature plays an important role in the Earth's climate system – particularly sea surface temperature because heat from ocean surface waters provides energy for storms and thereby influences weather patterns. Notably, the rate of ocean warming varies spatially and temporally, with some regions experiencing more pronounced changes than others. For instance, polar regions, particularly the Arctic and Antarctic, exhibit the most rapid warming due to the loss of sea ice and the melting of ice sheets and glaciers (Silvano et al. 2018; Euskirchen et al. 2013). Additionally,

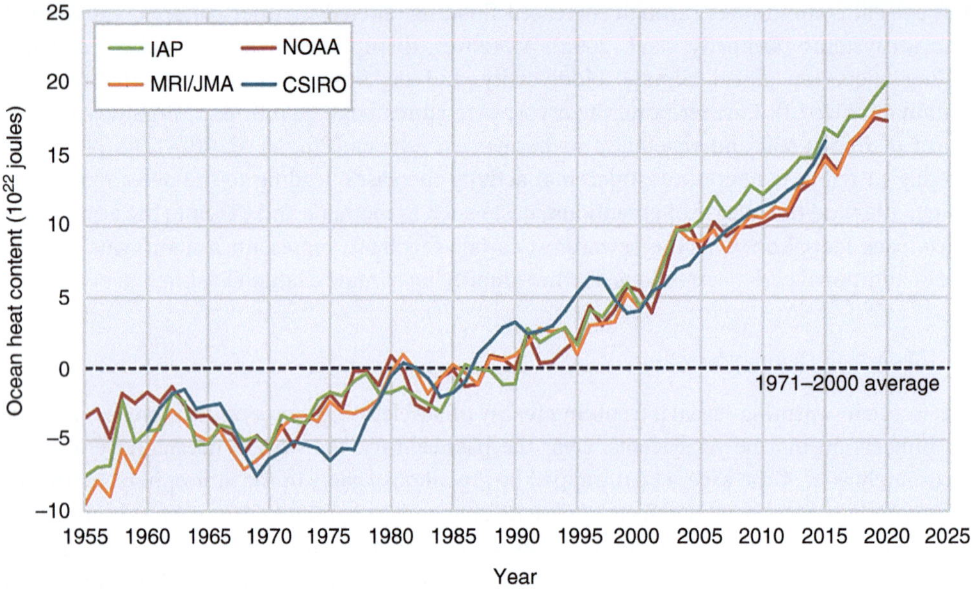

Figure 6.2 Heat Content in the Top 700m of the World's Oceans, 1955–2020 according to the government organizations in four countries: the United States' National Oceanic and Atmospheric Administration, Australia's Commonwealth Scientific and Industrial Research Organisation, China's Institute of Atmospheric Physics, and the Japan Meteorological Agency's Meteorological Research Institute. *Source:* https://www.epa.gov/climate-indicators/climate-change-indicators-ocean-heat/EPA/Public domain / Last accessed on 08-08-24.

coastal areas and shallow seas tend to warm more rapidly than deeper ocean regions (Silvano et al. 2018), amplifying the impacts on coastal ecosystems and communities.

The consequences of ocean warming extend beyond temperature increases alone, affecting ocean circulation patterns, marine biodiversity, and ecosystem functioning. These changes have profound implications for marine life, with shifts in species distributions, altered breeding and migration patterns, and increased risks of coral bleaching and ocean acidification. Furthermore, ocean warming exacerbates the intensity and frequency of extreme weather events, such as hurricanes and marine heat waves, posing additional risks to coastal communities and ecosystems.

6.1.3 Direct Impacts of Ocean Warming on the Ocean, Cryosphere, and Coastal Ecosystems

Ocean warming, propelled by human-induced climate change, has profound and direct impacts on the ocean, cryosphere, and coastal ecosystems. The escalating temperatures of the world's oceans trigger a cascade of changes that reverberate throughout these interconnected systems.

In the ocean, warming waters disrupt marine ecosystems, affecting species distribution, biodiversity, and ecosystem functioning. According to the IPCC (2019), global ocean temperatures have risen steadily, with the past decade being the warmest on record. This warming accelerates the loss of sea ice in polar regions and exacerbates the melting of glaciers and ice sheets. Arctic sea ice extent has declined rapidly, with summer minimums shrinking by approximately 13.1% per decade since 1979, according to satellite observations. The repercussions extend beyond polar regions, as warming oceans fuel more frequent and intense marine heatwaves, such as the "blob" in the North Pacific and the recent heatwave off the coast of Australia (Sen Gupta et al. 2020), which devastate marine ecosystems and disrupt fisheries.

Coastal ecosystems, including coral reefs, mangroves, and seagrass beds, are on the front lines of ocean warming's impacts. Coral bleaching events, driven by elevated sea surface temperatures, have become more frequent and severe, affecting over 75% of coral reefs during the global bleaching event of 2014–2017, according to National Oceanic and Atmospheric Administration (NOAA) (De et al. 2022). These events not only jeopardize the survival of coral reefs but also undermine the livelihoods of millions of people dependent on reef ecosystems for food, income, and coastal protection. Mangrove forests, essential for coastal resilience and biodiversity, face increasing threats from rising sea levels and extreme weather events. Accelerated rates of mangrove deforestation, estimated at approximately 1–2% per year (Richards and Friess 2016), further exacerbate coastal vulnerability. Additionally, seagrass meadows, critical for carbon sequestration and coastal protection, are declining due to warmer waters and habitat degradation.

The direct impacts of ocean warming on marine ecosystems reverberate throughout the broader environment, affecting climate systems, weather patterns, and human societies. Changes in ocean circulation and heat distribution influence regional climates, leading to shifts in precipitation patterns, altered storm tracks, and more frequent extreme weather events. These changes pose significant challenges for vulnerable coastal communities, exacerbating the risks of flooding, erosion, and saltwater intrusion. Furthermore, ocean warming threatens food security and economic livelihoods, particularly in developing countries reliant on marine resources for sustenance and income. Small-scale fisheries face increased uncertainty and variability due to shifting fish stocks and changing ocean conditions.

Addressing the direct impacts of ocean warming requires urgent and concerted action on multiple fronts. Mitigating greenhouse gas emissions is paramount to limit further warming and reduce the severity of future impacts. Additionally, enhancing the resilience of marine and coastal ecosystems through conservation measures, sustainable management practices, and ecosystem-based adaptation strategies is essential to safeguarding biodiversity and ecosystem services. Collaboration among governments, scientists, communities, and stakeholders is critical to developing and implementing effective solutions to mitigate and adapt to the impacts of ocean warming on the ocean, cryosphere, and coastal ecosystems. Failure to act decisively risks irreversible damage to marine ecosystems, coastal communities, and the broader environment, with far-reaching consequences for future generations.

6.1.4 Anthropocentric Ocean Warming and its Direct and Indirect Impacts on Human

Anthropocentric ocean warming, driven primarily by human-induced climate change, poses significant direct and indirect impacts on human populations worldwide. Directly, rising sea temperatures lead to the bleaching of coral reefs, disrupting marine ecosystems upon which millions depend for food, income, and protection from coastal hazards. This disruption reverberates through the entire marine food chain, affecting fish stocks and diminishing the livelihoods of countless communities reliant on fishing industries. Moreover, warmer oceans fuel more intense and frequent tropical storms, exacerbating risks to coastal populations through increased flooding, erosion, and infrastructure damage.

Indirectly, ocean warming amplifies sea level rise, threatening coastal cities and low-lying island nations with inundation and displacement. The resulting loss of land and infrastructure not only displaces millions but also amplifies social, economic, and political tensions as populations compete for dwindling resources and safe havens. Moreover, warmer oceans disrupt global weather patterns, leading to altered rainfall patterns, more frequent droughts, and extreme weather events, impacting agriculture, water resources, and human health. In addition to environmental consequences, anthropogenic ocean warming directly affects human health. Warmer waters facilitate the spread of waterborne diseases such as cholera and Vibrio infections, endangering communities reliant on marine resources for drinking water and sanitation. Furthermore, the proliferation of harmful algal blooms (HABs), fueled by higher sea temperatures, threatens seafood safety, risking human health through the ingestion of contaminated seafood.

Economically, the impacts of ocean warming are substantial and multifaceted. Declining fish stocks due to coral bleaching and habitat loss diminish not only the income of coastal communities but also global food security, given the critical role seafood plays in the diets of millions worldwide. Moreover, coastal infrastructure damage and increased insurance costs strain economies already grappling with the consequences of climate change, diverting resources away from essential social services and development initiatives.

Addressing the multifaceted challenges posed by anthropogenic ocean warming requires urgent and coordinated action on local, national, and international levels. Mitigation efforts must focus on reducing greenhouse gas emissions through the transition to renewable energy sources, sustainable land use practices, and the implementation of carbon pricing mechanisms. Additionally, adaptation strategies, including the protection and restoration of coastal ecosystems such as mangroves and salt marshes, can enhance the resilience of coastal communities to the impacts of ocean warming. Furthermore, international cooperation is essential to address the transboundary nature of ocean warming and its impacts. Collaborative initiatives are needed to promote sustainable fishery management, develop early warning systems for extreme weather events, and support vulnerable communities in adapting to a changing climate. By prioritizing the health of our oceans and the well-being of coastal communities, we can mitigate the impacts of anthropogenic ocean warming and build a more sustainable and resilient future for all.

6.2 Water Quality Deterioration in Oceans

6.2.1 Brief Overview of Water Quality Issues in Oceans

Recent research on water quality issues in oceans highlights various pollutants, their sources, and their impacts on human health and marine ecosystems (Boehm et al. 2017). The main pollutants affecting ocean water quality include nutrients, chemicals, industrial waste, and microplastics, each contributing to the deterioration of water quality in distinct ways (Feely et al. 2010; Karydis and Kitsiou 2013; Devlin et al. 2015; Weisberg et al. 2016).

To concisely address ocean water quality deterioration, attention can be primarily directed toward three key areas.

Nutrients and Chemicals: Excessive nutrients, such as nitrogen and phosphorus from agriculture runoff and sewage discharge, lead to eutrophication and HABs, adversely affecting marine life and human health by producing toxins that can contaminate seafood. Chemical pollutants, including pesticides, heavy metals like mercury and lead, and persistent organic pollutants (POPs), originate from industrial discharge and urban runoff, posing significant risks to marine ecosystems and human health through bioaccumulation and biomagnification processes (Lin et al. 2022).

Industrial Waste and Microplastics: Industrial waste, including plastics, radioactive waste, and oil spills, significantly contributes to physical harm to marine organisms, chemical pollution, and habitat destruction. Microplastics, resulting from the degradation of larger plastic items, have been identified as a major problem, with estimates suggesting that annually between 4 and

12 million tons of plastic enter the seas and oceans. These microplastics pose a threat to marine organisms and human health through ingestion and entanglement (Ghosh et al. 2021).

Impact on Human Health: Ocean pollution is known as a significant component of global pollution and is responsible for an estimated nine million premature deaths per year, with ocean pollutants like methylmercury and polychlorinated biphenyls (PCBs) known to damage developing brains in infants and increase risks for cardiovascular disease and dementia in adults. Other health risks include severe neurological impairment and rapid death from the ingestion of toxins produced by HABs, respiratory diseases from airborne HAB toxins, and gastrointestinal diseases from pathogenic marine bacteria (Landrigan et al. 2020).

6.2.2 Main Pollutants Affecting Water Quality in Oceans

The recent research on water quality issues in oceans highlights various pollutants, their sources, and impacts on human health and marine ecosystems. The main pollutants affecting ocean water quality include nutrients, chemicals, industrial waste, and microplastics, each contributing to the deterioration of water quality in distinct ways (Table 6.1) (Lin et al. 2022; Landrigan et al. 2020).

The deterioration of ocean water quality has significant implications for marine biodiversity, human health, and coastal economies. It harms marine life, disrupts food chains, and diminishes the recreational and economic value of coastal areas. Addressing this issue requires coordinated efforts to reduce pollution, conserve habitats, and promote sustainable coastal development, guided by international cooperation and public awareness of the importance of preserving ocean water quality.

6.2.3 Ecosystem Disruption

Ecosystem disruption due to water quality deterioration in ocean environments has been a subject of increasing concern among scientists and environmentalists. This deterioration is caused by a range of pollutants, including nutrients from agricultural runoff, chemicals from industrial discharges, plastics, and microplastics, as well as heavy metals and oil spills (Mostofa et al. 2013).

Excessive nutrients, particularly nitrogen and phosphorus, from agricultural runoff and sewage lead to eutrophication. This process results in the overgrowth of algae that blocks sunlight, depleting oxygen levels in the water and creating dead zones where aquatic life cannot survive (Wurtsbaugh et al. 2019). Eutrophication has been shown to cause significant disruptions in marine ecosystems, leading to the loss of biodiversity, alterations in species composition and dominance, and the decline of fish populations critical to human food supply (Ngatia et al. 2019).

Chemicals such as pesticides, pharmaceuticals, and industrial compounds (e.g. polychlorinated biphenyls (PCBs), dichlorodiphenyltrichloroethane (DDT), and dioxins) accumulate in the marine environment, where they can be toxic to marine life. These substances can alter hormonal systems in fish and marine mammals, leading to reproductive failures and developmental problems (Rios-Fuster et al. 2021). Bioaccumulation and biomagnification of these chemicals within the marine food web further exacerbate their impacts, potentially affecting human health through the consumption of contaminated seafood (Thanigaivel et al. 2023). Plastics and microplastics have become ubiquitous contaminants in the ocean, posing physical and chemical threats to marine life. Marine organisms, including fish, seabirds, and mammals, can ingest plastics, mistaking them for food, which can lead to intestinal blockages, starvation, and death. Moreover, plastics can leach toxic additives and adsorb other pollutants from the water, introducing these substances into the food web (Thushari and Senevirathna 2020).

Table 6.1 A concise overview of the major pollutants impacting ocean water quality, their origins, and their effects on marine ecosystems and human health.

Pollutant category	Type of the pollutant	Source	Impact on the ocean	Impact on human health	Sources
Nutrients	Nitrates	Oxidation of ammonia by nitrifying bacteria, Agricultural runoff, Sewage discharge, and Emissions from fossil fuel combustion	Eutrophication, Harmful Algal Blooms (HABs), Coral Reef Degradation	Humans can be exposed to the toxins produced by HABs through the consumption of contaminated seafood. Groundwater in coastal areas can be contaminated with nitrates. Nitrates in drinking water are linked to methemoglobinemia.	Prospero and Savoie (1989), Landrigan et al. (2020)
	Phosphate	River Runoff, Atmospheric Deposition, and Hydrothermal Activity	Eutrophication, Algal blooms, and Coral reef degradation	Airborne toxins from HABs can cause respiratory issues in humans.	Migon and Sandroni (1999), Martinez-Escobar and Mallela (2019)
Chemicals	Pesticides	Atmospheric deposition due to higher temperatures and changes in ice cover patterns, Riverine discharge, surface runoff, Interflow, and Pesticides have been detected in marine ice and the surface microlayer of seawater	Herbicides can negatively affect the population size of primary producers, leading to a decrease in biodiversity and alterations in ocean ecosystem structure and function, Bioaccumulation and Biomagnification.	Exposure to pesticides, such as methylmercury and PCBs, through the consumption of contaminated seafood, can generate Neurological disorders and generate endocrine and reproductive effects.	Bidleman et al. (2015), Mojiri et al. (2020)
	Heavy Metals and Metalloids	Heavy metals are transported via atmospheric particles and are deposited into ocean waters through dry deposition, Riverine Discharge, Agricultural runoff, Uban runoff	Bioaccumulation of heavy metals affects species at all levels of the food web from plankton to higher trophic levels, Leading to biodiversity loss and alterations in ecosystem structure and function	Exposure to these metals through the consumption of contaminated seafood can lead to various health issues, including neurological and developmental effects, endocrine and reproductive effects, liver failure, cardiovascular disease, and in severe cases, carcinogenesis.	Dauvin (2008), Briffa et al. (2020)

POPs	Ocean Dumping, Marine and biological transport, Industrial discharge, and atmospheric Deposition	POPs can disrupt the hormonal systems of marine wildlife, leading to altered reproductive cycles and growth patterns, Coral bleaching, and the degradation of coral reefs.	POPs can travel long distances through the atmosphere and water currents. Indigenous peoples and communities relying heavily on fish for their diet, especially in the Arctic and other remote areas, are particularly at risk due to the high levels of POPs found in marine species they consume.	Lohmann et al. (2021)
Microplastics	Shipping, fishing, and recreational boating contribute to the release of microplastics, Atmospheric Deposition, Sewage treatment, Fibers from synthetic textiles, Washed off during laundry processes	Ingestion by Marine Organisms, Bioaccumulation, and Biomagnification	Food chain contamination	Sharma and Chatterjee (2017)
Radioactive waste	Dispose into the ocean, including both liquid and solid wastes from nuclear-powered vessels and facilities, Accidental releases, Leakage from storage	Marine organisms exposed to radioactive materials may suffer from increased mortality, reduced reproductive success, genetic damage, and Coral Reef Damage.	Persistent radioactive contaminants in the marine environment pose long-term health risks to humans, especially for communities relying heavily on seafood for their diet.	Han (2023)
Oil spills	Offshore drilling Accidents, Shipwrecks, and collisions, operational discharges	Spills can devastate marine habitats such as coral reefs, mangroves, and seagrasses	Oil spills can contaminate seafood with polycyclic aromatic hydrocarbons (PAHs) and other harmful compounds. Respiratory and Skin Problems in humans.	Zhang et al. (2019)

Heavy metals such as mercury, lead, and cadmium, often released into the ocean through industrial discharges, mining, and atmospheric deposition, can have toxic effects on marine life. Mercury, for example, is converted into methylmercury in the water, a potent neurotoxin that affects the nervous system of fish and mammals, including humans who consume contaminated seafood. Heavy metals can cause a range of adverse effects, from behavioral changes and growth inhibition to reproductive failure and mortality in marine organisms (Butnariu 2022).

Oil spills are catastrophic events that can cause immediate and long-term damage to marine ecosystems. Oil coats the feathers of seabirds and the fur of marine mammals, reducing their insulation and buoyancy, leading to hypothermia or drowning (Magris and Giarrizzo 2020). In fish, oil exposure can affect heart function, impair sensory systems, and reduce reproductive success. The toxic components of oil, such as polycyclic aromatic hydrocarbons (PAHs), can persist in the environment, causing long-term ecological damage (Wallace et al. 2020).

6.3 Ocean Acidification

Emissions of anthropogenic carbon dioxide (CO_2) to the atmosphere and subsequent uptake by the ocean are changing ocean water chemistry, a process known as ocean acidification. As indicated by various research, the continuation of ocean acidification reflects the increase in CO_2 emissions. Although some species will be tolerant, ocean acidification will impact many marine organisms and processes, including the composition of communities and food webs. Remarkably, the ocean surface acidity has been stable for 800 000 years until the eighteenth century, between 8.2 and 8.3 (Zeeba and Ridgwell 2011). However, beginning with the industrial revolution, human activities, including combustion of fossil fuel emissions, production of cement, and changes in land use, have released increasing amounts of CO_2 into the atmosphere. In addition, forest cutting for agricultural purposes and fuel over the past 250 years has also contributed to higher CO_2 levels in the atmosphere since trees capture and store CO_2 *via* photosynthesis. The higher atmospheric CO_2 concentrations alter the Earth's climate but also impact the ocean's chemistry. Interestingly, only half of the CO_2 has remained in the atmosphere from the 10 Pg carbon released in 2010. As a result, the average ocean pH decreased to 8.1 (Turley and Gattuso 2012). However, depending on the CO_2 emissions, it is expected that the ocean pH could reach 7.7–7.9 in the year 2100 (Gattuso et al. 2011).

Due to human activities, the changes in seawater chemistry are referred to as anthropogenic ocean acidification. CO_2 is a weak acid, and when it is dissolved in seawater, it generates various changes in seawater chemistry (Gattuso and Hansson 2011; The Royal Society 2005). When the CO_2 dissolves in water, it reacts with water molecules and forms carbonic acid. Carbonic acid is further transformed into bicarbonate and carbonate ions. The dissolved carbon dioxide, carbonic acid, bicarbonate, and carbonate acid exist in balanced proportions in seawater (Figure 6.3). However, when a large amount of CO_2 is added to seawater, the balance could shift, and the carbonate ion concentration decreases as it is transformed into bicarbonate due to increasing acidity. The concentrations of bicarbonate ions ($[HCO_3^-]$) and the dissolved inorganic carbon are increased due to the dissolution of the CO_2 in seawater, and it lowers the pH and the carbonate ions concentration ($[CO_3^2-]$).

$$CO_2 + CO_3^{2-} \rightarrow H_2O + 2HCO_3^-$$

The concentration of protons ($[H^+]$) increases, which is proportionate to the ratio of the bicarbonate ion concentration to the carbonate ion concentration ($[HCO_3^-]/[CO_3^2-]$), pH decreases,

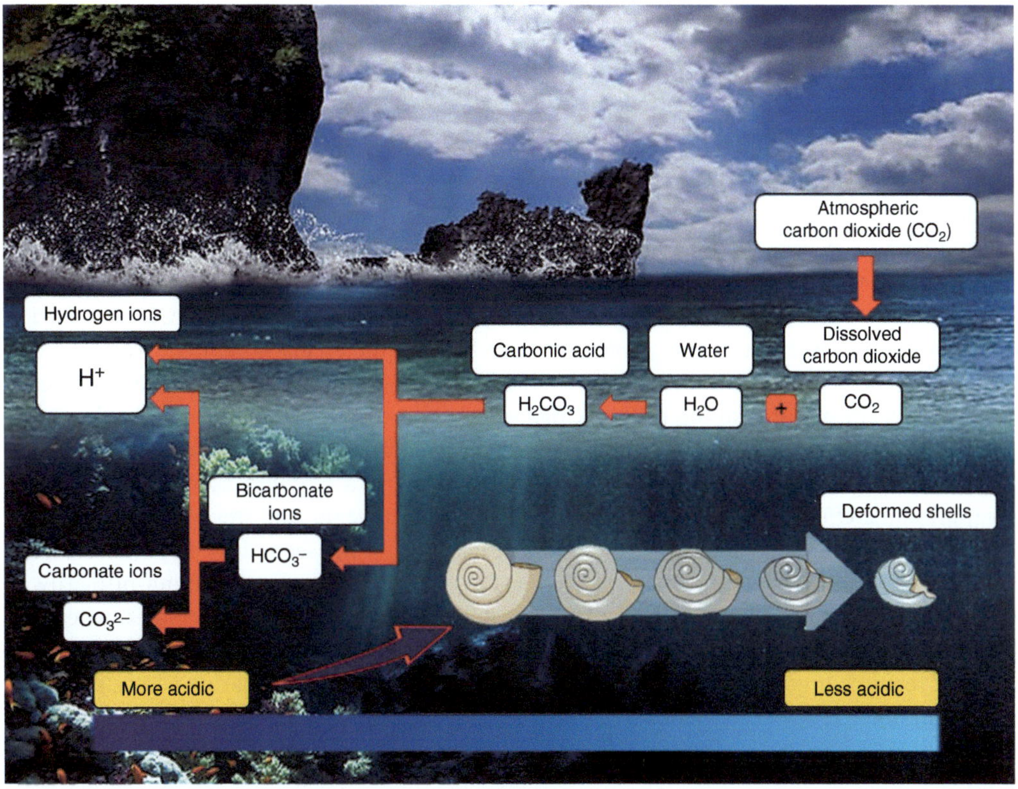

Figure 6.3 Schematic Diagram of Ocean Acidification; Carbon dioxide (CO_2) and water (H_2O) react with each other results in an increase in the hydrogen ions (H+) concentration; the additional changes include an increase in bicarbonate ions (HCO_3^-), and a great decrease in carbonate ions (CO_3^{2-}); carbonate ions (CO_3^{2-}) may modify the carbonate saturation state and these changes lead to acidification.

therefore the acidity of the water will be increased. The changes in pH play a key role in many physiological processes, including ion transportation, enzyme activity, and protein function (Gattuso and Hansson 2011). In addition, when one or several reactant(s) in a physiological process is a carbon species, such as photosynthesis and the precipitation and dissolution of calcium carbonate, the direct effects of ocean acidification could occur.

The rate of the dissolution of $CaCO_3$ minerals in the ocean is increased due to the decrease in carbonate ions. As described by Turley and Gattuso (2012), the saturation state of the seawater, which is denoted as Ω, is the degree of $CaCO_3$ saturation:

$$\Omega = \left[Ca^{2+}\right]\left[CO_3^{\;2-}\right] / K_{sp}$$

where $[Ca^{2+}]$ and $[CO_3^{\;2-}]$ are the *in situ* calcium and carbonate concentrations and K_{sp} is the solubility product for $CaCO_3$, respectively. Values of K_{sp} depend not only on the crystalline form of $CaCO_3$ but also on temperature and pressure, where the $CaCO_3$ is more soluble in cold water than in warm water unusually (Orr 2011).

The saturation status of seawater for a mineral can be defined as a measure of the ability of the mineral to form or dissolve (Birchenough et al. 2017). Most of the environments with high saturation states are more suitable for calcifying organisms such as plants and animals, which produce shells, plates, and $CaCO_3$ skeletons, as high saturation values reduce the energy required for calcification. Currently, the saturation state (Ω) is more than one for most of the surface where the seawater is supersaturated concerning $CaCO_3$ (28th reference: Future Biological). In contrast, most of the deep ocean is unsaturated, where the saturation state is lower than 1, as the changes in temperature, pressure, and the accumulation of biologically produced CO_2. Therefore, the unprotected $CaCO_3$ structures could be corrosive, and calcareous fossils are absent from sediments below the level at which the rate of $CaCO_3$ dissolution exceeds the rate of its supply (Orr 2011). However, it is difficult for marine calcifies such as mollusks, crustaceans, and reef-forming corals to build and maintain their skeletons and shells when the saturation state is less than 1, as the carbonate ions are likely to dissolve. As described by Birchenough et al., (2017), increased CO_2 dissolved in the seawater will lead to lower aragonite saturation levels in the oceans around the world.

Based on the different monitoring projects that have been conducted to monitor the changes in carbonate chemistry around the United Kingdom, in the North-East Atlantic and Arctic, and in the Southern Ocean, the datasets have shown a decline of pH values over the sampling period of 2008–2015 for the coastal time series, while also revealing a strong seasonally pattern. The hydrodynamic effects such as mixing/stratification, local currents, and freshwater inflows may affect these intra-annual changes. However, these changes mostly seem driven by two biological processes, including photosynthesis, which removes CO_2 from the water column primarily by phytoplankton, and respiration by zooplankton, bacteria, and benthic invertebrates, where the cycle is completed by returning it (Birchenough et al. 2017). In conclusion, addressing ocean acidification requires urgent action to mitigate CO_2 emissions and protect marine ecosystems from further harm.

6.3.1 The Impacts of Ocean Acidification on Marine Eco-systems

The largest natural reservoir of dissolved carbon is the ocean, and it holds an immense buffering capacity for changes in atmospheric CO_2 concentrations. The rapid increase of atmospheric CO_2 due to the Industrial Revolution has caused it to absorb increasingly larger amounts of CO_2 in the oceans and seas. The pre-existing chemical equilibrium of the sea has been disturbed by this process, and therefore, the chemical state of the sea changes and alters the ocean pH (Birchenough et al. 2017).

Marine primary producers are phylogenetically very diverse groups, from prokaryotes to angiosperms, and they differ widely based on their photosynthetic apparatus and the use of dissolved inorganic carbon for photosynthesis. Based on the evidence, ocean acidification acts differently across species, with some impacts on sensitive marine species and ecosystems with direct and indirect repercussions. Some species live in marine ecosystems that possess effective carbon-concentrating mechanisms and are less sensitive to increased CO_2 levels than those lacking them (Turley and Gattuso 2012). A variety of phytoplankton taxa such as diatoms, coccolithophores, cyanobacteria, and dinoflagellates have stimulation effects on elevated CO_2 for photosynthesis and carbon fixation (Riebesell and Tortell 2011). Changes in species morphology, ecology, and behavior are some of the direct effects, while indirect effects may be for processes or higher trophic groups, including wider food web effects and interactions within and between species (Birchenough et al. 2017; Williamson et al. 2017; Browman 2016).

Based on the previous evidence, crustaceans are less sensitive to the changes in pH when compared to the other groups, while the larval fishes are more sensitive to PCO_2 changes than some vertebrates (Wittmann and Pörtner 2013). Furthermore, marine flora has demonstrated some effects due to the combined effects of both ocean acidification and warming. The seagrasses and epiphytes are some species that can proliferate under lower pH conditions. However, these new conditions are not optimum for calcified algae, which causes some changes in their distribution (Birchenough et al. 2017). Further, ocean acidification may exacerbate the overall effects of "nuisance" species, which produce further degradation of ecosystem services. The new environmental conditions produced due to the ocean's carbonate chemistry are changing, so these new environmental conditions may lead to the introduction of new species. As an example, the increased temperature and the decreased pH are benefits for the harmful bloom-forming algae and invasive seaweeds that thrive in these modified/changed environments. In addition, these new environmental conditions may affect native fauna, cause mortality from displacement, and create modified habitats (Birchenough et al. 2017; Hall-Spencer and Allen 2015). These changes may have overall effects on ecosystem processes; however, these effects remain uncertain.

Results of the research conducted by UK Ocean Acidification and the Defra-funded placing ocean acidification in a wider fisheries context (PLACID) programs have demonstrated that certain groups of species, including mollusks, echinoderms, and corals, are generally more sensitive than others such as crustaceans and finfish. Further, the different life stages, including larvae, juveniles, and adults, have exhibited different vulnerability levels for some species (Kroeker et al. 2013; Wittmann and Pörtner 2013). A recent study conducted to determine the potential impacts of ocean acidification and warming on future fisheries has demonstrated that the projected standing stock biomasses may decrease by 10–60% and losses in relevant employment, including fisheries and associated industries maybe, 3–20% during 2020–2050 (Fernándes et al. 2016).

Experimental studies conducted in the United Kingdom have indicated that the cold-water coral *Lophelia pertusa,* which forms biodiverse and functionally important deep-water habitats in the North Atlantic and the Celtic Sea, is particularly vulnerable to ocean acidification (Guinotte et al. 2006). As described by Jackson et al. (2006), around 85% of known deep-sea cold-water coral reefs in the United Kingdom could be exposed to waters that are corrosive to them by 2060. In addition, the studies based on future climate projections exhibited a potential risk toward the horse mussel beds (*Modiolus modiolus*) by 2100 due to rising temperatures and ocean acidification (Gormley et al. 2013). Moreover, marine life uses carbonate from the seawater to build the skeleton and shells. Carbonate is less available when the seawater becomes more acidic. Shells and skeletons may dissolve under severe acidification conditions. Therefore, coldwater coral reefs and mussel beds are particularly vulnerable to acidification, highlighting the urgent need for mitigation measures. As carbonate availability diminishes in acidic seawater, the dissolution of shells and skeletons poses further threats to marine life (Figure 6.4). In conclusion, addressing ocean acidification requires collaborative efforts to mitigate CO_2 emissions and protect marine ecosystems from irreversible damage.

6.3.2 Mitigation Strategies for Ocean Acidification

Although previously the people and industry had limited knowledge of how ocean acidification contributed to the losses, nowadays, there are various active monitoring programs on pH and water conditions supported by active networks of industry, scientists, and policymakers for helping to minimize future losses and monitoring the early larvae developments. In the United States of America, they have an active model of science–industry–lawmaker collaboration, which helps the industry to adapt and respond to rapid changes in ocean chemistry (Birchenough et al. 2017). The requirement to continue supporting monitoring practices may lead to preventing the mortality of important marine species and the destruction of various marine ecosystems. As an example,

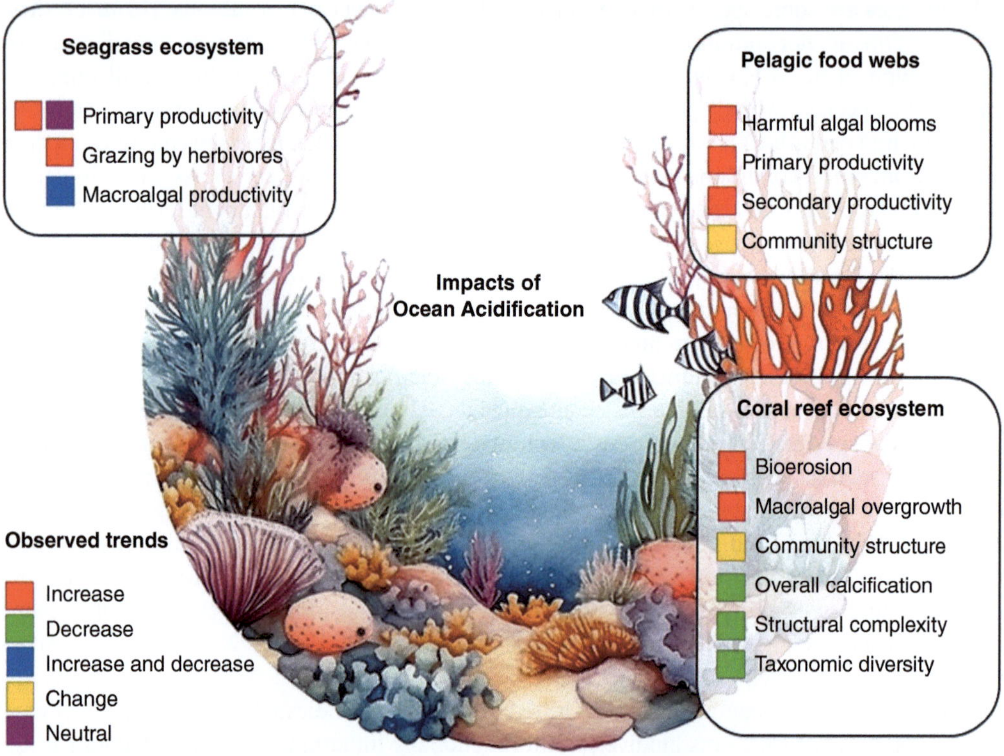

Figure 6.4 General trends in key community and ecosystem properties and processes in response to ocean acidification in seagrass meadows, coral reefs, other carbonate reef ecosystems, and pelagic food webs.

a recent review related to oyster biology has highlighted the key role of these reef-forming species in providing wider ecosystem services such as improving water quality, coastline protection, and food provision. These species could be likely affected by ocean acidification (Lemasson et al. 2017).

The ocean provides an enormous, often forgotten, diverse resource, while it provides 96% of the living spaces on Earth, providing diverse habitats for 80% of the planet's organisms. It produces around 50% of the oxygen in the atmosphere, which fixes around 50% of global primary production, and around 80% of world trade is carried across the ocean. Further, the ocean is an important waste recycler and an enormous carbon store that is substantially greater than on land or in the atmosphere. At present, approximately 30% of human-induced CO_2 emissions and around 90% of the heat generated by the Earth's recent warming have been absorbed by the ocean (Bindoff et al. 2007). Moreover, the ocean provides food and livelihoods for millions, and its coastal environments, including coral reefs, mangroves, salt marshes, seagrass, and kelp beds, play a crucial role in safeguarding shorelines and storing carbon (Duarte et al. 2005). The substantial changes in basic ocean chemistry associated with future ocean acidification are likely to have wide implications on marine life, leading to socioeconomic implications, notably concerning food security (Turley 2010; Turley and Gattuso 2012). As described by Brander et al. (2009), the yearly economic impact of coral reef loss caused by ocean acidification has been estimated to be US$500–870 billion by 2100, contingent upon the scenarios of CO_2 emissions. Polar, subpolar, and deep-sea ecosystems, along with shallow productive seas, including those connected to the upwelling of CO_2-enriched seawater, are also at risk due to the ocean acidification, and it will be most severe there and will commence within the next decade or two (Frommel et al. 2012; Nilsson et al. 2012; Dixson et al. 2009; Turley et al. 2010). As CO_2 levels continue to rise and equilibrate with ocean levels, alterations made in specific oceanic regions will ultimately impact the entire ocean. Therefore, mitigation efforts will necessitate significant modifications to ocean chemistry on a large scale.

Sequestering the carbon in the ocean floor by fertilizing certain ocean regions with iron is one mitigation approach suggested by researchers, which can limit the nutrients in these areas. As described by Pollard et al. (2009), the influxes of terrestrial iron led to rises in primary productivity in the Southern Ocean with high nutrients and low chlorophyll. Further, he described that the resulting phytoplankton blooms produce more carbon-containing molecules, which then travel through carbon flux and sink to be sequestrated on the sea floor. The recognition of this phenomenon has prompted suggestions from private companies to employ iron fertilization as a method of carbon offset within a worldwide carbon trading market.

The powdered limestone addition to the ocean water to react with CO_2 and form bicarbonate has also been proposed, as this would neutralize the acidity of added CO_2 (Rau and Caldiera 1999; Harvey 2007). In addition, this would lead the oceanic carbon equation toward carbonic acid and allow for more calcium carbonate to stay undissolved in the shells of marine life. As indicated by Rau et al. (2006), the ocean has a large and untapped ability to hold bicarbonate in the dissolved form if enough $CaCO_3$ is made available for the reaction. Therefore, this method is especially cheap and effective with no side effects.

Many approaches have been suggested to mitigate the accumulation of acidifying CO_2 in the world's oceans. Implementing strategies such as the addition of limestone or iron fertilization may offer potential solutions. However, given the global scale of the issue, any proposed solutions must also be global and are likely to have significant implications for ocean chemistry and biology. This calls for careful consideration, although some researchers question the extent of caution warranted regarding mitigation proposals, given the predicted accumulation of oceanic carbon and its potential to trigger drastic alterations in marine ecosystems (Jeffree 2009). Regardless of whether mitigation efforts are pursued, ocean chemistry is changing.

6.4 Plastic Pollution

6.4.1 Plastic Production, Usage, and Disposal in Oceans

Plastic is highly regarded for its versatility, durability, lightweight, and cost-effectiveness, making it widely utilized in various aspects of our everyday existence. Plastic has significantly influenced the progress of contemporary society and has benefited various industries, such as healthcare, science and technology, agriculture, packaging, transportation, and construction (Napper and Thompson 2020). The largest market demand for plastic is for single-use disposable packaging materials, with approximately half of all plastic production going toward single-use purposes (Xanthos and Walker 2017). Due to this high demand, plastic production has grown exponentially from 1.5 million tons in the 1950s, 322 million tons in 2016 (Carr 2017), excluding plastic fiber production, 359 million tons in 2018 (Plastics Europe 2019), to 400.3 million tons in 2022 (Plastics Europe 2023).

In addition to its growing production, owing to the exceptional durability and nonbiodegradability of plastic, its lifespan is projected to span from hundreds to thousands of years. Thus, the accumulation of plastics in a broader context of the environment poses a worldwide challenge in managing plastic waste. The marine environment, including seas and oceans, experiences the greatest impact from plastic accumulation. It is estimated that each year, a significant amount of plastic, ranging from 4 to 12 million tons, enters the seas and oceans (Figure 6.5). If this trend continues, it is projected that by 2050, the amount of plastic in these bodies of water will surpass the quantity of fish (Auta et al. 2017). Further, it is important to mention that plastics of all sizes, which are created and discarded on land due to human activities, have been widely observed in both terrestrial and nearby freshwater environments. Eventually, these plastics find their way into the marine environment (Li et al. 2017).

Marine plastic pollution, primarily caused by activities on land such as waste, spillage of virgin plastic pellets, litter flowing into storm drains and rivers, treated and untreated sewage effluent, and aerial deposition, can be categorized into three fundamental groups: first macroplastic, which consists of fishing nets, large pieces of Styrofoam, and parcels that have been lost or discarded from cargo ships; second microplastic, which refers to particles under 5 mm in diameter, including plastic nurdles and fragments that result from the degradation of larger plastic objects in marine ecosystems; and third, nano plastics either manufactured as such or the end state of microplastic degradation, invisible to the naked and more likely than microplastics to pass through biological membranes (Koelmans et al. 2015). In addition, shipping vessels, fishing and recreational boats, aquaculture, offshore oil industry, and tourism are also responsible maritime sources for plastic pollution. Although anthropogenic activities are the primary cause of marine plastic pollution, natural events such as floods, earthquakes, and tsunamis can result in large quantities of plastic debris unintentionally entering the ocean.

Approximately 4.8–12.7 million tons of plastic were introduced into the oceans from land in 2010 (Vince and Stoett 2018). Abundance assessments have projected that there are tens of millions of metric tons of plastic waste floating on the surfaces of oceans worldwide (Lebreton et al. 2018). Estimates for microplastic range from 15 to 51 trillion items, highlighting the global recognition of

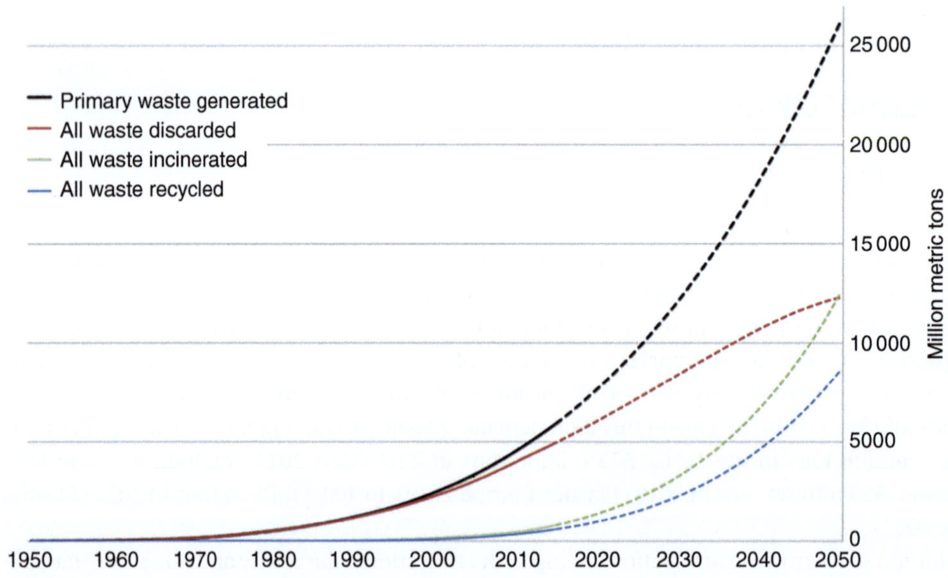

Figure 6.5 Cumulative plastic waste generation and disposal (in million metric tons). Solid lines show historical data from 1950 to 2015; dashed lines show projections of historical trends to 2050. *Source:* Geyer et al. (2017)/with permission of AAAS-American Association for the Advancement of Science.

plastic pollution (Vince and Stoett 2018). The longevity of plastic, large inputs into the ocean, and natural movement of the material *via* winds, currents, and eddies have made plastic a widespread and enduring pollutant throughout global coastal and marine environments, including in remote areas, such as the Arctic Ocean (Eriksen et al. 2020), Antarctic Ocean, and Southern Ocean, the Tibetan Plateau at 3000 m altitude, and the deep sea at greater than 1000 m in depth (Wang et al. 2019). The movement and distance at which objects travel are determined by the physical characteristics of the plastic object. The density, size, and shape of the polymer all affect the object's position in the water column and its ability to transition to another column.

Furthermore, Geyer et al. (2017) reported a total of 8300 million metric tons of virgin plastics had been produced by 2017. In 2015, a total of 6300 million metric tons of plastic waste was produced. Out of this, approximately 9% was recycled, 12% was incinerated, and the remaining 79% was either deposited in landfills or ended up in the natural environment. If the current patterns of production and waste management persist, approximately 12 000 million metric tons of plastic waste will be present in landfills or the natural environment by the year 2050 (Figure 6.5). Unsurprisingly, plastics constitute approximately 80% of the total marine debris.

6.4.2 Impacts of Plastic Debris on Marine Ecosystems

The wide distribution and varied properties of plastic debris give rise to numerous channels of interaction between humans and marine life. It is well known that plastic pollution can have physical and chemical effects on many trophic levels.

The most often observed interactions between marine plastic waste and wildlife are entanglement and ingestion (Kühn and van Franeker 2020). Certain shapes/types of plastic trash, such as ropes, bags, or circular plastic items like aluminum can six-pack rings, pose a higher risk for wildlife entanglement. Entanglement can result in tissue abrasion, strangling, decreased feeding efficiency, impaired growth and development, and ultimately death by drowning.

Plastic waste can cause further physical damage to marine environments and stationary bottom-dwelling creatures such as corals, seagrass, and mangroves by suffocating them and by being dragged across the ocean floor. Lost, abandoned, or discarded fishing nets, sometimes known as ghost nets, can ensnare fish and other marine creatures like rays and turtles for a long time. The environmental consequences could result in economic losses, mostly owing to the decline of fish species and the expenses incurred in replacing fishing gear.

Moreover, plastic debris is consumed by numerous marine species, ranging from zooplankton and fish to turtles, seabirds, whales, and other large marine mammals (Moore et al. 2020). The ingestion of plastic waste happens when an organism mistakenly consumes plastic debris, either because it visually resembles its prey, such as turtles mistaking plastic bags for jellyfish, or by smell; for example, some species of seabirds ingest microplastics after targeting zooplankton swarms. The impacts caused by ingesting plastic debris are often dependent on the size and shape of the plastics, an organism's feeding behavior, and feeding range within the water column. These impacts can range from modest effects to more severe consequences such as obstruction of the digestive tract, lower stomach capacity, internal injuries, changes in behavior, decreased swimming performance, poor reproduction, and oxidative stress (Foley et al. 2018). Furthermore, micro- and macroplastics serve as a substrate for the establishment of sessile organisms and microorganisms, including pathogens. The movement of colonized plastic debris can enhance the spread of organisms and facilitate the transportation of invasive species. The colonization of submerged plastic waste can modify the structure of the ecosystem by offering sessile benthic animals an alternate surface on which to settle and thrive.

Plastic waste offers a unique environment where bacteria can thrive. The term "plastisphere" refers to the distinct composition and classification of microorganisms that develop on the surface of marine plastic trash. This microbial community is notably different from the broader microbial population found on the surrounding surfaces (Feng et al. 2020). The plastisphere has been found to contain pathogenic Vibrio and *Escherichia coli* species, antibiotic-resistant bacteria, hazardous algal bloom species, and the bacteria Aeromonas salmonicida, which causes fish disease (Laverty et al. 2020; Moore et al. 2020). The physical processes that convert plastic objects into microplastics can provide the particles with a sizable specific surface area, which enables them to absorb pollutants from sediments or the water column through adsorption and function as sponges (Fred-Ahmadu et al. 2020). Metals and POPs are examples of contaminants that can be absorbed.

Chemical threats can also arise from the inclusion of chemical additives during the plastic production process, for example, stabilizers such as lead (Pb) and other metals, flame retardants, and plasticizers such as phthalates or bisphenol A. These may seep into the marine ecosystem when plastic weathers or, if ingested, enter its tissues and digestive systems (Gardon et al. 2020). Because the toxic compounds that are sorbed and leached are frequently persistent, plastic items can end up acting as bio-vectors or carriers of contaminants, which leads to bioaccumulation (Wang et al. 2020).

Similar to marine wildlife, marine plastic pollution affects humans through various channels. Humans can contact plastic waste through seafood, as microplastic waste has been discovered in invertebrates, crustaceans, and fish caught for human consumption (Walkinshaw et al. 2020). There is worry about the potential negative effects of ingesting micro- and nanoplastics on both marine wildlife and human health. These effects include the absorption of chemicals in the gastrointestinal tract, the toxicity caused by the release of plastic additives or absorbed contaminants, and the risks associated with microbial colonization (Wang et al. 2019).

6.4.3 Socioeconomic Effects of Marine Plastic Pollution

The economic costs linked to marine plastic pollution can manifest as either direct or indirect. It has adverse effects on various sectors such as tourism, fishing, shipping, aquaculture, recreation, and other ecosystem services. Ecosystem services refer to the advantages that individuals derive from nature, such as sustenance, carbon sequestration, and cultural benefits. Research indicates that plastic pollution has substantial effects on nearly all ecosystem services worldwide (Beaumont et al. 2019). In 2011, the economic costs of marine plastic were estimated to be between US$ 3300 and US$ 33,000 per ton per year, taking into account ecosystem service values and estimates of marine plastic abundance (Beaumont et al. 2019).

Beaches that are well-known or frequently visited and have a lot of litter can result in various economic costs, such as the need to spend money on cleaning up and the loss of revenue from tourism. The shipping, navy, coast guard, and fishing industries are adversely affected by the direct damage and entanglement of fishing gear in propellers. The fishing industry experiences economic losses from plastic debris, which has a detrimental effect on fish habitats and populations.

6.4.4 Initiatives on Marine Plastic Pollution Control and Prevention

Various techniques have been suggested to tackle the problem of plastic pollution. To control and avoid the buildup of plastic waste in marine and coastal habitats, it is crucial to have institutions at the global, regional, and national levels.

Considering the worldwide scope of marine plastic pollution, efforts have been made to address this issue, such as during the United Nations Environmental Assembly (UNEA) in December 2017. Member states of the United Nations Environment Program (UNEP) expressed their support for measures aimed at stopping the release of plastic waste and microplastics into the oceans. The measures encompassed the prevention of plastic waste, the promotion of reuse and recycling, and the reduction of needless plastic usage. Additionally, the importance of extended producer duties was emphasized. The Ad Hoc Open-Ended Expert Group on Marine Litter and Microplastics was created in 2017 to coordinate international efforts to address the issue of marine plastic litter and microplastics. The group's main objective is to explore various strategies to combat these pollutants, including the implementation of globally enforceable measures. In 2018, the Marine Environment Protection Committee of the International Marine Organization (IMO) approved the IMO Action Plan. This plan is designed to tackle the issue of marine plastic litter from ships by improving current rules and implementing new measures to reduce the amount of plastic waste in the ocean caused by ships. The UNEA adopted a resolution (UNEP/EA5/L23/REV.1) in March 2022 with the objective of eliminating plastic pollution and establishing a globally legally binding accord by 2024. Heads of State, Ministers of Environment, and other leaders from 175 nations have officially approved this significant agreement that covers the entire lifetime of plastic, from its origin to its impact on the ocean. Since 2020, the IMO and the Food and Agriculture Organization of the United Nations have collaborated on a worldwide initiative known as GloLitter to avoid and decrease the amount of plastic waste in the oceans originating from the shipping and fisheries industries.

In addition, nongovernment organizations also advocate for trash management education and widespread awareness, as well as the conversion of marine garbage into monetary value. The recommendation for reducing single-use plastic to address marine pollution involves conducting research and evaluation on the effectiveness of plastic bans. This will ensure that the policies enacted have a beneficial impact on the marine environment. Education initiatives are being implemented to limit the use of microbeads and plastic bags at their origin. Implementing take-back regulations and providing incentives for the return of plastic bottles, cans, and other waste in coastal areas can effectively mitigate plastic pollution caused by tourists. Furthermore, the installation of additional waste receptacles at popular tourist destinations would be a positive measure toward the collection and reduction of haphazard plastic discharge in the marine ecosystem.

6.4.5 Innovative Solutions for Marine Plastic Pollution Control and Prevention

A novel invention called the "plastic-eating drone" has recently been suggested as a means of removing plastic waste from the ocean. This method utilizes an independent device capable of pulling a net designed to capture and remove plastic waste from the ocean. Utilizing sonic transmitters could enable the detection of aquatic species to prevent their entanglement in the net (Boyle 2012). This method has been proposed for potential usage in both marine and freshwater environments.

An alternative and inventive device, known as "The Ocean Clean-up Array," can address the issue of plastic accumulation in ocean gyres by utilizing ocean currents' power. Platforms are affixed to the ocean floor and equipped with solid floating booms. The array is specifically engineered to enable marine animals to remain buoyant beneath the booms and to gather floating plastic along the booms at the surface water level (The Ocean Clean-Up 2014).

Furthermore, a novel method utilizing ellipsoid bodies has been devised to identify pixels of colored plastic waste on shorelines by analyzing images captured by a webcam. The methodology entails creating color references by utilizing a uniform color space (color space adopted by the International Commission on Illumination (CIE)) to identify plastic pixels and eliminate incorrectly identified pixels by applying a composite picture method. This technique has shown exceptional effectiveness in detecting plastic pixels of different hues compared to the prior method, which relied on the lightness values in the CIELUV color space. Webcam surveillance of plastic litter on beaches is efficient as it allows for remote quantification of plastic levels. Furthermore, the camera can be utilized as a valuable resource for systematically organizing beach clean-up efforts, in addition to its role in monitoring debris levels (Kataoka et al. 2012).

Biodegradation of marine plastic is also another innovative solution to control marine pollution, even though it is a complex phenomenon. The initial process of biodegradation commences with the development of a microbial biofilm on plastic, known as a plastisphere (Kirstein et al. 2019). This surface development contributes to the advancement of biodegradation. Exopolysaccharides produced by microorganisms enhance the adherence of biofilms to plastic surfaces. The polymer undergoes a process of conversion into a more basic form that is easily degradable.

Due to the large number of marine plastics, recyclability of the plastic is necessary. The weathering method, characterization method, and processing method are the methods generally used in the recycling of marine debris. Ronkay et al. (2021) have developed an artificial weathering method to investigate the recyclability of weathered and nonweathered polyethylene terephthalate. In addition, scientists are currently studying plastic-degrading enzymes as a possible remedy for decomposing specific forms of plastic trash. These enzymes break down the chemical bonds in plastic molecules, transforming them into simpler chemicals that microbes may biodegrade.

In addition, sea bins, interceptor boats, bio fences, drone surveillance, and autonomous underwater vehicles (AUVs) are several innovative technologies for clean-up and prevention of marine plastic pollution. The Seabin concept was launched in 2015 and is strategically placed at marinas, ports, and other aquatic environments to effectively gather and remove floating garbage, particularly plastic pollution. Sea bins function by employing a pump to extract water into the apparatus, ensuring suspended detritus and microplastics within a detachable filter bag and providing localized solutions for areas with high plastic pollution.

Interceptor vessels are strategically positioned in rivers and streams to gather and prevent plastic debris from reaching the ocean. These boats are equipped with conveyor belts or collection devices that efficiently gather plastic waste that is floating on the surface of the water. Biofences are another innovative solution that was developed to prevent the entry of plastic into marine ecosystems. They are buoyant barriers constructed from organic materials such as bamboo and other plant-derived fibers and designed to trap plastic waste in rivers and coastal regions. In recent years, biofences have been created and used in different places worldwide as a sustainable measure to capture plastic garbage in rivers and coastal areas.

AUVs are robotic devices designed to operate underwater and equipped with advanced sensors and cameras, enabling them to effectively navigate and gather information about the amounts and distribution of plastic pollution. This data is crucial in evaluating the extent to which plastic pollution affects marine ecosystems. Furthermore, drones equipped with cameras and sensors are also employed to observe and assess the presence of plastic pollution in oceans, offering vital data for scientific investigation and environmental remediation endeavors.

Circular economy initiatives also promote the recycling and responsible disposal of plastic waste to prevent it from entering the oceans. It creates closed-loop systems where plastic waste is recycled and reused, reducing the need for new plastic production and minimizing waste. For example, innovative packaging materials offer a sustainable alternative to traditional plastics, helping to reduce the amount of plastic waste entering the oceans and other ecosystems.

Moreover, the plastics industries should be responsible for the end-of-life of their products under requirements like those of electronic equipment producers. There is a suggestion to replace non-degradable or conventional materials with materials that have a high biodegradation rate, such as starch or pullulan. Biodegradable material is easily broken down by bacteria and fungi, which shortens the lifespan of these plastics in the natural environment. Furthermore, it is imperative for enterprises to contemplate the practice of recycling or enhancing the value of their plastic trash. In recent times, there has been an increase in awareness of tertiary recycling of plastic. This involves converting the plastic material into smaller molecules, which can be used as feedstock for new petrochemicals and plastics. For instance, polyethylene can be repurposed as a viable raw material for the manufacturing of gasoline (AL-Salem 2009). However, at the same time, the end user also has the responsibility to contribute to the prevention of plastic pollution by refusing single-use plastic products and demanding natural or recyclable products, as well as following the proper disposal methods.

6.5 Cross-Disciplinary Solutions for Interconnected Nature of Ocean Warming Acidification, Plastic Pollution, and Water Quality Deterioration

In confronting the complex and interrelated challenges of ocean warming, acidification, plastic pollution, and water quality deterioration, contemporary strategies demand interdisciplinary innovations integrating science, policy, and public awareness. Recent advancements underscore the urgency of collaborative, cross-disciplinary approaches to address these pressing issues effectively. In the scientific realm, cutting-edge technologies such as remote sensing, artificial intelligence, and genomics offer unprecedented opportunities for monitoring, modeling, and understanding ocean dynamics and ecosystem health. For instance, satellite-based remote sensing enables real-time tracking of ocean temperatures, currents, and surface pollution, providing valuable data for scientific research and environmental management. Similarly, advances in genomic sequencing allow researchers to study microbial communities in marine environments, shedding light on their role in nutrient cycling, carbon sequestration, and ecosystem resilience. By leveraging these innovative tools and techniques, interdisciplinary teams of scientists can enhance our understanding of ocean processes and inform evidence-based strategies for mitigating the impacts of climate change, pollution, and habitat degradation on marine ecosystems.

In the policy arena, recent efforts focus on fostering international cooperation, enhancing regulatory frameworks, and promoting sustainable governance mechanisms to address ocean degradation comprehensively. Multilateral initiatives such as the Paris Agreement and the United Nations Sustainable Development Goals provide frameworks for global collaboration on climate action, marine conservation, and sustainable development. Also, regional agreements and partnerships facilitate coordinated responses to transboundary issues like marine pollution and overfishing. Innovative policy instruments such as ecosystem-based management, marine spatial planning, and blue economy strategies promote holistic approaches to ocean governance that integrate environmental, social, and economic objectives. Moreover, emerging concepts such as ocean rights and nature-based solutions emphasize the intrinsic value of marine ecosystems and the importance of integrating indigenous knowledge and traditional practices into decision-making processes. By embracing innovative policy approaches and fostering inclusive, participatory governance processes, interdisciplinary initiatives can enhance the effectiveness and legitimacy of ocean conservation efforts at local, national, and global scales.

Public awareness and engagement are critical components of contemporary interdisciplinary approaches to ocean conservation, with recent efforts focusing on leveraging digital technologies, social media platforms, and behavioral insights to inspire action and promote stewardship. Innovative communication strategies such as immersive storytelling, virtual reality experiences, and interactive multimedia engage diverse audiences in understanding the importance of oceans and the interconnectedness of environmental issues. Citizen science projects empower individuals to contribute data, observations, and ideas to scientific research and monitoring efforts, fostering a sense of ownership and accountability for ocean health. Moreover, social marketing campaigns, online petitions, and youth-led movements mobilize public support for policy reforms, corporate accountability, and community-based conservation initiatives. By harnessing the power of digital communication tools and social networks, interdisciplinary approaches to public awareness can amplify voices for change, build alliances across sectors and generations, and catalyze a global movement for ocean stewardship and sustainability.

6.6 Conclusion

In conclusion, interdisciplinary solutions for ocean warming, acidification, plastic pollution, and water quality deterioration require innovative, updated approaches that integrate advances in science, policy, and public awareness. Recent developments in remote sensing, genomics, and artificial intelligence offer new opportunities for scientific research and monitoring, enabling more precise assessments of ocean health and ecosystem dynamics. Similarly, innovative policy instruments such as ecosystem-based management and blue economy strategies promote holistic approaches to ocean governance that balance environmental conservation with socioeconomic development. Finally, digital technologies, social media platforms, and citizen science initiatives empower individuals and communities to contribute to ocean conservation efforts, fostering a sense of collective responsibility and solidarity in addressing shared environmental challenges. By embracing interdisciplinary collaborations and embracing innovative approaches, By embracing interdisciplinary collaborations and innovative approaches, the health and resilience of our oceans can be safeguarded for future generations.

References

Al-Salem, S.M. (2009). Influence of natural and accelerated weathering on various formulations of linear low-density polyethylene (LLDPE) films. *Material and Design* 30 (5): 1729–1736. https://doi.org/10.1016/j.matdes.2008.07.049.

Auta, H.S., Emenike, C.U., and Fauziah, S.H. (2017). Distribution and importance of microplastics in the marine environment: a review of the sources, fate, effects, and potential solutions. *Environment International* 102: 165–176. https://doi.org/10.1016/j.envint.2017.02.013.

Barry, R.G. and Gan, T.Y. (2022). *The Global Cryosphere: Past, Present, and Future*. Cambridge University Press.

Beaumont, N.J., Aanesen, M., Austen, M.C. et al. (2019). Global ecological, social and economic impacts of marine plastic. *Marine Pollution Bulletin* 142: 189–195.

Benestad, R.E. (2006). *Earth's Climate*, 89–148. Berlin Heidelberg: Springer.

Bidleman, T., Agosta, K., Andersson, A. et al. (2015). Atmospheric pathways of chlorinated pesticides and natural bromoanisoles in the northern Baltic Sea and its catchment. *AMBIO* 44 (S3): 472–483. https://doi.org/10.1007/s13280-015-0666-4.

Bindoff, N.L., Willebrand, J., Artale, V. et al. (2007). Observations: oceanic climate change and sea level. *In Climate Change* 2007: 385–432.

Birchenough, S., Williamson, P., and Turley, C. (2017). Future of the Sea - Evidence Review. *Ocean Acidification* 3: 3–15.

Boehm, A.B., Ismail, N., Sassoubre, L., and Allan, E.A. (2017). Oceans in Peril: grand challenges in applied water quality research for the 21st century. *Environmental Engineering Science* 34 (1): 3–15. https://doi.org/10.1089/ees.2015.0252.

Boyle, R. (2012). Plastic-eating underwater drone could swallow the great pacific garbage patch. Popular Science, (Retrieved from http://www.popsci.com/technology/article/2012-07/plastic-eating-underwater-drone-could-swallow-greatpacificgarbage-patch).

Brander, L.M., Rehdanz, K., Tol, R.S.J., and van Beukering, P.J.H. (2009). The economic impact of ocean acidification on coral reefs. *ESRI Working Paper* 282: 1–33.

Briffa, J., Sinagra, E., and Blundell, R. (2020). Heavy metal pollution in the environment and their toxicological effects on humans. *Heliyon* 6 (9): e04691. https://doi.org/10.1016/j.heliyon.2020.e04691.

Browman, H.I. (2016). Applying organized scepticism to ocean acidification research. *ICES Journal of Marine Science* 73 (3): 529–536.

Butnariu, M. (2022). Heavy metals as pollutants in the aquatic Black Sea ecosystem. In: *Bacterial Fish Diseases* (ed. G.H. Dar, R.A. Bhat, H. Qadri, et al.), 31–57. Elsevier https://doi.org/10.1016/B978-0-323-85624-9.00003-8.

Carr, S.A. (2017). Sources and dispersive modes of micro-fibers in the environment. *Integrated Environmental Assessment and Management* 13 (3): 466–469. https://doi.org/10.1002/ieam.1916.

Chaturvedi, S. (2022). Maritime regionalism and 'Inclusive Development': opportunity and challenges before Bangladesh in anthropocene. *Journal of International Relations* 15 (1–2): 159–184.

Climate Change Indicators: Ocean Heat (n.d.). Available at: https://www.epa.gov/climate-indicators/climate-change-indicators-ocean-heat (accessed 15 March 2024).

Dauvin, J.-C. (2008). Effects of heavy metal contamination on the macrobenthic fauna in estuaries: the case of the Seine estuary. *Marine Pollution Bulletin* 57 (1–5): 160–169. https://doi.org/10.1016/j.marpolbul.2007.10.012.

De, K., Nanajkar, M., Arora, M. et al. (2022). Application of remotely sensed sea surface temperature for assessment of recurrent coral bleaching (2014–2019) impact on a marginal coral ecosystem. *Geocarto International* 37 (15): 4483–4508.

Denny, M. (2008). *How the Ocean Works: An Introduction to Oceanography*. Princeton University Press.

Devlin, M. et al. (2015). Water quality and river plume monitoring in the Great Barrier reef: an overview of methods based on ocean colour satellite data. *Remote Sensing* 7 (10): 12909–12941. https://doi.org/10.3390/rs71012909.

Dixson, D.L., Munday, P.L., and Jones, G.P. (2009). Ocean acidification disrupts the innate ability of fish to detect predator olfactory cues. *Ecological Letters* 12: 1–8.

Duarte, C.M., Middelburg, J.J., and Caraco, N. (2005). Major role of marine vegetation on the oceanic carbon cycle. *Biogeosciences* 2: 1–8.

Eriksen, M., Borgogno, F., Villarrubia-Gómez, P. et al. (2020). Mitigation strategies to reverse the rising trend of plastics in polar regions. *Environment International* 139: 105704.

Euskirchen, E.S., Goodstein, E.S., and Huntington, H.P. (2013). An estimated cost of lost climate regulation services caused by thawing of the Arctic cryosphere. *Ecological Applications* 23 (8): 1869–1880.

Feely, R.A., Alin, S.R., Newton, J. et al. (2010). The combined effects of ocean acidification, mixing, and respiration on pH and carbonate saturation in an urbanized estuary. *Estuarine, Coastal and Shelf Science* 88 (4): 442–449. https://doi.org/10.1016/j.ecss.2010.05.004.

Feng, L., He, L., Jiang, S. et al. (2020). Investigating the composition and distribution of microplastics surface biofilms in coral areas. *Chemosphere* 252: 126565. https://doi.org/10.1016/j.chemosphere.2020.126565.

Fernándes, J.A., Papathanasopoulou, E., Hattam, C. et al. (2016). Estimating the ecological, economic and social impacts of ocean acidification and warming on UK fisheries. *Fish and Fisheries* 18 (3): 389–411. https://doi.org/10.1111/faf.12183.

Foley, C.J., Feiner, Z.S., Malinich, T.D., and Hook, T.O. (2018). A meta-analysis of the effects of exposure to microplastics on fish and aquatic invertebrates. *Science of the Total Environment* 631-632: 550–559. https://doi.org/10.1016/j.scitotenv.2018.03.046.

Fountain, A.G., Campbell, J.L., Schuur, E.A. et al. (2012). The disappearing cryosphere: impacts and ecosystem responses to rapid cryosphere loss. *BioScience* 62 (4): 405–415.

Fred-Ahmadu, O.H., Bhagwat, G., Oluyoye, I. et al. (2020). Interaction of chemical contaminants with microplastics: principles and perspectives. *Science of the Total Environment* 706: 135978. https://doi.org/10.1016/j.scitotenv.2019.135978.

Frommel, A., Maneja, R., Lowe, D. et al. (2012). Severe tissue damage in Atlantic cod larvae under increasing ocean acidification. *National Climate Change* 2: 42–46.

Füssel, H.M. (2009). An updated assessment of the risks from climate change based on research published since the IPCC Fourth Assessment Report. *Climatic Change* 97 (3): 469–482.

Gardon, T., Huvet, A., Paul-Pont, I. et al. (2020). Toxic effects of leachates from plastic pearl-farming gear on embryo-larval development in the pearl oyster Pinctada margaritifera. *Water Research* 179: 115890. https://doi.org/10.1016/j.watres.2020.115890.

Gattuso, J. and Hansson, L. (2011). Ocean acidification: background and history. *In Ocean Acidification* 2011: 1–20.

Gattuso, J., Bijma, J., Gehlen, M., and Riebesell, U. (2011). Turley C: ocean acidification: knowns, unknowns and perspectives. *In Ocean Acidification* 2011: 291–311.

Geyer, R., Jambeck, J.R., and Law, K.L. (2017). Production, use, and fate of all plastics ever made. *Science Advances* 3: e1700782. https://doi.org/10.1126/sciadv.1700782.

Ghosh, S., Das, R., Bakshi, M. et al. (2021). Potentially toxic element and microplastic contamination in the river Hooghly: implications to better water quality management. *Journal of Earth System Science* 130 (4): 236. https://doi.org/10.1007/s12040-021-01733-9.

Gormley, K.S.G., Porter, J.S., Bell, M.C. et al. (2013). Predictive habitat modelling as a tool to assess the change in distribution and extent of an OSPAR priority habitat under an increased ocean temperature scenario: consequences for marine protected area networks and management. *PLoS One* 8 (7): e68263.

Gruber, N. (2011). Warming up, turning sour, losing breath: ocean biogeochemistry under global change. *Philosophical Transactions of the Royal Society A: Mathematical, Physical and Engineering Sciences* 369 (1943): 1980–1996.

Guinotte, J.M., Orr, J., Cairns, S. et al. (2006). Will human-induced changes in seawater chemistry alter the distribution of deep-sea scleractinian corals? *Frontiers in Ecology and the Environment* 4 (3): 141–146.

Hall-Spencer, J.M. and Allen, R. (2015). The impact of ocean acidification on nuisance species. *Research and Reports in Biodiversity Studies* 4: 33–46.

Han, X. (2023). Nuclear pollution and its effects on marine ecosystems. *Highlights in Science, Engineering and Technology* 69: 212–218. https://doi.org/10.54097/hset.v69i.11906.

Harvey, L.D.D. (2007). Mitigating the atmospheric CO_2 increase and ocean acidification by adding limestone powder to upwelling regions. *Journal of Geophysical Resources* 113: C04028.

Heinze, C., Meyer, S., Goris, N. et al. (2015). The ocean carbon sink–impacts, vulnerabilities and challenges. *Earth System Dynamics* 6 (1): 327–358.

IPCC (Intergovernmental Panel on Climate Change) (2019). Summary for policymakers. In: *IPCC Special report on the ocean and cryosphere in a changing climate* (ed. H.-O. Pörtner, D.C. Roberts, V. Masson-Delmotte, et al.). IPCC www.ipcc.ch/srocc/chapter/summary-for-policymakers.

Jeffree, R.A. (2009). Ocean acidification: a prognosis and treatment for this eclipsing issue in marine ecotoxicology. *Integrated Environmental Assessment and Management* 5: 173–175.

Jiao, N., Herndl, G.J., Hansell, D.A. et al. (2010). Microbial production of recalcitrant dissolved organic matter: long-term carbon storage in the global ocean. *Nature Reviews Microbiology* 8 (8): 593–599.

Johnson, G.C. and Lyman, J.M. (2020). Warming trends increasingly dominate global ocean. *Nature Climate Change* 10 (8): 757–761.

Karydis, M. and Kitsiou, D. (2013). Marine water quality monitoring: a review. *Marine Pollution Bulletin* 77 (1–2): 23–36. https://doi.org/10.1016/j.marpolbul.2013.09.012.

Kataoka, T., Hinata, H., and Kako, S. (2012). A new technique for detecting colored macro plastic debris on beaches using webcam images and CIELUV. *Marine Pollution Bulletin* 64: 1829–1836.

Kennedy, V.S., Twilley, R.R., Kleypas, J.A. et al. (2002). *Coastal and Marine Ecosystems & Global Climate Change*. Pew Center on Global Climate Change.

Kirstein, I.V., Wichels, A., Gullans, E. et al. (2019). The plastisphere–uncovering tightly attached plastic "specific" microorganisms. *PLoS One* 14 (4): e0215859.

Koelmans, A.A., Besseling, E., and Shim, W.J. (2015). Nanoplastics in the aquatic environment. Critical review. In: *Marine Anthropogenic Litter* (ed. M. Bergmann, L. Gutow, and M. Klages). Cham: Springer https://doi.org/10.1007/978-3-319-16510.

Kroeker, K.J., Kordas, R.L., Crim, R.N. et al. (2013). Impacts of ocean acidification on marine organisms: quantifying sensitivities and interaction with warming. *Global Change Biology* 19 (6): 1884–1896.

Kühn, S. and van Franeker, J.A. (2020). Quantitative overview of marine debris ingested by marine megafauna. *Marine Pollution Bulletin* 151: 110858. https://doi.org/10.1016/j.marpolbul.2019.110858.

Landrigan, P.J. et al. (2020). Human health and ocean pollution. *Annals of Global Health* 86 (1): 151. https://doi.org/10.5334/aogh.2831.

Laverty, A.L., Primpke, S., Lorenz, C. et al. (2020). Bacterial biofilms colonizing plastics in estuarine waters, with an emphasis on Vibrio spp. and their antibacterial resistance. *PLoS One* 15 (8): e0237704. https://doi.org/10.1371/journal.pone.0237704.

Lemasson, A.J., Fletcher, S., Hall-Spencer, J.M., and Knights, A.M. (2017). Linking the biological impacts of ocean acidification on oysters to changes in ecosystem services: a review. *Journal of Experimental Marine Biology and Ecology* 492: 49–62. https://doi.org/10.1016/j.jembe.2017.01.019.

Li, J., Liu, H., and Paul Chen, J. (2017). Micro plastics in freshwater systems: a review on occurrence, environmental effects, and methods for microplastics detection. *Water Research* 137: 362–374. https://doi.org/10.1016/j.watres.2017.12.056.

Lin, L., Yang, H., and Xu, X. (2022). Effects of water pollution on human health and disease heterogeneity: a review. *Frontiers in Environmental Science* 10: 880246. https://doi.org/10.3389/fenvs.2022.880246.

Lohmann, R., Markham, E., Klanova, J. et al. (2021). Trends of diverse POPs in air and water across the eestern Atlantic ocean: strong gradients in the ocean but not in the air. *Environmental Science & Technology* 55 (14): 9498–9507. https://doi.org/10.1021/acs.est.0c04611.

Magris, R.A. and Giarrizzo, T. (2020). Mysterious oil spill in the Atlantic Ocean threatens marine biodiversity and local people in Brazil. *Marine Pollution Bulletin* 153: 110961. https://doi.org/10.1016/j.marpolbul.2020.110961.

Martinez-Escobar, D.F. and Mallela, J. (2019). Assessing the impacts of phosphate mining on coral reef communities and reef development. *Science of the Total Environment* 692: 1257–1266. https://doi.org/10.1016/j.scitotenv.2019.07.139.

Maury, M. (2023). *The Physical Geography of the Sea*. BoD–Books on Demand.

Migon, C. and Sandroni, V. (1999). Phosphorus in rainwater: partitioning inputs and impact on the surface coastal ocean. *Limnology and Oceanography* 44 (4): 1160–1165. https://doi.org/10.4319/lo.1999.44.4.1160.

Mojiri, A., Zhou, J.L., Robinson, B. et al. (2020). Pesticides in aquatic environments and their removal by adsorption methods. *Chemosphere* 253: 126646. https://doi.org/10.1016/j.chemosphere.2020.126646.

Moore, R.C., Loseto, L., Noel, M. et al. (2020). Microplastics in beluga whales (*Delphinapterus leucas*) from the Eastern Beaufort Sea. *Marine Pollution Bulletin* 150: 110723. https://doi.org/10.1016/j.marpolbul.2019.110723.

Mostofa, K.M.G., Liu, C.-Q., Vione, D. et al. (2013). Sources, factors, mechanisms and possible solutions to pollutants in marine ecosystems. *Environmental Pollution* 182: 461–478. https://doi.org/10.1016/j.envpol.2013.08.005.

Napper, I.E. and Thompson, R.C. (2020). Plastic debris in the marine environment: history and future challenges. *Global Challenges* 4: 1900081. https://doi.org/10.1002/gch2.201900081.

Ngatia, L., Grace, J.M. III, Moriasi, D., and Taylor, R. (2019). Nitrogen and Phosphorus Eutrophication in Marine Ecosystems. In: *Monitoring of Marine Pollution* (ed. H.B. Fouzia). IntechOpen https://doi.org/10.5772/intechopen.81869.

Nilsson, G.E., Dixson, D.L., Domenici, P. et al. (2012). Near-future carbon dioxide levels alter fish behavior by interfering with neurotransmitter function. *National Climate Change* 2: 201–204.

Orr, J.C. (2011). Recent and future changes in ocean carbonate chemistry. *In Ocean Acidification* 2011: 41–66.

Parmentier, F.J.W., Christensen, T.R., Rysgaard, S. et al. (2017). A synthesis of the arctic terrestrial and marine carbon cycles under pressure from a dwindling cryosphere. *Ambio* 46: 53–69.

Ping, C.L., Jastrow, J.D., Jorgenson, M.T. et al. (2015). Permafrost soils and carbon cycling. *Soil* 1 (1): 147–171.

Plastics Europe (2019). Plastics—the facts 2019. An analysis of European plastic production, demand and waste data. https://plasticseurope.org/wp-content/uploads/2021/10/2019-Plastics-the-facts.pdf (accessed 7 February 2024).

Plastics Europe (2023). Plastics—the facts 2023. An analysis of European plastic production, demand and waste data. https://plasticseurope.org/knowledge-hub/plastics-the-fast-facts-2023/ (accessed 7 February 2024).

Pollard, R.T., Salter, I., Sanders, R.J. et al. (2009). Southern Ocean deep-water carbon export enhanced by natural iron fertilization. *Nature* 457: 577–580.

Prospero, J.M. and Savoie, D.L. (1989). Effect of continental sources on nitrate concentrations over the Pacific Ocean. *Nature* 339 (6227): 687–689. https://doi.org/10.1038/339687a0.

Qin, D., Ding, Y., Xiao, C. et al. (2018). Cryospheric science: research framework and disciplinary system. *National Science Review* 5 (2): 255–268.

Rahmstorf, S. (2006). Thermohaline ocean circulation. *Encyclopedia of quaternary sciences* 5: 4–9.

Rau, G.H. and Caldiera, K. (1999). Enhanced carbonate dissolution: a means of sequestering waste CO_2 as ocean bicarbonate. *Energy Conversion and Management* 40: 1803–1813.

Rau, G.H., Knauss, K.G., Caldiera, K., and Friedmann, J. (2006). Opportunities for low-cost CO_2 mitigation in electricity, oil, and cement production. *Proceedings of the Greenhouse Gas Technologies* 8: 1–4.

Richards, D.R. and Friess, D.A. (2016). Rates and drivers of mangrove deforestation in Southeast Asia, 2000–2012. *Proceedings of the National Academy of Sciences* 113 (2): 344–349.

Rios-Fuster, B., Alomar, C., Viñas, L. et al. (2021). Organochlorine pesticides (OCPs) and polychlorinated biphenyls (PCBs) occurrence in *Sparus aurata* exposed to microplastic enriched diets in aquaculture facilities. *Marine Pollution Bulletin* 173: 113030. https://doi.org/10.1016/j.marpolbul.2021.113030.

Ronkay, F., Molnar, B., Gere, D., and Czigany, T. (2021). Plastic waste from marine environment: demonstration of possible routes for recycling by different manufacturing technologies. *Waste Management* 119: 101–110.

Sen Gupta, A., Thomsen, M., Benthuysen, J.A. et al. (2020). Drivers and impacts of the most extreme marine heatwave events. *Scientific Reports* 10 (1): 19359.

Sharma, S. and Chatterjee, S. (2017). Microplastic pollution, a threat to marine ecosystem and human health: a short review. *Environmental Science and Pollution Research* 24 (27): 21530–21547. https://doi.org/10.1007/s11356-017-9910-8.

Silvano, A., Rintoul, S.R., Peña-Molino, B. et al. (2018). Freshening by glacial meltwater enhances melting of ice shelves and reduces formation of Antarctic bottom water. *Science Advances* 4 (4): 9467.

Thanigaivel, S., Vickram, S., Dey, N. et al. (2023). Ecological disturbances and abundance of anthropogenic pollutants in the aquatic ecosystem: critical review of impact assessment on the aquatic animals. *Chemosphere* 313: 137475. https://doi.org/10.1016/j.chemosphere.2022.137475.

The Ocean Clean-Up 2014. The Concept. http://www.theoceancleanup.com/the-concept.html. (accessed 7 February 2024).

Thushari, G.G.N. and Senevirathna, J.D.M. (2020). Plastic pollution in the marine environment. *Heliyon* 6 (8): e04709. https://doi.org/10.1016/j.heliyon.2020.e04709.

Turley, C. (2010). *UNEP Emerging Issues: Environmental Consequences of Ocean Acidification: A Threat to Food Security*. United Nations Environment Programme.

Turley, G. and Gattuso, J. (2012). Future biological and ecosystem impacts of ocean acidification and their socioeconomic-policy implications. *Current Opinion in Environmental Sustainability* 4: 278–286.

Turley, C., Eby, M., Ridgwell, A.J. et al. (2010). The societal challenge of ocean acidification. *Marine Pollution Bulletins* 60: 787–792.

Ussiri, D.A. and Lal, R. 2017. Introduction: climate overview. *Carbon Sequestration for Climate Change Mitigation and Adaptation*, pp.1–25 Springer.

Walkinshaw, C., Lindeque, P.K., Thompson, R. et al. (2020). Microplastics and seafood: lower trophic organisms at highest risk of contamination. *Ecotoxicology Environment Safety* 190: 110066. https://doi.org/10.1016/j.ecoenv.2019.110066.

Wallace, B., Stacy, B.A., Cuevas, E. et al. (2020). Oil spills and sea turtles: documented effects and considerations for response and assessment efforts. *Endangered Species Research* 41: 17–37. https://doi.org/10.3354/esr01009.

Wang, W., Gao, H., Jin, S. et al. (2019). The ecotoxicological effects of microplastics on aquatic food web, from primary producer to human: a review. *Ecotoxicology and Environmental Safety* 173: 110–117. https://doi.org/10.1016/j.ecoenv.2019.01.113.

Wang, Z., Dong, H., Wang, Y. et al. (2020). Effects of microplastics and their adsorption of cadmium as vectors on the cladoceran Moina monogolica Daday: implications for plastic- ingesting organisms. *Journal of Hazard Materials* 400: 123239. https://doi.org/10.1016/j.jhazmat.2020.123239.

Weisberg, S.B., Bednaršek, N., Feely, R.A. et al. (2016). Water quality criteria for an acidifying ocean: challenges and opportunities for improvement. *Ocean & Coastal Management* 126: 31–41. https://doi.org/10.1016/j.ocecoaman.2016.03.010.

Williamson, P., Turley, C., and Ostle, C. (2017). *Ocean Acidification*. MCCIP Science Review.

Winton, M. (2008). Sea ice-albedo feedback and nonlinear Arctic climate change. Arctic sea ice decline: observations, projections, mechanisms, and implications. *Geophysical Monograph* 180: 111–131.

Wittmann, A.C. and Pörtner, H.O. (2013). Sensitivities of extant animal taxa to ocean acidification. *Nature Climate Change* 3: 995–1001.

Wurtsbaugh, W.A., Paerl, H.W., and Dodds, W.K. (2019). Nutrients, eutrophication and harmful algal blooms along the freshwater to marine continuum. *WIREs Water* 6 (5): e1373. https://doi.org/10.1002/wat2.1373.

Xanthos, D. and Walker, T.R. (2017). International policies to reduce plastic marine pollution from single-use plastics (plastic bags and microbeads): a review. *Marine Pollution Bulletin*.

Zhang, B., Matchinski, E.J., Chen, B. et al. (2019). Marine Oil Spills—Oil Pollution, Sources and Effects. In: *World Seas: An Environmental Evaluation* (ed. C. Sheppard), 391–406. Elsevier https://doi.org/10.1016/B978-0-12-805052-1.00024-3.

7

Heavy Metal Pollution and Historical Legacies in Coastal–Marine Environment

Chamila Jinendra Nanayakkara[1], *Nimila Dushyantha*[2], *Pubudi Dilshara*[1], *Nadeera Batapola*[1], *Nalin Ratnayake*[1], *Ranjith Premasiri*[1], *and Bandara Abeysinghe*[1]

[1] *Department of Earth Resources Engineering, University of Moratuwa, Moratuwa, Sri Lanka*
[2] *Department of Applied Earth Sciences, Faculty of Applied Sciences, Uva Wellassa University, Badulla, Sri Lanka*

7.1 Introduction

Heavy metals are one of the major anthropogenic contaminants in coastal–marine environments and their inherent toxicity, persistence, and non-biodegradability can pose severe threats to ecological integrity and human health. The occurrence of the cadmium-caused Itai–itai disease and the mercury-caused Minamata disease in Japan are the best historical events that manifested the extensive consequences of heavy metal contamination in marine environments (Wang et al. 2013). Over the years, coastal–marine ecosystems have been subjected to numerous anthropogenic activities, such as industrial, agricultural, maritime transportation, and municipal wastes, leading to the introduction and accumulation of heavy metals in sediments, water column, and living organisms, even to the biomagnification through the food chain (Vodyanitskii 2013; Fang et al. 2019).

The majority of heavy metals entered into the coastal–marine environments via these anthropogenic sources are deposited in the sediment layer, leaving only a small proportion in the water column. Thus, coastal and marine sediments act as a chronicle of heavy metal contamination, and analyzing concentrations of heavy metals in successive sediment layers is crucial to trace the trajectory of these anthropogenic activities (Manahan 2002; Meena et al. 2018; Mishra et al. 2019). Furthermore, the impact of heavy metal pollution extends beyond mere contamination of sediments and water column. It impacts in the form of bioaccumulation and biomagnification within the bodies of various marine organisms, ultimately affecting the entire ecosystems (Saidon et al. 2024). As contaminants accumulate in organisms and progress up the food chain, they pose ecological consequences and public health concerns that demand sustainable remediation efforts (Kumar et al. 2019). Therefore, studying heavy metal contamination in coastal–marine sediments, water, and organisms would reveal the historical evolution of heavy metal pollution and its persistent challenges to ecosystems and human health.

Although coastal and marine heavy metal pollution has been a widely discussed subject among scientists, environmentalists, and policymakers, the effective prevention and management of anthropogenic stresses on these environments are still lacking. In this context, understanding the historical legacies of heavy metal pollution in coastal–marine environments will not only

Coastal and Marine Pollution: Source to Sink, Mitigation and Management, First Edition.
Edited by Meththika Vithanage, Sameera M. Samarasekara, Bryan D. James, and Christopher M. Reddy.

comprehend its evolution and enduring impact but also give insights for remedial and control measures to minimize the adverse effects of heavy metals on the coastal–marine ecosystems and human health. Therefore, by synthesizing historical perspectives with contemporary scientific understanding, this chapter aims to provide a holistic perspective on the challenges posed by heavy metal pollution in coastal–marine environments. Through an interdisciplinary approach, including historical analysis, scientific research, and policy considerations, it is planned to develop comprehensive mitigation strategies to address these challenges effectively.

7.2 Evolution of Heavy Metal Pollution in Coastal and Marine Environments

Heavy metals present in marine environments are derived from both natural and anthropogenic sources. However, concentrations naturally produced by the inherent composition of the earth's crusts are very low and lack the potential to induce any contamination. On the contrary, the evolution of heavy metal contamination in oceanic and coastal locales can be attributed predominantly to human activities (Nanayakkara et al. 2023). They include a multitude of anthropogenic routes such as dredging and coastal land reclamation, industrial and domestic waste disposal, agricultural runoff, petroleum exploration, refining, transportation, mining and mineral processing, and scrap metal recycling (Sharifuzzaman et al. 2016) (Figure 7.1). These activities have not only introduced new heavy metal contaminants to coastal environments but also mobilized previously

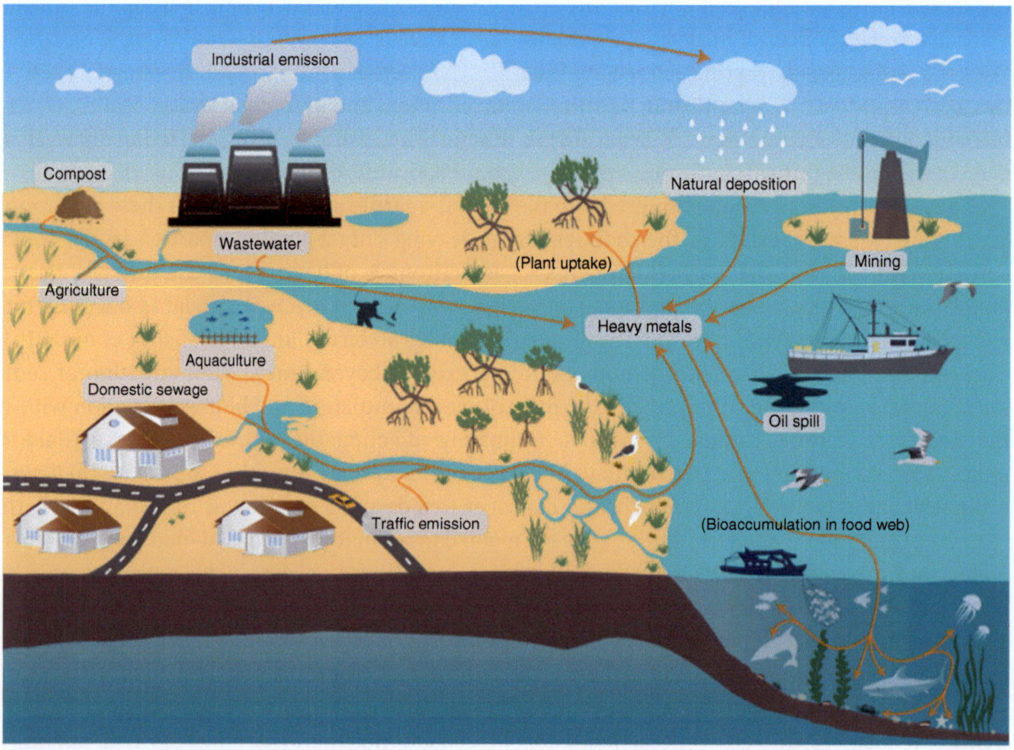

Figure 7.1 Anthropogenic sources of heavy metals in coastal and marine environments.
Source: Li et al. (2022).

deposited heavy metals within sediment layers, intensifying their tendency for incorporation into the food webs (Sharifuzzaman et al. 2016).

Owing to the mobile reactive form and solubility, heavy metals with an anthropogenic origin generally exhibit a high bioavailability (Ayangbenro and Babalola 2017). Furthermore, they persist in marine environments attributing to their nonbiodegradable and inert properties. After entering the marine environments, heavy metals get distributed between the water and sediment layers through the adsorption and desorption processes (Shah 2021a). Heavy metals present in marine environments have diverse forms: as ions in a free state, as complex compounds formed by metallic ions binding with various ligands, and as precipitated molecules of compounds. They are also adsorbed onto the surface of colloidal or suspended materials or suspended in the liquid phase. Among these forms, ionic species are detected to be more toxic and bioavailable. A majority of free ions get complexed through coprecipitation, hydrolysis, and adsorption, leading to their deposition in the sediment layers. Only a small fraction remains dissolved in the water column (Manahan 2002; Batapola et al. 2023). Therefore, sediment layers serve as a repository of historical records on heavy metal contamination in the littoral environment and can be analyzed to assess the extent of pollution. Table 7.1 succinctly presents the contribution of anthropogenic activities to the accumulation of heavy metals in coastal sediments based on previous research.

Table 7.1 Contribution of industrial activities for the accumulation of heavy metals in coastal sediments based on the previous research.

Anthropogenic activity	Location	Heavy metals concentrations in sediments (μg/g)									References
		As	Cd	Cr	Cu	Hg	Mn	Ni	Pb	Zn	
Dredging and land reclamation	Hong Kong estuarine	n.d.	1.5	80.0	65.0	1.0	—	40.0	75.0	200.0	Evans (1994)
	Victoria Harbour, Hong Kong	n.d.	32.0	80.0	11.0	—	196.0	14.0	—	67.0	Tanner et al. (2000)
Port	Port Klang, Malaysia	60.4	0.83	46.4	17.4	0.23	n.d.	11.4	59.4	51.1	Sany et al. (2013)
	Raysut Port, Oman	n.d.	0.20	31.2	0.83	0.002	n.d.	2.49	0.73	1.57	De Mora et al. (2004)
Desalination plants	Muscat, Oman	n.d.	0.6	95.1	10.0	n.d.	n.d.	299.2	1.7	13.9	Abdul-Wahab and Jupp (2009)
	Ras Tanajib, Saudi Arabia	n.d.	3.3	8.3	3.5	0.06	n.d.	12.1	14.8	7.4	Sadiq (2002)
Industrial activities	Jinzhou Bay, North China	396.5	n.d.	60.6	417	n.d.	n.d.	n.d.	753.2	6419	Zhang et al. (2008)
	Fujian coast, Southeast China	9.1	0.08	57.4	22	0.05	n.d.	27.4	37.1	96	Zhang (2008)
	Hong Kong, Southeast China	n.d.	0.3	48.9	119	0.19	n.d.	24.7	53.6	148	Zhou et al. (2007)
	Ulsan Bay, Korea	15.8	0.4	64.7	95.6	0.16	n.d.	32.2	90.7	361.9	Ra et al. (2014)
	Karachi Coast, Pakistan	n.d.	n.d.	96.8	n.d.	n.d.	n.d.	31.4	23.2	204.8	Mashiatullah et al. (2013)

n.d. – not determined.
Source: Sharifuzzaman et al. (2016).

7.3 Case Studies on Historical Legacies of Heavy Metal Contamination in Coastal and Marine Environments

The historical records on heavy metal contamination in coastal and marine regions span several decades and predominantly result from a myriad of anthropogenic activities. Among them, oil spills during the production and transportation phases and improper disposal of wastewater and other effluents from industries are very common and can be exemplified by various past incidents. These occurrences, together with their enduring and long-lasting impact on the coastal and marine ecosystems, are meticulously elaborated within this section.

7.3.1 Heavy Metal Contamination from Oil Spillages

Among the historical events that have caused heavy metal pollution in the oceanic environments, a notable proportion is accounted for the oil spills from maritime accidents. The ecological ramifications of heavy metals contamination from several major oil releases representing different coastal and marine settings in the global context are presented in Table 7.2, substantiated by previous research. Collectively, these facts conclude that there had been a considerable impact of heavy metal contamination on the ecological community worldwide following the maritime oil spillage.

7.3.2 Heavy Metal Contamination from Industrial Discharges

Over the past, improper disposal of industrial waste has caused heavy metal contamination in the littoral zone. Its direct impact on the inhabitant marine organisms and the repercussions on human beings via food chains are expounded in the ensuing description. This description involves case studies from several coastal localities illustrative of the highly heavy metal-contaminated regions of the world.

Mercury contamination in Minamata Bay, on the south-eastern coast of the Yatsushiro Sea, western Japan, serves as a paradigmatic illustration of heavy metal contamination in coastal environments through the introduction of industrial effluents. Between 1932 and 1968, a fertilizer company neighboring the Minamata harbor was found to have discharged waste liquids contaminated with organic and inorganic forms of Hg into Minamata Bay leading to an accumulation of approximately 250 tonnes of Hg therein (Nakata et al. 2008; Sarker 2021). Several studies conducted to monitor the Hg concentration in Minamata Bay sediments had recorded elevated values, as 29–713 ppm (dry weight) in 1963, 19–908 ppm (dry weight) in 1969, 8–253 ppm (dry weight) in 1970, and 14–586 ppm (dry weight) in 1971 which markedly transcended the permissible limit of 0.2 ppm recommended by the United States Environmental Protection Agency (US-EPA). In 1959, Hg levels in the bay-dwelling fish species were detected to be very high, with *Nibea schlegeli* recording 15 ppm (wet weight), *Venus japonica* registering 35.8 ppm (dry weight), and *Hormomya mutabilis* measuring 21.6 ppm (wet weight). These values conspicuously exceeded the 0.5 ppm permissible limit stipulated by the World Health Organization (WHO) implying the unsuitability of Minamata Bay fish for human consumption (Fujiki and Tajima 1992; Oppong et al. 2010).

Residents of the area were affected by Minamata disease due to methylmercury poisoning, after consuming the seafood harvested from Minamata Bay. The hair samples from the patients were detected to have pronounced concentrations of Hg, 705 ppm being the highest record in 1960. The postmortem examinations of deceased patients in 1959 reported elevated Hg content as 2.6–24.8 ppm (in the brain), 21.2–140.0 ppm (in kidneys), and 22.0–70.5 ppm (in the liver). Moreover, fetal methylmercury poisoning was common, given that the mothers were consuming contaminated seafood.

Table 7.2 Heavy metal contamination in coastal–marine areas from oil spills.

Incident	Cause	Area affected	Heavy metal contamination as evidenced by previous research
Exxon Valdez (1989)	Striking of Exxon Valdez oil tanker on Bligh Reef, near the coast of Alaska, adding 11 million gallons of Alaska North Slope crude oil into Prince William Sound (Griest et al. 1991; Mustafa et al. 2015).	Affected 783 km of coastal stretch, associated with Prince William Sound, and 1300 km surrounding the Gulf of Alaska. Also, a winter storm dispersed the oil slick to an extent of 2100 km within marine water (Mustafa et al. 2015).	Surficial sediment samples from Kachemak Bay, Alaska recorded an elevated As concentration (21.4 ± 1.8 mg/kg) substantially surpassing its international geochemical background shale value (8.1 mg/kg) (Apeti and Hartwell 2015; Olayinka-Olagunju et al. 2021). Significant amounts of heavy metals comprising Cr, Hg, Cd, Ni, V, Sb, As, Zn, Pb, Co, Mn, Se, Cu, and Sn were detected in environmental samples of marine organisms and sediments collected from the region affected by this oil spill (Mustafa et al. 2015).
Gulf War (1991)	Expulsion of about eight million crude oil barrels from two terminals, a refinery, and a tank field into the Persian Gulf (Hussain and Gondal 2008; Mustafa et al. 2015).	Covered 770 km along the shoreline stretching from the southern part of Kuwait to Abu Ali Island in Saudi Arabia (Hussain and Gondal 2008; Mustafa et al. 2015).	An analysis of the pre-spill (1990) and post-spill (1994) heavy metal content in gastropod snails (*Lunella coronatus*) revealed that their post-spill mean concentrations of Cu (11.24–55.00 μg/g), Ni (15.33–16.96 μg/g), Pb (0.37–0.57 μg/g), and Zn (28.86–486.61 μg/g) were notably higher than the corresponding pre-spill concentrations (0.35–0.67 μg/g for Cu, 1.50–4.50 μg/g for Ni, 0.16–2.98 μg/g for Pb, and 19.94–54.79 μg/g for Zn). Also, post-spill mean Cu concentration in bivalve pearl oysters (*Pinctada radiata*) (28.90–168.43 μg/g) exhibited a significant augmentation in contrast to pre-spill conditions (0.11–2.29 μg/g) (Bu-Olayan and Subrahmanyam 1997). The mean concentrations of Cu (123.8 μg/g) and Zn (206.0 μg/g) within their body organs of intertidal marine blue crabs (*Portunus pelagicus*), inhabiting the coastal regions of Kuwait had exceeded the corresponding concentrations of the global decapod crustacean's community (70.0 μg/g for Cu and 80.00 μg/g for Zn) (Al-Mohanna and Subrahmanyam 2001).
Prestige (2002)	Release of heavy fuel accounting for 60 000 tonnes from the tanker (Díez et al. 2007).	Affected the shores of Basque Country, Cantabria, Asturias, and Galicia (Díez et al. 2007).	An assessment of the heavy metal concentration in two seabird species (yellow-legged gulls and European shags) using the chick feathers, following this incident, showed that the Cu concentration in the samples (4.3–10 μg/g) was nearly twofold of the pre-spill level (1.5–3.6 μg/g). Correspondingly, Pb content in the feathers (1.0–1.4 μg/g) was approximately fourfold of the pre-spill level (0.1–0.4 μg/g) (Moreno et al. 2011). Hg concentrations (dry weight) in Atlantic puffins (1.774 ppm), common guillemot (1.208 ppm), and razorbills (2.275 ppm), seabirds from the Galician coast had reached a critical level capable of causing a toxic effect (Pérez-López et al. 2006).

(Continued)

Table 7.2 (Continued)

Incident	Cause	Area affected	Heavy metal contamination as evidenced by previous research
Deepwater Horizon (2010)	The explosion of a drilling platform at the Macondo exploration site leading to the expulsion of approximately 686 000 metric tonnes of crude oil into the gulf (Mustafa et al. 2015).	Generated an oil layer spanning over a distance of 1773 km along the shoreline. The weathering process of spilled oil generated lumps of tar that were deposited along the Alabama coast (Mustafa et al. 2015).	The Mean tissue concentrations of Ni (15.9 ± 3.5 ppm) and Cr (12.8 ± 2.6 ppm) within the tissues of the whales in the Gulf of Mexico after this incident were markedly higher than that of the global whale community (for Ni, 2.4 ± 0.4 ppm, for Cr 9.3 ± 1.0 ppm) (Wise et al. 2014)
Bohai Sea (2011)	The release of a significant amount of oil (723 barrels) and mineral-based drilling fluid (2620 barrels) into the sea from an inadvertent opening of a subsurface reservoir through natural geological fractures, induced by excessive pressure from water injection during production, at the off-shore oil field (Penglai 19-3) (Zhang et al. 2020b).	This incident had tainted an area of km^2 in the Bohai Sea (Zhang et al. 2020b).	The subsequent analysis of sediment samples from this area has proven moderate pollution in the subsurface strata, particularly with Cr as well as Ni, and to a lesser extent with Zn, Cu, Pb, and Co. It was also verified that this oil spill has contributed highly to the heavy metal contamination in the Bohai Sea with a comparatively minor influence from the riverine inputs and airborne dust (Zhang et al. 2020b). This catastrophe had a direct impact on a range of marine species with phytoplankton and fish included. A potential for advanced species, including mankind, to get affected by bioaccumulation and biomagnification of the contaminants from this incident via food chains and food webs was also recognized. Scientists have already expressed their concerns over the carcinogenic effect of this pollution on human beings (Guo et al. 2013).

The absence of an effective treatment for this malady during that period increased its severity. As of March 1992, 2252 patients had been officially diagnosed with Minamata disease and 1043 among them had died (Harada 1995). In this manner, Minamata Bay Hg contamination had exerted a profound impact on the ecosystem encompassing the human population as well.

The Gulf of Gabes, located in the southeastern expanse of Tunisia, is considered one of the most polluted areas in the Mediterranean Basin. Despite being a pivotal fishing ground, the environmental integrity of this gulf was found to be compromised by numerous contaminants featuring heavy metals due to the establishment of a large industrial complex in the year 1972. A considerable quantity ranging from 1000 to 13 000 tonnes of phosphogypsum, an inherent by-product of the phosphate fertilizer industry constituting heavy metals such as V, Cd, Zn, Cr, and Cu, was approximated to be released into the open sea daily without undergoing any pre-treatment (El Zrelli et al. 2015). Surficial sediments from the Gulf of Gabes had recorded elevated Cd (73.01 μg/g dry weight) and Zn (545.03 μg/g dry weight) content compared to similar marine settings worldwide. High contamination (Contamination Factor [CF] ≥ 6) concerning Zn and Cd was detected in the gulf with their CFs measuring 100.93 and 663.72, respectively (Hakanson 1980; El Zrelli et al. 2015).

Surficial sediments in the central region of this gulf demonstrated a Pollution Load Index (PLI) of 4.43, concluding that they are being polluted (PLI >1) with heavy metals (Tomlinson et al. 1980; El Zrelli et al. 2015). Based on the classification of mean geo-accumulation (I_{geo}) adapted by Muller (1969), surface sediment layer in the central coastal area exhibited a strong to extreme contamination with Pb (4.26) and extreme contamination with Cr and Zn (6.41 each) (El Zrelli et al. 2015). So, these three major pollution indices collectively verify that there has been a substantial environmental impact on the coastal environment of the Gulf of Gabes due to the unregulated waste disposal from the adjacent industrial complex.

A discernible impact of this contamination on the biosphere was visible because there was a local reduction in a seagrass species endemic to the Mediterranean Sea namely *Posidonia oceanica* after the phosphate fertilizer factory establishment in the Gabes industrial complex (Darmoul 1988). Local fishermen have noted a considerable number of marine fatalities, including turtles, cetaceans, mollusks, fish, cnidarians, and birds at different times of the year over the past few decades, particularly in the vicinity of the industrial complex. These ecological consequences can be predominantly ascribed to heavy metal-bearing phosphogypsum discharge into the seawater as verified by the marked surge in coastal halieutic population particularly at the industrial waste disposal sites upon the cessation of this waste stream (El Zrelli et al. 2015). Concentrations of Cd (16.11 µg/g), Zn (462.55 µg/g), and Cu (272.89 µg/g), in the edible gastropods (*Hexaplex trunculus*) sampled from the Gabes fishing harbor have transcended the European standards of 2 µg/g, (for Cd), 100 µg/g (for Zn), and 30 µg/g (for Cu) established by the Food and Agriculture Organization (FAO) (Lahbib et al. 2013). The exceedance of these regulatory limits has imperiled the food security of the local population posing a significant health risk given their heavy reliance upon the fisheries resources from the gulf.

The coastal–marine environments of the Arabian Gulf comprise one of the most anthropogenically stressed ecosystems in the world, and residents in the area vastly rely on marine resources for food, industry, and recreation. The heavy metal input into the gulf is highly diversified stemming from a multitude of sources including dredging and coastal land reclamation, effluents discharged from desalination plants, petroleum production, and various other industrial activities (Naser 2013).

The mean Pb content (10.8 µg/g wet weight) of grouper fish (*Epinephelus coioides*) collected from four major fishing locations in Bahrain, Arabian Gulf during 1990–1991 was detected to have surpassed existing international standards for fish (0.5 – 6 µg/g wet weight) (Al-Sayed et al. 1996). Al-Sayed et al. (1994) discovered that Cd (2.4 ± 0.39 µg/g) and Pb (7.6 ± 1.4 µg/g) content in tissues of pearl oysters (*Pinctada radiata*) sampled from pelagic waters in the shipping lane to Bahrain during 1991–1992 were over the WHO recommended values (2.0 µg/g for both Cd and Pb). It was inferred that the naval activities had disposed heavy metals into the surrounding water, which were bio-accumulated within oysters. In 1993, an analysis of shellfish and fish samples representing all fishing grounds around Bahrain island within the Arabian Gulf measured As levels of 3.61 and 1.7 µg/g (wet weight) respectively, which were higher than the permissible As limit (0.5 µg/g wet weight) adapted for Saudi Arabia and the United States (Madany et al. 1996). The Hg content in 20.6% of 330 fish samples collected from Kuwait Bay from 1996 to 1998 had exceeded the permissible limits (0.300 µg/g for methyl mercury [Me-Hg] and 0.500 µg/g for total mercury content) established by the WHO. The high Hg concentrations in fish tissues were attributed to the loss of Hg used in the electrolytic cells of a Cl_2 production plant into the environment (Al-Majed and Preston 2000).

In 2004, 140 samples of fish representing five distinct fish species were retrieved from the Iranian coastal zone adjacent to the Persian Gulf and analyzed for the presence of Hg in their liver and mussel tissues. According to the analysis results, 9% of the examined fish samples were categorized as polluted denoted by the Hg concentration in the range of 0.300 to 0.500 μg/g and 3% of the samples had Hg content above the maximum acceptable limit by WHO (0.5 μg/g) (Agah et al. 2010). All this evidence from previous research cumulatively indicates the retention of anthropogenic heavy metal input into the Arabian Gulf within the marine organisms. Given the widespread consumption of seafood among residents in the Arabian Gulf countries a substantial risk to human health exists from the pervasive heavy metal contamination in the gulf.

Santa Rosalía, a coastal town in the State of Baja California Sur, Mexico, adjoining the Gulf of California has had an over-century reputation for Cu mining and smelting activities. It was reported that wastes consisting of heavy metals, generated during the mining and mineral processing activities, were dumped into the adjacent local beaches (Rodríguez-Figueroa et al. 2009; Jonathan et al. 2016). Research by Jonathan et al. (2016) has revealed that the average concentrations of Cu (2748 mg/kg) and Zn (2599 mg/kg) had exceeded the Severe Effect Level concentrations established by US-EPA in 2001 (for Cu 110 mg/kg and Zn 820 mg/kg) indicating a level of contamination detrimental to a majority of sediment-inhibiting species (USEPA 2001). Considering the I_{geo} value, As, Zn, Cd, Sr, Co, Ni, Cu, and Cr were in the range from 4 to 6 implying a strong to extreme contamination with these metals. The association of the aforementioned elements validates the direct connection of their enrichment in the sediment layer to the Cu metal mining and processing industry (Jonathan et al. 2016). An investigation by Rodríguez-Figueroa et al. (2009) on the heavy metal composition in brown seaweeds (*Padina durvillaei*) sourced off the coast of Santa Rosalía showed that their Fe (2243 ± 2325 mg/kg) and Zn (63 ± 43 mg/kg) contents were higher than those measured by Carrillo Domínguez et al. (2002) (455 mg/kg for Fe and 11 mg/kg for Zn). This disparity strongly implies that these metals have bioaccumulated within the seaweed over time.

The environmental milieu of Morphou Bay located on the western coast of North Cyprus was significantly altered by the materials released from the Cu mining over a span of approximately 70 years. The residues resulting from Cu mining and processing as well as extraction of Au and Ag as secondary products from cyanide leaching were stored in large tailing ponds constructed with underdeveloped technology. This caused a considerable mine waste discharge into Morphou Bay along with the seasonal runoff, which carried the drainage from the tailing ponds as well as the surface flow of Cu mine wastes (Gokcekus et al. 2003). Therefore, a rigorous experimental study was undertaken to scrutinize the presence of heavy metal contaminants in the edible fish species from Morphou Bay. Accordingly, the mean Cd concentration in fish (0.13 μg/g) exceeded the standard Cd (0.05 μg/g) level stipulated by FAO in 2011. Thus the consumption of fish from this area can lead to health complications caused by high Cd intake (Karafistan and Gemikonakli 2019; Rizk et al. 2022).

7.4 Heavy Metals and Their Essentiality in the Environment

Most of the case studies elaborated in Section 7.3 present clear evidence for the bioaccumulation of heavy metals in marine organisms and their biomagnification through the food chain. Therefore, it is essential to study the physiological and biochemical functions of heavy metals in the coastal–marine environment, and how their excess or deficiency impacts the integrity of ecosystems and well-being of humans (Ali et al. 2019; Shah 2021). Heavy metals such as Fe, Mn, Ni, Co, Cu, Mo, and Zn categorized under micronutrients are essential for the synthesis of biomolecules such as carbohydrates, growth chemicals, nucleic acids, chlorophyll, and secondary metabolites. As part of the biochemical reactions, these metals play an important role in the catalytic activity of enzymes, accelerating the rates of the reaction. However, the essentiality of these metals may vary based on the different groups of organisms, such as plants, animals, and microorganisms. Despite these metals, heavy metals such as Pb, Hg, Cd, and As are considered nonessential to the environment and cause toxic effects on the organisms once they enter the body (Chalkiadaki et al. 2014; Ali et al. 2019).

7.4.1 Bioaccumulation of Heavy Metals in Coastal Ecosystems

Coastal ecosystems are major reservoirs of heavy metals as they are vulnerable to pollution by various anthropogenic sources. The runoff generated along the coast contains heavy metals from industrial effluents, oil spills, and solid waste dumping that can directly impact sediments and aquatic organisms (Mele et al. 2020; Nanayakkara et al. 2023). They are highly soluble in marine waters and readily consumed by aquatic organisms. These metals accumulate in the gills, tissues, and liver of fish and marine invertebrates (Danovaro et al. 2023). In addition, heavy metals are adsorbed by suspended particles and then settled on the seafloor during the sedimentation process. They are ingested by benthic organisms, such as bottom-feeding invertebrates, filter-feeding organisms, microbes, and microalgae (Kadokami et al. 2013; Danovaro et al. 2023). The gradual accumulation of heavy metals in tissues of organisms through the food chain is called bioaccumulation. Moreover, bioaccumulation can ultimately cause the progressive increase of heavy metals in the higher tropic levels, which is generally known as biomagnification (Mele et al. 2020; Danovaro et al. 2023) (Figure 7.2). These biological phenomena can cause severe health effects on living organisms.

7.4.2 Heavy Metals Bioaccumulation in Plankton

Plankton plays a major role in the food chains of aquatic ecosystems as primary consumers and producers. Phytoplankton is the primary producers in the aquatic environment that reinforce the productivity of the higher levels of the food chain (Rauf et al. 2019; Permana and Akbarsyah 2021),

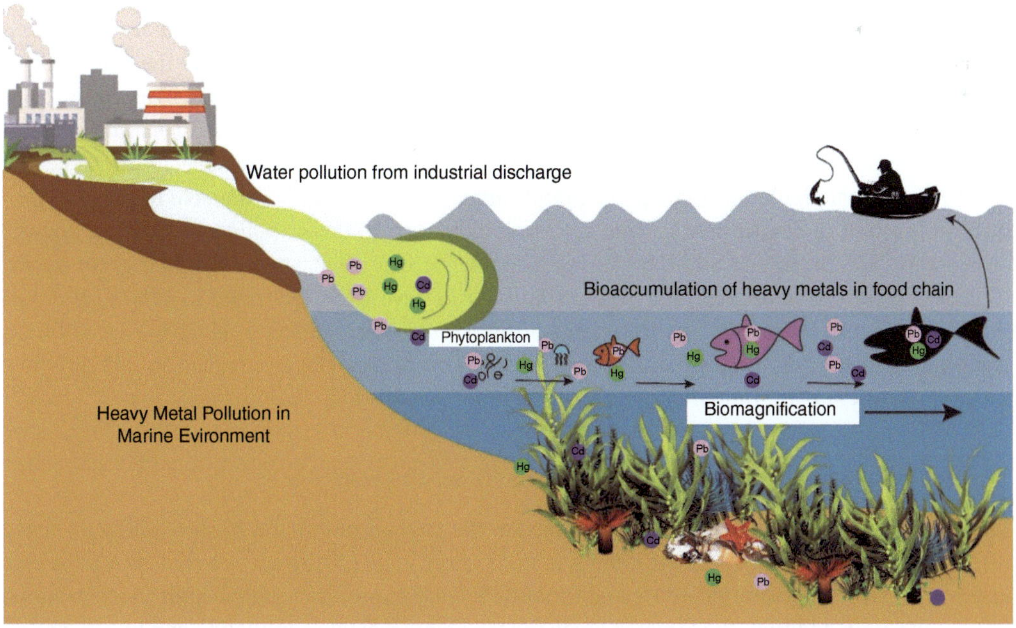

Figure 7.2 Bioaccumulation and biomagnification of heavy metals in the marine environment.

thus any threat to the phytoplankton population can lead to a trophic cascade and create significant stress on the environment. Therefore, both phytoplankton and zooplankton are considered ecologically sensitive species and often serve as indicators to determine heavy metal pollution (Permana and Akbarsyah 2021). According to Rauf et al. (2019), phytoplankton, such as *Amphora*, *Actinastrum*, *Staurastrum*, *Cymbella*, *Pediastrum*, *Spirulina*, and *Chroococcus*, and zooplanktons, such as *Brachionus*, *Filinia*, *Keratella*, and *Leptodora*, show considerable tolerance to high concentrations of Cd, Cr, Cu, and Co in water. Similarly, a study carried out in the Danube Delta, Romania, reported a high bioconcentration factor (accumulation levels of metal in the plankton samples compared to the concentration of water samples) in the plankton for Cd and Zn (Burada et al. 2014). Analysis of phytoplankton samples of Lake Manzala in Egypt found a high concentration of Zn (313–602 μg/g) compared to Cu, Cd, and Pb (63.35–131, 19.67–36.15, and 58.33–124 μg/g, respectively) (Bahnasawy et al. 2011).

7.4.3 Heavy Metal Uptake by Aquatic Plants and Algae

Aquatic plants accumulate heavy metals in their root and shoot parts through hemofiltration or phytofiltration, where heavy metals are taken up through absorption and adsorption processes. Heavy metals such as Fe, Zn, and Cu are nutritionally important to plants in trace levels. Bioaccumulation of such metals in plant biomass can significantly affect the productivity of the plant. For example, Pb accumulation in plants develops dwarf roots and dark green leaves, which are susceptible to falling and drooping. Furthermore, heavy metals affect the photosynthesis of plants by damaging the chloroplast, inactivating the photosynthetic enzymes, and altering the light reaction in the photosystem. In addition, the presence of heavy metals decreases carbon sequestration and inhibits root elongation and lignin synthesis in plants (Riyazuddin et al. 2021). Elevated concentrations of Cd, Cr, Pb, and Mn in the plant parts also increased the transpiration rates (Ankit et al. 2022). In *Sagittaria sagittifolia* plant, Cd concentration was higher in roots (510 μg/g) and caulis (548 μg/g) compared to the bulb (242 μg/g). Unlike Cd, other metals such as Cu and Pb were highly accumulated in the bulb (Hu et al. 2010). However, some aquatic plants such as *Lemna minor* (common duckweed) have shown the potential to remediate polluted wetlands, where the plant species survive and accumulate high concentrations of As in their biomass (Singh et al. 2016). In this context, *Salvinia auriculata, Limnobium laevigatum, Colocasia esculenta, L. minor, Typha latifolia*, and *Thelypteris palustris* plants are currently used in wastewater treatment (Muliyadi et al. 2023).

Similar to phytoplanktons and aquatic plants, algae perform an important role in primary production through photosynthesis. Algae act as bioindicators to detect pollution levels in aquatic ecosystems (Stevenson and Smol 2015). For example, brown algae, such as *Padina gymnospora*, *Cladophoropsis membranacea*, and *Hypnea hamulosa*, have been identified as bioindicators of metal pollution in the marine environments of the Persian Gulf (Bibak et al. 2020). In addition, algae are often used in the purification and remediation of water bodies contaminated by heavy metals and other organic and inorganic contaminants (Khalil et al. 2021).

7.4.4 Bioaccumulation of Heavy Metals in Shellfish and Fish

Metal bioaccumulation in fish mainly depends on the concentration and bioavailability of heavy metals in the water (Shah 2021). According to Bonsignore et al. (2018), species such as *Mullus* spp. and *Scorpaena porcus* accumulate Hg and Cr while *Squilla mantis* accumulate high concentrations of Cd and Cu. In addition, the targeted organ for the heavy metal accumulation can be also varied with specific organisms. For example, Ni accumulates significantly in the fish gills, whereas Cu, Cd, and Cr are accumulated in the muscles and liver (Zaynab et al. 2022). When comparing with metal accumulation levels in the body parts of *Liza aurata* species, elevated concentrations of Zn (62.92–99.8 μg/g) and Pb (6.90–10.27 μg/g) were found in fish gills compared to the muscles, which had concentrations of Zn and Pb (15.57–29.68 and 1.58–2.44 μg/g) (Bahnasawy et al. 2011). Bioaccumulation studies in Taihu Lake, China, with *Cyprinus carpio* species also exhibited high bioaccumulation levels in Pb > Cu > Cd > Cr during the summer season. Because seasonal variation affects the biotic factors and physicochemical properties in the lake, it influences the metal bioavailability (Rajeshkumar and Li 2018). Therefore, the specific preferences of the organism on the bioaccumulation process depend on their sensitivity toward the particular metal accumulation and the bioavailability of the metal. Some chemical and physical factors, such as lipid solubility, chemical speciation, kinetics for complexation, and participation coefficient among seawater and the particles, largely affect the metal accumulation. In addition, biological factors, such as biochemical and physiological adaptations of organisms and also influence on trophic interactions including herbivory, predation, and parasitism (Bonsignore et al. 2018; Moiseenko and Gashkina 2020; Shah 2021).

Bivalves accumulate a high capacity of heavy metals in their body as filter feeders. They consume food with contaminants in the water column, sediments, and suspended particles. The accumulation of heavy metals in bivalves is also affected by biological factors – such as age, size, physiological status, organisms, and gametogenesis – and abiotic factors – such as pH, temperature, salinity, and available contaminants (Zaynab et al. 2022). Heavy metals enter the fish mainly through metabolically active tissues, such as gills, and accumulate in muscles, gills, and the liver. However, the average concentrations of Cd, Hg, Pb, Cu, and Zn in the soft tissue parts of the crustaceans were higher compared to the muscle tissues of fish (Younis et al. 2015). During the study, *Chtenopteryx sicula* accumulated the highest Zn = 14.65 μg/g and lowest Cu = 0.99 μg/g levels, whereas *Panulirus penicillatus* showed the lowest Zn = 11.52 μg/g and highest Cu = 5.97 μg/g levels. Mollusks species such as *Ruditapes decussates*, *Mactra coralline*, and *Coralliophila meyendorffi* also have accumulated Pb > Cr > Co > Cd > Ni respectively in their bodies (Abdallah 2013). Because invertebrates such as crustaceans and mollusks are highly consumed by the fish in high trophic levels of the food web, elevated concentrations of heavy metals are accumulated in their body causing biomagnification.

7.4.5 Human Consumption and Health Effects

Heavy metals pose serious threats to humans, due to their persistence and high toxicity. They enter the human body through ingestion and affect the important organs and components in the digestive, cardiovascular, and central nervous systems (Bonsignore et al. 2018; Nanayakkara et al. 2023). Long-term accumulation of such metals in the human body causes nerve diseases, such as Parkinson's disease, depression, Alzheimer's disease, and dementia (Ungureanu and Mustatea 2022). Heavy metal toxicity often occurs when they produce reactive oxygen species (ROS) and cause oxidative damage to the organs, which ultimately causes severe health effects. For example, through the redox reactions in the biological metabolism, heavy metals, such as Ni and As produce free radicals, which cause oxidative damage to DNA and proteins (Fu and Xi 2020). In addition, these ROS can result in RNA affection, cancers in the skin, lungs, bladder, and bone mineralization in the bone system (Ungureanu and Mustatea 2022).

Humans are exposed to the heavy metals in the coastal ecosystems mainly through the ingestion of seafood. The consumption of heavy metals accumulated in bivalves, fish species, and crabs is detrimental to human health (Zaynab et al. 2022). Therefore, dietary reference values for these metals have been introduced by the Environmental Protection Agency (EPA), Agency for Toxic Substances and Disease Registry (ATSDRA), and Food and Drug Administration (FDA) organizations to reduce the effects on human health (Pb-0.26, As-0.3, Hg-0.3, Cd-1.0, and Cr-3.0 μg/kg/day) (Wong et al. 2022). The targeted organ and the health effects caused by the accumulation of heavy metals are indicated in Table 7.3.

Table 7.3 Toxic effects of heavy metals and their targeted organs in the human body.

Heavy metal	Toxic forms	Targeted organ/group of organs	Health effects	References
Pb	Pb(II)	The nervous system, reproductive system, liver, and kidney	- Attention deficit/hyperactivity disorder, Alzheimer's disease, amyotrophic lateral sclerosis, and Parkinson's disease - Alter the concentrations of sex hormones affecting the reproductive system - Affect on white blood count, hemoglobin, hematocrit concentrations, and serum urea, creatinine	Abd Elnabi et al. (2023)
As	- Mono-methylmalonic acid, di-methyl arsenic acid - AsH_3, As^{3+}, and As^{5+}	Lungs	- Effect functionality of cells, mitotic division, and respiration - Cause cardiovascular diseases and hypertension from arterial pressure	Ohiagu et al. (2022), Abd Elnabi et al. (2023)
Hg	Methyl mercury (MeHg)	Lungs, kidney, skin, and eyes	- Effect on child growth causes dyslipidemia in adults	Abd Elnabi et al. (2023)
Ni	Nickel sulfide, nickel oxide, and soluble nickel compounds	Lungs, liver, heart, brain, skin, and thyroid	- Heart defects in offspring when the mother is exposed to Ni during pregnancy - Immunological disorders due to the failure of apoptosis, Type I hypersensitivity - Skin diseases - Neurodegenerative diseases	Briffa et al. (2020), Ohiagu et al. (2022), Abd Elnabi et al. (2023)
Cd	Cd(II)	Lungs, kidney, and skeletal system	- Cause proliferation and metastasis of prostate cancer and basal breast tumors - Cause premature births, pregnancy loss, and reduced birth weight	Genchi et al. (2020), Abd Elnabi et al. (2023)
Cr	Cr(VI)	Lungs, digestive tract, and liver	- Affect on the fetal growth during pregnancy - Lung cancer - Impact on bronchial epithelial cells through the aberrant modulation of proteins such as apoptosis-related, cytoskeletal, and energy metabolism-associated proteins - Gastrointestinal ulceration - Liver and kidney damage	Ohiagu et al. (2022), Abd Elnabi et al. (2023)

Table 7.3 (Continued)

Heavy metal	Toxic forms	Targeted organ/group of organs	Health effects	References
Co	Co(II)	Respiratory tract and nervous system	- Respiratory diseases, such as bronchitis, rhinitis, fibrotic alterations in pulmonary tissue causing asthma, and lung fibrosis - Neural system disorders such as memory loss, bilateral nerve deafness, neuropathies, and optic atrophy - Hypertension and reversible electrocardiographic alterations - Skin diseases	Briffa et al. (2020), Abd Elnabi et al. (2023)
Mn	Mn(II)	Nervous system	- Hypertension - Cause behavioral changes such as dullness, weakness, anxiety, tremor, and lack of expressions	Briffa et al. (2020)
Zn	Zn(II)	Bone and muscles, liver, gastrointestinal tract, lung, brain, skin, pancreas	- Impaired immune functions and Neutropenia - Pancreatic complications - Acute illnesses such as abdominal cramps, vomiting, and nausea	Plum et al. (2010), Briffa et al. (2020)

7.5 Prevention and Management of Heavy Metal Pollution in Coastal and Marine Environments

Heavy metal pollution in coastal and marine environments has become a pressing global concern and is a result of continuous neglect and negligence of anthropogenic activities in these environments. Therefore, stringent prevention and control measures are required to mitigate the adverse impacts of coastal and marine heavy metal pollution (Zhang et al. 2020a; Aliko et al. 2022). Heavy metal remediation strategies can be both short and long term. Short-term measures include formulating quality guidelines and standards, enforcing existing national and regional environmental regulations and laws, and conducting holistic environmental monitoring programs while long-term actions include applying Environmental Impact Assessment (EIA) on existing and proposed projects or activities in the coastal or marine environments that may contribute to heavy metal pollution, advancing scientific research, and using the latest technological approaches and techniques in prevention and mitigation of heavy metal pollution (Naser 2013) (Figure 7.3).

7.5.1 Environmental Regulations and Legislations

Implementing environmental regulations and legislation can prevent and manage heavy metal pollution in coastal and marine environments, and it is often used in conjunction with other compatible approaches. Environmental regulation refers to the limitations or responsibilities enforced

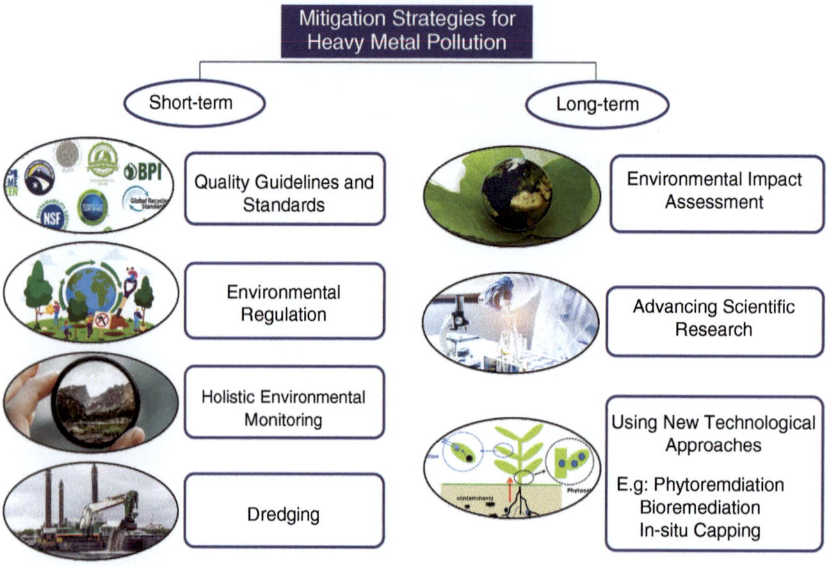

Figure 7.3 Schematic presentation of different types of mitigation strategies for the prevention and management of heavy metal pollution in coastal and marine environments.

upon individuals, corporations, and other entities to prevent and control environmental damage or to improve degraded environments. Environmental legislations include national and international laws and agreements established by parliaments or responsible organizations for the protection of the environment from pollution (McManus 2009).

The maritime sector has been identified as a prominent contributor to heavy metal pollution in marine environments. Therefore, implementing policies that regulate the maintenance of ships and engine boats – emphasizing the use of cleaner fuels and technologies, proper ballast water management, and waste handling – is critical (Lakshmi et al. 2021). Moreover, tourism has a significant impact on coastal and marine heavy metal pollution, thus governments should enforce sustainable tourism policies such as establishing carrying capacities and encouraging eco-friendly tourism to protect ecologically sensitive areas. In addition, industries operating in coastal areas need to be closely monitored for their pollution control measures and impose standard quality guidelines for heavy metal emissions (Meraj et al. 2022; Bhuyan et al. 2023). For example, shipbreaking industries in Bangladesh have severely degraded the biodiversity of its coastal zones, leading to increased attention on implementing environmentally friendly practices in the shipbreaking industry (Chowdhury et al. 2015). Enforcing stricter environmental regulations can reduce heavy metal pollution in coastal and marine environments. For instance, many Asian countries such as Indonesia, Malaysia, the Philippines, Singapore, Thailand, and Vietnam have accelerated the imposition of strict environmental regulations since the 1990s, resulting in a decline in the levels of heavy metal pollutants in coastal bays in the Philippines and Indonesia (Hosono et al. 2010).

7.5.2 Environmental Impact Assessment

Conducting EIA on proposed industries or projects in coastal and marine environments is critical, which is a systematic process of identifying, predicting, evaluating, and mitigating the negative environmental impacts of a project. EIA is helpful in altering and modifying the suggested process, design, or technology so that the environmental damage is minimal (Glasson and Therivel 2013). Moreover, EIA acts as an environmental management tool, controlling heavy metal emissions from industrial processes such as seawater desalination, sewage treatment, and oil refining and minimizing the mobilization of heavy metals due to dredging and reclamation activities in coastal and marine environments. Because changes in heavy metal concentrations in contaminated sites can adversely affect their ecological systems, holistic spatial and temporal monitoring of heavy metal contents is required. Such an integrated monitoring approach can elucidate the heavy metal pollution and bioaccumulation patterns in the environment, leading to the implementation of better prevention and remediation strategies (Bhuyan et al. 2023).

7.5.3 Dredging

Dredging has been widely used to control heavy metal pollution in coastal and marine environments by removing contaminated sediments. This method reduces the heavy metal contamination in water as it exposes the less polluted lower sediment layers, minimizing the exchange of heavy metals between the water column and sediments (Liu et al. 2015). However, dredging can cause more adverse effects to the surrounding environment rather than heavy metal pollution control. The most common disturbances that occurred while dredging include notorious intrusion of oxygen, resuspension of solids, and perturbation of sediments as well as possible variations of pH and temperature. These fluctuations in environmental conditions can promote the mobilization of heavy metals to the water column and aquatic organisms (Ferrans et al. 2021). Moreover, dredged sediments are mostly disposed of in landfills, posing threats such as the production of methane (greenhouse gas) and the leaching of heavy metals into the surroundings and groundwater (Depountis et al. 2009; Mymrin et al. 2017). Therefore, dredging must be carried out under vigilant supervision so that it does not pose more harmful threats to the ecological integrity of the environment.

7.5.4 In Situ Capping

In situ capping is an efficient and low-cost mitigation strategy for heavy metal pollution in coastal and marine environments compared to conventional approaches such as dredging. In this strategy, the contaminated sediment is capped by a layer of sand, clean sediment, or other active ingredients (Kumar et al. 2020). The use of chemically reactive materials called active capping segregates contaminants to reduce their mobility, toxicity, and bioavailability, leading to the immobilization of contaminants in sediments. For instance, active materials such as activated carbon, bio-char, surface-modified black carbon, and iron sulfide minerals have been used to successfully remediate heavy metal contaminants such as mercury (Ting and Hsi 2019).

7.5.5 Phytoremediation

Currently, increasing attention is being given to the utilization of technological approaches to eliminate or reduce heavy metal pollution in coastal and marine environments. Phytoremediation is such a cheap, eco-environmentally friendly strategy to prevent and manage heavy metal pollution, where plants are used to clean up the contaminated sites (Kumar et al. 2020). There are four different techniques used in phytoremediation, namely phytoextraction, rhizofiltration, phytostabilization, and phytovolatilization. Phytoextraction is one of the best methods to remove heavy metals from contaminated soils and sediments, where plants uptake pollutants by roots and translocate them to their shoots. In this method, heavy metals are generally removed by harvesting the plants. Rhizofiltration is the use of both terrestrial and aquatic plants to absorb and concentrate heavy metals from polluted aqueous sources in their roots. Moreover, plants are used to reduce the mobility or bioavailability of heavy metals in the environment, which is known as phytostabilization. This can effectively prevent heavy metals from migrating to groundwater or bio-magnifying via food chains. Phytovolatilization refers to the uptake of heavy metals by plants, transforming them into volatilized compounds, and then transpiring into the atmosphere (Trivedi and Ansari 2015).

Various types of plants are used for the phytoremediation of heavy metals from the coastal and marine environments. Common water hyacinth plants (*Eichhornia crassipes*) can remove several heavy metals from coastal areas. For instance, a study conducted in the coastal area of Ondo state in Nigeria showed that *E. crassipes* can accumulate high levels of Cr, Cd, Pb, and As, particularly in their roots and shoots. The use of halophytes to remediate polluted coastal areas such as lagoons and bays is also a promising solution to the increasing levels of heavy metal pollution (Lone et al. 2008; Agunbiade et al. 2009; Trivedi and Ansari 2015). Furthermore, mangroves (*Rhizophora mucronata*) have a high tolerance and are significantly adaptive to the presence of heavy metals, thus can be used for phytoremediation (Pahalawattaarachchi et al. 2009).

7.5.6 Bioremediation

Bioremediation is an emerging mitigation strategy for heavy metal pollution in coastal and marine environments, where specific metabolic activities of bacteria, fungi, yeast, microalgae, and microbial mats are used to purify ecosystems. The metabolic activities of these micro-organisms can mineralize or transform contaminants into less harmful substances, thus this method can be considered nondestructive and also economical (Aliko et al. 2022). Bioremediation utilizes two main strategies: (i) biostimulation where indigenous microbial populations are stimulated through the addition of specific nutrients such as air, organic substrates, or other electron donors/acceptors; and (ii) bioaugmentation, where viable microbial populations are introduced to assist populations that already exist in nature. Both these techniques can be used to remediate pollutants from the environment but abiotic and biotic factors that affect the biodegradation process need to be considered when applying either of them (Dash et al. 2013).

Different kinds of heavy metal-tolerant micro-organisms found in coastal and marine environments can be used for heavy metal bioremediation. For example, marine bacteria such as *Enterobacter cloacae* are capable of chelating heavy metals through the secretion of exopolysaccharides. Moreover, other purple nonsulfur marine bacteria such as *Rhodobium marinum* and *Rhodobacter sphaeroides* have been identified for their ability to metabolize heavy metals, including Cu, Zn, Cd, and Pb, through bioadsorption (Pramanik et al. 2015).

7.6 Conclusion

In summary, heavy metal pollution poses a persistent and complicated challenge in coastal–marine environments, driven by historical legacies and ongoing anthropogenic activities. The intricate interplay between essential and nonessential heavy metals highlights the complexity of their roles in ecological systems, necessitating a good understanding of their dynamics and impacts. Bioaccumulation and biomagnification processes further exacerbate the ecological consequences, highlighting the urgent need for effective mitigation strategies. By synthesizing historical perspectives with contemporary scientific understanding, stakeholders can develop holistic approaches to address heavy metal pollution, including environmental regulations, impact assessments, remediation techniques, and sustainable management practices. This comprehensive approach, guided by interdisciplinary collaboration and informed decision-making, holds the key to preserving the health and resilience of coastal–marine ecosystems for current and future generations.

7.7 Future Directives

Looking ahead, addressing heavy metal pollution in coastal–marine environments requires a proactive and adaptive approach that integrates emerging technologies, interdisciplinary research, and global cooperation. Future directives should prioritize the development and implementation of advanced monitoring systems capable of detecting and tracking heavy metal contamination in real time, enhancing the ability to respond swiftly to environmental threats. In addition, there is an essential need for further research to elucidate the long-term effects of heavy metal pollution on marine biodiversity and ecosystem functioning, facilitating the design of targeted conservation and restoration strategies. Embracing innovative remediation techniques, such as nanotechnology-based approaches and genetically engineered microorganisms, holds promise for enhancing the efficiency and sustainability of pollution mitigation efforts. Furthermore, fostering international collaborations and partnerships is essential for sharing knowledge, resources, and best practices, particularly in regions facing disproportionate environmental burdens. Education and public awareness campaigns play a vital role in promoting environmental stewardship and empowering communities to advocate for policy changes and sustainable practices. By embracing these future directives and leveraging the collective expertise of scientists, policymakers, industry stakeholders, and local communities, we can work toward a cleaner, healthier future for coastal–marine environments, safeguarding these invaluable ecosystems for generations to come.

Acknowledgment

This research was financially supported by the Senate Research Committee (SRC) of the University of Moratuwa, Sri Lanka, under Grant no. SRC/LT/2017/28.

References

Abd Elnabi, M.K., Elkaliny, N.E., Elyazied, M.M. et al. (2023). Toxicity of heavy metals and recent advances in their removal: a review. *Toxics* 11 (7): 580.

Abdallah, M.A.M. (2013). Bioaccumulation of heavy metals in mollusca species and assessment of potential risks to human health. *Bulletin of Environmental Contamination and Toxicology* 90: 552–557.

Abdul-Wahab, S.A. and Jupp, B.P. (2009). Levels of heavy metals in subtidal sediments in the vicinity of thermal power/desalination plants: a case study. *Desalination* 244 (1–3): 261–282.

Agah, H., Leermakers, M., Gao, Y. et al. (2010). Mercury accumulation in fish species from the Persian Gulf and in human hair from fishermen. *Environmental Monitoring and Assessment* 169: 203–216.

Agunbiade, F.O., Olu-Owolabi, B.I., and Adebowale, K.O. (2009). Phytoremediation potential of *Eichornia crassipes* in metal-contaminated coastal water. *Bioresource Technology* 100 (19): 4521–4526.

Ali, H., Khan, E., and Ilahi, I. (2019). Environmental chemistry and ecotoxicology of hazardous heavy metals: environmental persistence, toxicity, and bioaccumulation. *Journal of Chemistry* 2019: 6730305.

Aliko, V., Multisanti, C.R., Turani, B., and Faggio, C. (2022). Get rid of marine pollution: bioremediation an innovative, attractive, and successful cleaning strategy. *Sustainability* 14 (18): 11784.

Al-Majed, N.B. and Preston, M.R. (2000). An assessment of the total and methyl mercury content of zooplankton and fish tissue collected from Kuwait territorial waters. *Marine Pollution Bulletin* 40 (4): 298–307.

Al-Mohanna, S.Y. and Subrahmanyam, M.N.V. (2001). Flux of heavy metal accumulation in various organs of the intertidal marine blue crab, *Portunus pelagicus* (L.) from the Kuwait coast after the Gulf War. *Environment International* 27 (4): 321–326.

Al-Sayed, H.A., Mahasneh, A.M., and Al-Saad, J. (1994). Variations of trace metal concentrations in seawater and pearl oyster *Pinctada radiata* from Bahrain (Arabian Gulf). *Marine Pollution Bulletin* 28 (6): 370–374.

Al-Sayed, H.A., Al-Saad, J., Madany, I.M., and Al-Hooti, D. (1996). Heavy metals in the grouper fish *Epinephelus coioides* from the coast of Bahrain: an assessment of monthly and spatial trends. *International Journal of Environmental Studies* 50 (3–4): 237–246.

Ankit, Bauddh, K., and Korstad, J. (2022). Phycoremediation: use of algae to sequester heavy metals. *Hydrobiology* 1 (3): 288–303.

Apeti, D.A. and Hartwell, I.S. (2015). Baseline assessment of heavy metal concentrations in surficial sediment from Kachemak Bay, Alaska. *Environmental Monitoring and Assessment* 187: 1–11.

Ayangbenro, A.S. and Babalola, O.O. (2017). A new strategy for heavy metal polluted environments: a review of microbial biosorbents. *International Journal of Environmental Research and Public Health* 14 (1): 94.

Bahnasawy, M., Khidr, A.A., and Dheina, N. (2011). Assessment of heavy metal concentrations in water, plankton, and fish of Lake Manzala, Egypt. *Turkish Journal of Zoology* 35 (2): 271–280.

Batapola, N.M., Dushyantha, N.P., Ratnayake, N.P. et al. (2023). Risk assessment of heavy metals in the freshwater lake sediments around Eppawala phosphate deposit, Sri Lanka. *Journal of the National Science Foundation of Sri Lanka* 51 (4): 573–587.

Bhuyan, M.S., Haider, S.M.B., Meraj, G. et al. (2023). Assessment of heavy metal contamination in beach sediments of Eastern St. Martin's Island, Bangladesh: implications for environmental and human health risks. *Water* 15 (13): 2494.

Bibak, M., Sattari, M., Tahmasebi, S. et al. (2020). Marine macro-algae as a bio-indicator of heavy metal pollution in the marine environments. *Persian Gulf Indian Journal of Geo Marine Sciences* 49 (3): 357–363.

Bonsignore, M., Manta, D.S., Mirto, S. et al. (2018). Bioaccumulation of heavy metals in fish, crustaceans, molluscs and echinoderms from the Tuscany coast. *Ecotoxicology and Environmental Safety* 162: 554–562.

Briffa, J., Sinagra, E., and Blundell, R. (2020). Heavy metal pollution in the environment and their toxicological effects on humans. *Heliyon* 6 (9): e04691.

Bu-Olayan, A.H. and Subrahmanyam, M.N.V. (1997). Accumulation of copper, nickel, lead and zinc by snail, *Lunella coronatus* and pearl oyster, *Pinctada radiata* from the Kuwait coast before and after the gulf war oil spill. *Science of the Total Environment* 197 (1–3): 161–165.

Burada, A., Ţopa, C.M., Georgescu, L.P. et al. (2014). Heavy metals accumulation in plankton and water of four aquatic complexes from Danube Delta area. *Aquaculture, Aquarium, Conservation & Legislation* 7 (4): 301–310.

Carrillo Domínguez, S., Casas Valdez, M., Ramos Ramos, F. et al. (2002). Algas marinas de baja california Sur, México: Valor nutrimental. *Archivos Latinoamericanos de Nutrición* 52 (4): 400–405.

Chalkiadaki, O., Dassenakis, M., and Lydakis-Simantiris, N. (2014). Bioconcentration of Cd and Ni in various tissues of two marine bivalves living in different habitats and exposed to heavily polluted seawater. *Chemistry and Ecology* 30 (8): 726–742.

Chowdhury, P.A., Ali, M.M. and Shahjahan, A.T.M. (2015). Impacts of ship breaking industries on environment and socio-economic condition of Bangladesh-A Case Study of Sitakunda Shitolpur Ship Breaking Yard, Chittagong. Doctoral dissertation, Chittagong University of Engineering & Technology.

Danovaro, R., Cocozza di Montanara, A., Corinaldesi, C. et al. (2023). Bioaccumulation and biomagnification of heavy metals in marine micro-predators. *Communications Biology* 6 (1): 1206.

Darmoul, B. (1988). Pollution dans le golfe de Gabès (Tunisie): bilan de six années de surveillance (1976–1981). *Bulletin Marine and Freshwater Science* 15: 61–84.

Dash, H.R., Mangwani, N., Chakraborty, J. et al. (2013). Marine bacteria: potential candidates for enhanced bioremediation. *Applied Microbiology and Biotechnology* 97 (2): 561–571.

De Mora, S., Fowler, S.W., Wyse, E., and Azemard, S. (2004). Distribution of heavy metals in marine bivalves, fish and coastal sediments in the Gulf and Gulf of Oman. *Marine Pollution Bulletin* 49 (5–6): 410–424.

Depountis, N., Koukis, G., and Sabatakakis, N. (2009). Environmental problems associated with the development and operation of a lined and unlined landfill site: a case study demonstrating two landfill sites in Patra, Greece. *Environmental Geology* 56: 1251–1258.

Díez, S., Jover, E., Bayona, J.M., and Albaigés, J. (2007). Prestige oil spill. III. Fate of a heavy oil in the marine environment. *Environmental Science & Technology* 41 (9): 3075–3082.

El Zrelli, R., Courjault-Radé, P., Rabaoui, L. et al. (2015). Heavy metal contamination and ecological risk assessment in the surface sediments of the coastal area surrounding the industrial complex of Gabes city, Gulf of Gabes, SE Tunisia. *Marine Pollution Bulletin* 101 (2): 922–929.

Evans, N.C. (1994). *Effects of Dredging and Dumping on the Marine Environment of Hong Kong*. Terra et Aqua.

Fang, T., Yang, K., Lu, W. et al. (2019). An overview of heavy metal pollution in Chaohu Lake, China: enrichment, distribution, speciation, and associated risk under natural and anthropogenic changes. *Environmental Science and Pollution Research* 26: 29585–29596.

Ferrans, L., Jani, Y., Burlakovs, J. et al. (2021). Chemical speciation of metals from marine sediments: assessment of potential pollution risk while dredging, a case study in southern Sweden. *Chemosphere* 263: 128105.

Fu, Z. and Xi, S. (2020). The effects of heavy metals on human metabolism. *Toxicology Mechanisms and Methods* 30 (3): 167–176.

Fujiki, M. and Tajima, S. (1992). The pollution of Minamata Bay by mercury. *Water Science and Technology* 25 (11): 133–140.

Genchi, G., Sinicropi, M.S., Lauria, G. et al. (2020). The effects of cadmium toxicity. *International Journal of Environmental Research and Public Health* 17 (11): 3782.

Glasson, J. and Therivel, R. (2013). *Introduction to Environmental Impact Assessment*. Routledge.

Gokcekus, H., Kabdasli, S., Kabdasli, I. et al. (2003). Pollution of coastal region impacted by acid mine drainage in Morphou Bay, Northern Cyprus. *Journal of Environmental Science and Health, Part A* 38 (8): 1445–1457.

Griest, W.H., Ho, C.H., Guérin, M.R., and Tyndall, R.L. (1991). Chemical comparison of weathered spilled oil and Exxon/Valdez hold oil from an occupational health standpoint (No. CONF-910659-6). Oak Ridge National Lab., TN (USA).

Guo, J., Liu, X., and Xie, Q. (2013). Characteristics of the Bohai Sea oil spill and its impact on the Bohai Sea ecosystem. *Chinese Science Bulletin* 58: 2276–2281.

Hakanson, L. (1980). An ecological risk index for aquatic pollution control. A sedimentological approach. *Water Research* 14 (8): 975–1001.

Harada, M. (1995). Minamata disease: methylmercury poisoning in Japan caused by environmental pollution. *Critical Reviews in Toxicology* 25 (1): 1–24.

Hosono, T., Su, C.C., Siringan, F. et al. (2010). Effects of environmental regulations on heavy metal pollution decline in core sediments from Manila Bay. *Marine Pollution Bulletin* 60 (5): 780–785.

Hu, J., Zheng, A., Pei, D., and Shi, G. (2010). Bioaccumulation and chemical forms of cadmium, copper and lead in aquatic plants. *Brazilian Archives of Biology and Technology* 53: 235–240.

Hussain, T. and Gondal, M.A. (2008). Monitoring and assessment of toxic metals in Gulf War oil spill contaminated soil using laser-induced breakdown spectroscopy. *Environmental Monitoring and Assessment* 136: 391–399.

Jonathan, M.P., Shumilin, E., Rodríguez-Figueroa, G.M. et al. (2016). Potential toxicity of chemical elements in beach sediments near Santa Rosalia copper mine, Baja California Peninsula, Mexico. *Estuarine, Coastal and Shelf Science* 180: 91–96.5.

Kadokami, K., Li, X., Pan, S. et al. (2013). Screening analysis of hundreds of sediment pollutants and evaluation of their effects on benthic organisms in Dokai Bay, Japan. *Chemosphere* 90 (2): 721–728.

Karafistan, A. and Gemikonakli, E. (2019). Evaluation of contaminants in fish affected by mining in Morphou Bay, Cyprus, using statistical and artificial neural network methods. *Mine Water and the Environment* 38: 178–186.

Khalil, S., Mahnashi, M.H., Hussain, M. et al. (2021). Exploration and determination of algal role as bioindicator to evaluate water quality–Probing fresh water algae. *Saudi Journal of Biological Sciences* 28 (10): 5728–5737.

Kumar, S., Prasad, S., Yadav, K.K. et al. (2019). Hazardous heavy metals contamination of vegetables and food chain: role of sustainable remediation approaches – a review. *Environmental Research* 179: 108792.

Kumar, V., Sharma, A., Pandita, S. et al. (2020). A review of ecological risk assessment and associated health risks with heavy metals in sediment from India. *International Journal of Sediment Research* 35 (5): 516–526.

Lahbib, Y., Mleiki, A., Marigomez, I., and El Menif, N.T. (2013). Copper, zinc, and cadmium body concentrations in *Hexaplex trunculus* collected from the Tunisian coast. *Environmental Monitoring and Assessment* 185: 8967–8975.

Lakshmi, E., Priya, M., and Achari, V.S. (2021). An overview on the treatment of ballast water in ships. *Ocean & Coastal Management* 199: 105296.

Li, C., Wang, H., Liao, X. et al. (2022). Heavy metal pollution in coastal wetlands: a systematic review of studies globally over the past three decades. *Journal of Hazardous Materials* 424: 127312.

Liu, C., Shao, S., Shen, Q. et al. (2015). Use of multi-objective dredging for remediation of contaminated sediments: a case study of a typical heavily polluted confluence area in China. *Environmental Science and Pollution Research* 22: 17839–17849.

Lone, M.I., He, Z.L., Stoffella, P.J., and Yang, X.E. (2008). Phytoremediation of heavy metal polluted soils and water: progresses and perspectives. *Journal of Zhejiang University Science B* 9: 210–220.

Madany, I.M., Wahab, A.A.A., and Al-Alawi, Z. (1996). Trace metals concentrations in marine organisms from the coastal areas of Bahrain, Arabian Gulf. *Water, Air, and Soil Pollution* 91: 233–248.

Manahan, S.E. (2002). *Toxicological Chemistry and Biochemistry*. CRC Press.

Mashiatullah, A., Chaudhary, M.Z., Ahmad, N. et al. (2013). Metal pollution and ecological risk assessment in marine sediments of Karachi Coast, Pakistan. *Environmental Monitoring and Assessment* 185: 1555–1565.

McManus, P. (2009). Environmental regulation. In: *International Encyclopedia of Human Geography* (ed. R. Kitchin and N. Thrift), 546–552. Elsevier, Oxford https://www.sciencedirect.com/topics/earth-and-planetary-sciences/environmental-regulation; https://doi.org/10.1016/B978-008044910-4.00154-1.

Meena, R.A.A., Sathishkumar, P., Ameen, F. et al. (2018). Heavy metal pollution in immobile and mobile components of lentic ecosystems – a review. *Environmental Science and Pollution Research* 25: 4134–4148.

Mele, B.H., Russo, L., Crocetta, F. et al. (2020). Ecological assessment of anthropogenic impact in marine ecosystems: the case of Bagnoli Bay. *Marine Environmental Research* 158: 104953.

Meraj, G., Singh, S.K., Kanga, S., and Islam, M.N. (2022). Modeling on comparison of ecosystem services concepts, tools, methods and their ecological-economic implications: a review. *Modeling Earth Systems and Environment* 8 (1): 15–34.

Mishra, S., Bharagava, R.N., More, N. et al. (2019). Heavy metal contamination: an alarming threat to environment and human health. In: *Environmental Biotechnology: For Sustainable Future* (ed. R.C. Sobti, N.K. Arora, and R. Kothari), 103–125. Springer.

Moiseenko, T.I. and Gashkina, N.A. (2020). Distribution and bioaccumulation of heavy metals (Hg, Cd and Pb) in fish: influence of the aquatic environment and climate. *Environmental Research Letters* 15 (11): 115013.

Moreno, R., Jover, L., Diez, C., and Sanpera, C. (2011). Seabird feathers as monitors of the levels and persistence of heavy metal pollution after the Prestige oil spill. *Environmental Pollution* 159 (10): 2454–2460.

Muliyadi, M., Purwanto, P., Sumiyati, S., and Mussadun, M. (2023). Bioaccumulation of heavy metals using aquatic plants in wastewater. *International Journal of Public Health* 12 (2): 690–698.

Muller, G.M.M.G.M.G.M.G.P. (1969). Index of geoaccumulation in sediments of the Rhine River. *Geojournal* 2: 108–118.

Mustafa, A.D., Juahir, H., Yunus, K. et al. (2015). Oil spill related heavy metal: a review. *Malaysian Journal of Analytical Sciences* 19 (6): 1348–1360.

Mymrin, V., Stella, J.C., Scremim, C.B. et al. (2017). Utilization of sediments dredged from marine ports as a principal component of composite material. *Journal of Cleaner Production* 142: 4041–4049.

Nakata, H., Shimada, H., Yoshimoto, M. et al. (2008). Concentrations and distribution of mercury and other heavy metals in surface sediments of the Yatsushiro Sea including Minamata Bay, Japan. *Bulletin of Environmental Contamination and Toxicology* 80: 78–84.

Nanayakkara, C.J., Dushyantha, N.P., Ratnayake, N.P. et al. (2023). Risk evaluation and remedial measures for heavy metal contamination in lagoonal sediments of the Negombo Lagoon, Sri Lanka after the X-Press Pearl maritime disaster. *Regional Studies in Marine Science* 67: 103200.

Naser, H.A. (2013). Assessment and management of heavy metal pollution in the marine environment of the Arabian Gulf: a review. *Marine Pollution Bulletin* 72 (1): 6–13.

Ohiagu, F.O., Chikezie, P.C., Ahaneku, C.C., and Chikezie, C.M. (2022). Human exposure to heavy metals: toxicity mechanisms and health implications. *Material Science & Engineering International Journal* 6 (2): 78–87.

Olayinka-Olagunju, J.O., Dosumu, A.A., and Olatunji-Ojo, A.M. (2021). Bioaccumulation of heavy metals in pelagic and benthic fishes of Ogbese River, Ondo State, South-Western Nigeria. *Water, Air, & Soil Pollution* 232: 1–19.

Oppong, S.O.B., Voegborlo, R.B., Agorku, S.E., and Adimado, A.A. (2010). Total mercury in fish, sediments and soil from the River Pra Basin, Southwestern Ghana. *Bulletin of Environmental Contamination and Toxicology* 85: 324–329.

Pahalawattaarachchi, V., Purushothaman, C.S., and Vennila, A. (2009). Metal phytoremediation potential of *Rhizophora mucronata* (Lam.). *Indian Journal of Geo-Marine Sciences* 38 (2): 178–183.

Pérez-López, M., Cid, F., Oropesa, A.L. et al. (2006). Heavy metal and arsenic content in seabirds affected by the Prestige oil spill on the Galician coast (NW Spain). *Science of the Total Environment* 359 (1–3): 209–220.

Permana, R. and Akbarsyah, N. (2021). Phytoplankton susceptibility towards toxic heavy metal cadmium: mechanism and its recent updates. *World News of Natural Sciences* 38: 83–97.

Plum, L.M., Rink, L., and Haase, H. (2010). The essential toxin: impact of zinc on human health. *International Journal of Environmental Research and Public Health* 7 (4): 1342–1365.

Pramanik, A., Basak, P., Sengupta, S., and Bhattacharyya, M. (2015). Marine microbial diversity: potential functions to degrade xenobiotics. In: *Biodegradation and Bioremediation*, Environmental Science and Engineering, vol. 8 (ed. A. Singh and O.P. Ward), 305–318. Houston, TX: Studium Press LLC.

Ra, K., Kim, J.K., Hong, S.H. et al. (2014). Assessment of pollution and ecological risk of heavy metals in the surface sediments of Ulsan Bay, Korea. *Ocean Science Journal* 49: 279–289.

Rajeshkumar, S. and Li, X. (2018). Bioaccumulation of heavy metals in fish species from the Meiliang Bay, Taihu Lake, China. *Toxicology Reports* 5: 288–295.

Rauf, A., Javed, M., and Jabeen, G. (2019). Uptake and accumulation of heavy metals in water and planktonic biomass of the river Ravi, Pakistan. *Turkish Journal of Fisheries and Aquatic Sciences* 19 (10): 857–864.

Riyazuddin, R., Nisha, N., Ejaz, B. et al. (2021). A comprehensive review on the heavy metal toxicity and sequestration in plants. *Biomolecules* 12 (1): 43.

Rizk, R., Juzsakova, T., Ali, M.B. et al. (2022). Comprehensive environmental assessment of heavy metal contamination of surface water, sediments and Nile Tilapia in Lake Nasser, Egypt. *Journal of King Saud University-Science* 34 (1): 101748.

Rodríguez-Figueroa, G.M., Shumilin, E., and Sánchez-Rodríguez, I. (2009). Heavy metal pollution monitoring using the brown seaweed *Padina durvillaei* in the coastal zone of the Santa Rosalía mining region, Baja California Peninsula, Mexico. *Journal of Applied Phycology* 21: 19–26.

Sadiq, M. (2002). Metal contamination in sediments from a desalination plant effluent outfall area. *Science of the Total Environment* 287 (1–2): 37–44.

Saidon, N.B., Szabó, R., Budai, P., and Lehel, J. (2024). Trophic transfer and biomagnification potential of environmental contaminants (heavy metals) in aquatic ecosystems. *Environmental Pollution* 340: 122815.

Sany, S.B.T., Salleh, A., Sulaiman, A.H. et al. (2013). Heavy metal contamination in water and sediment of the Port Klang coastal area, Selangor, Malaysia. *Environmental Earth Sciences* 69 (6): 2013–2025.

Sarker, A. (2021). Ecological perspectives on water, food, and health security linkages: the Minamata case in Japan. *Environmental Science and Pollution Research* 28 (25): 32177–32189.

Shah, S.B. (2021). *Heavy Metals in Scleractinian Corals*. Springer International Publishing.

Sharifuzzaman, S.M., Rahman, H., Ashekuzzaman, S.M. et al. (2016). Heavy metals accumulation in coastal sediments. In: *Environmental Remediation Technologies for Metal-Contaminated Soils* (ed. H. Hasegawa, I.M.M. Rahman, and M.A. Rahman), 21–42. Springer.

Singh, N.K., Raghubanshi, A.S., Upadhyay, A.K., and Rai, U.N. (2016). Arsenic and other heavy metal accumulation in plants and algae growing naturally in contaminated area of West Bengal, India. *Ecotoxicology and Environmental Safety* 130: 224–233.

Stevenson, R.J. and Smol, J.P. (2015). Use of algae in ecological assessments. In: *Freshwater Algae of North America*, 2e (ed. J.D. Wehr, R.G. Sheath, and J.P. Kociolek), 921–962. Academic Press.

Tanner, P.A., Leong, L.S., and Pan, S.M. (2000). Contamination of heavy metals in marine sediment cores from Victoria Harbour, Hong Kong. *Marine Pollution Bulletin* 40 (9): 769–779.

Ting, Y. and Hsi, H.C. (2019). Iron sulfide minerals as potential active capping materials for mercury-contaminated sediment remediation: a mini review. *Sustainability* 11 (6): 1747.

Tomlinson, D.L., Wilson, J.G., Harris, C.R., and Jeffrey, D.W. (1980). Problems in the assessment of heavy-metal levels in estuaries and the formation of a pollution index. *Helgoländer meeresuntersuchungen* 33: 566–575.

Trivedi, S. and Ansari, A.A. (2015). Molecular mechanisms in the phytoremediation of heavy metals from coastal waters. In: *Phytoremediation: Management of Environmental Contaminants*, vol. 2 (ed. L. Newman, A.A. Ansari, S.S. Gill, et al.), 219–231. Springer.

Ungureanu, E.L. and Mustatea, G. (2022). Toxicity of heavy metals. In: *Environmental Impact and Remediation of Heavy Metals* (ed. H. Saleh and A.I. Hassan). IntechOpen. doi: 10.5772/intechopen.102441.

USEPA. (2001). The Role of Screening-Level Risk Assessments and Refining Contaminants of Concern in Baseline Ecological Risk Assessments (Intermittent Bulletin EPA 540/F-01/014; ECO Update). United States Environmental Protection Agency. https://www.epa.gov/sites/default/files/2015-09/documents/slera0601.pdf (accessed 12 December 2023).

Vodyanitskii, Y.N. (2013). Contamination of soils with heavy metals and metalloids and its ecological hazard (analytic review). *Eurasian Soil Science* 46: 793–801.

Wang, S.L., Xu, X.R., Sun, Y.X. et al. (2013). Heavy metal pollution in coastal areas of South China: a review. *Marine Pollution Bulletin* 76 (1–2): 7–15.

Wise, J.P. Jr., Wise, J.T., Wise, C.F. et al. (2014). Concentrations of the genotoxic metals, chromium and nickel, in whales, tar balls, oil slicks, and released oil from the gulf of Mexico in the immediate aftermath of the deepwater horizon oil crisis: is genotoxic metal exposure part of the deepwater horizon legacy? *Environmental Science & Technology* 48 (5): 2997–3006.

Wong, C., Roberts, S.M., and Saab, I.N. (2022). Review of regulatory reference values and background levels for heavy metals in the human diet. *Regulatory Toxicology and Pharmacology* 130: 105122.

Younis, A.M., Amin, H.F., Alkaladi, A., and Mosleh, Y.Y. (2015). Bioaccumulation of heavy metals in fish, squids and crustaceans from the Red Sea, Jeddah Coast, Saudi Arabia. *Open Journal of Marine Science* 5 (04): 369.

Zaynab, M., Al-Yahyai, R., Ameen, A. et al. (2022). Health and environmental effects of heavy metals. *Journal of King Saud University-Science* 34 (1): 101653.

Zhang, K.B. (2008). Characteristics of the elemental abundance and content of near-shore neritic sediment in Fujian Province. *Geol Fujian* 4: 402–408.

Zhang, Y.F., Wang, L.J., Huo, C.L., and Guan, D.M. (2008). Assessment on heavy metals pollution in surface sediments in Jinzhou Bay. *Marine Environmental Science* 2: 178–181.

Zhang, M., Sun, X., and Xu, J. (2020a). Heavy metal pollution in the East China Sea: a review. *Marine Pollution Bulletin* 159: 111473.

Zhang, S., Guo, H., Zhang, S. et al. (2020b). Are oil spills an important source of heavy metal contamination in the Bohai Sea, China? *Environmental Science and Pollution Research* 27: 3449–3461.

Zhou, F., Guo, H., and Hao, Z. (2007). Spatial distribution of heavy metals in Hong Kong's marine sediments and their human impacts: a GIS-based chemometric approach. *Marine Pollution Bulletin.* 54 (9): 1372–1384.

8

Marine Pollution due to Mariculture and Fishing Operations in Sri Lanka: Impacts and Mitigation Strategies

Menuka Udugama[1], Lahiru Udayanga[2], Suchima Gonapinuwala[3], and Dileepa de Croos[3]

[1] *Department of Agribusiness Management, Faculty of Agriculture and Plantation Management, Wayamba University of Sri Lanka, Gonawila (NWP), Sri Lanka*
[2] *Department of Biosystems Technology, Faculty of Agriculture and Plantation Management, Wayamba University of Sri Lanka, Gonawila (NWP), Sri Lanka*
[3] *Department of Aquaculture and Fisheries, Faculty of Livestock, Fisheries and Nutrition, Wayamba University of Sri Lanka, Gonawila (NWP), Sri Lanka*

8.1 Mariculture in Sri Lanka

With the advancement of science and technology, health services, and quality of life, the global human population is increasing rapidly. According to the estimates, it is expected to reach up to nine billion by 2050. Therefore, ensuring food security and sustainability have become prime concerns. Fisheries and aquaculture play pivotal roles in ensuring food security by providing a safe source of animal protein, while contributing largely to poverty alleviation due to the provision of livelihood opportunities. The global fish production from capture fisheries has remained around 90 million tons per year in the 1990s, which has rapidly increased up to 130 million tons per year over the last two to three decades (Jensen et al. 2007). According to FAO, the global fish production has reached up to 179 million tons in 2018, with an estimated total value of USD 401 billion, mainly due to the increasing world population and increased per capita consumption (Lichna et al. 2023). Therefore, natural fish stocks are being overexploited, leading the global fish production from capture fisheries to exceed the sustainable capacity of marine ecosystems to provide them (Jensen et al. 2007). Therefore, ensuring a sustainable increase in aquaculture production to meet the increasing global demand has become a major concern.

In general, fisheries refer to any activity including the cultivation, harvesting, processing, and management of fish and other aquatic resources within the marine, coastal, and inland waters. Basically, fisheries are composed of marine fisheries, coastal fisheries, inland fisheries, aquaculture, and fish processing. Aquaculture plays a key role in fisheries operations. A form of aquaculture, which includes the managed cultivation of aquatic species for a variety of purposes, including food, pharmaceuticals, cosmetics, culture media, and jewelry in marine waters, has been defined as mariculture (Lounas et al. 2020). In a broader sense, at least one phase of growth of the cultured organism is maintained under the control of humans in mariculture, in their natural environment, ponds, or raceways. The history of mariculture dates back to more than four millennia in time, when grey mullets were cultivated in lagoons in ancient Rome. Further, the Chinese have maintained oyster cultures even around 206 BCE to 220 CE. Countries like Japan, China, France, Italy, and

Coastal and Marine Pollution: Source to Sink, Mitigation and Management, First Edition.
Edited by Meththika Vithanage, Sameera M. Samarasekara, Bryan D. James, and Christopher M. Reddy.

the United Kingdom have a rich history in mariculture activities (Lovatelli et al. 2013). Within the last three decades, mariculture has shown an exponential growth, resulting a significant expansion at the global level. At present, South and Southeast Asian countries, the South American continent, North America, the European Union, and Mediterranean countries are actively engaged in mariculture (Lounas et al. 2020). Modern mariculture maneuvers range from small-scale ponds and hatcheries managed by families to large-scale commercial operations.

Sri Lanka is blessed with an extended coastline of 1770 km, along with exclusive commercial fishing rights over 517000 km^2 of ocean and 21500 km^2 of territorial waters (Rifath and Jemziya 2023). The diverse array of ecosystems present in the Sri Lankan coastal belt, such as lagoons, bays, and salt marshes, has provided perfect conditions for mariculture practices (Bournazel et al. 2015). In the Sri Lankan context, the agriculture, forestry, and fishing sectors have denoted 5.8% (686412 million LKR) and 6% (669802 million LKR) contribution to the gross domestic production (GDP) in 2019 and 2020, respectively (Central Bank of Sri Lanka (CBSL 2021)). In 2021, the agriculture and fisheries sectors have catered for 27% (2213015 people) of the labor force in Sri Lanka (Department of Census and Statistics (DCS 2022)). Based on the relative contribution to the GDP, paddy (11.7%), marine fishing and marine aquaculture (11.3%), spices and other pharmaceutical crops (10.2%), and fruits (9.6%) could be recognized as the leading subsectors contributing to the overall income earned by this sector. As a whole, the fisheries sector has reported a 1.3 to 1.4% contribution to the GDP of Sri Lanka in 2021 (Central Bank of Sri Lanka (CBSL 2021)).

Since the capture fisheries have reached their maximum sustainable limits, many countries including Sri Lanka are attempting to promote aquaculture to cater to the increasing global demand for fish-based products. Therefore, a variety of mariculture practices have been introduced and practiced globally to promote mariculture, while ensuring sustainability. Aquaculture was initially introduced in the mid-1980s in Sri Lanka as Black tiger shrimp (*Penaeus monodon*) farming along the North-Western coast from Chilaw to Puttalam lagoon. With the occurrence of white spot disease in *P. monodon* culture, Pacific white shrimp (*Litopenaeus vannamei*) was introduced to recover the shrimp farming sector (Jayasinghe and Amarasinghe 2021). The coastal aquaculture production in Sri Lanka has been estimated to increase from 20000 tons (USD 100 million) in 2020 to 60000 tons (USD 450 million) by 2025, according to the proposed aquaculture and culture-based fisheries sector development plan (2021–2025) of the National Aquaculture Development Authority of Sri Lanka (NAQDA). Therefore, strategic actions are being taken to promote mariculture as the prime seafood source by 2030 in Sri Lanka (Jayasinghe and Amarasinghe 2021).

Currently, shrimp culture remains as the major mariculture activity in Sri Lanka, accounting for 40 to 50% of total aquaculture exports in Sri Lanka (Senarath and Visvanathan 2001). Other mariculture practices in Sri Lanka include sea bass culture, sea cucumber breeding and farming, seaweed farming, milkfish breeding and farming, mud crab breeding and farming, and oyster farming (Jayasinghe and Amarasinghe 2021). Sea bass cage culture was started by NAQDA in 2018 under a Public–Private Partnership (PPP) approach in Negombo Lagoon, Mannar, Puttalam, Batticaloa, Galle, and Trincomalee. Sea cucumber farming in pens and ponds was also started by NAQDA in 2018 under PPP in Mannar, Kilinochchi, and Jaffna. Seaweed farming was facilitated by NAQDA in 2018 with community participation in the northern sea. Farming of *Kappaphycus alvarezii* is done using long lines as off-bottom, floating, and raft cultures. Milkfish culture in Sri Lanka dates back to the 1970s, which was initiated using wild-collected fry in the Puttalam district. Milkfish are used as bait for tuna. A hatchery was started by the private sector to produce mud crab (*Scylla serrata*) crablets in Chilaw. Oyster farming (*Crassostrea madrasensis*) was started in 2000 on a very small scale at Bolgoda Lake, south of Colombo (NAQDA 2021). The proposed aquaculture and culture-based fisheries sector development plan (2021–2025) of the NAQDA includes establishment of aquaculture industrial parks and crab cities in Hambantota, Batticaloa, and Mannar districts with expected production of 9075 tons. With the introduction of these mariculture practices, now the coastal aquaculture in Sri Lanka has extended to the northern and eastern provinces which was previously concentrated mainly in the north-western province (Jayasinghe and Amarasinghe 2021). With this expansion, targeted coastal mariculture production only from the northern province is 24 560 tons, including 20 450 tons of shrimps, 58 tons of sea bass, 135 tons of mud crabs, 2 500 tons of sea cucumber, 1 340 tons of seaweed, and 80 tons of milkfish (Jayasinghe and Amarasinghe 2021; Wu 1995).

Despite the numerical advantages provided by mariculture, at an industrial level, it is posing significant threats to marine and coastal biology due to certain operational practices. Degradation and destruction of natural habitats, overexploitation of natural stocks, addition of pollutants (organic and inorganic), addition of nutrients and antibiotics, intentional or unintentional releases of alien species, and transmission of diseases are few major impacts caused by mariculture and fishing operations. In Sri Lanka, the contribution of sea-based sources for marine pollution has been estimated as 10%, whereas it is 90% from land-based sources. Further, Sri Lanka has been ranked as the fifth worst in marine trash management, warranting the necessity of implementing strategies to manage marine pollution. Understanding the sources of pollutants from fisheries and aquaculture and their impacts on the marine environments is crucial for implementing effective management strategies to mitigate their impact on marine environments. This chapter provides a comprehensive overview on the causes of marine pollution in Sri Lanka caused by mariculture and fishing operations, along with their impacts and potential strategies for mitigation.

8.2 Marine Pollution by Aquaculture and Fisheries Practices and Impacts

Marine pollution is defined as "any detrimental alteration of the marine environment caused by the intentional or accidental release of dangerous or toxic substances, such as industrial, commercial and urban waste water, or energy causing harm to living resources, hazards to human health, hindrances to marine activities including fishing, impairment of the quality of sea water and reduction of amenities" by the United Nations Environment Programme (2021) and Chopin et al. (2001). Sri Lanka is at the 173rd place in the ocean health index (Chopin et al. 2001) and has been ranked fifth worst in marine trash management. In Sri Lanka, the contribution of sea-based sources for marine pollution is 10%, whereas it is 90% from land-based sources (Manage et al. 2022). Marine waters provide invaluable resources to human societies, supporting fisheries and aquaculture activities that contribute significantly to global food security and economic prosperity. However, these activities can also pose threats to marine ecosystems through the introduction of pollutants. Therefore, numerous studies are being conducted to characterize the major sources of pollutants from fisheries and aquaculture, while understanding their impacts on the marine environments, which is of essence for implementing effective management strategies to mitigate their impacts on marine environments.

8.3 Nutrient Pollution and Impacts from Fisheries and Mariculture Practices

More than 20 million tons of fish, which is similar to 25% of the world's fish catch is discarded annually (Rifath and Jemziya 2023; Central Bank of Sri Lanka (CBSL) 2021) mainly as by-catch or processing discards. In Sri Lanka, by-catch is either dumped into the sea or brought to the land to sell at cheaper prices (Rifath and Jemziya 2023). By-products from on-board processing are disposed to the sea, and those from processing factories are used for value-added product processing, dumped on land, or disposed to the sea. Around 50% of fish is discarded as by-products during processing, which include heads, skin, bones, fins, and viscera (Rifath and Jemziya 2023). Wastewater from cleaning the equipment, utensils, and surfaces onboard is also added into the sea. In addition to by-catch and processing by-products, a considerable amount of discards occur due to poor onboard fish handling practices, poor storage facilities, improper fishing gear, and lag time between catching and marketing (Rifath and Jemziya 2023).

Accumulation of these organic wastes in the sea pollutes the marine environment. High organic waste loading increases the Biological Oxygen Demand (BOD) and depletes the dissolved oxygen content in water (Rifath and Jemziya 2023), which refers to the amount of oxygen required by bacteria and other micro-organisms to decompose organic matter under aerobic conditions at a specified temperature. Anaerobic decomposition of organic matter produces hydrogen sulfide, ammonia, and methane, which are toxic to marine organisms (Islam et al. 2004). Sea birds such as seagulls and scavenger fish species are often used to feed upon this waste leading to their population growth, which can affect the ecosystem balance (Rifath and Jemziya 2023; Votier et al. 2004). A recent study has reported that this waste is unhealthy for gannets, which impede the development of their young (Grémillet et al. 2008). High nutrient loading leads to eutrophication (Rifath and Jemziya 2023) causing increased phytoplankton growth and toxic algal blooms (Rifath and Jemziya 2023; Bonsdorff et al. 1997). This affects the fish and benthic faunal diversity also (Rifath and Jemziya 2023; Bonsdorff et al. 1997).

Another significant contributor of organic pollutants related to fisheries is fishery harbours (FAO 2020). Release of untreated sewage including fish processing by-products and food waste in harbour basins and adjacent waters causes higher BOD levels (FAO 2020). Harbours also lack waste-treatment facilities. A recent study by Weerasekara et al. (2015) has analyzed the BOD levels in Galle, Hambantota, and Hikkaduwa fishery harbours located in the southern Sri Lanka in 2014 and reported average BOD levels of 7.10 ± 0.43, 9.17 ± 0.25, and 9.67 ± 0.70 mg/l, respectively. These values are higher than the environmental quality standard set by the Central Environmental Authority (CEA) of Sri Lanka (<4 mg/l) and the recommended level of primary water quality criteria for class SW-IV harbour water (<3 mg/l). Since the capture fisheries have reached their maximum sustainable limits, aquaculture should be developed to meet the global demand. Many approaches have been taken to develop mariculture practices and they are now being developed in Sri Lanka.

In Sri Lanka, better management practices (BMPs) have been introduced to minimize the adverse impacts of shrimp farming on the environment and on the farms themselves (FAO 2020). Although the effluent treatment in a sedimentation tank is recommended before discharging to the surrounding water bodies under BMPs, they are not practiced by farmers at the optimum level. Nowadays, all shrimp farms are operated at semi-intensive level in Sri Lanka to control diseases. However, differences in effluent quality are observed due to different feed, water, and disease management practices. Feed and fertilizer are the main sources of nutrients and organic matter inputs in the shrimp culture ponds. Some farmers use trash fish, mollusk meat, or eggs as supplementary feed. Excess amounts of feed and fecal matter are released with shrimp farm effluents and end up adding large amounts of nutrients, such as nitrogen, phosphorus, and sulfur compounds, organic carbon and suspended particles into the receiving water bodies (Jayasinghe and Amarasinghe 2021; Senarath and Visvanathan 2001; Deepananda 2011).

Other mariculture practices in Sri Lanka include sea bass culture, sea cucumber breeding and farming, seaweed farming, milkfish breeding and farming, mud crab breeding and farming, and oyster farming (Jayasinghe and Amarasinghe 2021). Similar to shrimp farming, feed, and fertilizer are the major nutrient sources and organic matter inputs in these mariculture systems also. Generally, 1 to 38% of feed wastage occurs in fish farming depending on the species, culture method, hydrography of culture site, feed type, and feed management practices (Grémillet et al. 2008; Deepananda 2011). Generally, 80–88% of C, 52–95% of N, and 85% of P available in feed are lost to the marine environment through waste feed, fecal matter, and respiration (Deepananda 2011). In addition, excess feed and fish metabolic products accumulate in the culture system. However, wild fish and other species eat these waste feed and reduce the accumulation in surrounding waters by 80% (Deepananda 2011).

Decomposition of a high organic load demands a high amount of oxygen, reducing the dissolved oxygen content in water and creating anaerobic conditions. Oxygen depletion causes ecological stress on aquatic organisms (Senarath and Visvanathan 2001) and could act as a critical factor in the collapse of fish populations (Manage et al. 2022). Toxic metabolic end products, such as nitrites, sulfides, and ammonia, also accumulate in marine water (Jayasinghe and Amarasinghe 2021). Nutrient loading in waters due to increased P, organic C, inorganic N (ammonia and urea), and total C can increase phytoplankton biomass and primary production. The N:P ratio in waste feed and fish fecal matter is 7:1 (the Redfield ratio) (Arulananthan et al. 1995), which is ideal for phytoplankton growth. The authors of Manage et al. (2022) have found a positive correlation between the growth of phytoplankton and dinoflagellates with N and P. Nitrogen is important for the somatic development of algae cells, whereas P is important for algal growth and metabolic activities (Muñiz et al. 2018; Dissanayake et al. 2021). Increased nutrient loadings also affect the growth and diversity of zooplankton and bacterial communities (Ackefors and Enell 1990), macroalgae, epiphytes (Deepananda 2011; Ackefors and Enell 1990), and benthic fauna (Deepananda 2011).

High amounts of nitrogen and phosphorus cause eutrophication in the receiving water bodies when it exceeds the carrying capacity of water (Brooks et al. 2002; Jayasinghe et al. 2013; Ocean Health Index 2020). Eutrophication causes microscopic toxic algal blooms which are harmful to aquatic fauna, aquatic plants, birds, other wildlife, and humans (Manage et al. 2022). Toxicity depends on the salinity, temperature, and pH (Ariyawansa 2013). These algal blooms increase the water turbidity, obstruct light penetration, and deoxygenate the underlying water, thereby reducing the natural productivity of marine waters. This is a threat to the near-shore fisheries (Manage et al. 2022). Unionized ammonia also has toxic effects on marine biota and promotes algal blooms (Caruso 2016). Fish fecal matter is a rich source of ammonia (Islam et al. 2004).

The extent of these impacts varies with the carrying capacity of receiving water bodies (Senarath and Visvanathan 2001). Carrying capacity is based on the balance between the quantity of waste and the capacity of microorganisms in the receiving water bodies to assimilate them. Carrying capacity is affected by the changes in the physicochemical parameters of water such as pH, salinity, suspended solids, toxins, and chemicals. Although it is expected that the natural system dilutes these pollutants, the residence time of these pollutants in coastal water bodies of Sri Lanka is relatively high (Arulananthan et al. 1995; Corea et al. 1998; Jayasinghe et al. 2013). In Sri Lanka, shrimp farm effluents are released to the Dutch canal which is a semi-enclosed coastal water system having a low flushing rate. Water pollution due to nutrient enrichment is further exacerbated due to low flushing rate and other contributing factors, such as hydrological changes, salinization of soil and water, and siltation (Jayasinghe and Amarasinghe 2021; Senarath and Visvanathan 2001; Niroshana et al. 2013; Weerasekara et al. 2015). These conditions become severe during harvesting due to the release of entire pond content. Further, water stagnation in Dutch Canal during dry seasons impairs the carrying capacity (Senarath and Visvanathan 2001).

These impacts are more significant on sea bottom sediments than marine water (Wu 1995). Generally, for every ton of fish or shrimp produced in mariculture, more than 500 kg of organic matter is produced which contains nearly 200 kg of carbon (Jayasinghe and Amarasinghe 2021). Studies have shown that nearly 23% of carbon (C), 21% of nitrogen (N), and 53% of phosphorous (P) in feed inputs accumulate in sediments (Wu 1995). This causes increased sediment oxygen demand (SOD), anaerobic conditions, anoxic or hypoxic conditions, reducing conditions, and production of toxic gases such as methane, ammonia, and hydrogen sulfide, thereby adversely affecting benthic biota (Jegatheesan et al. 2006). Increased SOD causes anoxic or hypoxic conditions in bottom waters. However, these impacts are confined to localized areas, normally within 1–1.5 km of the farm, and not over large areas (Wu 1995). Low dispersal of wasted food and fecal materials and accumulation of organic matter and nutrients in the sediments may be the reasons for this localized impact (Wu 1995). These conditions adversely affect cultured organisms as well (Deepananda 2011; Croomey et al. 2000; Brooks et al. 2002).

In addition to the above-mentioned impacts, mariculture practices pose threats to marine and coastal biological diversity due to large-scale destruction and degradation of natural habitats, intentional or unintentional releases of alien or modified organisms, transmission of disease to wild stocks, and displacement of local and indigenous communities (Seymour and Bergheim 1991).

8.4 Oil Pollution and Impacts from Fisheries and Mariculture Practices

Fisheries activities significantly contribute to oil pollution in marine waters. According to a study by Manage et al. (2022), fishery harbors are the main source of oil and grease in west-south coastal waters in Sri Lanka. This study included 65 sampling sites from Negombo to Mirissa. According to the findings, 62% of oil and grease found in marine waters were from fishery harbors followed by 16% from industrial wastewater, 13% from canals, and 9% from sewage discharges (USEPA), and the authors of Senarath and Visvanathan (2001), Jayawardhane et al. (2018), Jayasinghe and Jayasekara (1999), and Wu (2001) analyzed oil and grease levels in Galle, Hambantota, and Hikkaduwa fishery harbors from southern Sri Lanka in 2014. Oil and grease levels in Galle, Hambantota, and Hikkaduwa fishery harbors varied between 11.6 and 24.8, 10 and 29, and 11.5 and 15.0 mg/l, respectively. All these values were higher than the recommended levels of primary water quality criteria for class SW-IV harbor water (10 mg/l) (Weerasekara et al. 2015). Identified reasons for the higher content of oil and grease in these harbors were engine oil emissions from fishing vessels and accidental oil spills during vessel refueling. In addition to these identified sources at these harbors, common sources may include leakages of transferring fuel from vessels, bilge water from vessels contaminated with oil, and waste oil and fuel residues from vessels (Jayawardhane et al. 2018; Muñiz et al. 2018; Dissanayake et al. 2021; Chopin et al. 2001).

Organic waste, oil, and some other chemical compounds increase the chemical oxygen demand (COD) in harbor basins. COD is the amount of oxygen required to chemically oxidize organic compounds in water (Weerasekara et al. 2015). Weerasekara et al. (2015) reported COD values of 926.5 ± 93.5, 840.2 ± 104.8, and 719.51 ± 61.63 mg/l from Galle, Hambantota, and Hikkaduwa fishery harbors, respectively. The authors of Manage et al. (2022) reported much higher COD levels in fishery harbors along the south-west coast as 607 ± 4.5, 4622.64 ± 4.9, 5769.96 ± 9.9, 4622.64 ± 6.7, and 4622.64 ± 6.0 mg/l from Dikovita, Beruwala, Ambalangoda, Galle, and Mirissa, respectively. However, according to Manage et al. (2022), the reasons for these higher COD levels were effluents from nearby industries. All these values were higher than the acceptable limit of 250 mg/l.

Coating of oil and grease in aquatic animals and plants (Arulananthan et al. 1995), clogging of fish gills (Corea et al. 1998), reduced oxygen level, and anoxic conditions due to poor oxygen exchange between atmosphere and surface water layer due to the presence of surface oil layer

(Weerasekara et al. 2015), water toxicity and producing toxic substances, and generating rancid odors are some of the severe adverse effects of oil and grease in marine waters (Manage et al. 2022). Some hydrocarbon compounds of oil and grease are toxic to aquatic organisms even at low concentrations ((California Stormwater Quality Association 2003) in EJF (2004)).

8.5 Chemical Pollution and Impacts from Fisheries and Mariculture Practices

Chemical pollution is a significant threat to the marine environment (Bournazel et al. 2015; United States Environmental Protection Agency 2021). Persistent organic pollutants (POPs) are a group of chemicals which pose a significant threat to the marine environment. Some examples of POPs are polycyclic aromatic hydrocarbons (PAHs), polybrominated diphenyl ethers (PBDEs), organochlorine pesticides (OCPs) (for instance dichlorodiphenyltrichloroethane (DDTs), hexachlorocyclohexanes (HCHs), and chlordane), and polychlorinated biphenyls (PCBs) (Qiu et al. 2020). POPs are hydrophobic, so they are rapidly absorbed by suspended particulate matter and phytoplankton. Some POPs are precipitated and accumulated in sediments (Cornelissen et al. 2005; Qiu et al. 2017). Marine sediments are considered the largest sink of many POPs (Dachs et al. 2002). POPs bioaccumulate through the food web and pose a threat to marine organisms and humans (Dassanayake 1994). POPs are toxic, and many of them are mutagenic and carcinogenic (Qiu et al. 2020). In a study by Qiu et al. (2020) on POPs in sediments of the Eastern Indian Ocean, the highest PAH level was recorded near Colombo, Sri Lanka. The major source of PAHs, for example, naphthalene, is crude oil from ships. According to the estimates, around 50% of the global shipping container traffic and 70% of petroleum products are crossing the Indian Ocean, which can add PAHs in the marine environment in high amounts (Primavera 2006). Plastics are another source of POPs in the marine environment. Fragmentation and degradation of plastics release POPs into the marine sediments. Especially, dechlorane plus (DP) and PBDEs are only blended in the plastic polymer structure without covalent bonding, so they can be easily leached out (Qiu et al. 2020).

Tributyltin (TBT) is used in antifouling paints for netting material and vessels, and leaches into marine water and sediments. The authors of Manage et al. (2022) reported higher TBT levels in waters from the Dikovita fishery harbor (241 ng/l) compared to Galle, Mirissa, Ambalangoda, and Beruwala fishery harbors which ranged from 51 to 124 ng/l. TBT contamination can be due to boat painting, repair, and construction activities which are point sources (Manage et al. 2022). TBT levels of all sampling locations exceeded the no observed effective level (NOEL) of less than 1 ng/l recommended by the WHO as the threshold (Manage et al. 2022). TBT is bioaccumulated and biomagnified through the food chain resulting in high levels in marine fauna. Consumption of seafood contaminated with TBT is a significant health risk to humans. On the other hand, TBT can cause severe reproductive dysfunction such as imposex, which leads to sterile marine populations and biodiversity loss (Manage et al. 2022). TBT contamination and imposex have been occurring in cultured fish which have been reared in net pens treated with TBT (Wu 1995). Bandara et al. (2021) found a relationship between reproductive dysfunction and the levels of TBT in the western and southern coastal waters of Sri Lanka.

Antibiotics are used in shrimp farming and in some other mariculture practices to treat some diseases. Excessive use of antibiotics encourages the development of antibiotic-resistant bacterial strains (Qiu et al. 2020; Cornelissen et al. 2005). Previous studies have reported oxytetracycline-resistant bacteria in marine sediment near fish farms even 13 months after medication (Wu 1995).

Trace amounts of oxytetracycline residues were found in oysters near the farms, in which the amounts exceeded the US Food and Drug Administration limit for commercial seafood which is 0.1 μg/g (Wu 1995). However, this effect was localized. In contrast to oxytetracycline, furazolidone is degraded by microbes in sediments, therefore its antibacterial activity is not found (Wu 1995). Studies have reported the inhibition of sulfate reduction in sediments where antibiotics have been used in cage culture (Wu 1995). The adverse effects of antibiotics are more persistent in anoxic sediments due to longer half-life than toxic sediments. Ivermectin which is used to treat sea-lice infection in farmed salmon fish has caused toxicity in marine benthos such as polychaetes. Ivermectin has a long half-life of more than 100 days and it is highly toxic (Wu 1995).

8.6 Plastic Pollution and Impacts from Fisheries and Mariculture Practices

Plastics are synthetic and extremely durable organic polymers (Cornelissen et al. 2005; Dachs et al. 2002). Plastic production is the highest in Asia (50%) followed by Europe (19%) and North America (18%) (Qiu et al. 2017). According to a global assessment of marine litter and plastic pollution, 14% of plastics in beaches were of fisheries and aquaculture origin, such as nets and cage structures, lines, floating raft structures, poles, bags, and plastic sheeting. The same assessment reported that 15% of plastic debris found on the sea floor was of fisheries and aquaculture origin (Kuroda et al., 2020). In 2017, Sri Lanka was ranked the fifth highest that released plastic and polythene waste into the ocean (Centre for Environmental Justice 2021). Plastic waste generation in Sri Lanka is estimated at 1.59 MMT/year and mismanaged plastic is estimated at 5% (Centre for Environmental Justice 2021). The authors of Wurl et al. (2006) found that plastics comprise 80% of marine debris found on beaches of Sri Lanka which included plastic bags, bottles, and other plastic packaging materials.

Fisheries and mariculture practices release plastic debris in large quantities to the marine environment. The majority of plastic debris that originate from mariculture activities are from damaged installations and equipment due to regular wear and tear, catastrophic weather events, accidents, inadequate waste management, poor maintenance, and improper installation (Kuroda et al., 2020). Plastic debris that originates from fisheries activities includes discarded fishing gear, plastic bottles (Rangel-Buitrago et al. 2019; da Silva Videla and de Araujo 2021), synthetic nylon ropes with plastic net pieces, food wrappers, Styrofoam food containers, pieces of hard plastic, synthetic slippers, and buoys (Jung et al. 2018; Feld et al. 2022). Plastic wastes originating from ships are also accumulated in the marine environment (Woodall et al. 2014). Plastic particles can be classified according to size such as nanoplastic (less than 0.1 mm) (Zhang 2017), microplastic (less than 5 mm), mesoplastic (5–20 mm), macroplastic (>20 mm), and megaplastic (>100 mm) (Duis and Coors 2016). Among these different types, microplastics are given a significant concern due to health and environmental concerns associated with them. Microplastics are categorized as primary and secondary microplastics. Primary microplastics are manufactured for a specific purpose with a specific size. Secondary microplastics are formed by the degradation of larger pieces of plastic, such as meso-, macro-, and megaplastics into smaller pieces over time through physical (mechanical), chemical, and biological processes (Athapaththu et al. 2020).

Plastics are also categorized according to their source, namely, land-based sources and ocean-based sources. Land-based plastic pollutants include litter originating from streets and beaches. Ocean-based plastic pollutants include damaged and discarded fishing gear and plastic waste originating from ships (Pieper et al. 2019; Kosior and Crescenzi 2020). It is important to study the fate of these plastic debris in marine waters before discussing the impacts. Plastic debris accumulated on beaches are washed into the sea during tides and storm periods (De Vos et al. 2014). Floating plastic debris are distributed to rocky or sandy beaches by high-energy surface currents and onshore wave actions. High-density plastic debris sink along the shoreline. Large plastic debris such as macro- and mesoplastics are fragmented into secondary microplastics by physical (mechanical), chemical, and biological processes (De Vos et al. 2014). Physical fragmentation due to mechanical abrasion is prominent during inter-monsoonal and southwest monsoonal seasons due to strong winds with high-energy wave actions (Hardesty et al. 2017) on sandy and rocky surfaces in near-shore areas. In Sri Lanka, this process is significant due to frequent hydrodynamic factors such as tides, winds, waves, thermohaline gradients, and unique upwelling pattern in the southern sea (De Vos et al. 2014). Photodegradation upon exposure to UV radiation from sunlight and hydrolysis causes the fragmentation of polymer structure in plastic resulting in the formation of smaller plastic particles (Halsband and Herzke 2019; Moore 2008). Also, due to the wave actions near shore, sea surface water reaches the upper layer of beach sediments and carries low-density microplastics into coastal waters. This way the microplastic density in surface waters gets increased. Plastics and other marine debris may also float toward Sri Lanka beaches from outside (Jung et al. 2018).

The spatial and temporal distribution of microplastics depends on climate, coastal transport patterns (Zhang et al. 2017) such as ocean current circulation and wind direction, and anthropogenic activities (Athapaththu et al. 2020). Microplastics is transported and distributed over long distances by ocean currents and wind due to their smaller sizes and lower density (Auta et al. 2017) and they are mainly present on sea surface waters (Athapaththu et al. 2020). Due to the durability of plastics, they remain in coastal and marine waters and gradually accumulate in ocean ecosystems (Hidalgo-Ruz et al. 2012; Lusher et al. 2014).

In a study by Athapaththu et al. (2020) about plastics in the surface water of the southern coastal belt of Sri Lanka, microplastics were found as the dominant plastic type in all 12 study sites, and mesoplastics and macroplastics were found only in 5 sites. Their study sites were Hikkaduwa, Galle Fort, Dodanduwa, Unawatuna, Weligama, Mirissa, Dikwella, Tangalle, Polhena, Rekawa, Godawaya, and Hambantota, which involve recreational activities, tourism, harbor operations, and commercial fishing operations. According to their findings, the microplastic density was 10.52–23.46 items per m^3 and had recorded an intermediate lower risk level compared to previously published data on surface waters from other regions like Africa, Norway, the Middle East, Europe, Korea, Indonesia, Baltic Sea, North and North-East Atlantic Ocean, California, Japan, and China. However, 46.16% of the microplastics were less than 1 mm, and 0.4–0.6 mm size range was the highly abundant category. Similarly, Athawuda et al. (2020) recorded a lower availability of mesoplastics and macroplastics compared to microplastics along the west coast, off Colombo, Sri Lanka. However, the microplastic amount recorded in the southern coastal belt of Sri Lanka by Athapaththu et al. (2020) was less compared to the west coast, off Colombo by Athawuda et al. (2020). Koongolla et al. (2018) recorded a microplastic density of 0–29 items per m^3 in the southern coastal waters. Prabodha et al. (2020) recorded microplastic levels of 40.06 items per m^3 in surface water and 1.33 items/100 g in sediments in the Madu-Ganga estuary of the Southern Province, Sri Lanka. In contrast, a significantly lower microplastic concentration of 0.85 items per m^3 was recorded in the off north-western region of Sri Lanka by Weerakoon et al. (2019).

According to the findings of Sandaruwan et al. (2023), 13 types of marine debris were identified along the coast of Kandakuliya, Sri Lanka. These 13 types were hard plastic, soft plastic, plastic straps, metal, glass, rubber, cloth, timber, foam, paper, fishing, polythene, and pharmaceutical. Soft plastic debris accounted for the highest proportion of marine debris (40%), followed by fishing practices discards (39%), hard plastic (10%), and foam (4%) implying the contribution of petroleum-based synthetic products for the pollution. This study identified oil cans, toothbrushes, plastic spoons, toothpaste, combs, and shampoo bottles as soft plastic debris. PVC pipes, bottle cup lids, lighters, food containers, and beverage bottles were identified as hard plastic debris. Improper disposal of wear and tear from fishing gear such as long lines, gill nets, hand lines, Styrofoam buoys in fishing nets, and plastic bottles which are used to pull the nets in landing sites are the significant sources of petroleum-based synthetic products that end up in the oceans and beaches (Sandaruwan et al. 2023). They have classified marine debris as whole and fragmented, in which soft plastic accounted for the highest percentage of whole debris and fishing practices discards accounted for the highest percentage of fragmented debris. Similar to the findings by Sandaruwan et al. (2023), Jang et al. (2018), in Sandaruwan et al. 2023 reported plastics as the major type of marine debris (93.3 % of large debris and 98.7 % of small debris in terms of quantity) in 22 coastal beaches in Sri Lanka.

Different sources of microplastics are also identified according to their color. Athapaththu et al. (2020) found that the most abundant color of microplastics in the southern coastal waters of Sri Lanka was blue. Fisheries-related blue color plastic sources were fishing lines, fishing nets, ropes, and some other fishing materials (Athapaththu et al. 2020). According to Koongolla et al. (2018), the most abundant colors of microplastics in the southern coastal water were blue and green. Athapaththu et al. (2020) identified high-density polyethylene (HDPE), polypropylene (PP), and polystyrene (PS) as major polymer types of plastic samples collected from the southern coastal belt. Among different types of plastic debris, only the fisheries related debris are discussed as follows. Most fishing gear are made of HDPE or PP (Plastics Europe 2019). Fishing ropes are made of nylon, polyester, PP, and polyethylene (Corbett 2009). Floating ropes are made of PP, while buoyancy and Styrofoam are made of PS (Athapaththu et al. 2020).

Plastic debris are also categorized according to their shape, such as filaments, films, foams, fragments, and rod-shaped items. In the study of Athapaththu et al. (2020), macro-, meso-, and microfilaments were identified as major plastic categories on the southern coast of Sri Lanka. The potential filament-shaped plastic pollutants related to fisheries activities could be monofilament lines used in fishing gear (Athapaththu et al. 2020), monofilament fibers resulting from fibers of trawls (Suaria et al. 2016), and microfilaments from mending fishing nets and discarding old nets at harbors and ports (De Witte et al. 2014).

Microplastics are easily ingested by marine and coastal biota due to their small size (Nor and Obbard 2014). Ingestion of small microplastic particles by filter feeders (Cole et al. 2013; Foekema et al. 2013) and fish species (Prabodha et al. 2020) have been reported. Ingestion can be due to several reasons, such as the abundance in the ecosystems, color selectivity in feeding, and size and color resemblance to natural food particles (Cole et al. 2013; Verlis et al. 2013; Wright et al. 2013. Ingestion can be either active or passive. Plastic particles can block the food passage in the digestive tract (Lusher et al. 2013). Microplastics contain toxic chemicals such as polymer raw materials and additives (Do Sul and Costa 2014). In addition, they can absorb toxic pollutants on their surfaces (Rillig 2012). Therefore, there is a high risk of toxicological effects of ingestion, for example, digestive toxicity (Campanale et al. 2020; Oztekin and Bat 2017; Rillig 2012). Accumulation of toxic chemicals such as trace metals, persistent organic pollutants (POPs), and stabilizers along the food webs causes toxicity in higher trophic levels (Farrell and Nelson 2013). Consumption of contaminated seafood can cause health problems in humans, for example, disrupting thyroid and sex hormones (Auta et al. 2017; Ogunola et al. 2018; Smith et al. 2018).

Accumulation of plastics causes adverse effects on fragile aquatic ecosystems like seawater surfaces, sea beds, coastlines, and river basins, especially in the intertidal zones (Schonmayr 2017).

Opportunistic colonizers such as bacteria, Archaea, algae, and fungi attach to macro- and mega-plastics floating on the sea surface. They are transported from their native habitat to remote areas through surface drifting, become colonized as alien species, and threaten the equilibrium of native ecosystems that accumulate on the seafloor sediments and can modify the respective bottom ecosystems (De Vos et al., 2014).

8.7 Policy Framework Governing the Mariculture Sector in Sri Lanka

Sri Lanka holds the jurisdiction over 90000 sq. miles of ocean, including the sovereign rights to explore, exploit, conserve, and manage its natural resources as the Exclusive Economic Zone (EEZ). The Marine Zones Law of 1976 provides a declaration of the territorial sea and defines various Maritime Zones of the country. In 1964, the Water Resources Board Act was established to clarify the responsibilities of managing these water resources, although it focused mainly on water streams on land. Complementary, the National Water Supply and Drainage Act of 1974 (amended in 1992) established the National Water Supply and Drainage Board (NWS&DB) to develop and operate an efficient water supply, sewerage systems, and supply water in bulk to local authorities. The basis for environmental regulations was laid out with the National Environment Act of 1980 (amended in 1998, 2000, and 2005), which aimed to preserve the environment and reduce pollution. The Act also established the Central Environmental Authority (CEA) in 1981 as the primary state agency responsible for implementing and formulating policies and strategies to manage and protect the environment.

The continuous practice of illegal fishing methods is a key reason for pollution. Section 27 of the Fisheries and Aquatic Resources Act No. 2 of 1996 strictly prohibits using or attempting to use any poisonous, explosive, or stupefying substance or any other harmful substance for the purpose of poisoning, killing, stunning, or disabling any aquatic resources – including fish. However, blast fishing, which is a form of causing a blast in the water using an explosive such as dynamite, is still carried out in some coastal and maritime zones in Sri Lanka and is even reported to be increasing despite the prohibitions. This method of fishing not only kills non-target species of fish but also causes severe damage to marine ecosystems. The five prohibited methods of fishing in the Indian ocean per the 1996 act and subsequent amendments and regulations ban the use of push nets, harpooning, and moxi, trammel, gill, or monofilament nets. However, these are still practiced within the fishing communities all over the country. It is also observed that the use of trawl nets which was banned in Sri Lanka in 2017 due to the destruction they caused to all marine life still operates in the northern coastal zones. Purse seining is also carried out even though it was banned in 2010. In this method, a "Laila net" combined with either explosives or lights is used to encircle a large group of fish. When entire groups of fish are caught, it affects the ecological balance in marine environments. Therefore, it is evident that even though there are rules and regulations that prohibit destructive methods of fishing, they are not properly enforced.

Most of the abovementioned forbidden fishing methods ultimately result in overfishing, which can harm marine biodiversity in many ways. For instance, overfishing can remove either one or a group of marine species which can impact several trophic levels. It can also result in the by-catch and mortality of nontarget species. When the herbivorous species are so reduced, there is a possibility of algal dominance which can also cause significant damages to the marine ecosystems. Worsening the issues concerning illegal fishing practices and overfishing are the conflicts among the Sri Lankan and Indian fishermen in the maritime zones of Sri Lanka which have been a constant debate between the two countries.

The Sri Lankan fishermen argue that the Indian fishermen practice unsustainable methods of fishing in Sri Lankan waters which caused irreversible damage to the marine ecosystems. The territorial seas are not divided by a straight line and it is therefore challenging to decide whether the Indian fishermen deliberately invade our waters or do so by mistake. However, the responsibility not to damage marine ecosystems is an undeniable duty of any party. Being an island situated on a busy international shipping route, thousands of ships and boats roam in the territorial waters of Sri Lanka.

The marine environment is primarily protected by the Marine Pollution Prevention Act No. 35 of 2008, which gives comprehensive directions on preventing, reducing, controlling, and managing marine pollution in the maritime and coastal zones of Sri Lanka. Section 26 of the Act charges a criminal liability for discharging or the escaping of oil, harmful substances, or pollutants into the territorial waters of Sri Lanka. Civil liability for the same actions is provided in Section 34 of the 2008 Act. Per Section 41(2), any person who has caused marine or environmental pollution, failed to report such pollution, and failed to fulfill the necessary actions specified in the Act in the event of such marine pollution is liable to a fine not exceeding Rs. 3 million. The Act also establishes the Marine Environment Protection Authority (MEPA) who are bestowed with many authoritative functions to prevent, control, and reduce marine and coastal pollution. However, it is debatable as to what extent these regulations have succeeded in preventing or controlling marine pollution.

Yet another method of marine pollution is when the coastline and inland communities add plastic and garbage to the coastline and freshwater resources, and they get carried away into the marine waters. Such debris can be extremely harmful to marine ecosystems and creatures including fish, turtles, seals, and marine birds. Also, the plastic which is dumped into the ocean disrupts the microscopic marine algae called phytoplankton from removing carbon dioxide (CO_2) in the atmosphere. Therefore, polluting the marine ecosystems not only deteriorates the existing marine ecosystems but also exacerbates climate change. Invasive organisms or vessels which may carry these organisms which have traveled many international waters can be destructive for the Sri Lankan marine ecosystems. This is recognized as a threat to marine ecosystems all over the world. Usually, these organisms can be introduced to foreign ecosystems through natural phenomena such as being carried through ocean currents or winds or being attached to a whale or a log. An organism can move from one marine ecosystem to another through commercial marine shipping in two methods, namely, ballast water (water drawn and stored in specialized tanks) and biofouling (live organisms such as algae and microbes attach themselves to the hull). If such an organism is introduced to a marine ecosystem where its natural predators are not present, then these organisms can thrive and start harming the new environment in many ways. Therefore, there is a growing threat that needs immediate attention from the administrative authorities, to protect the marine ecosystems in Sri Lanka.

8.8 Potential Mitigation Measures, Approaches and Way Forward

Mitigating pollution from mariculture, the farming of marine organisms, is crucial for maintaining the health of marine ecosystems. One of the key mitigatory strategies is choosing appropriate locations for mariculture operations which can minimize environmental impact. Selecting sites with strong water currents can help disperse waste, preventing accumulation in sensitive areas. Integrated multi-trophic aquaculture which involves cultivating multiple species together in the same area is considered a potential and productive way forward. By utilizing organisms that can absorb excess nutrients or pollutants produced by others, these systems can well reduce environmental impact.

Implementing efficient effluent treatment systems can remove excess nutrients, organic matter, and other pollutants from mariculture wastewater before it is discharged into the surrounding environment. Recirculating aquaculture systems technology allows for the recycling and treatment of water within closed systems, minimizing the release of pollutants into the surrounding environment.

A key challenge faced by Sri Lanka is the use of banned fishing practices which are still carried out within many fishing communities. One major reason for this is the lack of proper monitoring systems and official authorities to enforce laws. It is necessary to appoint sufficient officials who can regularly patrol areas identified with most such occurrences of such illegal practices. These officials can represent the departments directly involved in marine environmental conservation such as the MEPA and the Coast Conservation and Coastal Resource Management Department and other related divisional and local authorities. It is necessary to ensure that they can discharge their duties, avoiding any personal conflicts and benefits. Government intervention and enforcement are imperative going forward. The government can implement regulations and standards to control mariculture practices, including waste discharge limits, site management requirements, and monitoring programs to ensure compliance. Educating mariculture operators and awareness about best practices and the environmental consequences of pollution can encourage responsible management and reduce the likelihood of harmful practices. Research and development of innovative technologies, such as biofilters, algae-based nutrient removal systems, and sensor networks for monitoring water quality, can help improve pollution mitigation in mariculture. Collaboration among mariculture operators, government agencies, environmental organizations, and local communities is essential for developing effective pollution mitigation strategies and ensuring their successful implementation.

Accordingly, contemporary laws which are practiced in Sri Lanka must be duly amended without further delay. There is an essential requirement of revising all the fines and penalties specified for the offenders in the applicable regulatory provisions. It is also important to impose provisions that would help the offenders to uplift their socioeconomic conditions. For instance, they must be encouraged to willingly discontinue the current livelihoods and practices which harm the marine environment and be provided with opportunities to engage in alternative business practices, to educate, train, and improve their capacity, and to obtain social security protection with special attention to gender equality. One such alternative employment opportunity is introducing replanting schemes such as restoring corals and planting mangrove forests which can not only help the communities but also regenerate the deteriorating ecosystems in the country. There are several such community-based and community-involved replanting projects initiated by state and private organizations.

However, environmental restoration projects must not be limited to a certain area but involve communities of different localities and even districts. Another recommendation is creating awareness of the protection of marine ecosystems among all involved stakeholders such as coastline and inland communities, state actors, and civil society organizations. It was evident that on many occasions, people do not acknowledge the value of ecosystems and only contemplate the short-term economic profits and their convenience. Therefore, communities can be educated through local-, municipal-, and urban-level programs on how to protect the environment and discontinue practices which cause environmental implications. Government officials who are involved in field activities and directly associated with the protection of the marine environment can be educated through national campaigns, seminars, and conferences on the necessity of safeguarding the ecosystems, disregarding their personal benefits. It is also important to promote awareness among high state officials, particularly state ministers, secretaries, etc., on the long-term effects of environmental degradation and the necessity of protecting the marine environment of Sri Lanka. Improving communication and co-operation among the state and private sectors to execute, improvise, and implement conservation and restoration plans and procedures is also necessary. The private sector must not only be able to initiate new plans to protect the marine environment but also criticize any governmental actions that would harm marine ecosystems.

At the same time, the government sector must be able to regulate private organizations and ensure that they do not misuse their funds. It is also necessary to build collaborations with international organizations in addressing concerns that are new to Sri Lanka, such as the introduction of invasive organisms through commercial shipping. Building regional collaborations is also essential in order to address fishing conflicts. It is necessary that the governments declare illegal fishing practices which are common in the region, and thereby, enter into mutual agreements to prevent such practices without engaging in blame games. Most importantly, the state and private sectors must encourage researchers, scientists, and academics to conduct exhaustive research to identify the risks and potentially vulnerable areas in marine ecosystems. This will be helpful in recognizing the illegal practices which harm the environment in specific areas and identifying invasive organisms and the resilience levels of marine ecosystems, community needs of those areas, and thereby finding applicable solutions. This intellectual analysis and statistics would be supportive in drafting the amendments and new regulations. Thereby, the above-discussed recommendations can be incorporated into regulations, policies, action plans, and strategies to protect and improve the marine ecosystems which are affected by harmful human activities. By implementing the right economic incentives along with mitigatory strategies, it is possible to mitigate pollution from mariculture and promote sustainable practices that will ensure the long term conservation of marine ecosystems.

References

Ackefors, H. and Enell, M. (1990). Discharge of nutrients from Swedish fish farming to adjacent sea areas. *Ambio* 19: 28–35.

Ariyawansa, S. (2013). Utilization of fish waste in Sri Lanka: an overview. *Journal of Fisheries Science and Technology* 116–120.

Arulananthan, K., Rydberg, L., Cederlöf, U., and Wiyeratne, E.M.S. (1995). Water exchange in a hypersaline tropical estuary, the Puttalam lagoon, Sri Lanka. *Ambio* 24: 438–443.

Athapaththu, A.M.A.I.K., Thushari, G.G.N., Dias, P.C.B. et al. (2020). Plastics in surface water of southern coastal belt of Sri Lanka (Northern Indian Ocean): distribution and characterization by FTIR. *Marine Pollution Bulletin* 161: 111750.

Athawuda, A.M.G.A.D., Jayasiri, H.B., Thushari, G.G.N., and Guruge, K.P.G.K.P. (2020). Quantification and morphological characterization of plastic litter (0.30–100 mm) in surface waters of off Colombo, west coast of Sri Lanka. *Environmental Monitoring and Assessment* 192: 1–17.

Auta, H.S., Emenike, C.U., and Fauziah, S.H. (2017). Distribution and importance of microplastics in the marine environment: a review of the sources, fate, effects, and potential solutions. *Environment International* 102: 165–176.

Bandara, K.R.V., Chinthaka, S.D.M., Yasawardene, S.G., and Manage, P.M. (2021). Modified, optimized method of determination of Tributyltin (TBT) contamination in coastal water, sediment and biota in Sri Lanka. *Marine Pollution Bulletin* 166: 112202.

Bonsdorff, E., Blomqvist, E.M., Mattila, J., and Norkko, A. (1997). Coastal eutrophication: causes, consequences and perspectives in the archipelago areas of the northern Baltic Sea. *Estuarine, Coastal and Shelf Science* 44: 63–72.

Bournazel, J., Kumara, M.P., Jayatissa, L.P. et al. (2015). The impacts of shrimp farming on land-use and carbon storage around Puttalam lagoon, Sri Lanka. *Ocean & Coastal Management* 113: 18–28.

Brooks, K.M., Mahnken, C., and Nash, C. (2002). Environmental effects associated with marine net pen waste with emphasis on salmon farming in the Pacific, Northwest. In: *Responsible Marine Aquaculture* (ed. R.R. Stickney and J.P. McVey), 149–203. Cambridge: CAB International.

Campanale, C., Massarelli, C., Savino, I. et al. (2020). A detailed review study on potential effects of microplastics and additives of concern on human health. *International Journal of Environmental Research and Public Health* 17 (4): 1212.

Caruso, G. (2016). Fishery wastes and by-products: a resource to be valorised. *Journal of Fisheries Sciences* 9 (4): 80–83.

Central Bank of Sri Lanka (CBSL) (2021). *Annual Report 2020*. Colombo, Sri Lanka: Central Bank of Sri Lanka. Central Bank of Sri Lanka (CBSL).

Centre for Environmental Justice (2021). Plastic Waste Management Country Situation Report - Sri Lanka. https://ipen.org/sites/default/files/documents/plastic_waste_management_in_sri_lanka.pdf (accessed 25 August 2024).

Chopin, T., Buscmann, A.H., Halling, C. et al. (2001). Seaweeds in sustainable integrated aquaculture. *Journal of Phycology* 37: 975–986; Aure, J. and Stigebrandt, A. (1990). Quantitative estimates of the eutrophication effects of fish farming on fjords. *Aquaculture* 90: 135–156.

Cole, M., Lindeque, P., Fileman, E. et al. (2013). Microplastic ingestion by zooplankton. *Environmental Science & Technology* 47 (12): 6646–6655.

Corbett, G.P. (2009). *Fire Engineering's Handbook for Firefighter I and II*. Fire Engineering Books.

Corea, A.S.L.E., Johnstone, R., Jayasinghe, J.M.P.K. et al. (1998). Self-pollution: a major threat to the prawn farming industry in Sri Lanka. *Ambio* 27 (8): 666–668.

Cornelissen, G., Gustafsson, Ö., Bucheli, T. et al. (2005). Extensive sorption of organic compounds to black carbon, coal, and kerogen in sediments and soils: mechanisms and consequences for distribution, bioaccumulation, and biodegradation. *Environmental Science and Technology* 39: 6881–6895.

Croomey, C.J., Nickell, T.D., and Black, K.D. (2000). DEPOMOD-modeling the deposition and biological effects of waste solids from marine cage farms. *Aquaculture* 214: 211–239.

Dachs, J., Lohmann, R., Ockenden, W.A. et al. (2002). Oceanic biogeochemical controls on global dynamics of persistent organic pollutants. *Environmental Science and Technology* 36: 4229–4237.

Dassanayake, H. (1994). An environmental assessment of the bay of Bengal region. *Bay of Bengal Programme* 209–225.

De Vos, A., Pattiaratchi, C.B., and Wijeratne, E.M.S. (2014). Surface circulation and upwelling patterns around Sri Lanka. *Biogeosciences* 11: 5909–5930. https://doi.org/10.5194/bg-11-5909-2014.

De Witte, B., Devriese, L., Bekaert, K. et al. (2014). Quality assessment of the blue mussel (*Mytilus edulis*): comparison between commercial and wild types. *Marine Pollution Bulletin* 85 (1): 146–155.

Deepananda, K.H.M.A. (2011). Mariculture, present trends and future prospects for Sri Lanka. In: *Conditions for entrepreneurship in Sri Lanka: A Handbook* (ed. B.W. Åmo). https://www.gbv.de/dms/zbw/665367155.pdf.

Department of Census and Statistics (DCS) (2022). *Annual Report of National Accounts - 2022*. Colombo, Sri Lanka: Department of Census and Statistics.

Dissanayake, D.A.S.J., Wickramasinghe, D.D., and Manage, P.M. (2021). Harmful diatoms and dinoflagellates in the Indian Ocean: a study from Southern coast of Sri Lanka. *Ukrainian Journal of Ecology* 11 (1): 279–285.

Duis, K. and Coors, A. (2016). Microplastics in the aquatic and terrestrial environment: sources (with a specific focus on personal care products), fate and effects. *Environmental Sciences Europe* 28: 1–25. https://doi.org/10.1186/s12302-015-0069-y.

EJF (2004). *Farming the Sea, Costing the Earth: Why We Must Green the Blue Revolution*. London, UK: Environmental Justice Foundation.

FAO (2020). The state of world fisheries and aquaculture – Sustainability in action. https://openknowledge.fao.org/server/api/core/bitstreams/170b89c1-7946-4f4d-914a-fc56e54769de/content. Accessed on 05 June 2024.

Farrell, P. and Nelson, K. (2013). Trophic level transfer of microplastic: *Mytilus edulis* (L.) to *Carcinus maenas* (L.). *Environmental Pollution* 177: 1–3.

Feld, L., da Silva, V.H., and Strand, J. (2022). Characterization of foamed plastic litter on Danish reference beaches–pollution assessment and multivariate exploratory analysis. *Marine Pollution Bulletin* 180: 113774.

Foekema, E.M., De Gruijter, C., Mergia, M.T. et al. (2013). Plastic in North Sea fish. *Environmental Science & Technology* 47 (15): 8818–8824.

Grémillet, D., Pichegru, L., Kuntz, G. et al. (2008). A junk-food hypothesis for gannets feeding on fishery waste. *Proceedings of the Royal Society B: Biological Sciences* 275 (1639): 1149–1156.

Halsband, C. and Herzke, D. (2019). Plastic litter in the European Arctic: what do we know? *Emerging Contaminants* 5: 308–318.

Hardesty, B.D., Harari, J., Isobe, A. et al. (2017). Using Numerical Model Simulations to improve the understanding of micro-plastic distribution and pathways in the marine environment. *Frontiers in Marine Science* 4: 1–9. https://doi.org/10.3389/fmars.2017.00030.

Hidalgo-Ruz, V., Gutow, L., Thompson, R.C., and Thiel, M. (2012). Microplastics in the marine environment: a review of the methods used for identification and quantification. *Environmental Science & Technology* 46 (6): 3060–3075.

Islam, M.S., Khan, S., and Tanaka, M. (2004). Waste loading in shrimp and fish processing effluents: potential source of hazards to the coastal and nearshore environments. *Marine Pollution Bulletin* 49 (1–2): 103–110.

Jang, Y.C., Ranatunga, R.R.M.K.P., Mok, J.Y. et al. (2018). Composition and abundance of marine debris stranded on the beaches of Sri Lanka: results from the first island-wide survey. *Marine Pollution Bulletin* 128: 126–131.

Jayasinghe, J.M.P.K. and Amarasinghe, U.S. (2021). Coastal aquaculture in Sri Lanka: opportunities and environmental challenges. *Journal of the Indian Society of Coastal Agricultural Research* 39 (2): 1–14.

Jayasinghe, J.M.P.K. and Jayasekara, A.M. (1999). Aquaculture research needs, research priorities and research capacities in Sri Lanka. *Sri Lanka Journal of Aquatic Sciences* 4: 1–11.

Jayasinghe, J.M.P.K., Gunaratne, T.V.N.M., and Sadaruwan, K.R.P. (2013). *Shrimp Culture in Sri Lanka: Better Management Interventions*. Makandura, Sri Lanka: Wayamba University of Sri Lanka 143p.

Jayawardhane, J.K.P.C., Manage, P.M., and Weerasekara, K.A.W.S. (2018). *Identification of Harmful Marine Microalgae with Special Reference to Physico-Chemical Aspects of Coastal Waters in Western Province*. Sri Lanka. In Proceedings of International Forestry and Environment Symposium (Vol. 23).

Jegatheesan, V., Zeng, C., Shu, L. et al. (2006). Technological advances in aquaculture farms for minimal effluent discharge to oceans. *Journal of Cleaner Production* https://doi.org/10.1016/j.jclepro.2006.07.043.

Jensen, O., Wroldsen, A.S., Lader, P.F. et al. (2007). Finite element analysis of tensegrity structures in offshore aquaculture installations. *Aquacultural Engineering* 36 (3): 272–284.

Jung, M.R., Horgen, F.D., Orski, S.V. et al. (2018). Validation of ATR FT-IR to identify polymers of plastic marine debris, including those ingested by marine organisms. *Marine Pollution Bulletin* 127: 704–716.

Koongolla, J.B., Andrady, A.L., Kumara, P.T.P., and Gangabadage, C.S. (2018). Evidence of microplastics pollution in coastal beaches and waters in southern Sri Lanka. *Marine Pollution Bulletin* 137: 277–284.

Kosior, E. and Crescenzi, I. (2020). Solutions to the plastic waste problem on land and in the oceans. In: *Plastic Waste and Recycling*, 415–446. Academic Press.

Kuroda, M., Uchida, K., Tokai, T. et al. (2020). The current state of marine debris on the seafloor in offshore area around Japan. *Marine Pollution Bulletin* 161 (1): 111670.

Lichna, A.I., Bezyk, K.I., and Kudelina, O.Y. (2023). Analysis of FAO data on the global fisheries and aquaculture production volume. *Водні біоресурси* 1 (13): 188–197.

Lounas, R., Kasmi, H., Chernai, S. et al. (2020). Towards sustainable mariculture: some global trends. *Thalassas: An International Journal of Marine Sciences* 36: 447–456.

Lovatelli, A., Aguilar-Manjarrez, J., and Soto, D., 2013. Expanding mariculture farther offshore. Technical, environmental, spatial and governance challenges. FAO Technical Workshop, Orbetello, Italy, 22–25 March 2010 (No. 24, pp. viii+-314).

Lusher, A.L., Mchugh, M., and Thompson, R.C. (2013). Occurrence of microplastics in the gastrointestinal tract of pelagic and demersal fish from the English Channel. *Marine Pollution Bulletin* 67 (1-2): 94–99.

Lusher, A.L., Burke, A., O'Connor, I., and Officer, R. (2014). Microplastic pollution in the Northeast Atlantic Ocean: validated and opportunistic sampling. *Marine Pollution Bulletin* 88 (1-2): 325–333.

Manage, P.M., Liyanage, G.Y., Abinaiyan, I. et al. (2022). Pollution levels in Sri Lanka's west-south coastal waters: making progress toward a cleaner environment. *Regional Studies in Marine Science* 51: 102193.

Moore, C.J. (2008). Synthetic polymers in the marine environment: a rapidly increasing, long-term threat. *Environmental Research* 108 (2): 131–139.

Muñiz, O., Rodríguez, J.G., Revilla, M. et al. (2018). Seasonal variations of phytoplankton community in relation to environmental factors in an oligotrophic area of the European Atlantic coast (southeastern Bay of Biscay). *Regional Studies in Marine Science* 17: 59–72.

National Aquaculture Development Authority of Sri Lanka (NAQDA) (2021). *Aquaculture and Culture Based Fisheries Sector Development Plan-2021-2025*. Colombo, Sri Lanka: National Aquaculture Development Authority of Sri Lanka.

Niroshana, K.H.H., Asanthi, H.B., and Kumara, P.B.T.P. (2013). An assessment of water quality and pollution in Puranawella fishery harbour, Dewinuwara, Sri Lanka. *Journal of the University of Ruhuna* 1 (1): 23–30.

Nor, N.H.M. and Obbard, J.P. (2014). Microplastics in Singapore's coastal mangrove ecosystems. *Marine Pollution Bulletin* 79 (1-2): 278–283.

Ocean Health Index (2020). Ocean health index. http://www.oceanhealthindex.org/region-scores/annual-scores-and-rankings.

Ogunola, O.S., Onada, O.A., and Falaye, A.E. (2018). Mitigation measures to avert the impacts of plastics and microplastics in the marine environment (a review). *Environmental Science and Pollution Research* 25: 9293–9310.

Oztekin, A. and Bat, L. (2017). Microlitter pollution in sea water: a preliminary study from Sinop Sarikum Coast of the southern Black Sea. *Turkish Journal of Fisheries and Aquatic Sciences* 17: 1387–1395.

Pieper, C., Amaral-Zettler, L., Law, K.L. et al. (2019). Application of Matrix Scoring Techniques to evaluate marine debris sources in the remote islands of the Azores Archipelago. *Environmental Pollution* 249: 666–675.

Plastics Europe (2019). https://www.plasticseurope.org/en/resources/publications/1804-plastics-facts-2019 (accessed 20 October 2024).

Prabodha, M.W.K., Wijethunga, H.N.S., Silva, A.P.R. et al. (2020). Screening of plastic pollution effects in Madu-ganga estuarine ecosystem in Southern Province, Sri Lanka: an approach toward the coastal zone management. *Proceedings of International Research Conference - IRCUWU2020.*

Primavera, J.H. (2006). Overcoming the impacts of aquaculture on the coastal zone. Ocean. *Coastal Management* 49: 531e545.

Qiu, Y.W., Zeng, E.Y., Qiu, H.L. et al. (2017). Bioconcentration of polybrominated diphenyl ethers and organochlorine pesticides in algae is an important contaminant route to higher trophic levels. *Science of the Total Environment* 579: 1885–1893.

Qiu, Y.W., Wang, D.X., and Zhang, G. (2020). Assessment of persistent organic pollutants (POPs) in sediments of the Eastern Indian Ocean. *Science of the Total Environment* 710: 136335.

Rangel-Buitrago, N., Gracia, A., Velez-Mendoza, A. et al. (2019). Where did this refuse come from? Marine anthropogenic litter on a remote island of the Colombian Caribbean sea. *Marine Pollution Bulletin* 149: 110611.

Rifath, M.R.A. and Jemziya, M.B.F. (2023). *Fish Waste Valorization to Active Food Packaging: Towards Blue Economy of Sri Lanka-Multisectoral Approaches to Accelerate Economic Transformation in the Face of Crisis in Sri Lanka Ed. D.M.S.B. Dissanayaka*. Sri Lanka: National Science and Technology Commission.

Rillig, M.C. (2012). Microplastic in terrestrial ecosystems and the soil? *Environmental Science & Technology* 46: 6453–6454.

Sandaruwan, R.D.C., Bellanthudawa, B.K.A., Perera, I.J.J.U.N. et al. (2023). Index based approach for assessment of abundance of marine debris and status of marine pollution in Kandakuliya, Kalpitiya, Sri Lanka. *Marine Pollution Bulletin* 197: 115724.

Schönmayr, D. (2017). *Automotive Recycling, Plastics, and Sustainability: The Recycling Renaissance*, 184. Springer International Publishing.

Senarath, U. and Visvanathan, C. (2001). Environmental issues in brackish water shrimp aquaculture in Sri Lanka. *Environmental Management* 27: 335–348.

Seymour, E.A. and Bergheim, A. (1991). Towards a reduction of pollution from intensive aquaculture with reference to the farming of salmonids in Norway. *Aquaculture Engineering*. 10: 73–88.

da Silva Videla, E. and de Araujo, F.V. (2021). Marine debris on the Brazilian coast: which advances in the last decade? A literature review. *Ocean & Coastal Management* 199: 105400.

Smith, M., Love, D.C., Rochman, C.M., and Neff, R.A. (2018). Microplastics in seafood and the implications for human health. *Current Environmental Health Reports* 5: 375–386.

Suaria, G., Avio, C.G., Mineo, A. et al. (2016). The Mediterranean Plastic Soup: synthetic polymers in Mediterranean surface waters. *Scientific Reports* 6 (1): 37551.

do Sul, J.A.I. and Costa, M.F. (2014). The present and future of microplastic pollution in the marine environment. *Environmental Pollution* 185: 352–364.

United Nations Environment Programme (2021). From Pollution to Solution: a global assessment of marine litter and plastic pollution. Nairobi. https://www.grida.no/resources/15029 (accessed 22 August 2024).

United States Environmental Protection Agency (2021). Vegetable oils and animal fats. https://www.epa.gov/emergency-response/vegetable-oilsand-animal-fats.

Verlis, K.M., Campbell, M.L., and Wilson, S.P. (2013). Ingestion of marine debris plastic by the wedge-tailed shearwater Ardenna pacifica in the Great Barrier Reef, Australia. *Marine Pollution Bulletin* 72 (1): 244–249.

Votier, S.C., Furness, R.W., Bearhop, S. et al. (2004). Changes in fisheries discard rates and seabird communities. *Nature* 427 (6976): 727–730.

Weerakoon, W.R.W.M.A.P., Grøsvik, B.E., Dalpadado, P. et al. (2019). Enumeration of Microplastics in Sri Lankan Waters: Preliminary Findings from the RV Dr. Fridtjof Nansen Ecosystem Survey, 2018 Scientific Sessions of NARA, 26 July 2019. vol. 15. NARA Auditorium, Colombo.

Weerasekara, K.A.W.S., Jayampathi, O.M.M.D., Hettige, N.D. et al. (2015). Assessment of water pollution status of selected fishery harbours located in the southern province of Sri Lanka. *Journal of Environmental Professionals Sri Lanka* 4 (2): 48–49.

Woodall, L.C., Sanchez-Vidal, A., Canals, M. et al. (2014). The deep sea is a major sink for microplastic debris. *Royal Society Open Science* 1: https://doi.org/10.1098/rsos.140317.

Wright, S.L., Thompson, R.C., and Galloway, T.S. (2013). The physical impacts of microplastics on marine organisms: a review. *Environmental Pollution* 178: 483–492.

Wu, R.S.S. (1995). The environmental impact of marine fish culture: towards a sustainable future. *Marine Pollution Bulletin* 4 (12): 159–166.

Wu, R.S.S. (2001). Environmental impacts of marine fish farming and their mitigation. In: *Responsible Aquaculture Development in Southeast Asia. Proceedings of the Seminar-Workshop on Aquaculture Development in Southeast Asia organized by the SEAFDEC Aquaculture Department, 12–14 October 1999, Iloilo City, Philippines*, 157–172. Aquaculture Department. Southeast Asian Fisheries Development Center.

Wurl, O., Potter, J.R., Obbard, J.P., and Durville, C. (2006). Persistent organic pollutants in the equatorial atmosphere over the open Indian Ocean. *Environmental Science and Technology 2006* (40): 1454–1461.

Zhang, H. (2017). Transport of microplastics in coastal seas. *Estuarine, Coastal and Shelf Science* 199: 74–86.

Zhang, W., Zhang, S., Wang, J. et al. (2017). Microplastic pollution in the surface waters of the Bohai Sea, China. *Environmental Pollution* 231: 541–548.

9

Marine Macro-litter: Sources, Abundance, Impacts, and Solutions

Minh-Ky Nguyen

Faculty of Environment and Natural Resources, Nong Lam University, Ho Chi Minh City, 700000, Viet Nam

9.1 Introduction

As any persistent, marine litter is recognized as processed or manufactured solid material abandoned or disposed of in the coastal and marine ecosystems (Programme et al., 2005; Galgani et al. 2013). Marine litter has led to an ecological risk for marine ecosystems around the globe (De-la-Torre et al. 2021; Nguyen et al. 2023c). It is frequently determined in all worldwide marine environments and reaches up to 90% plastic debris (Baxter et al. 2022; Al Nahian et al. 2022b). Marine macro-litter accumulates on the sea surface, the ocean, within the water bodies, along the beaches and ocean floor, etc. (Galgani et al. 2015; Goodman et al. 2021; Van-Giang et al. 2024). Once these items are introduced to marine ecosystems, plastic items continue to decompose into smaller-sized debris and could be conveyed long-transport from their polluted sources. Consequently, marine litter has been recognized in all beaches and oceans around the globe, adversely influencing marine wildlife, fishing, tourism, and shipping (Beaumont et al. 2019; Rakib et al. 2022). The widespread occurrence and persistent impact of marine macro-litter is of critical concern regarding environmental and human health (Engler 2012; Le et al. 2023b). Marine macro-litter, mostly especially plastic items, can be consumed by marine forms and degrade into (micro) plastics, and even (nano)plastics, releasing microcontaminants that pose harmful threats to ecological, organisms, and human health (Andrady 2011; Kumar et al. 2022; Al Nahian et al. 2022a; Nguyen et al. 2022a).

Pollutants can enter the marine environmental matrices through both land-based sources, such as drainage networks and river run-off, and sea-based activities, for example, fishing and shipping sectors. Growing evidence emphasizes the significant consequences of marine macro-litter on various economic activities, including recreation/tourism, aquaculture and fisheries, and maritime navigation. Additionally, marine litter poses challenges to the services and infrastructure that support local communities (including individuals) and businesses (McIlgorm et al. 2011; Schneider et al. 2018). The important issue of marine macro-litter arises from inadequate waste management practices and low public awareness, leading to the potential impacts of improper waste disposal (Jeftic et al. 2009; Andrady 2011; Chinh et al. 2021). Marine macro-litter poses significant challenges as it results in detrimental influences on the ecosystems, encompassing

Coastal and Marine Pollution: Source to Sink, Mitigation and Management, First Edition.
Edited by Meththika Vithanage, Sameera M. Samarasekara, Bryan D. James, and Christopher M. Reddy.

ecological, economic, and social damage (Andrady 2011; Cole et al. 2011). Inadequate waste management practices, particularly regarding plastics, contribute significantly to coastal littering in developing and developed countries (Jeftic et al. 2009; Jayasiri et al. 2013; Nguyen et al. 2023c). Therefore, it is crucial to identify their sources, abundance, impacts, and solutions toward sustainable development goals (SDGs). Conducting a baseline study to explore the abundance, occurrence, and distribution of marine macro-litter in oceans could enhance conservation efforts and aid in the strategic development of effective mitigation (Al Nahian et al. 2022b). The present attempt focuses on marine macro-litter pollution to provide an in-depth understanding of marine litter. This chapter aimed to (i) examine the abundance, density, and occurrence of marine macro-litter; (ii) investigate ecological impacts; and (iii) provide management approaches toward sustainable marine ecosystems.

9.2 Sources, Fate, and Transport of Marine Macro-litter

Generally, two important sources of marine macro-litter can be categorized: land based and ocean based (Figure 9.1). Marine litter is an anthropogenic hazard as a consequence of human activities and behavior, and the rise in marine macro-litter is closely associated with the development of the plastic industry. Marine litter originating from land enters the oceans through estuarine drainage, run-off, sewage systems, or wind, and especially rivers, whereas marine litter from sea-based sources includes trade activities, cruise ships, aquaculture and fishing activities, research and military vessels, offshore petroleum platforms, etc. (Galgani et al. 2015). For instance, aquaculture practices in marine environments significantly contribute to marine plastic pollution (Skirtun et al. 2022). In the ocean's depths, areas with high levels of marine litter were found to be correlated with the busiest shipping lanes, pointing to an additional source of litter input into the basin (Pasquini et al. 2016). Accidental loss or discharge of fishing gear, which comprises a notable portion of marine plastic litter originating from ships, contributes significantly to ocean pollution (Hwang 2020). Every year, incidents involving floating materials entangled in marine environments

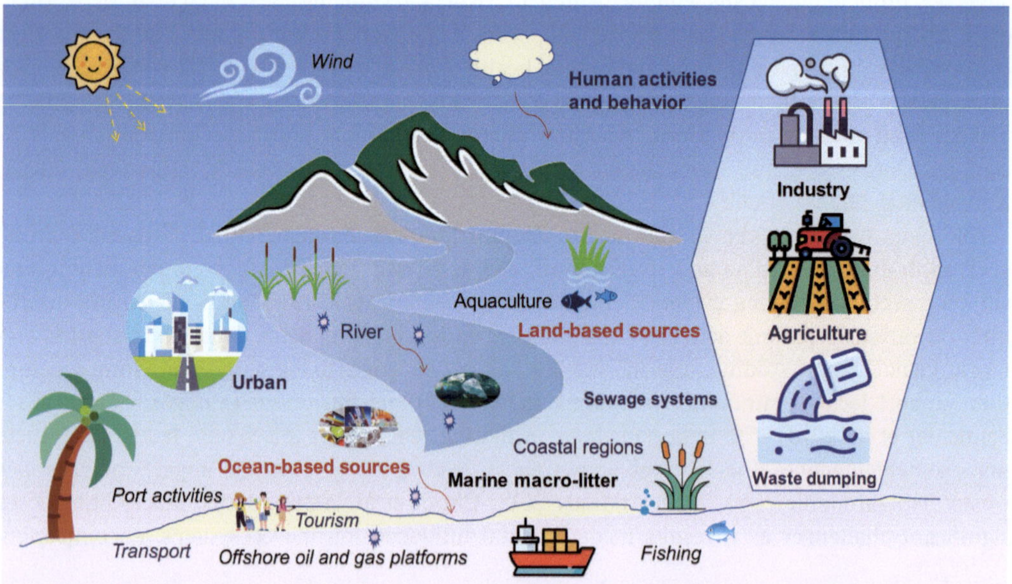

Figure 9.1 Sources, fate, and transport of marine macro-litter in the ecosystems.

occur frequently, with a significant proportion attributed to the presence of large marine plastic litter (Kim et al. 2023).

Marine macro-debris is any durable artificial material that discharges the marine environment through river run-off and/or is dumped directly into the seas. Estimated findings showed a range from 4.8×10^6 to 12.7×10^6 tons of plastic waste entered the oceans due to mismanaged land based (Jambeck et al. 2015). Mixed packaging, domestic waste, and fisheries activities were identified as the dominant marine litter sources. A study examined by García-Rivera et al. (2017) focused on evaluating the composition, occurrence/distribution, and sources of marine macro-litter in the Southeast Mediterranean (Spain). The study found that glass and plastics were predominantly located near the coast. Additionally, fishing activity accounted for 29.16% of the marine litter source observed in the area. In the study by Al Nahian et al. (2022b), local markets, domestic/household waste, tourism, hotels, and fishing sectors were identified as the sources of marine macro-litter. Also, significant mixed sources such as sewage and sanitary-related waste, illegal dumping, and non-sourced items also contribute to marine litter. In short, the promoted coastal population density and their lifestyle changes, the intensive socioeconomic aspects, for example, recreation, tourism, fisheries/aquaculture, and maritime transport, associated with the riverine inputs from various drainage basins, have caused an increase of litter inputs to the coastal regions and marine environment.

Regarding the fate and transport of marine macro-litter in the ecosystems, that is, marine macro-litter and plastic debris are carried both within and between environmental compartments, with transport processes playing a crucial role in determining the residence time of litter debris within a specific environmental compartment. These transport processes ultimately dictate the accumulation of hotspots of marine macro-litter. Key transport vectors linking the terrestrial compartment to the marine compartment encompass factors such as river systems, surface runoffs, and wind (Schwarz et al. 2019; Woods et al. 2021). Surface currents are primarily driven by the wind flowing over the sea surface, and this mechanism is accountable for transporting low-density floating litter (debris) (Kim et al. 2015). The prevalent beach litter along this coastline mainly comprised rubber debris and single-use plastics (SUPs) that had been carried through the nearby areas (Mishra et al. 2023). Moreover, the characteristics of litter/plastic fragments, including shape, size, and density, play a role in determining their ability to be transported. Fragmentation processes lead to the breakdown of macro-plastic fragments into micro-plastic fragments (Ky et al. 2023). This debris has the potential to sink to the ocean floor, wash up on beaches and shorelines, or decompose into microplastics (NOAA 2016; Pogojeva et al. 2021). Additionally, for certain polymer types, micro-plastics have the potential to further fragment into nano-plastic fragments. However, a notable portion can persistently remain afloat at the sea surface for extended periods and may be transported over long distances.

9.3 Abundance and Distribution of Marine Macro-litter

In recent studies, there has been a growing focus on quantifying marine debris due to the alarming accumulation rates of human-made items in marine ecosystems. Numerous local studies have already been carried out, with a primary emphasis on sandy beaches, aiming to quantify the amount of debris that drifted ashore. The composition of marine litter primarily consists of plastics, wood, cigarette butts, rubber, glass, paper, clothing, and metals (Özden et al. 2021). Several research studies have examined human-made plastic debris' quantity and geographical spread in beach and marine environments (Ribeiro et al. 2021). A comparable study was examined along the central Caribbean coast of Colombia, focusing on beach areas (Rangel-Buitrago et al. 2021). The results revealed that 75% of the beaches were categorized as having a "very high abundance" of plastic debris, while 17% were recognized as "high abundance," and the remaining 8% were categorized as having a "moderate abundance" of plastic debris (Rangel-Buitrago et al. 2021). In the Spanish Southeast Mediterranean, an investigation revealed that, by weight, plastic constituted 75.9% of the beached debris, while metal and glass were also present (García-Rivera et al. 2017). Nevertheless, data from middle- and low-income nations are lacking despite the frequency of plastic litter items along their coastal regions (Haarr et al. 2022). For instance, Vietnam is examined as one of the top nations discharging plastic waste into the oceans (Meijer et al. 2021).

Table 9.1 presents worldwide macro-litter density recognized from various studies. A high density of macro-litter items was reported on the North Sea coast, Belgium (6,429 items 100 m^{-1}), and plastics were the dominant type (95%) (Van Cauwenberghe et al. 2013). The mean litter density was measured to be 0.64 ± 0.09 items m^{-2} in Turkey (ErtaŞ et al. 2022). Plastic items constituted a significant portion of the recorded items in nearly all countries within the Adriatic-Ionian region,

Table 9.1 Global macro litter density reported from locations.

Location	Environment	Litter density	Remarks	References
Turkey	Aegean Sea	0.64 ± 0.09 items m^{-2}	Sources of marine litter were mixed domestic, packaging, and fisheries activities	ErtaŞ et al. (2022)
Japan	Sea of Japan	3.41 items m^{-2}	The most dominant type of macro-litter was plastic, which included 40–80% of the total debris	Kusui and Noda (2003)
India	Indian coast	$0.3 \pm 0.4 - 0.475 \pm 0.51$ items m^{-2}	Single-use plastics are predominant	(Mishra et al. (2023)
Nigeria	Several rivers	3487 items $area^{-1}$	Clean coast index varied from "very clean" to "extremely dirty"	Ebere et al. (2019)
Canada	Arctic Canada and West Greenland	1.0 ± 1.7 items m^{-2}	Local and long-range sources	Mallory et al. (2021)
Chile	Southeast Pacific	1.8 items m^{-2}	Major litter types include plastics, glass, and cigarette butts	Bravo et al. (2009)
Cyprus	Eastern Mediterranean	59 556 items $area^{-1}$	The top four most abundant plastics: plastic bags > plastic pieces > drinks containers > lids, caps	Özden et al. (2021)
Colombia	Caribbean and Pacific coast	$35 \pm 15 - 81 \pm 23$ items 100 m^{-1}	The most frequent (macro) plastic items involving packaging and containers	Garcés-Ordóñez et al. (2020)
Korea	Nakdong River Estuary	1.03 items m^{-2}	Plastic debris was most abundant	Lee et al. (2013)
Belgium	North Sea coast	6 429 items 100 m^{-1}	Plastics were the dominant type (95%)	Van Cauwenberghe et al. (2013)
Germany	The Isle of Rügen	318 items 100 m^{-1}	Plastic debris was the most common material (83%)	Hengstmann et al. (2017)

Note: Along the surveyed transect of 100 m, all macro-litter was recorded (items 100 m^{-1}).

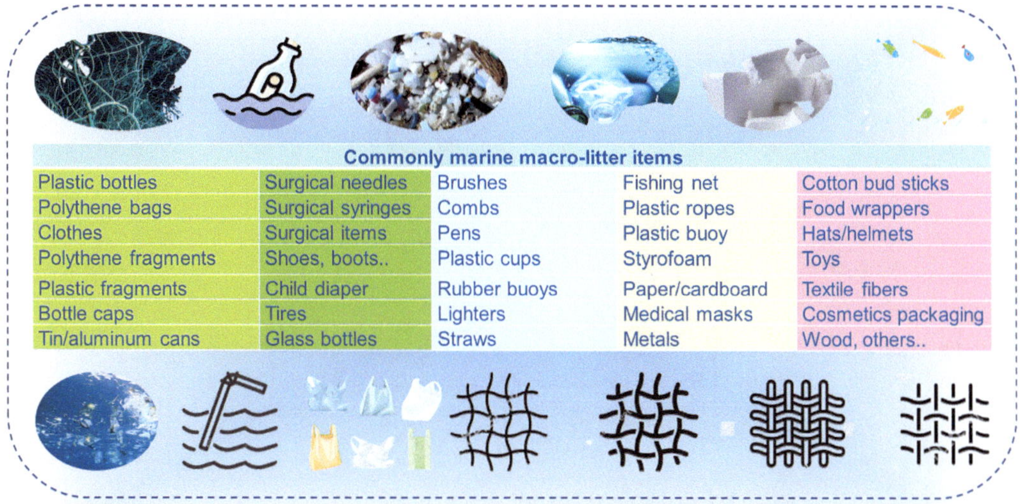

Figure 9.2 Examples of macro-litter items commonly found in marine environments.

ranging from 74% to 92% (except for Albania, where plastics made up 54.3% of the total), meanwhile at the regional level, the overall presence of plastic debris reached as high as 91% (Vlachogianni et al. 2018). Marine litter refers to items that are manufactured or utilized by humans and are either intentionally discarded or inadvertently lost at sea or along the coastlines. In many instances, plastic-made bottles, cotton bud sticks, bottle caps, straws, plastic buoys, and crisp packets were found on monitored beaches (Wenneker and Oosterbaan 2010; Al Nahian et al. 2022b). Figure 9.2 illustrates commonly recorded marine litter items, that is, plastic bottles, polythene bags, polyethylene terephthalate (PET)-plastic caps, styrofoam, chips, food wrappers, straws, clothes, plastic cups, sandals, fragments, rubber buoys, and fishing nets.

Recently, plastic-based medical face masks have played a crucial role in safeguarding individuals during the COVID-19 pandemic, providing significant protection (Ammendolia et al. 2021). However, medical-originated items, including tampons, and syringes, often find their way into aquatic ecosystems primarily through sewage systems. These masks have been haphazardly discarded into natural environments, resulting in their accumulation along coastlines. Thus, identification of the essential sources and distribution of litter is necessary in order to reduce beach macro-litter contamination and find sustainable approaches to the litter pollution issue.

9.4 Impacts of Marine Macro-litter

Marine contamination is determined as a worldwide emerging environmental problem related to a range of ecological risks, as well as economic costs and social impacts (Villarrubia-Gómez et al. 2018; Beaumont et al. 2019; Minh-Ky et al. 2024). Marine macro-debris and its effects can be recognized on any global beaches, sea/ocean, and seafloor. Estimates illustrated that many million man-made tons of macro-litter are introduced to oceans yearly, which are mostly plastic items, durable and hard materials (Law et al. 2020). Several investigations have revealed that marine macro-debris impacts sea turtles, marine mammals, and seabirds (Battisti et al. 2019b). The impact generally occurs through entanglement, entrapment, ingestion, and threat to ecosystems (Figure 9.3) (Kühn and Van Franeker 2020; Høiberg et al. 2022). For instance, marine animals

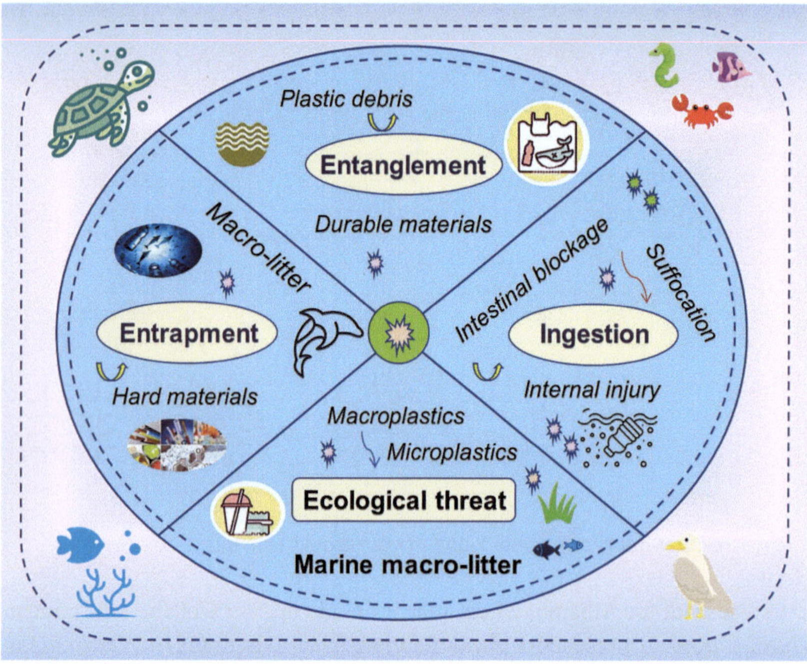

Figure 9.3 Impacts of marine macro-litter.

have been recognized to interact with and become entangled in plastic items for many years ago (Høiberg et al. 2022). Ingestion of (macro)plastics/litter could cause intestinal blockage, suffocation, and internal injury (Nelms et al. 2016).

Potential entanglement effect levels were recognized to be higher in ocean currents, along coastal regions, and in semi-enclosed oceans, concentrating plastic litter hotspots. Marine animals include mammals and seabirds are affected by marine debris, with a significant impact arising from plastic waste and debris associated with fishing practices. For example, during the period from 2016 to 2020, a total of 151 animals were documented to have been directly affected or observed engaging with marine debris within a protected natural area located in Peru (Ayala et al. 2022). Residual debris from ropes, fishing gear, plastic bags, and unidentified materials were recognized. Also, over 350 species have been known as noticed entangled in (macro)plastic items (Kühn and Van Franeker 2020). They can cause a risk to aquatic biodiversity, which is associated with stressors and leads to threaten their population. In another study, the prevailing interactions primarily involve ingestion, entanglement, and varied plastic applications, such as plastic serving as nesting material and providing refuge or shelter (Blettler and Mitchell 2021). Birds constituted the predominant taxonomic group in reported incidents, accounting for 72.2% of encounters, equal to 30 species. The prevailing form of interaction between plastic and fauna was the utilization of plastic as nesting material, which constituted 42% of the total cases (Blettler and Mitchell 2021). Following closely, the second most common encounter was entanglement, comprising 33% of the incidents. A significant majority of the entanglement instances resulted in fatal outcomes for the affected organisms, accounting for nearly 60% of the cases.

Due to its accumulation in the marine, this unavoidably impacts ecological communities and ecosystem functions (Villarrubia-Gómez et al. 2018; Nguyen et al. 2023a; Le et al. 2024).

Remarkably, some categories of macro-litter can cause detrimental impacts on life, acting with various mechanisms, including entanglement, suffocation, and poisoning (Battisti et al. 2019a; Blettler and Mitchell 2021). The extensive danger posed by large plastic debris entanglement to marine animals of significant size is a well-documented concern (Gall and Thompson 2015; Wilcox et al. 2016). Entanglement acts as a critical mechanism affecting several vertebrates (e.g. reptiles), mammals, and birds (Staffieri et al. 2019; Blettler and Mitchell 2021). Direct interaction with woven plastic components made from polypropylene had a detrimental impact on the well-being of the branched coral species *P. cylindrica* (Mueller et al. 2022). The engagement between wildlife and plastic waste can result in substantial harm to both individual animals and entire populations. However, the effects of marine litter are influenced by a range of factors, including their distribution, duration of exposure, the size of the litter, and the specific type of organism involved (Arabi et al. 2022).

Marine debris, a significant portion of which comprises plastic materials, poses a substantial environmental challenge (Galgani et al. 2015). Plastic contains additives (e.g. plasticizers, flame-retardants, and dyes), added during the process of production, which could leach into the ecosystems (Teuten et al. 2009; Rochman et al. 2014). Biota could potentially receive substantial transfers of chemical additives from plastic litter due to high contents (Napper and Thompson 2020; Nguyen et al. 2023b). Plastic comes in a variety of sizes and forms, and its effects on organisms can present in diverse ways, impacting multiple trophic levels and leading to ecological consequences (Browne et al. 2015). Furthermore, the degradation of larger plastic items is a significant contributor to the abundance of micro-plastic particles found within the marine ecosystem (Galloway et al. 2017; Le et al. 2023a). Over extended periods, the degradation of macroplastics results in the formation of micro- and nano-plastics, which organisms might unintentionally ingest. In addition to macroplastics, various other forms of marine debris also degrade into micro- and nano-sized particles, leading to their accumulation within the ecosystem. Marine organisms can bioaccumulate harmful substances (micro/nano-particles and other chemicals) through the entire food web, highlighting the potential ecological risks and impacting human health (Tuan Tran et al. 2022; Nguyen et al. 2023c).

Further, anthropogenic marine litter poses a global risk, demonstrating various impacts on economic and social aspects. Marine litter can substantially change coastal ecosystem services, leading to multiple social and economic influences at various levels (Herrera et al. 2023). The presence of marine debris presents a critical danger to ecological habitats associated with their organisms (Nguyen et al. 2022b; Nama et al. 2023). One of the important challenges to reefs is the common occurrence of marine items, which damages their habitats (de Carvalho-Souza et al. 2018; Nama et al. 2023). Degradation of marine ecosystems could have a detrimental impact on society (i.e. affecting the well-being and health of local residents and visitors) and the economy (i.e. losses in the tourism, fisheries, and shipping activities) (Watkins et al. 2015; UNEP 2017). Also, the prevalence of marine litter within marine ecosystems exerts a noteworthy influence on commercial fishing and industries, resulting in consequential socio-ecological outcomes (Lusher et al. 2017). Commercial shrimpers often encounter marine macro-debris entangled in their nets, leading to economic repercussions (Rodolfich et al. 2023). Marine litter could also be accountable for causing damage to fishing vessels and equipment, as well as potentially contaminating the caught seafood (Thompson et al. 2004; Andrady 2011). Especially in the European Union (EU), these economic impacts of marine litter have been approximated to result in a cost of nearly €62 million (Mouat et al. 2010). The effects of marine macro-litter highlighted the necessity of management approaches to protect the marine environment.

9.5 Solution Approaches Toward Sustainable Marine Ecosystem

In recent periods, there has been a growing interest among researchers and policymakers in addressing marine macro-litter, recognizing the urgent need for significant reduction efforts. Critical strategies for mitigating litter pollution can be proposed, including the implementation of technical standards and regulations, fostering behavior change, and reducing the usage of SUPs (Löhr et al. 2017; Lau et al. 2020; Calero et al. 2021). Achieving this goal requires concerted efforts and a multifaceted approach to address the issue at both individual and systemic levels (Figure 9.4). To combat plastic pollution, it is crucial to take meaningful actions such as avoiding or refusing nonessential plastic items and reducing reliance on SUPs (Walker and Xanthos 2018; Nguyen et al. 2023c). Embracing alternative materials, such as bioplastics (Roy Chong et al. 2022), and following the critical principles of the 4Rs "*Reduce–Reuse–Recycle–Recover*" – are essential strategies (Nižetić et al. 2019; Klemeš et al. 2021). The utilization of bio-based polymers can contribute to achieving lower carbon footprints and aligning with SDGs (Karan et al. 2019; Löhr and Van Belleghem 2020; Ahsan et al. 2023). Additionally, policy to encourage or regulate manufacturers toward environmentally sustainable practices (Klemeš et al. 2021). By adopting these approaches, it is possible to make meaningful progress in mitigating litter pollution and promoting more sustainable practices.

With a view to prioritize the effort and concentrate on the most significant culprits, analyzing the sources of marine macro-litter is the first stage in developing solutions to minimize marine litter (García-Rivera et al. 2017; Nguyen et al. 2023c). To effectively combat litter contamination, it is essential to possess information regarding the geographic sources of litter in coastal regions. Also, machine learning shows potential for explaining marine litter quantities and predicting plastic pollution (Kaandorp et al. 2021; Tran et al. 2023). Conducting regular litter surveys and analyzing the findings in relation to local weather conditions and coastal geomorphology are crucial steps in this process. Incentivizing establishments that utilize organic materials by offering reduced tax payments can be a driving force for plastic waste reduction (Varkey et al. 2021). In line with this approach, the EU Strategy on Plastics, recently adopted, aims to accelerate the evolution

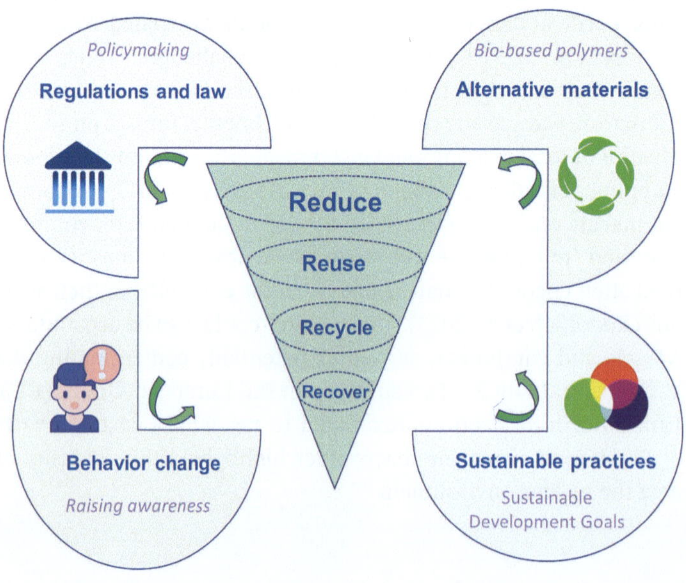

Figure 9.4 Solution approaches toward sustainable marine ecosystem.

toward a circular economy for the plastic industry and significantly reduce the consumption and impact of SUP items.

Thus, the reduction of litter contamination, especially plastic, is a major global challenge that necessitates comprehensive actions. Policymaking plays a crucial role in this regard, including the establishment of new regulations or laws to enforce marine macro-litter reduction (Chen 2015; Nguyen et al. 2023c). Moreover, it is crucial to develop initiatives that raise awareness among all types of vessels, local communities, and tourists to maintain clean seas (Bettencourt et al. 2023). By adopting these measures, significant progress can be made in addressing the issue of litter pollution in coastal areas.

9.6 Conclusions and Recommendations

There has been an increase in marine debris due to the impact of human-made activities on marine ecosystems. Marine litter originates from various sources, including both sea-based and land-based activities. Notably, land-based sources, particularly those from shoreline activities, agriculture, recreational and tourism activities, as well as medical-related waste, have been identified as significant contributors to litter. This highlights the need for targeted management solutions.

Marine litter, predominantly composed of plastic items, poses a significant ecological challenge. The global environmental effects of marine macro-litter are well documented. This impact generally occurs through entanglement, entrapment, and ingestion and poses a threat to ecosystems. The detrimental effects of litter debris on seabirds, marine mammals, and sea turtles have been recognized for decades. These environmental influences extend beyond individual organisms to affect entire populations and ecosystems. Therefore, mitigating litter pollution can be achieved through various measures, including the implementation of technical standards and regulations, fostering behavior change, and reducing marine litter sources, such as SUPs.

While the impact of marine macro-litter on living organisms has been extensively studied in various regions worldwide, it is vital to acknowledge the lack of research on this subject in the specific region under consideration. Further studies are necessary to comprehensively examine the impacts of marine macro-litter on living organisms in this area. Conducting such research is crucial for enhancing our understanding and implementing appropriate measures to mitigate the effects of marine litter on marine wildlife.

Additionally, there is a need for further research to improve our comprehension of the temporal and spatial patterns of marine litter. This understanding is essential for developing effective mitigation strategies and enhancing waste management practices. To address this issue, reducing plastic consumption emerges as a crucial step. This can be achieved by minimizing unnecessary packaging and labeling, increasing awareness, and promoting education about the impacts of plastic waste. Where feasible, prioritizing recycling efforts is also essential for effective marine litter management. Furthermore, it is crucial to explore and promote environmentally friendly alternatives to plastics that do not adversely affect the environment.

References

Ahsan, W.A., Hussain, A., Lin, C., and Nguyen, M.K. (2023). Biodegradation of different types of bioplastics through composting—a recent trend in green recycling. *Catalysts* 13 (2): 294.

Al Nahian, S., Rakib, M.R.J., Haider, S.M.B. et al. (2022a). Occurrence, spatial distribution, and risk assessment of microplastics in surface water and sediments of Saint Martin Island in the Bay of Bengal. *Marine Pollution Bulletin* 179: 113720.

Al Nahian, S., Rakib, M.R.J., Haider, S.M.B. et al. (2022b). Baseline marine litter abundance and distribution on Saint Martin Island, Bay of Bengal, Bangladesh. *Marine Pollution Bulletin* 183: 114091.

Ammendolia, J., Saturno, J., Brooks, A.L. et al. (2021). An emerging source of plastic pollution: environmental presence of plastic personal protective equipment (PPE) debris related to COVID-19 in a metropolitan city. *Environmental Pollution* 269: 116160.

Andrady, A.L. (2011). Microplastics in the marine environment. *Marine Pollution Bulletin* 62 (8): 1596–1605.

Arabi, S., Neehaul, Y., and Sparks, C. (2022). Impacts and threats of marine litter in African seas. In: *The African Marine Litter Outlook*, 91–136. Springer International Publishing Cham.

Ayala, F., Castillo-Morales, K., and Cárdenas-Alayza, S. (2022). Impact of marine debris recorded in a sympatric colony of otariids in the south coast of Peru. *Marine Pollution Bulletin* 174: 113281.

Battisti, C., Kroha, S., Kozhuharova, E. et al. (2019a). Fishing lines and fish hooks as neglected marine litter: first data on chemical composition, densities, and biological entrapment from a Mediterranean beach. *Environmental Science and Pollution Research* 26: 1000–1007.

Battisti, C., Staffieri, E., Poeta, G. et al. (2019b). Interactions between anthropogenic litter and birds: a global review with a 'black-list'of species. *Marine Pollution Bulletin* 138: 93–114.

Baxter, L., Lucas, Z., and Walker, T.R. (2022). Evaluating Canada's single-use plastic mitigation policies via brand audit and beach cleanup data to reduce plastic pollution. *Marine Pollution Bulletin* 176: 113460.

Beaumont, N.J., Aanesen, M., Austen, M.C. et al. (2019). Global ecological, social and economic impacts of marine plastic. *Marine Pollution Bulletin* 142: 189–195.

Bettencourt, S., Freitas, D.N., Lucas, C. et al. (2023). Marine litter education: from awareness to action. *Marine Pollution Bulletin* 192: 114963.

Blettler, M.C.M. and Mitchell, C. (2021). Dangerous traps: macroplastic encounters affecting freshwater and terrestrial wildlife. *Science of the Total Environment* 798: 149317.

Bravo, M., de los Ángeles Gallardo, M., Luna-Jorquera, G. et al. (2009). Anthropogenic debris on beaches in the SE Pacific (Chile): results from a national survey supported by volunteers. *Marine Pollution Bulletin* 58 (11): 1718–1726.

Browne, M.A., Underwood, A.J., Chapman, M.G. et al. (2015). Linking effects of anthropogenic debris to ecological impacts. *Proceedings of the Royal Society B: Biological Sciences* 282 (1807): 20142929.

Calero, M., Godoy, V., Quesada, L., and Martín-Lara, M.Á. (2021). Green strategies for microplastics reduction. *Current Opinion in Green and Sustainable Chemistry* 28: 100442.

de Carvalho-Souza, G.F., Llope, M., Tinôco, M.S. et al. (2018). Marine litter disrupts ecological processes in reef systems. *Marine Pollution Bulletin* 133: 464–471.

Chen, C.-L. (2015). Regulation and management of marine litter. *Marine Anthropogenic Litter* 395–428.

Chinh, P.C., Hung, N.T.Q., Ky, N.M. et al. (2021). Willingness to Pay for improving household solid waste management in Vietnam. *Applied Environmental Research* 43 (2): 1–14.

Cole, M., Lindeque, P., Halsband, C., and Galloway, T.S. (2011). Microplastics as contaminants in the marine environment: a review. *Marine Pollution Bulletin* 62 (12): 2588–2597.

De-la-Torre, G.E., Rakib, M.R.J., Pizarro-Ortega, C.I., and Dioses-Salinas, D.C. (2021). Occurrence of personal protective equipment (PPE) associated with the COVID-19 pandemic along the coast of Lima, Peru. *Science of the Total Environment* 774: 145774.

Ebere, E.C., Wirnkor, V.A., Ngozi, V.E., and Chukwuemeka, I.S. (2019). Macrodebris and microplastics pollution in Nigeria: first report on abundance, distribution and composition. *Environmental Analysis, Health and Toxicology* 34 (4): e2019012.

Engler, R.E. (2012). The complex interaction between marine debris and toxic chemicals in the ocean. *Environmental Science & Technology* 46 (22): 12302–12315.

ErtaŞ, A., Ribeiro, V.V., Castro, Í.B., and Sayim, F. (2022). Composition, sources, abundance and seasonality of marine litter in the Çakalburnu lagoon coast of Aegean Sea. *Journal of Coastal Conservation* 26 (2): 8.

Galgani, F., Hanke, G., Werner, S., Oosterbaan, L., Nilsson, P., Fleet, D., Kinsey, S., Thompson, R., Van-Franeker, J., Vlachogianni, T. 2013. Monitoring guidance for marine litter in European seas: A guidance document within the Common Implementation Strategy for the Marine Strategy Framework Directive. Joint Research Centre, Institute for Environment and Sustainability, European Commission, Luxembourg, 128pp, https://doi.org/10.2788/99475.

Galgani, F., Hanke, G., and Maes, T. (2015). Global distribution, composition and abundance of marine litter. *Marine Anthropogenic Litter* 29–56.

Gall, S.C. and Thompson, R.C. (2015). The impact of debris on marine life. *Marine Pollution Bulletin* 92 (1–2): 170–179.

Galloway, T.S., Cole, M., and Lewis, C. (2017). Interactions of microplastic debris throughout the marine ecosystem. *Nature Ecology & Evolution* 1 (5): 0116.

Garcés-Ordóñez, O., Espinosa, L.F., Cardoso, R.P. et al. (2020). Plastic litter pollution along sandy beaches in the Caribbean and Pacific coast of Colombia. *Environmental Pollution* 267: 115495.

García-Rivera, S., Lizaso, J.L.S., and Millán, J.M.B. (2017). Composition, spatial distribution and sources of macro-marine litter on the Gulf of Alicante seafloor (Spanish Mediterranean). *Marine Pollution Bulletin* 121 (1): 249–259.

Goodman, A.J., McIntyre, J., Smith, A. et al. (2021). Retrieval of abandoned, lost, and discarded fishing gear in Southwest Nova Scotia, Canada: preliminary environmental and economic impacts to the commercial lobster industry. *Marine Pollution Bulletin* 171: 112766.

Haarr, M.L., Falk-Andersson, J., and Fabres, J. (2022). Global marine litter research 2015–2020: geographical and methodological trends. *Science of the Total Environment* 153162.

Hengstmann, E., Gräwe, D., Tamminga, M., and Fischer, E.K. (2017). Marine litter abundance and distribution on beaches on the Isle of Rügen considering the influence of exposition, morphology and recreational activities. *Marine Pollution Bulletin* 115 (1): 297–306.

Herrera, M., Pita, P., Castelo, D. et al. (2023). Public perceptions of marine litter and impacts on coastal ecosystem services in Galicia (Spain). *Marine Policy* 155: 105742.

Høiberg, M.A., Woods, J.S., and Verones, F. (2022). Global distribution of potential impact hotspots for marine plastic debris entanglement. *Ecological Indicators* 135: 108509.

Hwang, D.-J. (2020). The IMO action plan to address marine plastic litter from ships and its follow-up timeline. *Journal of International Maritime Safety, Environmental Affairs, and Shipping* 4 (2): 32–39.

Jambeck, J.R., Geyer, R., Wilcox, C. et al. (2015). Plastic waste inputs from land into the ocean. *Science* 347 (6223): 768–771.

Jayasiri, H.B., Purushothaman, C.S., and Vennila, A. (2013). Plastic litter accumulation on high-water strandline of urban beaches in Mumbai, India. *Environmental Monitoring and Assessment* 185: 7709–7719.

Jeftic, L., Sheavly, S., Adler, E., and Meith, N. (2009). *Marine Litter: A Global Challenge*. UNEP.

Kaandorp, M.L.A., Ypma, S.L., Boonstra, M. et al. (2021). Using machine learning and beach cleanup data to explain litter quantities along the Dutch North Sea coast. *Ocean Science Discussions* 2021: 1–33.

Karan, H., Funk, C., Grabert, M. et al. (2019). Green bioplastics as part of a circular bioeconomy. *Trends in Plant Science* 24 (3): 237–249.

Kim, I.-S., Chae, D.-H., Kim, S.-K. et al. (2015). Factors influencing the spatial variation of microplastics on high-tidal coastal beaches in Korea. *Archives of Environmental Contamination and Toxicology* 69: 299–309.

Kim, B., Lee, H., and Kim, T. (2023). Possibility analysis of complex marine accidents caused by large marine plastic litter in the Republic of Korea. *Marine Policy* 152: 105591.

Klemeš, J.J., Fan, Y.V., and Jiang, P. (2021). Plastics: friends or foes? The circularity and plastic waste footprint. *Energy Sources, Part A: Recovery, Utilization, and Environmental Effects* 43 (13): 1549–1565.

Kühn, S. and Van Franeker, J.A. (2020). Quantitative overview of marine debris ingested by marine megafauna. *Marine Pollution Bulletin* 151: 110858.

Kumar, R., Sinha, R., Rakib, M.R.J. et al. (2022). Microplastics pollution load in Sundarbans delta of Bay of Bengal. *Journal of Hazardous Materials Advances* 7: 100099.

Kusui, T. and Noda, M. (2003). International survey on the distribution of stranded and buried litter on beaches along the Sea of Japan. *Marine Pollution Bulletin* 47 (1–6): 175–179.

Ky, N.M., Rakib, M.R.J., Lin, C. et al. (2023). A comprehensive review on ecological effects of microplastic pollution: an interaction with pollutants in the ecosystems and future perspectives. *TrAC Trends in Analytical Chemistry* 168: 117294.

Lau, W.W.Y., Shiran, Y., Bailey, R.M. et al. (2020). Evaluating scenarios toward zero plastic pollution. *Science* 369 (6510): 1455–1461.

Law, K.L., Starr, N., Siegler, T.R. et al. (2020). The United States' contribution of plastic waste to land and ocean. *Science Advances* 6 (44): eabd0288.

Le, V.-G., Nguyen, M.-K., Nguyen, H.-L. et al. (2023a). A comprehensive review of micro- and nano-plastics in the atmosphere: occurrence, fate, toxicity, and strategies for risk reduction. *Science of the Total Environment* 904: 166649.

Le, V.-R., Nguyen, M.-K., Nguyen, H.-L. et al. (2023b). Organic composts as a vehicle for the entry of microplastics into the environment: a comprehensive review. *Science of the Total Environment* 892: 164758.

Le, V.-G., Nguyen, M.-K., Nguyen, H.-L. et al. (2024). Ecotoxicological response of algae to contaminants in aquatic environments: a review. *Environmental Chemistry Letters* 22: 919–939.

Lee, J., Hong, S., Song, Y.K. et al. (2013). Relationships among the abundances of plastic debris in different size classes on beaches in South Korea. *Marine Pollution Bulletin* 77 (1-2): 349–354.

Löhr, A.J. and Van Belleghem, F. (2020). Sustainable development goals to reduce and prevent marine litter. *Life Below Water* 1–12.

Löhr, A., Savelli, H., Beunen, R. et al. (2017). Solutions for global marine litter pollution. *Current Opinion in Environmental Sustainability* 28: 90–99.

Lusher, A., Hollman, P., and Mendoza-Hill, J. (2017). *Microplastics in Fisheries and Aquaculture: Status of Knowledge on Their Occurrence and Implications for Aquatic Organisms and Food Safety*. FAO.

Mallory, M.L., Baak, J., Gjerdrum, C. et al. (2021). Anthropogenic litter in marine waters and coastlines of Arctic Canada and West Greenland. *Science of the Total Environment* 783: 146971.

McIlgorm, A., Campbell, H.F., and Rule, M.J. (2011). The economic cost and control of marine debris damage in the Asia-Pacific region. *Ocean & Coastal Management* 54 (9): 643–651.

Meijer, L.J.J., Van Emmerik, T., Van Der Ent, R. et al. (2021). More than 1000 rivers account for 80% of global riverine plastic emissions into the ocean. *Science Advances* 7 (18): eaaz5803.

Minh-Ky, N., Rakib, M.R.J., Nguyen, H.-L. et al. (2024). A mini-review on plasticrusts: occurrence, current trends, potential threats, and recommendations for coastal sustainability. *Environmental Monitoring and Assessment* 196 (2): 137.

Mishra, P., Kaviarasan, T., Sambandam, M. et al. (2023). Assessment of national beach litter composition, sources, and management along the Indian coast – a citizen science approach. *Marine Pollution Bulletin* 186: 114405.

Mouat, J., Lozano, R.L., and Bateson, H. (2010). *Economic Impacts of Marine Litter*. Kommunenes Internasjonale Miljøorganisasjon.

Mueller, J.S., Bill, N., Reinach, M.S. et al. (2022). A comprehensive approach to assess marine macro litter pollution and its impacts on corals in the Bangka Strait, North Sulawesi, Indonesia. *Marine Pollution Bulletin* 175: 113369.

Nama, S., Shanmughan, A., Nayak, B.B. et al. (2023). Impacts of marine debris on coral reef ecosystem: a review for conservation and ecological monitoring of the coral reef ecosystem. *Marine Pollution Bulletin* 189: 114755.

Napper, I.E. and Thompson, R.C. (2020). Plastic debris in the marine environment: history and future challenges. *Global Challenges* 4 (6): 1900081.

Nelms, S.E., Duncan, E.M., Broderick, A.C. et al. (2016). Plastic and marine turtles: a review and call for research. *ICES Journal of Marine Science* 73 (2): 165–181.

Nguyen, M.K., Hadi, M., Lin, C. et al. (2022a). Microplastics in sewage sludge: distribution, toxicity, identification methods, and engineered technologies. *Chemosphere* 308: 136455.

Nguyen, M.K., Lin, C., Hung, N.T.Q. et al. (2022b). Occurrence and distribution of microplastics in peatland areas: a case study in long an province of the Mekong Delta, Vietnam. *Science of the Total Environment* 844: 157066.

Nguyen, M.-K., Lin, C., Nguyen, H.-L. et al. (2023a). Ecotoxicity of micro- and nanoplastics on aquatic algae: facts, challenges, and future opportunities. *Journal of Environmental Management* 346: 118982.

Nguyen, M.-K., Lin, C., Nguyen, H.-L. et al. (2023b). Emergence of microplastics in the aquatic ecosystem and their potential effects on health risks: the insights into Vietnam. *Journal of Environmental Management* 344: 118499.

Nguyen, M.K., Lin, C., Quang Hung, N.T. et al. (2023c). Investigation of ecological risk of microplastics in peatland areas: a case study in Vietnam. *Environmental Research* 220: 115190.

Nižetić, S., Djilali, N., Papadopoulos, A., and Rodrigues, J.J.P.C. (2019). Smart technologies for promotion of energy efficiency, utilization of sustainable resources and waste management. *Journal of Cleaner Production* 231: 565–591.

NOAA. 2016. Report on Modeling Oceanic Transport of Floating Marine Debris, National, Oceanic Atmospheric Administration Marine Debris, Program, Silver Spring, MD.

Özden, Ö., Yıldırım, S., Fuller, W.J., and Godley, B.J. (2021). Anthropogenic marine litter on the north coast of Cyprus: insights into marine pollution in the eastern Mediterranean. *Marine Pollution Bulletin* 165: 112167.

Pasquini, G., Ronchi, F., Strafella, P. et al. (2016). Seabed litter composition, distribution and sources in the Northern and Central Adriatic Sea (Mediterranean). *Waste Management* 58: 41–51.

Pogojeva, M., Zhdanov, I., Berezina, A. et al. (2021). Distribution of floating marine macro-litter in relation to oceanographic characteristics in the Russian Arctic Seas. *Marine Pollution Bulletin* 166: 112201.

Programme, U.R.S., Plan, U.M.A., Secretariat of the Basel Convention on the Control of Transboundary Movements of Hazardous, W., Their, D., Office, U.G.C., Intergovernmental Oceanographic, C. (2005). Marine litter: an analytical overview.

Rakib, M.R.J., Ertaş, A., Walker, T.R. et al. (2022). Macro marine litter survey of sandy beaches along the Cox's Bazar Coast of Bay of Bengal, Bangladesh: land-based sources of solid litter pollution. *Marine Pollution Bulletin* 174: 113246.

Rangel-Buitrago, N., Mendoza, A.V., Mantilla-Barbosa, E. et al. (2021). Plastic pollution on the Colombian central Caribbean beaches. *Marine Pollution Bulletin* 162: 111837.

Ribeiro, V.V., Harayashiki, C.A.Y., Ertaş, A., and Castro, Í.B. (2021). Anthropogenic litter composition and distribution along a chemical contamination gradient at Santos Estuarine System—Brazil. *Regional Studies in Marine Science* 46: 101902.

Rochman, C.M., Lewison, R.L., Eriksen, M. et al. (2014). Polybrominated diphenyl ethers (PBDEs) in fish tissue may be an indicator of plastic contamination in marine habitats. *Science of the Total Environment* 476: 622–633.

Rodolfich, A., Sparks, E., Posadas, B. et al. (2023). The distribution and direct impacts of marine debris on the commercial shrimping industry. *Marine Pollution Bulletin* 186: 114417.

Roy Chong, J.W., Tan, X., Khoo, K.S. et al. (2022). Microalgae-based bioplastics: future solution towards mitigation of plastic wastes. *Environmental Research* 206: 112620.

Schneider, F., Parsons, S., Clift, S. et al. (2018). Collected marine litter—a growing waste challenge. *Marine Pollution Bulletin* 128: 162–174.

Schwarz, A.E., Ligthart, T.N., Boukris, E., and Van Harmelen, T. (2019). Sources, transport, and accumulation of different types of plastic litter in aquatic environments: a review study. *Marine Pollution Bulletin* 143: 92–100.

Skirtun, M., Sandra, M., Strietman, W.J. et al. (2022). Plastic pollution pathways from marine aquaculture practices and potential solutions for the North-East Atlantic region. *Marine Pollution Bulletin* 174: 113178.

Staffieri, E., de Lucia, G.A., Camedda, A. et al. (2019). Pressure and impact of anthropogenic litter on marine and estuarine reptiles: an updated "blacklist" highlighting gaps of evidence. *Environmental Science and Pollution Research* 26 (2): 1238–1249.

Teuten, E.L., Saquing, J.M., Knappe, D.R.U. et al. (2009). Transport and release of chemicals from plastics to the environment and to wildlife. *Philosophical Transactions of the Royal Society B: Biological Sciences* 364 (1526): 2027–2045.

Thompson, R.C., Olsen, Y., Mitchell, R.P. et al. (2004). Lost at sea: where is all the plastic? *Science* 304 (5672): 838–838.

Tran, H.-T., Hadi, M., Nguyen, T.T.H. et al. (2023). Machine learning approaches for predicting microplastic pollution in peatland areas. *Marine Pollution Bulletin* 194: 115417.

Tuan Tran, H., Lin, C., Bui, X.-T. et al. (2022). Phthalates in the environment: characteristics, fate and transport, and advanced wastewater treatment technologies. *Bioresource Technology* 344: 126249.

UNEP (2017). *Marine Litter Socio Economic Study*. United Nations Environment Programme.

Van Cauwenberghe, L., Claessens, M., Vandegehuchte, M.B. et al. (2013). Assessment of marine debris on the Belgian Continental Shelf. *Marine Pollution Bulletin* 73 (1): 161–169.

Van-Giang, L., Nguyen, H.-L., Nguyen, M.-K. et al. (2024). Marine macro-litter sources and ecological impact: a review. *Environmental Chemistry Letters* 22: 1257–1273.

Varkey, P.S., Walker, T.R., and Saunders, S.J. (2021). Identifying barriers to reducing single-use plastic use in a coastal metropolitan city in Canada. *Ocean & Coastal Management* 210: 105663.

Villarrubia-Gómez, P., Cornell, S.E., and Fabres, J. (2018). Marine plastic pollution as a planetary boundary threat–The drifting piece in the sustainability puzzle. *Marine Policy* 96: 213–220.

Vlachogianni, T., Fortibuoni, T., Ronchi, F. et al. (2018). Marine litter on the beaches of the Adriatic and Ionian Seas: an assessment of their abundance, composition and sources. *Marine Pollution Bulletin* 131: 745–756.

Walker, T.R. and Xanthos, D. (2018). A call for Canada to move toward zero plastic waste by reducing and recycling single-use plastics. *Resources Conservation and Recycling* 133: 99–100.

Watkins, E., Brink, P., Withana, S. et al. (2015). *Marine Litter: Socio-economic Study: Scoping Report.* Institute for European Environmental Policy: *London, UK*.

Wenneker, B., Oosterbaan, L. (2010). Guideline for Monitoring Marine Litter on the Beaches in the OSPAR Maritime Area. Edition 1.0.

Wilcox, C., Mallos, N.J., Leonard, G.H. et al. (2016). Using expert elicitation to estimate the impacts of plastic pollution on marine wildlife. *Marine Policy* 65: 107–114.

Woods, J.S., Verones, F., Jolliet, O. et al. (2021). A framework for the assessment of marine litter impacts in life cycle impact assessment. *Ecological Indicators* 129: 107918.

10

Nuclear and Radioactive Marine Pollution and Monitoring of Radioactivity in Oceans

R.M.N. Priyanga Rathnayake[1] *and Dulanjalee Rajapaksha*[2]

[1] *Radiation Protection and Technical Services Division, Sri Lanka Atomic Energy Board, Wellampitiya, Sri Lanka*
[2] *School of Natural Sciences, University of Tasmania, Hobart, Australia*

10.1 Introduction

Marine ecosystems are vital for regulating climate and biodiversity and providing food, resources, and livelihoods for millions of people worldwide. Contamination of the oceans with radioactive and nuclear materials, posing a potential harm to marine ecosystems as well as human health, is called radioactive and nuclear marine pollution.

Radiation is in different forms, and it is everywhere on the earth. This chapter focuses only on ionizing radiation in marine ecosystems and their impacts. The benefits and importance of ionizing radiation to the mankind are well understood and this led to the deployment of nuclear science in the wider arena of applications in different sectors including energy, medicine, industrial, agriculture, research and developments, space science, etc. The bombing of Hiroshima and Nagasaki (1945), nuclear weapon tests (1950s–1970s), and nuclear accidents such as 1979 Three-Mile-Island accident, 1986 Chernobyl accident, and 1999 Tokai Mura accident persuaded the scientific community to investigate the harmful impacts of ionizing radiation on the humans and environments. Most importantly Fukushima Accident (2011) can be considered an eye-opening incident for large-scale marine pollution by radioactive and nuclear materials.

Human beings are not at all sensitive to ionizing radiation. Hence, it is essential to deploy different types of detectors to measure the levels of radioactivity. All decisions and interventions must be based on the measured radioactivity in any object, body, or system including oceans. This highlighted the importance of establishment of baseline radioactivity in different seas and marine-originated products and continuous, real-time monitoring of marine radioactivity.

10.2 Sources of Radioactivity in Marine Ecosystems

Radioactivity in marine ecosystems is caused by naturally occurring radioactive elements and by manmade radionuclides (Figure 10.1).

Coastal and Marine Pollution: Source to Sink, Mitigation and Management, First Edition.
Edited by Meththika Vithanage, Sameera M. Samarasekara, Bryan D. James, and Christopher M. Reddy.

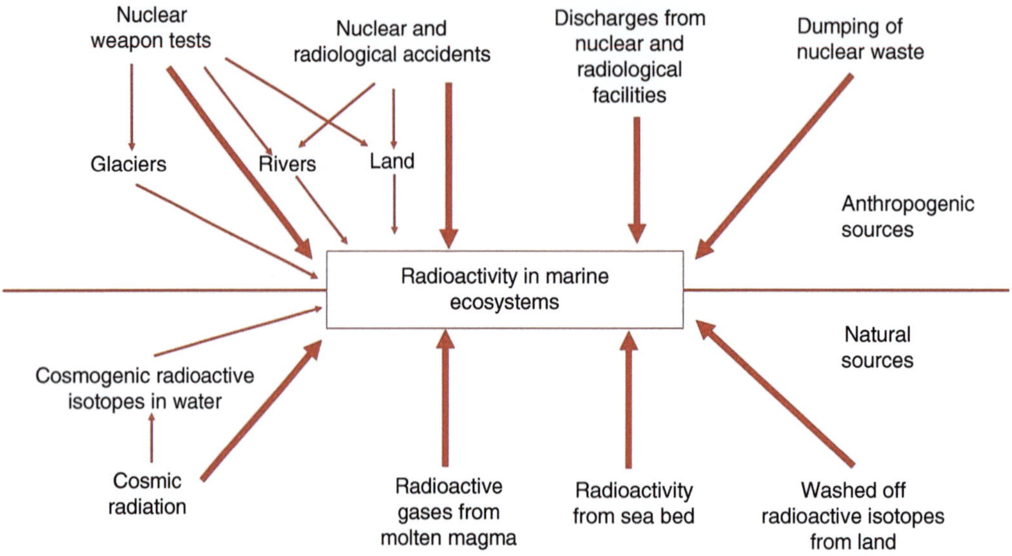

Figure 10.1 Sources of radioactivity in marine ecosystems.

10.2.1 Natural Radioactivity of Marine Ecosystems

Radioactivity was born with the birth of the universe. Life on the earth evolved under the positive influence of radiation. Almost everything on the earth is radioactive and the earth is continuously exposed to cosmic radiation. Even though it is not well discussed, the human body is radioactive. It is estimated that every second, about 9000 unstable nuclei decay in the human body emitting α, β, or γ rays (Grupen 2010).

Hence, naturally, the seawater, living beings in marine ecosystem, and seabed are radioactive. This is an important fact regarding radioactive pollution control as the presence of radiation does not necessarily mean that the sea is contaminated. The natural radioactivity of seawater is derived from a combination of decay of primordial radionuclides, the presence of radon in groundwater, radioactive isotopes in soil and rocks, and natural discharges from geological processes within the sea and shorelines. Some radioactive elements are washed out from terrestrial systems to the seas with river waters. Most of the heavy radioactive minerals are deposited in shorelines like thorium-containing monazite deposits in the east coast of Sri Lanka, one more example in the world.

The thorium series, the uranium series, and the actinium series are the three extended series of naturally occurring radioactive elements. When closely observing the different members of each series, there is characteristically a gaseous member, a very long half-life member and always the end product is the stable isotope of lead (^{206}Pb). Due to the differences in solubilities and/or half-life of compounds found in nature, the only members of the uranium series found in marine water in measurable amounts are uranium-238 (^{238}U) and radium-226 (^{226}Ra) (Love 1951). Attributed to higher solubility of potassium salts and radioactive isotope ^{40}K is the predominant, naturally occurring radioisotope in seawater (Draganić et al. 1991). Radioactive gases released by molten magma could have been added to radioactivity in seawater.

Cosmic rays are an important source of natural radioactivity and are composed of gamma rays, high-energy charged particles, and X-rays. Cosmic rays originated in space. On an average, cosmic radiation contributed 0.39 mSv (13%) out of a total of 3 mSv per year of natural background radiation exposure to human population. Cosmic rays contributed to several nuclear reactions in the earth's atmosphere resulting in radioactive isotopes such as ^{3}H, ^{10}Be, and ^{14}C. The intensity of the cosmic rays is greatly reduced at sea level due to atmospheric nuclear interactions. However, cosmic interactions occur within seawater creating isotopes such as ^{36}Cl (Lal 1999).

The radionuclides such as ^{238}U, ^{232}Th, and ^{40}K significantly contribute to natural radioactivity in both terrestrial and marine crusts (Al-Mur and Gad 2022). Levels of these radionuclides are often higher in seafood than in food produced on land, but they vary widely between species. However, ^{210}Po and 137Cs are contributing the most to the radiation dose from seafood (Aarkrog et al. 1995).

As a result of these radioactive elements, sediments, surface water, groundwater, rocks, soil, and biota in the world's oceans are naturally radioactive. Geographical contribution of radioactivity in seawater is quite similar and predominantly characterized by the naturally existing isotope ^{40}K in seawater (Solomon 1988).

10.2.2 Anthropogenic Radioactivity in Marine Ecosystems

The discovery of X-rays in 1895 by Wilhelm Roentgen, the discovery of radioactivity by Henri Becquerel, and the pioneering research work of Marie Curie and Pierre Curie opened the world to a new world of ionizing radiation. Soon after the discovery, wider applications sprung out around the globe. The increased use of radioactive and nuclear materials caused intentional, unintentional, or accidental addition of radioactivity to the environment. Several human activities have contributed to radioactive pollution of marine ecosystems.

Different sources contribute to this contamination such as dumping of nuclear waste in oceans, direct discharges of radioactive effluents into rivers/oceans, nuclear weapons testing, lost radioactive materials, and nuclear accidents. The main manmade sources are discussed below with examples.

10.2.3 Fallouts from Nuclear Weapon Tests

The end of World War II opened the door to the atomic era and most nations became nuclear powers during 1945–1964 (Prăvălie 2014). During this period, a large number of nuclear weapon tests were done, resulting in the contamination of both the marine environment and the atmosphere. This activity stands as the primary human source releasing radioisotopes into marine ecosystems, including ^{137}Cs, ^{14}C, and ^{90}Sr. Atmospheric nuclear weapons tests conducted during the period of 1945–1980 are the major contributor to the dispersed artificial radioactivity in the environment.

The Marshall Islands, Novaya Zemlya, Christmas Island, Lop Nop, French Polynesia, the Nevada test site, Semipalatinsk, and Johnson Atoll are key nuclear weapons test sites that have contributed to the contamination of the marine environment with radionuclides (Livingston and Povinec 2000).

The United States tested 66 nuclear devices between 1946 and 1958 in Enewetak and Bikini Atolls causing radioactive contamination of the aquatic environment of the atolls by fission products, unspent radioactive nuclear fuel, and activated products. Still, the sediments of these environments contain tens of terra becquerels (TBq) of the transuranic and activated products and some long-lived fission products (Robinson and Noshkin 1999). Over the concerns on contamination of seas by weapons tests, measurements of selected artificial radionuclides were first conducted in the 1960s in the Pacific Ocean. The study discovered that fallout radioactivity is not limited to the surface waters, but it has penetrated the deep ocean.

A global survey called The Geochemical Ocean Sections Study (GEOSECS) was conducted to survey the distributions of chemical, radiochemical tracers, and isotopes in the ocean and their deep thermohaline circulation. GEOSECS surveys were taken in the Atlantic Ocean (1972–1973), in the Pacific Ocean (1973–1974), and in the Indian Ocean (1977–1978). The survey provided the first comprehensive data on the lateral and vertical distributions of ^{90}Sr, Pu, and ^{137}Cs isotopes in studied oceans (Hamilton et al. 1996).

When considering the distribution of fall-out radioactivity, approximately 24% in the Southern hemisphere and 76% arrived in the Northern hemisphere were minimal at the equator and poles and maximal received at mid-latitudes (Livingston and Povinec 2000). According to the study of Hamilton et al. (1996), by 1990, the Pacific Ocean alone received 147 PBq of ^{137}Cs, 98 PBq of ^{90}Sr, and 3.8 PBq of Pu isotopes. The recorded ^{90}Sr was about one-third of the global total fallout. Further, the studies have shown that the measured total global deposit of long-lived fission products, such as ^{90}Sr and ^{137}Cs, agrees with the estimated fission yields (Hamilton 2005).

Having understood the impact of fallout radioactivity, Partial Nuclear Test-Ban Treaty was signed in 1963, by the United States, the Soviet Union, and the United Kingdom. This treaty banned nuclear weapon tests in the atmosphere, outer space, and underwater. It allowed only underground tests. In 1996, the United Nations General Assembly adopted the Comprehensive Nuclear Test Ban Treaty (CTBT) prohibiting all nuclear test explosions including underground tests. Hence, the fall out is not furthermore a source for radioactive and nuclear pollution. As a result of banning of tests, levels of fallout radionuclides, ^{90}Sr, Pu, and ^{137}Cs isotopes in the surface waters of the Pacific Ocean have reduced significantly and now are homogeneously spread (Hamilton et al. 1996).

10.2.4 Dumping of Radioactive Waste into the Oceans

A few decades ago, radioactive waste was dumped into the sea (IAEA 1995). The dumping of radioactive waste into the oceans has been a significant environmental concern due to its potential to contaminate marine ecosystems and pose risks to human health. While there have been instances of such dumping in the past, international regulations and agreements now prohibit or strictly regulate such practices.

The Mayak Production Association in Russia has a history of dumping radioactive waste into the Techa River and Lake Karachay, resulting in significant environmental contamination and health issues for nearby communities. The Soviet Union disposed of several decommissioned nuclear submarines, some containing radioactive waste, into the Kara Sea and other remote regions of the Arctic Ocean (Hamilton 1994). This has raised concerns about the potential long-term contamination of Arctic ecosystems. While not as large scale as some other examples, there have been instances of various countries dumping low-level radioactive waste into the ocean, mostly before stricter regulations were established.

As reported (Livingston and Povinec 2000), a very high amount of radioactive material has been disposed to the Pacific Ocean (1.4 PBq), Atlantic Ocean (45 PBq), and the Arctic Ocean (38 PBq) and adjacent seas. These nuclear wastes consisted of low-level materials enclosed in containers (found at disposing sites in the Pacific Oceans and Atlantic), although there were also complete reactor assemblies containing spent nuclear fuel (disposed of in bays of Novaya Zemlya in the Kara Sea) or reactor assemblies devoid of fuel (located in the Sea of Japan). Some of the nuclear wastes disposed to oceans were in liquid form such as disposal in the Sea of Japan and the Barents Sea. This was practiced in the Sea of Japan as late as 1993 (Livingston and Povinec 2000).

It is important to note that many of these practices have been largely discontinued or significantly reduced due to international agreements such as the 1972 Convention on the Prevention of Marine Pollution by Dumping of Wastes and Other Matter (London Convention) and the 1992 Convention for the Protection of the Marine Environment of the North-East Atlantic

(OSPAR Convention), which are aimed to prevent marine pollution by prohibiting or regulating the dumping of hazardous substances, including radioactive waste, into the oceans. However, the legacy of past dumping activities continues to have environmental implications in affected regions. Efforts to monitor and mitigate these impacts remain ongoing.

10.2.5 Discharges from Nuclear and Radiation Facilities and Accidents

Regular activities at fuel reprocessing facilities, nuclear power plants, medical facilities, and research centers may lead to the discharge of zero or minimal amounts of radioactive substances into the water. However, worldwide, these are the facilities under the highest level of regulatory control. Hence, for any such facility, radiation regulations are applied to minimize or control discharges and reduce the damage caused to the environment including the marine environment. As highlighted by Guéguéniat et al. (1996), nuclear power plants located near the sea cause very little environmental damage directly or indirectly.

Accidents happen in nuclear facilities such as nuclear fuel reprocessing plants, nuclear power plants, nuclear submarines, and nuclear-powered vessels that can release considerable amounts of radioactive and nuclear materials into the oceans. Here are some specific examples of the nuclear accidents that caused the impact of radioactive and nuclear marine pollution.

10.2.6 Chernobyl Nuclear Powerplant Accident

The Chernobyl Nuclear Power Plant accident happened on 26 April 1986, in the city of Pripyat, part of the Soviet Union, in the Ukrainian Soviet Socialist Republic (Present Ukraine). This accident remains alone as the worst nuclear disaster in history. During a late-night safety test, reactor number 4 experienced a catastrophic explosion due to a combination of operator errors, design flaws, communication failures, and safety system failures. The steam explosion caused the reactor's core to burst and discharge an immense number of radioactive materials into the atmosphere. The radioactive fallout spread over large parts of the Soviet Union and Europe, contaminating vast areas with dangerous levels of radiation. The city of Pripyat and surrounding areas were evacuated, and a 30-km exclusion zone was established around the plant to limit access to the highly contaminated areas. The immediate aftermath saw emergency responders and plant workers exposed to high levels of radiation, resulting in acute radiation sickness and fatalities. Long-term health effects, including increased prevalence of cancer and other illnesses, have been identified among those exposed to radiation due to the accident.

As the Chernobyl plant was not closer to the sea, there was no direct discharge of radioactivity into the oceans. However, the fallout radioactivity has created some marine pollution. The radionuclides found in significant quantities in the marine environments following the Chernobyl accident were ^{137}Cs, ^{134}Cs, ^{106}Ru, ^{103}Ru, and ^{m110}Ag. ^{137}Cs and ^{134}Cs were found in Ukraine shoreline seawaters and the concentrations were extreme in areas nearby to those where deposition over land was the highest. However, the measured levels were well below the established public exposure levels and the levels declined within a shorter period.

A considerable portion of the North Sea has been thoroughly surveyed, along with the Norwegian Sea. Evidence of Chernobyl radio cesium was detected as far north as 70°N and in many of these regions, including the northern North Sea. As Cs and Ru radionuclides concentrate in seaweeds and shellfish, the peak isotopic values were observed within a few days of the arrival of fallout over the United Kingdom (Holm 1995). After the accident, much attention was paid to the Baltic Sea, as Chernobyl fallout was detected in the adjacent regions (Povinec et al. 1996). The effects of fallouts were more prominent in the marine ecosystems of the Baltic Sea compared to the Black Sea or the Mediterranean Sea (Ilus 2007).

The Chernobyl disaster forced many nations to start environmental radioactivity measurements. Many countries have drafted laws and started radioactivity measurements of import and export food items including fish and fisheries products.

10.2.7 Fukushima Power Plant Accident

The Fukushima Daiichi Nuclear Power Plant was in the Fukushima Prefecture on the eastern coast of Japan. The Fukushima Daiichi nuclear disaster happened on March 11th of 2011, following a massive earthquake followed by a tsunami off the coast of Japan. The earthquake knocked out the external power supply to the facility and a tsunami wave flooded the backup generators, leading to the failure of the cooling systems in three of the plant's nuclear power reactors. Without proper cooling, the reactor cores overheated, leading to meltdowns, where the nuclear fuel rods melted through their containment vessels causing hydrogen explosions.

This disaster led to the discharge of large amounts of radioactive materials into the atmosphere and Pacific Ocean. This marked the most extensive "individual" unintentional discharge of anthropogenic radioactivity into the marine environment from civilian nuclear facilities (Beresford and Batlle 2012).

In contrast to Chernobyl, there was no significant explosive release of reactor core materials to the environment. Hence, the damage due to atmospheric fallout is smaller and limited primarily to radioactive gases and fission products such as cesium, which become volatile at elevated temperatures. The largest environmental impact was created by the fuel rod materials and nonvolatile activated products released when the acidic waters and corrosive brines were used to cool the reactors. The water interacted with the damaged fuel rods and carried radioactive materials into the ocean and ground. The full magnitude of the release and spectrum of isotopes released have not been well documented. As emergency cooling continued, radionuclides were kept on discharging directly into the Pacific Ocean during the initial months after the accident. Nearly 80% of the release to the Pacific Ocean occurred between 11 March and 8 April 2011. Major radionuclides released into the marine environment include ^{131}I,^{132}I, ^{134}Cs, ^{136}Cs,^{137}Cs, ^{129}Te, ^{129m}Te, and^{132}Te. However, due to the decay, short half-life radioisotopes had largely disappeared and ^{134}CS and ^{137}Cs became the dominant radionuclides at the end of May 2011 (UNSCEAR 2014).

Significant information on the level of several radioisotopes of Cs and I in the sea near the water release point was available shortly after the disaster (Garnier-Laplace et al. 2011). According to Buesseler et al. (2011), by July 2011, the radioactivity levels of ^{137}Cs at discharge points had exceeded those measured in 2010 in the coastal waters off Japan by more than 10000 times. Radioactivity concentrations of ^{137}Cs in June and July 2011, at the Dai-ni site (10 km from the Fukushima site) and Iwasawa Beach (16 km from the accident site), were of 4000–10000 Bq/m^3, respectively (Sanial et al. 2017). These values are equivalent to permissible drinking water limits for ^{137}Cs by the United States (7400 Bq/m^3) and the World Health Organization (10000 Bq/m^3).

A discussion was created on the risk of consumption of seafood and biological uptake from the sea closure to the Fukushima facility and many people avoided marine product consumption immediately after the accident. However, dose calculations suggested minimal effect on humans or marine biota due to direct exposure to surrounding ocean waters (Buesseler et al. 2011). Many countries have started radioactivity measurements in local and imported fish and fisheries products. Such activities were largely abandoned within a period of two to three years after the accident because no significantly higher levels were found.

The Fukushima disaster was eye opening for the potential marine radioactive pollution. Several countries have initiated marine radioactivity measurements after the Fukushima accident with the help of the International Atomic Energy Agency. For example, in Sri Lanka, the first comprehensive monitoring program around the coast was conducted in 2012 after the Fukushima nuclear power plant accident. ^{137}Cs was the only detectable gamma emitter in seawater samples and it implies that downward trend during the period from 1962 to 2019 (Madhusha et al. 2021). No detection of ^{134}Cs, which has only two years half-life, confirmed that global fallout is the only source of anthropogenic radionuclides in the Sri Lankan coastal environment and no fresh input of ^{137}Cs from the Fukushima nuclear power plant accidents to the Sri Lankan marine ecosystem.

10.2.7.1 Fukushima Water Release

More than 1.3 million cubic meters of water were sprayed into overheated reactor cores and spent fuel pools to prevent further damage. Thus, groundwater and rainwater entering the reactor buildings mixed with radioactive debris result in water contamination with 64 radionuclide elements (International Atomic Energy Agency (IAEA) 2023). The highly radioactive water was collected by the owner of the Fukushima Plant, Tokyo Electric Power Company (TEPCO), and securely stored on-site in more than 1000 specialized tanks to prevent its release into the environment. TEPCO faced challenges in storing large quantities of contaminated water and reaching the occupational exposure limit for the workers on the site. Considering these difficulties, TEPCO deployed a method to remove most of the radionuclides in the contaminated water, called advanced liquid processing system (ALPS), which reduces the radionuclides level to the acceptable safety limits (IAEA 2023).

Cesium and strontium comprise the highest amounts of contaminants in the water. Before being treated by the ALPS system, strontium and cesium are reduced by the devices called KURION and SARRY filtering systems (IAEA 2023) (Figure 10.2).

The ALPS system essentially operates as a filtration and pumping mechanism, employing ion exchange to remove 62 radionuclides from the contaminated water. However, ALPS is not effective in removing tritium which is a mildly radioactive isotope of hydrogen and carbon-14. The treated water is stored in large tanks of nearly 1,000 m^3 each. TEPCO and IAEA have conducted a

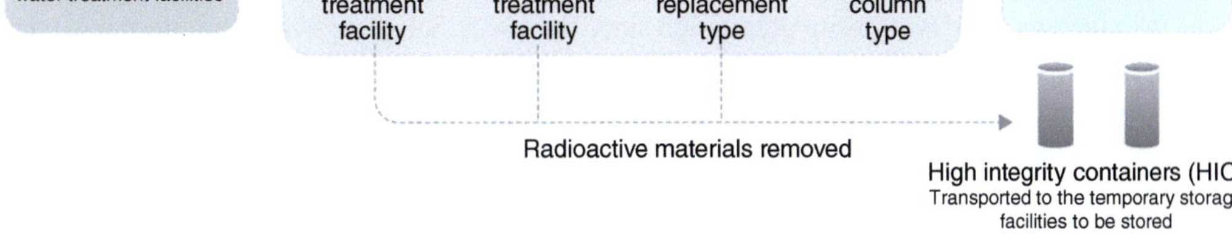

Figure 10.2 Details of water treatment process of contaminated water at Fukushima power plant to remove radionuclides. *Source:* IAEA (2023).

monitoring program to measure the radionuclide levels in treated water in various stages and at the discharging points.

After a decade of the disaster, in April 2021, the Government of Japan announced its policy on the discharge of treated water using the ALPS system into the ocean. The decision to release the treated water into the ocean has been met with controversy and concern from various stakeholders, including residents, fishermen, environmental groups, and neighboring countries. While proponents argue that the release is necessary to free up space at the Fukushima plant and to prevent further environmental damage from potential leaks or accidents, opponents raise fears about the potential impact on marine life, public health, and the livelihoods of fishing communities. Conversely, the Japanese government had a long-term plan for this, but this was a debated issue among many countries as well as the Japanese public (Wong 2023).

Environmental activists and some scientists have opposed the plan emphasizing that it is important to conduct a comprehensive study on radiological and ecological impact assessment for marine ecosystems. China has been the most vocal, condemning Japan for violating international moral and legal obligations (Wong 2023). Thousands of Korean people came to Seoul and protested, forcing the country's government to take appropriate action against this. As a response, TEPCO agreed to make available online real-time data on the radioactivity levels of water, on an online platform assigned to explaining the treatment and release processes in multiple languages (Wong 2023). After a two-year study, the IAEA issued a report saying there is a negligible radiological impact on people and the environment by releasing the treated water (Wong 2023). Japan commenced releasing treated radioactive contaminated water from the inactivated Fukushima Daiichi Nuclear Power Station into the Pacific Ocean, 12 years after the Fukushima accident in 2023. This has created a global concern and nearby countries have enacted many measures against contamination of their populations from this water.

IAEA conducted a continuous monitoring program with the Government of Japan before and after the discharge of ALPS-treated water into the ocean. The IAEA final report declared that treated water discharging reliable with the international safety regulations and the radioactivity level of treated water in the safe limits. And there are no environmental and health risks associated with the treated water in Japan and neighboring countries (IAEA 2023). The major radioactive isotope present in the released water is tritium (^{3}H). The IAEA had confirmed that the level of tritium in the water discharged was far below the safety limit of 1500 Bq/l. This level is lower than the limit of the World Health Organization for drinking water, which is 10 000 Bq/l.

It should be highlighted that the Fukushima accident prompted increased scrutiny of nuclear safety measures globally. As a result, several safety measures were upgraded particularly in passive safety systems. Further, the lessons learned have promoted new-generation power reactors (Gen III+) with advanced safety features. Hence, the likelihood of a similar accident is very small.

10.2.7.2 Discharges from the Nuclear Reprocessing Plants

Sellafield, formerly known as Windscale, has been involved in nuclear fuel reprocessing, nuclear waste management, and plutonium production since the 1940s. Sellafield has experienced few incidents over the years, including leaks, spills, and safety breaches, prompting scrutiny and criticism from environmental groups and the public. One of the most notable incidents at the site occurred in 1957 when the Windscale fire broke out in one of the reactors, resulting in the release of radioactive material into the atmosphere. Sellafield plant has been releasing radioactive materials into the Irish Sea for decades. This caused elevated levels of radioactivity in local fish and shellfish in Irish Sea.

Similarly, a significant number of radio isotopes have been discharged into the ocean by fuel reprocessing plants at several locations. The Bhabha Atomic Research Centre (BARC) at Trombay, near Mumbai, India, has been involved in nuclear research and reprocessing activities. Discharges from these activities, including liquid effluents containing radioactive isotopes, have been released into the nearby Arabian Sea, raising environmental concerns. The Tarapur Atomic Power Station, located on the west coast of India in Maharashtra, is one of India's oldest nuclear power plants. It has a reprocessing plant associated with it, which has also led to the discharge of radioactive isotopes into the Arabian Sea. The La Hague nuclear reprocessing plant is located on the northwest coast of France. It is one of the largest nuclear reprocessing facilities in the world. Discharges from this facility into the English Channel have been a subject of international concern due to the potential environmental impact on marine ecosystems.

These facilities have been subject to regulatory oversight, and efforts have been made to minimize the environmental impact of their operations. However, concerns about the release of radioactive isotopes into the oceans persist, and ongoing monitoring is necessary to assess their impact on marine life and ecosystems.

In addition to the above-mentioned anthropogenic sources, accidents in nuclear submarines and nuclear-powered ships, accidents during transport of nuclear materials over the sea, and nuclear war can add significant amount of radioactive and nuclear materials to the marine ecosystems.

10.3 Important Radionuclides to the Marine Ecosystems, Their Chemistry, and Behaviors

When considering the radionuclides that affect the radioactive pollution in the marine environments, more attention is focused on anthropogenic radionuclides. Cesium (^{137}Cs) and strontium (^{90}Sr), with half-lives of 30.05 and 28 years respectively (Bu et al. 2016), are identified as the most significant contributors to the radioactivity levels in marine environments due to their prolonged presence and persistence. ^{137}Cs and ^{134}Cs are produced as a by-product of fission reactions involving ^{235}U or ^{239}Pu. These two isotopes can enter marine ecosystems through nuclear accidents, industrial discharges, and the waste from nuclear facilities. Considering the short half-life of 2.06 years, ^{134}Cs acts as a distinct marker of the impact of nuclear accidents.

Cesium is an alkali metal, readily loses its outermost electron to form a positively charged ion (Cs^+). The distribution coefficient K_d (ratio between solid and liquid activities, l/kg) values give a better understanding of the interaction between radioisotopes and solid matrices such as marine sediments (Smodiš et al. 2003). The K_d values for radio cesium are decreasing from 6.66×10^3–1.35×10^5 in freshwater (Tomczak et al. 2019) to 4.5×10^2–2×10^3 in seawater (IAEA 2004), which is lower than to some transuranic elements (plutonium, americium, and curium) and many other important radionuclides (IAEA 1985). It emphasizes that retention and sorption of radio cesium in environmental systems is relatively less pronounced compared to many other radionuclides, especially certain transuranic elements.

The radioactive isotopes ^{129}I and ^{131}I are by-products of nuclear fission found in the wastes generated during the reprocessing of nuclear fuel with half-life 15.7 million years and 8 days respectively (IAEA 1985). Iodine is present in the aqueous solution as a simple anion as iodate (IO_3^-) or iodide (I^-), which is very poorly adsorbed in sedimentary materials and very soluble in seawater (IAEA 1985) (Zhang 1993).

The behavior of the tritium (^{3}H) molecule resembles that of the normal (H) water molecule. However, there is a slight difference in evaporation and condensation of tritiated water molecule from that of the normal water molecule due to the mass difference.

^{90}Sr is a fission product of nuclear reactors and weapons. It is a pure β-emitting radionuclide decaying to yttrium-90 (IAEA 1985). The physical half-life of ^{90}Sr is 28.79 years and a biological half-life of about 13 years (Uesugi et al. 2018). It has been identified as one of the most hazardous artificial radionuclides contributing for radiological pollution due to radiotoxicity and long half-life (Kim et al. 2023). The behavior of ^{90}Sr in the marine environment is identical with the stable counterpart of ^{89}Sr and follows exactly the similar chemical properties of calcium (Kim et al. 2023). ^{90}Sr is highly soluble in seawater and it easily accumulates in mammalian bones causing bone cancer due to radiation impact (Uesugi et al. 2018).

^{210}Po poses no external hazard but it does pose a hazard if ingested, inhaled, or if it enters the body through an opening in the skin. ^{226}Ra is taken by plants from the soil and transferred up the food chain to humans. Chemically like to calcium, radium is partially absorbed across the small intestine upon ingestion, with the majority deposited in bone, where 70–95% of the total ingested body radium is typically found.

Anthropogenic radionuclides contribute less than 0.3% of the dose to humans from marine food chains. ^{239,240}Pu, ^{137}Cs, and ^{90}Sr are the most significantly contributing three anthropogenic radioisotopes to humans via seafood consumption (IAEA-TECDOC-1429 2005). Furthermore, natural radionuclides, ^{210}Po and ^{210}Pb, are the major contributors for the total radiological dose from the marine foods (Holm 1995).

The physical half-life, often simply referred to as the half-life, is a term used in nuclear physics to describe the time it takes for half of the atoms in a radioactive substance to decay into a different element or isotope through radioactive decay. The biological half-life refers to the time it takes for half of a substance to be eliminated from the body through biological processes such as metabolism or excretion. The effective half-life refers to the time it takes for the concentration or activity of a substance in the body to decrease by half, considering both its physical decay and biological elimination processes (Table 10.1).

10.4 Impact of Radioactive and Nuclear Pollution on Marine Ecosystem

The effects of radioactive and nuclear marine pollution on marine life are different. Few of the potential impacts are given below.

10.4.1 Mutations and Genetic Damage

Radiation can harm the DNA of marine organisms, leading to mutations and birth defects. Radiation can indeed pose significant risks to marine organisms by damaging their DNA, potentially resulting in mutations and birth defects. Ionizing radiation emitted from sources such as nuclear facilities, radioactive waste, or fallout from nuclear accidents can penetrate marine environments and interact with the genetic material of marine organisms. This interaction can cause breaks in the DNA strands, alterations in genetic sequences, and other forms of damage that compromise the integrity and stability of the genome. The IAEA report from the Chernobyl Forum Expert Group on the Environment has revealed an increase in sex-linked recessive lethal mutations in natural populations of *Drosophila melanogaster* following the Chernobyl accident (IAEA 2006). Mutations level increased during the period of 1986–1987 among flies residing in areas tainted with radiation, initially experiencing exposure rates of 2 mGy/h or higher. In the subsequent two years, mutation frequencies gradually returned to their normal levels. Similar changes can be expected in marine organisms too. Kolar and Gugleta

Table 10.1 Important radionuclides in marine ecosystems and their characteristics.

Element	Availability	Half-life	Mode of decay	Biological half life
^{137}Cs	Byproducts of nuclear fission in weapons testing nuclear reactors	30.07 yr	Beta and gamma radiation	110 d
^{134}Cs		2.1 yr	Beta and gamma radiation	70 d
^{131}I		8.1 d	Strong gamma emitter	120 d
^{90}Sr	Fission by-product resulting from nuclear weapons testing and accidents, and from the discharge of radioactive waste	28 yr	Beta emitter	18 000 d
^{40}K	Naturally occurring and widely distributed in nature and is present in all plant, sediment, and animal tissues	1.3 billion yr	Beta and gamma radiation	16 d
^{210}Po	Naturally occurring primordial nuclide	138 d	Alpha radiation	60 d
^{226}Ra	Naturally occurring and can be found in various geological formations, particularly in uranium and thorium ores	1600 yr	Alpha and gamma radiation	
^{238}U	Naturally occurring primordial nuclide	4.468 billion yr	Alpha and gamma radiation	
^{3}H	Naturally occurring	12.33 yr	A very low-energy beta emitter	10–12 d
^{239}Pu	Primarily produced through nuclear reactors as a byproduct of uranium fission. Uranium-238 (^{238}U) absorbs a neutron in a reactor, becoming Uranium-239, which quickly decays into neptunium-239 (^{239}Np), and then further captures another neutron, transforming into plutonium-239 (^{239}Pu). While trace amounts of plutonium isotopes, including Pu-241, can be found in the environment due to nuclear testing, accidents, or the decay of other radioactive elements	24 110 yr	Alpha radiation	Bone, 50 yr; liver, 20 yr; gonads, considered permanent
^{240}Pu		6563 yr	Alpha radiation	
^{241}Pu		14.35 yr	Beta radiation	

IAEA-TECDOC-1429 (2005), DOH (2022), and Rahola and Suomela (1975).

(2019) have reported embryonic defects and embryonic death of fish. However, the impact depends not only on the radiation level but also on different factors including isotope, species, individual stages of growth, and tropic relationships.

10.4.2 Bioaccumulation

Pollution of marine ecosystems leads to the accumulation of radioactive isotopes in seawater, marine sediments, algae, and marine fauna. It has recorded bioaccumulation of radioactive isotopes in phytoplankton. This accumulation is visible and magnified through the food chains.

Isotopes of radioactive Iodine were found to be concentrated in microalgae and mollusks followed by fish (Kryshev et al. 2012). The studies conducted after the Fukushima accidents have confirmed the same pattern. The bioaccumulation of radioactive isotopes can result in harmful effects in physiology, reproduction, growth, and development. Further, damage to the genetic materials can be caused. Higher concentrations might cause the mortality of the organism. As marine organisms are an important and major player in food webs and food chains of humans, the risk of these detrimental effects is high to the humans if contaminated seafood is consumed (Uddin et al. 2015).

Radio cesium is a radioisotope which is biomagnified through planktonic food chains. A study conducted by Zhao et al. (2001) quantified the uptake of radio cesium by marine piscivorous fish, mangrove snapper (*Lutjanus argentimaculatus*). The uptake was reported to be dependent on cesium concentration, reduced salinity, and prey type. Further, the study has revealed immediate transport of absorbed cesium radioisotopes to the muscles. A joint research conducted by IAEA after the Chernobyl accident has studied the fate of long-lived radionuclides ^{137}Cs and ^{90}Sr in the Black Sea and across the Mediterranean, Atlantic, and Antarctic Oceans. Results showed that besides radioactive decay and distance from the accident site, the main factors controlling long-term changes in surface water concentration are the vertical mixing of the water masses and the water exchange via the Bosporus and Gibraltar Straits connecting the Black Sea, Mediterranean, and Atlantic basins (Gulin et al. 2012).

Temperature can be affected by the uptake of radioactive isotopes. As per Nakahara et al. (1977), uptake and release of radioactive ^{65}Zn by mackerel (*Scomber japonicus*) were increased with water temperature. Temperature dependence of ^{137}Cs concentration by blackfish (*Girella punctata*) was found to be most significant at 20–25 °C. Accumulation of ^{60}Co in midgut gland of clams (*Meretrix lamarkii*) was also affected by temperature increase.

Fish are considered the most radiosensitive aquatic organisms. In the first month after the Chernobyl accident, the ^{137}Cs concentrations were the highest in prey fish, whereas a few years later, the highest concentrations were recorded in predatory fish such as perch and pike (Lerebours et al. 2018).

As reported by Bezhenar et al. (2016), elevated concentrations of ^{137}Cs have been found in sediments, benthic organisms, and demersal fishes in the coastal zone around the Fukushima accident site. The bottom invertebrates ingest contaminated organic parts in sediments have adsorbed radionuclides which have migrated through the food chain (Garnier-Laplace et al. 2011). As humans are at the top of food chains, high risk is there for bioaccumulation of radioactive isotopes.

10.4.3 Psychological Impact

In general, people have a negative attitude about radioactivity due to nuclear bombs and nuclear accidents. They have a common concern on potential radioactive pollution from different nuclear and radiological facility. It seems that many communities around the globe have incorrect and exaggerated assessments of the effects of radioactivity in marine ecosystems. These attitudes and perceptions can largely disturb the fisheries and economy of fishing communities even at a minor radioactive release (Heldal 2001).

Several discussions were created on the risk of consumption of seafood and biological uptake from the seas closure to the Fukushima facility (WHO/FAO, 2011). Many people in the region including Japanese avoided marine products consumption immediately after the accident (Miyata & Wakamatsu, 2018). Several countries have started radioactivity measurements in local and imported fish and fisheries products. As the scientific data proved minimal effect on humans or to marine biota, regular consumption was started back (Batlle J, 2016). So, the psychological panic and refusal of seafood was largely limited to several weeks.

When the conditioned contaminated water is released to the Pacific Ocean in 2023 from Fukushima, widespread psychological concern was observed in the neighbouring countries as well as in Japan (Shalvey & Trotter, 2023). This has led to a significant reduction of consumption of fish and other marine-derived foods. Some neighbouring countries stopped the importation of fish and fisheries products from Japan causing a considerable economic loss and fisheries communities were affected. Despite the assurance given by IAEA and the Government of Japan on the level of radioactivity is within the permissible limits, all most all the countries around Japan started a vigorous radioactivity measurement of fish and fisheries products. The data from different countries revealed no significant changes in the radiation levels or contaminations; the concern faded within three months.

10.5 Monitoring of Radioactivity in Oceans and Marine-Originated Products

There are naturally occurring and anthropogenic radionuclides in marine environments. Therefore, it is important to identify the level of different radioisotopes in marine ecosystems around us. Monitoring radioactivity helps in safeguarding marine ecosystems from potential contamination, preserving biodiversity, and preventing adverse impacts on marine life. Marine-originated products, such as seafood, are essential components of many diets worldwide. Understanding of the radioactivity levels in seafood ensures the safety of these products for human consumption, minimizing the risk of exposure to harmful levels of radiation. Further, radioactive contaminants can enter the marine environment through various pathways, including nuclear accidents, industrial discharges, and natural processes. Many countries have established regulations and guidelines for permissible levels of radioactivity in marine environments and marine products. Regular monitoring ensures compliance with these regulations, thereby maintaining public health and safety standards.

A comprehensive database enables authorities to assess the safety of marine-derived products for human consumption, thereby protecting public health from potential radiation hazards. In the event of nuclear accidents, industrial spills, or other radiological incidents, having a pre-existing database of baseline radioactivity levels in the marine environment is crucial for detecting abnormal spikes in radiation levels. This facilitates prompt emergency response measures to minimize environmental and public health risks. Further, a comprehensive database provides the scientific evidence needed to establish protective measures, set regulatory limits, and develop strategies for mitigating potential risks to the marine environment and human health. Two such databases are discussed below.

10.5.1 Marine Radioactivity Information System (MARIS)

The MARIS serves as a critical tool for managing and understanding radioactivity in marine environments. MARIS is maintained and developed by the IAEA Radiometric Laboratory Monaco. It contains marine radioactivity data in sediments, seawater, biota, and suspended matters collected from scientific articles, research studies, and radiological surveys around the world. As reported by the IAEA, more than half a million marine radioactivity measurements in this database represent more than 100 different radionuclides or radionuclide ratios.

MARIS is an online and downloadable database for users since 2005. This information is currently used for radiological assessments, and it helps to investigate marine processes such as water-mass transport, carbon cycling, and sedimentation rate. Some important features of MARIS include the following (Figure 10.3).

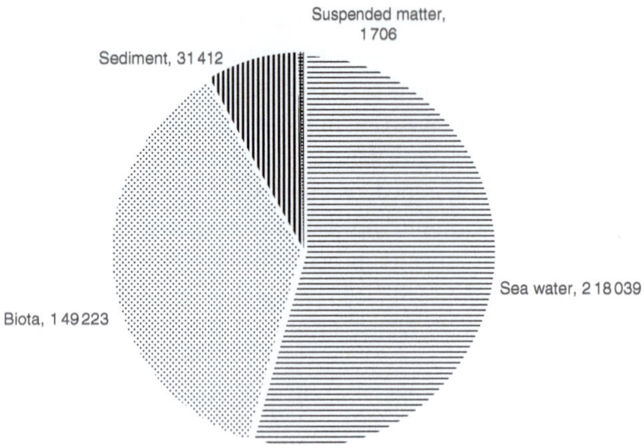

Figure 10.3 Details of the data available in MARIS database up to May 2024.

10.5.1.1 Data Centralization

MARIS collects, organizes, and centralizes data related to marine radioactivity from various sources, including research institutions, regulatory bodies, and monitoring programs. This centralized repository facilitates easy access to comprehensive information for researchers, policymakers, and the public.

10.5.1.2 Monitoring and Early Warning

MARIS enables continuous monitoring of radioactivity levels in oceans, coastal areas, and marine-derived products. By analyzing trends and anomalies in the data, MARIS can provide early warning of potential radiological hazards, such as contamination events or unusual spikes in radioactivity levels.

10.5.2 Risk Assessment and Management

MARIS supports risk assessment and management efforts by providing valuable insights into the sources, pathways, and behavior of radioactive substances in marine ecosystems. This information helps policymakers and environmental agencies make informed decisions to mitigate risks and protect marine environments and human health.

10.5.3 International Collaboration

MARIS facilitates collaboration and information exchange among countries and international organizations involved in monitoring and managing marine radioactivity. By promoting data sharing, harmonizing methodologies, and fostering cooperation, MARIS strengthens global efforts to address cross-border challenges related to marine radioactivity.

10.5.4 Public Awareness and Education

MARIS plays a crucial role in raising public awareness and understanding of marine radioactivity issues. Through accessible data portals, educational resources, and outreach activities, MARIS helps to educate stakeholders about the importance of monitoring, the potential risks associated with marine radioactivity, and the measures taken to ensure environmental and public safety.

10.5.5 Policy Support and Decision-Making

MARIS provides scientific evidence and data-driven insights to support the development of policies, regulations, and strategies for managing marine radioactivity. Policymakers can rely on MARIS to make evidence-based decisions that balance environmental protection, public health, and socioeconomic considerations.

Overall, the MARIS serves as a valuable platform for enhancing scientific understanding, promoting international cooperation, and guiding effective management of radioactivity in marine environments. Its role in monitoring, data sharing, risk assessment, and public engagement is indispensable for safeguarding marine ecosystems and ensuring sustainable use of marine resources.

10.5.6 Asia and Pacific Marine Radioactivity Database (ASPAMARD)

ASPAMARD is a radionuclide measurements database that includes sediment, seawater, suspended matter, and biota originates that from a range of sources including published research studies, technical reports, and scientific papers within Asia Pacific Institutes of IAEA Member States. This database was launched in 2011 as a joint project among the IAEA, 10 IAEA member states, and 3 nonmember states.

ASPAMARD is maintained by the Philippines under the supervision of the IAEA Marine Environment Laboratories in Monaco collaboratively with Asia Pacific IAEA member states. The ASPAMARD database provides comprehensive and up-to-date data and information on current and historical marine radioactivity levels in the region. As a result of the Fukushima nuclear power plant accident, Japan and neighboring countries conduct monitoring program to measure the radioactive levels in seawater, marine sediments, biota, and suspended particles. All data were initially collected by ASPAMARD and also submitted to MARIS. ASPAMARD radioactive measurement data is finally submitted to the IAEA global marine database (MARIS) run by the IAEA. This ASPARMD database improves the exchange of data gathered from marine measurements undertaken by the Asia Pacific region countries, as well as the information about the radiological and environmental impact assessments of the radioisotopes and support for risk assessments to marine biota and potentially to seafood safety.

10.6 Preventive Measures on Radioactive and Nuclear Marine Pollution

Preventive measures for radioactive and nuclear marine pollution are essential to minimize the risks associated with accidental releases, intentional acts, or routine operations involving radioactive materials. These measures typically involve a combination of regulatory frameworks, technical standards, monitoring systems, and emergency preparedness strategies. Some key preventive measures include the following.

10.6.1 Stricter Regulations on the Disposal of Radioactive Waste

International agreements and national regulations are crucial to minimize the risk of radiation hazards. Stricter regulations on the disposal of radioactive waste constitute a pivotal preventive measure in mitigating radioactive and nuclear marine pollution. By imposing stringent guidelines and standards governing the handling, storage, transportation, and final disposal of radioactive materials, these regulations aim to minimize environmental and public health risks. They establish clear criteria for classifying and categorizing radioactive waste, ensuring that appropriate disposal methods and safety requirements are implemented. Moreover, these regulations mandate licensing and permitting processes, requiring facilities to undergo rigorous safety assessments and environmental impact evaluations before obtaining permits for waste management activities. By promoting waste minimization, enhancing packaging and transportation protocols, and emphasizing site selection and characterization, stricter regulations on radioactive waste disposal work to prevent accidents, contain radioactive contaminants, and safeguard marine ecosystems from the adverse effects of radioactive pollution.

10.6.2 Improved Safety Standards for Nuclear Facilities

Improved safety standards for nuclear facilities serve as crucial preventive measures against radioactive and nuclear marine pollution by boosting the resilience and reliability of nuclear infrastructure. These standards encompass robust design criteria, advanced safety systems, and stringent operational procedures aimed at minimizing the risk of accidents and the release of radioactive materials into marine environments. By emphasizing comprehensive risk assessments, emergency preparedness, and continuous monitoring, improved safety standards enhance the ability of nuclear facilities to detect, mitigate, and respond effectively to potential incidents, thereby reducing the likelihood of radioactive contamination of marine ecosystems.

10.6.3 Monitoring and Research

Ongoing monitoring of radioactive levels in the marine environment is essential to assess the extent of contamination and identify potential risks. Monitoring and research play pivotal roles in preventing radioactive and nuclear marine pollution by providing essential data and insights for informed decision-making and proactive intervention. Continuous monitoring of marine environments allows for the early detection of abnormal radiation levels, enabling prompt responses to potential contamination incidents and minimizing their impact on marine ecosystems. Additionally, research efforts contribute to understanding the behavior, transport, and fate of radioactive substances in marine environments, facilitating the development of predictive models and risk assessments. Identification of sources of contamination, assessment of exposure pathways, evaluation of ecological risks, monitoring and research efforts pave the path for the implementation of preventive measures, such as improved safety standards, stricter regulations, and emergency preparedness plans, thus mitigating the likelihood and severity of radioactive and nuclear marine pollution. Furthermore, ongoing monitoring and research foster collaboration among scientists, policymakers, and stakeholders, promoting transparency, accountability, and innovation in addressing the complex challenges of radioactive pollution in marine ecosystems.

10.6.4 Public Education and Awareness

Public education and awareness are integral components in mitigating radiation hazards as they empower individuals and communities with the knowledge and understanding needed to protect themselves and the environment. By raising awareness about the sources, risks, and effects of radiation, public education initiatives foster a culture of safety and preparedness, encouraging proactive measures to minimize exposure and mitigate potential hazards.

By implementing these preventive measures, stakeholders can reduce the likelihood of radioactive and nuclear pollution in marine environments and minimize the associated risks to human health, marine ecosystems, and coastal communities. By taking these steps, we can work toward a better future where the oceans are free from radioactive and nuclear contamination.

References

Aarkrog, A., Baxter, M.S., Bettencourt, A.O. et al. (1995). A comparison of doses from Cs-137 and Po-210 in marine food: a major international study. *Journal of Environmental Radioactivity* 31 (1): 69–90.

Al-Mur, B. and Gad, A. (2022). Radiation hazard from natural radioactivity in the marine sediment of Jeddah Coast, Red Sea, Saudi Arabia. *Marine Science and Engineering* 10 (8): 1145.

Batlle J, V. I. (2016). Impact of the Fukushima accident on marine biota, five years later. *Integr Environ Assess Manag*, 654–658. doi:10.1002/ieam.1825. PMID: 27447852

Beresford, N., Vives i Batlle, J., & Vandenhove, H. (2012). One year on - Some thoughts about the ecological fallout from Fukushima. *The Scientist*, 26 (3), 29–29.

Bezhenar, R., Jung, K.T., Maderich, V. et al. (2016). Transfer of radiocaesium from contaminated bottom sediments to marine organisms through benthic food chains in post-Fukushima and post-Chernobyl periods. *Biogeosciences* 13 (10): 3021–3034.

Bu, W., Zheng, J., Liu, X. et al. (2016). Mass spectrometry for the determination of fission products 135Cs, 137Cs and 90Sr: a review of methodology and applications. *Spectrochimica Acta Part B* 1219: 65–75.

Buesseler, K., Aoyama, M., and Fukasawa, M. (2011). Impacts of the Fukushima nuclear power plants on marine radioactivity. *Environmental Science & Technology* 45 (23): 9931–9935.

DOH (2022). Washington State Department of Health. https://doh.wa.gov/community-and-environment/radiation

Draganić, I.G., Bjergbakke, E., Draganić, Z.D., and Sehested, K. (1991). Decomposition of ocean waters by potassium-40 radiation 3800 Ma ago as a source of oxygen and oxidizing species. *Precambrian Research* 52 (3–4): 337–345.

Garnier-Laplace, J., Beaugelin-Seiller, K., and Hinton, T.G. (2011). Fukushima wildlife dose reconstruction signals ecological consequences. *Environmental Science & Technology* 45 (12): 5077–5078.

Grupen, C. (2010). *Introduction to Radiation Protection. Practical Knowledge for Handling Radioactive Sources*. Heidelberg: Springer.

Guéguéniat, P., Herrmann, J., Kershaw, P.J. et al. (1996). Artificial radioactivity in the English Channel and the North Sea. In: *Radionuclides in the Oceans* (ed. P. Guéguéniat, P. Germain, and H. Métivier), 121–154. Les Ulis: EDP Sciences.

Gulin, S., Egorov, V., Polikarpov, G. et al. (2012). General trends in radioactive contamination of the marine environment from the Black Sea to Antarctic Ocean. In: *The Lessons of Chernobyl: 25 Years Later* (ed. E.B. Burlakova and V.I. Naydich). Hauppauge, New York, USA: Nova Science Publishers, Inc.

Hamilton, T.F. (1994). Radiometric investigations of Kara Sea sediments and preliminary radiological assessment related to dumping of radioactive wastes in the Arctic Seas. *Journal of Environmental Radioactivity* 25 (1–2): 113–134.

Hamilton, T.F. (2005). Chapter 2 Linking legacies of the Cold War to arrival of anthropogenic radionuclides in the oceans through the 20th century. *Radioactivity in the Environment* 6: 23–78.

Hamilton, T.F., Millies-Lacrox, J.C., and Hong, G.H. (1996). 137Cs(90Sr) and Pu isotopes in the Pacific Ocean sources & trends. Cherbourg-Octeville (France). www.osti.gov.

Heldal, H.E. (2001). *Radioactivity in Norwegian Waters: Distribution in Seawater and Sediments, and Uptake in Marine Organisms*. Norway: University of Bergen.

Holm, E. (1995). Plutonium in the Baltic Sea. *Applied Radiation and Isotopes* 46 (11): 1225–1229.

IAEA (1985). *Behaviour of Radionuclide Released into the Coastal Water*. Vienna: IAEA.

IAEA (1995). *Sources of Radioactivity in the Marine Environment and Their Relative Contributions to Overall Dose Assessment from Marine Radioactivity (MARDOS)*. Vienna: IAEA (International Atomic Energy Agency).

IAEA (2004). *Sediment Distribution Coefficients and Concentration Factors for Biota in the Marine Environment*. Vienna: IAEA.

IAEA (2005). *Worldwide Marine Radioactivity Studies (WOMARS)*. Vienna: IAEA.

IAEA (2006). *Environmental Consequences of the Chernobyl Accident and their Remediation: Twenty Years of Experience*. Vienna: IAEA.

IAEA (2023). *IAEA Comprehensive Report on the Safety Review of the ALPS-Treated Water at the Fukushima Daiichi Nuclear Power Station*. Vienna: IAEA.

IAEA (n.d.). Marine Radioactivity Information System (MARIS). https://www.iaea.org/resources/databases/marine-radioactivity-information-system-maris

Ilus, E. (2007). The Chernobyl accident and the Baltic Sea. *Boreal Environment Research* 12: 1–10.

Kim, G., Choi, S.-D., Lim, J.-M., and Kim, H. (2023). Strontium-90 levels in seawater southeast of Jeju Island during 2021–2023. *Marine Pollution Bulletin* 193.

Kolar, M.V. and Gugleta, M. (2019). The consequences of disposal and leakage of radioactive materials on various species of marine and freshwater fish. *International Journal of Fisheries and Aquatic Studies* 7 (6): 185–189.

Kryshev, I., Kryshev, A., and Sazykina, T. (2012). Dynamics of radiation exposure to marine biota in the area of the Fukushima NPP in March–May 2011. *Journal of Environmental Radioactivity* 114: 157–161.

Lal, D. (1999). An overview of five decades of studies of cosmic ray produced nuclides in oceans. *Science of the Total Environment, September* 237–238: 3–13.

Lerebours, A., Gudkov, D., Nagorskaya, L. et al. (2018). Impact of environmental radiation on the health and reproductive status of fish from Chernobyl. *Environmental Science & Technology* 52 (16): 9442–9450.

Livingston, H.D. and Povinec, P.P. (2000). Anthropogenic marine radioactivity. *Ocean & Coastal Management* 43 (8-9): 689–712.

Love, S.K. (1951). Natural radioactivity of water. *Industrial & Engineering Chemistry* 43 (7): 1541–1544.

Madhusha, R.K.D., Weerakkody, T.L., Weerasinghe, W.P.P.C. et al. (2021). Anthropogenic 137Cs distribution and radiological risk assessment in the Coastal Environments of Sri Lanka. *Global Scientific Journal* 9 (8): 2951–2960.

Miyata, T., & Wakamatsu, H. (2018). Who refuses safe but stigmatized marine products due to concern about radioactive contamination? *Fish Sci*, 1119–1133. doi:https://doi.org/10.1007/s12562-018-1250-1

Nakahara, M., Koyanagi, T., and Saiki, M. (1977). Temperature effect on the concentration of radionuclides by marine organisms. *Journal of Radiation Research* 18 (2): 122–131.

Povinec, P., Fowler, S., and Baxter, M. (1996). Chernobyl & the marine environment: the radiological impact in context. *IAEA Bulletin* 38 (1): 18–22.

Prăvălie, R. (2014). Nuclear weapons tests and environmental consequences: a global perspective. *PMID* 729–744.

Rahola, T. and Suomela, M. (1975). On biological half-life of potassium in man. *Annals of Clinical Research* 7: 62–65.

Robinson, W.L. and Noshkin, V.E. (1999). Radionuclide characterization and associated dose from long-lived radionuclides in close-in fallout delivered to the marine environment at Bikini and Enewetak Atolls. *Science of the Total Environment* 237–238 (1): 311–327.

Sanial, V., Buesseler, K.O., Charette, M.A., and Nagao, S. (2017). Unexpected source of Fukushima-derived radiocesium to the coastal ocean of Japan. *Environmental Sciences* 114 (42): 11092–11096.

Shalvey, K., & Trotter, A. (2023). *Japan begins releasing Fukushima's treated radioactive water into Pacific, prompting strong rebuke from China*. Sydney: ABC News.

Smodiš, B., Annareddy, V., and Rossbach, M. (2003). *Collection and Preparation of Bottom Sediment Samples for Analysis of Radionuclides and Trace Elements, IAEA-TECDOC-1360*. Vienna, Austria: IAEA.

Solomon, K. (1988). Sources of radioactivity in the ocean environment: from low level waste to nuclear powered submarines. *Journal of Hazardous Materials* 18 (3): 255–262.

Tomczak, W., Boyer, P., Krimissa, M., and Radakovitch, O. (2019). K_d distributions in freshwater systems as a function of material type, mass-volume ratio, dissolved organic carbon and pH. *Applied Geochemistry* 105 (2019): 68–77.

Uddin, S. et al. (2015). Radioactivity in the Kuwait marine environment – baseline measurements and review. *Marine Pollution Bulletin* 100 (2): 651–661.

Uesugi, M., Watanabe, R., Sakai, H., and Yokoyama, A. (2018). Rapid method for determination of 90Sr in seawater by liquid scintillation counting with an extractive scintillator. *The International Journal of Pure and Applied Analytical Chemistry* 178: 339–347.

UNSCEAR (2014). *Vienna*. Austria: United Nations Scientific Committee on the Effects of Atomic Radiation.

WHO/FAO. (2011). *Impact on Seafood Safety of the Nuclear Accident in Japan*. Rome: World Health Organization/Food and Agriculture Organization.

Wong, T. (2023). BBC. https://www.bbc.com/news/world-asia-66106162.

Zhang, L. (1993). *The Biogeochemistry of Iodine in Seawater*. Old Dominion University.

Zhao, X., Wang, W.-X., Yu, K.N., and Lam, P.K.S. (2001). Biomagnification of radiocesium in a marine piscivorous fish. *Inter-Research Science Publisher* 227–237.

11

GIS-Based Gray Water Footprint (GWF) Assessment of Coastal and Marine Pollution: A Minimization Approach

Pelin Soyertaş Yapıcıoğlu[1], *Abdullah İzzeddin Karabulut*[2], *and Mehmet İrfan Yeşilnacar*[1*]

[1] *Environmental Engineering Department, Harran University, Sanliurfa, Turkey*
[2] *Graduate School of Natural and Applied Sciences, Remote Sensing and Geographic Information Systems, 100/2000 CoHE, Harran University, Sanliurfa, Turkey*

11.1 Introduction

Microplastics (MPs) are becoming an increasing concern since they pose threats to the aquatic environment (Sun et al. 2019). MPs are the minor particles of plastics which are smaller than 5 mm (EU 2019, 2021). They are recalcitrant substances and very hard to remove and dispose of from nature. A raising quantity of MPs is in the marine ecosystem, including in the sea, ocean, surface water, wastewater, and groundwater and also in atmosphere and soil, as well as in agricultural areas and potable water (Sun et al. 2022; Sun et al. 2023; Wang et al. 2023a,b).

MP pollution has been regarded as one of the major threats to marine and coastal areas. MP pollution has been increasing in recent years, owing to many reasons such as random disposal and insufficient solid waste management systems (Pan et al. 2019; Sun et al. 2019; Dharmadasa et al. 2021). Wastewater discharge is the main cause of coastal and marine pollution. Wastewater could contain major quantities of MPs, and the degree of MPs contamination should be measured. Wastewater treatment plants potentially have played a crucial role in releasing MPs to the marine ecosystem (Sun et al. 2019).

Gray water footprint (GWF) is an indicator of the minimum volume of freshwater needed in order to dilute pollutant loads depending on the water quality standards. This chapter includes an overview of the GWF assessment of coastal and marine pollution using a geographic information system (GIS)-based analysis methodology. A GIS-based GWF approach has been recommended in this study. GWF is a determination methodology to quantify and measure the degree of pollution of water supplies.

GIS-supported mapping could be used as a tool to better describe marine and coastal pollution. This chapter has represented the significance of GIS-based GWF indicator in detail, especially for coastal areas. This study mainly aimed toward the integration of GWF and GIS tools to define and categorize the marine and coastal pollution.

Coastal and Marine Pollution: Source to Sink, Mitigation and Management, First Edition.
Edited by Meththika Vithanage, Sameera M. Samarasekara, Bryan D. James, and Christopher M. Reddy.

11.2 Gray Water Footprint of Coastal and Marine Pollution

Water footprint network (WFN) developed a calculation tool named as the GWF in order to estimate the water contamination stages of each pollutant parameter (Yapıcıoğlu and Yeşilnacar 2022). Many researchers have concentrated on GWF of surface and groundwater resources. Aldaya et al. (2020) used GWF for diffuse nitrogen pollution for groundwater and surface water resources in Spain. Serio et al. (2018) conducted a study to determine the correlation between nitrate pollution and agricultural activities by applying the GWF tool. Miglietta et al. (2017) reported the GWF of groundwater in Italy for the pollution resulting from mercury, vanadium, and ammonium. Yapıcıoğlu and Yeşilnacar (2022) investigated the GWF assessment of groundwater resources in southeastern Turkey in terms of the effect of recharge.

GWF is an approach for calculating the lowest volume of fresh water needed for diluting contaminant loads with respect to the water quality standards (Hoekstra et al. 2011; Morera et al. 2016). Also, it is an indicator of the assimilation capacity of water supplies (Cazcarro et al. 2016). Increasing the spatial resolution and analyzing the spatial variables affecting the GWF could be enabled using the GIS (Marini et al. 2015; Cazcarro et al. 2016; Shtull-Trauring et al. 2016). Furthermore, the impacts of spatial variables such as climatic factors and/or temporal factors such as seasonal variations in climatic factors could be assessed with high-resolution data by applying GIS (Shtull-Trauring et al. 2016).

Various environmental indicators were used to evaluate the water and sediment quality. GWF is an indicator tool to monitor the effect of pollutants which will be used for the determination of level of marine pollution. It enables to preserve the water supplies from various pollution. GWF assessment of marine and coastal areas should be done which have water scarcity problem. Several determination methods have been developed to indicate and monitor the contamination degree of marine pollution (Li et al. 2024). GWF is an indicator term for monitoring the effect of contaminants on the marine environment (WFN 2014; Yapıcıoğlu and Yeşilnacar 2021, 2022).

11.3 Integration of GIS and GWF Methods

A GIS, which is described as a system of hardware, software, and methods covering the processes of managing, processing, analyzing, modeling, and displaying the data located in the space, is designed to solve complex planning and management problems (Shtull-Trauring et al. 2016). GIS is an information technology system that stores all digital data in one place, and it quickly and efficiently retrieves and analyzes data in color-coded map format (Bajjali 2018). The basis of GIS is "where" and "what". GIS is concerned with the absolute and relative location ("where") of spatial features and their properties and attributes ("what"). The locations of important spatial objects such as rivers and streams can be recorded (Bolstad 2016). GIS is a tool for making and using spatial information. GIS has many definitions. Some of these definitions are:

GIS software integrates several tools that enable to study data related to a particular location in the world. GIS has many components designed to work with mapping data (Lan et al. 2015; Bolstad 2016). In the application of GIS tools, the conceptual model plays a crucial role in the selection of layers to be involved in creating a favorability map, as well as in scoring and weighting the selected layers. The score and weight in each tier are determined using a statistical approach or predicted based on expert opinion, which are called "data-driven" or "knowledge-driven" models, respectively. In data-driven modeling, layers are integrated by applying methodologies such as

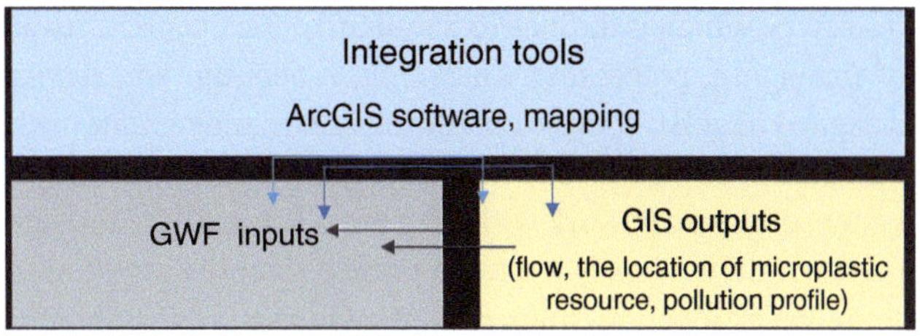

Figure 11.1 Conceptual framework of GIS-based GWF approach.

weight of evidence or logistic regression, while information-driven models generally use fuzzy logic, Boolean operators and index overlay (IO), and GIS methods (Trumpy et al. 2015).

GIS has been generally used to describe spatial relationships among various thematic information in each area to identify suitable marine areas (Zhang et al. 2020a). Also, a simple classification of marine and coastal areas could be performed using GIS. The map of possible estimated marine areas can be prepared because of resource exploration studies on ArcGIS program (Zhang et al. 2020b). From this point of view, GWF and GIS tools could integrate to define the pollution degrees of marine areas. The GIS tool could facilitate the determination of contamination levels in marine and coastal areas. ArcGIS program is one of the software which could be used for the integration of GIS technology and GWF approaches. Before discharging wastewater into the marine area, GIS technology could be used for the determination of pollution potential of the marine area which is the discharging point. The GWF approach could be integrated with GIS using mapping and classification systems. This study has presented the benefits of using high-resolution data with the help of the GIS tool to determine the effects of spatial factors (such as pollution points and flow) on the GWF tool. Flow is an important parameter to estimate GWF (Eq. (11.1)) (Yapıcıoğlu and Yeşilnacar 2021). Equation (11.1) adapted by Yapıcıoğlu and Yeşilnacar (2021) represents the calculation of GWF for marine area. The variables are Q, m_g, m_{max}, and m_{nat}. The flow was estimated using GIS technology. According to inverse distance weighted (IDW) interpolation, Q has been estimated in this study. Also, the pollutant concentrations have been mapped using GIS.

$$\text{GWF} = \frac{(Q \times m_g)}{(m_{max} - m_{nat})} \tag{11.1}$$

Q shows the discharging wastewater flow (volume/time) and m_g is defined as the concentration of MP concentration in wastewater. m_{max} shows the allowable highest concentration of MP in wastewater and m_{nat} represents the natural concentration of pollutants. Q has been ensured using the GIS method.

The proposed and used model of GIS and GWF is given in Figure 11.1. As seen from Figure 11.1, GIS outputs could be used as the input of GWF. Especially flow and the areas of high pollution load could be determined using GIS technology. These data could be used for the calculation of GWF.

11.4 GIS Solutions for Marine and Coastal Pollution Management

GISs have the advantage of being able to classify areas that are particularly vulnerable to contaminants by incorporating the scientific knowledge of coastal aquaculture activities into a geospatial model. GIS modeling appropriately classifies the vulnerability of coastal regions (Navas et al. 2011).

GIS is a computer-based system that assists in the storage, collection, analysis, distribution, and output of spatial data and information. When used well, GIS can help us live healthier and safer lives. GIS, which is defined as a system of hardware, software, and methods covering the processes of managing, processing, analyzing, modeling, and displaying the data located in the space, is designed to solve complex planning and management problems. GISs are various computer tools that allow people to work with data related to a particular location in the world. GIS is a database system designed to work with map data (Price 2010; Töreyen et al. 2010). The purpose of geographic information in GISs is to offer a professional information system technology where all people can share the production, management, analysis, and geographical data from distributed databases on the network.

11.5 3D Analyst

It provides users with effective surface modeling, visualization, and analysis. By using 3D Analyst, slope, aspect, and visibility analyses can be performed; vector raster transformations can be performed; and the space can be modeled in 3D. 3D Analyst has the ability to create and analyze surface data represented in leica absolute scanner (LAS) dataset formats, terrain, raster, and triangulated irregular network (TIN).

11.5.1 Spatial Analyst

Spatial Analyst has powerful spatial modeling and analytics that allow you to create, map, and query cell-based raster data. With Spatial Analyst, the user can perform distance, density analyses and interpolate and perform overlay analyses. Spatial Analyst applications include deriving distances from points, polylines, or polygons; calculating population density from quantities measured at specific points; reclassifying existing data into compliance classes; or generating slope, aspect, or hillshade outputs from elevation data (ESRI 2022). Some spatial analyses find suitable locations; interpolate data values for a study area based on samples; identify the best path between locations; derive new information from existing data, small neighborhoods, or predetermined zones; perform statistical analysis based on the local environment; and perform distance and cost-of-travel analyses.

11.5.2 Tracking Analyst

Tracking Analyst is used to map objects that change or move over time. With Tracking Analyst, temporal analyses of the past and the future can be performed.

11.5.3 Network Analyst

Network Analyst provides realistic modeling of road networks and performs route analyses on these models. Finding the shortest route, closest service, cost analysis, and fleet management analysis is performed with Network Analyst.

11.5.4 Geostatistical Analyst

Geostatistical Analyst offers statistical methods for analyzing and mapping continuous data. Geostatistical methods are widely used to explain spatial changes in events and complete missing data. In particular, the IDW interpolation technique is used to determine cell values of unsampled points based on values from known sample points. Kriging interpolation uses data from known nearby points to estimate the optimal values of the data at other points.

There is limited research on the integration of GIS and GWF in various topics. This method can ensure a visual pollution profile. A GIS-based analysis method, such as the IDW method, can be used to assess the GWF of coastal and marine pollution. It has been used to analyze and map the collected data. Detailed maps can provide estimates of land use and pollution flow and quantity, which can be used to determine the GWF. There are numerous entries of recorded literature on GIS-based analysis of marine pollution (Chybicki et al. 2008; Hamre et al. 2009; Kulawiak et al. 2010; Navas et al. 2011; Fustes et al. 2014; Abo-Taleb et al. 2016; Song et al. 2017; Gu and Gao 2019). The GIS-supported publications on marine pollution in the literature are summarized as follows.

GIS plays a pivotal role in contemporary environmental management, particularly in the context of marine ecosystems. Over the years, numerous studies have showcased the diverse applications of GIS in monitoring, assessing, and managing various environmental issues, ranging from water pollution to oil spills and ecosystem health. One significant contribution comes from Chybicki et al. (2008), who introduced a sophisticated GIS designed for near real-time monitoring of water pollution. This system, accessible remotely and integrated with pollution sensors, provides invaluable support to emergency management personnel by offering up-to-date location data and facilitating prompt decision-making. Expanding on this concept, Hamre et al. (2009) discussed the development of interoperable monitoring and forecasting services within the FP6 InterRisk project. By adhering to established web-GIS standards and INSPIRE recommendations, the project aimed to enhance risk management during natural hazards and industrial accidents across European marine and coastal areas. The successful demonstration of the pilot system highlighted the efficacy of open-source tools in improving information accessibility and service delivery. Moreover, Kulawiak et al. (2010) focused on implementing a web-based GIS for oil spill monitoring and forecasting services within the MARCOAST project. Through the integration of various web-GIS technologies, authenticated users can remotely access oil spill forecasting results, thus bolstering efforts in oil spill management and environmental monitoring. In the realm of aquaculture planning and environmental management, Navas et al. (2011) employed hydrodynamic and GIS models to assess the suitability of aquaculture sites in an Irish fjord. By integrating environmental vulnerability models, the study identified areas conducive to aquaculture development, offering valuable insights for sustainable marine resource management. Furthermore, Fustes et al. (2014) presented a novel tool for marine pollution detection and localization, integrating remote sensing, GIS, and cloud computing technologies. Advanced segmentation algorithms facilitated the identification of pollution areas in synthetic aperture radar (SAR) images, underscoring the potential of cloud computing for scaling algorithms and facilitating user communication. Additionally, Lan et al. (2015) outlined the construction of the Marine Oil Spill Risk Mapping (MOSRM) model, which assesses the risk of oil spills in marine environments. Validated through a case study in Dalian, China, the model categorized coastal zones into risk levels and proposed corresponding countermeasures, aiding in effective risk management strategies.

Abo-Taleb et al. (2016) underscored the importance of GIS-based decision support systems in assessing the vulnerability of coastal zones, particularly in Egypt's El Mex Bay. The study advocated for Integrated Coastal Zone Management (ICZM) approaches to balance development with environmental protection, emphasizing the significance of strategic planning in coastal areas. Moreover, Song et al. (2017) evaluated the health of Laizhou Bay's marine ecosystem using GIS-based monitoring data analysis. The assessment revealed environmental issues such as water eutrophication and heavy metal pollution, highlighting the need for targeted management strategies to ensure the long-term sustainability of the ecosystem. Lastly, Gu and Gao (2019) introduced a novel approach for identifying anthropogenic sources of heavy metals in Hong Kong marine sediments. Through a combination of enrichment factor (EF) determination, Network Management Systems (NMS), and GIS analysis, the study identified historical anthropogenic impacts on heavy metal distribution, emphasizing the necessity of effective pollution control measures.

11.6 Conclusions

Wastewater discharge is a major cause of coastal and marine pollution. This chapter includes an overview of the GWF assessment of coastal and marine pollution using GIS-based analysis methodology which has been developed in order to define the degree of contamination. GIS-supported mapping could make it easier to determine marine pollution. This chapter has presented the importance of GIS-based GWF indicator for coastal areas. This study mainly focused on the integration of GWF and GIS tools to determine marine and coastal pollution. Experimental and computational studies should be carried out to validate the model for coastal and marine areas in terms of MP pollution. In conclusion, many studies collectively underscore the transformative potential of GIS in advancing environmental monitoring, risk assessment, and management strategies in marine ecosystems, thereby contributing to the conservation and sustainable management of marine resources on a global scale. GIS gets easier to define the pollution area which creates the maximum GWF in the marine ecosystem. Hence, wastewater discharging has been planned to the points which have the low pollution risk. In the future, it will be a perspective to select the wastewater discharging areas that have lower pollution.

References

Abo-Taleb, H.A., Ezz, S.M.A., Aziz, N.E.A. et al. (2016). Detecting marine environmental pollution by biological beacons and GIS program. *Journal of Fisheries Sciences* 10 (4): 69.

Aeschlimann, M., Li, G., Kanji, Z.A., and Mitrano, D.M. (2022). Potential impacts of atmospheric microplastics and nanoplastics on cloud formation processes. *Nature Geoscience* 15 (12): 967–975.

Aldaya, M.M., Rodriguez, C.I., Fernandez-Poulussen, A. et al. (2020). Grey water footprint as an indicator for diffuse nitrogen pollution: the case of Navarra, Spain. *Science of the Total Environment* 698: 134338. https://doi.org/10.1016/j.scitotenv.2019.134338.

Bajjali, W. (2018). *ArcGIS for Environmental and Water Issues*. Springer https://link.springer.com/book/10.1007/978-3-319-61158-7.

Bolstad, P. (2016). *GIS Fundamentals: A First Text on Geographic Information Systems*. Livonia, USA: Eider (Press Minnesota).

Cazcarro, I., Duarte, R., and Sánchez-Chóliz, J. (2016). Downscaling the grey water footprints of production and consumption. *Journal of Cleaner Production* 132: 171–183. https://doi.org/10.1016/j.jclepro.2015.07.113.

Chybicki, A., Kulawiak, M., Lubniewski, Z., Dabrowski, J., Luba, M., Moszynski, M., & Stepnowski, A. (2008). GIS for remote sensing, analysis and visualisation of marine pollution and other marine ecosystem components. *2008 1st International Conference on Information Technology*, 18–21 May 2008 (pp. 1–4). Gdansk, Poland. IEEE. https://doi.org/10.1109/INFTECH.2008.4621628

Dharmadasa, W.S., Andrady, A.L., Kumara, P.T.P. et al. (2021). Microplastic pollution in marine protected areas of Southern Sri Lanka. *Marine Pollution Bulletin* 168: 112462.

Dong, Y.M., Bao, Q.L., Gao, M.L. et al. (2022). A novel mechanism study of microplastic and as co-contamination on indica rice (*Oryza sativa* L.). *Journal of Hazardous Materials* 421: 12.

ESRI (2022). What is the ArcGIS Spatial Analyst extension?https://desktop.arcgis.com/en/arcmap/latest/extensions/spatial-analyst/what-is-the-spatial-analyst-extension.htm. (accessed at 20 January 2024)

European (EU) Commission (2019). EU Green Deal. Communication from the Commission to the European Parliament, the 533 European Council, the Council, the European Economic and Social Committee and the 534 Committee of the Regions–The European Green Deal.

European (EU) Commission, European Green Deal (2021). Fit for 55 Package: The EU's Fit for 55 Package–Key Takeaways From Bernstein Renewables, Chemical, Airlines & Energy Teams.

Fustes, D., Cantorna, D., Dafonte, C. et al. (2014). A cloud-integrated web platform for marine monitoring using GIS and remote sensing. Application to oil spill detection through SAR images. *Future Generation Computer Systems* 34: 155–160. https://doi.org/10.1016/j.future.2013.09.020.

Gu, Y. G., & Gao, Y. P. (2019). An unconstrained ordination-and GIS-based approach for identifying anthropogenic sources of heavy metal pollution in marine sediments. *Marine Pollution Bulletin*, 146, 100–105. https://doi.org/10.1016/j.marpolbul.2019.06.008

Hamre, T., Krasemann, H., Groom, S., Dunne, D., Breitbach, G., Hackett, B.,. . . & Sandven, S. (2009). Interoperable web GIS services for marine pollution monitoring and forecasting. *Journal of Coastal Conservation*, 13, 1–13. https://doi.org/10.1007/s11852-009-0046-y

Hoekstra, A.Y., Chapagain, A.K., Aldaya, M.M., and Mekonnen, M.M. (2011). *The Water Footprint Assessment Manual*. London/Washington, DC: Earthscan.

Kulawiak, M., Prospathopoulos, A., Perivoliotis, L., Kioroglou, S., & Stepnowski, A. (2010). Interactive visualization of marine pollution monitoring and forecasting data via a Web-based GIS. *Computers & Geosciences*, 36(8), 1069–1080. https://doi.org/10.1016/j.cageo.2010.02.008

Lan, D., Liang, B., Bao, C., Ma, M., Xu, Y., & Yu, C. (2015). Marine oil spill risk mapping for accidental pollution and its application in a coastal city. *Marine Pollution Bulletin*, 96(1–2), 220–225. https://doi.org/10.1016/j.marpolbul.2015.05.023

Li, K., Du, L., Qin, C. et al. (2024). Microplastic pollution as an environmental risk exacerbating the greenhouse effect and climate change: a review. *Carbon Research* 3 (1): 9.

Marini, E., Papadopoulou, M.P., and Tsoukala, V. (2015). Water footprint assessment in river basin scale using GIS technology. In: *E-Proceedings of 9th World Congress of EWRA "Water Resources Management in a Changing World: Challenges and Opportunities"*. Istanbul, Turkey: (accessed 8 January 2024).

Miglietta, P.P., Toma, P., Fanizzi, F.P. et al. (2017). A grey water footprint assessment of groundwater chemical pollution: Case study in Salento (Southern Italy). *Sustainability* 9 (5): 799. https://doi.org/10.3390/su9050799.

Morera, S., Corominas, L., Poch, M. et al. (2016). Water footprint assessment in wastewater treatment plants. *Journal of Cleaner Production* 112: 4741–4748. https://doi.org/10.1016/j.jclepro.2015.05.102.

Navas, J.M., Telfer, T.C., and Ross, L.G. (2011). Spatial modeling of environmental vulnerability of marine finfish aquaculture using GIS-based neuro-fuzzy techniques. *Marine Pollution Bulletin* 62 (8): 1786–1799. https://doi.org/10.1016/j.marpolbul.2011.05.019.

Pan, Z., Liu, Q., Sun, Y. et al. (2019). Environmental implications of microplastic pollution in the Northwestern Pacific Ocean. *Marine Pollution Bulletin* 146: 215–224.

Price, M.H. (2010). *Mastering ArcGIS*. New York, NY, USA: McGraw-Hill.

Serio, F., Miglietta, P.P., Lamastra, L. et al. (2018). Groundwater nitrate contamination and agricultural land use: a grey water footprint perspective in Southern Apulia Region (Italy). *Science of the Total Environment* 645: 1425–1431. https://doi.org/10.1016/j.scitotenv.2018.07.241.

Shtull-Trauring, E., Aviani, I., Avisar, D., and Bernstein, N. (2016). Integrating high resolution water footprint and GIS for promoting water efficiency in the agricultural sector: a case study of plantation crops in the Jordan Valley. *Frontiers in Plant Science* 7: 1877. https://doi.org/10.3389/fpls.2016.01877.

Song, D., Gao, Z., Zhang, H., Xu, F., Zheng, X., Ai, J.,. . . & Zhang, H. (2017). GIS-based health assessment of the marine ecosystem in Laizhou Bay, China, *Marine Pollution Bulletin*, 125(1–2), 242–249. doi:https://doi.org/10.1016/j.marpolbul.2017.08.027

Sun, J., Dai, X., Wang, Q. et al. (2019). Microplastics in wastewater treatment plants: detection, occurrence and removal. *Water research* 152: 21–37.

Sun, Y., Li, X., Cao, N. et al. (2022). Biodegradable microplastics enhance soil microbial network complexity and ecological stochasticity. *Journal Hazardous Materials* 439: 129610.

Sun, X., Tao, R., Xu, D. et al. (2023). Role of polyamide microplastic in altering microbial consortium and carbon and nitrogen cycles in a simulated agricultural soil microcosm. *Chemosphere* 312: 137155.

Töreyen, G., Özdemir, İ., and Kurt, T. (2010). *Arcgıs 10 Desktop Application Document*. Ankara: Islem Geographic Information Systems Engineering and Education.

Trumpy, E., Donato, A., Gianelli, G. et al. (2015). Data integration and favourability maps for exploring geothermal systems in Sicily, southern Italy. *Geothermics* 56: 1–16. https://doi.org/10.1016/j.geothermics.2015.03.004.

Tsang, Y.Y., Mak, C.W., Liebich, C. et al. (2017). Microplastic pollution in the marine waters and sediments of Hong Kong. *Marine Pollution Bulletin* 115 (1–2): 20–28.

Wang, Z., Hu, X., Kang, W. et al. (2023a). Interactions between dissolved organic matter and the microbial community are modified by microplastics and heat waves. *Journal of Hazardous Materials* 448: 130868–130868.

Wang, Z., Hu, X., Qu, Q. et al. (2023b). Dual regulatory effects of microplastics and heat waves on river microbial carbon metabolism. *Journal of Hazardous Materials* 441: 129879.

Water Footprint Network, WFN (2014). Available from: http://http://waterfootprint.org/ (accessed 03 Jan 2024).

Yapicioğlu, P. and Yeşilnacar, M.İ. (2021). Grey water footprint assessment of geothermal water resources in the southeastern Anatolia region. *Turkish Journal of Earth Sciences* 30 (SI-2): 1200–1207. https://doi.org/10.3906/yer-2105-22.

Yapıcıoğlu, P. and Yeşilnacar, M.İ. (2022). Grey water footprint assessment of groundwater resources in southeastern Turkey: effect of recharge. *Water Supply* 22 (1): 615–627. https://doi.org/10.2166/ws.2021.247.

Zhai, X., Zhang, X.H., and Yu, M. (2023). Microbial colonization and degradation of marine microplastics in the plastisphere: a review. *Frontiers in Microbiology* 14: 1127308.

Zhang, B., Yang, X., Chen, L. et al. (2020a). Microplastics in soils: a review of possible sources, analytical methods and ecological impacts. *Journal of Chemical Technology & Biotechnology* 95 (8): 2052–2068.

Zhang, X., Chen, J., and Li, J. (2020b). The removal of microplastics in the wastewater treatment process and their potential impact on anaerobic digestion due to pollutants association. *Chemosphere* 251: 126360.

12

Underestimated Threats: Personal Care Products (PCPs) in Marine and Coastal Environments

Madhuni Madhushika Wijesooriya[1], Meththika Vithanage[2], and Hasintha Wijesekara[3]

[1] *Department of Biological and Chemical Engineering, Aarhus University, Aarhus C, Denmark*
[2] *Ecosphere Resilience Research Center, Faculty of Applied Sciences, University of Sri Jayewardenepura, Nugegoda, Sri Lanka*
[3] *Department of Natural Resources, Sabaragamuwa University, Belihuloya, Sri Lanka*

12.1 Introduction

Since the twentieth century, the annual global production and consumption of personal care products (PCPs) have grown at an exponential rate (Paul et al. 2023; Statista 2024). PCPs are a diverse range of compounds, from body lotions to toothpaste, hair care products, cleaning agents, and, beyond, including fragrances, sunscreens, and cosmetics (Brausch and Rand 2011; Yang et al. 2020; Khalid and Abdollahi 2021). These include cosmetic and medicinal products that serve aesthetic and therapeutic needs (Khalid and Abdollahi 2021). For instance, in the review of Montes-Grajales et al. (2017), 72 PCPs were recognized as emerging contaminants across 30 countries, with concentrations varying from 0.029 ng/l to 7.811×10^6 ng/l. The most often reported categories included sunscreens, antiseptics, and fragrances (Figure 12.1). The consumption patterns of PCPs are directly linked to socioeconomic status and lifestyle, driven by the growing demands of modern living, which prioritize hygiene, beauty, and health (Khalid and Abdollahi 2021). Thus, the increased market demand has prompted competition, leading to the continuous innovation and introduction of new products and brands (Łopaciuk and Łoboda 2013; Khalid and Abdollahi 2021). Despite their short-term application to the human body, the environmental consequences of this consumption trend are profound. PCPs persist in the environment due to continuous discharge from daily use, often in low concentrations. Hence, PCPs are considered pseudo-persistent pollutants and are ubiquitously detected in various environmental compartments including the marine and coastal environments such as in marine organisms, seawater, corals, and sediments (Pedrouzo et al. 2011; Liang et al. 2013; Beretta et al. 2014; Groz et al. 2014; Martínez Bueno et al. 2014; Downs et al. 2016; Yang et al. 2020).

Almost all the PCPs are externally applied to the human body and subsequently released into the environment without undergoing metabolic changes, often as fragments of complex molecules washed off during use (Brausch and Rand 2011). Furthermore, despite being designed to nourish the human body, these products' formulations and packaging can also harm the marine ecosystem. Consequently, most of the chemical compounds present in these PCPs persist in the environment as biologically active, bioaccumulate substances (Montes-Grajales et al. 2017) that interfere with

Coastal and Marine Pollution: Source to Sink, Mitigation and Management, First Edition.
Edited by Meththika Vithanage, Sameera M. Samarasekara, Bryan D. James, and Christopher M. Reddy.

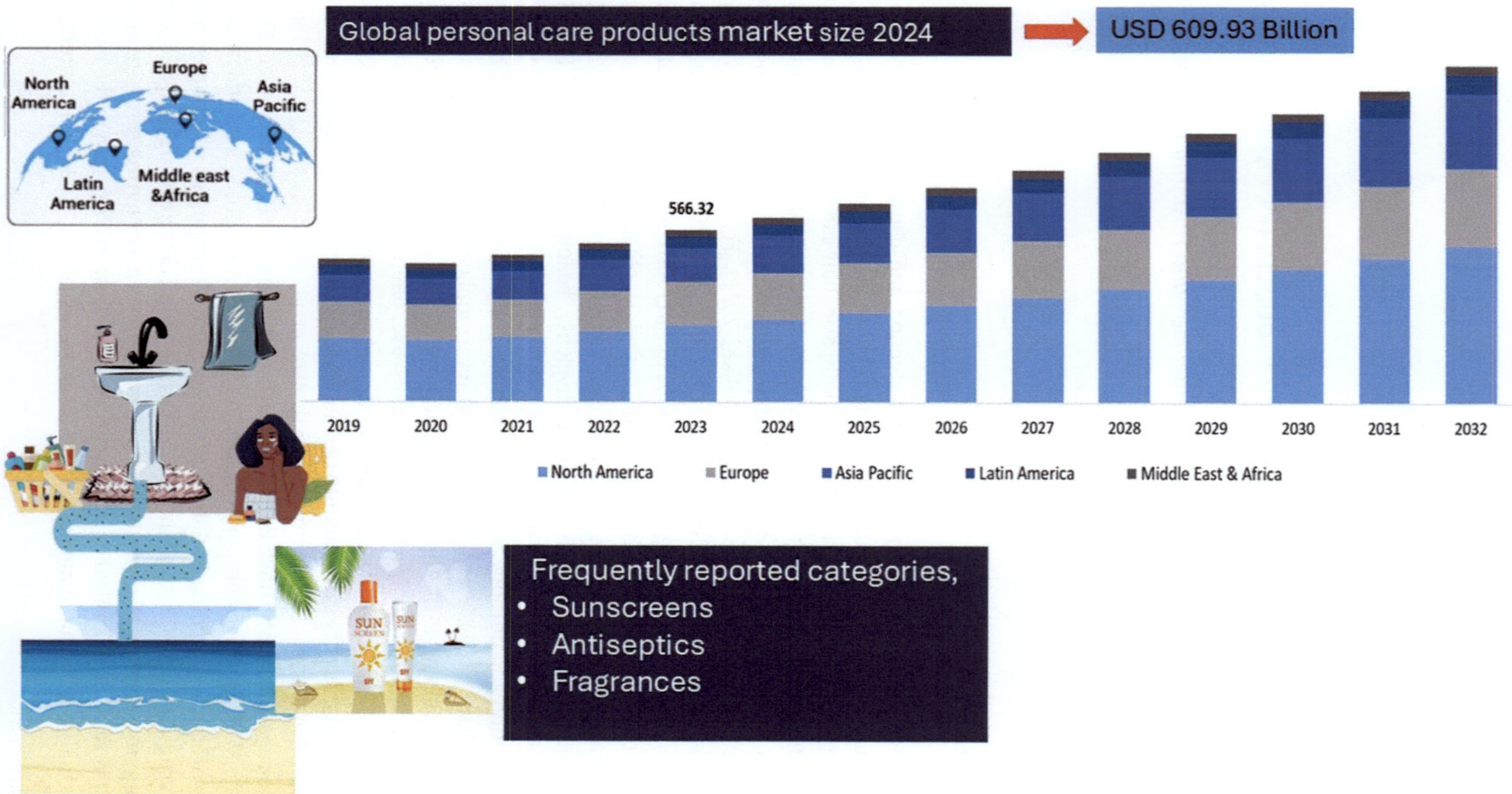

Figure 12.1 The global trend of the personal care products (PCPs) market concerning skin, hair, lip, oral, and body care, men's grooming, color cosmetics, bath and shower, and fragrance (*Source:* Adapted from https://www.polarismarketresearch.com/industry-analysis/personal-care-products-market).

the physiology of non-targeted organisms including fish, shrimps, algae, coral, and plankton leading to either acute or chronic toxicity (Jyoti and Sinha 2023). The widespread use of sunscreen products by tourists has inadvertently led to the release of these products into seawater, posing a significant threat to tropical corals by inducing extensive and rapid bleaching (Corinaldesi et al. 2018). The inefficiencies in wastewater treatment processes lead to insufficient removal of mid-polar to polar PCP residues, including sunscreen components. This results in the contamination of surface water, subsequently extending to the marine environment (Pai et al. 2020). Indeed, PCPs containing microplastics or other potentially harmful compounds are often found in the environment as complex mixtures. Even if the concentrations of individual compounds are low, the combined effect of these mixtures, known as the cocktail effect, can result in either enhancement or weakening of ecotoxicological effects (Chen et al. 2021). Consequently, these pollutants diffuse into seawater, circulate, and accumulate in sediments where they may interact synergistically or antagonistically, thereby amplifying or attenuating their environmental impact (Khalid and Abdollahi 2021; Wydro et al. 2023).

Despite the widespread presence of PCPs in seawater, sediments, and marine organisms, there remains limited evidence regarding their toxicity, necessitating further research to assess the associated hazards and risks. Challenges in evaluating PCP contamination in marine environments arise from factors such as dilution, diffusion, and complex hydrodynamics, leading to fewer studies conducted in coastal zones (Arpin-Pont et al. 2016). Nevertheless, in the detection of PCPs in marine organisms, Klosterhaus et al. (2013) highlight the importance of assessing contamination ranges and identifying vulnerable marine sites to enhance the understanding of marine organism exposure. Moreover, PCP contamination poses a risk to both aquatic ecosystems and human health due to their potential toxicities. This emphasizes the need for effective management strategies, particularly due to the rapid expansion of tourism in coastal areas. Therefore, this chapter aims to provide an overview of the impacts of PCP contamination in marine and coastal environments, highlighting the urgent need for interdisciplinary research and integrated approaches to mitigate PCP pollution.

12.2 Sources and Environmental Fate of PCPs in Marine and Coastal Environments

The extensive usage of PCPs, coupled with improper disposal, everyday activities, recreational activities, and inadequate treatment of domestic and industrial wastewater, leads to widespread distribution of PCPs and their associated compounds in marine and coastal environments (Montes-Grajales et al. 2017). The lifecycle of PCPs from manufacturing facilities to retailer outlets and their eventual distribution in the marine environment involves multiple point and nonpoint sources, including surface runoff. The widespread occurrence of PCPs in marine environments is attributed to outdoor recreational activities and tourism. Further, sewage effluents discharged from wastewater treatment plants (WWTPs) are often due to incomplete or inefficient removal processes (Blair et al. 2013). Notably, the inefficiency of many WWTPs in removing fragrance compounds exacerbates the problem, leading to their discharge into rivers and streams (Bridges 2002). For instance, Nakata et al. (2012) have reported the detection of synthetic musks in mussels from Asia-Pacific coastal waters indicating fragrance contamination in the marine environment. However, compared to pharmaceutical products, only a few studies have highlighted the detection of active ingredients in PCPs and their transforming metabolites in surface water, groundwater, outdoor recreational areas such as swimming pools and ponds, and beaches (Lu et al. 2017;

Ma et al. 2018; Chakraborty et al. 2019). For instance, a study conducted in Portugal during the summer season revealed the presence of UV filters (UVFs) in beach sand, with concentrations ranging from 0.030 to 373 ng/g dry weight (Capela et al. 2019). Further, UVFs have been detected in marine sediments of the Baltic Sea and Chinese Yellow Sea (Apel et al. 2018a,b). Further, there is a high possibility of PCPs that may exist in sewage sludge (biosolids) from WWTPs used for agriculture purposes entering the sea via stormwater runoff (Díaz-Cruz et al. 2009).

Once released into the environment, PCPs including their biologically active and inactive ingredients exhibit various fates in different environmental compartments (Figure 12.2), including persistence and bioaccumulation in sediments or dispersion with water (Hawthorne et al. 2005; Girijan and Kumar 2020; Khalid and Abdollahi 2021). Upon entering WWTPs, PCPs may either undergo partial degradation, leading to residual concentrations and potential degradation end-products and intermediates in the discharge, or resist degradation altogether. This contributes to the presence of PCPs in surface water sources. Additionally, subsurface water flows can facilitate the interaction and transfer of these contaminants to groundwater. Direct disposal of solid waste from households and treatment plants into soil or landfills causes the potential for dispersing PCPs into groundwater or surface water via leachate and runoff. Nevertheless, the fate and distribution of PCPs and their associated chemicals significantly depend on their chemical properties such as Henry's constant and dissociation constant values (Girijan and Kumar 2020). For example, monoethanolamine has been found to persist in soil for decades due to its strong binding capacity and is less likely to migrate into groundwater (Hawthorne et al. 2005). Further, significant concentrations of specific chemicals in PCPs such as nonylphenol (26.5 ng/l) and octylphenol (113 ng/l) were detected in Sail Rock Beach and Boli River estuary of Taiwan, respectively (Kung et al. 2018). Nevertheless, there is limited information regarding the fate and distribution pattern of certain chemical constituents, such as ethanolamines and glycol ethers, in different environmental compartments.

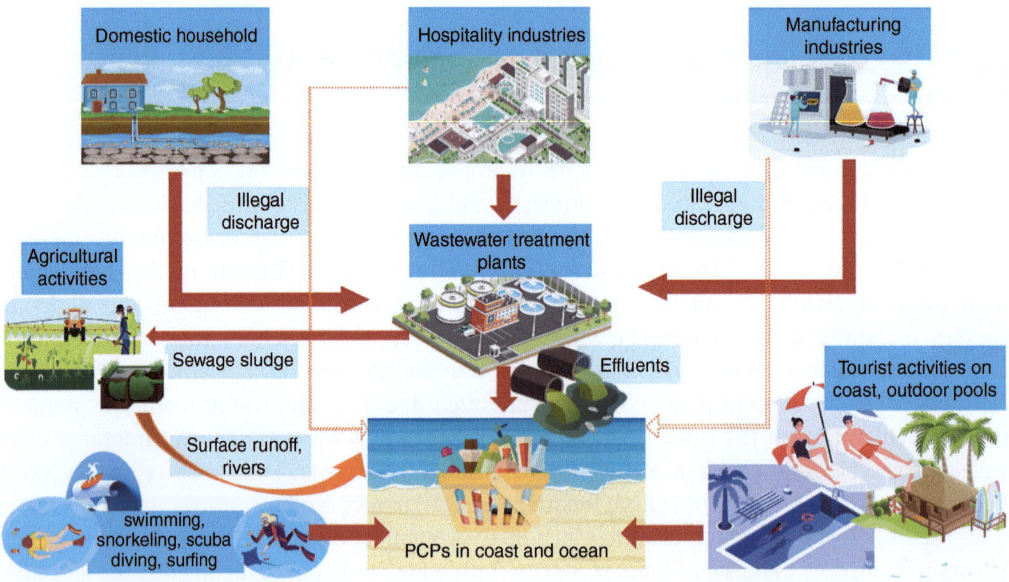

Figure 12.2 Various pathways through which PCPs are transported into coastal and marine environments from different sources.

The global distribution of PCP contamination suggests diverse usages and sources across different regions and countries (Table 12.1). According to previous studies, there exists a significant variation in PCP contamination levels among different regions. It has been identified that metabolites of PCPs have been detected in coastal waters at varying concentrations, ranging from 0.01 to 6800 ng/l (Gaw et al. 2014); nevertheless, these levels are not sufficient to predict the concentration experienced by marine organisms (Granek et al. 2016). Another finding showed that Chinese coastal waters have higher concentrations of PCPs, ranging from 0.880 to 1194 ng/l, whereas Korean coastal waters exhibit lower levels, ranging from 9.91 to 422 ng/l (Yang et al. 2020). The average concentration of triclosan detected in the Gulf of Urabá, Colombia, was in the range from 0.10 to 0.79 μg/l (Pemberthy et al. 2020). As revealed by Bayen et al. (2013), the reasons for these observed differences across regions or countries were mainly due to the coastal hydrodynamics. The study further emphasized the effects of the tidal stage on dilution and diffusion of contaminants. Additionally, seasonal variations dry or wet seasons, and recreational activities also impact the spatial and temporal distribution of PCPs in seawater and sediments (Arpin-Pont et al. 2016).

Besides the distribution of PCPs in seawater columns, beach sand, top layer of sediments, as well as estuaries, aquatic sediments possess a remarkable capacity to accumulate substances, driven by their high sorption capacity or affinity for binding, akin to potentially toxic elements (PTEs) (Wang and Wang 2016). Thus, the excess PCPs are unable to persist in the water for extended durations, as they partition between the water column and the underlying soil bed through physical and chemical processes influenced by factors such as soil pH and sedimentation affinity (Ammeri et al. 2020). Furthermore, PCP accumulation in sediments is influenced by various factors such as sediment texture, mineralogical composition, and particle size (Billah et al. 2017). Moreover, the binding strength of contaminant to sediment also relies on the fractionation of pollutants within the solid matrix, including exchangeable proportions, oxidizable and reducible fractions, and residuals (Ferraro et al. 2023). Concerning bioavailability, similar to PTEs, the bioavailability of PCPs in marine sediment may vary depending on environmental and redox conditions that potentially lead to changes in chemical speciation (Ferraro et al. 2023).

Table 12.1 Occurrence of PCPs in different compartments of the coastal and marine environment.

Location	Active ingredients of PCPs	Detected concentration	References
Tamar Estuary coastal water (Plymouth, UK)	Musk: Galaxolide® (HHCB) and Tonalide® (AHTN)	Detected concentrations of HHCB and AHTN in sewage effluents were 987–2098 ng/l and 55–159 ng/l, respectively, and in coastal water were 5–30 ng/l and 3–15 ng/l, respectively.	Sumner et al (2010)
Coastal water (US)	UV filter: benzophenone 3	2013 ng/l	Bratkovics and Sapozhnikova (2011)
	UV filter: octocrylene	1409 ng/l	
Coastal bathing zone (Mediterranean area)	UV filter: ethylhexyl methoxycinnamate	9 ng/g (dry weight: d.w)	Amine et al. (2012)
	Octocrylene	78 ng/g (d.w)	
	UV filter: ethylhexyl dimethyl PABA	9 ng/g (d.w)	
Sediment (Estuary, US)	Triclosan	400 ng/g (d.w)	Cantwell et al. (2010)

The mechanisms potentially responsible for removing PCPs from coastal marine environments include sorption, biodegradation, and photodegradation (Ilyas and Hullebusch 2020), but further investigation is necessary to fully comprehend the fate of these compounds in marine ecosystems.

12.3 Active Ingredients of Concern in PCPs

The main categories of PCPs contain various compounds such as disinfectants and cleansing agents (antimicrobials), fragrances (musks), insect repellents (DEET), preservatives (parabens), UVFs (oxybenzone and methyl benzylidene camphor), surfactants (alkylphenol polyethoxylates), volatile organic compounds (cyclosiloxanes and glycol ether), and plasticizers (phthalates and bisphenols) where their toxicity varies depending on the mode and quantity of exposure (Brausch and Rand 2011; Khalid and Abdollahi 2021; Jyoti and Sinha 2023). Many of these components are lipophilic, persistent, and tend to bioaccumulate in aquatic organisms and sediments (Rainieri et al. 2017; Chavoshani et al. 2020).

12.3.1 Disinfectants and Cleansing Agents

The widely used antimicrobials in PCPs are triclocarban, triclosan, chloroprene, 1,4-dichlorobenzene, ortho-phenylphenol, and diazolidinyl urea (Capkin et al. 2017; Musee 2018). Triclosan is the most used synthetic antibacterial agent, with a structural similarity to thyroid hormone (Hwang et al. 2017). Despite the different modes of action, the antibacterial properties of both triclosan and triclocarban are similar. Further, the transforming products of triclosan and triclocarban such as methyl-triclosan and chlorinated phenols, respectively, are more toxic than the original compounds (Chevillot et al. 2018; Jyoti and Sinha 2023).

12.3.2 Preservatives

The commonly used preservatives in PCPs are parabens and benzalkonium chloride (Bilal et al. 2020; Al-Halaseh et al. 2022). Parabens are an active component in more than 22 000 cosmetic products due to their wide range of antimicrobial properties and high stability over time (Andersen 2008; Błędzka et al. 2014). The most widely detected paraben analogs include butylparaben, methylparaben, ethylparaben, and propylparaben (Bolujoko et al. 2021).

12.3.3 UV Filters

Among the UVFs, benzophenone 4 (BP-4) exhibited higher concentrations compared to other compounds, attributed to its greater polarity and solubility. Compounds such as BP-3, ES, and octocrylene were detected at high concentrations due to their widespread use in cosmetics (Ramos et al. 2016). Oxybenzone is another active component in over 3000 PCPs including sunscreens.

12.3.4 Fragrances

Even though there is a substantial demand for fragrance-containing products such as perfumes, deodorants, and detergents, the concentration of musk is only 2% (Martínez-Girón et al. 2010). The synthetic aromatics used in PCPs include nitro-, alicyclic-, polycyclic-, and macrocyclic musk which contain chemicals such as aldehydes, carboxylic acids, ketones, amines, esters, and dioxanes (Patel et al. 2021).

12.3.5 Micro(nano)plastics, Microbeads, and Nano-additives

PCPs are a direct source of generating secondary microplastics, often termed 'microbeads' (Anderson et al. 2016; Hernandez et al. 2017). These microbeads are included in PCPs such as toothpaste, cleansers, and cosmetic products to enhance abrasiveness and decorative elements (Zitko and Hanlon 1991; Leslie 2014). The sizes of these microbeads/microplastics range from a few microns to a thousand microns, enabling them to easily pass through water filtration systems (Vesilind 2003) and enter rivers, and ultimately, the sea (Ziajahromi et al. 2017). It is estimated that facial/body cleansers or exfoliants could release between 4594 and 94 500 microbeads during a single application (Napper et al. 2015). Among the released microbeads, polyethylene (PE) microbeads are frequently detected in certain cleansing products along with alumina and sodium tetraborate decahydrate particles (Fendall and Sewell 2009).

Marine environments are particularly affected by the substantial presence of MPs, which heavily contaminate not only the sea surface and water column but also accumulate in sediments over the long term regardless of whether the areas are remote or protected (Cincinelli et al. 2019). Cosmetics have been identified as a possible source of microplastics in the marine environment (Möhlenkamp et al. 2018). A study indicates that microbeads and other micro(nano)plastics introduced through cosmetics and PCPs represent a small proportion of marine microplastic pollution in terms of weight (Cheung and Fok 2016; Guerranti et al. 2019). For instance, in Europe, cosmetic items account for between 0.1% and 4.1% of the plastic that enters the marine environment annually (Duis and Coors 2016; Hirst and Bennett 2017). Nevertheless, the significance of this emission lies in its abundance in terms of the number of particles, especially considering their small sizes (Guerranti et al. 2019).

Apart from microbeads, nano-additives are used in PCPs to enhance the effectiveness of the formulation. For instance, silver nanoparticles are included in some therapeutic shampoos and creams for their antimicrobial properties (Lorenz et al. 2011; McGillicuddy et al. 2017).

12.4 Consequences to Marine and Coastal Ecosystems

The health of our oceans is a matter of growing concern as human activities continue to exert unprecedented pressure on marine ecosystems (Boonstra et al. 2015). Within this context, the detrimental effects caused by PCPs on marine and coastal environments are significant, affecting organisms ranging from tiny creatures to the entire ecosystem (Table 12.2).

12.4.1 Effects on Marine Species and Corals

Marine organisms undergo a range of natural and anthropogenic stressors, such as fluctuations in temperature and pH, along with exposure to long-lasting, emerging contaminants. Thus, it is necessary to identify and understand the emerging threats they face from rising concentrations of such contaminants to manage species and ecosystems. Concerning PCP contamination, there is a substantial research gap in understanding the impacts of PCPs on various marine taxa at cellular and subcellular levels, including marine mammals, sea turtles, seabirds, and floating and suspended macro/microalgae (Figure 12.3). Furthermore, there are limited studies on benthic fauna that are vulnerable to exposure to high concentrations of PCPs. Besides that, the existing literature does not provide insights into how the presence and concentration of PCPs and their additives in coastal water and marine sediment align with the exposure levels encountered by

Table 12.2 Impacts of active ingredients of PCPs on non-targeted marine species.

PCPs	Active ingredient	Targeted species	Threshold level and toxic effects	References
Sunscreens	4-MBC, benzophenone-3 (BP-3), BP-4, and EHMC	Unicellular microalgae (*Isochrysis galbana*), Mediterranean mussel (*Mytilus galloprovincialis*), sea urchin (*Paracentrotus lividus*), and coastal zooplankton (*Siriella armata*)	UV filter concentrations in beach water ranged between 35 and 164 ng/l. Toxicity thresholds (μg/l), *S. armata* > *P. Lividus* > *M. galloprovincialis* BP-3 selectively toxic to *Isochrysis galbana*	Paredes et al. (2014)
	ZnO	*Acropora* spp.	Bleaching of stony corals (impact on symbiotic algae, *Symbiodinium* sp.)	Corinaldesi et al. (2018)
	4-MBC, BP-4, and octocrylene (OC)	Mussel (*Mytilus galloprovincialis*)	Bioaccumulated in tissues with BCF values in the range of 1000–2000 l/kg	Vidal-Liñán et al. (2018)
	Oxybenzone (OBZ), BP-3	Eukaryotic (*Chlorella* sp.) and prokaryotic algae (*Arthrospira* sp.)	Inhibited the growth and chlorophyll synthesis of both algae at a concentration of 22.8 ng/l	Zhong et al. (2019)
	Not specified (tested commercially available sunscreens in lotion, gel, and milk spray form)	Estuarine shrimp (*Palaemon varians*)	Avoided the highest concentrations of sunscreens in lotion and gel form. Avoidance caused an immediate population decline. Homogenization in the medium has reduced mortality.	Araújo et al. (2020)
	ZnO	Sea urchin (*Strongylocentrotus purpuratus*)	Malformations (skeletal abnormality, stage arrest, axis determination disruption)	Cunningham et al. (2020)
Fragrances	Polycyclic musk tonalide	Zebrafish (*Danio rerio*) juvenile stages	Long exposure (28 days) showed changes in antioxidant and detoxifying enzyme activities (adaptive responses to tonalide toxicity)	Blahova et al. (2018)
	Polycyclic musk compounds commercially available (Galaxolide and tonalide)	Clam (*Ruditapes philippinarum*)	Concentrations: 0.005, 0.05, 0.5, 5.0, and 50.0 μg/l of HHCB and AHTN-induced biotransformation enzymes at environmental concentrations, genotoxicity, oxidative stress	Ehiguese et al. (2020)
	Musk Galaxolide® (HHCB)	Clam (*Ruditapes philippinarum*)	Induce change at cellular level at 500 ng/l concentration (modulations of SOD and CAT activities)	Rilievo et al. (2021)

Table 12.2 (Continued)

PCPs	Active ingredient	Targeted species	Threshold level and toxic effects	References
Antimicrobials	Triclosan	Mediterranean mussel (*Mytilus galloprovincialis*)	Exposed concentration was 300 μg/l. Accumulated 881 ng/g (dry weight) at the end of 28 days	Gatidou et al. (2010)
		Dunaliella tertiolecta	Susceptibility to triclosan ranked as *Dunaliella* ~ *Scenedesmus* > *Chlorella* ~ *Chlamydomonas*	Bi et al. (2018)
Insect repellent	*N,N*-Diethyl-meta-toluamide (DEET)	Mussels (*Perna perna*)	Cytotoxicity at environmental concentration of 0.001–1.0 mg/l	de Assis Martini et al. (2017)
Plasticizers	Bisphenol A	Yellowfin seabream (*Acanthopagrus latus*)	Decreased RBC, hematocrit, hemoglobin, and lymphocytes when increasing BPA concentration (1, 5, 10, and 50 μg/g BPA)	Yaghoobi et al. (2017)
Not specified	UV filter: BP-3 Disinfectant: triclosan Fragrance: tonalide	Marine microalga *Tetraselmis suecica*	BP-3 increased growth rate, metabolic activity, and chlorophyll *a* fluorescence from 10 pg/cell. Growth and esterase activity decreased in cells exposed to triclosan and tonalide. All treated cells remained viable, but membrane potential changed. BP-3 and AHTN caused a significant depolarization of cells from 10 to 1.6 pg/cell	Seoane et al. (2017)

marine organisms. Therefore, uncertainty persists regarding whether the anticipated or unexpected effects of PCPs on marine organisms will present a greater risk, unlike the known detrimental impacts caused by pharmaceuticals. Furthermore, profound implications affecting marine populations, communities, and ecosystems need to be explored considering organism–community interactions.

The detrimental effects caused by various compounds in sunscreens, specifically UVFs and preservatives, range from single-celled to multicellular organisms (Danovaro and Corinaldesi 2003; Fent et al. 2010; Groz et al. 2014; Paredes et al. 2014). Previous studies have identified substances such as butylparaben, ethylhexyl methoxycinnamate, and benzophenone-3 that pose a threat to coral reefs and marine planktons (Danovaro and Corinaldesi 2003; Danovaro et al. 2008; Sánchez-Quiles and Tovar-Sánchez 2015). Fent et al. (2010) identified that preservatives such as parabens and certain UVFs such as oxybenzone, octocrylene, or ethylhexyl triazone can be toxic to marine invertebrates and fish through the risk assessment study. The lipophilic properties of sunscreens and the included additives tend to bioaccumulate in aquatic

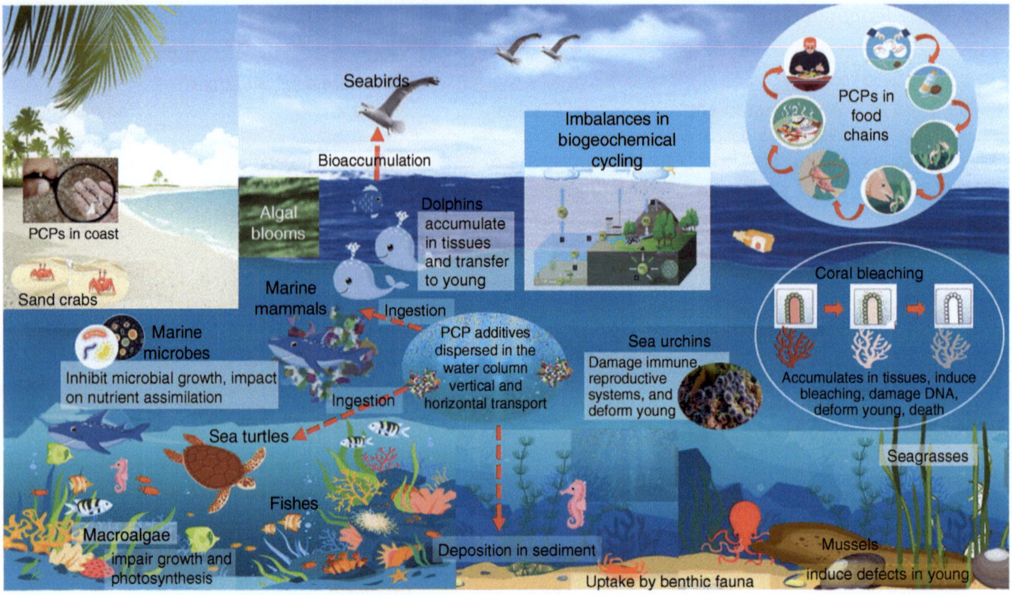

Figure 12.3 Effects of PCPs on the functioning and well-being of the marine and coastal ecosystem. *Source:* Sansert/Adobe Stock; Eduardo/Adobe Stock.

organisms (Giokas et al. 2007), causing effects similar to those of other xenobiotics (Danovaro et al. 2008). Oxybenzone (OBZ) has detrimental impacts on pelagic phytoplanktons, rodents, fish, invertebrates, and algae at the cellular and physiological levels (Mao et al. 2017; Seoane et al. 2017; Zhong et al. 2019). Further, OBZ adversely affects corals disturbing genomic integrity, bleaching, and altering the symbiotic relationship between coral host and algae (Downs et al. 2016). Marcellini (2017) found that benzophenone-3, homosalate, and preservatives in a particular sunscreen cause damage to the early development stages of *Paracentrotus lividus* by altering both embryo development and larvae structure. Benzotriazole UV stabilizers (BUVSs) are a group of compounds found in a range of PCPs such as sunscreens, hair dyes, makeup formulations, and soaps. Previous studies have discovered various BUVSs accumulated in marine organisms such as clams, oysters, gastropods, snarks, coastal birds (Nakata et al. 2009; Li et al. 2023), and marine mammals (Gago-Ferrero et al. 2013; Alonso et al. 2015) representing their adverse effects.

The usage of parabens in the cosmetic industry is expected to rise by 8% over the next six years, causing a threat of parabens contamination in aquatic environments in the future (Gago-Ferrero et al. 2013; Bilal et al. 2020). Xue et al. (2015) have detected high-concentration levels of paraben and its associated metabolites, such as 4-hydroxybenzoic acid (4-HB) and methylparaben in liver tissues of dolphins and sea otters from coastal waters of the United States. Furthermore, these compounds were detected in liver tissues of polar bears from remote regions, Chukchi Sea and the Beaufort Sea in trace levels, indicating that these substances have spread widely throughout the ocean. The lipophilic nature of parabens facilitates bioaccumulation in aquatic organisms, with chlorinated paraben exerting higher levels of toxicity compared to other paraben derivates (Bilal et al. 2020; Arfaeinia et al. 2022).

The rising demand for recreational activities in coral reefs, such as swimming, snorkeling, and scuba diving, poses a threat to coral reef ecosystems due to the extensive use of PCPs containing organic UVFs (Danovaro et al. 2008; Lamb et al. 2014). One of the well-known effects of UVFs is the bleaching and mortality of coral reefs (Richardson and Ternes 2014). A study conducted by Tsui et al. (2017) revealed the presence of organic UVFs in seawater, sediment, and five coral species collected from the eastern Pearl River Estuary of the South China Sea, along with the exclusive accumulation of UVFs in soft coral tissues. Another study has indicated that corals contaminated by UVFs are at levels surpassing threshold values known to induce larval deformities and mortality (Mitchelmore et al. 2019).

The persistence of MPs and microbeads in marine environments poses a significant threat to various marine organisms due to their small size and bioavailability. Zooplanktons, turtles, crabs, lobsters, worms, sea urchins, and fish are particularly vulnerable as they can ingest these nondegradable particles. The effects caused by ingestion of MPs include intestinal blockage, inhibition of enzymes released from the stomach, imbalances in steroid hormones, ovulation delay, and infertility. A previous study has shown that the gastrointestinal tract can absorb MPs and microbeads inadvertently consumed from toothpaste (Waring et al. 2018).

Studies have shown that nano-additives such as titanium dioxides (TiO_2) and zinc oxides (ZnO) can have toxicological effects on a wide range of taxa, including bacteria, plants, fish, and corals (Dedman et al. 2021; Shinde et al. 2023). For example, Dedman et al. (2021) found that environmentally relevant concentrations ranging from 1 μg/l to 100 mg/l of titanium dioxide nanoparticles ($n TiO_2$) pose negligible risk to marine microbial species, including the dominant cyanobacterium *Prochlorococcus*. However, a population decline was observed at high-concentration levels of $n TiO_2$. Moreover, Corinaldesi et al. (2018) have demonstrated that inorganic UVFs, such as ZnO nanoparticles, induce negative effects on corals, warranting further investigation.

12.4.2 PCP Accumulation in Marine Sediments and Seashore

Marine sediments serve as the ultimate sink for many organic pollutants, including PCPs. The accumulation of PCPs in sedimentary environments is a growing concern due to their potential ecological impacts and bioaccumulation potential (Wydro et al. 2023). Thus, it poses significant impacts on sediment-bound organisms, including shrimps, Mollusca, and other benthic fauna, causing bioaccumulation and biomagnification along the marine food chain. Consequently, the consumption of seafood contaminated with PCPs and associated toxins presents far-reaching implications for human health (Zhang et al. 2016). Gomez et al. (2012) have highlighted the substantial bioaccumulation potential of UVFs, musks, and antimicrobial agents such as methyl-triclosan in marine organisms. However, the bioaccumulation potential of PCPs across tropic levels is a cumulative effect of multiple factors such as chemical constituents of PCPs, environmental factors, exposure routes, and metabolic routes of organisms. For instance, the bioaccumulation level of UVFs in fish was comparable to the levels of PCBs and DDT, with bioconcentration factors (BCFs) exceeding 5000 (Brausch and Rand 2011). Furthermore, microplastics and microbeads released from PCPs are initially present in the water column and adhere to microalgae, increasing their density. Subsequently, they sink through the water column and settle in marine sediments, where they may persist for prolonged durations (Isobe 2016). However, the mere presence of PCPs in aquatic environments does not necessarily indicate an immediate risk of exposure, as these compounds require binding agents before they can increase internal tissue concentrations and induce acute toxicity. Therefore, as emphasized in the review of Prichard and Granek (2016), it is crucial to collect bioaccumulation data from marine ecosystems to understand the effects of PCPs on organisms across different trophic and taxonomic levels, as well as to assess their exposure risk.

12.4.3 Marine Eutrophication and Algal Blooms

Coastal and marine eutrophication occurs when excessive nutrients, such as nitrogen and phosphorus, enter marine environments, leading to accelerated algal growth and subsequent ecological imbalances (Malone and Newton 2020). While agricultural runoff and industrial discharges have been recognized as primary contributors to eutrophication, emerging research suggests that PCPs may also play a substantial role in this phenomenon. Nevertheless, the link between PCPs and marine eutrophication is complex. Some examples of eutrophication-driven chemicals commonly included in standard PCPs are ammonium chloride, methylisothiazolinone, parabens, phthalates, triclosan, sodium laureth sulfate, and sodium lauryl sulfate (Pennington et al. 2004; EPA 2020). Besides that, a variety of PCPs contain phosphorous-derived compounds, as mentioned by Tjandraatmadja et al. (2010) commonly found in dishwasher detergents and toothpaste formulations. The widespread use of such PCPs results in the introduction of phosphorous compounds into wastewater systems during everyday hygiene practices (Richards et al. 2015; Chong et al. 2019). Despite efforts in sewage treatment, not all facilities are equipped to adequately remove these compounds. As a result, a portion of these phosphorous compounds ended up in marine environments, contributing to nutrient loading in the ocean. Further, a study done by Ma and Chen (2018) has detected nitrogen and phosphorous-containing compounds in wastewater effluents of the cosmetic industry.

12.4.4 Marine Microbiome and Ocean Biogeochemical Cycling

Marine microbes are vital contributors to marine ecosystem function and biochemical cycling (Bar-On et al. 2018). Their impact extends globally, as they regulate CO_2 and CH_4 levels and are fundamental in decomposing organic matter within the ocean (Glöckner et al. 2012). Further, they play a key role in symbiotic relationships with many organisms such as algae, seaweeds, corals, and sponges (Lozano et al. 2020). Despite their vital importance, our understanding of marine microbial diversity remains limited, highlighting the need for further research to comprehensively assess their environmental impact, especially in the context of PCPs (Callieri et al. 2018). A recent review by Świacka et al. (2023) summarized the impacts of pharmaceuticals on marine microbes, highlighting how these substances can disrupt normal metabolic functions, alter microbial abundance and composition, and affect nutrient cycling. However, the impacts caused by PCPs on sediment-dwelling and water column microbial communities are either similar to or in contrast with pharmaceuticals, potentially resulting in profound and lasting effects that disrupt natural balances and diversity.

While most of the antimicrobial agents found in PCPs such as triclosan and triclocarban may not exert a major influence on the human microbiome, their effects on coastal and marine microbiomes remain largely unknown (Poole et al. 2016). A few studies have investigated the specific impacts of PCPs on microbial processes. For example, some microbes are capable of degrading toxic compounds found in PCPs, as highlighted in a recent review by Paul et al. (2023). However, numerous PCPs, ranging from deodorants and detergents to toothpaste and soaps, contain antimicrobial agents such as 1,4-dichlorobenzene, ortho-phenylphenol, triclocarban, and triclosan, which are known to inhibit microbial growth (Allmyr et al. 2006; Dodson et al. 2012). For instance, studies have shown that triclosan exposure can lead to the development of antimicrobial-resistant strains of bacteria (Braoudaki and Hilton 2004) which can have cascading effects on microbial communities, decomposition and mineralization processes, and biogeochemical cycles. The study of Lopez et al. (2021) revealed a significant inhibitory impact of triclosan, a common antimicrobial agent, on the nitrification process in WWTPs. Conversely, the impact of the insect-repellent ingredient

N,N-diethyl-*m*-toluamide (DEET) was observed to be moderate. Further, triclosan exposure has been found to significantly affect microalgae species, with chronic toxicity values ranging from 1.4 to 3.5 μg/l (DeLorenzo et al. 2008; Franz et al. 2008). Moreover, according to Yang et al. (2020), *Microcystis aeruginosa* and *Selenastrum capricornutum* have also exerted high sensitivity to antimicrobials in PCPs. A recent study by Chen et al. (2023) found that triclosan negatively impacted microalgal–bacterial symbiosis systems, affecting microbial growth and nutrient assimilation capacity, and ultimately impacting chemical oxygen demand, ammonium-nitrogen, and phosphorus removal efficiencies. A recent review by Pereira et al. (2023) has elaborated on the effects of parabens on the structure and metabolism of microorganisms. Further, the study indicated that parabens have the potential to cause varying impacts on the thickness and formation of microbial biofilms and increase the risk of spreading antimicrobial resistance (AMR). Similarly, another review by Caioni et al. (2023) also highlighted the antimicrobial resistance potential of parabens, triclosan, and triclocarban in PCPs and the underlying mechanisms of AMR through the zebrafish model.

In addition, it is known that microplastics (MPs) exert a significant influence on microbial oxidation-reduction reactions, thereby affecting nutrient balance (Rogers et al. 2020; da Fonseca et al., 2022). For instance, Rong et al. (2021) have shown that the presence of polyethylene (PE) can alter the local microbial community compositions. Therefore, given that a significant portion of the microbeads present in PCPs are composed of PE, comprising approximately 93% of the total MPs (Fendall and Sewell 2009; Gouin et al. 2015), there exists the potential for these microplastics to adversely affect microbial communities and disrupt nutrient cycles.

12.5 Risks to Humans via Seafood Consumption

As discussed in the above section, PCPs, and their associated additives tend to accumulate in the edible tissues of commercial fish and shellfish, which are commonly consumed as seafood. Further, microbeads, micro(nano)plastics, and nano-additives have the potential to serve as vectors for microbial pathogens, biocides, and trace metals in the contaminated environment (Curren et al. 2020). The global consumption of seafood has increased rapidly exceeding the population growth rate (Canton 2021). This suggests a higher chance of significant exposure pathways to PCPs in the human body through seafood consumption, despite their external applications (Xie et al. 2019). Since PCPs are not intended for internal use, the presence of these compounds, their metabolites, and derivatives may pose a toxicity to internal systems and functions of the human body. So far, studies have detected several PCPs that have accumulated in fish plasma and bile (Yao et al. 2018; Yao et al. 2019). However, there is a lack of comprehensive data regarding the bioaccumulation levels of active components of PCPs across the food chain as well as their toxicity and toxicokinetic properties in the human body.

12.6 Prevention and Mitigation Strategies

Despite many attempts, current mitigation and prevention strategies are struggling to keep control of the increasing amounts of PCPs entering the marine and coastal environment via various routes. The application of remediation strategies in these environments is not feasible, as contaminants are diffuse and widely distributed, extending from coast to polar regions through sea waves and ocean circulation. Therefore, the most effective approach is prevention and mitigation before they reach coastal and marine environments. Most of the PCPs enter the coast and seawater due to inefficiencies in WWTPs (Jyoti and Sinha 2023). Therefore, actions should be taken to improve the efficiency of WWTPs to capture and retain PCPs and their associated additives (Miller et al. 2016). Remediation strategies should be investigated to remove PCPs from WWTPs at the industrial scale. Further, it is necessary to implement strategies to collect and properly dispose of leftovers of PCPs from households and tourism sectors (Miller et al. 2016).

There is an urgent need to promote green chemistry approaches to develop low-cost alternatives (Paul et al. 2023) that are less or no toxic, for example, resistant to endocrine-disrupting activities. For instance, biosurfactants derived from *Lactobacillus paracasei* are an alternative option to replace petroleum-based surfactants used in shampoo, toothpaste, body lotions, and cosmetics (Ferreira et al. 2017; Huynh et al. 2020).

Eco-labeling is another market-driven and product-specific approach adopted to manufacture ecologically compatible PCPs, for instance prioritizing nature-based products (Klaschka et al. 2007). In 2021, the European Commission updated the European Union Ecolabel criteria for cosmetics including rinse-off and leave-on cosmetics. Consequently, all cosmetic products, such as shampoos and shower gels, are required to apply the EU Ecolabel which aims to minimize products' impact on the environment, fostering a toxic-free environment and circular economy (EU 2018). Nonetheless, there is no ban on the use of microbeads in cosmetics in some parts of Europe. A recent review by Khalid and Abdollahi (2021) has provided some recommendations to ensure that packages of PCPs are clearly labeled, thereby enhancing consumer safety. So far, there is no law or regulation either at the national or international level (Jyoti and Sinha 2023). Thus, it is essential to implement common regulations across regions, as the impacts caused are diffused and widespread across the globe. For instance, parabens-containing products have been produced in China in higher volumes: methyl- and ethyl-parabens at 500 tons/month and propyl-paraben at over 10 000 tons/month, without restrictions on their use in cosmetic and pharmaceutical industries (Melinda and Bao 2011). However, the EU has banned the use of many analogs of parabens in PCPs since 2014 (EU 2014). Another example of unclear rules and regulations which has caused loopholes in handling products by different authorities and across regions is that in the United States, manufacturers are not required to mention the safety of cosmeceutical products before marketing (Pandey et al. 2019). Furthermore, improper labeling fails to mention active ingredients in formulations, and their correct quantities have caused problems with consumers' awareness of the content of PCPs they are consuming. This could be due to the absence of strict rules in legislation for manufacturing and marketing (Dodson et al. 2012) regarding human safety and exposure to hazardous ingredients in PCPs. In addition, the rules, and regulations relevant to PCPs are different from region to region based on the class to which they belong such as drugs, PCPs, cosmeceuticals, or other regulatory products. For instance, in the United States, many cosmeceuticals such as sunscreens, antidandruff shampoos, and antiperspirants are categorized as drugs whereas in Europe these are considered cosmeceuticals (Pandey et al. 2019). These differences affect the sales of PCPs in different regions. Another instance is that, despite banning the use of phthalate esters by the European Chemicals Agency, manufacturers have substituted them with several plasticizers that might also have toxic effects (Hammel et al. 2019). Consequently, these limitations and loopholes in rules and regulations have resulted in unnecessary chemical exposure for consumers (Steinemann 2009), as well as unintentional damage to non-targeted organisms due to contamination of the environment. Therefore, addressing these failures requires careful monitoring of manufacturing and selling PCPs that contain harmful substances. This includes checking standards and quality control measures and establishing common criteria for defining and categorizing PCPs while enforcing strict regulations worldwide.

Public awareness plays a major role in preventing and mitigating anthropogenic menace. However, a qualitative analysis conducted by Anderson et al. (2016) showed that beauticians and consumers are unaware of the negative impacts of microplastics on PCPs. Thus, the most efficient approach involves eliminating synthetic microbeads from PCP formulations (Cheung and Fok 2016). It can be achieved by changing consumer behavior through awareness and implementing a legislative ban on the use of microplastic beads in cosmetics and other PCPs (Anderson et al. 2016; Hirst and Bennett 2017), not only for microbeads but also for other hazardous additives in PCPs. The recent report of the World Economic Forum has mentioned the importance of consumer behavior in the use and disposal of PCPs, especially focusing on listing the products' impact on the environment in the packaging (World Economic Forum 2023). Nevertheless, due to a lack of evidence-based data, further research is needed to identify controlled and safe chemicals for use in PCPs, enabling the implementation of rules and regulations to regulate manufacturers (Hubinger and Havery 2006).

Apart from the above facts, it is necessary to consider that many of the proposed and implemented strategies concerning the banning of toxic PCP ingredients and the promotion of environmentally and economically favorable products in the recent decade (2015–2025) are not immediate measures and are planned to be achieved by 2030. This extended timeline for adhering to new strategies is reasonable concerning industrial and business goals. However, it presents a considerable period during which significant damage to the environment can occur due to the release of numerous harmful PCP contaminants that might not have been considered.

12.7 Conclusions and Future Remarks

Every second, more and more PCPs and their transforming products enter the marine and coastal environments. Thus, there is an urgent need for research to investigate both acute and chronic toxicities experienced by organisms exposed to their environmental concentrations. Further, the effects caused by a cocktail of contaminants need to be explored. The research gaps on impacts extending beyond individual species to entire ecosystems need to be explored. Moreover, understanding how interspecific interactions may either mediate or exacerbate the effects of PCP exposure is important. Additionally, it is crucial to explore how toxicity is transferred across trophic levels, posing a potential threat to human health through seafood consumption.

When assessing these impacts on laboratory scale, it is vital to simulate real environmental scenarios by considering physiochemical parameters such as salinity, temperature, and pH of seawater and sediments. Furthermore, it is necessary to assess the impacts of PCPs on microbial communities and study changes in community composition and distribution, their mechanisms, and alternative pathways relevant to biogeochemical cycles through advanced omics tools. The scientific findings of such research should be disseminated to the public to raise their awareness and formulate policies and regulations at a global level. One of the targets for sustainable development is the preservation of the ocean and marine resources which must be accomplished by 2030. In this context, policies, rules, and regulations relevant to PCPs should be taken into consideration.

References

Al-Halaseh, L.K., Al-Adaileh, S., Mbaideen, A. et al. (2022). Implication of parabens in cosmetics and cosmeceuticals: advantages and limitations. *Journal of Cosmetic Dermatology* 21: 3265–3271.

Allmyr, M., Adolfsson-Erici, M., McLachlan, M.S., and Sandborgh-Englund, G. (2006). Triclosan in plasma and milk from Swedish nursing mothers and their exposure via personal care products. *Science of the Total Environment* 372: 87–93.

Alonso, M.B., Feo, M.L., Corcellas, C. et al. (2015). Toxic heritage: maternal transfer of pyrethroid insecticides and sunscreen agents in dolphins from Brazil. *Environmental Pollution* 207: 391–402.

Amine, H., Gomez, E., Halwani, J. et al. (2012). UV filters, ethylhexyl methoxycinnamate, octocrylene and ethylhexyl dimethyl PABA from untreated wastewater in sediment from eastern Mediterranean river transition and coastal zones. *Marine Pollution Bulletin* 64: 2435–2442.

Ammeri, R.W., Hidri, Y., and Abdenaceur, H. (2020). Effect of PCP pesticide contamination on soil quality. In: *Soil Contamination-Threats and Sustainable Solutions* (ed. L.L. Marcelo and S. Soloneski), 37–58. IntechOpen.

Andersen, F.A. (2008). Final amended report on the safety assessment of methylparaben, ethylparaben, propylparaben, isopropylparaben, butylparaben, isobutylparaben, and benzylparaben as used in cosmetic products. *International Journal of Toxicology* 27: 1–82.

Anderson, A., Grose, J., Pahl, S. et al. (2016). Microplastics in personal care products: Exploring perceptions of environmentalists, beauticians and students. *Marine Pollution Bulletin* 113: 454–460.

Apel, C., Joerss, H., and Ebinghaus, R. (2018a). Environmental occurrence and hazard of organic UV stabilizers and UV filters in the sediment of European North and Baltic Seas. *Chemosphere* 212: 254–261.

Apel, C., Tang, J., and Ebinghaus, R. (2018b). Environmental occurrence and distribution of organic UV stabilizers and UV filters in the sediment of Chinese Bohai and Yellow Seas. *Environmental Pollution* 235: 85–94.

Araújo, C.V., Rodríguez-Romero, A., Fernández, M. et al. (2020). Repellency and mortality effects of sunscreens on the shrimp *Palaemon varians*: toxicity dependent on exposure method. *Chemosphere* 257: 127190.

Arfaeinia, H., Asadgol, Z., Ramavandi, B. et al. (2022). Monitoring and eco-toxicity effect of paraben-based pollutants in sediments/seawater, north of the Persian Gulf. *Environmental Geochemistry and Health* 44: 4499–4521.

Arpin-Pont, L., Bueno, M.J.M., Gomez, E., and Fenet, H. (2016). Occurrence of PPCPs in the marine environment: a review. *Environmental Science and Pollution Research* 23: 4978–4991.

de Assis Martini, G., Rogero, S.O., and Rogero, J.R. (2017). Cytotoxic effects caused by N, N-diethyl-meta-toluamide and radiation in *Perna perna* mussels. *Ecotoxicology and Environmental Contamination* 12: 33–37.

Bar-On, Y.M., Phillips, R., and Milo, R. (2018). The biomass distribution on Earth. *Proceedings of the National Academy of Sciences of the United States of America* 115: 6506–6511.

Bayen, S., Zhang, H., Desai, M.M. et al. (2013). Occurrence and distribution of pharmaceutically active and endocrine disrupting compounds in Singapore's marine environment: influence of hydrodynamics and physical–chemical properties. *Environmental Pollution* 182: 1–8.

Beretta, M., Britto, V., Tavares, T.M. et al. (2014). Occurrence of pharmaceutical and personal care products (PPCPs) in marine sediments in the Todos os Santos Bay and the north coast of Salvador, Bahia, Brazil. *Journal of Soils and Sediments* 14: 1278–1286.

Bi, R., Zeng, X., Mu, L. et al. (2018). Sensitivities of seven algal species to triclosan, fluoxetine and their mixtures. *Scientific Reports* 8: 15361.

Bilal, M., Mehmood, S., and Iqbal, H. (2020). The beast of beauty: environmental and health concerns of toxic components in cosmetics. *Cosmetics* 7: 13.

Billah, M.M., Mustafa Kamal, A.H., Idris, M.H., and Ismail, J. (2017). Mangrove macroalgae as biomonitors of heavy metal contamination in a tropical estuary, Malaysia. *Water, Air, & Soil Pollution* 228: 347.

Blahova, J., Divisova, L., Plhalova, L. et al. (2018). Multibiomarker responses of juvenile stages of zebrafish (*Danio rerio*) to subchronic exposure to polycyclic musk tonalide. *Archives of Environmental Contamination and Toxicology* 74: 568–576.

Blair, B.D., Crago, J.P., Hedman, C.J. et al. (2013). Evaluation of a model for the removal of pharmaceuticals, personal care products, and hormones from wastewater. *Science of the Total Environment* 444: 515–521.

Błędzka, D., Gromadzińska, J., and Wąsowicz, W. (2014). Parabens. From environmental studies to human health. *Environment International* 67: 27–42.

Bolujoko, N.B., Unuabonah, E.I., Alfred, M.O. et al. (2021). Toxicity and removal of parabens from water: a critical review. *Science of the Total Environment* 792: 148092.

Boonstra, W.J., Ottosen, K.M., Ferreira, A. et al. (2015). What are the major global threats and impacts in marine environments? Investigating the contours of a shared perception among marine scientists from the bottom-up. *Marine Policy* 60: 197–201.

Braoudaki, M. and Hilton, A.C. (2004). Low level of cross-resistance between triclosan and antibiotics in *Escherichia coli* K-12 and *E. coli* O55 compared to *E. coli* O157. *FEMS Microbiology Letters* 235: 305–309.

Bratkovics, S. and Sapozhnikova, Y. (2011). Determination of seven commonly used organic UV filters in fresh and saline waters by liquid chromatography-tandem mass spectrometry. *Analytical Methods* 3: 2943–2950.

Brausch, J.M. and Rand, G.M. (2011). A review of personal care products in the aquatic environment: environmental concentrations and toxicity. *Chemosphere* 82: 1518–1532.

Bridges, B. (2002). Fragrance: emerging health and environmental concerns. *Flavour and Fragrance Journal* 17: 361–371.

Caioni, G., Benedetti, E., Perugini, M. et al. (2023). Personal care products as a contributing factor to antimicrobial resistance: current state and novel approach to investigation. *Antibiotics* 12: 724.

Callieri, C., Eckert, E.M., Cesare, A., and Bertoni, F. (2018). Microbial communities. *Encyclopedia Ecology* 1: 126–134.

Canton, H. (2021). Food and agriculture organization of the United Nations—FAO. In: *The Europa Directory of International Organizations 2021*, 297–305. Europa Publications, Routledge https://doi.org/10.4324/9781003179900.

Cantwell, M.G., Wilson, B.A., Zhu, J. et al. (2010). Temporal trends of triclosan contamination in dated sediment cores from four urbanized estuaries: evidence of preservation and accumulation. *Chemosphere* 78: 347–352.

Capela, D., Vila, M., Llompart, M. et al. (2019). Footprints in the sand – assessing the seasonal trends of volatile methylsiloxanes and UV-filters. *Marine Pollution Bulletin* 140: 9–16.

Capkin, E., Ozcelep, T., Kayis, S., and Altinok, I. (2017). Antimicrobial agents, triclosan, chloroxylenol, methylisothiazolinone and borax, used in cleaning had genotoxic and histopathologic effects on rainbow trout. *Chemosphere* 182: 720–729.

Chakraborty, P., Mukhopadhyay, M., Sampath, S. et al. (2019). Organic micropollutants in the surface riverine sediment along the lower stretch of the transboundary river Ganga: occurrences, sources and ecological risk assessment. *Environmental Pollution* 249: 1071–1080.

Chavoshani, A., Hashemi, M., Amin, M.M., and Ameta, S.C. (2020). Personal care products as an endocrine disrupting compound in the aquatic environment. In: *Micropollutants and Challenges* (ed. A. Chavoshani, M. Hashemi, M.M. Amin, and S.C. Ameta), 91–144. Amsterdam, The Netherlands: Elsevier.

Chen, H., Chen, W., Guo, H. et al. (2021). Pharmaceuticals and personal care products in the seawater around a typical subtropical tourist city of China and associated ecological risk. *Environmental Science and Pollution Research* 28: 22716–22728.

Chen, S.L., Wang, T.Y., Tang, C.C. et al. (2023). Revealing mechanisms of triclosan on the removal and distribution of nitrogen and phosphorus in microalgal-bacterial symbiosis system. *Environmental Pollution* 337: 122539.

Cheung, P.K. and Fok, L. (2016). Evidence of microbeads from personal care product contaminating the sea. *Marine Pollution Bulletin* 109: 582–585.

Chevillot, F., Guyot, M., Desrosiers, M. et al. (2018). Accumulation and sublethal effects of triclosan and its transformation product methyl-triclosan in the earthworm *Eisenia andrei* exposed to environmental concentrations in an artificial soil. *Environmental Toxicology and Chemistry* 37: 1940–1948.

Chong, J., Dominish, E., Tjandraatmadja, G. et al. (2019). *Ecosystem Impacts of Phosphorus and Surfactants in Consumer Products.* Institute for Sustainable Futures, University of Technology Sydney for Stewart Investors.

Cincinelli, A., Martellini, T., Guerranti, C. et al. (2019). A potpourri of microplastics in the sea surface and water column of the Mediterranean Sea. *TrAC Trends in Analytical Chemistry* 110: 321–326.

Corinaldesi, C., Marcellini, F., Nepote, E. et al. (2018). Impact of inorganic UV filters contained in sunscreen products on tropical stony corals (*Acropora* spp.). *Science of the Total Environment* 637: 1279–1285.

Cunningham, B., Torres-Duarte, C., Cherr, G., and Adams, N. (2020). Effects of three zinc-containing sunscreens on development of purple sea urchin (*Strongylocentrotus purpuratus*) embryos. *Aquatic Toxicology* 218: 105355.

Curren, E., Leaw, C.P., Lim, P.T., and Leong, S.C.Y. (2020). Evidence of marine microplastics in commercially harvested seafood. *Frontiers in Bioengineering and Biotechnology* 8: 562760.

Danovaro, R. and Corinaldesi, C. (2003). Sunscreen products increase virus production through prophage induction in marine bacterioplankton. *Microbial Ecology* 45: 109–118.

Danovaro, R., Bongiorni, L., Corinaldesi, C. et al. (2008). Sunscreens cause coral bleaching by promoting viral infections. *Environmental Health Perspectives* 116: 441–447.

Dedman, C.J., King, A.M., Christie-Oleza, J.A., and Davies, G.-L. (2021). Environmentally relevant concentrations of titanium dioxide nanoparticles pose negligible risk to marine microbes. *Environmental Science: Nano* 8: 1236–1255.

DeLorenzo, M., Keller, J., Arthur, C. et al. (2008). Toxicity of the antimicrobial compound triclosan and formation of the metabolite methyl-triclosan in estuarine systems. *Environmental Toxicology: An International Journal* 23: 224–232.

Díaz-Cruz, M.S., García-Galán, M.J., Guerra, P. et al. (2009). Analysis of selected emerging contaminants in sewage sludge. *TrAC Trends in Analytical Chemistry* 28: 1263–1275.

Dodson, R.E., Nishioka, M., Standley, L.J. et al. (2012). Endocrine disruptors and asthma-associated chemicals in consumer products. *Environmental Health Perspectives* 120: 935–943.

Downs, C., Kramarsky-Winter, E., Segal, R. et al. (2016). Toxicopathological effects of the sunscreen UV filter, oxybenzone (benzophenone-3), on coral planulae and cultured primary cells and its environmental contamination in Hawaii and the US Virgin Islands. *Archives of Environmental Contamination and Toxicology* 70: 265–288.

Duis, K. and Coors, A. (2016). Microplastics in the aquatic and terrestrial environment: sources (with a specific focus on personal care products), fate and effects. *Environmental Sciences Europe* 28: 1–25.

Ehiguese, F.O., Alam, M.R., Pintado-Herrera, M.G. et al. (2020). Potential of environmental concentrations of the musks galaxolide and tonalide to induce oxidative stress and genotoxicity in the marine environment. *Marine Environmental Research* 160: 105019.

EPA (2020). Contaminants of emerging concern including pharmaceuticals and personal care products. https://www.epa.gov/wqc/contaminants-emerging-concern-including-pharmaceuticals-and-personal-care-products#:~:text=Contaminants%20of%20emerging%20concern%20(CECs,an%20impact%20on%20aquatic%20life. (accessed 15 March 2024).

EU (2018). Environmental label regulation. https://ec.europa.eu/environment/ecolabel/documents/Factsheet_EU_Ecolabel_Cosmetic_product.pdf. (accessed Mar. 20, 2024).

EU Commission Regulation (2014). no. 1004/2014. https://www.legislation.gov.uk/eur/2014/1004/introduction. (accessed Apr. 16, 2024).

Fendall, L.S. and Sewell, M.A. (2009). Contributing to marine pollution by washing your face: microplastics in facial cleansers. *Marine Pollution Bulletin* 58: 1225–1228.

Fent, K., Kunz, P.Y., Zenker, A., and Rapp, M. (2010). A tentative environmental risk assessment of the UV-filters 3-(4-methylbenzylidene-camphor), 2-ethyl-hexyl-4-trimethoxycinnamate, benzophenone-3, benzophenone-4 and 3-benzylidene camphor. *Marine Environmental Research* 69: S4–S6.

Ferraro, A., Parisi, A., Barbone, E. et al. (2023). Characterizing contaminants distribution in marine-coastal sediments through multivariate and nonparametric statistical analyses: a complementary strategy supporting environmental monitoring and control. *Environmental Monitoring and Assessment* 195: 59.

Ferreira, A., Vecino, X., Ferreira, D. et al. (2017). Novel cosmetic formulations containing a biosurfactant from *Lactobacillus paracasei*. *Colloids and Surfaces B: Biointerfaces* 155: 522–529.

da Fonseca, E.M., Gaylarde, C., Baptista Neto, J.A. et al. (2022). Microbial interactions with particulate and floating pollutants in the oceans: a review. *Micro* MDPI: 257–276.

Franz, S., Altenburger, R., Heilmeier, H., and Schmitt-Jansen, M. (2008). What contributes to the sensitivity of microalgae to triclosan? *Aquatic Toxicology* 90: 102–108.

Gago-Ferrero, P., Alonso, M.B., Bertozzi, C.P. et al. (2013). First determination of UV filters in marine mammals. Octocrylene levels in Franciscana dolphins. *Environmental Science & Technology* 47: 5619–5625.

Gatidou, G., Vassalou, E., and Thomaidis, N.S. (2010). Bioconcentration of selected endocrine disrupting compounds in the Mediterranean mussel, *Mytilus galloprovincialis*. *Marine Pollution Bulletin* 60: 2111–2116.

Gaw, S., Thomas, K.V., and Hutchinson, T.H. (2014). Sources, impacts and trends of pharmaceuticals in the marine and coastal environment. *Philosophical Transactions of the Royal Society, B: Biological Sciences* 369: 20130572.

Giokas, D.L., Salvador, A., and Chisvert, A. (2007). UV filters: from sunscreens to human body and the environment. *TrAC Trends in Analytical Chemistry* 26: 360–374.

Girijan, S. and Kumar, M. (2020). Microbial degradation of pharmaceuticals and personal care products from wastewater. *Microbial Bioremediation & Biodegradation* 173–201.

Glöckner, F.O., Gasol, J.M., McDonough, N., and Calewaert, J.B. (2012). Marine microbial diversity and its role in ecosystem functioning and environmental change. In: (ed. J.B. Calewaert and N. McDonough), 1–80. Ostend, Belgium: Marine Board European Scientific Foundation.

Gomez, E., Bachelot, M., Boillot, C. et al. (2012). Bioconcentration of two pharmaceuticals (benzodiazepines) and two personal care products (UV filters) in marine mussels (*Mytilus galloprovincialis*) under controlled laboratory conditions. *Environmental Science and Pollution Research* 19: 2561–2569.

Gouin, T., Avalos, J., Brunning, I. et al. (2015). Use of micro-plastic beads in cosmetic products in Europe and their estimated emissions to the North Sea environment. *SOFW Journal* 141: 40–46.

Granek, E.F., Conn, K.E., Nilsen, E.B. et al. (2016). Spatial and temporal variability of contaminants within estuarine sediments and native Olympia oysters: a contrast between a developed and an undeveloped estuary. *Science of the Total Environment* 557: 869–879.

Groz, M.P., Bueno, M.M., Rosain, D. et al. (2014). Detection of emerging contaminants (UV filters, UV stabilizers and musks) in marine mussels from Portuguese coast by QuEChERS extraction and GC–MS/MS. *Science of the Total Environment* 493: 162–169.

Guerranti, C., Martellini, T., Perra, G. et al. (2019). Microplastics in cosmetics: environmental issues and needs for global bans. *Environmental Toxicology and Pharmacology* 68: 75–79.

Hammel, S.C., Levasseur, J.L., Hoffman, K. et al. (2019). Children's exposure to phthalates and non-phthalate plasticizers in the home: the TESIE study. *Environment International* 132: 105061.

Hawthorne, S.B., Kubátová, A., Gallagher, J.R. et al. (2005). Persistence and biodegradation of monoethanolamine and 2-propanolamine at an abandoned industrial site. *Environmental Science & Technology* 39: 3639–3645.

Hernandez, L.M., Yousefi, N., and Tufenkji, N. (2017). Are there nanoplastics in your personal care products? *Environmental Science & Technology Letters* 4: 280–285.

Hirst, D. and Bennett, O. (2017). *Microbeads and Microplastics in Cosmetic and Personal Care Products.* Parliament UK: The House of Commons Library.

Hubinger, J.C. and Havery, D.C. (2006). Analysis of consumer cosmetic products for phthalate esters. *Journal of Cosmetic Science* 57: 127–137.

Huynh, A., Maktabi, B., Reddy, C.M. et al. (2020). Evaluation of alkenones, a renewably sourced, plant-derived wax as a structuring agent for lipsticks. *International Journal of Cosmetic Science* 42: 146–155.

Hwang, J., Suh, S.S., Park, M. et al. (2017). Differential gene expression patterns during embryonic development of sea urchin exposed to triclosan. *Environmental Toxicology* 32: 426–433.

Ilyas, H. and van Hullebusch, E.D. (2020). A review on the occurrence, fate and removal of steroidal hormones during treatment with different types of constructed wetlands. *Journal of Environmental Chemical Engineering* 8 (3): 103793.

Isobe, A. (2016). Percentage of microbeads in pelagic microplastics within Japanese coastal waters. *Marine Pollution Bulletin* 110: 432–437.

Jyoti, D. and Sinha, R. (2023). Physiological impact of personal care product constituents on non-target aquatic organisms. *Science of the Total Environment* 905: 167229.

Khalid, M. and Abdollahi, M. (2021). Environmental distribution of personal care products and their effects on human health. *Iranian Journal of Pharmaceutical Research: IJPR* 20: 216.

Klaschka, U., Liebig, M., and Knacker, T. (2007). Eco-Labelling of Shampoos, Shower Gels and Foam Baths. *Environmental Science and Pollution Research-International* 14: 24–29.

Klosterhaus, S.L., Grace, R., Hamilton, M.C., and Yee, D. (2013). Method validation and reconnaissance of pharmaceuticals, personal care products, and alkylphenols in surface waters, sediments, and mussels in an urban estuary. *Environment International* 54: 92–99.

Kung, T.A., Lee, S.H., Yang, T.C., and Wang, W.H. (2018). Survey of selected personal care products in surface water of coral reefs in Kenting National Park, Taiwan. *Science of the Total Environment* 635: 1302–1307.

Lamb, J.B., True, J.D., Piromvaragorn, S., and Willis, B.L. (2014). Scuba diving damage and intensity of tourist activities increases coral disease prevalence. *Biological Conservation* 178: 88–96.

Leslie, H. (2014). Review of microplastics in cosmetics. *IVM Institute for Environmental Studies* 476: 1–33.

Li, Q., Wang, P., Wang, C. et al. (2023). Benzotriazole UV stabilizer-induced genotoxicity in freshwater benthic clams: a survey on apoptosis, oxidative stress, histopathology and transcriptomics. *Science of the Total Environment* 857: 159055.

Liang, X., Chen, B., Nie, X. et al. (2013). The distribution and partitioning of common antibiotics in water and sediment of the Pearl River Estuary, South China. *Chemosphere* 92: 1410–1416.

Łopaciuk, A. and Łoboda, M. (2013). Global beauty industry trends in the 21st century. *Management, Knowledge and Learning International Conference* 19–21.

Lopez, C., Nnorom, M.-A., Tsang, Y.F., and Knapp, C.W. (2021). Pharmaceuticals and personal care products'(PPCPs) impact on enriched nitrifying cultures. *Environmental Science and Pollution Research* 28: 60968–60980.

Lorenz, C., Von Goetz, N., Scheringer, M. et al. (2011). Potential exposure of German consumers to engineered nanoparticles in cosmetics and personal care products. *Nanotoxicology* 5: 12–29.

Lozano, C., Matallana-Surget, S., Givens, J. et al. (2020). Toxicity of UV filters on marine bacteria: combined effects with damaging solar radiation. *Science of the Total Environment* 722: 137803.

Lu, J., Mao, H., Li, H. et al. (2017). Occurrence of and human exposure to parabens, benzophenones, benzotriazoles, triclosan and triclocarban in outdoor swimming pool water in Changsha, China. *Science of the Total Environment* 605: 1064–1069.

Ma, G. and Chen, J. (2018). Nitrogen and phosphorus pollutants in cosmetics wastewater and its treatment process of a certain brand. *IOP Conference Series: Earth and Environmental Science* 113: 012051.

Ma, X., Wan, Y., Wu, M. et al. (2018). Occurrence of benzophenones, parabens and triclosan in the Yangtze River of China, and the implications for human exposure. *Chemosphere* 213: 517–525.

Malone, T.C. and Newton, A. (2020). The globalization of cultural eutrophication in the coastal ocean: causes and consequences. *Frontiers in Marine Science* 7: 670.

Mao, F., He, Y., Kushmaro, A., and Gin, K.Y.-H. (2017). Effects of benzophenone-3 on the green alga *Chlamydomonas reinhardtii* and the cyanobacterium *Microcystis aeruginosa*. *Aquatic Toxicology* 193: 1–8.

Marcellini, F. (2017). Development of innovative approaches for assessing the impact of personal care products and other micro-contaminants in the marine environment. https://iris.univpm.it/retrieve/e18b8790-6728-d302-e053-1705fe0a27c8/tesi_marcellini.pdf.

Martínez Bueno, M., Boillot, C., Munaron, D. et al. (2014). Occurrence of venlafaxine residues and its metabolites in marine mussels at trace levels: development of analytical method and a monitoring program. *Analytical and Bioanalytical Chemistry* 406: 601–610.

Martínez-Girón, A.B., Crego, A.L., González, M.J., and Marina, M.L. (2010). Enantiomeric separation of chiral polycyclic musks by capillary electrophoresis: application to the analysis of cosmetic samples. *Journal of Chromatography A* 1217: 1157–1165.

McGillicuddy, E., Murray, I., Kavanagh, S. et al. (2017). Silver nanoparticles in the environment: sources, detection and ecotoxicology. *Science of the Total Environment* 575: 231–246.

Melinda, M. and Bao, L. (2011). Standards for uses of food additives - part I. China. *In:* China, Ministry of Health of the People's Republic of China (ed.). https://www.dgav.pt/wp-content/uploads/2020/12/China-GB-2760_2011.pdf. (accessed Mar. 15, 2024).

Miller, E.L., Nason, S.L., Karthikeyan, K., and Pedersen, J.A. (2016). Root uptake of pharmaceuticals and personal care product ingredients. *Environmental Science & Technology* 50: 525–541.

Mitchelmore, C.L., He, K., Gonsior, M. et al. (2019). Occurrence and distribution of UV-filters and other anthropogenic contaminants in coastal surface water, sediment, and coral tissue from Hawaii. *Science of the Total Environment* 670: 398–410.

Möhlenkamp, P., Purser, A., and Thomsen, L. (2018). Plastic microbeads from cosmetic products: an experimental study of their hydrodynamic behaviour, vertical transport and resuspension in phytoplankton and sediment aggregates. *Elementa: Science of the Anthropocene* 6: 61.

Montes-Grajales, D., Fennix-Agudelo, M., and Miranda-Castro, W. (2017). Occurrence of personal care products as emerging chemicals of concern in water resources: a review. *Science of the Total Environment* 595: 601–614.

Musee, N. (2018). Environmental risk assessment of triclosan and triclocarban from personal care products in South Africa. *Environmental Pollution* 242: 827–838.

Nakata, H., Murata, S., and Filatreau, J. (2009). Occurrence and concentrations of benzotriazole UV stabilizers in marine organisms and sediments from the Ariake Sea, Japan. *Environmental Science & Technology* 43: 6920–6926.

Nakata, H., Shinohara, R.-I., Nakazawa, Y. et al. (2012). Asia–Pacific mussel watch for emerging pollutants: distribution of synthetic musks and benzotriazole UV stabilizers in Asian and US coastal waters. *Marine Pollution Bulletin* 64: 2211–2218.

Napper, I.E., Bakir, A., Rowland, S.J., and Thompson, R.C. (2015). Characterisation, quantity and sorptive properties of microplastics extracted from cosmetics. *Marine Pollution Bulletin* 99: 178–185.

Pai, C.-W., Leong, D., Chen, C.-Y., and Wang, G.-S. (2020). Occurrences of pharmaceuticals and personal care products in the drinking water of Taiwan and their removal in conventional water treatment processes. *Chemosphere* 256: 127002.

Pandey, A., Jatana, G.K., and Sonthalia, S. (2019). Cosmeceuticals. In: *StatPearls [Internet]*. Treasure Island (FL): StatPearls Publishing: 2024. PMID: 3133494.

Paredes, E., Pérez, S., Rodil, R. et al. (2014). Ecotoxicological evaluation of four UV filters using marine organisms from different trophic levels *Isochrysis galbana*, *Mytilus galloprovincialis*, *Paracentrotus lividus*, and *Siriella armata*. *Chemosphere* 104: 44–50.

Patel, S., Homaei, A., and Sharifian, S. (2021). Need of the hour: to raise awareness on vicious fragrances and synthetic musks. *Environment, Development and Sustainability* 23: 4764–4781.

Paul, S., Paul, S., Bhattacharjee, S. et al. (2023). Ecological problems associated with personal care products and the role of microbes in aquatic environment. *Environmental Quality Management* 00: 1–8.

Pedrouzo, M., Borrull, F., Marcé, R.M., and Pocurull, E. (2011). Analytical methods for personal-care products in environmental waters. *TrAC Trends in Analytical Chemistry* 30: 749–760.

Pemberthy, D., Padilla, Y., Echeverri, A., and Peñuela, G.A. (2020). Monitoring pharmaceuticals and personal care products in water and fish from the Gulf of Urabá, Colombia. *Heliyon* 6: e04215.

Pennington, D.W., Potting, J., Finnveden, G. et al. (2004). Life cycle assessment Part 2: current impact assessment practice. *Environment International* 30: 721–739.

Pereira, A.R., Gomes, I.B., and Simões, M. (2023). Parabens as environmental contaminants of aquatic systems affecting water quality and microbial dynamics. *Science of the Total Environment* 905: 167332.

Poole, A.C., Pischel, L., Ley, C. et al. (2016). Crossover control study of the effect of personal care products containing triclosan on the microbiome. *Msphere* 1: e00056–e00015.

Prichard, E. and Granek, E.F. (2016). Effects of pharmaceuticals and personal care products on marine organisms: from single-species studies to an ecosystem-based approach. *Environmental Science and Pollution Research* 23: 22365–22384.

Rainieri, S., Barranco, A., Primec, M., and Langerholc, T. (2017). Occurrence and toxicity of musks and UV filters in the marine environment. *Food and Chemical Toxicology* 104: 57–68.

Ramos, S., Homem, V., Alves, A., and Santos, L. (2016). A review of organic UV-filters in wastewater treatment plants. *Environment International* 86: 24–44.

Richards, S., Paterson, E., Withers, P.J., and Stutter, M. (2015). The contribution of household chemicals to environmental discharges via effluents: combining chemical and behavioural data. *Journal of Environmental Management* 150: 427–434.

Richardson, S.D. and Ternes, T.A. (2014). Water analysis: emerging contaminants and current issues. *Analytical Chemistry* 86: 2813–2848.

Rilievo, G., Fabrello, J., Roverso, M. et al. (2021). Effects of the fragrance galaxolide on the biomarker responses of the clam *Ruditapes philippinarum*. *Journal of Marine Science and Engineering* 9: 509.

Rogers, K.L., Carreres-Calabuig, J.A., Gorokhova, E., and Posth, N.R. (2020). Micro-by-micro interactions: how microorganisms influence the fate of marine microplastics. *Limnology and Oceanography Letters* 5: 18–36.

Rong, L., Zhao, L., Zhao, L. et al. (2021). LDPE microplastics affect soil microbial communities and nitrogen cycling. *Science of the Total Environment* 773: 145640.

Sánchez-Quiles, D. and Tovar-Sánchez, A. (2015). Are sunscreens a new environmental risk associated with coastal tourism? *Environment International* 83: 158–170.

Seoane, M., Esperanza, M., Rioboo, C. et al. (2017). Flow cytometric assay to assess short-term effects of personal care products on the marine microalga *Tetraselmis suecica*. *Chemosphere* 171: 339–347.

Shinde, S.A., More, P.R., and Ingle, A.P. (2023). Hazardous effects of nanomaterials on aquatic life. *Nanotechnology in Agriculture and Agroecosystems* 423–450.

Statista (2024). Beauty & Personal Care – Worldwide. https://www.statista.com/outlook/cmo/beauty-personal-care/worldwide#:~:text=The%20Beauty%20%26%20Personal%20Care%20market,US%24282.80bn%20in%202024. (accessed July 08, 2024).

Steinemann, A.C. (2009). Fragranced consumer products and undisclosed ingredients. *Environmental Impact Assessment Review* 29: 32–38.

Sumner, N.R., Guitart, C., Fuentes, G., and Readman, J.W. (2010). Inputs and distributions of synthetic musk fragrances in an estuarine and coastal environment; a case study. *Environmental Pollution* 158: 215–222.

Świacka, K., Maculewicz, J., Kowalska, D., and Grace, M.R. (2023). Do pharmaceuticals affect microbial communities in aquatic environments? A review. *Frontiers in Environmental Science* 10: 2657.

Tjandraatmadja, G., Pollard, C., Sheedy, C., and Gozukara, Y. (2010). Sources of contaminants in domestic wastewater: nutrients and additional elements from household products. *Water for a Healthy Country Flagship Report: CSIRO, Canberra*. https://webcat.niwa.co.nz/documents/Tjandraatmadja2010.pdf. (accessed July 08, 2024).

Tsui, M.M., Lam, J.C., Ng, T. et al. (2017). Occurrence, distribution, and fate of organic UV filters in coral communities. *Environmental Science & Technology* 51: 4182–4190.

Vesilind, P. (2003). *Wastewater Treatment Plant Design*. IWA Publishing.

Vidal-Liñán, L., Villaverde-de-Sáa, E., Rodil, R. et al. (2018). Bioaccumulation of UV filters in *Mytilus galloprovincialis* mussel. *Chemosphere* 190: 267–271.

Wang, J. and Wang, S. (2016). Removal of pharmaceuticals and personal care products (PPCPs) from wastewater: a review. *Journal of Environmental Management* 182: 620–640.

Waring, R.H., Harris, R., and Mitchell, S. (2018). Plastic contamination of the food chain: a threat to human health? *Maturitas* 115: 64–68.

World Economic Forum, O. W. (2023). Nature positive: role of the household and personal care products sector. *Part of the World Economic Forum's Sector Transitions to Nature Positive Report Series.*

Wydro, U., Wołejko, E., Luarasi, L. et al. (2023). A review on pharmaceuticals and personal care products residues in the aquatic environment and possibilities for their remediation. *Sustainability* 16: 169.

Xie, H., Hao, H., Xu, N. et al. (2019). Pharmaceuticals and personal care products in water, sediments, aquatic organisms, and fish feeds in the Pearl River Delta: Occurrence, distribution, potential sources, and health risk assessment. *Science of the Total Environment* 659: 230–239.

Xue, J., Sasaki, N., Elangovan, M. et al. (2015). Elevated accumulation of parabens and their metabolites in marine mammals from the United States coastal waters. *Environmental Science & Technology* 49: 12071–12079.

Yaghoobi, Z., Safahieh, A., Ronagh, M.T. et al. (2017). Hematological changes in yellowfin seabream (*Acanthopagrus latus*) following chronic exposure to bisphenol A. *Comparative Clinical Pathology* 26: 1305–1313.

Yang, L., Zhou, Y., Shi, B. et al. (2020). Anthropogenic impacts on the contamination of pharmaceuticals and personal care products (PPCPs) in the coastal environments of the Yellow and Bohai seas. *Environment International* 135: 105306.

Yao, L., Lv, Y.-Z., Zhang, L.-J. et al. (2018). Determination of 24 personal care products in fish bile using hybrid solvent precipitation and dispersive solid phase extraction cleanup with ultrahigh performance liquid chromatography-tandem mass spectrometry and gas chromatography-mass spectrometry. *Journal of Chromatography A* 1551: 29–40.

Yao, L., Lv, Y.-Z., Zhang, L.-J. et al. (2019). Bioaccumulation and risks of 24 personal care products in plasma of wild fish from the Yangtze River, China. *Science of the Total Environment* 665: 810–819.

Zhang, W., Wang, W.-X., and Zhang, L. (2016). Comparison of bioavailability and biotransformation of inorganic and organic arsenic to two marine fish. *Environmental Science & Technology* 50: 2413–2423.

Zhong, X., Downs, C.A., Che, X. et al. (2019). The toxicological effects of oxybenzone, an active ingredient in suncream personal care products, on prokaryotic alga *Arthrospira* sp. and eukaryotic alga *Chlorella* sp. *Aquatic Toxicology* 216: 105295.

Ziajahromi, S., Neale, P.A., Rintoul, L., and Leusch, F.D. (2017). Wastewater treatment plants as a pathway for microplastics: development of a new approach to sample wastewater-based microplastics. *Water Research* 112: 93–99.

Zitko, V. and Hanlon, M. (1991). Another source of pollution by plastics: skin cleaners with plastic scrubbers. *Marine Pollution Bulletin* 22: 41–42.

13

Monitoring and Assessment of Coastal and Marine Pollution: Methods and Technologies

G.K.M.R. Peter[1], S.M. Dassanayake[2], C.L. Jayawardena[1], and I. Mahakalanda[2]

[1] *Department of Earth Resources Engineering, University of Moratuwa, Moratuwa, 10400, Sri Lanka*
[2] *Department of Decision Sciences, University of Moratuwa, Moratuwa, 10400, Sri Lanka*

13.1 Introduction

Coastal and marine ecosystems are vital components of our environment experiencing rapidly increasing pollution in recent decades (Hafeez et al. 2018). Growing human activity has taken a toll on these environments to the extremes, and the consequences could be unbearable. The coastal regions have been recorded as the most densely populated areas globally, which continues to increase (Mahrad et al. 2020). Marine pollutants usually include (Figure 13.1) a variety of plastics, oil spills, toxic chemicals, heavy metals (HMs), radioactive waste, and sewage from both industrial and domestic sources (Hafeez et al. 2018). Stressors like overfishing, rising ocean temperatures, ocean acidification, and sea level rise further exacerbate the conditions (Álvarez-Romero et al. 2013). This upward trend can profoundly impact both the environment and socioeconomic conditions (Islam and Tanaka 2004; Karydis and Kitsiou 2013; Zheng and DiGiacomo 2017). Discharge of HMs, oil spills, sediment pollution, eutrophication, and microplastic pollution are identified as the most significant contributors to marine pollution (Anthony et al. 2023). Moreover, land-based activities, such as agriculture, industry, and waste disposal, contribute to approximately 80% of marine pollution (Halpern et al. 2008; Butchart et al. 2010; Álvarez-Romero et al. 2013; Reichelt-Brushett 2023). Therefore, it is crucial to minimize these activities and aim to establish a healthy coastal and marine ecosystem.

Preservation and conservation of sustainable coastal and marine ecosystems demand continuous and accurate monitoring of the influx of pollutants into the marine environments (Hafeez et al. 2018; Collard and Ask 2021; Anthony et al. 2023). Typically, the conventional ocean pollution monitoring approaches include sampling water, sediments, and marine organisms to analyze physical and chemical properties in laboratories or using underwater sensors to record specific elements of interest and biological monitoring to extract changes in marine organisms. More robust techniques include remote sensing (RS) methodologies such as satellite data with spectral and spatial signatures, aerial images, and sonar and laser systems, which have facilitated both coastal and marine observations with increased precision of results and location. The latest technological advancements amalgamated with artificial intelligence (AI) and machine learning (ML) have enabled RS techniques to reflect real-time data visualizations and conventional approaches to reach molecular-level investigations (Kannangara et al. 2022).

Coastal and Marine Pollution: Source to Sink, Mitigation and Management, First Edition.
Edited by Meththika Vithanage, Sameera M. Samarasekara, Bryan D. James, and Christopher M. Reddy.

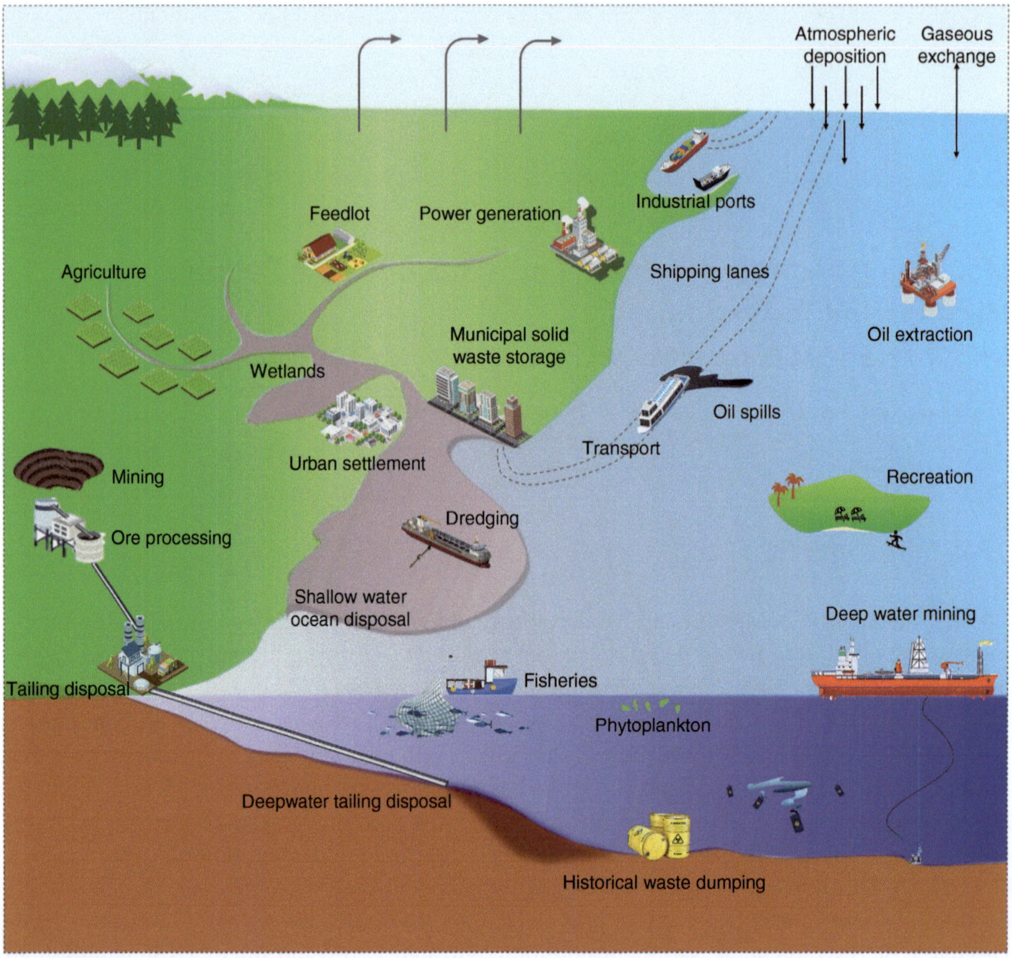

Figure 13.1 Varieties and sources of pollution affecting coastal and marine ecosystems.

Furthermore, sensor networks have been established from floating platforms to the sea floor level facilitated by underwater cameras capable of transmitting real-time visual information.

These sophisticated systems gather data on crucial parameters including temperature, conductivity, pH, salinity, dissolved oxygen, chlorophyll fluorescence, turbidity, and CDOM (Hafeez et al. 2018). Recent studies have identified several emerging technologies for monitoring marine pollution encompassing cloud-native platforms, decision intelligence, robotic process automation, augmented reality, and virtual reality (Anthony et al. 2023). Among the above, ML and AI have notably emerged as key technologies by significantly enhancing image processing capabilities beyond microscopic, spectroscopic, and biomonitoring techniques (Lorencin et al. 2019; Kylili et al. 2020; Politikos et al. 2021). These innovative technologies hold promise for improving further by complementing the current monitoring methods and further research will be essential to fully comprehend their potential applications (Concejcão et al. 2021).

In this chapter, we compile methods and technologies utilized to monitor and assess pollution in marine and coastal settings with attention to diverse techniques currently in practice and emerging systems. A bibliometric analysis courtesy of the VOSviewer indicates the intricate associations and visualization of the key scientific literature and RS footprint in this domain. With that realization, we evaluate the applicability of those monitoring methods in different environmental conditions based on their performance through a comparative analysis to extract relative effectiveness. Moreover, this chapter will expand our understanding of pollution monitoring in marine and coastal environments and enhance our comprehension of current practices with valuable insights on future adaptations and advancements in strategic environmental monitoring.

13.2 Bibliographic Review

A bibliometric analysis was conducted to uncover trends and identify the primary types of pollution discussed in key publications, in the most affected countries and regions. The analysis revealed the current state of research, gaps or emerging areas of interest, and methods and technologies for monitoring and assessing coastal and marine pollution. The process of gathering and analyzing bibliometric data involves several steps. The Scopus database was selected as the main source for collecting relevant academic papers. To find the most pertinent studies on marine and coastal pollution and the methods used to monitor them, the search terms contained "(marine OR coastal) AND pollution AND (monitoring OR assessment) AND (methods OR techniques)." The retrieved bibliometric data were filtered based on the relevance to the scope of marine pollution monitoring.

The co-occurrence of the keywords was used to examine their relevance in the literature (Figure 13.2). Each bubble in Figure 13.2 represents a keyword, with its size indicating how frequently it appears. The thickness of the lines between bubbles indicates how often those keywords appear together. Bubbles that are closer together are more similar and have similar colors. The strength of the connections between keywords reflects the frequency with which they appear together in publications (Wang et al. 2022; Anthony et al. 2023).

Overall bibliometric analysis conducted from 2019 to 2024 (Figure 13.2a) shows the primary types of marine pollution as plastic pollution, HM pollution, sediment pollution, and oil spill pollution. At the end of the period (Figure 13.2b), the analysis highlights microplastic as the main pollution type among the others. Moreover, eutrophication also has become a dominant keyword revealing it has become an emerging threat to marine ecosystems eventually. After identifying major pollution types, a country-specific analysis was done to understand mostly responsible countries for coastal and marine pollution. This analysis as in Figure 13.3 identified three major countries such as China followed by the United States and Italy.

According to the analysis shown in Figure 13.3, all three countries are mainly affected by microplastic pollution. As the next major common pollution types, HMs and oil spills could be identified. In addition to that, the United States shows harmful algal blooms (HABs) as a main marine pollution event. Most of the pollution incidents have happened in or near the Bohai Sea and East China Sea since they appear as major keywords related to marine and coastal pollution in China. While HMs are more prominent in the East China Sea area, the Bohai Sea is particularly prominent in oil spill events. The Gulf of Mexico and the Mediterranean Sea can be identified as closely related areas with pollution events in the United States and Italy, respectively. The highly urbanized coastal lines, population density, industrial activities, coastal activities, and poor regulations and enforcement might be the main reasons for extensive pollution within these regions. However, several methods are being used to understand and evaluate pollutant behavior.

Laboratory methods have been utilized mainly for monitoring and assessing marine pollution in the overall analysis. But at the end of the period, from 2023 to 2024 (Figure 13.2c), RS has come forward as an emerging technique over traditional laboratory methods. This trend can be seen in

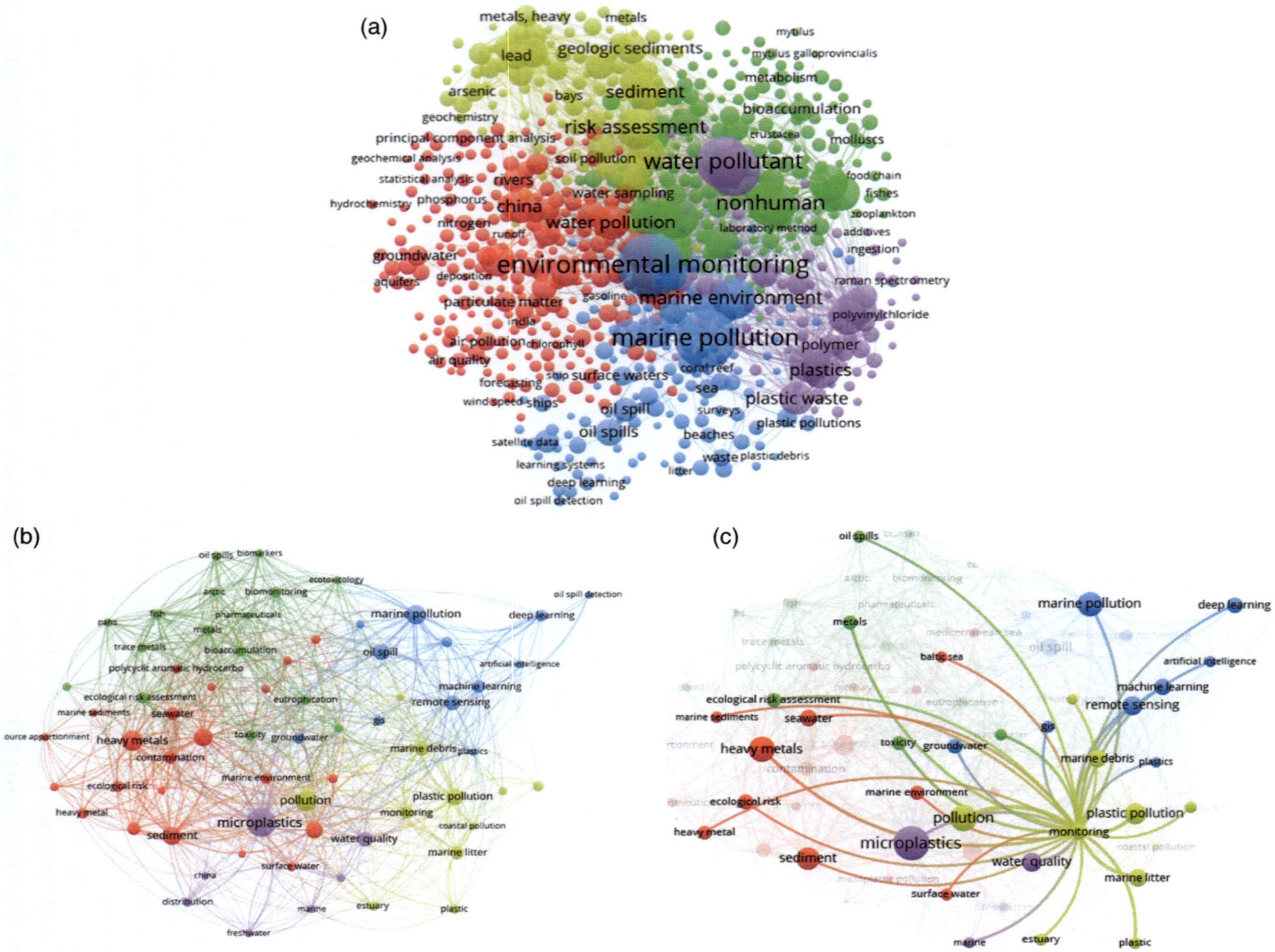

Figure 13.2 (a) Bibliometric analysis for the period of 2019–2024. (b) Bibliometric analysis for the recent period of 2023–2024. (c) Directly correlated keywords with "monitoring" in the bibliometric analysis for the period of 2023–2024.

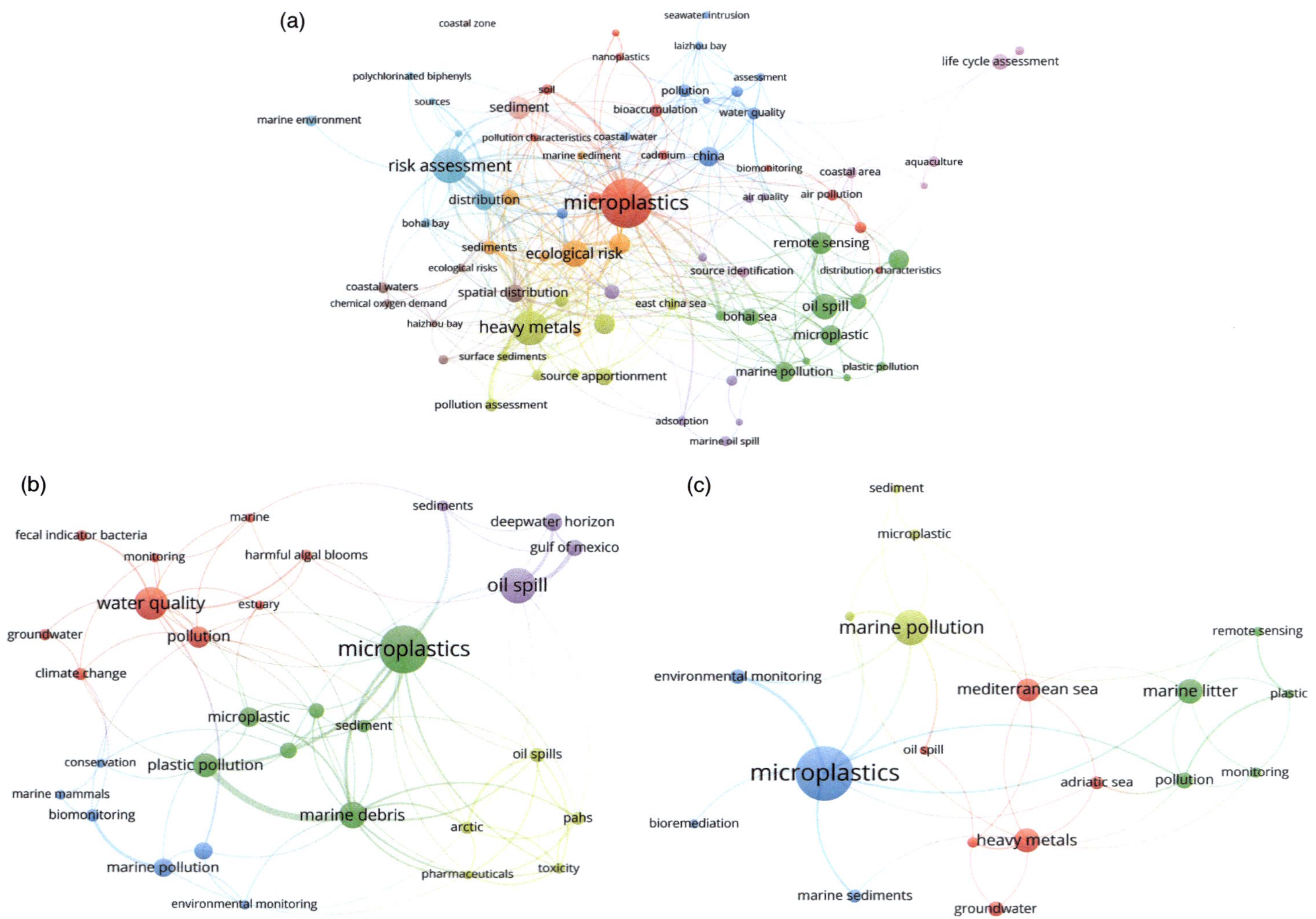

Figure 13.3 Geographical distribution of the keywords: (a) China, (b) the United States, and (c) Italy.

both overall analysis and country-wise analysis (Figure 13.3). A close connection can be seen between RS, oil spills, and plastic debris. This proves RS technology is widely used for monitoring these two pollution types. On the other hand, less connectivity can be seen between RS and HMs. Along with RS techniques, ML, and deep learning (DL) methods, subcategories of AI have been employed.

This analysis identified the major coastal and marine pressures as plastic pollution, HMs, oil spills, and eutrophication. China, the United States, and Italy lead as the most responsible countries for marine pollution due to several reasons as mentioned. Even though laboratory methods have been widely employed as the prominent monitoring method for the past few years, RS has been introduced to the marine monitoring sector recently. The applicability of emerging technologies like AI coupling with RS will be more useful in the future as monitoring technologies.

13.3 Remote Sensing in Coastal and Marine Pollution Monitoring

Recent progress in RS technologies offers promising opportunities for mapping and evaluating the stressors affecting coastal and marine environments (Mahrad et al. 2020). Efficient monitoring of coastal and marine pollution, encompassing occurrences like oil spills, algal blooms, and suspended sediment contamination requires real-time or near-real-time measurements. These measurements are indispensable for prompt responses, wide coverage, enhanced efficiency, and the ability to capture multiband imaging (Wang et al. 2022). The integration of RS technology facilitates these measurements (Hafeez et al. 2018), offering a macro-observational method that overcomes limitations associated with traditional approaches, such as prolonged time requirements, restricted spatial reach, and the necessity for skilled personnel. RS efficiently detects temporal and spatial variations in various physical parameters on the ocean surface with accuracy (Collard and Ask 2021; Wang et al. 2022).

A thorough comprehension of marine dynamics is needed for proficient monitoring of marine pollutants. These dynamics include factors like ocean current direction and intensity, surface wind direction and speed, sea surface temperature (SST), and sea surface salinity (SSS). This understanding facilitates precise evaluation and forecasting of pollutant dispersion patterns and their effects on marine ecosystems. In recent years, RS technology has undergone significant advancements. At present, there are several main types of satellite and airborne sensors as described in Figure 13.4, capable of capturing detailed data about marine dynamics across extensive oceanic areas (Hafeez et al. 2018).

Airborne and spaceborne (Hafeez et al. 2018) are the major two platforms in RS. Airborne platforms operate at relatively low altitudes, ranging from a few hundred meters to a few kilometers above the surface, resulting in data with higher levels of detail. Spaceborne platforms are located in outer space and the sensors in these platforms have relatively low resolution compared to airborne sensors.

Mahrad et al. (2020) have outlined six types of airborne and spaceborne RS platforms utilized for managing coastal and marine environments as follows.

1) Satellite remote sensing
2) Aerial remote sensing (ARS) or manned aircraft
3) Unmanned aerial vehicles (UAVs)
4) Unmanned surface vehicles (USVs)
5) Unmanned underwater vehicles (UUVs)
6) Static sensors (SS) as ground methods

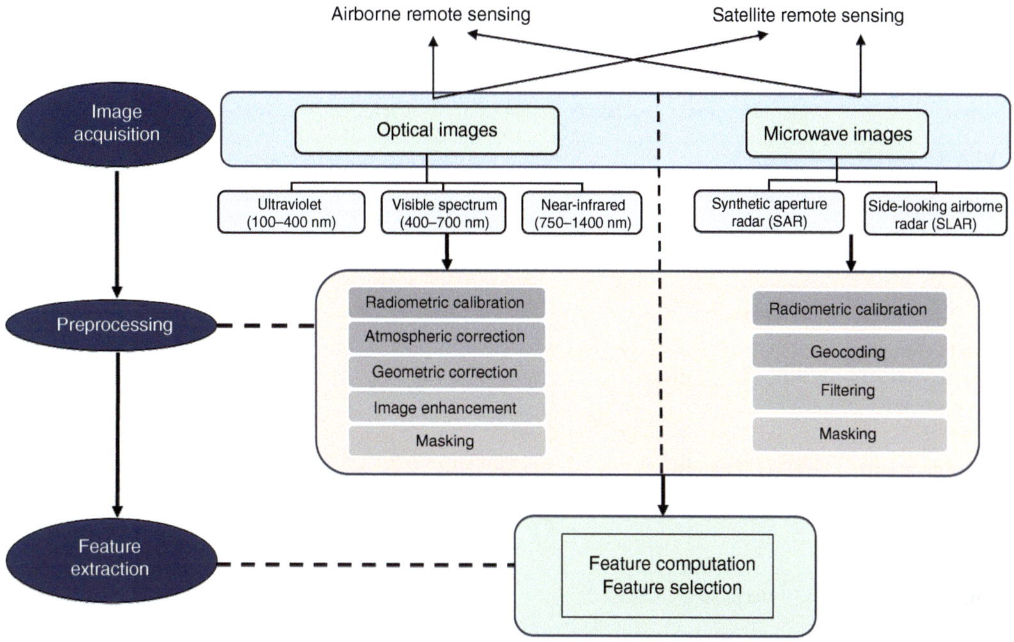

Figure 13.4 Image acquisition using airborne and satellite RS technologies *Source:* Al-Ruzouq et al. (2020) MDPI/CC BY 4.0.

Tracking marine pollutants needs to identify the features correctly from the captured images. Whatever the pollutant types, extracting targeted features needs an effective image classification procedure. This process can be described in Figure 13.4. Image acquisition is the first step toward image processing in RS. In this step, sensors can capture optical or microwave images. This can be done using both platforms. Optical images normally lie in the wavelength range between 100 and 1400 nm. In that range, there are mainly three spectral regions: ultraviolet, visible, and near-infrared (IR).

Microwave images can be categorized mainly into two: synthetic aperture radar (SAR) images and side-looking airborne radar (SLAR) images. SAR can measure patterns of sea surface roughness, providing valuable data for determining sea surface wind conditions. Laser fluorometry involves the use of laser light to stimulate molecules, making it a widely used method for detecting oil and chemical spills. Active spaceborne sensors, such as SAR and laser fluorometry, find various applications including the monitoring of sea surface currents, oil spills, biogenic films like algal blooms, and the dispersion of river plumes (Hafeez et al. 2018).

After successful image acquisition, preprocessing needs to be done. The way it is done for optical images and microwave images is a little different from each other. For optical images, necessary corrections such as radiometric, atmospheric, and geometric should be done in the first place. Then images are enhanced followed by masking finally. For microwave images, after doing radiometric correction, geocoding and filtering are implemented before masking. Then the features can be extracted from both optical and microwave images by feature computation and selection.

This study analyzed different airborne and spaceborne sensors used for marine water quality monitoring purposes. While satellite sensors could be identified as the widely used method for monitoring purposes for the past few decades, several studies revealed some efforts made to use airborne sensors as well but with limited amounts. Table 13.1 presents such airborne sensors

Table 13.1 Hyperspectral airborne sensors in water quality assessment.

Sensor	Manufacturer	Number of bands	Spectral range (nm)	Spatial resolution (m)	Studied parameter
Airborne visible infrared imaging spectrometer (AVIRIS)	NASA Jet Propulsion Lab	224	400–2500	17	Bottom albedo, water absorption, backscattering coefficients, Chl-a, CDOM, TSS
HyMap	Earth Search Sciences Inc.	128	400–2500	3–10	Heavy metals
Portable remote imaging spectrometer (PRISM)	NASA Jet Propulsion Lab	-	350–1050, SWIR band (1240 and 1640)	0.3	Sediment, CDOM, chlorophyll fluorescence, turbidity, Chl-a, and dissolved organic carbon
Airborne prism experiment (APEX)	VITO (Belgium)	313	VIS and NIR (380–970), SWIR (970–2500)	2–5	Chlorophyll fluorescence, SS

Source: Hafeez et al. 2018/IntechOpen/CC BY 3.0.

used for studying different water quality parameters. The results show how much resolution can be achieved by these airborne sensors.

Since using airborne sensors still has some major drawbacks such as limited spatial coverage and time limitations, spaceborne sensors come forward to overcome such limitations. Optical spaceborne sensors primarily employed for marine monitoring typically operate in sun-synchronous orbits. Notably, the Geostationary Ocean Color Imager (GOCI) sensor, tailored for marine monitoring, is positioned in a geostationary orbit. These sensors offer spatial coverage spanning tens to hundreds of kilometers and provide temporal frequency ranging from hourly to weekly monitoring intervals.

The routine collection of satellite data offers the opportunity for synoptic-scale analysis, encompassing extensive geographic areas over extended periods. Moreover, these datasets are often available as open-source resources, readily accessible, particularly when acquired by governmental bodies and institutions. This accessibility facilitates the availability of a substantial volume of data for analysis and dissemination purposes. For the past few decades, there have been several studies conducted using satellites to retrieve marine water quality parameters as summarized in Table 13.2.

The introduction of Hyperion in 2000 helped bridge the gap between high-resolution airborne sensors and low-resolution satellite sensors. Brando and Dekker conducted a study in 2003 to evaluate Hyperion's capabilities across various water targets in Eastern Australia, particularly Moreton Bay in southern Queensland. Their primary aim was to assess estuarine and coastal water quality using satellite hyperspectral RS data. By focusing on Moreton Bay, they found that, after applying preprocessing and noise reduction techniques, Hyperion imagery could effectively map optical water quality concentrations of colored dissolved organic matter, chlorophyll, and total suspended matter simultaneously in the complex waters of estuarine and coastal systems.

While using airborne or satellite data separately, there could be identified some effort of combining in situ data with earth observation data. Copernicus program by the European Space Agency has made substantial strides in merging both in situ and satellite observations. Arabi et al. (2020) conducted a study focusing on this integration to enhance water quality monitoring in coastal

Table 13.2 Satellite sensors mostly used to retrieve marine water quality parameters.

Satellite sensor	Launch date	Spectral bands (nm)	Spatial resolution (m)	Swath width (km)	Marine parameter accessed
Satellite sensors with moderate spatial resolution					
Landsats 4 and 5 TM	1 March 1984	5 (450–1750),1 (2080–2350), and 1 (1040–1250)	30–120	185	Chl-a, SS Secchi depth
Landsat 7 ETM+	15 April 1999	6 (450–1750),1 Pan (520–900),1 (2090–2350), and 1 (2090–2350), and 1 (1040–1250)	15–30–60	183	Chl-a, SS, Secchi depth, turbidity
Terra Aster	18 December 1999	3 (520–860), 6 (1600–2430), and 5 TIR (8125–11 650)	15–30–90	60	Chl-a
EO-I ALI	November 2000	(443–2350)	30	—	Turbidity, SS
EO-1 Hyperion	1 November 2000	242 (350–2570)	30	7.5	Chl-a, SS, CDOM SS
PROBA CHRIS	22 October 2001	19 (400–105)	18–36	14	Chl-a, phycocyanin, behenic macroalgae
HICO	10 September 2009	128 (350–1080)	100	45-50	Chl-a, turbidity, CDOM, SS
Landsat 8 OLI/ TIRS	11 February 2013	1 cirrus cloud detection (1360–1380), 5 (430–880), 1 Pan (500–680), 2 (1570–2290), 2 TIRS (10 600–12 510)	15–30–100	170	ChL-a, SS, turbidity, TN, TP
Sentinel-2 MSI	23 June 2015	8 (490–865),1 (443) coastal aerosol, and 3 (1375–2190)	30–120	290	Chl-a, CDOM, DOC, SS
Satellite sensors for regional coverage					
Orb View 2 SeaWiFS	1 August 1997	8 (402–885)	1130	2806	Chl-a
Terra, Aqua MODIS	18 December 1999	2 (620–876), 5 (459–2155), 29 (405–877), and thermal	250–500–100	2330	Chl-a, CDOM SS, turbidity, TP
ENVISAT-1 MERIS	1 March 2002	15 (390–1040)	300–1200	1150	Chl-a, SS
GOCI	26 June 2010	8 (412–865)	500	2500	Chl-a, SS, turbidity
Suomi NPP VIIRS	28 October 2011	5 bands (640–1145), 16 bands (412–12, 013), DNB (500–900)	375–750	3060	Chl-a
Sentinel-3 OLCI	16 February 2016	21 (400–1020)	300	1270	Chl-a, SS, CDOM, and Secchi depth

Source: Hafeez et al. 2018/IntechOpen/CC BY 3.0.

regions. They investigated a 15-year diurnal variation of water constituent concentrations (WCCs) by combining satellite and in situ data, employing radiative transfer (RT) modeling techniques. Their study centered on the Dutch Wadden Sea revealed a strong agreement between WCC values obtained from in situ measurements and those derived from satellite observations. However, it is essential to note that only a limited number of studies have explored this integrated approach thus far.

These studies showed the applicability of RS for monitoring coastal and marine pollution monitoring as an emerging technology. Airborne and satellite sensors are the main two types of optical and microwave image-capturing methods. Selection of the sensor depends on the requirement of the study and satellite sensors can be seen as the widely used sensors over airborne sensors. There have been several efforts to try innovative methods to improve the quality and the resolution of the data. Moreover, this study intends to analyze how the RS technique has been used for monitoring major pollution types identified earlier.

13.3.1 Monitoring Plastic Pollution

Plastic pollution is an immensely pressing issue that has concerned scientists and environmentalists (Kikaki et al. 2020; Mahrad et al. 2020; Baruah et al. 2022) due to its adverse chemical, biological, and ecological impacts on the marine ecosystem. It is estimated that 4.8–12.7 million metric tons of plastic are dumped in the sea every year due to the increased use of plastic in industry and daily life (Hafeez et al. 2018). Moreover, Kikaki et al. (2020) have presented the largest plastics accumulator (Great Pacific Garbage Patch) as sufficient evidence for marine plastic debris which is constantly growing.

Degraded plastic particles are commonly carried by rain into water bodies such as rivers, lakes, and sewers, eventually reaching the oceans. These plastics accumulate in marine species, even in deep-sea environments (Kikaki et al. 2020). Microplastics have emerged as significant global concerns due to their unpredictable threats to the biosphere (Hurley et al. 2018). The widespread use of plastic-containing products and inadequate disposal management contribute to the buildup of microplastics in water bodies and aquatic ecosystems (Baruah et al. 2022).

There is a notable absence of long-term and large-scale monitoring efforts (Hafeez et al. 2018), and regular monitoring of microplastics is recommended due to their pervasive presence in the environment (Grbic et al. 2019). Detecting microplastics is typically challenging due to their diverse shapes, sizes, and properties like transparency and translucency, posing difficulties in experimental conditions (Asamoah et al. 2019). Analytical techniques have played a vital role in detecting and analyzing microplastics in the environment before the introduction of RS (Qiu et al. 2016). Commonly employed techniques for microplastic detection include microscopy, Fourier-transform IR spectroscopy (FTIR), Raman spectroscopy, and thermal analysis. Microscopy methods such as light microscopy, stereomicroscopy, fluorescent microscopy, and scanning electron microscopy (SEM) are frequently used for microplastic detection (Baruah et al. 2022).

Recent studies have explored the use of RS for marine plastic detection thoroughly. One study focused on the rapid detection of microplastics in seawater using hyperspectral imaging (HSI) technology (Shan et al. 2019). Satellite RS, according to Mahrad et al. (2020) and Kikaki et al. (2020), could provide valuable information for detecting plastic debris on the ocean surface at high spatial and temporal resolutions. Garaba et al. (2018) utilized aerial remote sensing (ARS) hyperspectral shortwave IR images to distinguish marine litter from surrounding seawater using polymer identification techniques. Goddijn-Murphy et al. (2018) proposed a spectral model concept for a hyperspectral RS algorithm that calculates sea surface plastic litter, suggesting its potential visibility with spectral light reflectance measurements. Work is ongoing to develop a comprehensive short-wave IR (SWIR) spectral signature library for oceanic plastics. Kikaki et al. (2020) utilized multispectral satellite observations over the Caribbean Sea to monitor the sources and trajectories of marine plastic debris, determining the speed of plastic litter movement with ocean currents. This study demonstrated that satellite RS is a valuable and cost-effective tool for monitoring plastic debris, aiding decision-making processes.

Monitoring plastic pollution is essential as the major pressure on marine and coastal ecosystems. At present, RS has been able to cover a vast polluted area over traditional laboratory methods. Even though some larger plastic debris can be easily tracked down, microplastic detection needs some conventional laboratory methods. Several studies have identified this could be achieved using hyperspectral images. In the sense of cost-effectiveness and the larger area coverage, most studies agree with using satellite sensors over airborne sensors as currently plastic pollutants have been spread to a larger area.

13.3.2 Monitoring Oil Spills

Oil pollution remains a significant concern for coastal environments, posing serious threats due to its sensitivity to both natural occurrences and human activities (Li et al. 2017; Naz et al. 2021; D'Affonseca et al. 2023). Oil spills typically result from human activities. This releases liquid petroleum hydrocarbons into the environment during various stages of oil exploration, production, transportation, and storage (Al-Ruzouq et al. 2020). Estimates suggest that annually, approximately 706 million gallons of oil are spilled into the ocean. These can be stemmed from natural seepage and anthropogenic incidents such as ship accidents and oil extraction platform malpractices (Li et al. 2017). Human-related sources account for nearly 45% of global ocean oil pollution, with oil transportation contributing 5% (Lu 2003; Naz et al. 2021).

The detection of oil slicks on sea surfaces has garnered significant attention recently (Song et al. 2018). Traditional monitoring methods, such as aerial or field investigations, often fall short in swiftly and efficiently identifying oil spill areas, leading to delays in response efforts. Accurate detection and prediction of oil spills and their trajectories are crucial for fisheries, wildlife preservation, resolving liability disputes, and effective resource management (Al-Ruzouq et al. 2020). Therefore, RS imagery has become a valuable tool for the efficient detection and treatment of oil spills (Zhao et al. 2022).

Four common airborne sensors employed for spill surveillance include IR, SLAR, microwave radiometer, and laser fluorosensor. Airborne hyperspectral sensors, featuring fine spatial resolution, are adept at capturing detailed spectral variations. Consequently, they aid in selecting the appropriate spectral region for studying specific water quality parameters, designing satellite sensors, and refining existing algorithms. Certain airborne hyperspectral sensors are especially beneficial for analyzing coastal and river water quality (Hafeez et al. 2018).

Recent research has focused on SAR image processing techniques as SAR data from microwave satellites are the preferred choice due to their weather independence and availability (Fan et al. 2015; Al-Ruzouq et al. 2020). SAR image processing can be categorized into physical mechanism-based classification, statistical characteristics, and image processing methods (Fan et al. 2015). The goal of oil spill detection typically involves four stages: image preprocessing, dark area detection, parameter extraction, and image classification (Naz et al. 2021).

Several major oil spill incidents worldwide have prompted extensive research efforts to monitor and assess their impacts. Naz et al. (2021) conducted a study utilizing Sentinel-1 SAR imagery to detect marine oil spills over the Indian Ocean, modeling spill trajectories and weathering effects, demonstrating the potential of SAR imagery for real-time detection. Similarly, Fan et al. (2015) employed SAR imagery to detect the 2010 Dalian Xingang oil spill in China, showcasing the effectiveness of a support vector machine (SVM) for image segmentation.

Al-Ruzouq et al. (2020) reviewed sensor applications in oil spill monitoring, noting a growing use of multispectral data from various satellite platforms, including MODIS, Sentinel-2, Landsat, and others. Hyperspectral systems offer high-resolution data but require advanced processing and analysis, limiting their real-time monitoring capabilities. Integration of multisource passive RS satellites can enhance oil spill detection and mapping efficiency despite their limitations (Yang et al. 2021; Zhao et al. 2022).

Both optical and microwave images can be used to detect marine oil spills. However, optical images limit their usage due to difficulty in separately identifying oil spills from the lookalike. Multisource passive RS sheds light on this issue through several efforts made. Many researchers have suggested to use SAR imagery from active satellite sensors. Recent studies showed many positive responses to using SAR imagery in monitoring oil spills. On the other hand, satellite RS is far better in usage since its ability to quickly respond and track the trajectories.

13.3.3 Monitoring Heavy Metal Pollution

HMs present in sediments pose a significant threat to marine ecosystems due to their biotoxicity, wide sources, persistence, and inability to degrade. Exceeding certain thresholds, HMs can induce toxic effects, particularly nonessential metals like Cd and As, even at low concentrations, affecting organism growth and development. These metals originate from both natural geological processes and human activities, with anthropogenic sources, including smelting, mining, agriculture, aquaculture, and industrial waste, being predominant. Once released into aquatic environments, HMs can accumulate in marine organisms and magnify the food chain, leading to elevated levels in predators and ultimately affecting human health (Zhang et al. 2020).

Traditional approaches to monitoring HM pollution typically rely on analytical techniques, with less emphasis on RS methods in coastal and marine environments. El-Alfy et al. (2019) investigated HM toxicity in Egypt's Rosseta area by integrating chemical analysis with RS to pinpoint potential pollution sources. Similarly, Chen et al. (2010) assessed the feasibility of using RS to estimate HM concentrations in coastal waters. Their findings suggested that HM distribution in estuarine waters is influenced by factors such as suspended sediments, phytoplankton, and CDOM (Colored Dissolved Organic Matter), which affect water-leaving radiance or RS reflectance. This indirect relationship implies the potential of RS to infer HM concentrations based on changes in reflectance.

Even though HMs mixed into marine systems pose a great threat, the lack of RS techniques applicable to detecting them makes them less effective in monitoring. Airborne hyperspectral sensors could be used as a solution to this, but this method needs high computational power. Therefore, more research will be needed to use satellite RS for monitoring HMs in the future as HM pollution is getting expanded more in the future at the current rate.

13.3.4 Monitoring Harmful Algal Blooms/Red Tides

Most algae found in coastal and open ocean environments are harmless, forming the foundation of the marine food chain. However, their rapid proliferation, known as algal blooms, can pose significant threats to the marine ecosystem. Excessive algae growth leads to the accumulation of biomass, which can block sunlight and create oxygen-depleted conditions, endangering marine life.

Notably, prolonged blooms of green macroalgae, such as Enteromorpha prolifera in the Yellow Sea, have been observed to disrupt marine ecosystems and coastal activities (Hu 2009). Algal blooms can persist for varying durations, ranging from a few days to several weeks, and can span from small areas to extensive regions (Hafeez et al. 2018).

Given the limitations of traditional ship-based sampling and analysis methods in both spatial coverage and temporal frequency, the detection, monitoring, and prediction of HABs are crucial tasks. RS technology offers a promising solution by providing synoptic information over large areas with relatively high frequency. Early studies in the 1970s and 1980s demonstrated a correlation between chlorophyll concentration and natural fluorescence using airborne sensors and in situ surveys (Hu et al. 2005). Satellite ocean color sensors like Sea-viewing Wide Field-of-view Sensors (SeaWiFS) are valuable tools for estimating phytoplankton biomass globally, particularly during episodic blooms (Ahn and Shanmugam 2006).

Algal blooms are closely linked to nutrient-rich coastal regions and continental shelves, with equatorial regions experiencing moderate chlorophyll-a concentrations due to the upwelling of nutrient-rich waters (Blondeau-Patissier et al. 2014). The bibliometric analysis done before confirmed this by reporting China, the United States, and Italy as the most pressing countries as they contained similar conditions. Detection of algal blooms has been facilitated by innovative approaches such as the Floating Algae Index (FAI) developed by Hu (2009), which utilizes medium-resolution data from MODIS satellites. FAI offers advantages over traditional indices like NDVI and EVI by being less sensitive to environmental and observational conditions, thereby enabling detection through thin clouds. SAR data also play a role in detecting large algal blooms, especially in cloudy weather conditions where they may appear as areas of low backscatter compared to surrounding water surfaces (Hafeez et al. 2018).

HABs are a recurring phenomenon on the West Florida shelf (WFS), primarily occurring between late fall and early spring. Studies by Hu et al. (2005) using MODIS fluorescence data and by Ahn and Shanmugam (2006) for optically complex coastal waters of northeast Asia have developed indices like the Floating Algae Index (FAI) and the Red Tide Index (RI) to identify potential HAB areas, demonstrating the utility of RS in HAB detection and monitoring.

Even though this phenomenon does not occur always, when there algal blooms, it spreads so fast and poses significant threats. Therefore, monitoring the growing HABs is necessary. Optical satellite sensors are widely used for this purpose and in the presence of severe weather conditions active RS also can be used. For the decision-making process, it is essential to make use of gathered data by assessing them. The current status of a particular pollution type and future direction can be easily identified by a thorough analysis of them as follows.

13.4 Marine Pollution Assessment

Recent research emphasizes the significance of Earth Observation (EO) data and tools, such as satellite surveillance, in bolstering efforts to manage coastal areas. Ensuring the sustainability of marine ecosystems relies heavily on the proficient management of coastal resources and aquaculture, with a primary focus on assessing and predicting water quality. Advances in satellite technology have significantly improved the capacity for monitoring water quality on a global scale, enabling comprehensive planning and management strategies for coastal and marine environments. The fusion of RS methodologies with emerging technologies like the Internet of Things (IoT), EO Big Data, cloud computing, and AI offers robust solutions to the challenges encountered in marine ecosystems.

ML, a subset of AI, empowers machines to discern and comprehend the connections between inputs and outputs from a comprehensive set of representative training data. This enables the construction of predictive and empirical classification models without presupposing any data distribution. ML adeptly tackles specific issues, even in scenarios where the theoretical understanding remains insufficient, leveraging vast datasets. Given the proliferation of high-dimensional remotely sensed data and the intricacies of pattern recognition tasks, ML techniques find application across a wide spectrum of earth observation domains, including oceanography and environmental monitoring. ML models navigate complex classification tasks by iteratively analyzing candidate solutions from provided training data and features, without explicit programming directives.

After feature extraction, several ML algorithms are used to train and classify data to get into a usable status. Common ML approaches aligned with either pixel-based analysis or object-based analysis before image segmentation. In the final stage, there are common ML algorithms in use for image processing such as artificial neural networks (ANNs), SVM, decision trees (DTs), neural networks (NNs), and embedded learning (EL).

Yet, traditional ML approaches encounter constraints in handling large datasets and capturing the nuanced spatiotemporal variations in water quality parameters. DL emerges as a promising avenue to overcome these challenges. Autonomously extracting intricate features from extensive datasets without manual intervention, DL furnishes advanced analytical capabilities for understanding marine environments (Figure 13.5).

Drawing inspiration from the human brain's structure and function, DL algorithms consist of distinct deep neural networks (DNNs) that automatically discern complex discriminative features from vast datasets in a hierarchical manner, enabling information extraction.

A case study on HABs detection and prediction in Florida's coastal waters integrated RS data with convolutional neural networks (CNNs), long short-term memory (LSTM) components, along

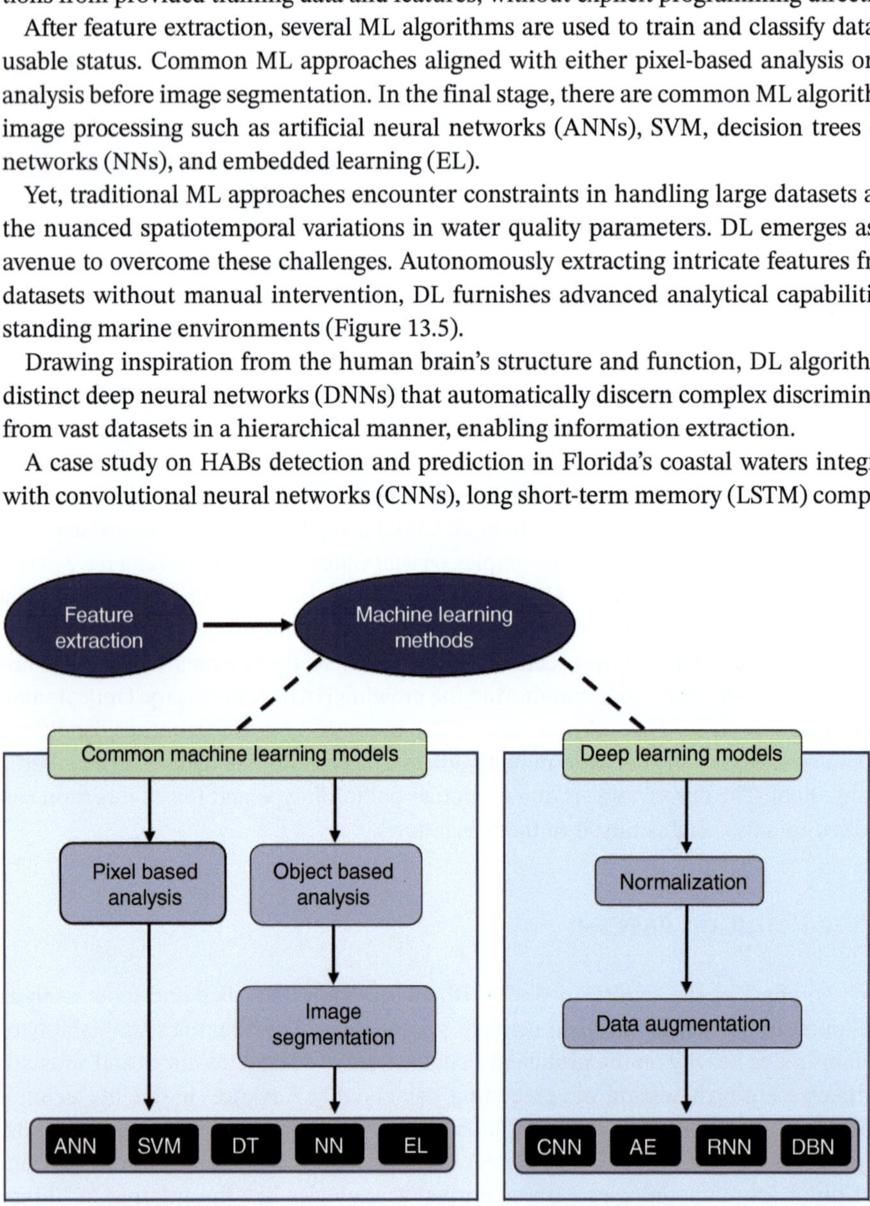

Figure 13.5 Machine learning and deep learning frameworks on image processing. *Source:* Al-Ruzouq et al. (2020)MDPI/CC BY 4.0.

with random forest (RF) and SVM classification methods. While yielding high accuracy, these sophisticated methods may not be feasible when data availability is limited. In such instances, empirical statistical models serve as simpler yet successful alternatives for HAB forecasting.

Several studies have addressed oil spill assessments, with one proposed method involving semantic segmentation using deep CNNs, demonstrating efficiency. Additionally, a publicly accessible SAR image dataset serves as a benchmark for future oil spill detection approaches. Similarly, a study employing a CNN architecture detected oil spills in China's Bohai Bay, reflecting efforts across various marine pollution events.

As the coastal and marine pollution assessment techniques, RS has actively engaged with AI for the past few years. Some drawbacks of using conventional ML techniques have directed the processes toward DL methods. The analysis made by this study reveals the emerging trends of using DL basically for marine pollution assessment. Yet some challenges need to be addressed through thorough research in the future.

Figure 13.6 illustrates the global spread of several coastal and marine pollution events reported from the past until the present. All these incidents have caused severe damage to the marine ecosystems. Among them, one famous incident was chosen for a case study. This occurred within the Sri Lankan marine region in the recent past and RS techniques were utilized by several studies. This study presents one of them to gain valuable insights into how RS has been utilized for monitoring the damage caused by the event.

In recent times, numerous incidents of oil spills have been recorded in the Indian Ocean, primarily due to the heavy transportation of oil. Being the third-largest ocean globally and interconnected with the Pacific Ocean via major shipping routes, the Indian Ocean is rich in seabed oil reservoirs, owing to extensive oil extraction activities in its northwestern coastal countries (Naz et al. 2021).

One notable event occurred in September 2020, when two major fires erupted on a large crude oil tanker named "New Diamond" within Sri Lanka's maritime zones. Sri Lanka, an island nation with vast sea areas in the Indian Ocean, is particularly susceptible to pollution from vessels. SAR data from Sentinel-1 was employed to detect potential marine pollution caused by the "New Diamond" tanker (Mahindapala 2020).

The fire initially broke out on 3rd September 2020 on the "New Diamond," a 330-meter tanker carrying crude oil, approximately 38 nautical miles off Sangamankanda on Sri Lanka's east coast. Despite the fire being extinguished two days later, a subsequent fire emerged, with both fires finally being put out by 10 September 2020. According to the disaster management team, the crude oil storage remained secure, and the tanker was situated around 50 nautical miles off the east coast.

SAR sensors operating in vertical-vertical (VV) polarization mode (vertical signal transmission and reception) were utilized for oil spill detection. VV polarization is sensitive to sea surface roughness, with oil spills causing a reduction in radar backscatter, resulting in darker areas on the sea surface within SAR images. Sentinel-1 GRD DV data, comprising dual VV+VH polarization, processed by the European Space Agency (ESA), were utilized for this study.

Sentinel Application Platform (SNAP) algorithms were employed to process SAR data, visualize results, and identify potential oil spill spots, indicated by darker regions within SAR images. The Adaptive Threshold algorithm was applied to detect these dark spots, indicative of oil spills. The analysis revealed a strong correlation between SAR data processing and visual images, successfully identifying the "New Diamond" tanker and associated oil spills. The timely acquisition of Sentinel-1 SAR data, typically within a few hours, is crucial for prompt response to oil spill incidents, covering territorial seas, contiguous zones, and exclusive economic zones, facilitating efficient detection of vessels and oil spills in these regions.

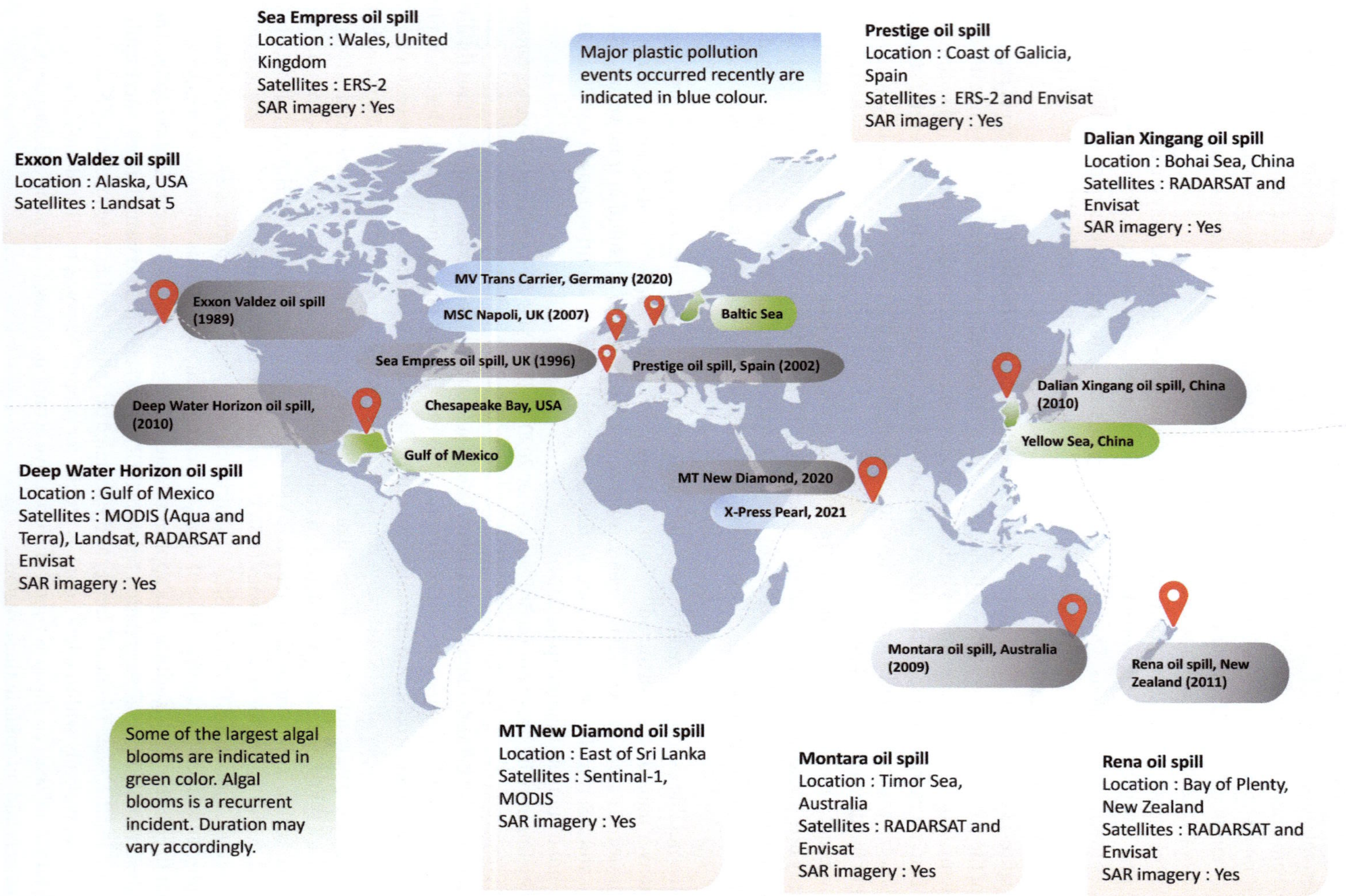

Figure 13.6 Several alarming instances of marine pollution have come to light globally.

13.5 Discussion

China, the United States, and Italy are major contributors to global plastic pollution, damaging coastal and marine ecosystems. High population density, rapid urbanization, and industrialization exacerbate the problem. These countries' extensive coastal regions also make them highly susceptible to pollution. RS finds extensive application in monitoring oil spills, plastic pollution, and eutrophication through real-time or near-real-time monitoring. Facilitating expansive coverage, high efficiency, and multiband imaging capabilities. RS has been able to overcome the limitations of traditional methods such as time constraints, spatial restrictions, and the need for specialized personnel. Most studies have primarily utilized satellite data over airborne sensor data. Remote sensors require careful selection based on user requirements, including study area, revisit time, and, crucially, the sensor's capabilities such as spatial resolution, spectral availability, and radiometric quality.

For oil spill detection, SAR has been extensively employed due to its independence from weather conditions, prompt response capabilities, wide coverage area, and accessibility. Oil spills typically induce changes in the roughness of the sea surface, which SAR can readily identify. Conversely, optical sensors may encounter challenges in distinguishing oil spills from similar natural features. Hence, SAR imagery stands out as the preferred method for effective oil spill monitoring. However, in the context of detecting plastic debris, optical sensors emerge as the most suitable method. Unlike oil spills, plastic debris does not alter the sea surface roughness, making it challenging for SAR. Instead, the distinctive spectral signature of plastics can be easily discerned by optical sensors, which are adept at covering larger areas. Nonetheless, regardless of the data type employed, processing for both oil and plastic debris typically involves two main stages: data preprocessing and classification of floating elements into categories such as plastic litter, oil spills, or lookalikes (Topouzelis et al. 2021). As shown in Figure 13.2, however, a strict upsurge in microplastic pollution, among other pollutants, challenges the use of typical RS approaches for pollution assessments.

For detecting HABs, optical satellite sensors are widely utilized. HABs often occur in nutrient-rich, densely populated coastal regions, transitioning somewhat from closed to open ocean, as depicted in Figure 13.6. While optical sensors are the primary choice, SAR satellite data is sometimes utilized when high cloud cover or adverse weather conditions limit the effectiveness of optical sensors in monitoring HABs.

Detecting HMs using RS proves challenging. HMs, being diluted in seawater, do not exist independently like oil spills or plastic debris. Consequently, distinguishing HMs from other elements poses difficulties. RS functions as a qualitative technique, incapable of precisely quantifying HM quantities or concentrations. Although limited efforts have been made to detect HMs using RS, these typically involve attempts to predict HM concentrations by leveraging other parameters measurable through RS. As a result, extracting reliable information regarding HM concentration from remotely sensed data necessitates sophisticated algorithms and integration with other environmental data sources such as water samples or oceanographic information.

Following the feature extraction stage, the subsequent image processing phase now employs AI, with ML standing out as the primary method currently employed. However, traditional ML techniques encounter challenges when handling large datasets for assessment. To address this limitation, DL has emerged as a promising alternative. Mimicking the functioning of the human brain, DL excels at processing complex datasets without necessitating manual feature engineering. Among DL models, CNNs are particularly noteworthy for their high effectiveness. Moreover, these models exhibit the capability to integrate data from multiple sources, including satellites, underwater sensors, and environmental measurements. Furthermore, DL models possess the advantage of handling real-time or near-real-time data, proving invaluable for monitoring critical incidents such as oil spills, which demand quick responses. Advances in DL methodologies indicate ongoing progress, with future research likely to focus on leveraging DL for marine pollution monitoring and assessment.

13.6 Concluding Remarks

In recent decades, the use of RS technology has significantly improved our understanding and monitoring of coastal and marine pollution. The trend is now shifting toward using satellite data and DL methods for detecting oil spills, plastic debris, and HABs. However, RS techniques face challenges in detecting HMs due to their diluted nature in seawater. Future research should focus on developing sophisticated algorithms and integrating data from various sources, such as water samples and oceanographic data to improve the accuracy and reliability of HM detection through RS. Additionally, there is ongoing research on DL methods, particularly CNNs, which show promise for real-time monitoring capabilities and addressing the challenges of handling large datasets. Overall, future research should aim to bridge existing gaps in RS capabilities, enhance the integration of ML and DL methodologies, and explore innovative approaches to effectively monitor and mitigate marine pollution. By doing so, we can contribute to the preservation and sustainable management of marine ecosystems worldwide.

References

Ahn, Y.-H. and Shanmugam, P. (2006). Detecting the red tide algal blooms from satellite ocean color observations in optically complex Northeast-Asia Coastal waters. *Remote Sensing of Environment* 103: 419–437.

Al-Ruzouq, R., Gibril, M.B.A., Shanableh, A. et al. (2020). Sensors, features, and machine learning for oil spill detection and monitoring: a review. *Remote Sensing* 12: 3338.

Álvarez-Romero, J.G., Devlin, M., da Silva, E.T. et al. (2013). A novel approach to model exposure of coastal-marine ecosystems to riverine flood plumes based on remote sensing techniques. *Journal of Environmental Management* 119: 194–207.

Anthony, D., Siriwardana, H., Ashvini, S. et al. (2023). Trends in marine pollution mitigation technologies: Scientometric analysis of published literature (1990-2022). *Regional Studies in Marine Science* 103156. https://doi.org/10.1016/j.rsma.2023.103156.

Arabi, B., Salama, M., Pitarch, J. et al. (2020). Integration of multi-sensor satellite images and in-situ measurements for long-term water quality monitoring in coastal areas. *Remote Sensing of Environment* 239: 111632.

Asamoah, B.O., Kanyathare, B., Roussey, M., and Peiponen, K.-E. (2019). A prototype of a portable optical sensor for the detection of transparent and translucent microplastics in freshwater. *Chemosphere* 231: 161–167.

Baruah, A., Sharma, A., Sharma, S., and Nagraik, R. (2022). An insight into different microplastic detection methods. *International Journal of Environmental Science and Technology* 19: 5721–5730.

Blondeau-Patissier, D., Gower, J.F., Dekker, A.G. et al. (2014). A review of ocean color remote sensing methods and statistical techniques for the detection, mapping and analysis of phytoplankton blooms in coastal and open oceans. *Progress in Oceanography* 123: 123–144.

Butchart, S.H., Walpole, M., Collen, B. et al. (2010). Global biodiversity: indicators of recent declines. *Science* 328: 1164–1168.

Chen, C., Liu, F., He, Q., and Shi, H. (2010) The possibility on estimation of concentration of heavy metals in coastal waters from remote sensing data. In: 2010 IEEE International Geoscience and Remote Sensing Symposium, pp. 4216–4219

Collard, F. and Ask, A. (2021). Plastic ingestion by Arctic fauna: a review. *Science of the Total Environment* 786: 147462.

Conceįcão, M.R.A., de Mendonça, L.F.F., Lentini, C.A.D. et al. (2021). SAR oil Spill detection system through random forest classifiers. *Remote Sensing* 13: 2044.

D'Affonseca, F.M., Reis, F.A.G.V., Correa, C.V.S. et al. (2023). Environmental sensitivity index maps to manage oil spill risks: a review and perspectives. *Ocean & Coastal Management* 239: 106590.

El-Alfy, M.A., Hasballah, A.F., El-Hamid, H.T.A., and El-Zeiny, A.M. (2019). Toxicity assessment of heavy metals and organochlorine pesticides in freshwater and marine environments, Rosetta area, Egypt using multiple approaches. *Sustainable Environment Research* 29: 19.

Fan, J., Zhang, F., Zhao, D., and Wang, J. (2015). Oil spill monitoring based on SAR remote sensing imagery. *Aquatic Procedia* 3: 112–118.

Garaba, S.P., Aitken, J., Slat, B. et al. (2018). Sensing ocean plastics with an airborne hyperspectral shortwave infrared imager. *Environmental Science & Technology* 52: 11699–11707.

Goddijn-Murphy, L., Peters, S., Sebille, E.V. et al. (2018). Concept for a hyperspectral remote sensing algorithm for floating marine macro plastics. *Marine Pollution Bulletin* 126: 255–262.

Grbic, J., Nguyen, B., Guo, E. et al. (2019). Magnetic extraction of microplastics from environmental samples. *Environmental Science & Technology Letters* 6: 68–72.

Hafeez, S., Wong, M.S., Abbas, S. et al. (2018). Detection and monitoring of marine pollution using remote sensing technologies. In: *Monitoring of Marine Pollution* (ed. H.B. Fouzia). https://doi.org/10.5772/intechopen.81657.

Halpern, B.S., Walbridge, S., Selkoe, K.A. et al. (2008). A global map of human impact on marine ecosystems. *Science* 319: 948–952.

Hu, C. (2009). A novel ocean color index to detect floating algae in the global oceans. *Remote Sensing of Environment* 113: 2118–2129.

Hu, C., Muller-Karger, F.E., Taylor, C.J. et al. (2005). Red tide detection and tracing using MODIS fluorescence data: a regional example in SW Florida coastal waters. *Remote Sensing of Environment* 97: 311–321.

Hurley, R., Woodward, J., and Rothwell, J.J. (2018). Microplastic contamination of river beds significantly reduced by catchment-wide flooding. *Nature Geoscience* 11: 251–257.

Islam, M.S. and Tanaka, M. (2004). Impacts of pollution on coastal and marine ecosystems including coastal and marine fisheries and approach for management: a review and synthesis. *Marine Pollution Bulletin* 48: 624–649.

Kannangara, K., Shoukie, M., Nayomi, M., et al. (2022). Spatiotemporal growth dynamics of invasive plant distribution in Bolgoda Lake, Sri Lanka: a GIS based approach. In: 2022 Moratuwa Engineering Research Conference (MERCon). pp. 1–6

Karydis, M. and Kitsiou, D. (2013). Marine water quality monitoring: a review. *Marine Pollution Bulletin* 77: 23–36.

Kikaki, A., Karantzalos, K., Power, C.A., and Raitsos, D.E. (2020). Remotely sensing the source and transport of marine plastic debris in Bay Islands of Honduras (Caribbean Sea). *Remote Sensing* 12: 1727.

Kylili, K., Hadjistassou, C., and Artusi, A. (2020). An intelligent way for discerning plastics at the shorelines and the seas. *Environmental Science and Pollution Research* 27: 42631–42643.

Li, Y., Cui, C., Liu, Z. et al. (2017). Detection and monitoring of oil spills using moderate/high-resolution remote sensing images. *Archives of Environmental Contamination and Toxicology* 73: 154–169.

Lorencin, I., Anđelić, N., Mrzljak, V., and Car, Z. (2019). Marine objects recognition using convolutional neural networks. *NĂSE MORE: znanstveni \vcasopis za more i pomorstvo* 66: 112–119.

Lu, J. (2003). Marine oil spill detection, statistics and mapping with ERS SAR imagery in south-east Asia. *International Journal of Remote Sensing* 24: 3013–3032. https://doi.org/10.1080/014311 60110076216.

Mahindapala, W.K.M. (2020). Oil spill detection in the east of Sri Lanka with Sentinel-1 SAR. In: E3S Web of Conferences, p. 02013

Mahrad, B.E., Newton, A., Icely, J.D. et al. (2020). Contribution of remote sensing technologies to a holistic coastal and marine environmental management framework: a review. *Remote Sensing* 12: 2313.

Naz, S., Iqbal, M.F., Mahmood, I., and Allam, M. (2021). Marine oil spill detection using synthetic aperture radar over Indian Ocean. *Marine Pollution Bulletin* 162: 111921.

Politikos, D.V., Fakiris, E., Davvetas, A. et al. (2021). Automatic detection of seafloor marine litter using towed camera images and deep learning. *Marine Pollution Bulletin* 164: 111974.

Qiu, Q., Tan, Z., Wang, J. et al. (2016). Extraction, enumeration and identification methods for monitoring microplastics in the environment. *Estuarine, Coastal and Shelf Science* 176: 102–109.

Reichelt-Brushett, A. (2023). *Marine Pollution--Monitoring, Management and Mitigation.* Springer Nature.

Shan, J., Zhao, J., Zhang, Y. et al. (2019). Simple and rapid detection of microplastics in seawater using hyperspectral imaging technology. *Analytica Chimica Acta* 1050: 161–168.

Song, D., Wang, B., Chen, W. et al. (2018). An efficient marine oil spillage identification scheme based on an improved active contour model using fully polarimetric SAR imagery. *IEEE Access* 6: 67959–67981.

Topouzelis, K., Papageorgiou, D., Suaria, G., and Aliani, S. (2021). Floating marine litter detection algorithms and techniques using optical remote sensing data: a review. *Marine Pollution Bulletin* 170: 112675.

Wang, Q., Wang, J., Xue, M., and Zhang, X. (2022). Characteristics and trends of ocean remote sensing research from 1990 to 2020: a bibliometric network analysis and its implications. *Journal of Marine Science and Engineering* 10: 373.

Yang, Z., Chen, Z., Lee, K. et al. (2021). Decision support tools for oil spill response (OSR-DSTs): approaches, challenges, and future research perspectives. *Marine Pollution Bulletin* 167: 112313.

Zhang, M., Sun, X., and Xu, J. (2020). Heavy metal pollution in the East China Sea: a review. *Marine Pollution Bulletin* 159: 111473.

Zhao, D., Tan, B., Zhang, H., and Deng, R. (2022). Monitoring marine oil spills in hyperspectral and multispectral remote sensing data by the spectral gene extraction (SGE) method. *Sustainability* 14: 13696.

Zheng, G. and DiGiacomo, P.M. (2017). Uncertainties and applications of satellite-derived coastal water quality products. *Progress in Oceanography* 159: 45–72.

14

Toxicological Techniques for Coastal and Marine Pollution Monitoring

Siddihalu Lakshitha Madunil and Jithya Wijesinghe

Department of Indigenous Medical Resources, Faculty of Indigenous Health Sciences and Technology, Gampaha Wickramarachchi University of Indigenous Medicine, Yakkala, Sri Lanka

14.1 Introduction

Coastal and marine environments are considered the ultimate dumping sites for a wide range of pollutants that have been inadvertently or purposefully released into the environment by human activity in the past. However, population growth has a significant impact on coastal and marine pollution. Pollutants can arise from diverse origins, encompassing land-based sources like urban runoff, industrial discharges, and agricultural runoff. Additionally, shipping and port activities contribute through the discharge of ballast water, accidental oil spills, and operational discharges from vessels (Jahan and Strezov 2019), introducing pollutants such as petroleum hydrocarbons, halogenated hydrocarbons, heavy metals, radionuclides, plastic debris, and fertilizers into coastal environments (Čulin and Bielić 2016). Moreover, offshore activities, including oil and gas exploration, maritime mining, and drilling operations, pose a threat with the potential for spills and discharges, further exacerbating the issue of coastal and marine pollution (Zhou et al. 2022). Coastal and marine pollution represent a pressing challenge with significant implications for ecosystems, biodiversity, and human well-being. However, the mitigation of pollution associated with global consequences is complicated due to various types and originates from diverse sources of pollutants. Thus, to maintain a healthy marine ecosystem, it is necessary to monitor their pollution closely (Häder et al. 2020).

Monitoring marine pollution is of paramount importance, serving critical roles in environmental preservation and safeguarding human health. It helps identify sources of pollution, track the dispersion of contaminants, and evaluate the effectiveness of mitigation measures. It enables the identification of contaminated seafood, harmful algal blooms, and polluted coastal areas, helping to mitigate risks associated with consuming contaminated products or engaging in recreational activities in affected areas. A comprehensive monitoring framework incorporating advanced technologies and international collaboration is crucial for informed decision-making and the development of effective strategies to address the multifaceted challenges posed by marine and coastal pollution (Bean et al. 2017).

The evolution of toxicological techniques for monitoring coastal and marine pollution showcases significant advancements and diversification. In the early twentieth century, the inception of marine

Coastal and Marine Pollution: Source to Sink, Mitigation and Management, First Edition.
Edited by Meththika Vithanage, Sameera M. Samarasekara, Bryan D. James, and Christopher M. Reddy.

monitoring primarily arose in response to public concerns about the sustainability of fishing activities (Kennelly and Broadhurst 2002). In the early stages of marine monitoring, traditional methods focused on physical and chemical parameters to assess pollution levels (Karydis and Kitsiou 2013). However, recognizing the limitations of these approaches, the field has evolved, integrating toxicological techniques that directly measure how contaminants affect marine life biologically (Jin et al. 2015). This shift toward bioassessment reflects a more holistic understanding of the impact of pollutants on ecological systems. Initially rooted in bioassays and acute toxicity tests, these techniques have progressively advanced to include a broader spectrum of biomarkers, genomic analyses, and sophisticated molecular tools (Collings and Vaidya 2008). Biomarkers, such as enzyme activities or genetic responses in organisms, serve as indicators of exposure and biological effects caused by pollutants (Lionetto et al. 2019). Bioassays involve exposing living organisms to environmental samples and measuring their responses to assess the toxicity of pollutants (Kokkali and van Delft 2014). Molecular tools, such as DNA-based analyses, provide insights into the genetic responses of organisms to pollution stress (Šrut 2021). This evolution reflects a growing awareness of the intricate nature of marine pollution and the necessity for refined, sensitive approaches to identify and quantify contaminants. Integrating historical knowledge with modern advancements in toxicological techniques significantly contributes to our ability to monitor and address the impact of pollution on coastal and marine environments. This chapter describes toxicological techniques used for monitoring coastal and marine pollution, like toxicity tests, bioindicators, bioassays, biomarker analysis, biosensors, chemical analysis, histopathology, ecotoxicological surveys, and toxicogenomics.

None of the substances is considered non-poison, but their dose determines whether they behave as poison or not. Therefore, toxicity can be used as a tool to monitor any foreign substances entering the environment. Nowadays, toxicological techniques have widely been applied for monitoring marine and coastal pollution (Auffan et al. 2014). A comprehensive toxicological assessment for environmental monitoring combines knowledge from various fields, including chemistry, biology, ecology, pharmacology, epidemiology, and toxicology. Most of the toxicological studies are performed under laboratory conditions, thereby deviating the results from the real environmental conditions (Wolska et al. 2007). When applying toxicological techniques, they are only useful for contaminants of interest as well as their responses. Therefore, the additional contaminants present in the coastal and marine ecosystems may not be calculated. The toxicity of pollutants cannot be directly measured in the presence of different mixtures because different substances have different modes of uptake and, thereby, different consumption rates. When various pollutants coexist, their ultimate impact may manifest as a combined effect, including synergistic or antagonistic interactions (Vieira and Morgado 2020). As a result of that, toxicity may not be directly connected to the concentration of the pollutants. The bioavailability of different pollutants depends on the other pollutants remaining in the environment. The matrix decides whether the toxic compounds are available in a free or bound form. Moreover, the toxicity of a substance depends not only on the species but also on age, health, and nutrients, which are key parameters even for the same species. Further, some species show an adaptation to a specific substance by living with it for a long time (Nebbia 2012). Additionally, some species have special detoxification mechanisms for particular substances, that is, some hyperaccumulator plant species for heavy metals (Sharma et al. 2021). Environmental factors such as temperature, pH, salinity, alkalinity, conductivity, turbidity, dissolved oxygen, suspended sediment, and other pollutants are of utmost importance for toxicity (Chariton et al. 2010a). Further, toxicity depends on the life stage of the organism, the duration of the exposure, and the route of administration (Wang et al. 2018b; de Almeida Rodrigues et al. 2022).

Toxicological techniques play a crucial role in monitoring coastal and marine pollution, helping to assess the impact of contaminants on aquatic ecosystems. These techniques involve identifying, quantifying, and evaluating toxic substances and their effects on marine organisms. The rate of absorption, distribution, metabolism, and excretion determines the toxicity of an organism.

Therefore, understanding the factors mentioned above helps to use organisms as monitors of pollution. A living organism shows changes in some features, such as its concentration in tissues, physiochemical, morphological, or behavioral, as well as population density, when exposed to pollutants (Connell et al. 1999). Thus, those features can be used as indicators of pollution.

14.2 Toxicity Testing

Toxicity testing is a critical component of monitoring coastal and marine pollution. These tests help us comprehend how contaminants affect aquatic life. Toxicity tests involve exposing organisms to known concentrations of pollutants. The goal is to observe and quantify adverse effects on the organisms. From the 1940s onward (Zbinden 1993), the evolution of organism toxicity testing has been spurred by the imperative to grasp the consequences of chemicals on living entities.

14.2.1 Acute Toxicity Tests

Short-term assessments, known as acute toxicity tests, are typically conducted within a timeframe of 24–96 hours. These tests aim to evaluate the immediate impact of a substance on an organism. Acute toxic effects refer to the immediate and severe adverse impacts that exposure to high levels of toxic substances can have on living organisms. These effects typically manifest shortly after exposure and can include symptoms such as rapid mortality, physiological distress, and behavioral abnormalities. The severity of acute toxicity depends on factors like the type of toxic substance, concentration, and duration of exposure. Understanding and assessing acute toxic effects are crucial for evaluating the immediate risks posed by pollutants, guiding regulatory measures, and implementing effective strategies to prevent or mitigate these harmful outcomes in both aquatic and terrestrial ecosystems.

14.2.2 Chronic Toxicity Tests

Tests conducted over an extended period, ranging from weeks to months, are employed to assess sublethal effects. These tests observe parameters such as growth, reproduction, and other endpoints, offering insights into the risks associated with chronic exposure. Chronic toxic effects characterize the long-term and often delayed consequences of continuous exposure to low levels of toxic substances over an extended period of time. Unlike acute toxicity, chronic effects may not be immediately apparent but can lead to persistent health issues and cumulative damage to organisms. These effects may include developmental abnormalities, reproductive impairments, and chronic diseases, impacting the overall health and fitness of individuals and populations. Assessing chronic toxic effects is crucial for understanding the prolonged risks associated with pollutants, guiding regulatory frameworks, and developing sustainable strategies to safeguard ecosystems and human health from the enduring impacts of chronic exposure to toxic substances. Therefore, toxicity tests can be categorized as either acute or chronic, depending on the duration of exposure, enabling a thorough evaluation of short-term and long-term impacts. In designing standardized toxicity tests, it is essential to establish clearly defined and easily measurable endpoints such as lethality and sub-lethality.

14.2.3 Whole Effluent Toxicity (WET) Tests

WET tests are designed to assess the toxicity of entire effluents, such as wastewater discharges. These tests utilize bioindicators, such as fish and invertebrates, to evaluate the overall toxicity of a sample. The selection of appropriate organisms for these tests is crucial, as different species can exhibit varying sensitivities to pollutants. Commonly used test species include *Daphnia magna*, also known as the water flea, for freshwater tests and Mysid shrimp for marine tests. Fish species, such as zebrafish, are often used for both acute and chronic tests because of their high sensitivity to a broad spectrum of pollutants. These tests provide a comprehensive view of the potential environmental impact of effluents, assisting in the formulation of efficient waste disposal plans and regulatory policies.

14.3 Endpoints and Biomarkers

Assessing the adverse impacts of substances on organisms relies heavily on toxicity endpoints and biomarkers. The endpoints differ depending on the type of test performed. For example, mortality is a frequently observed outcome in acute tests, serving as a straightforward and immediate indicator of the lethal effects of a substance. Growth inhibition is frequently employed as an endpoint in chronic tests to assess the enduring effects of a substance on an organism's development. Biomarkers, such as enzyme activity, act as indicators of stress in an organism, providing information about the physiological responses caused by exposure to a substance. Molecular endpoints, such as DNA damage, offer a more comprehensive comprehension of the potential genetic effects of a substance. The various endpoints and biomarkers together contribute to a thorough comprehension of toxicity, aiding in the creation of efficient strategies for managing and reducing the risks linked to exposure to harmful substances (Figure 14.1).

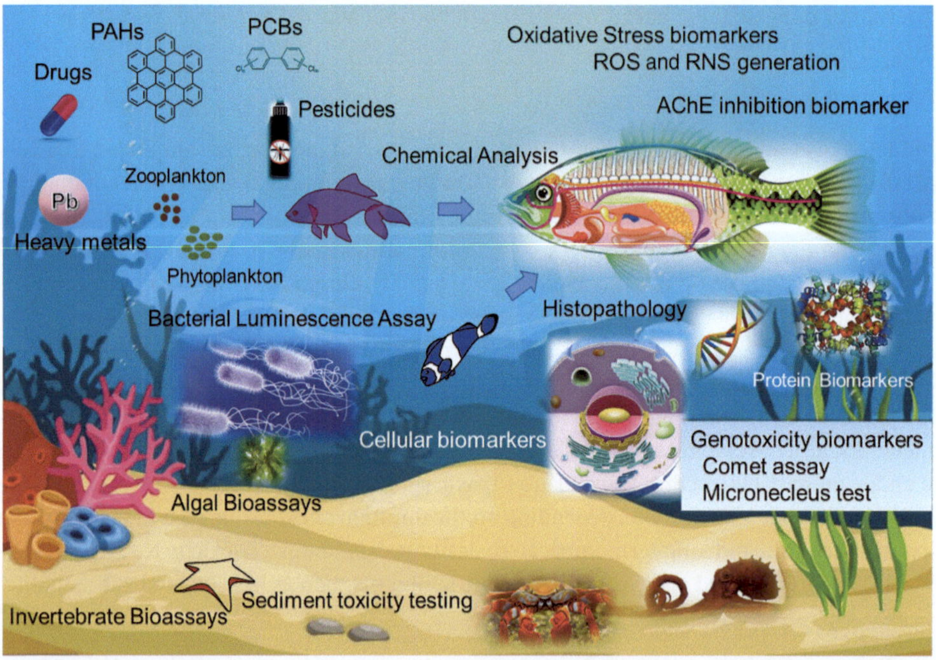

Figure 14.1 Diverse impacts of contaminants on organisms can be discerned through population and intra-organism changes, which serve as pollution indicators. Contaminants accumulate to different extents in different species, leading to diverse sensitivity levels. *Source:* studiographicmh/Adobe Stock; Peter Wilton/Wikimedia commons; FMSTUDIO/Adobe Stock Photos.

14.4 Sediment Toxicity Tests

Sediments consist of a combination of rocks, minerals, and biogenic matter. As they naturally migrate from seashores to oceans through coastal erosion, water, ice, and wind, sediments serve as potential sources for various contaminants, such as PAHs and PCBs (Maletić et al. 2019). As a result, the concentration of pollutants, especially hydrophobic substances, may be higher in sediments than in marine water (Voice et al. 1983). The toxic effects of sediments present a heightened risk to benthic organisms (Campana et al. 2012) due to their close association with the sediment environment, particularly impacting bottom-feeding organisms such as crabs, prawns, and certain fish species. In particular, suspended sediments can pose a significant threat to filter-feeding species (Simpson and Spadaro 2016), as the dissolution of toxicants in the pore spaces of sediments may lead to the exposure of a substantial quantity of toxic substances to animal gills. Another risk involves the direct ingestion of sediments, leading to the potential bioaccumulation of contaminants by aquatic organisms. It is noted that the availability and toxicity of sediment depend on the organism's activities, such as digestion and complexation.

Sediment toxicity tests are assessments conducted to evaluate the potential toxicity of contaminants in sediments to aquatic organisms. Sediments can accumulate various pollutants, including heavy metals, organic chemicals, and other substances, posing risks to benthic organisms and potentially impacting the broader aquatic ecosystem (Burton 2010). In sediment toxicity tests, a designated biological endpoint, such as mortality or reduced growth, serves as a measurable parameter to evaluate the organism's response to contaminants present in the sediments. Moreover, amphipods, bacteria, sea urchins, and microalgae have commonly been employed as species in toxicity tests (Byrne and O'halloran 2001).

A study revealed that sea urchin embryo-larval development and the growth rate of microalgae were identified as the most sensitive bioassays for evaluating contaminants such as PAHs, metals, detergents, and pharmaceutical products in bulk sediment (Khosrovyan et al. 2013). Additionally, amphipod mortality and the Microtox solid-phase test were identified as suitable bioassays for assessing the quality of bulk sediment in marine environmental evaluations (Maranho et al. 2015). Monitoring contamination in deep-sea sediments involves the use of multiple species and endpoints to assess the response to contaminants, particularly focusing on elements such as Cr and Ni. An approach utilizing multiple species and diverse endpoints would be well-suited for the monitoring of contamination in deep-sea sediments. However, the selected endpoint should elucidate the dynamics of contaminants in marine sediments at considerable depths, especially as no direct relationships were observed between the outcomes of chemical and ecotoxicological characterizations (McCready et al. 2006). It is noted that mercury levels in mollusks exhibit an inverse relationship with mercury levels in sediment. In contrast, the levels of other metals (Cd, Pb, Cu, and Zn) in organisms displayed positive associations with the levels of those metals in water or sediment (Taylor et al. 2012). Several studies have examined the condition of sediment toxicity in tropical and temperate regions. However, these studies have found that additional testing methods must be developed to assess the impact on ecologically significant species. Consequently, evaluating sediment toxicity is difficult due to the difficulty of determining the level of exposure or dosage. Additionally, the presence of different sediment types affects the extent to which contaminants can be absorbed, and this availability will differ depending on the specific compounds or complexes being examined (Adams and Stauber 2008; Chariton et al. 2010a). These tests can be used individually or in combination to provide a comprehensive understanding of the effects of marine pollution on living organisms. They play a crucial role in environmental monitoring and management efforts to protect marine ecosystems.

14.5 Impacts of Pollution on Different Marine Species

Marine pollution, arising from sources such as plastic waste, chemical runoff, oil spills, and heavy metals, presents a substantial danger to marine organisms. It affects a wide range of organisms, from microscopic plankton to whales, disrupting ecosystems, harming animal health, and affecting biodiversity. Recent studies have highlighted the multifaceted impacts of marine pollution on various marine species, illustrating the urgent need for concerted conservation efforts (Abreo et al. 2015).

The foundation of the marine food web, plankton, is directly affected by pollution, with microplastics being ingested by zooplankton, leading to physical harm and the potential transfer of toxic substances (Cole et al. 2013). Pollution, particularly from chemical runoffs and oil spills, can cause bleaching and diseases in coral reefs, critically endangering these biodiversity hotspots. Exposure to pollutants can result in developmental issues, reproductive failure, and increased mortality in fish populations. Heavy metals and chemicals can accumulate in fish tissues, posing health risks to predators, including humans (Garai et al. 2021). Marine mammals are vulnerable to pollution through ingestion of contaminated prey, entanglement in debris, and direct exposure to toxic substances, which can lead to health problems and decreased reproductive success (Khoshnood 2017). Plastic ingestion is a significant issue for seabirds, leading to intestinal blockage, malnutrition, and even death (Wilcox et al. 2016). Chemical pollutants can also affect seabird health and breeding success. The review by Galloway et al. (2017) underscores the pervasive nature of marine pollution and its detrimental effects across the marine ecosystem. By focusing on microplastics, the study sheds light on the interconnectedness of marine life and the direct and indirect pathways through which pollutants impact species and ecosystems. The findings call for a global response to reduce pollution sources and protect marine biodiversity, highlighting the importance of research, policy changes, and public awareness in tackling this environmental challenge.

The extent of exposure for marine species depends on the chemical characteristics of contaminants in diverse environmental conditions. Furthermore, it is influenced by the physiology of a specific species and its interactions with the surrounding environment. The behavior, activity, physiological processes, and lifespan of a species determine the absorption, storage, and elimination of pollutants. Thus, the same species exhibits different toxicity levels. Organisms with minimal or restricted movement are suitable for biomonitoring or bioindication, considering important factors such as bioaccumulation and biomagnification (Hodkinson and Jackson 2005).

14.6 Biomonitors and Bioindicators

The term biomonitors should be distinct from bioindicators. A biomonitor uses organisms or their parts to offer quantitative insights into environmental quality, while a bioindicator conveys information about the quality of the environment or specific parts of it. As a result, while a biomonitor is fundamentally a bioindicator, not all bioindicators meet the criteria to be classified as biomonitors. Furthermore, bioindicators reveal the presence of pollutants by displaying typical symptoms or measurable responses, providing insights into the marine environmental state through their observed reactions to stressors. For example, the abundance of specific aquatic macroinvertebrates can signal water quality. Some bioindicators are benthic organisms, aquatic macroinvertebrates, phytoplankton, zooplankton, macrophytes (seagrasses and algae), fish, shellfish, and animal toxins (Figure 14.1).

14.6.1 Benthic Organisms

They inhabit the seafloor and are sensitive indicators of environmental conditions. The group includes bacteria, microalgae, crustaceans, foraminifera, and benthic macroinvertebrates (Islam et al. 2024). For instance, the presence and diversity of bacteria serve as indicators of seawater quality, while variations in microalgae populations signify nutrient levels and pollution (Kriwy and Uthicke 2011). Additionally, the abundance and diversity of crustaceans respond to contaminants, and foraminifera are employed to assess sediment quality. Benthic macroinvertebrates are indicative of ecosystem health (Covich et al. 1999).

14.6.2 Aquatic Macroinvertebrates

Aquatic macroinvertebrates, such as insects, crustaceans, mollusks, and worms, are tiny organisms without a backbone that inhabit aquatic environments. The widespread presence, diverse species, and high sensitivity of marine invertebrates to pollutants make them valuable bioindicators for monitoring the health of marine and coastal ecosystems (Sumudumali and Jayawardana 2021). Further, they can reflect the presence of contaminants in sediment (Costa et al. 2021).

14.6.3 Phytoplankton and Zooplankton

They are microscopic organisms that live in aquatic environments and form the base of the marine food web (Figure 14.1). They are also important indicators of water quality and environmental changes, as they respond to various pollutants and stressors in different ways. Pollution affects phytoplankton and zooplankton in multiple ways, with significant consequences for the marine ecosystem. These impacts include changes in the number and diversity, as well as interference with vital functions such as growth, reproduction, and survival (D'Costa et al. 2017b). Furthermore, pollution causes the accumulation of harmful substances in the plankton's tissues and cells (Wurl and Obbard 2004). Pollution alters the trophic interactions and energy transfer in aquatic ecosystems, changing the fragile balance of species interactions and affecting the structure and function of plankton communities (Telesh 2004). Further, pollution makes phytoplankton and zooplankton more susceptible to diseases and parasites, which affect their health and resilience (Lafferty and Kuris 1999). This higher vulnerability can intensify the decline in population health, disturbing the ecological balance (Worm et al. 2006).

The occurrence of harmful algal blooms (HABs) and hypoxia is heightened by pollution (Kazmi et al. 2022). Excessive nutrient input from pollution promotes the expansion of toxic algae, which creates HABs (Gilbert et al. 2018). The decomposition of these algal blooms can lower oxygen levels in the water, causing hypoxia, which harms the survival of plankton (Breitburg et al. 2018). Monitoring phytoplankton and zooplankton can yield valuable insights into the conditions and changes of marine and coastal ecosystems, along with the potential consequences of pollution on human well-being and livelihoods. For instance, monitoring coastal and marine pollution can involve assessing parameters like biomass, abundance, diversity, productivity, respiration, nutrient cycling, dynamics, distribution, and plankton transport. Moreover, concerns related to plankton, including eutrophication and HABs, offer valuable insights into pollution dynamics (Heisler et al. 2008).

14.6.4 Macrophytes

Macrophytes refer to aquatic plants that grow in or around water, including seagrasses and algae (De and Roy 2019). By mirroring the environmental state and fluctuations of their habitats, they serve as bioindicators of marine and coastal pollution. Being widely distributed and easy to sample and identify is an advantage of using macrophytes as bioindicators. With extended life spans, they can absorb the effects of pollution over time, enabling them to accumulate and tolerate various pollutants in their tissues and cells (Milan et al. 2006). Hence, identifying sources of pollutants and pathways becomes more straightforward. They can respond to different types of pollution, such as nutrients, metals, organic compounds, and acidification (Polechońska and Klink 2023). Further, they have the potential to affect the physical, chemical, and biological properties of both water and sediments. Therefore, macrophytes are vital for aquatic ecosystems, as they produce organic matter and create habitats for other organisms. Seagrasses, such as *Cymodocea nodosa* and *Zostera marina*, act as indicators for eutrophication, sedimentation, and habitat degradation (Burkholder et al. 2007; Yang et al. 2018). Macroalgae, represented by species like *Caulerpa prolifera* and *Ulva lactuca*, serve as indicators for nutrient enrichment, organic pollution, and hypoxia (Valiela et al. 1997). Mangroves, such as *Avicennia marina* and *Rhizophora mucronata* can indicate heavy metal contamination, oil spills, and salinity changes (Pan and Wang 2012). These plants have different mechanisms to cope with stress, such as detoxification, compartmentalization, and the excretion of pollutants (Hall 2002). They can also provide ecological benefits, such as stabilizing sediments, filtering nutrients, and providing habitats for other organisms (Madsen et al. 2001).

14.6.5 Fish and Shellfish

They are aquatic animals that can be used as bioindicators for marine and coastal pollution monitoring (Van der Oost et al. 2003). When exposed to numerous pollutants through water, sediment, and food sources, these organisms can accumulate diverse contaminants within their tissues and organs. Indeed, fish exhibit a wide range of responses when faced with stress induced by pollutants. These responses include physiological, biochemical, and behavioral adaptations. For instance, they may display heightened aggression or avoidance behaviors, and their feeding patterns may change. They can mirror the impacts of pollution across various levels, ranging from individual organisms to entire populations and communities. Fish species, including herring, flounder, and cod, serve as indicators of eutrophication, heavy metal contamination, and oil spills (Authman et al. 2015). Additionally, shellfish, such as mussels, oysters, and scallops, can serve as indicators of nutrient enrichment, organic pollution, and hypoxia (Viarengo and Canesi 1991). Further, crustaceans, such as shrimp, crab, and lobster, can serve as indicators of pesticide exposure, endocrine disruption, and habitat degradation (Le Moullac and Haffner 2000).

14.6.6 Animal Toxins

Some marine organisms, including dinoflagellates, urolophus, trachinus, pterois, scorpaena, synanceja, and cyanobacteria, have developed the ability to produce toxins as a defense mechanism in response to pollution (Selander et al. 2006; Vasconcelos et al. 2010). Indeed, these toxins can be small molecular compounds or larger entities such as proteins. This adaptation allows them to survive in environments that have been adversely affected by various pollutants. These toxins can serve as bioindicators, providing valuable information about the health of the marine ecosystem. It also helps identify potential risks to human health, especially for communities relying on these marine resources (Tevell Åberg et al. 2013). Aquatic organisms are known to produce a range of toxins, such as domoic acid, kainic acid, saxitoxin, tetrodotoxin, and polyether polyketides. These toxins, each with their own unique structures and effects, are essential for the survival of these organisms in polluted environments (Chekan et al. 2020).

14.7 Bioassays

A bioassay, also known as a biotest, is a conventional method used in toxicology to evaluate the effects of substances, like environmental pollutants, on living organisms. These assessments are usually conducted under controlled conditions in a lab, but sometimes, they may be carried out in the field. Both living (biotic) and nonliving (abiotic) factors are taken into account during these tests. Therefore, bioassays involve cell or tissue culture to examine enzyme activity, physiological responses, and evaluations involving various organisms such as microorganisms, plants, and animals. These assessments can be conducted on a single species or multiple species within representative ecological systems, like microcosms and mesocosms (Benton et al. 2007). Bioassays are versatile tools employed to evaluate the impact of substances on biological entities or systems. They use specific biomarkers and, on occasion, particular biosensors. They can be utilized in both bioindication and biomonitoring processes. The selection of particular parameters is contingent upon the objectives and emphasis of the individual assay. They provide a complete understanding of the environmental impact, and thus, they are widely employed in coastal and marine pollution monitoring. These bioassays can be used individually or in combination to provide a comprehensive assessment of the ecological impact of marine pollution. It is important to note that bioassays should be complemented with chemical analyses to identify specific pollutants and their concentrations to obtain a thorough understanding of the pollution status in a given marine environment (Farré and Barceló 2003; Van der Oost et al. 2003; Torres et al. 2008). Here are some commonly used bioassays for detecting marine pollution:

14.7.1 Algal Bioassays

Algal bioassays are assessments that use algae, a group of photosynthetic microorganisms, to evaluate the potential toxicity of substances in water. These bioassays are valuable tools in marine pollution monitoring as they provide insights into the effects of pollutants on primary producers and, consequently, on the overall health of aquatic ecosystems. A microalgal growth inhibition test is essential to assess the impact of contaminants on primary producers. The general method is the exposure of marine microalgae cultures to water samples and measurement of the growth inhibition compared to a control group (Nyholm and Källqvist 1989). *Phaeodactylum tricornutum* can be chosen as the experimental algae due to its physiological importance, simplicity, and measurement speed (Bowler and Falciatore 2019). While the growth of algae remained unaffected by variations in the size of polyethylene (PE) microplastics (PE-MPs), it is noteworthy that a considerable quantity of MPs has been identified in seawater (Feng et al. 2020). Another standard assessment of the acute toxicity of MPs on bacteria (*Pseudomonas putida*), microalgae (*Scenedesmus* sp.), and yeast (*Saccharomyces cerevisiae*) has indicated that yeast shows susceptibility to MPs (Miloloža et al. 2022). The conventional ecotoxicological endpoints, such as the inhibition of microalgae (*Phaeodactylum tricornutum*) growth and bacteria (*Vibrio fischeri*) bioluminescence, lack the necessary sensitivity to adequately assess the potential effects of MPs on marine unicellular organisms. It is noted that MPs may not pose a threat to organisms in the same way that nanoplastics do (Gambardella et al. 2019). Moreover, the size range of MPs from 1 to 500 μm was not deemed problematic; however, particles smaller than 1 μm raised concerns due to their capacity to permeate through the pores of the cell wall, potentially resulting in adverse effects (Zhang et al. 2017). The microalgae *Chlorella vulgaris* was used in a growth-inhibition test to assess the impact of nickel oxide nanoparticles (NiONPs). The test revealed that the EC50 values (the term "Effective Concentration 50" refers to the concentration of a harmful substance that causes a specific adverse effect in 50% of the tested individuals) were less than 45 mg/l after 72 and 120 hours of exposure. Furthermore, they exhibited significant adverse effects, such as cytoplasmolysis and rupture of the cytomembrane, which can be attributed to the stress induced by NiONPs (Gong et al. 2011).

It was found that *Nitzschia closterium* exhibited greater sensitivity to Al compared to Ga and V, with reported IC50 (inhibitory concentration 50, which represents the concentration at which a substance inhibits a biological process or response by 50%) values for the species being 190, 19 640, and 42 000 μg/l for Al, Ga, and V, respectively (Harford et al. 2011) (Table 14.1).

14.7.2 Invertebrate Bioassays

Small invertebrates have an essential role in assessing the effects of substances and environmental conditions, making them crucial for environmental monitoring (Impellitteri et al. 2022; Chowdhury et al. 2023). Furthermore, their short lifespan allows for the assessment of generative and cross-generational effects, enhancing the comprehensive evaluation of environmental impacts (Sánchez and Tarazona 2002). Invertebrates, such as insects, crustaceans, mollusks, and worms, have been widely used due to their ecological importance and sensitivity to environmental changes (Prather et al. 2013).

Species selection is critical; *Tigriopus japonicus*, a copepod species, exhibits tolerance to specific stressors such as UV lights, γ-rays, and certain metals. Consequently, it is imperative to utilize more sensitive species for precise monitoring (Lee et al. 2007). *Paracyclopina nana* has found extensive application in environmental monitoring owing to its biological features (small size, reproductive characteristics, and ease of cultivation), wide distribution, and high abundance. Additionally, it has consistently demonstrated responsiveness to various chemicals (Dahms et al. 2016). In contrast to other species, *Paracyclopina nana* exhibits heightened sensitivity to Cd, Cu, UVB rays, and γ radiation (Won and Lee 2014; Won et al. 2015). The Daphnia immobilization test provides an evaluation of acute toxicity in water samples. Daphnia (water fleas) are exposed to the sample and observe immobilization or mortality rates (Zhu et al. 2009). Some research has been carried out to identify the effects of Fe^{3+}, Cr^{3+}, Ni^{2+}, and Cu^{2+} in wastewater from the electroplating industry using Daphnia magna (Yi et al. 2011).

In a study, sediment contamination emerged as a notable stressor affecting the composition of benthic macroinvertebrate communities; nevertheless, the presence of nutrients was discovered to

Table 14.1 Some species commonly employed as bioassays for monitoring specific contaminants.

Bioassay	Species	Contaminants
Algal bioassays	*Phaeodactylum tricornutum*	Polyethylene microplastics
	Scenedesmus sp.	Microplastics
	Chlorella vulgaris	Nickel oxide nanoparticles
	Nitzschia closterium	Al, Ga, and V
Invertebrate bioassays	*Tigriopus japonicus*	UV lights, γ-rays, and metals
	Paracyclopina nana	Cd, Cu, UVB rays, and γ-rays
	Daphnia magna	Fe^{3+}, Cr^{3+}, Ni^{2+}, and Cu^{2+}
Fish bioassays	*Oryzias melastigma* *Oryzias latipes*	Anilines, triclosan, fenitrothion, and pyriproxyfen

potentially mitigate the adverse effects on *Chironomus riparius* (De Lange et al. 2005; Lagauzère et al. 2009). Certain invertebrates are suitable for use as bioassays due to their widely known culture expertise, which makes them easier to utilize in lab settings. Additionally, their vulnerability to common stressors and basic genome information reinforces their role as essential organisms (Table 14.1).

14.7.3 Fish Bioassays

Fish bioassays are experimental assessments conducted using fish species to evaluate the potential impact of substances or environmental conditions on aquatic ecosystems. Fish exhibit diverse adaptive and behavioral responses when exposed to a polluted environment or stress, reflecting either an instinctive escape response or an adaptive adjustment to the prevailing conditions (Sharma et al. 2019).

In an in vivo acute toxicity bioassay involving marine fish *Oryzias melastigma* and freshwater fish *Oryzias latipes*, conducted for chlorine-substituted anilines, triclosan, fenitrothion, and pyriproxyfen, they displayed comparable toxicity responses to 3-chloroaniline and fenitrothion. However, *O. latipes* exhibited greater resilience than *O. melastigma* when exposed to the other three compounds. Therefore, *O. melastigma* is suitable for conducting acute toxicity bioassays in marine water (Horie and Takahashi 2021).

Further, a fish embryo toxicity test is vital to evaluate the effects of pollutants on fish's early life stages. The test method is used to expose fish embryos to water samples and observe for developmental abnormalities, hatching success, and mortality. For example, several studies have demonstrated the sensitive response of freshwater fish embryos, specifically *Danio rerio* (zebrafish), to different compositions of wastewater (Nagel 2002; Embry et al. 2010). Larval fish tests are conducted to evaluate the impact of pollutants on the growth and survival of fish larvae. Additionally, bioaccumulation studies aid in quantifying the accumulation of contaminants in fish tissues, offering insights into exposure levels (Van der Oost et al. 2003). These tests are designed to measure various responses in fish, providing valuable information about the health and quality of water bodies (Figure 14.1). Fish bioassays often involve exposing fish to specific concentrations of chemicals, pollutants, or other stressors and observing their physiological, behavioral, or reproductive responses (Table 14.1).

14.7.4 Bacterial Luminescence Assay

Bioluminescence is a natural process that involves the emission of light through an oxidative reaction facilitated by the interaction of luciferases and luciferins in living organisms. Moreover, these compounds are present in diverse marine organisms such as bacteria, fungi, and insects, leading to the emission of light within the wavelength range of 400–700 nm. Thus, various organisms generate a spectrum of colors, with blue being the most prevalent among the bioluminescent emissions (Vega-Corrales and Marín-Vindas 2021). As bioluminescence is an enzymatic process, exposure to toxic pollutants can lead to changes in light intensity by inhibiting the enzymes involved in the bioluminescent reaction. Hence, the change in light intensity serves as a quantifiable measure of toxicity in the context of bioluminescence assays. Bioluminescence-based assays, owing to their convenience, have found extensive application in monitoring pollution due to their high sensitivity and precision (Perin et al. 2022). Moreover, this technique has been employed in diverse applications, including the assessment of contamination generated by metals, pesticides, and drugs in natural water bodies (Wang et al. 2009; Van De Merwe and Leusch 2015; Banerjee et al. 2022). Another study demonstrates the utilization of luminescent bacteria to assess the effects of human activities on coastal waters (Baums et al. 2007). Indeed, luminescent bacteria offer an intriguing capability to evaluate both unknown environmental samples and unidentified substances (Menz et al. 2013).

Another study highlighted the necessity for further research to precisely determine the minimum concentration of metals and other potential toxins required to inhibit both the luminescence and growth of bacteria (Strotmann et al. 2020).

14.8 Biomarker Analysis

Biomarkers, which also fall under biomonitors or bioindicators, are measurable endpoints at the suborganismic level, including genetic, enzymatic, physiological, and morphological characteristics. They indicate structural or functional changes that occur due to exposure to pollutants. Biomarkers such as concentrations of contaminants in tissues, molecular-level changes, cellular-level changes, and histological-level changes are measurable endpoints for biomonitoring. It is significant to note that various halogenated hydrocarbons can cause an increase in the activity of Phase I enzymes, particularly cytochrome P450 enzymes. These enzymes are crucial in the initial metabolism of foreign substances, known as xenobiotics. The change in their activity can be used as a valuable biomarker, signaling exposure to specific environmental contaminants. Evaluating these biomarkers in conjunction with their impact on the overall organism, encompassing factors like physiology, behavior, reproduction, growth, and survival, offers a comprehensive perspective on the organism's reaction to environmental stressors. Biological alterations induced by pollutants can be effective biomarkers for contamination monitoring (Lionetto et al. 2019). These alterations should have the potential to be measured in terms of pollutant toxicity and the degree of exposure. Thus, biomarkers serve as measurable endpoints that define the potential effects of contaminants on organisms in the environment (Turesky and Lu 2020). These indicators are commonly utilized in laboratory-based ecotoxicological testing to assess the impact of pollutants on living organisms. Biomarkers such as concentrations of contaminants in tissues, molecular level changes, cellular level changes, and histological level changes, along with the assessment of effects on the entire organism including physiology, behavior, reproduction, growth, and existence comprise comprehensive indicators for biomonitoring (Torres et al. 2008; Hampel et al. 2016).

14.8.1 Acetylcholinesterase (AChE) Inhibition Biomarker

The AChE enzyme plays a vital role in the nervous system by disintegrating the neurotransmitter acetylcholine. Blocking this enzyme leads to an accumulation of acetylcholine, which disrupts neurotransmission. It is noted that the presence of pollutants is positively associated with the inhibition of AChE activity (Fulton and Key 2001). As a result of the neurotoxicity posed by pollutants creating harmful effects on animals, lowering their well-being and survivability, inhibition of AChE is a significant biomarker with particular ecological significance (Lionetto et al. 2013). Exposure to dioxins, PCBs, herbicides, pyrethroids, plasticizers, flame retardants, heavy metals, PAHs, carbamate, and organophosphorus pesticides can all cause AChE inhibition (Fu et al. 2018) (Figure 14.1).

These substances can directly bind to AChE, thereby suppressing its enzymatic activity. The primary mechanism of action observed was the inhibition of AChE's catalytic activity; however, pollutants like dioxins induce a decrease in the biosynthesis of mRNA (Fu et al. 2018). The research findings demonstrate a significant reduction in AChE activity in mussels' gills and digestive glands, and this decrease in activity is directly associated with an accumulation of metals in these organs (Perić et al. 2017; Zhou et al. 2021). Similarly, native mussels from industrialized areas in various regions also exhibited comparable findings, indicating a connection between the suppression of this enzyme and heavy metal pollution (Magni et al. 2006). While pesticides such as organophosphates and carbamates are known for inducing AChE inhibition, it is noteworthy that herbicides like glyphosate and propanil may exhibit the potential to enhance AChE activity (Moraes et al. 2007; Samanta et al. 2014). Indeed, additional research is necessary to clarify the processes involved in neurotoxicity caused by pollutants. Therefore, comprehensive knowledge of neurotoxicity will enhance our understanding and could lead to more effective prevention and treatment strategies.

14.8.2 Aldehyde Dehydrogenase (ALDH) Inhibition Biomarker

Aldehyde dehydrogenase oxidizes acetic aldehyde to acetic acid. Because the reduction of NAD to NADH is coupled to this reaction, the enzyme activity can be measured by the increasing extinction of NADH. Alterations in ALDH activity can act as a signal of exposure to certain pollutants, providing essential data on the well-being and adaptive responses of marine species when faced with environmental pressures. Thus, ALDH emerges as a noteworthy biomarker within the realm of marine pollution monitoring (Sharma et al. 2023).

14.8.3 Protein Biomarkers

Metallothioneins (MTs) are petite proteins characterized by a high concentration of cysteine, resilience to heat, and the absence of aromatic amino acids. Additionally, they feature cysteine residues that exhibit a strong affinity for binding to metal ions. The thiol (–SH) groups present in cysteine residues play a crucial role in enabling MTs to efficiently detoxify metals through chelation. Moreover, they are widely recognized as antioxidants. Considering that metallothionein (MT) formation rises notably as the accumulation of metals in an organism increases, the induction of MT becomes a promising biomarker (Bakiu et al. 2013; Hemmadi 2016).

Glutathione (GSH), a tripeptide, functions as an antioxidant, aiding in the neutralization of reactive oxygen species (ROS) and free radicals. GSH plays a crucial role in cellular antioxidant defense mechanisms by efficiently inhibiting protein oxidation through the presence of its thiol (SH) group (Cnubben et al. 2001). The reduction in GSH levels indicates the activation of the detoxification mechanism in response to free radicals induced by foreign chemical substances (Lin and Yang 2007). A significant reduction in GSH levels was noted in brown mussels (*Perna perna*) when subjected to elevated concentrations of trace metals. This observation aligns with findings from diverse studies conducted on different mussel species (Regoli and Principato 1995). Similarly, there is a strong correlation between trace metal contamination levels and the reduction in GSH levels in various organisms, particularly metals and metalloids (Srikanth et al. 2013; Hernández et al. 2015). The most robust results can be achieved by concurrently assessing MT, GSH, and malondialdehyde (MDA). Additionally, it is important to acknowledge that biomarkers may exhibit variability based on environmental conditions and seasonal changes (Del Rio et al. 2005).

14.8.4 Genotoxicity Biomarkers

Substances or factors that have the potential to cause damage to DNA are referred to as genotoxic agents that can perform various processes such as directly binding to DNA, interfering with DNA replication, interfering with DNA repair, inducing ROS, forming DNA adducts, and inducing chromosomal aberration. Most of the pollutants found in coastal and marine environments are genotoxic substances. These pollutants can originate from various sources, including industrial discharges, agricultural runoff, shipping activities, oil spills, and urban wastewater. The genotoxic effects of these pollutants can pose significant risks to marine life and may have broader ecological consequences. Some examples of genotoxic marine and coastal pollutants are heavy metals, polycyclic aromatic hydrocarbons (PAHs), certain pesticides, polychlorinated biphenyls (PCBs), brominated flame retardants, oil and petroleum products, microplastics, radiative substances, pharmaceuticals, and personal care products (Figure 14.1).

The detection of those genotoxic agents is difficult in marine water. Therefore, the easiest way is to measure the DNA damage. There are several techniques, such as the comet assay (CA) and micronucleus (MN) test, available for the detection of DNA damage. MN formation is a permanent change, whereas the comet assay measures transient changes due to DNA damage.

During cell division, MN formation is an indicator of exposure to genotoxic agents that can create aneugenic and clastogenic effects. In a comet assay, the damage can be estimated from the relative extent of the comet head (undamaged DNA) and tail (broken DNA strand fragments). The longer the tail, the more DNA damage is present. The MN test serves as a sensitive method for identifying chromosomal DNA damage arising from either chromosomal breakage during cell division or missegregation of chromosomes during mitosis. MNs serve as an index reflecting the cumulative genetic damage sustained by cells throughout their lifespan. This enables the detection of integrated responses to complex mixtures of contaminants, notably high levels of PAHs, PCBs, organochlorinated compounds, and metals (Pb and Cd) (Bolognesi and Hayashi 2011). Note that in field studies, the occurrence of MNs has been linked to oxidative stress induced by metal pollution, environmental thermal stress, and various factors, including salinity, pH, and food availability. These factors influence cell turnover rates, consequently modulating the extent of MN formation (Janssen et al. 1993).

The CA offers a swift, straightforward approach and exhibits a consistent linear response to a range of exposures, encompassing UV radiation, oxidized bases, damage from alkylation (Collins et al. 1996; Azqueta et al. 2011), and petroleum oil (Lewis et al. 2010). CA has been utilized to evaluate DNA damage, revealing that the extent of damage is dependent on the dosage of pollutants (Mahmood et al. 2019). The application of comet and micronucleus tests on fish, such as *Epinephelus chlorostigma* and *Scamberomorus commerson*, serves as an early warning system for monitoring the overall health of the sea (Mahboob et al. 2022). Additionally, both of these tests have been utilized to evaluate the genotoxic effects induced by trace metals and petroleum hydrocarbons on edible fish (D'Costa et al. 2017a). Moreover, the comet assay has been employed as a biomarker for invertebrates to assess organic pollution (Livingstone et al. 2000). Genotoxicity biomarkers are commonly employed in extensive biomonitoring activities, employing standardized procedures specifically designed to assess DNA damage in hemocytes and gill cells of mussels (Katsumiti et al. 2014). As a result of the vulnerability of DNA to oxidative stress, DNA damage can be used as a biomarker for pollution.

14.8.5 Oxidative Stress Biomarkers

Oxidative stress can be used as a biomarker to measure the capacity of the antioxidant defense system of an organism. The capacity to neutralize the free radicals by the antioxidant defense system of an organism is directly related to oxidative stress, which occurs when there is an imbalance between the free radicals and the antioxidant defense system. Most marine and coastal environmental contaminants, such as heavy metals, pesticides, PAH, and radioactive substances, are responsible for oxidative stress. For instance, metals have the potential to induce the production of ROS and reactive nitrogen species (RNS) by interfering with metal-related processes and generating free radicals (Figure 14.1). The activity of GSH, glutathione *S*-transferase (GST), GSH-peroxidase, and GSH-catalase helps protect from oxidative stress. Moreover, the activities of catalase (CAT), superoxide dismutase (SOD), glutathione peroxidase (GPx), GST, GSH, thiobarbituric acid reactive substances (TBARS), lipid peroxidation (LPO), maleic dialdehyde (MDA), and ROS can be classified as biomarkers of oxidative stress. Interestingly, the monitoring of oxidative stress biomarkers also offers direct evidence of perturbations in biological systems, thereby serving as a valuable tool for assessing marine and coastal pollution (Marrocco et al. 2017; Halappanavar and Mallach 2021).

Enzymes like SOD and CAT finely regulate cellular oxidative balance by converting superoxide ($O_2^{\cdot}-$) to oxygen (O_2) and hydrogen peroxide (H_2O_2) to water, contributing to effective antioxidant defense (Wang et al. 2018a; Sies and Jones 2020). Furthermore, the enzymatic activity of GPx is crucial in reducing H_2O_2 and lipid peroxides, utilizing reduced GSH to produce oxidized glutathione (GSSG), thereby preventing these reactive species from causing cellular damage. Glutathione reductase (GR) facilitates the conversion of GSSG to GSH by utilizing NADPH as an electron donor. Subsequently, NADPH is regenerated through the action of glycolytic enzymes such as glucose-6-phosphate dehydrogenase (G6PDH). Therefore, the activities of SOD and GPx are among the most frequently used biomarkers to assess oxidative stress (Lubos et al. 2011). Studies on fish and sea worms revealed a notable increase in ROS, SOD, and CAT levels upon exposure to PE and polystyrene (PS) (Missawi et al. 2020; Cocci et al. 2022; Lombardo et al. 2022).

Certain enzymes, such as glutathione-S-transferases (GST), indirectly function as antioxidants by participating in the detoxification of reactive compounds. A moderate rise in both GST and CAT activity was noted at the site, which was contaminated with high levels of trace metals. Elevated concentrations of trace metals have the potential to stimulate antioxidant enzymes (Jomova and Valko 2011). Various studies, including those on *Perna perna* by Saenz et al. (2010) and on *Mytilus galloprovincialis* by Mejdoub et al. (2017), have reported elevated levels of GST and CAT in marine organisms inhabiting polluted sites when compared to non-polluted areas. Numerous researchers have studied the stimulation of these enzymes in aquatic invertebrates, including *Paracentrotus lividus*, *Venerupis philippinarum*, and *Ruditapes clams* when exposed to trace metal pollution (Amri et al. 2017).

Phospholipid hydroperoxide GPx (PHGPx) plays a vital role as an antioxidant enzyme, directly counteracting intracellular phospholipid hydroperoxides, which are a specific type of ROS (Imai and Nakagawa 2003). This enzyme potentially contributes to metal-modulated protection against LPO (Lubos et al. 2011). Moreover, both chronic and acute exposure to metals can notably elevate the levels of CAT, PHGPx, and GSH (Atli and Canli 2007).

MDA has been the most extensively investigated among oxidative stress endpoints (Toto et al. 2022). An increase in the concentration of MDA is often observed due to the peroxidation process, causing damage to the lipid compartment of cells and is frequently associated with neurotoxic effects. Indeed, the MDA content can function as an early indicator of the extent of exposure, reflecting the damage caused by radical species to the cell membrane and serving as a valuable biomarker for stress induced by chemical pollutants (Del Rio et al. 2005). For example, a study revealed a notable observation where a significant increase in MDA content was exclusively found in the digestive gland and not in the gills of *Perna perna*, implying that the digestive gland accumulates higher levels of trace metals than the gills. Notably, multiple studies provide compelling evidence of a strong correlation between trace metal accumulation and the rise in MDA concentration spanning across various organisms (Viarengo et al. 1989; Sellami et al. 2014). Elevated levels of MDA are observed following exposure to micro-sized polystyrene (PS) (Cheng et al. 2022; Maity et al. 2023). Sea worms exhibited a notable increase in MDA levels compared to controls following exposure to polystyrene (PS), PE, and polypropylene (PP) (Missawi et al. 2020). Additionally, the interaction of MDA with nucleic acids has the potential to form DNA adducts, which may result in mutations capable of advancing into cancer (Marnett 1999).

14.8.6 Cellular Biomarkers

Lysosomal membrane stability (LMS) can be employed as a biomarker to assess the impact of marine and coastal pollution on the health of aquatic organisms. To utilize LMS as a biomarker in marine and coastal pollution studies, researchers typically employ techniques such as fluorometric assays or vital dyes to measure changes in lysosomal integrity. Combining LMS assessments with other biomarkers and ecological indicators enhances the ability to evaluate the overall health of marine ecosystems and the efficacy of pollution management efforts. The decline of antioxidant defense can lead to harm to the cellular membrane, lipids, DNA, and proteins (Figure 14.1). Measuring the activities of antioxidant enzymes and determining potential damages can be helpful in identifying oxidative stress caused by acidification and pollutants (Richardson et al. 2008). Following a 96-hour exposure of mussels (*Perna perna*) to different concentrations of crack-cocaine combined with varying pH levels, the results likely indicated notable effects on biomarkers, including LMS, LPO, and DNA strand breaks. More precisely, the findings showed that decreased pH levels and higher concentrations of crack-cocaine resulted in reduced stability of lysosomal membranes, increased LPO levels, and more DNA strand breaks in the mussels. Both a decrease in pH and exposure to crack cocaine have harmful effects on mussels, as indicated by changes in these biomarker responses (da Silva Souza et al. 2021). These findings suggest possible dangers to mussel populations in environments where these stressors exist. Furthermore, mussels exhibited an inclination toward higher levels of mucus secretion, nitrogen excretion, and oxygen consumption rates when exposed to elevated concentrations of copper in the short term. This suggests that the mussels' physiological processes were activated in response to increased levels of copper in the short term. Consequently, increased levels of copper can affect different physiological processes in mussels, which may have an impact on their overall well-being and performance (Vernon and Jha 2019). Over 24 days, *Perna perna* (mussels) experienced a reduction in oxygen uptake when exposed to copper (Cu) and zinc (Zn). In contrast, mussels that were exposed to mercury (Hg) exhibited a contrasting reaction, as their oxygen uptake increased during prolonged exposure. Furthermore, the filtration rate of the mussels exhibited fluctuations when exposed to metals, occasionally remaining consistent while at other times displaying variability. Therefore, long-term exposure to copper (Cu), zinc (Zn), and mercury (Hg) can have varying impacts on physiological functions in mussels, such as oxygen intake and filtration speed. These findings emphasize the intricate and diverse reactions of mussels to various metal exposures over prolonged durations (Anandraj et al. 2002). Thus, the use of morphological and physiological biomarkers is restricted due to their significant variability among populations and individuals (Anandraj et al. 2002; Madeira et al. 2019).

14.9 Biosensors

Absolutely, biosensors are indeed a method of biomonitoring. A biosensor functions as an analytical device, generating a signal directly correlated with the concentration of specific target substances. This is achieved by combining a selective biological component, whether it be an enzyme, antibody, membrane, organelle, cell, or tissue, with a physical transducer, such as a potentiometric or amperometric electrode or an optical/optoelectronic receiver. For example, the Toxiguard bacterial taximeter and the EuCyanobacterial electrode. Biosensors are valuable due to their capacity to deliver swift, uninterrupted, and exceptionally precise measurements. They necessitate minimal reagent usage for calibration, exhibit rapid response times, and can quantify nonpolar molecules that are beyond the estimation capabilities of conventional devices. Biosensors have the capability to detect the initial reactions of living organisms to environmental stressors. They offer insights into harm and disease in marine ecosystems resulting from pollution, whether it is anthropogenic or naturally occurring. The application of biosensors spans from detecting toxins to evaluating the overall well-being of marine ecosystems. They have the capability to identify both general and specific toxicity resulting from various pollutants. Innovative biosensors, integrating nano-biosensors and molecularly imprinted polymers, offer a comprehensive approach to overcoming challenges in the development and deployment of marine biosensors.

14.9.1 Enzyme-Based Biosensors

These biosensors use enzymes as the biological sensing element. They are highly specific and can detect a wide range of pollutants, including heavy metals and organic compounds. Enzyme-based biosensors play a crucial role in monitoring pollution in marine and coastal environments, given their high specificity and sensitivity. The Toxiguard bacterial taximeter, designed to measure indicators associated with endocrine effects, assesses parameters like the Vitellogenin (VTG) level in male fish (Matozzo et al. 2008). Indeed, these biosensors typically exhibit less susceptibility to interference from redox-active compounds (Narayanasamy et al. 2024). These instances highlight the diverse uses of enzyme-based biosensors in tracking marine and coastal pollution (Li et al. 2023).

14.9.2 Immunosensors

These immunosensors offer immediate, remote, and in situ evaluations for environmental pollution surveillance and preemptive alerts, utilizing the principles of bio-recognition and signal transduction. These sensors are primarily classified into three types: electrochemical, optical, and mass-based sensors. Each type is unique in its physicochemical properties and functioning principles. They have arisen as a compelling technology for monitoring pollution in marine and coastal environments. As an example, optical immunosensors, encompassing fiber-optic, surface plasmon resonance, and Raman spectroscopy-based and FTIR-based biosensors, have demonstrated substantial promise in the identification and surveillance of diverse pollutants within marine ecosystems.

Progress in sensor technologies has transformed the approach to monitoring marine environments, providing in-depth perspectives on extensive regions, remarkable spatial and temporal resolution, and continuous time-series coverage spanning numerous years to decades.

While precision challenges arise due to the dynamic nature of coastal zones, the integration of these sensors into numerical models can facilitate thorough 4D monitoring of the ocean and precise forecasting. Adopting this comprehensive monitoring approach for marine pollution is vital to safeguarding the sustainability of marine ecosystems (Mehrotra 2016).

14.9.3 DNA-Based Biosensors

They have emerged as a promising toxicological technique for monitoring marine and coastal pollution. These biosensors have been employed to detect pathogenic microorganisms and heavy metals in water. The swift progress in molecular biology, sequencing technology, materials science, electronics, and chemical science has significantly enhanced the development and utilization of aptasensors. Aptamer-based biosensors have substantially decreased the detection threshold for marine toxins from nanomolar to femtomolar levels, providing elevated sensitivity and specificity. They offer convenience, time efficiency, cost-effectiveness, and high-throughput capabilities. Hence, biosensors based on DNA exhibit significant promise in pushing forward the realm of monitoring marine and coastal pollution (Kim et al. 2016).

14.9.4 Whole-Cell Biosensors

These biosensors, designed as living cells that react to specific substances, offer a unique method for identifying a broad spectrum of marine toxins. Whole-cell biosensors have the capability to offer real-time data on both chemical and biological parameters. Whole-cell biosensors possess the potential to monitor various marine parameters, enabling the assessment of activity impacts and the implementation of remediation measures. This underscores their significant promise in advancing marine and coastal pollution monitoring (Justino et al. 2017).

14.9.5 Optical Biosensors

Optical biosensors, employing fluorescence as the predominant technique, exhibit the capability to identify diverse pollutants such as pesticides, pharmaceuticals, microorganisms, toxins, and heavy metals. This technique measures various parameters, including intensity, decay time, quenching efficiency, anisotropy, and quenching (Yoo and Lee 2016).

14.9.6 Electrochemical Biosensors

Electrochemical biosensors exhibit notable selectivity and sensitivity toward specified targets, with an emphasis on potential enhancements in contaminant diagnostics and monitoring. Their application extends to the analysis of diverse contaminants, encompassing organophosphorus, toxic hydrocarbons, organic compounds, heavy metals, as well as toxic anions and cations.

Marine pollution biosensors must possess full automation, resistance to biofouling, high resistance to rusting and physical damage, low power consumption, sensitivity to measure pollutants in tiny amounts, and the ability to facilitate precise calibration. A wide range of biosensors have been developed for analyzing diverse marine pollutants, such as eutrophication, antibiofouling agents, algal toxins, trace metals, and organic substances. The identification of organic pollutants in marine water involves the use of numerous microorganisms. This process requires sampling, laboratory testing, and analysis. It is a tedious and laborious procedure that requires the collection of samples at specific locations and subsequent preparation. The cost of this procedure depends on its complexity. Therefore, biosensor and bioassay technology have been developed to enhance the efficiency of monitoring marine pollutants (Justino et al. 2015).

14.10 Chemical Analysis

14.10.1 Sample Collection Techniques

The meticulous collection of samples is crucial in chemical analysis, ensuring accurate measurements and providing insights into pollution levels. This precision is essential for obtaining a comprehensive understanding of the types and concentrations of chemicals in the environment. Two water sampling methods, passive and active techniques, are available for precise environmental analysis. Passive samplers, such as the Polar Organic Chemical Integrative Sampler (POCIS), offer the advantage of time-weighted average concentration measurement over extended periods, while active sampling techniques are essential for capturing episodic pollution events. Techniques available for extracting pollutants from sediment matrices include Soxhlet extraction, pressurized liquid extraction, and microwave-assisted extraction. The collection and preparation of biota samples from various organisms, including fish, bivalves, and algae. Given the complexities associated with measuring the bioaccumulation and biomagnification of pharmaceuticals in food webs, it is crucial to carefully select target species and tissues. By continually refining these techniques, the scientific community can enhance its ability to monitor pollutants, understand their impacts, and develop effective strategies for mitigating pollution in aquatic ecosystems.

14.10.2 Analytical Techniques

Certain physical and chemical parameters, such as pH, turbidity, biochemical oxygen demand (BOD), and chemical oxygen demand (COD), serve as key indicators for the assessment of marine pollution (Borja et al. 2008). These parameters can provide insights into water quality, and advanced techniques can be employed for accurate analysis of both inorganic and organic pollutants. Moreover, various techniques are available for measuring the concentrations of heavy metals, including mercury, lead, and cadmium, which can accumulate in marine organisms. Atomic absorption spectrophotometry (AAS) and inductively coupled plasma optical emission spectrometry (ICP-OES) techniques facilitate the detection of over 70 metals and metalloids in seawater, achieving sensitivity at the μg/l (parts per billion) level. Inductively coupled plasma mass spectrometry (ICP-MS) is a modern and reliable technique for determining almost all elements and their isotopes. The detection limits can reach up to ng/l (parts per trillion) level due to the sensitivity of the MS detector. Ion chromatography (IC) is a technique utilized for the determination of both anions and cations by employing a suitable column, with the retention time (RT) serving as the parameter to identify compounds in relation to their respective standards. This is suitable for the determination of nitrate, phosphate, sulfate, fluoride, and chloride in seawater. The determination of organic pollutants such as carbamates, organophosphates, pyrethroids, PAHs, PCBs, and oils is determined by gas chromatography and liquid chromatography (LC) with different detectors like ultraviolet-visible, fluorescence, flame ionization detector, electron capture detector, nitrogen–phosphorus detector and MS. Further, tandem mass spectrometry, high-resolution mass spectrometry (HRMS), and laser ionization time-of-flight mass spectrometry (LI-TOFMS) are emerging technologies for the detection of chlorinated pesticides, oils, PCBs, PAHs, and pharmaceuticals in complex matrices (Magi and Di Carro 2018; Cariou et al. 2020; Madunil et al. 2020; Madunil et al. 2022).

The identification of microplastics (MPs) is commonly conducted through various techniques, including Fourier transform infrared spectroscopy (FTIR), Raman spectroscopy, and pyrolysis-gas chromatography coupled to mass spectrometry (Pyr-GC-MS) (Lenz et al. 2015; Khalik et al. 2018; Peters et al. 2018). Less frequently applied, high-performance liquid chromatography (HPLC) with UV detection and Orbitrap-MS techniques have been utilized in the analysis of MPs. In coastal areas, the evaluation of pollutants absorbed by microplastics, such as pesticides, pharmaceuticals, PAHs, PCBs, and metals, was performed through a comprehensive approach involving LC-MS/MS, GC-TOFMS, and ICP-MS techniques (Elseblani et al. 2023).

14.10.3 Tissue Analysis

A comprehensive tissue analysis was conducted to biomonitor polycyclic aromatic compounds (PACs) and trace metals in mussels (*Mytilus* sp.) (King et al. 2023). Various techniques have been employed for the extraction of PACs from soft tissues, including accelerated solvent extraction (ASE) and dispersive solid phase extraction (dSPE) (Idowu et al. 2018; Xia et al. 2021). Following the extraction process, the measurement of PACs has been carried out using GC-MS/MS with multiple reaction monitoring (MRM) modes (Idowu et al. 2018; Xia et al. 2021; King et al. 2023). The limit of detection (LOD) for both ASE and dSPE using the GC-MS/MS method was in the range of pg/μl, representing parts per billion sensitivity. In the analysis of total mercury (Hg), tissue pools were homogenized, and freeze-dried tissue samples were directly subjected to analysis using a mercury analyzer (Hebert and Popp 2018; Idowu et al. 2018; Xia et al. 2021; King et al. 2023). Tissue samples, subjected to microwave digestion, were analyzed for trace metals that are closely associated with marine pollution, employing inductively coupled plasma mass spectrometry (ICP-MS) (Hebert and Popp 2018; Idowu et al. 2018; Thomas et al. 2021; Xia et al. 2021; King et al. 2023). The detection limits for trace metals were expressed in μg/g. Apart from the metal concentration at a specific marine site, factors such as metal speciation and complexation can also influence the bioavailability of metals. Moreover, physiological conditions and the sizes of organisms exert an inevitable influence on their absorption of metals by mussels (Cossa 1989; Casas et al. 2008; Bruland et al. 2014).

14.11 Histopathology

Histopathology, involving the microscopic examination of tissues to detect disease or abnormalities, serves as a crucial tool for monitoring marine pollution. By analyzing changes at the cellular level in marine organisms, scientists can infer the presence and potential impact of environmental contaminants. This approach is instrumental in detecting sub-lethal effects of pollutants that may not immediately manifest at the organism or population level but can significantly affect marine biodiversity and ecosystem health over time. Histopathology provides detailed insights into the physiological stress and pathological changes induced by pollutants in marine organisms (Salamat and Zarie 2016). It is particularly useful for identifying specific organ damage caused by exposure to contaminants like heavy metals, hydrocarbons, pesticides, and plastics (Vasanthi et al. 2013) and, also, understanding the mechanisms of toxicity and disease progression within marine organisms, further, assessing the overall health of marine ecosystems and the impact of human activities (Stentiford et al. 2003).

Marine species, including fish, invertebrates, and mammals, are often the subjects of histopathological studies due to their exposure to various environmental stressors. Histopathological analysis of these organisms can reveal tissue-specific responses to pollution, such as gill and liver damage in fish, neurological effects in marine mammals, and reproductive disorders in marine invertebrates (Ruiz-Picos et al. 2015). The gills and the liver are primary targets for aquatic toxicants due to their roles in respiration and detoxification, respectively (Shah et al. 2020). Further, liver and kidney tissues showed significant alterations, including necrosis and inflammation, suggesting that heavy metals can severely impact common loon chicks' health (Vinodhini and Narayanan 2009). Similar results can be observed in the gills, liver, and kidneys of fish, indicating that significant impacts occur when exposed to heavy metals (Javed et al. 2017). Another study found that winter flounder exposed to PAH-contaminated sediments displayed a range of liver lesions, including hepatic necrosis and fibrosis (Murchelano and Wolke 1991). Note that long-term exposure to plastic debris could result in the bioaccumulation of metals in marine organisms, potentially causing histopathological changes, such as inflammation and tissue damage (Sıkdokur et al. 2020; Guerrera et al. 2021). Exposure to fluoxetine, an antidepressant, led to histopathological alterations in fish, including changes in gonadal structure and function (Duarte et al. 2019). Therefore, histopathological examination can reveal alterations like hyperplasia, necrosis, and fibrosis, indicating exposure to toxic substances (Bhuvaneshwari et al. 2015). The accumulation of neurotoxic pollutants, such as mercury and POPs, can lead to behavioral changes and brain lesions, detectable through histopathological analysis. Exposure to endocrine-disrupting chemicals (EDCs) can cause reproductive anomalies in marine invertebrates, including shell thickening in mollusks and intersex conditions in crustaceans, which can be identified through tissue examination (Zala and Penn 2004).

Histopathological biomarkers are used to assess the health status of marine organisms and ecosystems. These biomarkers include cellular alterations (e.g. LMS, micronucleus formation), tissue changes (e.g. hepatic and renal histopathology), and the appearance of specific pathologies associated with contaminant exposure (Cajaraville et al. 2000). The use of histopathological biomarkers in conjunction with chemical analysis of pollutants provides a comprehensive understanding of the environmental impact of marine pollution. Numerous case studies across the world have utilized histopathology to monitor the health of marine environments affected by oil spills, industrial discharge, agricultural runoff, and other sources of pollution (Salamat and Zarie 2016).

Interestingly, the potential impact of pharmaceuticals on marine species at the histological level is significant. These studies often focus on sentinel species, which are indicators of ecosystem health due to their sensitivity to environmental changes. Histopathological analysis of these species has revealed various pathological changes, including those related to endocrine disruption, immune system effects, and carcinogenesis, highlighting the utility of this approach in detecting the subtle effects of pollution. Recent advancements in histopathological techniques, including digital image analysis and molecular histopathology, have enhanced the sensitivity and specificity of pollution monitoring. These techniques allow for quantitative analysis of tissue damage, the identification of molecular pathways affected by pollutants, and the correlation of histopathological changes with environmental levels of contaminants. This progress has improved the accuracy and efficiency of diagnosing the impacts of marine pollution (Smits and Fernie 2013).

Neoplastic (cancerous) and non-neoplastic diseases are observed in fish populations exposed to toxic chemicals in their environment. By employing histopathological techniques, the researchers were able to correlate the presence of specific pollutants, such as PAHs and chlorinated hydrocarbons, with increased incidences of liver tumors and other pathological changes in fish (Myers et al. 2003; Vogelbein and Unger 2006).

14.12 Ecotoxicological Surveys

Ecotoxicological surveys have become essential tools for monitoring coastal and marine pollution. These surveys involve the assessment of contaminants and their effects on organisms within the marine ecosystem. The increasing population and industrial activities in coastal areas have contributed to a rise in marine pollution, emphasizing the importance of ecotoxicological surveys for assessing the health of these ecosystems.

These surveys utilize various techniques, including selective biosensors, bioassays, bioaccumulation modeling, biomarkers, and gene expression analysis. These methods empower scientists to assess the cellular, subcellular, and molecular effects of low-level contaminants on diverse organisms (Sepúlveda et al. 2011). Information gathered from these surveys can be assimilated into computational models, allowing for comprehensive 4D surveillance of the ocean and predictive studies. This aids in assessing alterations resulting from overuse, changes in habitat, runoff from watersheds, emissions of pollutants, and even climate change (Khan et al. 2019). Moreover, ecotoxicological surveys play a crucial role in detecting HABs, heavy metal contamination, and bioaccumulation and assessing physicochemical variables and biological indicators in coastal ecosystems. Hence, ecotoxicological surveys have significant potential to propel the progress of monitoring marine and coastal pollution.

Coastal zones function as transitional areas connecting terrestrial and marine environments. These zones are marked by higher population density, increased rates of population growth and urbanization compared to the hinterland, and the concentration of economic assets and critical infrastructure. Consequently, safeguarding the health of these ecosystems is crucial, and ecotoxicological surveys play a pivotal role in this endeavor.

These surveys offer invaluable insights that can inform policy decisions and guide conservation efforts. With advancing technology, it is anticipated that these assessments will become increasingly precise and comprehensive, further strengthening our capacity to protect and preserve the invaluable marine and coastal ecosystems.

14.13 Toxicogenomics

Toxicogenomics, which investigates changes in gene expression in response to toxic substances, presents an innovative approach for monitoring and assessing the impact of pollution in coastal and marine environments. Conventional monitoring techniques typically emphasize chemical analyses, whereas toxicogenomics introduces a molecular dimension to evaluating the impacts of pollution. Through the examination of alterations in gene expression patterns, toxicogenomics enables a more nuanced comprehension of biological responses to contaminants (Sturchio et al. 2008). The field encompasses various technologies, such as genome sequence analysis, proteomics, metabolomics, and bioinformatics, to identify changes in genetic expression, protein synthesis, and the generation of metabolites caused by toxic agents (Stierum et al. 2005). Additionally, it investigates how these changes affect phenotypic expressions at the cellular, tissue, and organismal levels. This information enhances our comprehension of how biological pathways react to harmful substances.

Key components of toxicogenomics for marine pollution monitoring are transcriptomic profiling, biomarker discovery, and pathway analysis. Toxicogenomic studies involve transcriptomic profiling to identify genes that are differentially expressed in response to pollutants. High-throughput sequencing technologies, such as RNA-Seq, enable a comprehensive analysis of the transcriptome (Pujolar et al. 2012; Reid and Whitehead 2016). Toxicogenomics facilitates the discovery of molecular biomarkers indicative of exposure to specific pollutants (Dix et al. 2007). Understanding the pathways affected by pollutants is crucial. Toxicogenomics allows for pathway analysis, revealing the interconnected network of genes involved in cellular responses to pollution (Chu et al. 2016).

14.14 Integrated Monitoring Techniques

The biomarkers of condition index (CI), MT, and AChE were combined to assess various locations, demonstrating the differences in biomarker responses between reference sites and those affected by pollution. By comparing biomarker responses across diverse sites, researchers can gain valuable insights into the effects of pollution in these locations (Mitrić and Ramšak 2021). A combination of the neutral red retention assay with the CA and MN biomarkers was used to assess the well-being of *Mytilus edulis* by Strubbia et al. (2019). However, the responses of the three biomarkers did not exhibit any significant correlation. Thus, a set of biomarkers is necessary to accurately assess pollution levels. A set of biomarkers can include physiological, biochemical, genetic, or other measurable characteristics that provide information about exposure to stressors or environmental contaminants, as well as overall health status. By using multiple biomarkers together, researchers can obtain a more comprehensive understanding of the biological responses and potential impacts of various stressors on organisms. Various elements can impede or alter biomarker responses, including the species, age, and reproductive stage of the organism, as well as environmental conditions or the existence of potential inhibitors. Moreover, the lack of correlation between biomarker responses could be attributed to differences in their sensitivities, particularly in cases of low contamination levels. For this type of situation, chemical analysis techniques would be more useful for measuring low-level pollutants in marine and coastal monitoring.

Integrated toxicological techniques for marine pollution monitoring involve the combination of various methods to comprehensively assess the impacts of pollutants on marine ecosystems. This approach typically includes the use of selective biosensors, bioassays, biomarker analysis, transcriptomic profiling, pathway analysis, and integration with other data sources, such as satellite remote sensing. By employing multiple techniques in tandem, researchers can obtain a thorough understanding of the complex interactions between pollutants and marine organisms, facilitating more effective monitoring and mitigation strategies. Note that we cannot completely examine marine and coastal pollution by using only a single toxicological technique. Therefore, combined monitoring techniques are pivotal.

14.15 Conclusion

Toxicological techniques for monitoring marine and coastal pollution employ diverse methodologies, including selective biosensors, bioassays, and tissue analysis. These methods facilitate the measurement of cellular, subcellular, and molecular effects caused by low-level contaminants on various organisms. Biomarker discovery is essential for identifying molecular indicators of exposure to specific pollutants, while understanding the pathways affected by pollutants is crucial. Toxicogenomics enables pathway analysis, unveiling the interconnected network of genes involved in cellular responses to pollution. Satellite-borne sensors enable effective quasi-global ocean monitoring, providing synoptic views of large areas with high spatial and temporal resolution. This sustained time-series data spans several years to decades, facilitating comprehensive analysis. Together, these methods play a pivotal role in fostering a holistic comprehension of the repercussions of marine and coastal pollution, thus assisting in the formulation of impactful mitigation strategies.

Future directions in toxicological monitoring of marine pollution involve significant advancements. The evolution of advanced sequencing methods, coupled with the integration of machine learning algorithms, holds the potential to transform transcriptomic profiling. This promises improved abilities to detect genes exhibiting diverse expression patterns in reaction to pollutants, offering deeper insights with heightened efficiency. Future research will focus on uncovering molecular biomarkers that signal exposure to particular pollutants, representing a crucial area of investigation. Integrating these biomarkers with diverse datasets, including satellite remote sensing information, offers the potential for a comprehensive grasp of pollution effects. An important aspect of future research lies in pathway analysis, which explores the interconnected network of genes engaged in cellular responses to pollution. Such analysis aids in comprehending toxicity mechanisms and the impacts of pollutants on marine organisms.

References

Abreo, N.A.S., Macusi, E.D., Cuenca, G.C. et al. (2015). Nutrient enrichment, sedimentation, heavy metals and plastic pollution in the marine environment and its implications on Philippine marine biodiversity: a review. *IAMURE International Journal of Ecology and Conservation* 15 (1): 111–168.

Adams, M.S. and Stauber, J.L. (2008). Marine whole sediment toxicity tests for use in temperate and tropical Australian environments: current status. *Australasian Journal of Ecotoxicology* 14 (2/3): 155–167.

de Almeida Rodrigues, P., Ferrari, R.G., Kato, L.S. et al. (2022). A systematic review on metal dynamics and marine toxicity risk assessment using crustaceans as bioindicators. *Biological Trace Element Research* 200: 881–903.

Amri, S., Samar, M.F., Sellem, F., and Ouali, K. (2017). Seasonal antioxidant responses in the sea urchin *Paracentrotus lividus* (Lamarck 1816) used as a bioindicator of the environmental contamination in the South-East Mediterranean. *Marine Pollution Bulletin* 122 (1-2): 392–402.

Anandraj, A., Marshall, D.J., Gregory, M.A., and McClurg, T.P. (2002). Metal accumulation, filtration and O(2) uptake rates in the mussel *Perna perna* (Mollusca: Bivalvia) exposed to Hg^{2+}, Cu^{2+} and Zn^{2+}. *Comparative Biochemistry and Physiology, Part C: Toxicology & Pharmacology* 132 (3): 355–363.

Atli, G. and Canli, M. (2007). Enzymatic responses to metal exposures in a freshwater fish *Oreochromis niloticus*. *Comparative Biochemistry and Physiology Part C: Toxicology & Pharmacology* 145 (2): 282–287.

Auffan, M., Tella, M., Santaella, C. et al. (2014). An adaptable mesocosm platform for performing integrated assessments of nanomaterial risk in complex environmental systems. *Scientific Reports* 4 (1): 5608.

Authman, M.M., Zaki, M.S., Khallaf, E.A., and Abbas, H.H. (2015). Use of fish as bio-indicator of the effects of heavy metals pollution. *Journal of Aquaculture Research & Development* 6 (4): 1–13.

Azqueta, A., Shaposhnikov, S., and Collins, A.R. (2011). DNA repair measured by the comet assay. In: *DNA Repair* (ed. I. Kruman), 615–636. InTech Open http://www.intechopen.com/books/dna-repair/dna-repair-measured-by-the-cometassay.

Bakiu, R., Santovito, G., Hoda, A. et al. (2013). Metallothionein (MT): a good biomarker in marine sentinel species like sea bream (*Sparus aurata*). *Albanian Journal of Agricultural Sciences* 12 (2): 247.

Banerjee, I., Paul, A. and Saha, S. (2022). Biochemical and molecular basis of bacterial bioluminescence and its utilization as a biosensor for the assessment of environmental toxicity. https://api.semanticscholar.org/CorpusID:252357299 (accessed 05 December 2023).

Baums, I.B., Goodwin, K.D., Kiesling, T. et al. (2007). Luminex detection of fecal indicators in river samples, marine recreational water, and beach sand. *Marine Pollution Bulletin* 54 (5): 521–536.

Bean, T.P., Greenwood, N., Beckett, R. et al. (2017). A review of the tools used for marine monitoring in the UK: combining historic and contemporary methods with modeling and socioeconomics to fulfill legislative needs and scientific ambitions. *Frontiers in Marine Science* 4: 263.

Benton, T.G., Solan, M., Travis, J.M., and Sait, S.M. (2007). Microcosm experiments can inform global ecological problems. *Trends in Ecology & Evolution* 22 (10): 516–521.

Bhuvaneshwari, R., Padmanaban, K., and Babu Rajendran, R. (2015). Histopathological alterations in muscle, liver and gill tissues of zebra fish Danio rerio due to environmentally relevant concentrations of organochlorine pesticides (OCPs) and heavy metals. *International Journal of Environmental Research* 9 (4): 1365–1372.

Bolognesi, C. and Hayashi, M. (2011). Micronucleus assay in aquatic animals. *Mutagenesis* 26 (1): 205–213. https://doi.org/10.1093/mutage/geq073.

Borja, A., Bricker, S.B., Dauer, D.M. et al. (2008). Overview of integrative tools and methods in assessing ecological integrity in estuarine and coastal systems worldwide. *Marine Pollution Bulletin* 56 (9): 1519–1537.

Bowler, C. and Falciatore, A., 2019. *Phaeodactylum tricornutum*. https://www.cell.com/trends/genetics/abstract/S0168-9525(19)30109-X (accessed 20 December 2023).

Breitburg, D., Levin, L.A., Oschlies, A. et al. (2018). Declining oxygen in the global ocean and coastal waters. *Science* 359 (6371): eaam7240.

Bruland, K.W., Middag, R., and Lohan, M.C. (2014). Controls of trace metals in seawater. In: *Treatise on Geochemistry*, vol. 8 (ed. H.D. Holland and K.K. Turekian), 19–51. Amsterdam, Netherlands: Elsevier https://doi.org/10.1016/B978-0-08-095975-7.00602-1.

Burkholder, J.M., Tomasko, D.A., and Touchette, B.W. (2007). Seagrasses and eutrophication. *Journal of Experimental Marine Biology and Ecology* 350 (1-2): 46–72.

Burton, G.A. Jr. (2010). Metal bioavailability and toxicity in sediments. *Critical Reviews in Environmental Science and Technology* 40 (9–10): 852–907. https://doi.org/10.1080/10643380802501567.

Byrne, P.A. and O'halloran, J. (2001). The role of bivalve molluscs as tools in estuarine sediment toxicity testing: a review. *Hydrobiologia* 465: 209–217.

Cajaraville, M.P., Bebianno, M.J., Blasco, J. et al. (2000). The use of biomarkers to assess the impact of pollution in coastal environments of the Iberian Peninsula: a practical approach. *Science of the Total Environment* 247 (2–3): 295–311.

Campana, O., Simpson, S.L., Spadaro, D.A., and Blasco, J. (2012). Sub-lethal effects of copper to benthic invertebrates explained by sediment properties and dietary exposure. *Environmental Science & Technology* 46 (12): 6835–6842.

Cariou, R., Méndez-Fernandez, P., Hutinet, S. et al. (2020). Nontargeted LC/ESI-HRMS detection of polyhalogenated compounds in marine mammals stranded on French Atlantic coasts. *ACS ES&T Water* 1 (2): 309–318.

Casas, S., Gonzalez, J.L., Andral, B., and Cossa, D. (2008). Relation between metal concentration in water and metal content of marine mussels (*Mytilus galloprovincialis*): impact of physiology. *Environmental Toxicology and Chemistry* 27: 1543–1552. https://doi.org/10.1897/07-418.1.

Chariton, A.A., Roach, A.C., Simpson, S.L., and Batley, G.E. (2010a). Influence of the choice of physical and chemistry variables on interpreting patterns of sediment contaminants and their relationships with estuarine macrobenthic communities. *Marine and Freshwater Research* 61: 1109–1122.

Chekan, J.R., Fallon, T.R., and Moore, B.S. (2020). Biosynthesis of marine toxins. *Current Opinion in Chemical Biology* 59: 119–129.

Cheng, W., Li, X., Zhou, Y. et al. (2022). Polystyrene microplastics induce hepatotoxicity and disrupt lipid metabolism in the liver organoids. *Science of the Total Environment* 806 (Pt 1): 150328.

Chowdhury, S., Dubey, V.K., Choudhury, S. et al. (2023). Insects as bioindicator: a hidden gem for environmental monitoring. *Frontiers in Environmental Science* 11: 1146052.

Chu, J.H., Hart, J.E., Chhabra, D. et al. (2016). Gene expression network analyses in response to air pollution exposures in the trucking industry. *Environmental Health* 15: 1–10.

Cnubben, N.H., Rietjens, I.M., Wortelboer, H. et al. (2001). The interplay of glutathione-related processes in antioxidant defense. *Environmental Toxicology and Pharmacology* 10 (4): 141–152.

Cocci, P., Gabrielli, S., Pastore, G. et al. (2022). Microplastics accumulation in gastrointestinal tracts of *Mullus barbatus* and *Merluccius merluccius* is associated with increased cytokine production and signaling. *Chemosphere* 307 (Pt 3): 135813.

Cole, M., Lindeque, P., Fileman, E. et al. (2013). Microplastic ingestion by zooplankton. *Environmental Science & Technology* 47 (12): 6646–6655.

Collings, F.B. and Vaidya, V.S. (2008). Novel technologies for the discovery and quantitation of biomarkers of toxicity. *Toxicology* 245 (3): 167–174.

Collins, A.R., Dusinska, M., Gedik, C.M., and Stetina, R. (1996). Oxidative damage to DNA: do we have a reliable biomarker? *Environmental Health Perspectives* 104 (S3): 465–469.

Connell, D., Lam, P., Richardson, B., and Wu, R. (1999). *Introduction to Ecotoxicology*, 180. Cornwall: Blackwell Science Ltd.

Cossa, D. (1989). A review of the use of *Mytilus* spp. as quantitative indicators of cadmium and mercury contamination in coastal waters. *Oceanologica Acta* 12: 417–432.

Costa, L.L., da Costa, M.F., and Zalmon, I.R. (2021). Macroinvertebrates as biomonitors of pollutants on natural sandy beaches: overview and meta-analysis. *Environmental Pollution* 275: 116629.

Covich, A.P., Palmer, M.A., and Crowl, T.A. (1999). The role of benthic invertebrate species in freshwater ecosystems: zoobenthic species influence energy flows and nutrient cycling. *Bioscience* 49 (2): 119–127.

Čulin, J. and Bielić, T. (2016). Plastic pollution from ships. *Pomorski zbornik* 51 (1): 57–66.

da Silva Souza, L., Bonnail, E., Maranho, L.A., Pusceddu, F.H., Cortez, F.S., Cesar, A., Ribeiro, D.A., Riba, I., de Souza Abessa, D.M., DelValls, Á., and Pereira, C.D.S. (2021). Sub-lethal combined effects of illicit drug and decreased pH on marine mussels: A short-time exposure to crack cocaine in CO2 enrichment scenarios. *Marine Pollution Bulletin* 171: 112735.

D'Costa, A., Shyama, S.K., and Kumar, M.P. (2017a). Bioaccumulation of trace metals and total petroleum and genotoxicity responses in an edible fish population as indicators of marine pollution. *Ecotoxicology and Environmental Safety* 142: 22–28.

D'Costa, P.M., D'Silva, M.S., and Naik, R.K. (2017b). Impact of pollution on phytoplankton and implications for marine econiches. *Marine Pollution and Microbial Remediation* 205–222.

Dahms, H.U., Won, E.J., Kim, H.S. et al. (2016). Potential of the small cyclopoid copepod *Paracyclopina nana* as an invertebrate model for ecotoxicity testing. *Aquatic Toxicology* 180: 282–294.

De Lange, H.J., De Haas, E.M., Maas, H., and Peeters, E.T.H.M. (2005). Contaminated sediments and bioassay responses of three macroinvertebrates, the midge larva *Chironomus riparius*, the water louse *Asellus aquaticus* and the mayfly nymph *Ephoron virgo*. *Chemosphere* 61 (11): 1700–1709.

De, M. and Roy, C. (2019). Diverse role of Macrophytes in aquatic ecosystems: a brief review Mitu De, Chayanika Roy 2, Suchismita Medda 3, Sulagna Roy 4 and Santi Ranjan Dey 2. *International Journal of Experimental Research and Review* 19: 40–48.

Del Rio, D., Stewart, A.J., and Pellegrini, N. (2005). A review of recent studies on malondialdehyde as toxic molecule and biological marker of oxidative stress. *Nutrition, Metabolism, and Cardiovascular Diseases* 15 (4): 316–328.

Dix, D.J., Houck, K.A., Martin, M.T. et al. (2007). The ToxCast program for prioritizing toxicity testing of environmental chemicals. *Toxicological Sciences* 95 (1): 5–12.

Duarte, I.A., Pais, M.P., Reis-Santos, P. et al. (2019). Biomarker and behavioural responses of an estuarine fish following acute exposure to fluoxetine. *Marine Environmental Research* 147: 24–31.

Elseblani, R., Cobo-Golpe, M., Godin, S. et al. (2023). Study of metal and organic contaminants transported by microplastics in the Lebanese coastal environment using ICP MS, GC-MS, and LC-MS. *Science of the Total Environment* 887: 164111.

Embry, M.R., Belanger, S.E., Braunbeck, T.A. et al. (2010). The fish embryo toxicity test as an animal alternative method in hazard and risk assessment and scientific research. *Aquatic Toxicology* 97 (2): 79–87.

Farré, M. and Barceló, D. (2003). Toxicity testing of wastewater and sewage sludge by biosensors, bioassays and chemical analysis. *TrAC Trends in Analytical Chemistry* 22 (5): 299–310.

Feng, Z., Zhang, T., Shi, H. et al. (2020). Microplastics in bloom-forming macroalgae: distribution, characteristics and impacts. *Journal of Hazardous Materials* 397: 122752.

Fu, H., Xia, Y., Chen, Y. et al. (2018). Acetylcholinesterase is a potential biomarker for a broad spectrum of organic environmental pollutants. *Environmental Science & Technology* 52 (15): 8065–8074.

Fulton, M.H. and Key, P.B. (2001). Acetylcholinesterase inhibition in estuarine fish and invertebrates as an indicator of organophosphorus insecticide exposure and effects. *Environmental Toxicology and Chemistry: An International Journal* 20 (1): 37–45.

Galloway, T.S., Cole, M., and Lewis, C. (2017). Interactions of microplastic debris throughout the marine ecosystem. *Nature Ecology & Evolution* 1 (5): 0116.

Gambardella, C., Piazza, V., Albentosa, M. et al. (2019). Microplastics do not affect standard ecotoxicological endpoints in marine unicellular organisms. *Marine Pollution Bulletin* 143: 140–143.

Garai, P., Banerjee, P., Mondal, P., and Saha, N.C. (2021). Effect of heavy metals on fishes: toxicity and bioaccumulation. *Journal of Clinical Toxicology* 18: 1–10.

Gilbert, P.M., Beusen, A.H., Harrison, J.A. et al. (2018). Changing land-, sea-, and airscapes: sources of nutrient pollution affecting habitat suitability for harmful algae. In: *Global Ecology and Oceanography of Harmful Algal Blooms*, 53–76.

Gong, N., Shao, K., Feng, W. et al. (2011). Biotoxicity of nickel oxide nanoparticles and bio-remediation by microalgae *Chlorella vulgaris*. *Chemosphere* 83 (4): 510–516.

Guerrera, M.C., Aragona, M., Porcino, C. et al. (2021). Micro and nano plastics distribution in fish as model organisms: histopathology, blood response and bioaccumulation in different organs. *Applied Sciences* 11 (13): 5768.

Häder, D.P., Banaszak, A.T., Villafañe, V.E. et al. (2020). Anthropogenic pollution of aquatic ecosystems: emerging problems with global implications. *Science of the Total Environment* 713: 136586.

Halappanavar, S. and Mallach, G. (2021). Adverse outcome pathways and in vitro toxicology strategies for microplastics hazard testing. *Current Opinion in Toxicology* 28: 52–61. https://www.sciencedirect.com/science/article/pii/S2468202021000450.

Hall, J.Á. (2002). Cellular mechanisms for heavy metal detoxification and tolerance. *Journal of Experimental Botany* 53 (366): 1–11.

Hampel, M., Blasco, J., and Martín Díaz, M.L. (2016). Biomarkers and effects. In: *Marine Ecotoxicology* (ed. J. Blasco, P.M. Chapman, O. Campana, and M. Hampel), 121–165. Academic Press.

Harford, A.J., Hogan, A.C., Tsang, J.J. et al. (2011). Effects of alumina refinery wastewater and signature metal constituents at the upper thermal tolerance of: 1. The tropical diatom *Nitzschia closterium*. *Marine Pollution Bulletin* 62 (3): 466–473.

Hebert, C.E. and Popp, B.N. (2018). Temporal trends in a biomagnifying contaminant: application of amino acid compound-specific stable nitrogen isotope analysis to the interpretation of bird mercury levels. *Environmental Toxicology and Chemistry* 37: 1458–1465.

Heisler, J., Glibert, P.M., Burkholder, J.M. et al. (2008). Eutrophication and harmful algal blooms: a scientific consensus. *Harmful Algae* 8 (1): 3–13.

Hemmadi, V. (2016). Metallothionein-A potential biomarker to assess the metal contamination in marine fishes-A review. *International Journal of Bioassays* 5 (4): 4961–4973.

Hernández, L.E., Sobrino-Plata, J., Montero-Palmero, M.B. et al. (2015). Contribution of glutathione to the control of cellular redox homeostasis under toxic metal and metalloid stress. *Journal of Experimental Botany* 66 (10): 2901–2911.

Hodkinson, I.D. and Jackson, J.K. (2005). Terrestrial and aquatic invertebrates as bioindicators for environmental monitoring, with particular reference to mountain ecosystems. *Environmental Management* 35: 649–666.

Horie, Y. and Takahashi, C. (2021). Development of an in vivo acute bioassay using the marine medaka *Oryzias melastigma*. *Environmental Monitoring and Assessment* 193 (11): 725.

Idowu, I., Francisco, O., Thomas, P.J. et al. (2018). Validation of a simultaneous method for determining polycyclic aromatic compounds and alkylated isomers in biota. *Rapid Communications in Mass Spectrometry* 32 (3): 277–287. https://doi.org/10.1002/rcm.8035.

Imai, H. and Nakagawa, Y. (2003). Biological significance of phospholipid hydroperoxide glutathione peroxidase (PHGPx, GPx4) in mammalian cells. *Free Radical Biology and Medicine* 34 (2): 145–169.

Impellitteri, F., Curpăn, A.S., Plăvan, G. et al. (2022). Hemocytes: a useful tool for assessing the toxicity of microplastics, heavy metals, and pesticides on aquatic invertebrates. *International Journal of Environmental Research and Public Health* 19 (24): 16830.

Islam, S.S., Samanta, S., Mahato, S. et al. 15 -(2024). Diversity of meiobenthic fauna in costal environment: As a bioindicator. In: *Advances in Pollution Research, Spatial Modeling of Environmental Pollution and Ecological Risk*Woodhead Publishing (ed. P.K. Shit, D.K. Datta, B. Bera, et al.), 275–299. ISBN 9780323952828, https://doi.org/10.1016/B978-0-323-95282-8.00029-8.

Jahan, S. and Strezov, V. (2019). Effect of shipping on the distribution of trace elements and petroleum hydrocarbons in the coastal basins of Australia: a review. *Marine and Freshwater Research* 71 (7): 794–809.

Janssen, Y.M., Van Houten, B., Borm, P.J., and Mossman, B.T. (1993). Cell and tissue responses to oxidative damage. *Laboratory Investigation* 69 (3): 261–274.

Javed, M., Ahmad, M.I., Usmani, N., and Ahmad, M. (2017). Multiple biomarker responses (serum biochemistry, oxidative stress, genotoxicity and histopathology) in *Channa punctatus* exposed to heavy metal loaded waste water. *Scientific Reports* 7 (1): 1675.

Jin, L., Gaus, C., and Escher, B.I. (2015). Bioanalytical approaches to understanding toxicological implications of mixtures of persistent organic pollutants in marine wildlife. *Comprehensive Analytical Chemistry* 67: 57–84.

Jomova, K. and Valko, M. (2011). Advances in metal-induced oxidative stress and human disease. *Toxicology* 283 (2–3): 65–87.

Justino, C.I., Freitas, A.C., Duarte, A.C., and Santos, T.A.R. (2015). Sensors and biosensors for monitoring marine contaminants. *Trends in Environmental Analytical Chemistry* 6: 21–30.

Justino, C.I., Duarte, A.C., and Rocha-Santos, T.A. (2017). Recent progress in biosensors for environmental monitoring: a review. *Sensors* 17 (12): 2918.

Karydis, M. and Kitsiou, D. (2013). Marine water quality monitoring: a review. *Marine Pollution Bulletin* 77 (1–2): 23–36.

Katsumiti, A., Gilliland, D., Arostegui, I., and Cajaraville, M.P. (2014). Cytotoxicity and cellular mechanisms involved in the toxicity of CdS quantum dots in hemocytes and gill cells of the mussel *Mytilus galloprovincialis*. *Aquatic Toxicology* 153: 39–52.

Kazmi, S.S.U.H., Yapa, N., Karunarathna, S.C., and Suwannarach, N. (2022). Perceived intensification in harmful algal blooms is a wave of cumulative threat to the aquatic ecosystems. *Biology* 11 (6): 852.

Kennelly, S.J. and Broadhurst, M.K. (2002). By-catch begone: changes in the philosophy of fishing technology. *Fish and Fisheries* 3 (4): 340–355.

Khalik, W.M.A.W.M., Ibrahim, Y.S., Anuar, S.T. et al. (2018). Microplastics analysis in Malaysian marine waters: a field study of Kuala Nerus and Kuantan. *Marine Pollution Bulletin* 135: 451–457.

Khan, K., Baderna, D., Cappelli, C. et al. (2019). Ecotoxicological QSAR modeling of organic compounds against fish: application of fragment based descriptors in feature analysis. *Aquatic Toxicology* 212: 162–174.

Khoshnood, Z. (2017). Effects of environmental pollution on fish: a short review. *Transylvanian Review of Systematical and Ecological Research* 19 (1): 49–60.

Khosrovyan, A., Rodríguez-Romero, A., Salamanca, M.J. et al. (2013). Comparative performances of eggs and embryos of sea urchin (*Paracentrotus lividus*) in toxicity bioassays used for assessment of marine sediment quality. *Marine Pollution Bulletin* 70 (1-2): 204–209.

Kim, Y.S., Raston, N.H.A., and Gu, M.B. (2016). Aptamer-based nanobiosensors. *Biosensors and Bioelectronics* 76: 2–19.

King, M.D., Elliott, J.E., Idowu, I. et al. (2023). Polycyclic aromatic compound and trace metal element residues in *Mytilus* mussels at marine wildlife hotspots on the Pacific coast of Canada. *Environmental Pollution* 316: 120624.

Kokkali, V. and van Delft, W. (2014). Overview of commercially available bioassays for assessing chemical toxicity in aqueous samples. *TrAC Trends in Analytical Chemistry* 61: 133–155.

Kriwy, P. and Uthicke, S. (2011). Microbial diversity in marine biofilms along a water quality gradient on the Great Barrier Reef. *Systematic and Applied Microbiology* 34 (2): 116–126.

Lafferty, K.D. and Kuris, A.M. (1999). How environmental stress affects the impacts of parasites. *Limnology and Oceanography* 44 (3part2): 925–931.

Lagauzère, S., Boyer, P., Stora, G., and Bonzom, J.M. (2009). Effects of uranium-contaminated sediments on the bioturbation activity of *Chironomus riparius* larvae (Insecta, Diptera) and *Tubifex tubifex* worms (Annelida, Tubificidae). *Chemosphere* 76 (3): 324–334.

Le Moullac, G. and Haffner, P. (2000). Environmental factors affecting immune responses in Crustacea. *Aquaculture* 191 (1–3): 121–131.

Lee, K.W., Raisuddin, S., Hwang, D.S. et al. (2007). Acute toxicities of trace metals and common xenobiotics to the marine copepod *Tigriopus japonicus*: evaluation of its use as a benchmark species for routine ecotoxicity tests in Western Pacific coastal regions. *Environmental Toxicology: An International Journal* 22 (5): 532–538.

Lenz, R., Enders, K., Stedmon, C.A. et al. (2015). A critical assessment of visual identification of marine microplastic using Raman spectroscopy for analysis improvement. *Marine Pollution Bulletin* 100 (1): 82–91.

Lewis, C., Guitart, C., Pook, C. et al. (2010). Integrated assessment of oil pollution using biological monitoring and chemical fingerprinting. *Environmental Toxicology and Chemistry* 29: 1358–1366.

Li, J., Chang, H., Zhang, N. et al. (2023). Recent advances in enzyme inhibition based-electrochemical biosensors for pharmaceutical and environmental analysis. *Talanta* 253: 124092.

Lin, T. and Yang, M.S. (2007). Benzo [a] pyrene-induced elevation of GSH level protects against oxidative stress and enhances xenobiotic detoxification in human HepG2 cells. *Toxicology* 235 (1-2): 1–10.

Lionetto, M.G., Caricato, R., Calisi, A. et al. (2013). Acetylcholinesterase as a biomarker in environmental and occupational medicine: new insights and future perspectives. *BioMed Research International* 2013 (1): 321213.

Lionetto, M.G., Caricato, R., and Giordano, M.E. (2019). Pollution biomarkers in environmental and human biomonitoring. *The Open Biomarkers Journal* 9 (1): 1–9.

Livingstone, D.R., Chipman, J.K., Lowe, D.M. et al. (2000). Development of biomarkers to detect the effects of organic pollution on aquatic invertebrates: recent molecular, genotoxic, cellular and immunological studies on the common mussel (*Mytilus edulis* L.) and other mytilids. *International Journal of Environment and Pollution* 13 (1–6): 56–91.

Lombardo, J., Solomando, A., Cohen-Sánchez, A. et al. (2022). Effects of human activity on markers of oxidative stress in the intestine of *Holothuria tubulosa*, with special reference to the presence of microplastics. *International Journal of Molecular Sciences* 23 (16): 9018.

Lubos, E., Loscalzo, J., and Handy, D.E. (2011). Glutathione Peroxidase-1 in health and disease: from molecular mechanisms to therapeutic opportunities. *Antioxidants & Redox Signaling* 15 (7): 1957–1997.

Madeira, D., Mendonça, V., Madeira, C. et al. (2019). Molecular assessment of wild populations in the marine realm: importance of taxonomic, seasonal and habitat patterns in environmental monitoring. *Science of the Total Environment* 654: 250–263.

Madsen, J.D., Chambers, P.A., James, W.F. et al. (2001). The interaction between water movement, sediment dynamics and submersed macrophytes. *Hydrobiologia* 444: 71–84.

Madunil, S.L., Imasaka, T., and Imasaka, T. (2020). Resonant and non-resonant femtosecond ionization mass spectrometry of organochlorine pesticides. *Analyst* 145 (3): 777–783.

Madunil, S.L., Imasaka, T., and Imasaka, T. (2022). Determination of barbiturates by femtosecond ionization mass spectrometry. *Analytical Chemistry* 94 (42): 14691–14698.

Magi, E. and Di Carro, M. (2018). Marine environment pollution: the contribution of mass spectrometry to the study of seawater. *Mass Spectrometry Reviews* 37 (4): 492–512.

Magni, P., De Falco, G., Falugi, C. et al. (2006). Genotoxicity biomarkers and acetylcholinesterase activity in natural populations of *Mytilus galloprovincialis* along a pollution gradient in the Gulf of Oristano (Sardinia, western Mediterranean). *Environmental Pollution* 142 (1): 65–72.

Mahboob, S., Ahmed, Z., Farooq Khan, M. et al. (2022). Ecogenotoxicological studies for an early toxicity screening and monitoring in *Epinephalus chlorostigma* and *Scamberomorus commerson*. *Saudi Journal of Biological Sciences* 29 (4): 2719–2726. https://doi.org/10.1016/j.sjbs.2021.12.064.

Mahmood, S.K., Mhawesh, A.A., and Abdullah, T.S. (2019). Comet assay for rapid detection of dna base damage in plants. *The Iraqi Journal of Agricultural Science* 50 (1): 495–499.

Maity, S., Guchhait, R., De, S., and Pramanick, K. (2023). High doses of nano-polystyrene aggravate the oxidative stress, DNA damage, and the cell death in onions. *Environmental Pollution* 316 (Pt 2): 120611.

Maletić, S.P., Beljin, J.M., Rončević, S.D. et al. (2019). State of the art and future challenges for polycyclic aromatic hydrocarbons is sediments: sources, fate, bioavailability and remediation techniques. *Journal of Hazardous Materials* 365: 467–482.

Maranho, L.A., Garrido-Pérez, M.C., DelValls, T.A., and Martín-Díaz, M.L. (2015). Suitability of standardized acute toxicity tests for marine sediment assessment: pharmaceutical contamination. *Water, Air, & Soil Pollution* 226: 1–14.

Marnett, L.J. (1999). Chemistry and biology of DNA damage by malondialdehyde. *IARC Scientific Publications* 150: 17–27.

Marrocco, I., Altieri, F., and Peluso, I. (2017). Measurement and clinical significance of biomarkers of oxidative stress in humans. *Oxidative Medicine and Cellular Longevity* 2017: 6501046. https://doi.org/10.1155/2017/6501046.

Matozzo, V., Gagné, F., Marin, M.G. et al. (2008). Vitellogenin as a biomarker of exposure to estrogenic compounds in aquatic invertebrates: a review. *Environment International* 34 (4): 531–545.

McCready, S., Birch, G.F., Long, E.R. et al. (2006). Relationships between toxicity and concentrations of chemical contaminants in sediments from Sydney Harbour, Australia, and vicinity. *Environmental Monitoring and Assessment* 120: 187–220.

Mehrotra, P. (2016). Biosensors and their applications – a review. *Journal of Oral Biology and Craniofacial Research* 6 (2): 153–159.

Mejdoub, Z., Fahde, A., Loutfi, M., and Kabine, M. (2017). Oxidative stress responses of the mussel *Mytilus galloprovincialis* exposed to emissary's pollution in coastal areas of Casablanca. *Ocean and Coastal Management* 136: 95–103.

Menz, J., Schneider, M., and Kümmerer, K. (2013). Toxicity testing with luminescent bacteria–characterization of an automated method for the combined assessment of acute and chronic effects. *Chemosphere* 93 (6): 990–996.

Milan, B., Slobodanka, P., Živko, S., and Borivoj, K. (2006). Macrophytes as phytoindicators and potential phytoremediators in aquatic ecosystems. In: *International Conference*. Klosterneuburg, Austria: International Austrian Committee Danube Research.

Miloloža, M., Bule, K., Prevarić, V. et al. (2022). Assessment of the influence of size and concentration on the ecotoxicity of microplastics to microalgae *Scenedesmus* sp., bacterium *Pseudomonas putida* and yeast *Saccharomyces cerevisiae*. *Polymers* 14 (6): 1246.

Missawi, O., Bousserrhine, N., Belbekhouche, S. et al. (2020). Abundance and distribution of small microplastics (≤ 3 μm) in sediments and seaworms from the southern Mediterranean coasts and characterisation of their potential harmful effects. *Environmental Pollution* 263 (Pt A): 114634.

Mitrić, M. and Ramšak, A. (2021). Sampling site specific biomarker responses in mediterranean mussels from the Adriatic Sea. *Bulletin of Environmental Contamination and Toxicology* 106: 310–317.

Moraes, B.S., Loro, V.L., Glusczak, L. et al. (2007). Effects of four rice herbicides on some metabolic and toxicology parameters of teleost fish (*Leporinus obtusidens*). *Chemosphere* 68: 1597–1601.

Murchelano, R.A. and Wolke, R.E. (1991). Neoplasms and nonneoplastic liver lesions in winter flounder, *Pseudopleuronectes americanus*, from Boston Harbor, Massachusetts. *Environmental Health Perspectives* 90: 17–26.

Myers, M.S., Johnson, L.L., and Collier, T.K. (2003). Establishing the causal relationship between polycyclic aromatic hydrocarbon (PAH) exposure and hepatic neoplasms and neoplasia-related liver lesions in English sole (*Pleuronectes vetulus*). *Human and Ecological Risk Assessment* 9 (1): 67–94.

Nagel, R. (2002). DarT: The embryo test with the Zebrafish Danio rerio – a general model in ecotoxicology and toxicology. *ALTEX* 19: 38–48.

Narayanasamy, D., Taufik, S., Azmi, A.F.M. et al. (2024). Modern technology advances of *Pseudomonas aeruginosa* based biosensor approach. *Biosensors and Bioelectronics* X: 100441.

Nebbia, C. (2012). Factors affecting chemical toxicity, Chapter 4. In: *Veterinary Toxicology*, 2e (ed. R.C. Gupta), 48–61. Academic Press ISBN: 9780123859266.

Nyholm, N. and Källqvist, T. (1989). Methods for growth inhibition toxicity tests with freshwater algae. *Environmental Toxicology and Chemistry: An International Journal* 8 (8): 689–703.

Pan, K. and Wang, W.X. (2012). Trace metal contamination in estuarine and coastal environments in China. *Science of the Total Environment* 421: 3–16.

Perić, L., Nerlović, V., Žurga, P. et al. (2017). Variations of biomarkers response in mussels *Mytilus galloprovincialis* to low, moderate and high concentrations of organic chemicals and metals. *Chemosphere* 174: 554–562.

Perin, L.S., Moraes, G.V., Galeazzo, G.A., and Oliveira, A.G. (2022). Bioluminescent dinoflagellates as a bioassay for toxicity assessment. *International Journal of Molecular Sciences* 23 (21): 13012.

Peters, C.A., Hendrickson, E., Minor, E.C. et al. (2018). Pyr-GC/MS analysis of microplastics extracted from the stomach content of benthivore fish from the Texas Gulf Coast. *Marine Pollution Bulletin* 137: 91–95.

Polechońska, L. and Klink, A. (2023). Macrophytes as passive bioindicators of trace element pollution in the aquatic environment. *Wiley Interdisciplinary Reviews: Water* 10 (2): e1630.

Prather, C.M., Pelini, S.L., Laws, A. et al. (2013). Invertebrates, ecosystem services and climate change. *Biological Reviews* 88 (2): 327–348.

Pujolar, J.M., Marino, I.A., Milan, M. et al. (2012). Surviving in a toxic world: transcriptomics and gene expression profiling in response to environmental pollution in the critically endangered European eel. *BMC Genomics* 13: 1–21.

Regoli, F. and Principato, G. (1995). Glutathione, glutathione-dependent and antioxidant enzymes in mussel, *Mytilus galloprovincialis*, exposed to metals under field and laboratory conditions: implications for the use of biochemical biomarkers. *Aquatic Toxicology* 31 (2): 143–164.

Reid, N.M. and Whitehead, A. (2016). Functional genomics to assess biological responses to marine pollution at physiological and evolutionary timescales: toward a vision of predictive ecotoxicology. *Briefings in Functional Genomics* 15 (5): 358–364.

Richardson, B.J., Mak, E., De Luca-Abbott, S.B., Martin, M., McClellan, K., and Lam, P.K. (2008). Antioxidant responses to polycyclic aromatic hydrocarbons and organochlorine pesticides in green-lipped mussels (Perna viridis): do mussels "integrate" biomarker responses?. *Marine Pollution Bulletin* 57(6-12): 503–514.

Ruiz-Picos, R.A., Sedeño-Díaz, J.E., and López-López, E. (2015). Histopathological indicators in fish for assessing environmental stress. In: *Environmental Indicators* (ed. R. Armon and O. Hänninen). Dordrecht: Springer https://doi.org/10.1007/978-94-017-9499-2_38.

Saenz, L.A., Seibert, E.L., Zanette, J. et al. (2010). Biochemical biomarkers and metals in *Perna perna* mussels from mariculture zones of Santa Catarina, Brazil. *Ecotoxicology and Environmental Safety* 73 (5): 796–804.

Salamat, N. and Zarie, M. (2016). Fish histopathology as a tool for use in marine environment monitoring: a review. *Comparative Clinical Pathology* 25: 1273–1278.

Samanta, P., Pal, S., Mukherjee, A.K., and Ghosh, A.R. (2014). Biochemical effects of glyphosate based herbicide, Excel Mera 71 on enzyme activities of acetylcholinesterase (AchE), lipid peroxidation (LPO), catalase (CAT), glutathione-S-transferase (GST) and protein content on teleostean fishes. *Ecotoxicology and Environmental Safety* 107: 120–125.

Sánchez, P. and Tarazona, J.V. (2002). Development of a multispecies system for testing reproductive effects on aquatic invertebrates. Experience with *Daphnia magna*, *Chironomus prasinus* and *Lymnaea peregra*. *Aquatic Toxicology* 60 (3–4): 249–256.

Selander, E., Thor, P., Toth, G., and Pavia, H. (2006). Copepods induce paralytic shellfish toxin production in marine dinoflagellates. *Proceedings of the Royal Society B: Biological Sciences* 273 (1594): 1673–1680.

Sellami, B., Louati, H., Dellali, M. et al. (2014). Effects of permethrin exposure on antioxidant enzymes and protein status in Mediterranean clams *Ruditapes decussatus*. *Environmental Science and Pollution Research* 21: 4461–4472.

Sepúlveda, M.S., Ralston-Hooper, K.J., Sanchez, B.C. et al. (2011). Use of proteomic and metabolomic techniques in ecotoxicological research. *General, Applied and Systems Toxicology* https://doi.org/10.1002/9780470744307.gat215.

Shah, N., Khan, A., Ali, R. et al. (2020). Monitoring bioaccumulation (in gills and muscle tissues), hematology, and genotoxic alteration in *Ctenopharyngodon idella* exposed to selected heavy metals. *BioMed Research International* 2020 (1): 6185231.

Sharma, M., Thakur, J., Verma, S., and Sharma, P. (2019). Behavioural responses in effect to chemical stress in fish: a review. *International Journal of Fisheries and Aquatic Studies* 7 (1): 01–05.

Sharma, P., Ngo, H.H., Khanal, S. et al. (2021). Efficiency of transporter genes and proteins in hyperaccumulator plants for metals tolerance in wastewater treatment: sustainable technique for metal detoxification. *Environmental Technology & Innovation* 23: 101725.

Sharma, S., Bhardwaj, A., Thakur, M. et al. (2023). Understanding microplastic pollution of marine ecosystem: a review. *Environmental Science and Pollution Research* https://doi.org/10.1007/s11356-023-28314-1.

Sies, H. and Jones, D.P. (2020). Reactive oxygen species (Ros) as pleiotropic physiological signalling agents. *Nature Reviews Molecular Cell Biology* 21 (7): 363–383.

Sıkdokur, E., Belivermiş, M., Sezer, N. et al. (2020). Effects of microplastics and mercury on manila clam Ruditapes philippinarum: feeding rate, immunomodulation, histopathology and oxidative stress. *Environmental Pollution* 262: 114247.

Simpson, S.L. and Spadaro, D.A. (2016). Bioavailability and chronic toxicity of metal sulfide minerals to benthic marine invertebrates: implications for deep sea exploration, mining and tailings disposal. *Environmental Science & Technology* 50 (7): 4061–4070.

Smits, J.E. and Fernie, K.J. (2013). Avian wildlife as sentinels of ecosystem health. *Comparative Immunology, Microbiology and Infectious Diseases* 36 (3): 333–342.

Srikanth, K., Pereira, E., Duarte, A.C., and Ahmad, I. (2013). Glutathione and its dependent enzymes' modulatory responses to toxic metals and metalloids in fish – a review. *Environmental Science and Pollution Research* 20: 2133–2149.

Šrut, M. (2021). Ecotoxicological epigenetics in invertebrates: Emerging tool for the evaluation of present and past pollution burden. *Chemosphere* 282: 131026.

Stentiford, G.D., Longshaw, M., Lyons, B.P. et al. (2003). Histopathological biomarkers in estuarine fish species for the assessment of biological effects of contaminants. *Marine Environmental Research* 55 (2): 137–159.

Stierum, R., Heijne, W., Kienhuis, A. et al. (2005). Toxicogenomics concepts and applications to study hepatic effects of food additives and chemicals. *Toxicology and Applied Pharmacology* 207 (2): 179–188.

Strotmann, U., Pastor Flores, D., Konrad, O., and Gendig, C. (2020). Bacterial toxicity testing: modification and evaluation of the luminescent bacteria test and the respiration inhibition test. *Processes* 8 (11): 1349.

Strubbia, S., Lyons, B.P., and Lee, R.J. (2019). Spatial and temporal variation of three biomarkers in *Mytilus edulis*. *Marine Pollution Bulletin* 138: 322–327.

Sturchio, E., Ficociello, B., Minoia, C. et al. (2008). Gene expression and environmental exposure to xenobiotics: overview and applications. *Giornale Italiano di Medicina del Lavoro ed Ergonomia* 30 (2): 101–114.

Sumudumali, R.G.I. and Jayawardana, J.M.C.K. (2021). A review of biological monitoring of aquatic ecosystems approaches: with special reference to macroinvertebrates and pesticide pollution. *Environmental Management* 67: 263–276. https://doi.org/10.1007/s00267-020-01423-0.

Taylor, D.L., Linehan, J.C., Murray, D.W., and Prell, W.L. (2012). Indicators of sediment and biotic mercury contamination in a southern New England estuary. *Marine Pollution Bulletin* 64 (4): 807–819.

Telesh, I.V. (2004). Plankton of the Baltic estuarine ecosystems with emphasis on Neva Estuary: a review of present knowledge and research perspectives. *Marine Pollution Bulletin* 49 (3): 206–219.

Tevell Åberg, A., Björnstad, K., and Hedeland, M. (2013). Mass spectrometric detection of protein-based toxins. *Biosecurity and Bioterrorism: Biodefense Strategy, Practice, and Science* 11 (S1): S215–S226.

Thomas, P.J., Newell, E.E., Eccles, K. et al. (2021). Co-exposures to trace elements and polycyclic aromatic compounds (PACs) impacts North American river otter (*Lontra canadensis*) baculum. *Chemosphere* 265: 128920. https://doi.org/10.1016/J.CHEMOSPHERE.2020.128920.

Torres, M.A., Barros, M.P., Campos, S.C. et al. (2008). Biochemical biomarkers in algae and marine pollution: a review. *Ecotoxicology and Environmental Safety* 71 (1): 1–15.

Toto, A., Wild, P., Graille, M. et al. (2022). Urinary Malondialdehyde (Mda) concentrations in the general population-a systematic literature review and meta-analysis. *Toxics* 10 (4).

Turesky, R.J. and Lu, K. (2020). Biomarkers of environmental toxicants: exposure and biological effects. *Toxics* 8 (2): 37.

Valiela, I., McClelland, J., Hauxwell, J. et al. (1997). Macroalgal blooms in shallow estuaries: controls and ecophysiological and ecosystem consequences. *Limnology and Oceanography* 42 (5part2): 1105–1118.

Van De Merwe, J.P. and Leusch, F.D. (2015). A sensitive and high throughput bacterial luminescence assay for assessing aquatic toxicity – the BLT-screen. *Environmental Science: Processes & Impacts* 17 (5): 947–955.

Van der Oost, R., Beyer, J., and Vermeulen, N.P. (2003). Fish bioaccumulation and biomarkers in environmental risk assessment: a review. *Environmental Toxicology and Pharmacology* 13 (2): 57–149.

Vasanthi, L.A., Revathi, P., Mini, J., and Munuswamy, N. (2013). Integrated use of histological and ultrastructural biomarkers in *Mugil cephalus* for assessing heavy metal pollution in Ennore estuary, Chennai. *Chemosphere* 91 (8): 1156–1164.

Vasconcelos, V., Azevedo, J., Silva, M., and Ramos, V. (2010). Effects of marine toxins on the reproduction and early stages development of aquatic organisms. *Marine Drugs* 8 (1): 59–79.

Vega-Corrales, L. and Marín-Vindas, C. (2021). Effect of metal concentration on growth and luminescence of luminous bacteria strains isolated from golfo de Nicoya, Costa Rica. *Journal of Marine and Coastal Sciences* 13 (1): 27–38.

Vernon, E.L. and Jha, A.N. (2019). Assessing relative sensitivity of marine and freshwater bivalves following exposure to copper: application of classical and novel genotoxicological biomarkers. *Mutation Research, Genetic Toxicology and Environmental Mutagenesis* 842: 60–71.

Viarengo, A. and Canesi, L. (1991). Mussels as biological indicators of pollution. *Aquaculture* 94 (2–3): 225–243.

Viarengo, A., Pertica, M., Canesi, L. et al. (1989). Lipid peroxidation and level of antioxidant compounds (GSH, vitamin E) in the digestive glands of mussels of three different age groups exposed to anaerobic and aerobic conditions. *Marine Environmental Research* 28 (1–4): 291–295.

Vieira, L.R. and Morgado, F. (2020). Fundamentals of toxicology and ecotoxicology: promoting education and lifelong learning to protect the environment and human health. In: *Quality Education. Encyclopedia of the UN Sustainable Development Goals* (ed. W. Leal Filho, A.M. Azul, L. Brandli, et al.). Cham: Springer https://doi.org/10.1007/978-3-319-69902-8_120-1.

Vinodhini, R. and Narayanan, M. (2009). Heavy metal induced histopathological alterations in selected organs of the *Cyprinus carpio* L.(Common Carp). *International Journal of Environmental Research* 3 (1): 95–100.

Vogelbein, W.K. and Unger, M.A. (2006). Liver carcinogenesis in a non-migratory fish: the association with polycyclic aromatic hydrocarbon exposure. *Bulletin of the European Association of Fish Pathologists* 26 (1): 11.

Voice, T.C., Rice, C.P., and Weber, W.J. (1983). Effect of solids concentration on the sorptive partitioning of hydrophobic pollutants in aquatic systems. *Environmental Science & Technology* 17 (9): 513–518.

Wang, X., Lin, J.M., Liu, M.L., and Cheng, X.L. (2009). Flow-based luminescence-sensing methods for environmental water analysis. *TrAC Trends in Analytical Chemistry* 28 (1): 75–87.

Wang, Y., Branicky, R., Noë, A., and Hekimi, S. (2018a). Superoxide dismutases: dual roles in controlling ROS damage and regulating ROS signaling. *Journal of Cell Biology* 217 (6): 1915–1928.

Wang, Y., Wang, S., Feng, X.N. et al. (2018b). The impact of exposure route for class-based compounds: a comparative approach of lethal toxicity data in rodent models. *Drug and Chemical Toxicology* 41 (1): 95–104. https://doi.org/10.1080/01480545.2017.1320405.

Wilcox, C., Mallos, N.J., Leonard, G.H. et al. (2016). Using expert elicitation to estimate the impacts of plastic pollution on marine wildlife. *Marine Policy* 65: 107–114.

Wolska, L., Sagajdakow, A., Kuczyńska, A., and Namieśnik, J. (2007). Application of ecotoxicological studies in integrated environmental monitoring: possibilities and problems. *TrAC Trends in Analytical Chemistry* 26 (4): 332–344.

Won, E.J. and Lee, J.S. (2014). Gamma radiation induces growth retardation, impaired egg production, and oxidative stress in the marine copepod *Paracyclopina nana*. *Aquatic Toxicology* 150: 17–26.

Won, E.J., Han, J., Lee, Y. et al. (2015). In vivo effects of UV radiation on multiple endpoints and expression profiles of DNA repair and heat shock protein (Hsp) genes in the cycloid copepod *Paracyclopina nana*. *Aquatic Toxicology* 165: 1–8.

Worm, B., Barbier, E.B., Beaumont, N. et al. (2006). Impacts of biodiversity loss on ocean ecosystem services. *Science* 314 (5800): 787–790.

Wurl, O. and Obbard, J.P. (2004). A review of pollutants in the sea-surface microlayer (SML): a unique habitat for marine organisms. *Marine Pollution Bulletin* 48 (11–12): 1016–1030.

Xia, Z., Idowu, I., Kerr, E. et al. (2021). New approaches to reduce sample processing times for the determination of polycyclic aromatic compounds in environmental samples. *Chemosphere* 274: 129738.

Yang, X., Zhang, P., Li, W. et al. (2018). Evaluation of four seagrass species as early warning indicators for nitrogen overloading: implications for eutrophic evaluation and ecosystem management. *Science of the Total Environment* 635: 1132–1143.

Yi, X., Kim, E., Jo, H.J. et al. (2011). A comparative study on toxicity identification of industrial effluents using *Daphnia magna*. *Bulletin of Environmental Contamination and Toxicology* 87: 319–323.

Yoo, S.M. and Lee, S.Y. (2016). Optical biosensors for the detection of pathogenic microorganisms. *Trends in Biotechnology* 34 (1): 7–25.

Zala, S.M. and Penn, D.J. (2004). Abnormal behaviours induced by chemical pollution: a review of the evidence and new challenges. *Animal Behaviour* 68 (4): 649–664.

Zbinden, G. (1993). The concept of multispecies testing in industrial toxicology. *Regulatory Toxicology and Pharmacology* 17 (1): 85–94.

Zhang, C., Chen, X., Wang, J., and Tan, L. (2017). Toxic effects of microplastic on marine microalgae *Skeletonema costatum*: interactions between microplastic and algae. *Environmental Pollution* 220: 1282–1288.

Zhou, L., Li, M., Zhong, Z. et al. (2021). Biochemical and metabolic responses of the deep-sea mussel *Bathymodiolus platifrons* to cadmium and copper exposure. *Aquatic Toxicology* 236: 105845.

Zhou, Q., Wang, S., Liu, J. et al. (2022). Geological evolution of offshore pollution and its long-term potential impacts on marine ecosystems. *Geoscience Frontiers* 13 (5): 101427.

Zhu, X., Zhu, L., Chen, Y., and Tian, S. (2009). Acute toxicities of six manufactured nanomaterial suspensions to *Daphnia magna*. *Journal of Nanoparticle Research* 11: 67–75.

15

Marine Sediment Remediation Through Tiered Risk Assessment Approach

Divya Pal and Jonas Gunnarsson

Department of Ecology, Environment and Plant Sciences (DEEP), Stockholm University, Svante Arrhenius Väg 20A 114 18, Stockholm, Sweden

15.1 Introduction

Sediment tends to accumulate both inorganic [i.e. metals or metal(loid)s] and hydrophobic organic or persistent organic pollutants (POPs) that pose to environmental contamination and risk to aquatic organisms. The POPs derived from housecleaning products, pharmaceuticals, pesticides, personal care products, and chemically transformed constituents from the aforementioned materials can create a serious threat to aquatic ecosystems. These chemicals find their way into the sediment either through urban runoff (including sewer overflow also) or industrial discharges (Reible et al. 2006; Leppänen et al. 2017). Furthermore, inorganic pollutants can also be geologically originated (Pal and Maiti 2018a).

The most important sinks in the aquatic environment for POPs and inorganic pollutants are the sediments. Approximately, more than 99% of total POPs and inorganic pollutants, either adsorb or make a surface complexation with sediment particles (Lee et al. 2003). However, alterations in aquatic physiochemical conditions can induce the release of bound pollutants from sediment into the overlying water, thereby turning the sediment into a secondary source of contaminants (Pal and Maiti 2018b; Pal and Maiti 2019). Due to their lipophilic characteristics and resistance to biotransformation, these pollutants are also prone to bioaccumulation and biomagnification within the food chain (Khim and Hong 2014). These chemicals can cause a wide range of toxic effects on benthic and aquatic fauna. For instance, the white Pacific shrimp (*Litopenaeus vannamei*), a prominent species in Asian aquaculture, has been notably impacted by the bioaccumulation of these contaminants, making it especially susceptible to pollutant uptake from sedimentary sources owing to its benthic feeding behavior (Varadharajan and Pushparajan 2013; Hidayati et al. 2021).

Thus, bioaccumulation of POPs and metals in aquatic organisms poses a dual threat to both the integrity of the aquatic ecosystem and the stability of the blue economy (Chiesa et al. 2019). In recent years, the aquaculture industry has witnessed remarkable expansion, driven by its ability to offer sustainable, safe, and dependable aquatic products (Costello et al. 2020). According to the Food and Agriculture Organization (FAO 2018), mariculture contributed 28.7 million tonnes (equivalent to 67.4 billion USD) to global aquaculture production in 2016 (Nobakht-Kolur et al. 2021), underscoring its pivotal role in meeting growing demands for seafood.

Coastal and Marine Pollution: Source to Sink, Mitigation and Management, First Edition.
Edited by Meththika Vithanage, Sameera M. Samarasekara, Bryan D. James, and Christopher M. Reddy.

Therefore, the presence of pollutants in sediment not only poses a threat to aquatic organisms but also carries a potential risk to human health and global blue economy.

Hence, it becomes imperative to both decrease ongoing contaminant emissions and to better managing already historically contaminated sediments in order to limit the accessibility of contaminants within the aquatic food chain. The predominant approaches for sediment remediation comprise two methods: (i) sediment dredging coupled with *ex situ* treatment or disposal, and (ii) *in situ* treatment or sediment capping. Dredging or removal of contaminated sediment is one of the most commonly used techniques in shallow areas, for example, in contaminated harbors, where it also serves a way for maintaining a good sailing depth. Over 200 million cubic meters of contaminated sediment are dredged annually in Europe alone (EMODnet 2015). However, managing dredged contaminated sediment poses a significant economic and environmental challenge, necessitating billions of dollars and considerable space for sediment handling (USEPA 2005; Ekman 2017). Moreover, the issue of contaminant resuspension during dredging exacerbates environmental concerns (Pal and Maiti 2020). The *in situ* method is broadly categorized into two main types: (i) conventional capping, that is, isolation capping, where the contaminated sediment is covered with a thick layer (30–50 cm) of sand or gravel and eventually geo-textiles as a barrier to prevent the contaminants from leaching through; and (ii) thin-layer capping (2–5 cm) with active sorbents, such as activated carbon, where the capping act as a semipermeable barrier to let through nutrients but keeping the contaminants sequestered in the sediment (SGI 2016). Utilizing thick layers of sand, silt, or clay for capping serves as an alternative to dredging, but it requires a significant amount of inert material for capping, which may have adverse effects on the seabed (Näslund et al. 2012). In thin-layer capping, materials with high sorption abilities such as activated carbon (AC) and calcareous materials are predominantly utilized (Ghosh et al. 2011; Pal and Maiti 2019; cöm et al. 2021). However, some recent research highlights the moderate adverse effects of thin-layer capping on benthic organisms. These effects encompass disrupted feeding behavior, reduced growth rates, and increased mortality rates among certain species (Trannum et al. 2021).

Samuelsson et al. (2017) reported the negative effect of AC+clay-treated sediment on *Amphiura filiformis* and suggested that utilization of powdered AC has significantly transformed the habitat, making it less conducive for macrofauna species. In another study, the application of powdered AC negatively impacted benthic organisms, with findings indicating that the biological response depends on the ingestibility of AC particles. Ingestible particles may induce starvation by reducing the availability of co-ingested food or by prompting the rejection of AC-treated sediment as a food source (Rämö et al. 2021).

To mitigate ecological and economic impacts, a thorough examination of pollutants in sediment and their effects on aquatic ecology and human health risks is crucial before implementing sediment treatment methods (Apitz et al. 2005). Environmental regulatory agencies in numerous countries prioritize accurate measurement of these parameters before sediment remediation (USEPA 1998; Chapman and Wang 2001; Sparrevik and Breedveld 2010). The sediment quality triadapproach integrates measurements of diverse chemicals, potential toxicities, and benthic community structures, providing a comprehensive method for assessment. While it directly evaluates sediment quality for all concerning chemicals, it does not establish a direct cause-and-effect relationship between individual chemical concentrations and adverse biological effects. This method could be enhanced by the lines of evidence approach, which explores physiological components within resident aquatic communities and their interactions (Khim et al. 2014).

Exploring biomagnification patterns and utilizing sediment toxicity identification and evaluation (TIE) techniques are common practices in assessing sediment conditions. To assess the adverse effects of these compounds on aquatic biota, neither biological tests nor chemical tests alone suffice. Integrated approaches are necessary to comprehend the ecological state of contaminated sediment. Thus, this chapter aims to discuss the detailed steps involved in the triad approach and its role in decision-making for sediment remediation.

15.2 Sediment Remediation Techniques

15.2.1 Active or Thin-Layer Capping

15.2.1.1 Remediation with Alkali-Activated Materials (AAMs)

AAMs result from activating alumina–silica-rich materials with an alkali solution. Aluminosilicate-rich ores like calcined kaolin, bentonite, or albite, along with industrial by-products such as fly ashes, metallurgical slags, or mine tailings, can be utilized in the production of AAMs (Provis 2018). Alkali-metal silicates, hydroxides, or carbonates dissolved in water are the preferred activators for preparing AAMs under ambient conditions of room temperature and atmospheric pressure. These traits imply that AAMs might provide improved energy efficiency and cost-effectiveness when compared to other remediation materials.

The presence of nanoscale pore structures on the surface of AAMs facilitates the adsorption of contaminants. Geopolymers, characterized by an amorphous structure and a Si/Al ratio > 1.5, exhibit rings consisting of at least eight Al or Si atoms. This distinct configuration facilitates a thorough ion-exchange process, effectively converting alkali ions such as Na^+ ions to NH_4^+ or Co^{2+} ions with unparalleled efficiency, thereby highlighting the materials' remarkable selectivity and adaptability. Furthermore, under high pH conditions, the deprotonation of existing silanol groups (Si-OH) to $Si\text{-}O^{2-}$ aids in the adsorption of metal cations (Habbaba and Plank 2012). Nevertheless, the adsorption of anionic contaminants is controlled by the layered double hydroxides present in AAMs, particularly in calcium (Ca)-rich AAMs with sufficient magnesium (Mg) content (Ke et al. 2017). The aforementioned attributes, including their adsorption and pH-regulating capabilities, render AAMs an interesting choice for use as an active capping material.

In a recent study, alkali-activated blast furnace slag AAMs, activated by sodium silicate and hydroxide, were utilized as a capping material for in situ remediation of Al, Zn, Cu, Cr, Ti, Fe, Sr, and Ni metals in the sediment. The results disclosed that in bioavailable fraction of sediment of Al, Zn, Ni, Cu, and Fe metals were decreased significantly than unamended sediments (Kutuniva et al. 2019). In a mesocosm study fly ash-based zeolite (FABZ), activated with sodium hydroxide tested for the immobilization of Cd metal from the sediment. The study indicated that the utilization of 10% FABZ resulted in the immobilization of approximately 80% of Cd, compared to untreated sediment. Moreover, there was an observed reduction of Cd around 91% in the bioavailable fraction of sediment (Prabhakar and Samadder 2020). Thus, AAMs could be used as a potential sorbent for immobilization of several metals in sediment and could be cost-effective too, but their capping efficiency on a long-term basis remains to be studied.

15.2.1.2 Carbonaceous Materials (CMs) for Capping

The utilization of CMs in active capping holds significant potential as a sustainable, economical, and durable approach for remediating contaminated sediment, including POPs. The most commonly used carbonaceous materials includes activated carbon (AC), biochar, bituminous coal-based AC, magnetic AC, and other modified CMs (Rakowska et al. 2012; Han et al. 2015; Nybom et al. 2016) (Figure 15.1). Within this category, ACs have always attracted the researchers, due to their strong sorption capabilities, which frequently exceed those of organic carbon (OC)

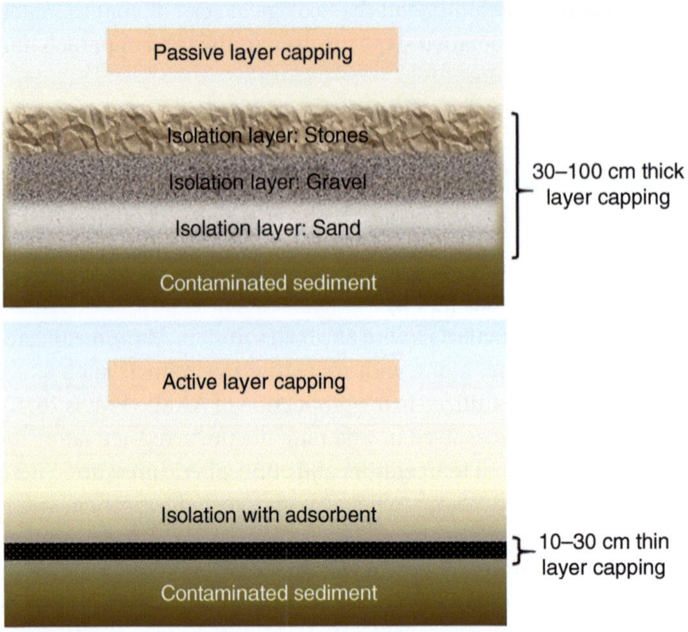

Figure 15.1 Diagram illustrating passive and active capping setups.

Table 15.1 Characteristics of different AC-derived products employed for in situ sediment remediation.

Article	Characteristics	Aim pollutant	Remediation mechanisms
BOS-100	GAC infused by zero valent iron (ZVI)	Chlorinated compounds	Abiotic reductive dechlorination
BOS-200®	Blend of nutrients, electron acceptors, and versatile bacteria	Petroleum Hydrocarbons	Aerobic and anaerobic biostimulation
CAT-100®	BOS-100® and reductive dichlorination bacterial strains	Chlorinated compounds	Abiotic and biotic reductive dechlorination
COGAC®	GAC or PAC mixed with calcium peroxide, and sodium persulfate	Chlorinated compounds and petroleum hydrocarbons	Chemical oxidation and aerobic and anaerobic biostimulation
PlumeStop®	Colloidal AC suspension with an organic stabilizer, co-applied with hydrogen or oxygen release compounds, and/or corresponding bacterial strains	Chlorinated compounds or petroleum hydrocarbons	Enhanced biotic reductive dechlorination for chlorinated solvents and aerobic biodegradation for petroleum hydrocarbons
Carbo-Iron®	Colloidal AC impregnated with ZVI	Chlorinated compounds	Abiotic reductive dichlorination

Soure: Adapted from USEPA (2018)/U.S. Environmental Protection Agency (EPA)/Public Domain.

by up to 100-fold (USEPA 2014). This characteristic renders them particularly adept at sorbing a wide range of contaminants, including POPs such as polyaromatic hydrocarbons (PAHs), polychlorinated biphenyls (PCBs), dioxins/furans, pesticides, and metals/metalloids, thereby reducing their bioavailability (listed in Table 15.1).

CMs can be produced through the thermochemical conversion of biomass in an oxygen-limited environment (IBI 2012). In thermochemical conversion, supplied energy initiates activation reactions within the biomass, prompting a decrease of the carbon atoms from the pore walls and thereby increasing the number of pores at active sites. This process generates a porous structure, ultimately resulting in an increase in total pore volume and a high specific surface area (Li et al. 2022). Based on their particle size characteristics, AC can be classified into two main categories: powdered activated carbon (PAC) and granular activated carbon (GAC). GAC typically exhibits a particle size range of 0.2–5 mm, while PAC particles are typically <0.188 mm in diameter, with high specific surface area generally >2200 m^2/g (Ghosh et al. 2011). Owing to their specific properties, the addition of 2–5% of CMs by the dry weight of sediment within the top 10–30 cm can significantly reduce the bioavailability of POPs and other contaminants in the aquatic environment, effectively (Ghosh et al. 2011). Multiple biodynamic models integrating species-specific physiological parameters have effectively elucidated the diminished uptake capacity of POPs and other contaminants in biota tissues and in pore water (Bonaglia et al. 2019; Pal and Maiti 2019; Yan et al. 2020).

15.2.1.3 Organo-clay Materials (OCMs) for Capping

The OCMs are composed of a modified clay matrix primarily comprising "montmorillonite," a benthonic clay imparting notable characteristics such as high surface area, significant cation exchange capacity (CEC), expansive swelling behavior, and elevated adsorption capabilities, particularly for heavy metals (Bortone et al. 2020). Owing to significant CEC, extensive surface area, fine particle size, thin flake structure, and pronounced swelling capacity, smectite clays are widely favored for the production of OCMs. During the synthesis of OCMs, surfactant molecules primarily experience adsorption within the interlayer space through ion exchange at low levels of surfactant loading. With higher loading, surfactants are adsorbed via both ion exchange and hydrophobic interactions (Xi et al. 2007; Sarkar et al. 2012). In contemporary practices, cationic surfactants, especially those based on ammonium, are widely utilized for clay modification. Additionally, commercially available modifiers such as alkyl carboxy (Sheng et al. 1999), pyridium (Chigwada et al. 2006), alkyl carbazole (Chigwada et al. 2006), and other reactive agents are also utilized in this process. The intercalation process not only shifts their affinity from hydrophilic to hydrophobic but also notably augments the basal spacing within the interlayers (Sarkar et al. 2012). The expanded basal spacing, coupled with the hydrophobic properties of the surface, renders organoclays effective agents for sequestering both organic and inorganic contaminants in various environments.

The adsorption of POPs by OCMs primarily entails the partitioning of organic compounds. In this process, the quaternary amine groups present on the surface of OCMs act as hydrophobic organic phases, enhancing the sorption of POPs to these groups and facilitating their absorption (Richards and Bouazza 2007). On the other hand, the adsorption of inorganic pollutants is facilitated by the formation of inner-sphere complexes with silanol and aluminol groups situated at the surface periphery of the clay particles. Moreover, modification of OCMs with anionic surfactants, such as sodium dodecyl sulfate, could improve the sorption of metal ions by encouraging the formation of surface-cationic complexes (Sarkar et al. 2012). Although the use of OCMs shows promise for sediment remediation, the widespread application of surfactants in their modification may result in secondary toxicity in aquatic ecosystem (Yoo et al. 2004).

15.2.1.4 Other Materials

Polonite® is a sorbent used as a filter to retain phosphate in wastewater treatment plants and also for sediment treatment. It originates from a sedimentary rock called opoka that undergoes thermal treatment at 900 °C, followed by crushing and sieving. As a result of thermal treatment, the $CaCO_3$

in opoka undergoes a transformation into CaO, offering great potential for the treatment of nutrient-rich sediment, particularly phosphorus (P). Polonite is typically manufactured within the 2–5.6 mm fraction, resulting in excellent hydraulic characteristics (Renman and Renman 2010). A study conducted at a lake sediment in Sweden used the polonite, an activated calcium silicate sorbent for the immobilization of phosphorus (P), organic contaminants, and heavy metals. The study concluded that the application of polonite has successfully reduced the phosphate ($PO_4^{3}-$) flux in anoxic conditions by almost 50% and a load of Cd and Zn was reduced by 67% and 89%, respectively, however, found polonite ineffective for organic contaminants remediation (Wikström et al. 2021).

Another two materials, limestone (LS) and steel slag (SS), hold great potential for the remediation of heavy metals from sediment due to their high calcium content. The surface area of LS and SS can be > 0.08 and 1.19, correspondingly (Park et al. 2019). The adsorption of heavy metals relies on two primary processes: precipitation and sorption. This interaction is heavily contingent upon pH levels. The addition of LS and SS elevates the pH levels, which promote the adsorption of cationic heavy metals onto metal oxides, leading to the formation of metal precipitates (Park et al. 2018).

15.2.2 Passive Layer Capping

In passive layer capping of sediment, inert materials such as sand and silt are majorly used as a barrier between contaminants and overlying water. This technique demonstrates high efficiency, particularly in cases where pollutants are tightly bound to solid components, and where the site-specific transport mechanisms at the operational level do not effectively promote rapid migration of contaminants through the material. Passive capping utilizing sand has been scientifically validated as an insulated barrier, effectively decreasing the dispersion of contaminated and unconsolidated muddy harbor sediments and facilitating the restoration of coastal ecosystems (Simpson et al. 2002; Bruhn et al. 2020). An experimental study has demonstrated that capping muddy sediment with a 10-cm layer of sand efficiently prevents contaminants from dispersing into the overlying water without inducing mixing (Flindt et al. 2021). In a separate large-scale restoration study carried out in Odense Fjord, Denmark, the application of a 10 cm thick sand cap over organic-rich muddy sediment resulted in a reduction of fine sediment particle mixing in the overlying water by 9–22%. Furthermore, this intervention induced a shift in the benthic community from low to high diversity, leading to an enhancement in ecosystem function (Oncken et al. 2022). However, utilization of sand capping may prove insufficient for sites with high contamination levels or where transport rates are accelerated by phenomena such as porewater upwelling or tidal pumping effects.

The utilization of dredged materials from nearby areas as a passive barrier has garnered significant attention due to its cost-effectiveness and alignment with the principles of circular economy. A research project undertaken in Ward Cove, Alaska, implemented a stratified layer of dredged material, measuring between 15 and 30 cm in thickness, upon sediment contaminated with ammonia and 4-methylphenol, aiming to facilitate the colonization of benthic macroinvertebrates. Subsequent long-term monitoring revealed that the application of dredged material onto contaminated sediment effectively lowered ammonia concentrations by over 10% and 4-methylphenol concentrations by more than 20% compared to benchmark values. Additionally, a significant increase in taxa richness, ranging from 3 to ≥20, was reported in the remediated sites (Becker et al. 2009). However, careful consideration must be given to potential risks associated with introducing additional contaminants and the challenges in handling fine-grained sediment.

15.2.3 Dredging and Other Methods for Sediment Remediation

The extraction of contaminated sediment through dredging has become a popular long-term solution for decreasing environmental risks from contaminated sediments. In contaminated hotspot areas, despite dredging efforts, residual contaminant often persists in the sediment surface due to incomplete excavation, mixing of more heavily contaminated deeper sediment with clean surficial sediment, and the resuspension of contaminated sediment into the water column. However, the effectiveness of sediment excavation heavily relies on the availability of technical infrastructure, water depth, high operational cost, and other physical constraints, which may restrict its applicability on a broader scale (Palermo and Hays 2013).

Sediment washing *ex situ* represents an alternative method for treating contaminated sediment, designed to effectively remove contaminants adhering to the surface of sediment particles to eliminate soluble pollutants (Akcil et al. 2015). This method can be used for removal of heavy metals, especially from dredged sediment (Wang et al. 2017; Pal and Hogland 2022). Metal removal can be accomplished through a two-step process: first, physical treatment involves washing sediment with water to remove solubilized metals. Second, chemical treatment entails the application of various additives, including nitric acid, hydrochloric acid, chelating agents such as EDTA and acetic acid, and surfactants like rhamnolipid, which enhance the efficacy of the sediment washing method (Löser et al. 2007; Peng et al. 2009). Sediment washing is especially effective for extracting heavy metals that are bound to sediment particles through weak associations with carbonates, oxides, and hydroxides (Akcil et al. 2015). Consequently, washing is unsuitable for sediment containing fine particles, as metals are firmly bound to sediment particles owing to the high availability of surface area (Peng et al. 2009). Numerous alternative methods exist, including electrochemical separation, thermal treatment of sediment, and biological treatment, among others. However, these methodologies are not addressed in this chapter as their focus is primarily on the discussion of risk assessment relating to contaminated sediments.

15.3 Risk Assessment-Based Strategy

The aforementioned sediment remediation methods highlight concerns regarding cost and time and also underscore the absence of assurances regarding potential future pollutant release. Consequently, risk assessment is integral to both remedial methods and feasibility for sites-specific investigation of contaminated sediment. It serves as a fundamental step in assessing the necessity for sediment remediation, which often driven by concerns about potential risks to human health and ecosystems, particularly through the consumption of aquatic food sources such as shrimp and fish. The primary goal of traditional sediment remediation methods was to lower contaminant concentrations in the sediment, often neglecting to consider the potential adverse effects associated with the selected remediation method itself. Expanding the decision framework to include factors like short- and long-term ecological impacts, residual effects, and efficacy allows for a more thorough assessment of possible less intrusive remedial measures, such as *in situ* treatments. Criteria for selecting an appropriate sediment remediation method should consider efficiency, benefits, adverse effects, and long-term cost implications to ensure the feasibility and sustainability of the chosen approach. However, the primary risk often stems from exposure to porewater concentrations rather than total sediment load. Hence, engaging in remediation practices without prior risk assessment may prove ineffective in achieving long-term mitigation objectives. The adoption of a risk-based approach to establish clean-up thresholds allows for the evaluation of site-specific conditions and the bioavailability of contaminants.

15.3.1 Ecological Risk Assessment (ERA)

ERA is a structured methodology involving the systematic collection, arrangement, and analysis of environmental data to quantitatively estimate the potential risk of contamination to ecosystems. The ERA has two stages: (i) the first stage aims at anticipating potential environmental impacts of hazardous substances and is often linked to regulatory processes, particularly within the European Union where it enables a prospective assessment of substances. This step is preferably conducted before the substance is released into the environment. (ii) The second stage involves estimating the effects of sites already contaminated, that is, a retrospective risk assessment by assessing changes in population dynamics or ecosystems. Thus, this phase can be described as an impact assessment rather than a risk assessment.

The ERA is typically conducted in tiers or stages, with each tier advancing toward more detailed or site-specific measurements in order to decrease the uncertainty in evaluating risks. In the context of sediment risk assessment, Tier 1 typically involves establishing remediation thresholds using preliminary screening criteria that are usually derived by comparing contaminated concentrations to background levels. The primary objective is to maintain sediment concentrations below the thresholds outlined by the two lowest screening levels, which are in accordance with the Environmental Quality Standards (EQS) set by the European Union. When Tier 2 and Tier 3 investigations are conducted, they incorporate more and more site-specific variables such as partition coefficients and sediment organic content. As a result, the findings of these investigations may prompt consideration of elevated remediation thresholds. However, insufficient data collection efforts by project stakeholders often hinder the effectiveness of this method, resulting in incomplete risk assessments. Subsequently, Tier 3 remediation thresholds are formulated based on the bioavailability and specific site conditions (Figure 15.2).

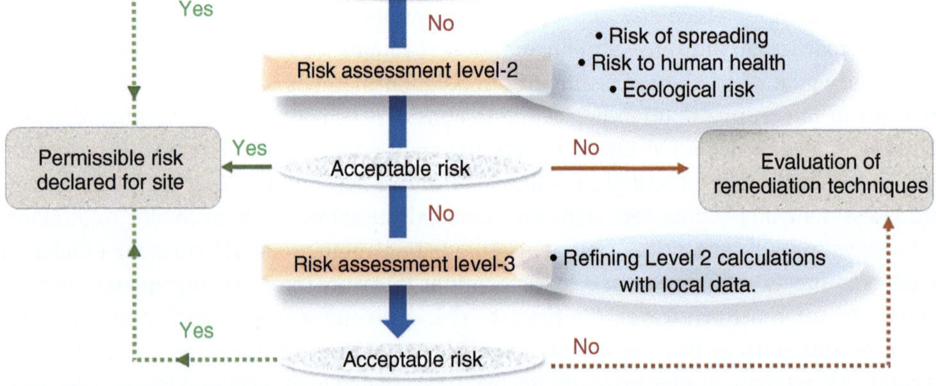

Figure 15.2 Outline of sediment risk assessment framework.

15.3.2 Tier-Based/TRIADTriad Approach for Sediment Toxicity Assessment

Assessing the distribution of POPs, metals, and other endocrine-disrupting chemicals (EDCs) in both sediment and the water column is crucial for understanding their behavior and impact within aquatic ecosystems. However, due to the intricate nature of environmental samples, fully comprehending or forecasting their potential environmental impacts solely through instrumental analysis has become increasingly challenging. Moreover, this methodology, while expensive and time-intensive, may yield outcomes that either overestimate or underestimate the risk posed by pollutants.

Utilizing the TRIADTriad or tier-based approach for sediment quality assessment based on three independent lines of evidence has proven to be a potent method, including evaluations of chemical concentrations, potential toxicity, and impacts on benthic community structure (Figure 15.2) (Chapman 1990). This approach offers a straightforward evaluation of sediment quality applicable to all relevant chemicals, although it does not establish a direct causal link between the concentrations of specific chemicals and hostile biological effects.

Assessing ecological risks posed by pollutants in the environment requires considering both pollutant concentrations, physicochemical characteristics and toxicity and ecosystem vulnerabilities. While obvious effects like contaminated sediment covering can occur, the accumulation of pollutants within organisms can disrupt metabolism and growth, highlighting the need for a comprehensive risk assessment approach. The triad approach evaluates ecological quality by combining chemical analyses with biological data and ecological observations (Chapman 1999). This validated method involves three line of evidences (LOEs): (i) environmental chemistry, (ii) toxicity testing using bioassays, and (iii) field ecology observations. Comparisons to reference samples help evaluate pollutant effects, although finding suitable reference sites can be challenging. Tailored toxicity tests and analyses are necessary depending on pollution types and environmental conditions. Operating on a tiered system, the triad approach balances effort and cost to avoid unnecessary expenses, providing a robust framework for assessing ecological risks in polluted sediments.

When evaluating contaminated sediments, it is vital to consider the combined impact of pollutants rather than focusing solely on individual concentrations. Even when present at levels below their No Observed Effect Concentrations (NOECs), pollutants can interact synergistically, resulting in compounded effects. Regulatory authorities have addressed this concern, issuing draft guidance outlining standardized methodologies for assessing the combined risks associated with exposure to multiple chemicals. These guidelines encompass assessments related to human health, animal health, and ecological risk, ensuring a comprehensive approach to evaluating the potential consequences of combined chemical exposures (Chapman and Hollert 2006).

15.3.2.1 Line of Evidence-1: Environmental Chemistry

The chemical component of the triad approach involves analyzing water, sediment, or soil to assess pollution levels, focusing not only on total concentrations but also on the bioavailable fraction of contaminants. Various techniques are employed for this purpose, such as persulfate oxidation for assessing PAHs accessibility and solid-phase extraction for POPs availability. Additionally, the bioavailable fraction of metals can be measured through sediment extraction with $CaCl_2$ followed by chemical analysis. Total pollutant concentrations are used to calculate the potentially affected fraction of species (PAF) at a site, with models like OMEGA used for freshwater environments (Wagelmans and Lieten 2017). These models can estimate PAF for single and multiple compounds based on total concentrations, and in some cases, PAF based on bioavailable fractions (Wagelmans and Lieten 2017). The same models can also assess bioaccumulation in the food chain using NOEC and LC_{50} data

(Wagelmans and Lieten 2017). Supportive analyses conducted in this part of the triad approach include pH, dissolved organic carbon (DOC), conductivity, and sediment particle distribution.

15.3.2.2 Line of Evidence-2: Toxicity

In the biological component of the triad approach, bioassays or biomarker tests are conducted to evaluate pollution impacts. Various model organisms and biomarker analysis are employed based on pollution types, environmental characteristics, and ecosystem considerations (Piva et al. 2011). Selection of the most sensitive model organism and representative species for the polluted site is crucial, with aquatic invertebrates, algae, and fishes typically used for aquatic samples. Utilizing a combination of organisms from different trophic levels provides a comprehensive assessment of risks within the entire system. Both chronic and acute tests, assessing long-term and short-term effects respectively, as well as functionality tests measuring processes like nitrification rate, are conducted (see Table 15.2). This table presents a selection of validated and standardized bioassays for assessing the acute and chronic effects of pollutants across different environments, although nonstandardized tests may be employed if validated and deemed beneficial.

15.3.2.3 Line of Evidence-3: Ecology

In the ecological line of evidence of the triad approach, biological data from polluted sites are evaluated for assessing the effects of contaminants on the environment. Monitoring of benthic organisms can be used to measure the abundance of species and the proportion of sensitive or tolerant taxonomic groups and calculate various diversity indexes in order to determine the ecological status, such as the benthic quality status (BQI) (Lamberson et al. 2018) and benthic response index (BRI) (Bay and Weisberg 2012). The ecological status can then be compared between the polluted site and a reference site. In marine habitats, these groups typically consist of polychaetes, crustaceans (like amphipods), mollusks, and echinoderms, while in freshwater environments, oligochaetes and insect larvae (such as chironomids) are generally included. Utilization of biotic integrity indices is prevalent in watersheds with well-documented benthic community compositions relative to habitat conditions. Analyses of amphipods and nematodes involve quantifying the total worm count per 100 g of sediment, species diversity, and abundance per feeding and life strategy groups. A Maturity Index (MI) can be calculated as the ratio of new colonies to persisters and serves as an indicator of benthic community health, with colonizers dominating in disturbed ecosystems. Additionally,

Table 15.2 Recommended bioindicator tests for sediment toxicity.

Toxicity test	Time (in days)	Target endpoints	Sediment required (in l)
Marine/estuarine waters			
Amphipod	10–28	Survival, growth, reproduction	1.5
Bivalve/Echinoderm larvae	2–5	Survival, development	0.5
Polychaete worm	20	Survival, growth	2.0
Freshwater			
Amphipod	10–28	Survival, growth, reproduction	1.5
Water flea	10–21	Survival, reproduction	0.5
Insect larvae	20	Survival, growth	1.5
Oligochaete worm	20	Survival, growth, reproduction	1.5

Source: Adapted from Chapman and McDonald (2005).

when assessing ecology, particular attention should be given to the surface sediments, as they host the majority of benthic organisms (Chapman et al. 2002). Efforts should be made to distinguish between potential impacts on benthic ecology stemming from contaminants and those resulting from other stressors, such as human activities or natural phenomena like unstable bed sediments or intermittent freshwater influxes. Moreover, the use of cutting-edge methods such as genetic barcoding may hold great promise for categorizing benthic populations.

15.3.3 Scaling, Weighting, and Integration of Results

Through the combination and integration of above-mentioned LOEs, the assessment of actual ecological risks is conducted. In essence, the findings underscore the imperative of combining data from diverse assays into a cohesive risk assessment paradigm, in accordance with the guidelines (Burton et al. 2002; Smith et al. 2002). Utilizing a standardized scaling method is crucial to ensure consistency across different LOEs while maintaining detailed quantitative data. The main goal is to maximize the quantitative utilization of test results and integrate them smoothly into decision-making processes, with tabulated decision matrices being the primary method for clear and quantitative information sharing (Burton et al. 2002). To ease the evaluation and merging of results from different tests within the triad approach, it is suggested to use a scale ranging from zero to one, covering the range from no impact to maximum influence. The translation of outcomes from individual parameters onto this scale depends on synthesizing available empirical evidence or achieving professional consensus. Distinct assays necessitate tailored methodologies; for example, growth assays may quantify impact via percentage inhibition, while ecological field assessments are anchored against the ecological baseline of reference locales and a hypothetical state denoting maximal perturbation (Jensen et al. 2006). The weight of evidence (WoE) framework aids in rectifying mismatches in scaling methodologies arising from erroneous postulates (Chapman et al. 2002). Following the normalization of outcomes into a unified metric, the collective response of a suite of methodologies, such as the chemical LOE, can be quantified by computing the geometric mean of the reciprocal effect (1 – effect). This analytical method places particular importance on results that show positive ecological effects, recognizing that biological assays, particularly in early stages, may miss detecting negative impacts due to their limited sensitivity (Chapman et al. 2002; Jensen et al. 2006).

Once the results from each test have been scaled, they can be combined within each LOE. This process aims to gather comprehensive information for contamination risk estimation. Various pieces of information are utilized for this assessment. For example, the species sensitivity distribution (SSD) approach considers the importance of all organisms, acknowledging their varying sensitivities to contamination. Additionally, estimates of effects based on different exposure scenarios may be combined to address species-specific differences in contaminant availability. In the next step, the different pieces of data from the three LOEs are merged to produce a singular risk value. This process also assesses the extent to which the three LOEs align in indicating the same level of risk, incorporating a measure of variance between them (Jensen et al. 2006). A significant difference in the results of the three LOEs could signal the need for further investigation.

15.4 Conclusions

This chapter emphasizes the crucial role of risk assessment in achieving successful sediment remediation. Among the various remediation techniques discussed, sediment thin-layer capping emerges as the most suitable option for containing contaminants within sediment phases, particularly in highly contaminated sites, albeit necessitating long-term monitoring. Furthermore, the use of inert materials for capping effectively manages both metal contamination and HOCs, especially in scenarios where groundwater upwelling is not significant. Integrating multiple lines of evidence using a step by step, that is, tiered-based risk assessment such as the triad method also provides a valuable tool for estimating the risks and the uncertainty associated with each LOE, thus providing data that can be used to better decide if and how a polluted site needs to be remediated, that is, facilitating the task for environmental management based on best available science and evidence of potential risks and highlighting uncertainties that may warrant more analyses before a decision can be taken. The tire-based multiple WoE approach also offers a practical means of considering bioavailability when assessing and managing ecological risks associated with contaminated sediments. The triad approach combines chemistry, toxicity, and ecology to assess ecological risks associated with pollution. By integrating these lines of evidence, both numerical and nonnumerical methods are employed to conduct a comprehensive evaluation of ecological risks. This holistic approach enables a thorough understanding of the impacts of contaminants on the environment, facilitating informed decision-making for effective pollution management and ecosystem protection. While promising, further research is necessary to fully validate its effectiveness and better comprehend the complex processes of sediment ageing, which influence contaminant availability and toxicity over time. Nevertheless, experts widely advocate for integrating bioavailability measures into ecological risk assessments, essential for enhancing the accuracy of site-specific evaluations. This transition away from relying solely on total concentration assessments signifies a significant advancement in effectively managing environmental risks associated with contaminated sediments. Moving forward, exploring innovative technologies and methodologies, along with continued research into sediment ageing processes and the refinement of bioavailability assessments, will be key in further improving remediation strategies and enhancing environmental protection measures.

Acknowledgment

The authors would like to take this opportunity to acknowledge the Department of Ecology, Environment and Plant Sciences (DEEP), Stockholm University and the Swedish Environmental Protection Agency, PROMISE Grant number 2020–0002; and the Swedish Geotechnical Institute, TUFFO program, Grant number 1.1–1602-0106 for the support they provided during the writing of this chapter.

References

Akcil, A., Erust, C., Ozdemiroglu, S. et al. (2015). A review of approaches and techniques used in aquatic contaminated sediments: metal removal and stabilization by chemical and biotechnological processes. *Journal of Cleaner Production* 86: 24–36.

Apitz, S.E., Davis, J.W., Finkelstein, K. et al. (2005). Assessing and managing contaminated sediments: part II, evaluating risk and monitoring sediment remedy effectiveness. *Integrated Environmental Assessment and Management: An International Journal* 1 (1): e1–e14.

Bay, S.M. and Weisberg, S.B. (2012). Framework for interpreting sediment quality triad data. *Integrated Environmental Assessment and Management* 8 (4): 589–596.

Becker, D.S., Sexton, J.E., Jacobs, L.A. et al. (2009). Biological responses to sediment remediation based on thin layer placement near a former pulp mill in Ward Cove, AK (USA). *Environmental Monitoring and Assessment* 154: 427–438.

Bonaglia, S., Rämö, R., Marzocchi, U. et al. (2019). Capping with activated carbon reduces nutrient fluxes, denitrification and meiofauna in contaminated sediments. *Water Research* 148: 515–525.

Bortone, I., Labianca, C., Todaro, F. et al. (2020). Experimental investigations and numerical modelling of in-situ reactive caps for PAH contaminated marine sediments. *Journal of Hazardous Materials* 387: 121724.

Bruhn, A., Flindt, M.R., Hasler, B., Krause-Jensen, D., Larsen, M.M., Maar, M., Petersen, J.K. and Timmermann, K., 2020. Marine virkemidler: beskrivelse af virkemidlernes effekter og status for vidensgrundlag.

Chapman, P.M. (1990). The sediment quality triad approach to determining pollution-induced degradation. *Science of the Total Environment* 97: 815–825.

Chapman, P.M. and Hollert, H. (2006). Should the sediment quality triad become a tetrad, a pentad, or possibly even a hexad? *Journal of Soils and Sediments* 6: 4–8.

Chapman, P.M. and Wang, F. (2001). Assessing sediment contamination in estuaries. *Environmental Toxicology and Chemistry: An International Journal* 20 (1): 3–22.

Chiesa, L.M., Nobile, M., Ceriani, F. et al. (2019). Risk characterisation from the presence of environmental contaminants and antibiotic residues in wild and farmed salmon from different FAO zones. *Food Additives & Contaminants: Part A* 36 (1): 152–162.

Chigwada, G., Wang, D., and Wilkie, C.A. (2006). Polystyrene nanocomposites based on quinolinium and pyridinium surfactants. *Polymer Degradation and Stability* 91 (4): 848–855.

Costello, C., Cao, L., Gelcich, S. et al. (2020). The future of food from the sea. *Nature* 588 (7836): 95–100.

EMODnet 2015. European Marine Observation and Data Network. https://emodnet.ec.europa.eu/en

FAO 2018. FAOSTAT- Food Balance Sheet. Consumption of Fish and Fishery Products. United Nations Food and Agriculture Organization, Rome. Food and Agriculture.

Flindt, M.R., Oncken, N.S., Kuusmesae, K. et al. (2021). Sand-capping of muddy sediment improves benthic light conditions and sediment anchoring capacity to support recovery of eelgrass (Zostera marina). *Journal of Sea Research* 181: 102177.

Ghosh, U., Luthy, R.G., Cornelissen, G., Werner, D. and Menzie, C.A., 2011. In-situ sorbent amendments: a new direction in contaminated sediment management

Habbaba, A. and Plank, J. (2012). Surface chemistry of ground granulated blast furnace slag in cement pore solution and its impact on the effectiveness of polycarboxylate superplasticizers. *Journal of the American Ceramic Society* 95 (2): 768–775.

Han, Z., Sani, B., Akkanen, J. et al. (2015). A critical evaluation of magnetic activated carbon's potential for the remediation of sediment impacted by polycyclic aromatic hydrocarbons. *Journal of Hazardous Materials* 286: 41–47.

Hidayati, N.V., Asia, L., Khabouchi, I. et al. (2021). Ecological risk assessment of persistent organic pollutants (POPs) in surface sediments from aquaculture system. *Chemosphere* 263: 128372.

Initiative, I. B (2012). *Standardized product definition and product testing guidelines for biochar that is used in soil.* IBI biochar Stand.

Jensen, J., Mesman, M., Bierkens, J., Loibner, A., Rutgers, M., Bogolte, T., Celis, R., Dirven-van Breemen, E.M., Erlacher, E., Ehlers, C. and Hartnik, T., 2006. Ecological risk assessment of contaminated land-Decision support for site specific investigations. Rijksinstituut voor Volksgezondheid en Milieu RIVM.

Ke, X., Bernal, S.A., and Provis, J.L. (2017). Uptake of chloride and carbonate by Mg-Al and Ca-Al layered double hydroxides in simulated pore solutions of alkali-activated slag cement. *Cement and Concrete Research* 100: 1–13.

Khim, J.S. and Hong, S. (2014). Assessment of trace pollutants in Korean coastal sediments using the triad approach: a review. *Science of the Total Environment* 470: 1450–1462.

Kutuniva, J., Mäkinen, J., Kauppila, T. et al. (2019). Geopolymers as active capping materials for in situ remediation of metal (loid)-contaminated lake sediments. *Journal of Environmental Chemical Engineering* 7 (1): 102852.

Lamberson, J.O., DeWitt, T.H., and Swartz, R.C. (2018). Assessment of sediment toxicity to marine benthos. In: *Sediment Toxicity Assessment* (ed. G.A. Burton), 183–211. CRC Press.

Li, S., Tan, X., Li, H. et al. (2022). Investigation on pore structure regulation of activated carbon derived from sargassum and its application in supercapacitor. *Scientific Reports* 12 (1): 10106.

Löser, C., Zehnsdorf, A., Hoffmann, P., and Seidel, H. (2007). Remediation of heavy metal polluted sediment by suspension and solid-bed leaching: estimate of metal removal efficiency. *Chemosphere* 66 (9): 1699–1705.

Näslund, J.S.G.S., Samuelsson, G.S., Gunnarsson, J.S. et al. (2012). Ecosystem effects of materials proposed for thin-layer capping of contaminated sediments. *Marine Ecology Progress Series* 449: 27–39.

Nobakht-Kolur, F., Zeinoddini, M., Aliakbari, T., and Bahram, M. (2021). Wave attenuation/build-up around and inside marine fouled floating aquaculture cages under regular wave regimes. *Journal of Ocean Engineering and Marine Energy* 7: 59–81.

Nybom, I., Abel, S., Waissi, G. et al. (2016). Effects of activated carbon on PCB bioaccumulation and biological responses of Chironomus riparius in full life cycle test. *Environmental Science & Technology* 50 (10): 5252–5260.

Oncken, N.S., Lange, T., Kristensen, E. et al. (2022). Sand-capping–a large-scale approach to restore organic-enriched estuarine sediments. *Marine Environmental Research* 173: 105534.

Pal, D. and Hogland, W. (2022). An overview and assessment of the existing technological options for management and resource recovery from beach wrack and dredged sediments: an environmental and economic perspective. *Journal of Environmental Management* 302: 113971.

Pal, D. and Maiti, S.K. (2019). Abatement of cadmium (Cd) contamination in sediment using tea waste biochar through meso-microcosm study. *Journal of Cleaner Production* 212: 986–996.

Pal, D. and Maiti, S.K. (2020). An approach to counter sediment toxicity by immobilization of heavy metals using waste fish scale derived biosorbent. *Ecotoxicology and Environmental Safety* 187: 109833.

Palermo, M. and Hays, D.F. (2013). Sediment dredging, treatment and disposal. In: *Processes, Assessment and Remediation of Contaminated Sediments* (ed. Y.-M. Cho, D. Werner, E.M.-L. Janssen, and R. Luthy), 365–391. New York, NY: Springer New York.

Park, S.J., Kang, K., Lee, C.G., and Choi, J.W. (2019). Remediation of metal-contaminated marine sediments using active capping with limestone, steel slag, and activated carbon: a laboratory experiment. *Environmental Technology* 40 (26): 3479–3491.

Peng, J.F., Song, Y.H., Yuan, P. et al. (2009). The remediation of heavy metals contaminated sediment. *Journal of Hazardous Materials* 161 (2-3): 633–640.

Piva, F., Ciaprini, F., Onorati, F. et al. (2011). Assessing sediment hazard through a weight of evidence approach with bioindicator organisms: a practical model to elaborate data from sediment chemistry, bioavailability, biomarkers and ecotoxicological bioassays. *Chemosphere* 83 (4): 475–485.

Prabhakar, R. and Samadder, S.R. (2020). Effective immobilization and reduction in bioavailability of Cd in a L. succinea growing in contaminated sediment by the application of alkali synthesized fly ash based zeolite (FABZ). *Microporous and Mesoporous Materials* 306: 110416.

Provis, J.L. (2018). *Alkali-activated materials. Cement and concrete research* 114: 40–48.

Rakowska, M.I., Kupryianchyk, D., Harmsen, J. et al. (2012). In situ remediation of contaminated sediments using carbonaceous materials. *Environmental Toxicology and Chemistry* 31 (4): 693–704.

Rämö, R.A., Honkanen, J., Nybom, I., and Gunnarsson, J.S. (2021). Biological effects of activated carbon on benthic macroinvertebrates are determined by particle size and ingestibility of activated carbon. *Environmental Toxicology and Chemistry* 40 (12): 3465–3477.

Renman, A. and Renman, G. (2010). Long-term phosphate removal by the calcium-silicate material Polonite in wastewater filtration systems. *Chemosphere* 79 (6): 659–664.

Richards, S. and Bouazza, A. (2007). Phenol adsorption in organo-modified basaltic clay and bentonite. *Applied Clay Science* 37 (1–2): 133–142.

Samuelsson, G.S., Raymond, C., Agrenius, S. et al. (2017). Response of marine benthic fauna to thin-layer capping with activated carbon in a large-scale field experiment in the Grenland fjords, Norway. *Environmental Science and Pollution Research* 24: 14218–14233.

Sarkar, B.I.N.O.Y., Xi, Y., Megharaj, M. et al. (2012). Bioreactive organoclay: a new technology for environmental remediation. *Critical Reviews in Environmental Science and Technology* 42 (5): 435–488.

Sheng, G., Boyd, S.A., and Xu, S. (1999). A dual function organoclay sorbent for lead and chlorobenzene. *Soil Science Society of America Journal* 63 (1): 73–78.

Simpson, S.L., Pryor, I.D., Mewburn, B.R. et al. (2002). Considerations for capping metal-contaminated sediments in dynamic estuarine environments. *Environmental Science & Technology* 36 (17): 3772–3778.

Sparrevik, M. and Breedveld, G.D. (2010). From ecological risk assessments to risk governance: evaluation of the Norwegian management system for contaminated sediments. *Integrated Environmental Assessment and Management: An International Journal* 6 (2): 240–248.

Swedish Geotechnical Institute (SGI) 2016. *In-situ* capping of contaminated sediments (SGI Publikation 30-6E). https://sgi.se/globalassets/publikationer/sgi-publikation/sgi-p30-6e.pdf

Trannum, H.C., Raymond, C., Næss, R. et al. (2021). Long-term response of marine benthic fauna to thin-layer capping with powdered activated carbon in the Grenland fjords, Norway. *Science of the Total Environment* 776: 145971.

U.S. Environmental Protection Agency (USEPA) 1998. EPA's contaminated sediment management strategy. Washington DC: USEPA. EPA-823-R-98-001.

U.S. Environmental Protection Agency (USEPA) 2014. OSWER Directive 9200.2-128FS - Use of Amendments for In Situ Remediation at Superfund Sediment Sites – February 2024.

U.S. Environmental Protection Agency (USEPA) 2018. Remedial Technology Fact Sheet – Activated Carbon Based Technology for In Situ Remediation

Varadharajan, D. and Pushparajan, N. (2013). Food and feeding habits of aquaculture candidate a potential Crustacean of Pacific white shrimp Litopenaeus Vannamei, South East Coast of India. *Journal of Aquaculture Research and Development* 4 (1): https://doi.org/10.4172/2155-9546.1000161.

Wagelmans, M. and Lieten, S. (2017). Protocols for ecological risk assessment using the triad approach. In: *Hydrocarbon and Lipid Microbiology Protocols: Pollution Mitigation and Waste Treatment Applications*, 65–79. Springer.

Wang, H., Liu, T., Tsang, D.C., and Feng, S. (2017). Transformation of heavy metal fraction distribution in contaminated river sediment treated by chemical-enhanced washing. *Journal of Soils and Sediments* 17: 1208–1218.

Wikström, J., Bonaglia, S., Ramo, R. et al. (2021). Sediment remediation with new composite sorbent amendments to sequester phosphorus, organic contaminants, and metals. *Environmental Science & Technology* 55 (17): 11937–11947.

Xi, Y., Zhou, Q., Frost, R.L., and He, H. (2007). Thermal stability of octadecyltrimethylammonium bromide modified montmorillonite organoclay. *Journal of Colloid and Interface Science* 311 (2): 347–353.

Yan, S., Rakowska, M., Shen, X. et al. (2020). Bioavailability assessment in activated carbon treated coastal sediment with in situ and ex situ porewater measurements. *Water Research* 185: 116259.

Yoo, J.Y., Choi, J., Lee, T., and Park, J.W. (2004). Organobentonite for sorption and degradation of phenol in the presence of heavy metals. *Water, Air, and Soil Pollution* 154: 225–237.

16

Biochar Application for Mitigation of Coastal and Marine Pollution: An Experimental Modeling

Pelin Soyertaş Yapıcıoğlu

Environmental Engineering Department, Engineering Faculty, Harran University, 63050, Sanliurfa, Turkey

16.1 Introduction

Microplastics (MPs) are minor pieces of plastics that are smaller than 5 millimeters (mm) (European Union [EU] 2019, EU 2021). They are recalcitrant, very mobile, and difficult in order to remove and dispose of from nature. An increasing amount of MPs is available in the flora and fauna, in the sea, ocean, surface water, wastewater, and groundwater and also in the atmosphere and soil, as well as in food and potable water (Wang et al. 2023a; Wang et al. 2023b; Sun et al. 2022; Sun et al. 2023). Many MPs do not biodegrade and tend to accumulate in the receiver bodies (Zhai et al. 2023; Zhang et al. 2020a, Zhang et al. 2020b). Biodegradability is a complex phenomenon, especially in the marine and coastal environment. There are raising concerns about the presence of MPs in various receiver media such as water, wastewater, soil, and atmosphere and their effect on the environment and biodiversity and human health (EU 2021). In 2023, the European Commission declared a restriction on MPs added to products and a suggestion for a regulation on prohibiting plastic pellet releases to the environment. These preventions will directly contribute to achieving the 30% minimization aim for MP emissions set out in the Zero Pollution Action Plan (EU 2019). The EU has aimed to decrease MP emissions by 30% by 2030. This would be obtained by minimizing plastic pollution (as these degrade into MPs), limiting the use of intentionally added MPs to products, and minimizing MP emissions (EU 2021).

MPs are available in various locations and also in the human body. The risks corresponding to higher stages of human exposure are a cause of great concern. Plastic pellets are available in water and soil, contaminating agricultural lands and ecosystems and biodiversity. Pellets are considered to be eaten by several types of marine and coastal species, and when ingested, they could lead to physicochemical hazards or death (Andrady 2011; Tsang et al. 2017; Pan et al. 2019, Markic et al. 2022). MP pollution has been considered one of the main threats to marine and coastal areas. MP pollution has increased nowadays, due to many reasons such as random disposal and insufficient solid waste management systems (Pan et al. 2019; Dharmadasa et al. 2021).

MPs not only pose a significant pollution problem to the environment but also have the potential to contribute to greenhouse gas emissions and climate change (Li et al. 2024). Due to these factors, marine pollution due to MPs could be reduced and taken under control using various technical

Coastal and Marine Pollution: Source to Sink, Mitigation and Management, First Edition.
Edited by Meththika Vithanage, Sameera M. Samarasekara, Bryan D. James, and Christopher M. Reddy.

methodologies. MPs have been transferred into the marine and/or coastal area directly or with the help of wastewater discharge into the receiver point. This study has concentrated on wastewater discharge among these components of pollution. MPs can be decreased significantly using various techniques, such as biological, chemical, and physical methods.

There are many remediation and treatment technologies such as altering process conditions, microalgal and microbial treatment methods, and the use of renewable energy resources in terms of water and/or wastewater treatment. Renewable energy resources have been used generally in the last decades (Yapıcıoğlu and Yeşilnacar 2022a, Zhang et al. 2021). Also, biomass energy could be a sufficient reduction technique owing to its many advantages. The advantages of biomass energy are that it is less expensive and feasible for application in water and wastewater treatment plants (Yapıcıoğlu and Yeşilnacar 2023). Among different kinds of biomass energy, biochar is generally applied for water and wastewater treatment processes in recent years (Yapıcıoğlu and Yeşilnacar 2023; Qambrani et al., 2017).

Biochar, which is a carbon-rich material produced by pyrolysis or carbonization of biomass wastes, is a significant matter as it could remove pollutants and be carried out as a carbon-negative technology (Yapıcıoğlu &Yeşilnacar 2022b, Qambrani et al. 2017, Kumar et al. 2023). Biochar is regarded as a carbon-rich material that could be generated using different organic and inorganic feedstocks (Qambrani et al., 2017, Yapıcıoğlu &Yeşilnacar 2022b). Biochar has gained increasing significance due to its stable nature, higher carbon context, ion exchange capacity, larger surface area, and higher uptake capacity (Wang and Wang 2019). Biochar has gained great popularity recently because of its significant role in environmental remedies (Qambrani et al., 2017). It is cheaper than other treatment techniques and can easily adsorb various pollutants in water and wastewater (Rajapaksha et al. 2016, Sadhu et al. 2021). From this perspective, biochar could adsorb MPs, which are the recalcitrant pollutants in the marine environment.

From this point of view, this study aimed to develop an experimental and computational model related to biochar application in order to reduce coastal and marine pollution in terms of MP contamination. This chapter has mainly suggested that biomass wastes resulting from marine ecosystems such as algal wastes would be used as biochar feedstock. Also, wastewater discharging has been focused on as the source of MP in the marine environment. Firstly, the MP removal mechanism of biochar has been defined. Then, a new pollution reduction index (PRI_{MPs}) based on biochar removal capacity and wastewater discharge has been developed in this chapter. This chapter presents an experimental and computational calculation tool corresponding to biochar application in order to mitigate coastal and marine pollution in terms of MP pollution.

16.2 Materials and Methods

16.2.1 Biochar Application for Coastal and Marine Pollution

Biochar is a porous carbonaceous material with a larger specific surface area, and it is derived by pyrolysis of limited oxygen media (Premarathna et al. 2019). Biochar is produced by pyrolysis of biomass, which are sewage sludge, agricultural crops, marine wastes, manures, and solid wastes (Ravindiran et al. 2024; Yuan et al. 2017; Yapıcıoğlu & Yeşilanacar 2022b; Sadhu et al. 2021). In this study, algal biochar has been recommended to remove MPs from wastewater. Algal biochar has been preferred because of its ability to decrease marine pollution. Marine algae could be used as biochar feedstock. *Ulva reticulata* is a marine alga; it is an aquatic biomass and exists in large amounts in coastal environments (Ravindiran et al. 2024).

This species could be used for biochar production. Also, algae have been known as a wastewater treatment technology. It is a greater effect to generate biochar as a raw material. In the literature, algal biochar has been used for many purposes (Chen et al. 2020). One purpose is MP removal. MPs have a long lifetime, a biologically persistent character, and high stability. Investigations on MP removal have demonstrated higher efficiency using activated biochar obtained from different land biomass.

The mechanisms of biochar to remove MPs could be classified as entanglement, physisorption, heteroaggregation, chemical bonding and electrostatic attraction, electrostatic attraction and formation of surface complexation, hydrophobic interaction and hydrogen bonding, electrostatic repulsion, and ionic competition between MPs and humic acids (Kumar et al. 2023). The adapted (Kumar et al. 2023) schematic diagram of MPs interactions with biochar is shown in Figure 16.1.

One of the main resources of MPs in marine environments is wastewater discharge. Wastewater could contain large quantities of MPs. Because of the high adsorption capacity of biochar and the low cost of the generation process, biochar application for wastewater treatment has gained great attention (Wang et al. 2020). Research on MPs in the influent and effluent of wastewater treatment plants has approved that the efficiency of combined conventional and advanced treatment processes to remove MPs is nearly above 90% (Wang et al. 2020; Conley et al. 2019). The biochar adsorption process is an innovative and efficient treatment technique for wastewater systems. In this study, algae-derived biochar adsorption process has been recommended as an MP removal method from wastewater mass. Figure 16.2 shows the recommended MP removal methodology and the main outlines of the chapter.

16.2.2 Experimental and Computational Modeling of Biochar Application

A new estimation tool has been developed for MP reduction in marine and coastal areas due to wastewater discharge. In the adsorption process, the pollutant parameter is MPs and the adsorbent is algae-derived biochar. The basic adsorption calculation term is given below (Eq. 16.1) (Metcalf & Eddy 2014):

$$\mathrm{qe} = ((\mathrm{Po} - \mathrm{Pe})V)/W \tag{16.1}$$

In Eq. (16.1), Po is the initial pollutant concentration (mM), Pe is the equilibrium pollutant concentration (mM), *V* is the wastewater volume (l), and *W* is the adsorbent dose (g). In this study,

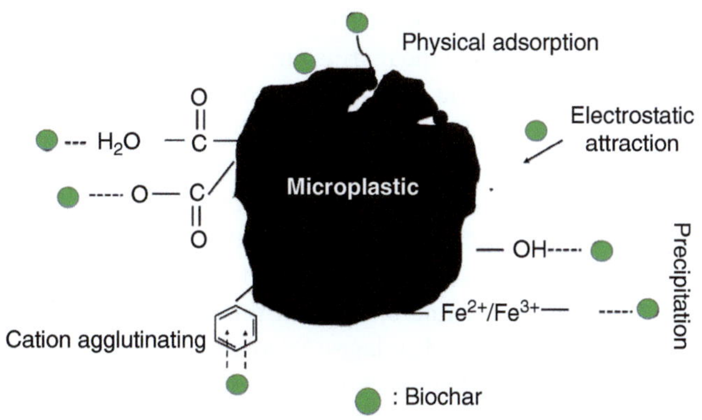

Figure 16.1 Schematic diagram of MP uptake by biochar.

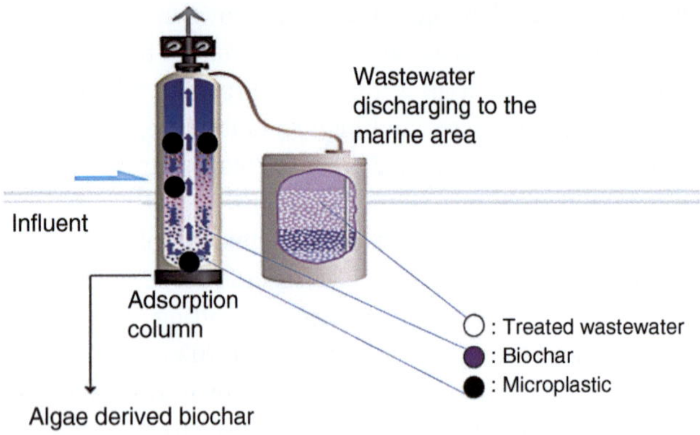

Figure 16.2 Conceptual framework of the MP reduction methodology using biochar.

a new adsorption calculation term based on this adsorption theory and experimental and computational studies has been developed. The derived equation is represented as Eq. (16.2):

$$\mathrm{RR}(\%) = (((\mathrm{MPi} - \mathrm{MPf})Q)\mathrm{ADB}) \times \mathrm{DR} \times h \tag{16.2}$$

In Eq. (16.2), RR is the removal rate (%), MPi and MPf are the initial and final concentrations of MP (particles/l), and Q is the discharged wastewater flow (m^3/d). Algae-derived biochar (ADB) is the dose of algae-derived biochar (g), DR is the dilution ratio (%) in the marine ecosystem (discharging point), and h is the discharging depth (m).

The validation of the recommended model has been performed using the Monte Carlo simulation. The simulation tool has been given in Eq. (16.3). In Eq. (16.3), C is the correspondence between RR and the MPi, MPf, Q, ADB, DR, and h. For each parameter, the simulation has been done to define the correlation:

$$\mathrm{RRmax} = \mathrm{Riskoutput}(\mathrm{Lognormal}) + \mathrm{RiskLognorm}(C, \mathrm{RR}) \tag{16.3}$$

16.3 Results and Discussion

Biochar is a porous carbonaceous material with a higher specific surface area produced by pyrolysis and consists of carbon and minerals. Adsorption is the major mechanism for biochar to remove contaminants. One of these contaminants could be MPs in the marine environment. In the literature, there are limited studies related to this topic. Previous studies have confirmed that various types of biochar could remove MPs (Siipola et al. 2020; Singh et al. 2021; Wang et al. 2022; Tong et al. 2020; Magid et al. 2021). MP removal efficiency is approximately 100% using biochar according to the previous studies. Table 16.1 shows the MP removal using various biomass-derived biochar. As shown in Table 16.1, a reduction in MP pollution is ~100%. From this perspective, it can be suggested that a unique biochar, which is algae-derived, could be used for MP removal from wastewater. There are several pyrolysis methods to generate biochar. In this study, microwave-assisted pyrolysis has been recommended due to its many advantages. Microwave-assisted pyrolysis has gained increasing importance in recent years due to its unique energy transfer mechanism, fast and equal heating at the molecular level, fast and selective heating with high energy efficiency, and easy control mechanisms (Sahoo and Remya 2022).

Table 16.1 The studies related to MP removal using biochar.

Reference	Biochar type	MP types	Removal efficiency
Siipola et al. (2020)	Pine and spruce bark	Polyethylene microbeads	~100%
Singh et al. (2021)	Modified biochar derived from Prosopis juliflora	Nanoplastics: 30– 1000 nm	~100%
Wang et al. (2022)	Peanut shell, biochar, and magnesium oxide (MgO)–peanut shell biochar	Polystyrene MPs (1 μm)	59.1–75.5%
Tong et al. (2020)	Cellulose-based biochar	Polystyrene (2 μm)	90%
Magid et al. (2021)	Corncob raw and oxidized biochar	Polystyrene nanoplastics	>90%

According to several studies, wastewater is a resource of MPs (Wang et al. 2020; Conley et al. 2019). From this point of view, wastewater discharging is a resource of MPs in the marine environment. This study has recommended to investigate the reduction potential of MP pollution originating from wastewater discharge using biochar application in marine ecosystems. This chapter mainly focused on pollution reduction using algae-derived biochar, which is a kind of novel biochar.

In this study, a new estimation tool has been developed for MP removal capacity in marine and coastal areas due to wastewater discharge using biochar adsorption. The main factors in adsorption have been considered. A sensitivity analysis has been performed to validate the recommended model using the Monte Carlo simulation with the help of @RISK software. One simulation and 1000 iterations have been carried out to validate the model. The inputs are MPi, MPf, Q, ADB, DR, and h. The desirable output is the maximum MP removal rate. According to the sensitivity analysis, the correspondence is meaningful in the range of 0.97–0.98 of R^2. The standard deviation has been nearly 0.04. The validation of the recommended model (Eq. 16.2) is represented in Figure 16.3 in detail.

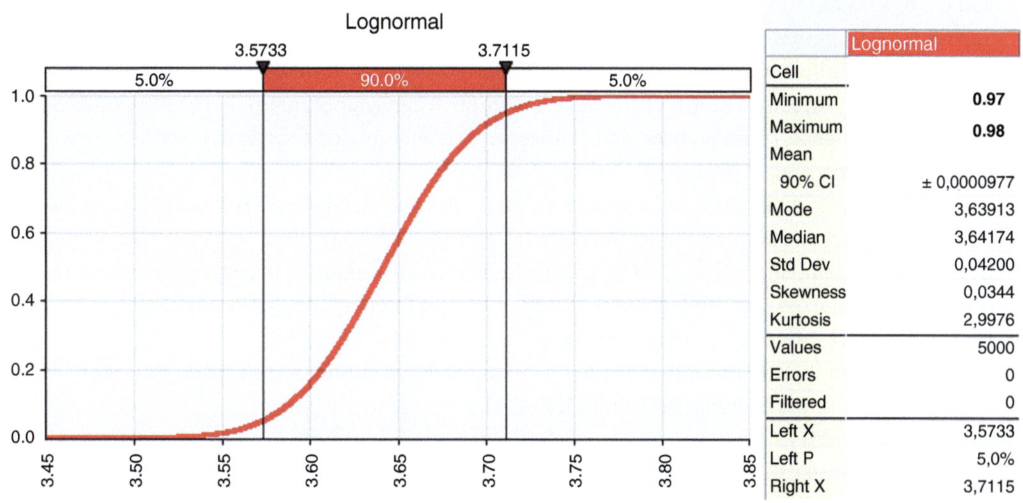

Figure 16.3 The validation of the recommended model.

16.4 Conclusion

MP management in the marine environment is a special issue, which has been an environmental challenge in recent years. Also, the EU Green Deal recommended a 30% reduction in MP pollution. Wastewater discharging is a component of MP accumulation in the marine environment. The biochar adsorption process is an innovative technique in order to remove MPs from wastewater. From this perspective, algae-derived biochar could be used for the adsorption of MPs. A new estimation index based on biochar adsorption and wastewater discharge has been developed, and the recommended model has been validated using the Monte Carlo simulation. The model has been validated in the values of 0.97 and 0.98 of R^2. The experimental studies in the marine environment on this topic should be improved.

References

Aeschlimann, M., Li, G., Kanji, Z.A., and Mitrano, D.M. (2022). Potential impacts of atmospheric microplastics and nanoplastics on cloud formation processes. *Nature Geoscience* 15 (12): 967–975.

Andrady, A.L. (2011). Microplastics in the marine environment. *Marine Pollution Bulletin* 62 (8): 1596–1605.

Chen, Y.D., Liu, F., Ren, N.Q., and Ho, S.H. (2020). Revolutions in algal biochar for different applications: state-of-the-art techniques and future scenarios. *Chinese Chemical Letters* 31 (10): 2591–2602.

Conley, K., Clum, A., Deepe, J. et al. (2019). Wastewater treatment plants as a source of microplastics to an urban estuary: removal efficiencies and loading per capita over one year. *Water Research X* 3: 100030.

Dharmadasa, W.S., Andrady, A.L., Kumara, P.T.P. et al. (2021). Microplastic pollution in marine protected areas of Southern Sri Lanka. *Marine Pollution Bulletin* 168: 112462.

Dong, Y.M., Bao, Q.L., Gao, M.L. et al. (2022). A novel mechanism study of microplastic and as co-contamination on indica rice (*Oryza sativa L.*). *Journal of Hazardous Materials* 421: 12.

European (EU) Commission (2019). Communication from the Commission to the European Parliament, the 533 European Council, the Council, the European Economic and Social Committee and the 534 Committee of the Regions–The European Green Deal.

European (EU) Commission (2021). European Green Deal 2021 Fit for 55 Package: The EU's Fit for 55 Package – Key Takeaways From Bernstein Renewables, Chemical, Airlines & Energy Teams.

Kumar, R., Verma, A., Rakib, M.R.J. et al. (2023). Adsorptive behavior of micro (nano) plastics through biochar: co-existence, consequences, and challenges in contaminated ecosystems. *Science of the Total Environment* 856: 159097.

Li, K., Du, L., Qin, C. et al. (2024). Microplastic pollution as an environmental risk exacerbating the greenhouse effect and climate change: a review. *Carbon Research* 3 (1): 9.

Magid, A.S.I.A., Islam, M.S., Chen, Y. et al. (2021). Enhanced adsorption of polystyrene nanoplastics (PSNPs) onto oxidized corncob biochar with high pyrolysis temperature. *Science of Total Environment* 784: 147115.

Markic, A., Bridson, J.H., Morton, P. et al. (2022). Microplastic pollution in the surface waters of Vava'u, Tonga. *Marine Pollution Bulletin* 185: 114243.

Metcalf and Eddy (2014). *Wastewater Engineering: Treatment and Resource Recovery*, 5e. McGraw-Hill International Editions.

Pan, Z., Liu, Q., Sun, Y. et al. (2019). Environmental implications of microplastic pollution in the Northwestern Pacific Ocean. *Marine Pollution Bulletin* 146: 215–224.

Premarathna, K.S.D., Rajapaksha, A.U., Adassoriya, N. et al. (2019). Clay-biochar composites for sorptive removal of tetracycline antibiotic in aqueous media. *Journal of Environmental Management* 238: 315–322.

Qambrani, N.A., Rahman, M.M., and Won, S. (2017). Biochar properties and eco-friendly applications for climate change mitigation, waste management, and wastewater treatment: a review. *Renewable and Sustainable Energy Reviews* 79: 255–273.

Rajapaksha, A.U., Chen, S.S., Tsang, D.C. et al. (2016). Engineered/designer biochar for contaminant removal/immobilization from soil and water: potential and implication of biochar modification. *Chemosphere* 148: 276–291.

Ravindiran, G., Rajamanickam, S., Ramalingam, M. et al. (2024). Conversion of seaweed waste to biochar for the removal of heavy metal ions from aqueous solution: a sustainable method to address eutrophication problem in water bodies. *Environmental Research* 241: 117551.

Sadhu, M., Bhattacharya, P., Vithanage, M., and Sudhakar, P.P. (2021). Adsorptive removal of fluoride using biochar–a potential application in drinking water treatment. *Separation and Purification Technology* 119106.

Sahoo, D. and Remya, N. (2022). Influence of operating parameters on the microwave pyrolysis of rice husk: biochar yield, energy yield, and property of biochar. *Biomass Conversion and Biorefinery* 12: 3447–3456.

Siipola, V., Pflugmacher, S., Romar, H. et al. (2020). Low-cost biochar adsorbents for water purification including microplastics removal. *Applied Science* 10: 788.

Singh, N., Khandelwal, N., Ganie, Z.A. et al. (2021). Eco-friendly magnetic biochar: an effective trap for nanoplastics of varying surface functionality and size in the aqueous environment. *Chemical Engineering Journal* 418: 129405.

Sun, Y., Li, X., Cao, N. et al. (2022). Biodegradable microplastics enhance soil microbial network complexity and ecological stochasticity. *Journal Hazardous Materials* 439: 129610.

Sun, X., Tao, R., Xu, D. et al. (2023). Role of polyamide microplastic in altering microbial consortium and carbon and nitrogen cycles in a simulated agricultural soil microcosm. *Chemosphere* 312: 137155.

Tong, M., He, L., Rong, H. et al. (2020). Transport behaviors of plastic particles in saturated quartz sand without and with biochar/Fe_3O_4-biochar amendment. *Water Research* 169: 115284.

Tsang, Y.Y., Mak, C.W., Liebich, C. et al. (2017). Microplastic pollution in the marine waters and sediments of Hong Kong. *Marine Pollution Bulletin* 115 (1–2): 20–28.

Wang, J. and Wang, S. (2019). Preparation, modification and environmental application of biochar: a review. *Journal of Cleaner Production* 227: 1002–1022.

Wang, Z., Sedighi, M., and Lea-Langton, A. (2020). Filtration of microplastic spheres by biochar: removal efficiency and immobilisation mechanisms. *Water Research* 184: 116165.

Wang, C., Wei, W., Zhang, Y.-T., and Ni, B.-J. (2022). Evaluating the role of biochar in mitigating the inhibition of polyethylene nanoplastics on anaerobic granular sludge. *Water Research* 221: 118855.

Wang, Z., Hu, X., Kang, W. et al. (2023a). Interactions between dissolved organic matter and the microbial community are modified by microplastics and heat waves. *Journal of Hazardous Materials* 448: 130868–130868.

Wang, Z., Hu, X., Qu, Q. et al. (2023b). Dual regulatory effects of microplastics and heat waves on river microbial carbon metabolism. *Journal of Hazardous Materials* 441: 129879.

Yapıcıoğlu, P. and Yeşilnacar, M.I. (2022a). Economic performance index assessment of an industrial wastewater treatment plant in terms of the European Green Deal: effect of greenhouse gas emissions. *Journal of Water and Climate Change* 13 (8): 3100–3118.

Yapıcıoğlu, P. and Yeşilnacar, M.İ. (2023). Energy cost optimization of groundwater treatment using biochar adsorption process: an experimental approach. *Water Supply* 23 (1): 14–33.

Yuan, Y., Bolan, N., Prévoteau, A. et al. (2017). Applications of biochar in redox-mediated reactions. *Bioresource Technology* 246: 271–281.

Zhai, X., Zhang, X.H., and Yu, M. (2023). Microbial colonization and degradation of marine microplastics in the plastisphere: a review. *Frontiers in Microbiology* 14.

Zhang, B., Yang, X., Chen, L. et al. (2020a). Microplastics in soils: a review of possible sources, analytical methods and ecological impacts. *Journal of Chemical Technology & Biotechnology* 95 (8): 2052–2068.

Zhang, X., Chen, J., and Li, J. (2020b). The removal of microplastics in the wastewater treatment process and their potential impact on anaerobic digestion due to pollutants association. *Chemosphere* 251: 126360.

Zhang, Q., Smith, K., Zhao, X. et al. (2021). Greenhouse gas emissions associated with urban water infrastructure: What we have learnt from China's practice. *Wiley Interdisciplinary Reviews: Water* 8 (4): e1529.

Yapıcıoğlu, P. and Yeşilnacar, M.İ. (2022b). Grey water footprint assessment of groundwater resources in southeastern Turkey: effect of recharge. *Water Supply* 22 (1): 615–627.

17

Coastal and Marine Plastic Pollution Monitoring and Control Using Remote Sensing (RS) and Artificial Intelligence (AI) Technologies

Iddamalgoda Jayawardenage Judi Udari Nisansala Perera[1,2,3], *Ranepura Dewage Charuka Sandaruwan*[4,5], *and Bellanthudawage Kushan Aravinda Bellanthudawa*[1,3,6]

[1] *Department of Agricultural Engineering and Environmental Technology, Faculty of Agriculture, University of Ruhuna, 81110, Matara, Southern Province, Sri Lanka*
[2] *South China Sea Institute of Oceanology, Chinese Academy of Sciences, Guangzhou, 510301, China*
[3] *University of Chinese Academy of Sciences, Beijing, 100049, China*
[4] *Sri Lanka Wildlife Conservation Society, Pussallayaya, Handungamuwa, 21536, Matale, Central Province, Sri Lanka*
[5] *Faculty of Graduate Studies, University of Kelaniya, Kelaniya 11600, Sri Lanka*
[6] *Research Center for Eco-Environmental Sciences, Chinese Academy of Sciences, Beijing, 100085, China*

17.1 Introduction

Marine debris is a ubiquitous and increasing threat to marine ecosystems, raising an alarm due to its widespread presence, durability, and persistence across all oceans globally (Napper and Thompson, 2020; Agamuthu et al., 2019). Plastics (i.e. synthetic polymers) are the most frequent type of marine litter, accounting for 60–95% of the global ocean marine litter (Abalansa et al., 2020). While reports of plastic litter date back to the 1960s, it has only recently emerged as a pressing global environmental concern (Pilapitiya and Rathnayake, 2024). If current trends in production and waste management continue, over 12 000 metric tons of plastic waste is expected to end up in landfills or the natural environment by 2050 and that mass will surpass the mass of the fish population (Geyer, 2020; Williams and Rangel-Buitrago 2022). Plastics exist in a variety of marine and coastal ecosystems, generating from both land-based and sea-based sources. In the past few decades, plastic pollution has not only increased in places near populated beaches, bays, gulfs, river plumes, and estuarine systems but also reached remote areas of the Earth and carried away with wind, waves, and tides (Ryan et al., 2020).

Plastic debris items in the oceans vary widely in terms of size, shape, or chemical composition. In particular, large plastic products and thousands of plastic items can break down into smaller pieces, called microplastics or nanoplastics, through a combination of photodegradation, oxidation, and mechanical abrasion (Pham et al., 2021). While microplastics are prevalent, larger plastic items, or macroplastics, can endure for extended periods in the ocean, particularly when shielded from UV radiation underwater or within sediments. These macroplastics constitute a significant portion of the global mass of floating debris, causing substantial harm to the marine environment and the well-being of marine organisms. Many types of plastic debris do not remain on the ocean's surface but instead sink to the depths, accumulating in vast patches called ocean gyres. The most

Coastal and Marine Pollution: Source to Sink, Mitigation and Management, First Edition.
Edited by Meththika Vithanage, Sameera M. Samarasekara, Bryan D. James, and Christopher M. Reddy.

prominent of these gyres is the Great Pacific Garbage Patch (GPGP) located halfway between Hawaii and California, United States, covering an expansive 1.6 million square kilometers (617 763 square miles). Over the past few decades, it has become evident that anthropogenic marine debris, particularly plastic, has profound impacts on marine organisms (Kühn and Van Franeker 2020). The detrimental effects of plastic litter on marine wildlife adversely affect biodiversity, primarily through ingestion and entanglement. Animal species, including fish, birds, and invertebrates, are primarily susceptible to the impact of microplastic litter and mesoplastic litter. Moreover, mesolitter and macrolitter can lead to the degradation of coastal habitats and adversely affect the quality of life in coastal communities. Simultaneously, microplastic litter can transfer through the seafood chains, posing a threat to human health. Additionally, plastic debris facilitates the transport of organisms through rafting and provides new habitats for colonization (Wright et al., 2020). Beyond the environmental consequences, plastic pollution poses challenges to maritime traffic and can lead to adverse socioeconomic impacts on fisheries, tourism, and human health. Consequently, addressing ocean pollution accelerated by plastic contamination is imperative, thus efforts are underway to tackle this issue through two overarching approaches: the removal of existing marine debris and the establishment of deterrence by monitoring and identifying sources. These comprehensive methods aim to understand, quantify, and mitigate the detrimental effects of plastic pollution and safeguard the health of our oceans.

17.2 Challenges of In Situ Visual Monitoring of Plastic Litter in Marine and Coastal Ecosystems

In situ visual census surveys, such as beachcombing, represent the most technically simplistic method for directly collecting data on mesoplastic litter or macroplastic litter, as these larger debris items are readily visible to the naked eye (Bhuyan et al., 2021). Other data acquisition systems involve the use of Manta nets or Neuston nets to collect and monitor floating microplastic litter in open sea environments, as well as along dunes and the seafloor (Pasquier et al., 2022). However, in situ collections, for instance, prove to be both costly and labor-intensive, demanding a significant amount of time (Politikos et al., 2023). Furthermore, their coverage is confined to relatively small areas, making them challenging to implement in remote regions. Another drawback that the scientists encounter is the lack of standardization across these methods. Notably, there exists a variability in net sizes, and however, information on particle sizes and individual piece dimensions is occasionally omitted (Madricardo et al., 2020). Studies often present their measurements either in particle count or by mass, contributing to a lack of consistency (Madricardo et al., 2020). Nonetheless, estimating microplastics with these methods is especially difficult and challenging, as identification is often done by visual inspection, which can lead to errors from observer bias, misidentification, or under-identification (Madricardo et al., 2020; Politikos et al., 2023). Therefore, there exists a substantial gap to overcome these aforementioned challenges in the effective and viable estimation of plastics and microplastic pollution levels in marine ecosystems.

17.3 Remote Sensing (RS) Approaches for Monitoring Marine and Coastal Environments

In the beginning, airplanes were used for visual observation of open ocean monitoring, but their usage was limited to favorable weather conditions and lacked the capability to consistently provide dedicated and continuous data feeds for a specific region over prolonged period (Veenstra and

Churnside, 2012). Nevertheless, the manual nature of these observations introduces challenges such as tedium, variability in results among different observers, and the absence of a permanent objective record for the identified objects (Veenstra and Churnside, 2012). In response to these challenges, satellites were introduced in 1972 for terrestrial monitoring and later in the late 1970s for marine spaces. These sensing technologies in satellites are designed to provide observations of global scope, continuous temporal coverage, and harmonized data collection and processing, thus potentially being ideal tools for global marine debris monitoring, filling the gaps between sparse in situ observations and providing, at the same time, a relatively uniform coverage over larger areas and long time periods (Salgado-Hernanz et al., 2021). Furthermore, development of diverse sensors adaptable to various platforms has expanded the practice of remote sensing (RS), despite each sensor having its own capabilities and limitations.

17.3.1 Sensor Types

RS platforms include satellites, aircraft, and unmanned aerial vehicles (UAVs, commonly equipped with both passive and active sensors capable of targeting the Earth's surface) (Yasin et al., 2020). The relevant passive sensors are exclusively imaging systems, such as video cameras, multispectral and hyperspectral cameras, and thermal imagers. Based on existing studies, the passive sensors approach involves the utilization of multispectral red-green-blue (RGB) composite natural color images within the visible spectrum (400–700 nm) (Salgado-Hernanz et al., 2021). Additionally, false color RGB combinations in the near-infrared (NIR) and shortwave infrared (SWIR) regions of the spectrum have been commonly utilized to discern discrete floating plastics from space (Sannigrahi et al., 2022). Similarly, some studies have revealed the employability of hyperspectral imagery to effectively identify plastics. Specifically, the most promising set of potential bands for plastic detection is situated in the NIR range (890–970 nm), along with three bands in the SWIR spectrum (1160–1250 nm, 1360–1440 nm, and 1680–1760 nm). Furthermore, peaks for the absorption of marine plastics are identified at 931, 1215, 1417, and 1732 nm (Garaba and Dierssen, 2018). On the other hand, active sensors include light detection and ranging (LiDAR) and radar systems in detection of plastics and marine debris (Sabatini et al., 2015). The working principle of these active sensors include by emitting signals and measuring the return time to detect the geographical position and characteristics of items, which makes them highly effective in detecting plastic and other marine debris.

17.3.2 RS Platforms

In the endeavor to assess plastic pollution in aquatic environments and along coastlines, various RS platforms have been successfully tested at close range, mid-range, and far distances (Gnann et al., 2022). These platforms encompass satellites, aircraft, balloons, ships, elevated points such as bridges, and UAVs or autonomous underwater vehicles (AUVs). The three main platforms for RS of marine plastic debris are proximal spectrometers, airborne including UAVs, and spaceborne data acquisition. In addition, nearshore debris can be monitored using in situ, vessel-based photographs, terrestrial photographs, video footages, and simple digital camera photographs (Ramos, 2022).

17.3.2.1 Airborne Platforms for Remote Detection of Plastic Debris in Marine and Coastal Environments

Data and imagery acquired from airborne platforms, such as airplanes, UAVs, balloons, and kites, have been employed for monitoring and detecting marine plastic litter (Veettil et al., 2022). In this context, various applications have emerged over the past years, including a combination of in situ mass measurements and balloon-assisted aerial photography for quantifying total litter mass. Additionally, techniques have been developed for the detection of colored macroplastic debris using sequential images captured by webcams. This involves the generation of color references using a uniform color space to identify plastic pixels (Kako et al., 2018). While webcam photography proves to be valuable for sequential monitoring of nearshore marine environments and offers cost-effectiveness, its implementation necessitates elevated ground or a hill for camera installation, a stable electricity supply, and a file transfer protocol for sending imagery to another location. Furthermore, it is subjected to restrictions imposed by land features and coastal landforms. Moreover, aerial imagery, encompassing optical (RGB), multispectral, and hyperspectral data, has been affirmed as a more practical and effective means of monitoring floating marine plastic litter when compared to observer-based monitoring. Recent studies have emphasized three key advantages of employing aerial photography over observer-based surveys for monitoring floating marine plastic litter (Veettil et al., 2022). First, aerial surveys provide higher accuracy by allowing precise designation of either a priori or determined a posteriori (knowledge based solely on experience or personal observation) from photographs, facilitating the estimation of the exact size of the target when the image ground sampling distance is known (Unger et al., 2021). Second, the use of aerial imagery reduces human error as it serves as a permanent record, minimizing the likelihood of inaccuracies, and permits data reanalysis at any time (Merlino et al., 2020). Third, the approach offers enhanced safety and cost-effectiveness, posing fewer risks to human safety and proving to be relatively economical (Maximenko et al., 2019). Despite not being the most cost-effective option, aerial photographs from airplanes can be employed for the accurate quantification of marine macro debris abundance through the application of appropriate georeferencing techniques.

Presently, the most widely used airborne platform for marine plastic litter monitoring is UAVs or drones, owing to their ease of operation and the capability to obtain high-resolution data in a cost-effective manner. Commonly utilized UAV optical data includes RGB or multispectral imagery, while hyperspectral data is predominantly acquired from airplanes. Especially, UAV technology has advanced significantly, with increasingly sophisticated hyperspectral sensors being mounted on drones in recent developments. Previous research has utilized UAV-based data to map and monitor the spatial distribution of marine litter on coastal dunes, an area that has received limited attention from scientists in comparison to sandy shores (Andriolo et al., 2020). UAV-based data offers several key advantages, including a high frequency of image acquisition, high spatial resolution, the capability to capture images below the clouds, and high mobility (Merlino et al., 2020). The accurate estimation of the spatial and temporal distribution of marine plastic debris across various zones of a beach, crucial for understanding the dynamics of debris accumulation over time, can be achieved through the utilization of high-resolution data obtained from UAVs. This is catered by the ability to acquire data at any desired time, enabling detailed investigations. One of the practical advantages of employing UAVs for mapping beached marine litter is the experiment's repeatability in a short time frame. However, it is important to note that partially buried objects pose a challenge as they are not easily detected through this method. Despite being costly, hyperspectral sensors can be employed to overcome the limitations associated with detecting partially buried objects. Yet, one of the primary challenges of UAVs lies in their restricted flight time and the expenses associated with repeated campaigns (Escobar-Sánchez et al., 2022).

17.3.2.2 Spaceborne Platforms for Remote Detection of Marine Plastic Litter in Marine and Coastal Environments

Spaceborne platforms possess the capability to gather data from various sensor types mounted on them, including optical (multispectral and hyperspectral), LiDAR, and radar sensors. Multispectral and hyperspectral data obtained from spaceborne platforms play a vital role in mapping marine plastic debris. Satellite RS stands as the primary technique for collecting high-quality, timely, and standardized optical imagery on a global synoptic scale. The essential criterion for detecting marine and shoreline debris from satellite data is achieving sufficient spatial resolution, and a noteworthy advantage lies in the ability for repeated monitoring, facilitated by data obtained from orbiting satellites (Veettil et al., 2022). Multispectral and hyperspectral data obtained from satellite sensors are particularly susceptible to weather conditions, such as the presence of cloud cover. Moreover, the initial cost of spaceborne RS is higher when contrasted with airborne data acquisition. In recent studies, Sentinel-2 multispectral data have been employed for mapping marine floating debris by calculating spectral indices. Additionally, spectral indices derived from Sentinel-2 data have been utilized for the detection of floating plastic patches (Sannigrahi et al., 2022). Unlike clear water, which efficiently absorbs NIR to SWIR radiation, floating materials such as macroalgae and macroplastics exhibit reflection in the near-infrared (Mukonza and Chiang, 2022). This characteristic is pivotal in spectrally differentiating aggregated materials floating on the ocean surface when observed from space. Furthermore, hyperspectral satellite data, exemplified by PRecursore IperSpettrale della Missione Applicativa (PRISMA), has been employed for the identification of floating marine plastic litter based on their distinctive spectral properties. Thus, it is clear that the trade-off between spatial and spectral resolution is an important consideration in selecting hyperspectral satellite data for mapping marine plastic litter. Besides, the quantification of beached plastic debris can also be accomplished through spectral analysis of optical satellite data (Acuña-Ruz et al., 2018). Even with very high-resolution imagery, the identification of individual plastic items on beaches is challenging without additional spectral information.

17.3.2.3 Underwater Platforms for Monitoring Subsurface and Submerged Marine Plastic Debris

Conducting autonomous surveying and identification of submerged plastic debris presents a nontrivial challenge. Conventional RS methods, for instance, satellite imagery or UAVs, are limited to surface-level observations. To address underwater object identification, the most effective approach involves utilizing sonar sensing technology. However, interpreting sonar images poses significant difficulties, particularly in the case of small objects. This challenge is exacerbated in the context of large field-of-view sensors like synthetic aperture sonar (SAS) and side-scan sonar (SCS) (Hansen et al., 2011). Among the various sonar technologies available, forward-looking sonars (FLSs) emerge as the most reliable option for the designated task. FLS systems are capable of generating high-resolution and high-frame-rate sonar images, catering detailed coverage down to 2.3 mm per pixel. This particular characteristic makes FLS ideal for the identification of small objects, especially when integrated into AUVs. This approach encompasses remotely operated vehicles (ROVs), and these AUVs have been employed to capture images for the detection and identification of nonbuoyant marine litter as well (Veettil et al., 2022).

17.3.2.4 LiDAR, Radar, and Sonar RS for Marine Debris Monitoring

While not as cost-effective as employing spaceborne multispectral or hyperspectral data, some recent studies have employed LiDAR data for quantifying marine debris abundance based on its shape. An advantage of laser scanners is their capability to estimate the volume of debris on beaches by identifying their three-dimensional structure (Szafarczyk and Toś 2022). Furthermore, LiDAR data can offer depth-resolved information through the water column, proving beneficial in detecting submerged litter in less turbid waters. However, key limitations in using LiDAR data for mapping marine plastic debris include the high cost of LiDAR sensors and a reduction in mapping accuracy as beach topography becomes more undulated, especially in the case of beached debris.

Spaceborne radar technology offers the capability to generate all-weather, all-day, wide-swath images of the sea, enabling accurate data acquisition even in the presence of cloud cover (Veettil et al., 2022). The remarkable flexibility of spaceborne SAR operations, combined with its wide-swath high-resolution imaging capabilities, constitutes a unique and powerful tool for ocean monitoring. Recently, some studies have applied X-band radar to monitor small garbage islands, specifically floating debris. However, the efficacy of this method was limited to discerning the target only under calm wind conditions. A handful of studies have explored the use of sonar data for mapping subsurface and submerged marine plastic debris, with floating LiDAR systems being employed for detecting submerged marine debris, including plastic and fiber materials. A significant advantage of using sonar data lies in the potential for high-resolution and high-frame images, making it suitable for the identification of small objects on a millimeter scale. Yet, the resolution of the data is contingent on sonar characteristics and the distance from the target. A notable drawback is the challenge of detecting deformable objects, a common occurrence in submerged debris due to water pressure. Besides, high-flow velocities in the marine environment can lead to noise and stripping in sonar images. Figure 17.1 depicts the overview of the different technologies that are suitable or have the potential to detect plastic on the seafloor.

Semiautomatic Recognition of Anthropogenic Macrodebris Using LiDAR Technology – Nanhui Beach, Shanghai, China This study was conducted at Nanhui Beach, Shanghai, China, and three additional field studies were carried out to assess marine debris classification at a beach in Beihai, China. In this study, LiDAR, an active surveying technology was employed to generate point-cloud datasets. A point-cloud refers to a collection of points resembling a cloud, representing a vector dataset distinct from a raster image. These datasets, serving as the digitized and discretized representation

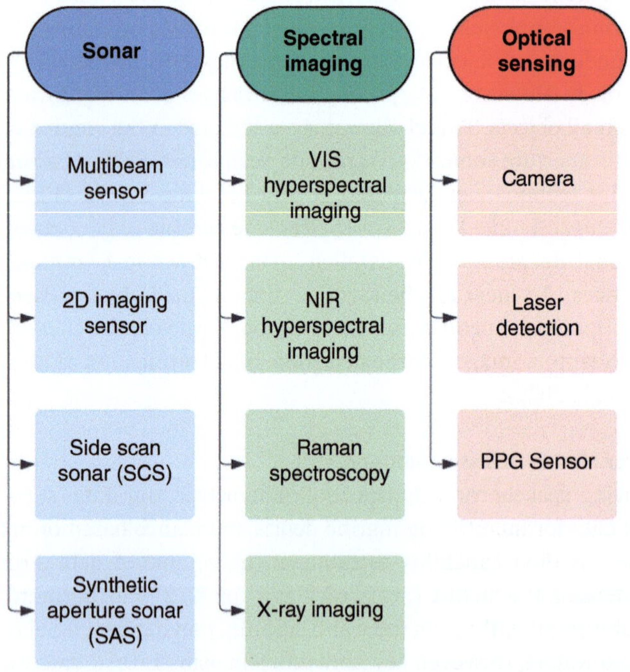

Figure 17.1 Conceptual overview of the different technologies that are suitable or have the potential to detect plastic on the seafloor.

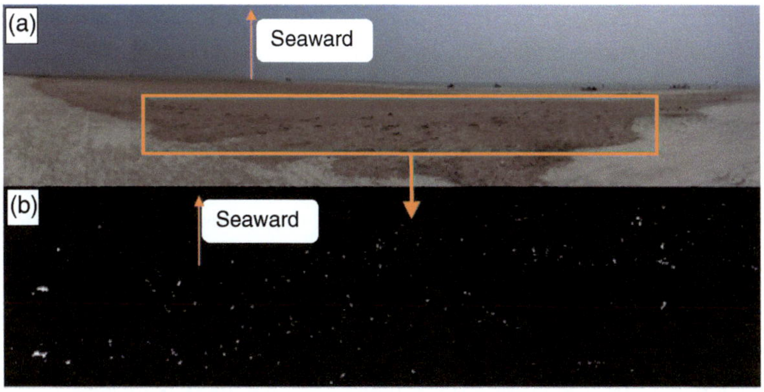

Figure 17.2 (a) Validation of the semiautomatic recognition method for marine debris on Beihai beach (the box represents the study area where the statistical data of the debris were manually obtained); (b) distribution of debris identified by LiDAR in the study area. *Source:* Reproduced from Ge et al., 2016/Springer Nature/CC BY 4.0.

of real objects through laser scanning, capture the spatial 3D features of the actual objects (Ge et al., 2016).

The results indicate that compared to manual methods, LiDAR identified significantly more debris for each type, except for glass debris. Despite challenges in identifying glass debris, the mean accuracy of LiDAR was found to be 75.4%. Yet, during validation experiments, successful detection of glass debris proved elusive. An interesting observation is that both glass and sand share SiO_2 as their major chemical component. This commonality may be a key factor contributing to the misclassification of glass, highlighting a limitation of the proposed method. The beach experiment results suggest that marine debris detection with LiDAR is characterized by low time consumption and high recognition efficiency compared to manual methods (Figure 17.2).

17.4 Methods for Mapping Marine Plastic Litter from Remote Sensing Data: Object-Based Image Analysis (OBIA) Using Machine Learning (ML) and Deep Learning (DL)

In recent years, AI has gained attention in marine litter research by providing automatic visual recognition and identification of marine plastics. Image processing methods employed for shoreline and marine plastic litter detection vary from manual photointerpretation to ML and object-based image analysis (OBIA) (Zaaboub et al., 2023). Particularly, when utilizing satellite data with lower resolution, compared to centimeter-level UAV imagery, manual mapping becomes less effective, necessitating the use of automatic or semiautomatic image processing algorithms. Moreover, the repeatability of automatic methods using different orthophotos provides an advantage over manually delineating beached marine litter and can serve as a viable substitute for manual mapping in beach clean-up operations (Kandrot and Holloway, 2020). Consequently, the adoption of ML has become increasingly prevalent, and the incorporation of in situ observations can enhance the development of training datasets. Accordingly, in ML, the data collected from various sources are utilized to train the machine to detect, predict, and analyze a target.

Furthermore, ML requires image data samples relevant to a specific task whose statistical structure can be analyzed in order to identify rules for automating the image classification, detection, or identification tasks. Similar to color space-based approaches, ML algorithms utilize representations or transformations of data for these rules. The model therefore searches for the best representations or transformations from a predefined space of possibilities. In the field of macroplastic or litter detection and identification, the most widely used ML algorithms include random forest (RF) or support vector machine (SVM) classifiers (Gonçalves et al., 2020a). However, ML can neither extract locally generated signatures that are present in a small percentage of the RS image nor can it extract long-time series data. Hence, this necessitates the need for new data mining algorithms that would eventually support the developing of new DL approaches. In contrast to traditional ML methods that necessitate domain-specific manually crafted feature extraction, DL techniques possess the capability to automatically discern and acquire latent data representations directly from raw data and provide end-to-end learning mechanisms.

Automated identification of marine debris shows several challenges. First, there is a wide array of marine debris types, each exhibiting significant intra-class variations. For example, plastic bottles, bags, and cups may possess distinct shapes but fall under the same category of marine debris (plastics) (Marin et al., 2021). Second, the detection of waste objects should be robust across various degradation phases, regardless of recording angles, water turbidity, or illumination conditions. Furthermore, the training of deep neural networks demands a substantial amount of annotated data. However, acquiring a sufficiently large and diverse dataset for image-based marine debris classification is often costly and not always feasible. To address the scarcity of labeled training data, a common and effective strategy is called transfer learning (Marin et al., 2021). This involves transferring knowledge from a source domain to the desired target domain. The concept revolves around leveraging features learned by a neural network pretrained on an extensive dataset, such as the ImageNet dataset, which comprises 1.2 million training images from 1000 classes. This approach proves beneficial when applying the pretrained network to a new task, such as the classification of marine debris. As an example, Kylili et al. (2019) used ImageNet data alongside artificial intelligence (AI) tools for image classification using a convolutional neural network (CNN), specifically the bottleneck method. This approach was utilized for detecting both shoreline and floating debris.

Politikos et al. (2021) used a region-based CNN for the automated detection of marine debris on the seafloor in Ermoupolis Bay, Syros Island, Greece. However, it is important to note that the accuracy of marine debris classification can be influenced by sunlight conditions and associated terrain shading effects (Fallati et al., 2019). Moreover, DL methods have also been successfully applied to AUV multispectral data for mapping underwater marine debris and sonar data for mapping submerged marine debris. The segmentation threshold method, when applied to UAV data, has been asserted as superior to the maximum-likelihood classification method in certain studies (Abdu and Noor, 2022). Nevertheless, it is crucial to note that the success of the threshold value applied may vary across different study sites and the spectral reflectance of various debris materials. It is noteworthy that a single sensor has limitations in detecting various types of plastic debris due to constraints in spectral resolution. Zielinski et al. (2022) emphasized the significance of developing adequate sensor types with multiple stressor assessment capabilities in remote areas, including metadata storage and assessment capacities. Hence, there's a need to develop more advanced sensor types that can assess multiple stressors in remote areas and have robust metadata storage and assessment capabilities.

17.4.1 AI Workflow for Marine Litter Detection

A typical AI workflow, when applied to a marine macrolitter dataset encompassing various data types and tasks, is illustrated in Figure 17.3. The initial step involves retrieving the litter dataset, which may comprise images, orthomosaics/orthophotos, or tabular data. ML algorithms exhibit

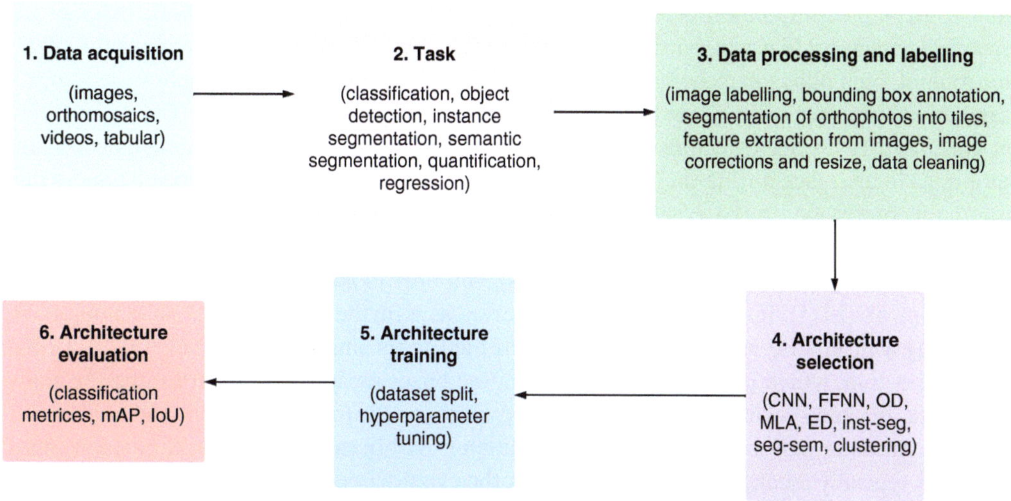

Figure 17.3 Generic view of AI workflow, as applied to a marine litter dataset for common data types and tasks (convolutional neural network (CNN); object detection architecture (OD); feedforward neural network (FNNN); machine learning algorithm (MLA); encoder–decoder architecture (ED); instance segmentation architecture (Inst-Seg); semantic segmentation architecture (Seg-Sem), classification metrics: accuracy, recall, precision, FI-score; object detection/segmentation metrics: mean average precision (mAP)). *Source:* Adapted from Politikos et al., 2023.

a versatile capability to handle diverse types of data, including but not limited to tabular data, images, videos, and documents. Second, the AI task to be addressed is defined. The primary tasks include classification, object detection, instance segmentation, semantic segmentation, quantification (via segmentation), and regression (Yang and Yu, 2021). Image classification refers to the process of determining the content or representation of an image. This field witnessed a significant advancement in computer vision with the introduction of DL. CNNs stand out as the most prominent type of networks tailored for image classification tasks. In ML, the two prevailing approaches commonly employed with real-world datasets are supervised learning, where the model is trained on labeled data, and unsupervised learning, which operates without labeled data (Saravanan and Sujatha, 2018). Besides, regression predictive modeling technique prediction is based on a set of independent variables, regarded as potential influencers of the target variable. The training process involves using a training set to train a regressor, and subsequently, the performance of the trained model is assessed using a test set. In unsupervised learning, the algorithm is presented with unlabeled data and tasked with autonomously identifying inherent structures within the dataset. Common unsupervised tasks include clustering (grouping data items based on similarities or differences), dimensionality reduction (eliminating nonessential variables in a dataset), association rule mining (identifying meaningful relationships between variables), outlier/anomaly detection (identifying data items deviating from normal behavior in a dataset), and density estimation (estimating a continuous density field from discretely sampled data items) (Politikos et al., 2023).

In contrast, supervised learning requires a labeled dataset for training the algorithm, allowing it to acquire experience. Following training, the algorithm is then employed to make predictions on new datasets. The two most prevalent supervised tasks in ML are classification and regression (Osisanwo et al., 2017). Classification involves assigning data to specific classes. Object detection, a task in computer vision, encompasses both classifying and localizing objects within an image. More precisely, object detection seeks to determine the presence of objects belonging to specified categories (e.g. bag, bottle, paper, can, and cup) and, if identified, to provide the spatial location of each object enclosed by a bounding box. Object detection approaches are typically categorized into two main types: two-stage detectors and one-stage detectors (Carranza-García et al., 2020). Two-stage detectors employ a sliding window to generate candidate object region proposals in an image during the initial stage, and subsequently, classify and localize them during the second stage. In the existing literature, an extensive array of object detection architectures is available, encompassing both two-stage detectors, such as R-CNN, and one-stage detectors, including single shot detectors (SSDs), you only look once (YOLO), and EfficientDet (Zagitov et al., 2024). Instance segmentation serves as a further extension of object detection, producing a pixel-wise mask for the object identified in an image (Wang et al., 2020). Unlike object detection, which yields rectangular bounding boxes incapable of determining the shape of objects, instance segmentation offers more quantitative information about the object, enabling the estimation of their surface area or perimeter. Within these networks, an instance mask branch is typically incorporated in parallel with the classification and bounding components of object detection architectures. Prominent DL models designed to address this aspect include Mask R-CNN, PANet, YOLACT, and many others. Object-based semantic segmentation stands out as another pivotal topic in computer vision, aiming to assign a class label to every pixel in the image (Ulku and Akagündüz 2022). This task is commonly denoted as "dense prediction." Semantic segmentation is also proved to be valuable in various applications, such as in autonomous driving for identifying frontal objects like roads, vehicles, pavements, and humans, as well as in medical image diagnostics. The literature introduces a range of DL architectures tailored for semantic segmentation, including fully convolutional networks, encoder–decoder models, attention-based models, recurrent neural network-based models, and dilated convolutional models. Notable architectures designed for semantic segmentation tasks include ReSeg, SegNet, and DeepLab (Garcia-Garcia et al. 2017).

In the third step, preprocessing the dataset is imperative to prepare the data for parsing by AI algorithms. Typical tasks during the data preprocessing phase include image labeling, bounding box annotation, cleaning of tabular data, feature extraction from images, extraction of orthophotos from orthomosaic images, segmentation of orthophotos into tiles, and enhancement of image quality (Politikos et al., 2023). Fourth, one or more algorithms (for comparison purposes) are chosen to address the specified task. The prevalent architectures used in recent studies include CNN, OD, FFNN, MLA, encoder–decoder architecture (EDA), Inst-Seg, Seg-Sem, and clustering (Politikos et al., 2023). Fifth, the dataset is partitioned into training and validation subsets. These subsets are utilized to train the AI algorithm(s) while fine-tuning hyperparameters. Then in the sixth step, a dataset unfamiliar to the training process, known as the test set (created during the dataset split), is utilized to assess the predictive performance of the trained algorithm using task-specific evaluation metrics. Assessing the performance of an ML–DL algorithm is crucial for gauging its predictive accuracy. Particularly in tasks such as image classification, where the system is expected to effectively differentiate between categories like "bottle" and "bag," the evaluation process is vital. Common metrics used for evaluating image classification tasks include accuracy, recall, precision, and F1-score. In the context of segmentation tasks, two prevalent evaluation metrics are employed. The first is mean average precision (mAP), and the second is intersection over union (IoU). These metrics provide quantitative measures to assess the accuracy and effectiveness of the segmentation performed by the algorithm.

17.4.1.1 Mapping Marine Litter with Unmanned Aerial Systems: Cabedelo Beach, Western Portuguese Coast

This study was carried out at Cabedelo Beach, a sandy coastal stretch situated on the western Portuguese coast facing the North Atlantic Ocean. The research focuses on the application and performance comparison of unmanned aerial system (UAS)-derived orthophoto analysis techniques, specifically manual image screening and ML methods. The findings indicate that manual image screening of the orthophoto is preferable when detailed characterization of marine litter, including type and material, is necessary. However, it is noted that the description of buried, decolored, or broken objects may be challenging or even impossible using this method. The comparison between the two automated techniques reveals that RF yielded better results than CNN. RF exhibited fewer false positives than CNN, and its marine litter density map demonstrated greater similarity to the one produced by manual screening.

While the manual procedure, involving the manual digitization of object contours in the training areas, can be laborious and time-consuming, CNN offers a more automated alternative, eliminating the need for manual feature selection. Both ML techniques, however, share a limitation in their inability to categorize marine litter items. Figure 17.4 shows the detection results, with details about the marine litter items correctly identified (TP), objects not detected (FN), and objects misled as marine litter items (FP) by the automated techniques, in comparison with the MS records (Gonçalves et al., 2020b).

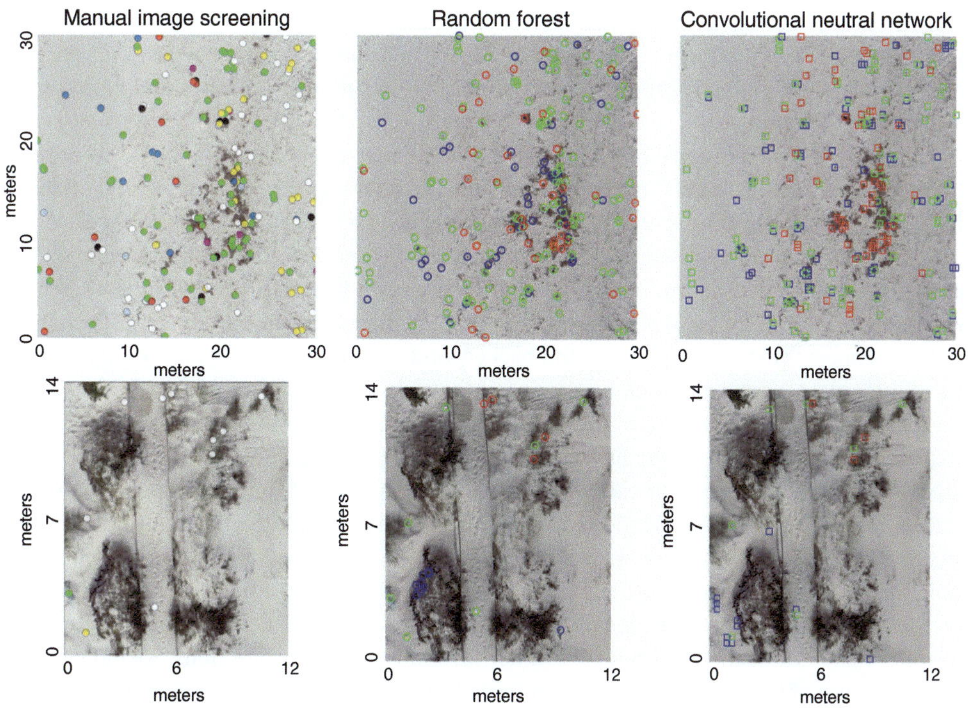

Figure 17.4 Comparison of the three detection techniques in the training areas. For the manual image screening technique, marine litter items are plotted as filled circles with the respective color recognized on the orthophoto. Regarding the automated techniques (random forest and convolutional neural network), green represents the true positive (TP), red the false negative (FN), and blue the false positive (FP) detections. *Source:* Reproduced from Gonçalves et al., 2020b/Elsevier.

17.5 Challenges Using RS/AI Technologies to Detect Marine Plastics

Identifying litter in real-world marine environments presents a considerable challenge for AI algorithms. The automated recognition of coastal and marine litter can be influenced by complex natural backgrounds, including vegetation, the presence of natural wood litter, and seashells interspersed among litter items. Additionally, identifying plastic debris through RS becomes more challenging due to deterioration, both for floating and beached plastics, as well as their movements in currents for floating debris. The processes of accretion and erosion along the coasts play a crucial role in the context of marine plastic debris, as a substantial portion of plastic litter goes unaccounted for when buried under the soil, making it nearly impossible to detect using RS approaches. The same problem arises in the case of submerged plastic litter buried under the seafloor. Thus, the easiness of detecting marine plastic debris decreases from beached debris to floating plastic to submerged plastic litter. Moreover, litter objects may exhibit substantial intraclass variability, such as items of the same type having different shapes (e.g. bags), and interclass similarity, where different types share similarities (e.g. cloth and plastic segments). These factors can significantly impact predictability in the identification process.

Detection and mapping of marine plastic litter with different sizes are the next challenge. Current photographic technology does not have enough capability to detect objects with sizes less than 10 × 10 cm and, hence, only mega, macroplastics, and mesoplastics higher than this size can be detected using photographic surveys. Some studies have attempted to address the detection of marine microplastics using hyperspectral data. Shan et al. (2019) used hyperspectral imagery to obtain information on microplastic organic particles, polymer types, and particle size. Nevertheless, cost-effective methods for estimating marine microplastics at a global scale are still under development.

In satellite images, the presence of high cloud coverage or erroneous annotations by experts in intricate scenarios (e.g. distinguishing between a ship and its wake) can lead to mispredictions of floating items. Additionally, the spatial resolution of satellite images has specific limitations. While smaller plastic targets can be discerned and identified under specific conditions, satellite data is primarily suited for the detection and monitoring of large-sized litter (Shen et al., 2024). Yet the false predictions can also arise when litter items share similar spectral signatures, especially in cases of low concentrations compared to the background signal. The acquisition of high-quality images of litter plays a crucial role in the success of the task. For example, the distance between a sampling system and an object can impact the accurate classification of both beached/dune and floating litter. At higher altitudes, identification becomes challenging due to various water surface conditions, beached landscapes, or variations in the color and size of targets (Papakonstantinou et al., 2016). Armitage et al. (2022) reported that the object detection of floating litter items can be influenced when they occupy a low pixel percentage within an image, proposing the option of cropping the image into smaller portions. Therefore, a proper orientation is vital for a camera to capture clear images of seafloor marine litter, preventing any loss in model accuracy. Additionally, aerial surveys should ideally be conducted under favorable weather conditions, including calm winds; the absence of fog, mist, and rain; and minimal sunlight reflections, to ensure the collection of high-quality data.

Remarkably, from the available literature, it is evident that the majority of studies on detecting and mapping marine plastic debris focus on either beached or floating objects, with relatively few accountings for submerged/underwater litter. Detecting floating marine debris is comparatively more manageable using data from various platforms, including airborne, spaceborne, or terrestrial sources (Martínez-Vicente et al., 2019). Nevertheless, achieving global-scale mapping of submerged plastic debris remains a challenging task. In the case of nearshore, shallow water submerged plastic, effective algorithms incorporating water column correction can be applied to multispectral data obtained from airborne or spaceborne sources. Unlike submerged vegetation, such as seagrasses, which exhibits a common factor influencing reflectance characteristics (e.g. the presence of chlorophyll), submerged plastic with varying chemical compositions (e.g. PET, foam) and differing reflectance properties cannot be addressed with a single image processing algorithm. This poses a significant challenge due to limitations in the spectral resolution of available RS data in the market.

RS holds considerable potential for mapping and monitoring marine plastic litter; however, the validation of approaches through standard data is imperative. Access to in situ data of sufficient quality would substantially enhance the accuracy of RS applications for marine and beached plastic litter. This improvement is crucial for transitioning studies from a regional scale to a global scale, as validation needs to be performed across different regions (Martínez-Vicente et al., 2019). In situ data plays a vital role in ML algorithms, serving as the necessary training dataset (Orth, 2021). Selecting in situ data for validating remotely sensed data presents a challenge due to the nonuniqueness of spectral properties in most samples, stemming from the various materials they comprise. The background in images (beached, floating, and submerged) is another significant factor influencing the detection and mapping of marine plastic litter (Veettil et al., 2022). ML techniques applied to high-resolution data have demonstrated success in homogeneous backgrounds (e.g. water and sand). However, their performance tends to diminish in areas with a rocky or gravel beach background. As a result, validating data from multiple sources with different spatial and spectral resolutions presents a challenge. For example, validating the quantification of floating plastic debris is complicated by the need for a universal mathematical equation to estimate debris volume, considering varying densities.

17.6 Conclusion

In conclusion, the detection and identification of marine and coastal plastics are essential to get an overall picture of the magnitude of socioeconomic and health-related damage that plastic pollution brings along. Traditional monitoring methods, reliant on manual surveys and visual counting, have proven insufficient, hampered by limitations in scale and efficiency, mainly to assess the plastic content in coastal and marine areas. The integration of RS technologies with AI-based analysis, including ML and DL, emerges as a transformative solution. This synergy not only offers a more automated and scalable approach but also holds the promise of unraveling intricate pathways and distribution patterns of plastics.

Whether implementing classification methods or more advanced techniques like AI and ML, existing technologies face certain constraints as described above. The most significant bottleneck, however, is the insufficient availability of high-quality training data for the development of effective detection algorithms. Therefore, obtaining high-quality in situ-validated spectral measurements, along with relevant ancillary metadata of marine litter, is essential. Furthermore, calibration and validation datasets also has become integral requirements for the future development of efficient RS detection methodologies.

This chapter emphasizes that these essential steps are crucial for achieving accurate, effective, and large-scale monitoring of marine and coastal plastics through RS and AI platforms. Embracing these advanced methodologies represents a critical step toward a more comprehensive understanding of the plastic pollution crisis, which is essential for implementing effective mitigation and conservation efforts in marine and coastal ecosystems.

References

Abalansa, S., El Mahrad, B., Vondolia, G.K. et al. (2020). The marine plastic litter issue: a social-economic analysis. *Sustainability* 12 (20): 8677.

Abdu, H. and Noor, M.H.M. (2022). A survey on waste detection and classification using deep learning. *IEEE Access* 10: 128151–128165.

Acuña-Ruz, T., Uribe, D., Taylor, R. et al. (2018). Anthropogenic marine debris over beaches: spectral characterization for remote sensing applications. *Remote Sensing of Environment* 217: 309–322.

Agamuthu, P., Mehran, S.B., Norkhairah, A., and Norkhairiyah, A. (2019). Marine debris: a review of impacts and global initiatives. *Waste Management & Research* 37 (10): 987–1002.

Andriolo, U., Gonçalves, G., Bessa, F., and Sobral, P. (2020). Mapping marine litter on coastal dunes with unmanned aerial systems: a showcase on the Atlantic Coast. *Science of the Total Environment* 736: 139632.

Armitage, S., Awty-Carroll, K., Clewley, D., and Martinez-Vicente, V. (2022). Detection and classification of floating plastic litter using a vessel-mounted video camera and deep learning. *Remote Sensing* 14 (14): 3425.

Bhuyan, M.S., Venkatramanan, S., Selvam, S. et al. (2021). Plastics in marine ecosystem: a review of their sources and pollution conduits. *Regional Studies in Marine Science* 41: 101539.

Carranza-García, M., Torres-Mateo, J., Lara-Benítez, P., and García-Gutiérrez, J. (2020). On the performance of one-stage and two-stage object detectors in autonomous vehicles using camera data. *Remote Sensing* 13 (1): 89.

Escobar-Sánchez, G., Markfort, G., Berghald, M. et al. (2022). Aerial and underwater drones for marine litter monitoring in shallow coastal waters: factors influencing item detection and cost-efficiency. *Environmental Monitoring and Assessment* 194 (12): 863.

Fallati, L., Polidori, A., Salvatore, C. et al. (2019). Anthropogenic marine debris assessment with unmanned aerial vehicle imagery and deep learning: a case study along the beaches of the Republic of Maldives. *Science of the Total Environment* 693: 133581.

Garaba, S.P. and Dierssen, H.M. (2018). An airborne remote sensing case study of synthetic hydrocarbon detection using short wave infrared absorption features identified from marine-harvested macro-and microplastics. *Remote Sensing of Environment* 205: 224–235.

Garcia-Garcia, A., Orts-Escolano, S., Oprea, S., Villena-Martinez, V., & Garcia-Rodriguez, J. (2017). A review on deep learning techniques applied to semantic segmentation. *arXiv preprint arXiv:1704.06857*.

Ge, Z., Shi, H., Mei, X. et al. (2016). Semi-automatic recognition of marine debris on beaches. *Scientific Reports* 6 (1): 25759.

Geyer, R. (2020). Production, use, and fate of synthetic polymers. In: *Plastic Waste and Recycling* (ed. T.M. Letcher), 13–32. Academic Press.

Gnann, N., Baschek, B., and Ternes, T.A. (2022). Close-range remote sensing-based detection and identification of macroplastics on water assisted by artificial intelligence: a review. *Water Research* 222: 118902.

Gonçalves, G., Andriolo, U., Gonçalves, L. et al. (2020a). Quantifying marine macro litter abundance on a sandy beach using unmanned aerial systems and object-oriented machine learning methods. *Remote Sensing* 12 (16): 2599.

Gonçalves, G., Andriolo, U., Pinto, L., and Duarte, D. (2020b). Mapping marine litter with unmanned aerial systems: a showcase comparison among manual image screening and machine learning techniques. *Marine Pollution Bulletin* 155: 111158.

Hansen, R.E., Callow, H.J., Sabo, T.O., and Synnes, S.A.V. (2011). Challenges in seafloor imaging and mapping with synthetic aperture sonar. *IEEE Transactions on geoscience and Remote Sensing* 49 (10): 3677–3687.

Kako, S.I., Isobe, A., Kataoka, T. et al. (2018). Sequential webcam monitoring and modeling of marine debris abundance. *Marine Pollution Bulletin* 132: 33–43.

Kandrot, S., and Holloway, P. (2020). Applications of drone technology for sustainable development of the coastal zone: A literature review. A Literature Review, Work Package 2: Technological Innovation for Sustainable Development Deliverable T2.1.1: Project report. In Sustainable Resilient Coast; The European Union: Brussels, Belgium, 2020.

Kühn, S. and Van Franeker, J.A. (2020). Quantitative overview of marine debris ingested by marine megafauna. *Marine Pollution Bulletin* 151: 110858.

Kylili, K., Kyriakides, I., Artusi, A., and Hadjistassou, C. (2019). Identifying floating plastic marine debris using a deep learning approach. *Environmental Science and Pollution Research* 26: 17091–17099.

Madricardo, F., Ghezzo, M., Nesto, N. et al. (2020). How to deal with seafloor marine litter: an overview of the state-of-the-art and future perspectives. *Frontiers in Marine Science* 7: 505134.

Marin, I., Mladenović, S., Gotovac, S., and Zaharija, G. (2021). Deep-feature-based approach to marine debris classification. *Applied Sciences* 11 (12): 5644.

Martínez-Vicente, V., Clark, J.R., Corradi, P. et al. (2019). Measuring marine plastic debris from space: initial assessment of observation requirements. *Remote Sensing* 11 (20): 2443.

Maximenko, N., Corradi, P., Law, K.L. et al. (2019). Toward the integrated marine debris observing system. *Frontiers in Marine Science* 6: 447.

Merlino, S., Paterni, M., Berton, A., and Massetti, L. (2020). Unmanned aerial vehicles for debris survey in coastal areas: long-term monitoring programme to study spatial and temporal accumulation of the dynamics of beached marine litter. *Remote Sensing* 12 (8): 1260.

Mukonza, S.S. and Chiang, J.L. (2022). Satellite sensors as an emerging technique for monitoring macro-and microplastics in aquatic ecosystems. *Water Emerging Contaminants & Nanoplastics* 1 (4): 17.

Napper, I.E. and Thompson, R.C. (2020). Plastic debris in the marine environment: history and future challenges. *Global Challenges* 4 (6): 1900081.

Orth, R. (2021). Global soil moisture data derived through machine learning trained with in-situ measurements. *Scientific Data* 8 (1): 1–14.

Osisanwo, F.Y., Akinsola, J.E.T., Awodele, O. et al. (2017). Supervised machine learning algorithms: classification and comparison. *International Journal of Computer Trends and Technology (IJCTT)* 48 (3): 128–138.

Papakonstantinou, A., Topouzelis, K., and Pavlogeorgatos, G. (2016). Coastline zones identification and 3D coastal mapping using UAV spatial data. *ISPRS International Journal of Geo-Information* 5 (6): 75.

Pasquier, G., Doyen, P., Kazour, M. et al. (2022). Manta net: the golden method for sampling surface water microplastics in aquatic environments. *Frontiers in Environmental Science* 10: 811112.

Pham, T.H., Do, H.T., Thi, L.A.P. et al. (2021). Global challenges in microplastics: from fundamental understanding to advanced degradations toward sustainable strategies. *Chemosphere* 267: 129275.

Pilapitiya, P.N.T., and Ratnayake, A.S. (2024). The world of plastic waste: A review. *Cleaner Materials* 11: 100220. https://doi.org/10.1016/j.clema.2024.100220

Politikos, D.V., Fakiris, E., Davvetas, A. et al. (2021). Automatic detection of seafloor marine litter using towed camera images and deep learning. *Marine Pollution Bulletin* 164: 111974.

Politikos, D.V., Adamopoulou, A., Petasis, G., and Galgani, F. (2023). Using artificial intelligence to support marine macrolitter research: a content analysis and an online database. *Ocean & Coastal Management* 233: 106466.

Ramos, E.A. (2022). Adapting small unmanned aerial systems for behavioral research with coastal marine mammals. Doctoral dissertation, City University of, New York. Rasee, A.I., Awual, E., Rehan, A.I., Hossain, M.S., Waliullah, R.M., Kubra, K.T., et al., 2023. Efficient separation, adsorption, and recovery of Samarium (III) ions using novel ligand-based composite adsorbent. Surface. Interfac. 41, 103276.

Ryan, P.G., Pichegru, L., Perolod, V., and Moloney, C.L. (2020). Monitoring marine plastics-will we know if we are making a difference? *South African Journal of Science* 116 (5–6): 1–9.

Sabatini, R., Richardson, M.A., Gardi, A., and Ramasamy, S. (2015). Airborne laser sensors and integrated systems. *Progress in Aerospace Sciences* 79: 15–63.

Salgado-Hernanz, P.M., Bauzà, J., Alomar, C. et al. (2021). Assessment of marine litter through remote sensing: recent approaches and future goals. *Marine Pollution Bulletin* 168: 112347.

Sannigrahi, S., Basu, B., Basu, A.S., and Pilla, F. (2022). Development of automated marine floating plastic detection system using Sentinel-2 imagery and machine learning models. *Marine Pollution Bulletin* 178: 113527.

Saravanan, R., and Sujatha, P. (2018). A state of art techniques on machine learning algorithms: a perspective of supervised learning approaches in data classification. In *2018 Second international conference on intelligent computing and control systems (ICICCS)*, pp. 945–949. IEEE.

Shan, J., Zhao, J., Zhang, Y. et al. (2019). Simple and rapid detection of microplastics in seawater using hyperspectral imaging technology. *Analytica Chimica Acta* 1050: 161–168.

Shen, A., Zhu, Y., Angelov, P., and Jiang, R. (2024). Marine debris detection in satellite surveillance using attention mechanisms. *IEEE Journal of Selected Topics in Applied Earth Observations and Remote Sensing* 17: 4320–4330. https://doi.org/10.1109/JSTARS.2024.3349489.

Szafarczyk, A. and Toś, C. (2022). The use of green laser in LiDAR bathymetry: state of the art and recent advancements. *Sensors* 23 (1): 292.

Ulku, I. and Akagündüz, E. (2022). A survey on deep learning-based architectures for semantic segmentation on 2d images. *Applied Artificial Intelligence* 36 (1): https://doi.org/10.1080/08839514.2022.2032924.

Unger, B., Herr, H., Viquerat, S. et al. (2021). Opportunistically collected data from aerial surveys reveal spatio-temporal distribution patterns of marine debris in German waters. *Environmental Science and Pollution Research* 28: 2893–2903.

Veenstra, T.S. and Churnside, J.H. (2012). Airborne sensors for detecting large marine debris at sea. *Marine Pollution Bulletin* 65 (1–3): 63–68.

Veettil, B.K., Quan, N.H., Hauser, L.T. et al. (2022). *Coastal and marine plastic litter monitoring using remote sensing: a review*, 108160. Coastal and Shelf Science: *Estuarine*.

Wang, S., Gong, Y., Xing, J. et al. (2020, April). Rdsnet: a new deep architecture forreciprocal object detection and instance segmentation. *Proceedings of the AAAI Conference on Artificial Intelligence* 34 (07): 12208–12215.

Williams, A.T. and Rangel-Buitrago, N. (2022). The past, present, and future of plastic pollution. *Marine Pollution Bulletin* 176: 113429.

Wright, R.J., Erni-Cassola, G., Zadjelovic, V. et al. (2020). Marine plastic debris: a new surface for microbial colonization. *Environmental Science & Technology* 54 (19): 11657–11672.

Yang, R. and Yu, Y. (2021). Artificial convolutional neural network in object detection and semantic segmentation for medical imaging analysis. *Frontiers in Oncology* 11: 638182.

Yasin, J.N., Mohamed, S.A., Haghbayan, M.H. et al. (2020). Unmanned aerial vehicles (uavs): collision avoidance systems and approaches. *IEEE Access* 8: 105139–105155.

Zaaboub, N., Guebsi, R., Chaouachi, R.S. et al. (2023). Using unmanned aerial vehicles (UAVs) and machine learning techniques for the assessment of Posidonia debris and marine (plastic) litter on coastal ecosystems. *Regional Studies in Marine Science* 67: 103185.

Zagitov, A., Chebotareva, E., Toschev, A., and Magid, E. (2024). Comparative analysis of neural network models performance on low-power devices for a real-time object detection task. *Computer* 48: 2.

Zielinski, S., Anfuso, G., Botero, C.M., and Milanes, C.B. (2022). Beach litter assessment: critical issues and the path forward. *Sustainability* 14 (19): 11994.

18

Policy and Governance Approaches for Coastal and Marine Pollution Management

Chandrasekara SSK

Department of Agricultural Engineering, Faculty of Agriculture, University of Peradeniya, Peradeniya, 20400, Sri Lanka

18.1 Introduction

The coastal and marine ecosystems consist of biotic and abiotic components, and their biological processes, which are primarily to maintain the biodiversity of the allied ecosystems and, secondarily, to support community needs. Ecosystem services provided by the coastal environment are the provisioning of (i) basic water and nutrient requirements for marine and estuarine flora and fauna including human beings, (ii) regulatory services such as ocean-atmospheric weather and climate systems, coastal erosion, and pollution-associated systems, (iii) cultural services in the communities allied to coast and marine environment, and (iv) nutrient recycling. Nevertheless, the limited resources – coast and marine resources are fragile and under threat because of the (i) adverse impact due to the growth of urbanization in the coastal regions, (ii) blooming economic activities such as industrialization, and (iii) unplanned tourism in the regions (Laknath et al., 2023).

The Government of Sri Lanka (GoSL) introduced various governance approaches to reduce and control coastal and marine pollution in the country. Narangoda et al. (2023) summarized that (i) enactment of specific rules and regulations, (ii) implementation of the enacted rules and regulations, and (iii) frequent amendments of the existing rules and regulations are a must to manage pollution in coastal and marine environments. Nevertheless, coastal and marine ecosystem pollution is at an alarming rate in Sri Lanka and is identified as one of the major problems in the coastal zones.

18.1.1 Coastal and Marine Resources of Sri Lanka At a Glance

Sri Lanka is an island nation blessed with coastal and marine resources due to its location in the (i) Indian Ocean and (ii) near to the land mass – India, which is also vulnerable to climate change-induced natural disasters as well as anthropogenic impacts such as frequent political invasions since the ancient era. Sri Lanka consists of flat coastal plains averaging an elevation of less than 30 m. The continental shelf underneath the sea extends outward from the island with a width of about 8–40 km and a depth of about 65 m below sea level. The continental shelf is narrower around the southern coast of Sri Lanka and, interestingly, it is wider on the northern coast of Sri Lanka while merging the shelf around India (Coastal Zone Management Plan, 2006). The coastline of Sri Lanka is approximately 1620 km, including shorelines of bays and inlets, but excluding lagoons

Coastal and Marine Pollution: Source to Sink, Mitigation and Management, First Edition.
Edited by Meththika Vithanage, Sameera M. Samarasekara, Bryan D. James, and Christopher M. Reddy.

(Coastal Zone Management Plan, 2006). The 103 rivers in the country form diverse estuaries around Sri Lanka and nourish the coastal and marine ecosystems, but are also a source of pollutant transport to the coastal and marine ecosystems. Sri Lanka owns four types of shorelines such as (i) bays and headlands, (ii) straight sandy shorelines, (iii) deltas and saline flats, and (iv) limestone coasts. Furthermore, coastal profile types consist of (i) sandy beach, (ii) sandy beach and low dunes, (iii) sandy beach and high dunes, (iv) sandy beach and beach rock, (v) sandy beach and reef, (vi) sits and barriers and, (vii) reefs or coral reefs (Coastal Zone Management Plan, 2006). The beach around the country is rich in mineral resources such as ilmenite and monazite bearing beach sands, silica sands, miocene limestone, kaolin, china clays, copper magnetite, and peat, which are economically important for industries (Savunclranayagam et al., 1993).

The coastal zone of Sri Lanka is diverse geographically and demographically. Sri Lanka is divided into 25 administrative districts/units, of which 14 are entailed with coastal boundaries and cover 28% of the total land. According to the Coast Conservation Act No. 57 of 1981, the coastal zone of Sri Lanka is delineated as the regions which have a limit of 300 m landward of the Mean High-Water Line (MHWL) and a limit of 2 km seaward of the Mean Low Water Line (MLWL). The coastal regions of Sri Lanka represent all the climatic zones i.e. the wet zone, the dry zone, and the intermediate zone of Sri Lanka. The seasonal moderate wave heights characterize the surrounding wave climate of Sri Lanka. The weak micro and mixed to semi-diurnal tidal currents govern tides in the marine regions of Sri Lanka. The monsoons and inter-monsoons govern the wind climate and wind speed of Sri Lanka. Nevertheless, cyclones are not frequent in Sri Lanka, but weather systems in the Bay of Bengal could influence the generation and traversing the cyclones over Sri Lanka (Coastal Zone Management Plan, 2006). Furthermore, the intensity and frequency of the cyclones traversing via the country would alter with the existing climate change and variabilities throughout Sri Lanka.

Since the ancient era, Sri Lanka has been a hub of major shipping routes in the world and the country's coastal regions have experienced diverse external trading and cultural influences. The coastal zone of Sri Lanka has been contributing to the social, cultural, and economic development of the island even before the arrival of the first King of the nation – Vijaya in 550 BCE. Human settlement patterns existed during the colonial era, making the high population density (i.e. nearly 25% of the total population) in the coastal regions of the island (Coastal Zone Management Plan, 2006). The colonial architecture, pre-historical monuments, and cultural diversity appeal to tourism in the coastal regions of Sri Lanka. Therefore, rapid urbanization with many development activities in the region damage the coastal and marine environment and ecosystems.

18.1.2 Diversity Paves for Pollution

The diversity of the coastal zone of Sri Lanka aids the way for many economic activities contributing to the gross domestic product (GDP) of Sri Lanka. Fish – the main animal protein provider of the nation and, marine and brackish water fisheries are the main sector of the coastal regions. The coastal regions of Sri Lanka are the core for industrial production, because of four major ports such as Colombo, Galle, *Trincomalee* – a natural harbor, and *Kankasanthurei* and country's fishery harbors. Furthermore, due to the accessibility of the resources via the harbors the Free Trade Zone industries are concentrated in the coastal regions of the country (Savunclranayagam et al., 1993). As mentioned previously, tourism is the fifth largest income generator of the country and tourism is encouraged in coastal regions due to this reason. According to the Coastal Zone and Coastal Resources Management Plan (CZMP) – 2018 (Sri Lanka Coastal Zone and Coastal Resources Management Plan – 2018), the coastal pollution of Sri Lanka due to (i) untreated or partially treated wastewater and toxic substances from industries, the tourism sector, and the shrimp farms, (ii) leachate from the soil waste dumping sites at coastal areas, (iii) raw sewage from communities and industries, and (iv) waste oil from fishing boats, ships i.e. Bilge and ballast water, coastal services stations and oil spills (Sri Lanka Coastal Zone and Coastal Resources Management Plan, 2018). Furthermore, runoff contributing from 103 rivers transports the point and non-point source pollutants from agriculture, as well as the waste from squatter settlements and river reservation encroachers. Moreover, beach and river sand mining, and sea coral mining for lime production are disastrous for the coastal and marine environments. The discharge of hot water and desalinized water from thermal power plants i.e. located in *Kerawalapitiya*, *Norachcholai*, and *Puttalam* threatens the quality of coastal and marine water.

18.1.3 Pollution from Ignorant Natural Disaster – Tsunami

Tsunami struck in 2004 was identified as the worst natural disaster in Sri Lankan history because of the damages caused by the event. Besides the other social and economic impacts, drinking water sources were polluted due to the debris of the Tsunami event and saltwater intrusion (Vithanage et al., 2011). Furthermore, ecosystems of coral reefs and mangroves were damaged extensively (Sri Lanka Coastal Zone and Coastal Resources Management Plan, 2018; Caldecott and Wickremasinghe, 2005).

18.1.4 Marine Plastic Pollution – Concealed Catastrophic Disaster

The land-based plastics are the major share of marine plastic pollution and the remaining (20%) are from the transboundary litter generated from marine activities. Further, microplastics are the major marine debris in Sri Lanka (Nijamdeen et al., 2020). Buoyancy facilitates the transportation of plastics and the area of distribution upsurges due to the prevailing winds, currents, and ocean tides. Nevertheless, marine plastic pollution is lightly considered in Sri Lanka due to the knowledge gap and lack of professional expertise for science-based solutions (Arulnayagam, 2020; Jang et al., 2018).

In recent years two major disastrous maritime accidents happened in the peripheral ocean of Sri Lanka, (i) leakage of the crude oil from the MT New Diamond ship fire on 3 September 2020, and (ii) release of chemicals i.e. heavy metals and plastic pollutants from MV X – Press Pearl fire on 20 May 2021, which had short-term losses as (i) threatening to fisheries industries, (ii) loss of properties, (iii) marine pollution and allied threats to marine ecosystems, (iv) impact on tourism, and most importantly (v) transboundary impacts due to movement of the pollutants (Perera et al., 2022; Arachchige et al., 2021). Nevertheless, the long-term environmental impact assessments were not conducted at the time of emergency for these incidents because the consequent long-term damages from the incidents are not as rapid as the incidents. Furthermore, Sri Lanka failed to act swiftly in these incidents because (i) less infrastructure and human resources to combat rapid maritime emergencies as observed (Arachchige et al., 2021), (ii) absence of adequate protocols and agreements to such rapid emergencies, (iii) COVID-19 operational and logistical constraints in 2021, and (iv) prevailing strong southwest monsoon and south-westerly winds (United Nations Environment Programme, and United Nations Office for the Coordination of Humanitarian Affairs, 2021).

18.2 Institutional Arrangement for the Coast and Marine Resources Management in Sri Lanka

18.2.1 Establishment of a Main Institution for Coastal Resources Development and Management

The institutions related to water resources development and management in Sri Lanka were established in different approaches such as establishment (i) before their relevant legislations, (ii) from legislation relevant to the institution, (iii) by merging several institutions and forming one institution, and (iv) by expanding the existing division into an individual institution as an example the Coast Conservation Department (CCD). Before 1963, coastal protection works were done by many institutions, and coastal erosion was considered one of the threatening problems that had to be controlled at that time. Therefore, the Coast Protection Engineering Unit was developed under the Colombo Port Commission in the year 1963. In 1978, the unit was transferred to the Ministry of Fisheries and Aquatic Resources and renamed as Coast Conservation Division. In 1981, the Parliament of Sri Lanka enacted the Coast Conservation Act No. 57 of 1981, and the Coast Conservation Division upgraded into CCD. The CCD had three divisions (i) coast protection, (ii) coastal research and design, and (iii) coastal resources management. The Coast Conservation Act No. 57 of 1981 was amended as Act No. 64 of 1988 and Act No. 49 of 2011 for better performance with the changes in the coastal and marine resources. The CCD was further upgraded in 2011 as the Coast Conservation and Coastal Resources Management Department (CC&CRMD).

The primary mandated responsibilities of the CC&CRMD are (i) acquisition and declaration of lands for coast and marine environment, (ii) planning, construction, maintenance, and development of water bodies in ocean and coastal zones, (iii) conservation of reservations of ocean and seashore, (iv) soil conservation in foreshore and sea bed, (v) pollution control in wetlands and ocean, (vi) water allocation for environmental and public water use, (vii) navigation in the oceans, (viii) integrating with other natural resources such as wildlife and fisheries, (ix) advising ministries and ministers, (x) inter-sectoral coordination with the relevant institutions, and (xi) conducting research and experiments.

18.2.2 Integration of Multidisciplinary Institutions to Support the Mandated Responsibilities

The diversity of the coastal and marine resources and their uses flagged the requirement for multidisciplinary institutional involvement. Therefore, the Coast Conservation Act No. 57 of 1981 mandated to establish Coast Conservation Advisory Council compromising (i) the secretary to the Ministry of the Minister in charge of the subject of Coast Conservation, (ii) a senior officer representing the Ministry subject of Tourism, (iii) a senior officer representing the Ministry subject of shipping, (iv) a senior officer representing the Ministry subject of Local Government, (v) a senior officer representing the Ministry subject of home affairs, (vi) a senior officer representing the Ministry subject of industries, (vii) the director of the CCD as the secretary to the council, (viii) the director representing aquatic resources, research, and development, (ix) the Land commissioner, (x) the General Manager of Urban Development Authority (UDA), (xi) the Director of irrigation, (xiii) the minister appointed academic staff of one of the Universities in Sri Lanka and, (xiv) a representative from the voluntary organization representing coastal environment and fishing industry (Coast Conservation Act No. 57, 1981).

18.2.3 Subsidiary/Secondary Institutions for Collaborative Coastal and Marine Environment Management

CC&CRMD collaboratively works with other institutions to administrate coastal and marine resources and allied consequences. Table 18.1 shows the collaborative institutions that have mandated subsidiary responsibilities for coastal and marine resources management and development. Furthermore, marine litter management in the country is vested primarily in the Marine Environment and Protection Authority (MEPA), Central Environmental Authority (CEA),

Table 18.1 The collaborative institutions for the coastal and marine resources management and development in Sri Lanka.

Institution	Responsibilities	Legislation
(MEPA)	Plan, construct, maintain, and develop water bodies in the ocean and coastal zone Pollution control in Ocean	Marine Pollution Prevention Act No. 35 of 2008
Land Commissioner Generals' Division (LCGD)	Declaration, management, and conservation of coastal reservations including seashores Soil conservation at the ocean i.e., foreshore and seabed Environmental and public water use	Land Development Ordinance No. 19 of 1935
CEA	Pollution control in wetlands and waste management Fishing and fisheries in territorial water Integrating with other natural resources Ex. Wildlife and fisheries	National Environmental Act No. 47 of 1980
Ministry of local governments PC	Implementation of policies and plans Coordinative administration with the central GoSL Navigation over inland waterways and tidal water Fishing and fisheries in territorial water and beyond	13th amendment to the constitution of the Democratic Socialist Republic of Sri Lanka Provincial Councils Act No. 42 of 1987
Department of Wildlife and Conservation (DoWC) Director of Fisheries	Fishing and fisheries in natural reserves Integrating with other natural resources Ex. Water resources Marine reserves protection and conducting Environmental Impact Assessments (EIA) before the development activities in the declared areas.	Fauna and Flora Protection Ordinance No. 02 of 1937 Amendment Act No. 49 of 1993 Fisheries Ordinance No. 24 of 1940
National Water Supply and Drainage Board (NWSDB), *Pradeshiya Sabha* (PS), Municipal Councils (MC), Urban Councils (UC), Water Resources Board (WRB)	Environmental and public water use Control of water pollution in the municipality-declared areas.	Urban Councils Act No. 22 of 1955 Water Resources Board Act No. 29 of 1964 National Water Supply and Drainage Board Law No. 02 of 1974 Pradeshiya Sabha Act No. 15 of 1987 Municipal Councils Ordinance No. 19 of 1987

(*Continued*)

Table 18.1 (Continued)

Institution	Responsibilities	Legislation
National Aquaculture Development Authority of Sri Lanka (NAQDA)	Integrating with other natural resources Ex. Wildlife and fisheries	National Aquaculture Development Authority Act No. 53 of 1998
Department of Agrarian Development (DoAD), Farmer Organizations (FO), Department of Agriculture (DoA), WRB, CEA, Irrigation Department (ID), LCGD, NWSDB, *Mahaweli* Authority of Sri Lanka (MASL)	Intersectoral coordination with other relevant institutions	Department of Agriculture Ordinance No. 37 of 1921 *Mahaweli* Authority of Sri Lanka Act No. 23 of 1979 Agrarian Services Act No. 58 of 1979 Agrarian Development Act No. 46 of 2000 The Irrigation Ordinance No. 32 of 1856 and 1946 National Environmental Act No. 47 of 1980 Water Resources Board Act No. 29 of 1964 Land Development Ordinance No. 19 of 1935 National Water Supply and Drainage Board Law No. 02 of 1974
Merchant Shipping Department (MSD) at the Ministry of Shipping	Regulate all aspects of merchant shipping except, marine pollution from the shipping industry.	The Merchant Shipping Act No: 52 of 1971
Port Authority under The Ministry of Port Development	Management of removal of ballast and waste into harbor (quote No. 60)	Sri Lanka Port Authority Act No. 15 of 1979, amended Act No. 7 of 1984, and Act No. 35 of 1984.

CC&CRMD and aligned institutions such as the Ministry of Local Government and Provincial Councils (PC), Ministry of Mahaweli Development and Environment, Ministry of Megapolis and Western Development, Ministry of Health, Nutrition, and Indigenous Medicine, and National Solid Waste Management Support Center contribute in planning, implanting and monitoring approaches for the marine litter management (Nijamdeen et al., 2020).

18.2.4 Enactment of Legislation for Provisioning of Responsibilities

The Coast Conservation Act No. 57 of 1981 is the first enactment enacted entitled for coastal resources in Sri Lanka. The act was amended in 1988 as Coast Conservation (Amendment) Act No. 64 of 1988. The main objectives of the act are to a) facilitate surveying of the coastal zones of Sri Lanka and to prepare the CZMP, b) regulate and control development activities within the coastal zone of Sri Lanka, c) implementing coast conservation approaches within the coastal zone of Sri Lanka, and d) making required amendments to the written laws (The Coast Conservation Act No. 57, 1981). This legislation is indirectly concerned with the water quality of the coastal resources but does not consider marine pollution. Therefore, Marine Pollution Prevention

Act No. 59 of 1981 later amended as Marine Pollution Prevention Act No. 35 of 2008 enacted to provide for the prevention, control, and reduction of pollution in the territorial waters of Sri Lanka or any other maritime zones, its foreshore, and the coastal zones (Marine Pollution Prevention Act No. 35, 2008). The primary objective of the legislation is to establish MEPA (i.e. MEPA initially named as Marine Pollution Prevention Authority in 1981). The most important responsibilities vested from the legislation but which are not identified in other legislations in Sri Lanka are (i) to prevent, reduce, control, and management of pollution arising out of ship-based and shore-based maritime activities, (ii) to provide adequate and effective reception facilities for harmful substances/pollutants, (iii) to recommend and adherence to all international conventions and relevant protocols dealing for marine pollution, (iv) to formulate and implement the National Oil Pollution Contingency Plan, and (v) to oversee, regulate and supervise the exploration of natural resources in marine environment such as petroleum (Marine Pollution Prevention Act No. 35, 2008). Nevertheless, obscure legislation – Maritime Zones Law No. 22 of 1976 enacted to declare pollution prevention areas in the ocean to maintain environmental equilibrium. This legislation vested the responsibilities for CCD and the His Excellency the President of Sri Lanka.

MARPOL 73/78 Convention or The International Convention for the Prevention of Pollution from Ships, 1973 as modified by the Protocol of 1978 is the main international convention declared by the International Maritime Organization (IMO) for the prevention of marine pollution from ships and Sri Lanka became a member in the convention in 1977. Became a member of the Indian Ocean Memorandum of Understanding on Port State Control in 1998 to strengthen the adoption of MARPOL 73/78. Nevertheless, Marine Pollution Prevention Act No: 59 of 1981 has not covered the intention of MARPOL 73/78 and two bills drafted as (i) Marine Pollution Prevention Bill to affect the convention, and (ii) Marine Environmental Protection Authority Bill to establish a general body to administrate and regulate convention and allied future laws to prevention marine pollution in Sri Lanka (Fernando, 2009).

Therefore, these legislations (i) Coast Conservation Act No. 57 of 1981, (ii) Marine Pollution Prevention Act No. 35 of 2008, and (iii) Maritime Zones Law No. 22 of 1976 cover all the required responsibilities for coastal and marine water pollution in Sri Lanka. Moreover, Nijamdeen et al. (2020) showed that an abundance of local programs, policies, and international conventions ratified for marine debris management in the country could mitigate land-based marine debris pollution.

18.3 The Necessity of Governance Approaches to Increase the Effectiveness of Coastal and Marine Pollution Control in Sri Lanka

It is identified that there are an adequate number of institutions related to coastal and marine resources management and development in Sri Lanka. Nevertheless, except MEPA and CC&CRMD, other institutions do not have direct mandated authority to control coastal and marine pollution, but the vested responsibilities to these institutions indirectly control the coastal and marine pollution. Furthermore, most of the responsibilities overlap among different institutions, and sectoral division of the mandated institutions hinders the management of coastal and marine resources (Nijamdeen et al., 2020; Rajasuriya, 2003; Lowry and Wickremeratne, 1988). Merging institutions which have the same responsibilities to control pollution is not advisable, since coastal and marine ecosystems in Sri Lanka are diverse. Therefore, strengthening the existing representations/integration from all the relevant sectors/institutions for a board of management in either CC&CRMD or MEPA or both would be advantageous. Furthermore, the worn-off mandated penalties applied for the offenses make loopholes for implementing prevention/mitigation strategies for coastal and marine resource pollution in the country (Kularatne, 2020). Therefore, frequent amendments to the existing legislation are a must to mitigate the diverse and rapidly changing coast and marine pollution in Sri Lanka. Moreover, prompt completion of the legal process which is inattentive in the country could trigger the pollution. Consequently, it is a must to strengthen the legal framework for the governance of coastal and marine pollution in Sri Lanka. Nevertheless, Nijamdeen et al. (2020) suggested (i) the introduction of integrated policies, laws, and plans covering regional and international corporations, and (ii) establishing a new legal framework and institutional structure for diverse marine debris management in the country.

CC&CRMD and MEPA are the pioneering institutions that were declared for pollution control in the coastal and marine environment. Nonetheless, frequent movement of CC&CRMD to different ministries could hinder the effective performance of the institute because of changes in its priorities. Table 18.2 shows the allocation of ministries to CC&CRMD from 1978 to 2022. These frequent changes could be due to political instability in the country and public awareness could mitigate the consequences, but it would not give rapid outcomes.

Empowering community-based organizations (CBOs) for coastal and marine pollution control could be an effective mechanism. CCD developed Special Area Management (SAM) plans to encourage community participation in decision-making and management of the common resources. Nonetheless, community participation in decision-making was limited because higher hierarchy officers neglected the decisions/suggestions made by the community in planning and implementing processes. Therefore, increasing the awareness of the community on controlling and mitigating

Table 18.2 Frequent movement of the CC&CRMD to different ministries.

Year	Mandated Ministry	Special notes
1978	Ministry of Fisheries	
Until 2010	Ministry of Fisheries and Aquatic Resource Development	
April, 2010	Ministry of Ports and Aviation	Extra Ordinary Gazette notification number 1651/20 dated 30/04/2010
November, 2010	Ministry of Defence	Extra Ordinary Gazette notification number 1681/3 dated 22/11/2010
October, 2011	Ministry of Defence and Urban Development	The name of the Ministry of Defence changed in to Ministry of Defence & Urban Development, Sri Lanka.
2015	Ministry of Mahaweli Development and Environment	Extra Ordinary Gazette notification number 1933/13 dated 21/09/2015
2019	Ministry of Environment and Wildlife Resources	Extraordinary Gazette Notification No. 2153/12 dated 10/12/2019
2020	State Ministry of Urban Development, Coast Conservation, Waste Disposal and Community Cleanliness	Extraordinary Gazette Notification No. 2187/27 dated 02/08/2020
2021	Ministry of Urban Development and Housing/State Ministry of Coastal Conservation and Lowland Development	Extraordinary Gazette Notification No. 2235/45 dated 07/07/2021
2022	Ministry of Urban Development and Housing	Extraordinary Gazette Notification No. 2289/43 dated 22/07/2022

Source: Coast Conservation and Coastal Resources Management Department, 2022.

coastal and marine pollution and empowering the CBOs to participate in decision-making in community-involved activities would be advantageous. Awareness and willingness towards changing the behavioral patterns/traits of the community on waste/pollutant generation would be a long-term but rapid solution for the coastal and marine pollution in Sri Lanka (Ratnayake, 2023). Furthermore, granting legal provisions to the decisions made by the communities/CBOs and effective incentives for the changes in behavioral patterns/traits would strengthen coastal and marine pollution control. Nevertheless, local communities in Sri Lanka still depend on the state institutions and approaches for implementation and enforcement, and in some instances, ineffective performances of the state institutions or approaches aggravate the problem (Perera and de Vos, 2007).

Increasing the awareness among the related officials would provide long-term benefits for pollutant control in the coastal and marine environments. Therefore, the motivation of the officials, aware them of the latest technologies to mitigate pollution and monitoring their performances would be beneficial.

The evidence of the research gap on the management of coastal and marine resources in Sri Lanka is identified especially in newly emerged microplastic pollution (Dharmadasa et al., 2021; Nijamdeen et al., 2020). Research and monitoring of the coastal and marine resources and environment is mainly done by the National Aquatic Research Development Agency (NARA) which is under the Ministry of Fisheries and Aquatic Resources. Simultaneously, MEPA, CC&CRMD, CEA, and some non-governmental organizations are facilitating research and spreading awareness on the protection of coastal and marine ecosystems in the country. Nevertheless, the deficiencies in data and resources hinder the effectiveness of coastal and marine pollution control. Furthermore, the non-cohesive research by academic and other research institutions failed to provide rigid answers for the coastal and marine pollution in Sri Lanka (Perera and de Vos, 2007). Therefore, collaborative research could be cooperative for the mitigation of coastal and marine pollution. Effective research and monitoring could be achievable by the successful dissemination of knowledge that is deficient in the coastal and marine pollution prevention context in Sri Lanka. Arulnayagam (2020) identified that the public prefers social media platforms as the best way for the dissemination of knowledge on marine pollution in Sri Lanka.

18.4 Conclusion and the Way Forward

The coastal and marine ecosystem/environment in Sri Lanka is immensely diverse and the diversity paves pollution from diverse sources/origin of the pollutants. Therefore, GoSL enacted legislation, established institutions, and introduced governance approaches for the prevention/mitigation of coastal and marine pollution in Sri Lanka. CCD presently known as CC&CRMD is the mandated institution for the coastal and marine resources development which indirectly focuses on controlling coastal and marine pollution in Sri Lanka. Furthermore, MEPA was established solemnly for the mitigation of marine pollution and most importantly the institution considered the conventions and protocols to be adhered to in Sri Lanka for the mitigation of pollution from ships and other marine pollutants. Furthermore, it governs the extraction of marine resources for petroleum production. Some individual institutions also partake in responsibilities but overlapping responsibilities would cause inefficiencies in controlling coastal and marine pollution. The Coast Conservation Act No. 57 of 1981 and its amendments, Marine Pollution Prevention Act No. 35 of 2008, and Maritime Zones Law No. 22 of 1976 are the major legislations related to coastal and marine resources in Sri Lanka. These legislations include almost all the required provisions for the control and mitigation of coastal and marine water pollution. However, governance approaches considering legal provisioning of (i) empowering CBOs on decision making at planning and implementing, (ii) awareness increasing on monitoring and mitigation of coastal and marine environments, and (iii) cohesive approaches for research and development towards mitigating coastal and marine pollution are suggested. Most importantly, frequent amendments for the legitimate governance and strengthening the legal framework for the implementation of the amended governance approaches including incorporating international maritime laws are desperately required because the existing legal framework for coastal and marine resources of the country is not updated according to the dynamic changes in environment, economic and social aspects in Sri Lanka.

References

Arachchige, U.S.P.R., Sathsara, K.L.T., Preethika, P. et al. (2021). The impact of shipping on marine environment – A study of Sri Lankan water ways. *International Journal of Scientific Engineering and Science* 5 (7): 30–38.

Arulnayagam, A. (2020). Public perception towards plastic pollution in the marine ecosystems of Sri Lanka. *American Journal of Marine Science* 8 (1): 6–13. https://doi.org/10.12691/marine-8-1-2.

Caldecott, J. and Wickremasinghe, W.R.M.S. (2005). Sri Lanka. Post-Tsunami Environmental Assessment. United Nations Environment Programme and Ministry of Environment & Natural Resources of Sri Lanka. DEP/0758/GE, Asian Disaster Task Force, United Nations Environment Programme, 48–52. Geneva.

Coast Conservation and Coastal Resources Management Department (2022). Annual Performance Report – 2022, Coast Conservation and Coastal Resources Management Department, Sri Lanka.

Coastal Zone Management Plan, Government of Sri Lanka. § 1(1) (2006).

Dharmadasa, W.L.S.S., Andrady, A.L., Kumara, P.B.T.P. et al. (2021). Microplastics pollution in Marine Protected Areas of Southern Sri Lanka. *Marine Pollution Bulletin* 168: 112462. https://doi.org/10.1016/j.marpolbul.2021.112462.

Fernando, S.A. (2009). Marine Pollution Prevention & Marine Environment Protection Authority [Two bills]. A Legislation Drafting Project submitted in partial fulfilment of the requirements for the award of the Degree of Master of Laws (LL.M.) at the IMO International Maritime Law Institute.

Jang, Y.C., Ranatunga, R.R.M.K.P., Mok, J.Y. et al. (2018). Composition and Abundance of Marine Debris Stranded on the Beaches of Sri Lanka: Results from the First Island-Wide Survey. *Marine Pollution Bulletin.* 128: 126–131. https://doi.org/10.1016/j.marpolbul.2018.01.018.

Kularatne, R.K.A. (2020). Sri Lanka's legal framework for marine pollution control: an evaluation of the Marine Pollution Prevention Act, No. 35 of 2008. *Journal of the Indian Ocean Region* 16 (3): 244–267. https://doi.org/10.1080/19480881.2020.1816364.

Laknath, D.P.C., Kumara, I.G.I.K., and Manamperi, T.U.S. (2023). Coastal pollution in Sri Lanka: perspectives on the current status, policy implementation, and institutional mechanism. In: *ICSBE 2022*, Lecture Notes in Civil Engineering, vol. 32 (ed. R. Dissanayake et al.), 771–783. Singapore: Springer https://doi.org/10.1007/978-981-99-3471-3_52.

Lowry, K. and Wickremeratne, H.J.M. (1988). Coastal area management in Sri Lanka. *Coastal Management, Ocean Year Book Online* 7 (1): 263–293. https://doi.org/10.1163/221160088X00174.

Marine Pollution Prevention Act No. 35, Government of Sri Lanka (2008). § L.D.-O. 11/2006.

Narangoda, S.R.C.N.K., Weerasekara, K.A.W.S., Amarathunga, A.A.D., and Hettige, N.D. (2023). *A review of marine litter pollution in coastal waters of Sri Lanka. Annual Scientific Session – 2023*, 79. Sri Lanka: National Aquatic Resources Research and Development Agency.

Nijamdeen, T.W.G.F., Atugoda, T., Kumara, P.B.T.P. et al. (2020). Status of Particulate Marine Plastics in Sri Lanka. Research gaps and policy needs. In: *Particulate Plastics in Terrestrial and Aquatic Environments*, 1e (ed. N.S. Bolan, M.B. Kirkham, C. Halsband, et al.), 298–322. CRC Press.

Perera, N. and de Vos, A. (2007). Marine protected areas in Sri Lanka. A review. *Journal of Environment Management* 40: 727–738. https://doi.org/10.1007/s00267-005-0154-x.

Perera, U.L.H.P., Subasinghe, H.C.S., Ratnayake, A.S. et al. (2022). Maritime pollution in the Indian Ocean after the MV X-Press Pearl accident. *Marine Pollution Bulletin* 185 (A): 114301. https://doi.org/10.1016/j.marpolbul.2022.114301.

Rajasuriya, A. (2003). Impact of fisheries on coral reef habitats in Sri Lanka. *Second International Tropical Marine Ecosystems Management Symposium (ITMEMS 2)*, Manilla, Philippines. March 24–27, 2003.

Ratnayake, A.S. (2023). Marine pollution after the MV X-Press Pearl maritime disaster: societal impacts based on research outcomes. *Societal Impacts* 1 (1–2): 100003. https://doi.org/10.1016/j.socimp.2023.100003.

Savunclranayagam, T., de Alwis, L., Joseph, L., and Siripala, N. (1993). The Economic Significance of the Coastal Region of Sri Lanka. CRMP Working Paper No. 6/1994. Coastal Conservation Department and the Coastal Resources Management Project, Sri Lanka.

Sri Lanka Coastal Zone and Coastal Resources Management Plan (2018). Government of Sri Lanka. § 12(1).

The Coast Conservation Act No. 57, Government of Sri Lanka. (1981). §§ 1372-4-2004 260-22-1983.

United Nations Environment Programme, and United Nations Office for the Coordination of Humanitarian Affairs (2021). X-Press Pearl maritime disaster: Sri Lanka. Report of the UN Environmental Advisory Mission. https://wedocs.unep.org/20.500.11822/36608.

Vithanage, M., Engesgaard, P., Villholth, K.G., and Jensen, K. H. (2011). The effects of the 2004 Tsunami on a coastal aquifer in Sri Lanka. *Groundwater* 50 (5): 170–174. "https://urldefense.com/v3/__https://doi.org/10.1111/j.1745-6584.2011.00893.x__;!!N11eV2iwtfs!rgUbSHIUnELckTc1r_YWx7l7aFzG_5VZSQrG3FFY67yV-jof3J4R178Qy-MEEsBD4bldnZ3mCHybAu75KrISI555R9g$" https://doi.org/10.1111/j.1745-6584.2011.00893.x

19

Impact of Climate Change on Marine Ecosystems

Leneka Terika Rhoden

Department of Life Sciences, Faculty of Science and Technology, The University of the West Indies, Mona, Jamaica

19.1 Introduction

Marine ecosystems are intricately sustained by the transfer of energy starting from primary producers, progressing through various levels of consumers, including apex predators like humans, and returning through decomposition processes (Halpern et al. 2019). This interconnected web of life, characterized by biological interactions such as predation, competition, and mutualism, determines the success of species within these communities. Collectively, these interactions shape ecosystem functions like nutrient cycling and productivity, which are vital for providing essential services to society such as fisheries, water purification, and recreational opportunities.

However, the integrity of marine ecosystems is increasingly threatened by human activities, notably climate change (Hoegh-Guldberg et al. 2018). Elevated atmospheric carbon dioxide (CO_2) levels stand out as a particularly critical concern due to their global and enduring impact on ecological dynamics. This rise in CO_2 leads to rising ocean temperatures and acidity, along with associated consequences such as sea level rise, alters precipitation patterns, and changes ocean circulation.

While marine ecosystems play a central role in the global biological landscape, understanding of how human-induced climate change is affecting them remains limited (Mann et al. 2008). Recent research suggests that the rapid increase in greenhouse gas concentrations is pushing oceanic systems toward conditions not witnessed in millions of years, posing a significant risk of irreversible ecological changes (IPCC 2021). The impacts of human-driven climate change on marine ecosystems include reduced productivity, shifts in food web dynamics, diminished populations of species that shape habitats, changes in species distribution patterns, and an increase in disease prevalence (Doney et al. 2020).

19.2 Implications of Anthropogenic Climate Change on Marine Ecosystems

Despite uncertainties regarding the exact spatial and temporal manifestations of these changes, it is evident that climate change is reshaping marine ecosystems. This ongoing transformation poses challenges and costs for societies globally.

Coastal and Marine Pollution: Source to Sink, Mitigation and Management, First Edition.
Edited by Meththika Vithanage, Sameera M. Samarasekara, Bryan D. James, and Christopher M. Reddy.

Human activities significantly impact the Earth's climate by emitting greenhouse gases such as carbon dioxide (CO_2), methane (CH_4), nitrous oxide (N_2O), tropospheric ozone, and chlorofluorocarbons, primarily through the burning of fossil fuels, industrial processes, agriculture, and land use changes (Solomon et al. 2007). These gases trap heat in the atmosphere, resulting in global surface warming. Additionally, aerosols, which have shorter lifespans, can contribute to either regional warming or cooling. The warming effect is further intensified by positive climate feedback loops, such as increased water vapor and reductions in sea ice.

Projections indicate that the Earth's average surface temperature could rise by approximately 1.8 to 4.0 °C during the twenty-first century, depending on various emission scenarios (Solomon et al. 2007). Recent data show that CO_2 emissions from fossil fuels have been on the rise, particularly due to rapid economic growth in developing nations (Le Quéré et al. 2009).

The repercussions of human-induced climate change are increasingly evident across the expanse of the world's oceans, from the tropics to the polar regions, and from the ocean's surface to its deepest depths. While impacts from climate change are inevitable, the extent of harm inflicted on marine organisms, along with the vital services they furnish, can be significantly mitigated through prompt reductions in greenhouse gas emissions. There are six main climate-related factors- namely, warming, acidification, deoxygenation, sea level rise and associated storm events, loss of sea ice, stratification, and alterations in nutrient supply – that can affect marine ecosystems such as coral reefs, kelp forests, seagrass meadows, rocky and sandy intertidal zones, salt marshes, estuaries, and mangroves.

The implications of six primary climate-related factors on marine ecosystems, as summarized in Table 19.1, are diverse and interconnected. Warming oceans cause shifts in species distribution and abundance, coral bleaching, and disruptions in ocean circulation and productivity. Acidification reduces carbonate ion availability, affecting calcifying organisms and marine food webs. Deoxygenation leads to hypoxic conditions, impacting aerobic organisms and ecosystem dynamics. Sea level rise and storms intensify coastal flooding and erosion, threatening coastal habitats and biodiversity. Loss of sea ice disrupts habitats and food webs in polar regions. Stratification reduces nutrient transport, affecting phytoplankton and marine food webs. Alterations in nutrient supply can disrupt primary productivity and species composition. These factors collectively highlight the complex interplay between climate change and marine ecosystem health.

Recent assessments by the *Intergovernmental Panel on Climate Change (IPCC)* highlight significant changes in the Earth's oceans. Since the 1950s, there has been a notable increase in upper-ocean heat content, accompanied by a rise in global sea-surface temperatures by about 0.4 °C (Levitus et al. 2009).

19.3 Linear Graph Showing Annual Global Sea-Surface Temperature Anomalies from 1880 to 2015

Figure 19.1 shows annual global sea-surface temperature anomalies from 1880 to 2015 with a linear trend. Source: ncdc.noaa.gov/cag/time-series/global/globe/ocean/ytd/12/1880-2016.

Changes in ocean salinity also reflect the warming trends, with fresher waters in low-salinity areas and saltier waters in high-salinity regions, driven by increased evaporation and precipitation rates due to warmer temperatures (Boyer et al. 2005; Bindoff et al. 2007; Durack and Wijffels 2010). These alterations in salinity contribute to vertical stratification in the water column, affecting ocean currents, mixing processes, and ventilation. The decline in sea ice extent is particularly evident in the Arctic and along the western Antarctic Peninsula, especially during the summer season

Table 19.1 Climate-related factors and their implications for marine ecosystems.

Climate-related factor	Implications for marine ecosystems	Case studies
Warming	1) Shifts in species distributions and abundance patterns. 2) Coral bleaching and mortality due to increased sea-surface temperatures. 3) Changes in ocean circulation and productivity.	The Great Barrier Reef has experienced several mass coral bleaching events due to increased sea-surface temperatures. In 2016 and 2017, consecutive bleaching events affected large portions of the reef, resulting in widespread coral mortality and ecosystem degradation (Hughes et al. 2018).
Acidification	1) Reduced availability of carbonate ions, impacting calcifying organisms like corals, shellfish, and some planktonic species. 2) Disruption of marine food webs and ecosystem functioning.	The Pacific Northwest oyster industry has been impacted by ocean acidification (Hare and Mantua 2000). In areas such as Oregon and Washington, corrosive waters have affected oyster larvae, leading to decreased survival rates and significant economic losses for oyster farmers (Waldbusser et al. 2015).
Deoxygenation	1) Expansion of oxygen minimum zones, leading to hypoxia and anoxia in coastal and deep-sea habitats. 2) Negative impacts on aerobic organisms, including fish and invertebrates, affecting ecosystem dynamics.	The Gulf of Mexico experiences an annual dead zone, largely caused by nutrient runoff from agricultural activities (Rabalais et al. 2010b). This runoff fuels algal blooms, which upon decomposition, deplete oxygen levels in the water, resulting in hypoxic conditions detrimental to marine life.
Sea level rise and storms	1) Increased coastal flooding and erosion, altering coastal habitats and ecosystems. 2) Intensification of storm events, causing damage to marine infrastructure and ecosystems.	Mangrove forests in countries like Indonesia and Bangladesh are threatened by sea level rise and intensified storm events (Alongi 2008). Coastal mangroves provide critical habitat for numerous species and act as natural buffers against storm surges, but rising sea levels and stronger storms are leading to habitat loss and increased vulnerability to coastal erosion.
Loss of sea ice	1) Habitat loss for ice-dependent species such as polar bears, seals, and some seabirds. 2) Altered light regimes, affecting primary productivity and marine food webs.	The Arctic region is experiencing rapid sea ice decline, particularly during the summer months. This loss of sea ice habitat affects species such as polar bears, which rely on sea ice platforms for hunting seals (Stroeve et al. 2012). Additionally, reduced sea ice alters light penetration into the water, impacting primary productivity and ecosystem dynamics.
Stratification	1) Reduced vertical mixing of ocean waters, affecting nutrient transport and distribution. 2) Changes in phytoplankton community composition and productivity.	The Baltic Sea experiences significant stratification due to its relatively low salinity and limited connection to the North Atlantic Ocean. This stratification hinders vertical mixing of water layers, leading to oxygen depletion in deeper waters and altered nutrient dynamics, which can affect phytoplankton productivity and biodiversity (Conley et al. 2009).

(Continued)

Table 19.1 (Continued)

Climate-related factor	Implications for marine ecosystems	Case studies
Alterations in nutrient supply	1) Shifts in nutrient availability and ratios, impacting primary productivity and ecosystem structure. 2) Changes in trophic interactions and species composition.	The Chesapeake Bay (United States) suffers from eutrophication due to nutrient runoff from agricultural and urban areas. Excess nutrients lead to algal blooms, which, upon decomposition, deplete oxygen levels in the water, causing fish kills and disrupting the bay's ecosystem balance (Kemp et al. 2005).

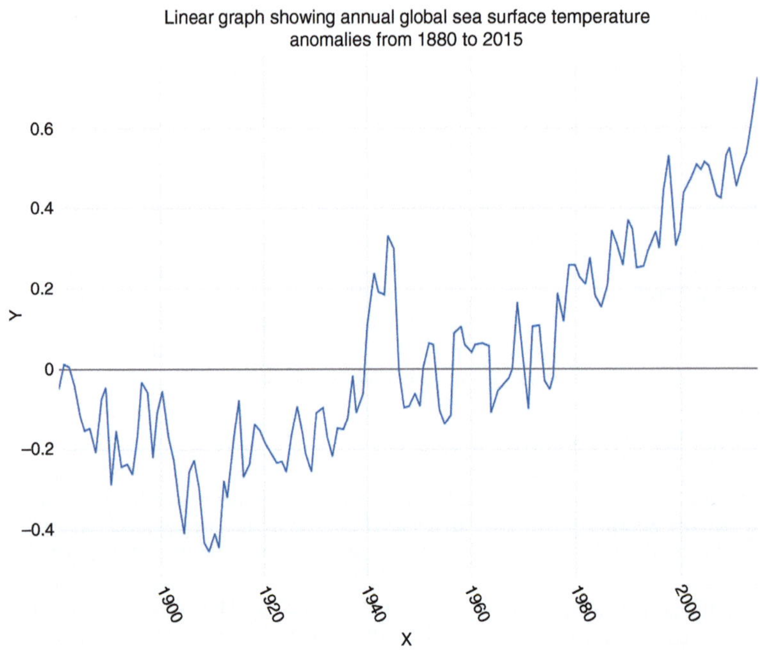

Figure 19.1 (*X axis representing years; Y axis representing temperature*) shows the warming of the ocean with the greatest ocean warming occurring in the Southern Hemisphere and contributing to the subsurface melting of Antarctic ice shelves. This warming is not uniform across all regions due to factors like ocean circulation patterns, wind changes, and natural climate phenomena such as El Niño/Southern Oscillation (ENSO) and the North Atlantic Oscillation (Ishii et al. 2006).

(Stroeve et al. 2007; Stammerjohn et al. 2008; Polyakov et al. 2010). Projections suggest that the Arctic could be devoid of sea ice during summer by the mid to late twenty-first century (Lawrence et al. 2008). Moreover, the thickness of Arctic ice is diminishing, with thin, first-year ice replacing thick, multiyear ice (Maslanik et al. 2007). Warming-induced thermal expansion and melting of landfast ice, including glaciers and ice caps/sheets, contribute to rising sea levels, which currently increase at a rate of approximately 3mm per year (Cazenave et al. 2008). However, local sea level changes can vary due to differences in upper-ocean warming, alterations in atmospheric and oceanic circulation patterns, and local geological factors like land uplift or subsidence.

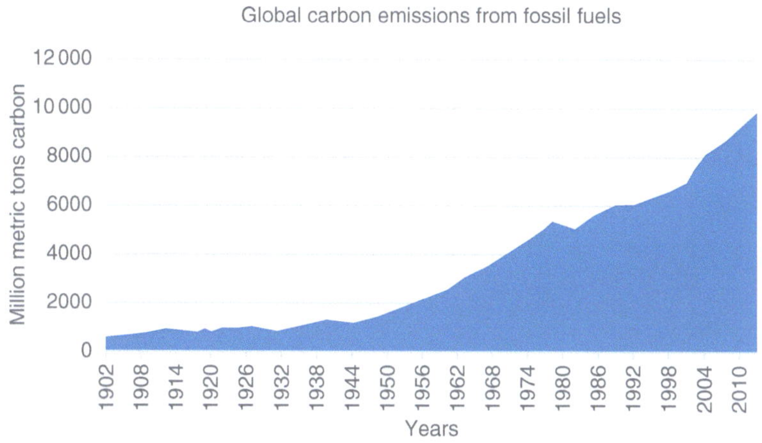

Figure 19.2 Trends in global emissions. *Source:* Boden et al. (2017).

There exists a consensus within the scientific community affirming the reality of contemporary global climate change, with much of this ongoing alteration attributed directly to human activities (Hansen et al. 2007; IPCC 2007a). Specifically, activities such as fossil fuel combustion, cement production, and alterations in land use have driven atmospheric carbon dioxide concentrations ($CO_2[_{atm}]$) from a preindustrial level of approximately 280parts per million (ppm) to 385ppm by 2008 (Meure et al. 2006). This increase in $CO_2[_{atm}]$ has been accompanied by annual increments surpassing 2ppm, a trajectory that surpasses even the direst projections outlined by the Intergovernmental Panel on Climate Change (IPCC 2007a), and illustrated in Figure 19.2.

A direct correlation exists between global temperature fluctuations and $CO_2[_{atm}]$ levels (IPCC 2007a). The increase in atmospheric CO_2, methane, and other greenhouse gases has engendered heightened radiative forcing in the lower atmosphere and Earth's surface, a phenomenon commonly referred to as the "greenhouse" effect, contributing to an approximate radiative forcing value of $3W/m^2$ (IPCC 2007a). This warming trend is unprecedented in at least the past 22000 years (Joos and Spahni 2008), exerting discernible physical ramifications on marine environments and their inhabitants.

19.4 Conceptual Framework

The conceptual diagram illustrating the complex interactions between human activities and climate variabilityis shown in Figure 19.3. Influenced by Rabalais et al. (2010b), the diagram reflects the intricate relationships shaping marine ecosystems, with evidence derived from paleological and archaeological records elucidating the consequences of environmental changes over varying timescales (Enghoff et al. 2007). Marine ecosystems, including fish stocks and fisheries, are subject to fluctuations driven by climate variability and anthropogenic activities, impacting ecosystem dynamics (Hamilton et al. 2003). The response of biological systems to climate change, whether physiological processes, individuals, populations, or entire ecosystems, depends on the magnitude of the climate change and the sensitivity of the system. Responses are often nonlinear, necessitating caution in assuming a linear response when predicting future climate impacts. The conceptual framework (Figure 19.3) covers two broad focus areas: biological responses and changes in the physical environment. Climate variability shows biological responses such as shifts in species distributions,

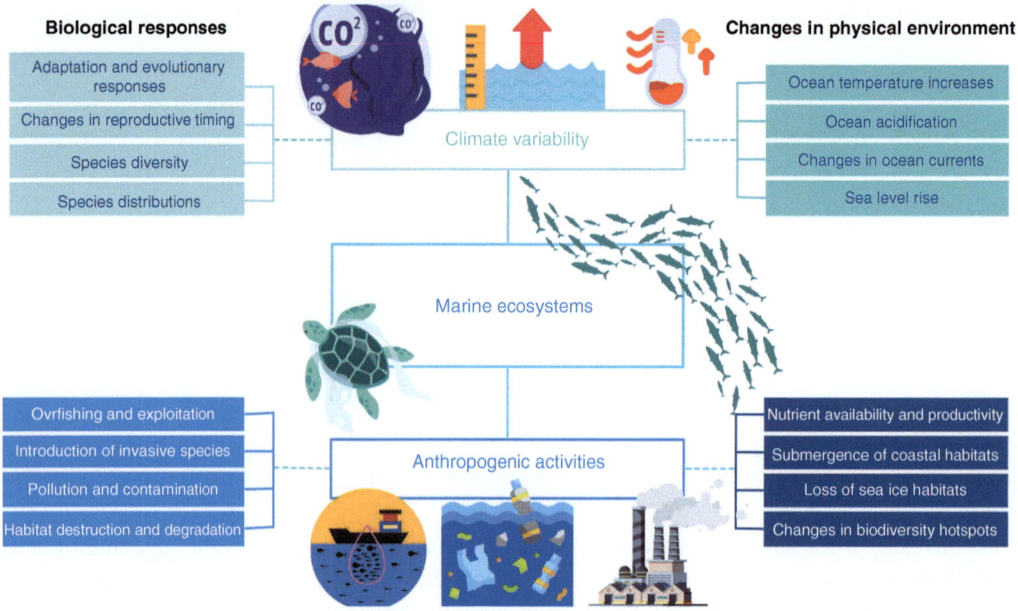

Figure 19.3 Conceptual framework exploring biological responses and changes in physical environment in marine ecosystems.

alterations in reproductive timing, changes in phenology, and adaptive evolutionary responses of marine organisms. These responses are linked to fluctuations in environmental conditions such as temperature, nutrient availability, and ocean chemistry. Concurrently, changes in the physical environment resulting from climate variability, including ocean temperature increases, ocean acidification, shifts in ocean currents, and sea level rise, exert effects on marine ecosystems by altering habitat suitability, nutrient dynamics, and ecosystem connectivity. Recent studies highlight contemporary impacts on biodiversity, such as a decline of over 13% in calcification rates of massive porite coral heads on the Great Barrier Reef since 1990, attributed to higher temperatures and lower pH levels (De'Ath et al. 2009). Furthermore, global biogeochemical models predict a prolonged decline in oxygen levels due to increased ocean stratification, with significant implications for marine life and ocean productivity over tens to hundreds of thousands of years (Shaffer et al. 2009).

Anthropogenic activities further exacerbate the pressures on marine ecosystems, eliciting both biological responses and changes in the physical environment. Overfishing and exploitation of marine resources lead to the depletion of fish populations, disrupting trophic interactions and ecosystem stability. The Black Sea underwent a significant ecological shift following prolonged heavy fishing, a surge in jellyfish populations, and eutrophication (Daskalov et al. 2007). Introduction of invasive species by human activities alters community composition, competes with native species, and disrupts ecosystem functioning. Pollution and contamination from anthropogenic sources, such as plastics, heavy metals, and chemicals, accumulate in marine environments, detrimentally affecting the health of marine organisms and ecosystems. Habitat destruction and degradation arising from coastal development, dredging, and unsustainable fishing practices degrade marine habitats, reducing biodiversity and compromising ecosystem resilience.

Anthropogenic inputs of nutrients, stemming from sources like agricultural runoff and sewage discharge, fuel eutrophication processes within marine ecosystems. This leads to alterations in nutrient availability and productivity, fostering harmful algal blooms and oxygen depletion.

Coastal development, coupled with sea level rise attributed to urbanization and climate change, exacerbates the submergence of critical habitats such as mangroves and salt marshes. This loss of coastal habitats amplifies habitat degradation, coastal erosion, and susceptibility to storm surges, further jeopardizing marine biodiversity and ecosystem resilience. Climate change-induced loss of sea ice, accelerated by anthropogenic greenhouse gas emissions, impacts polar ecosystems by altering habitats of ice-dependent species like polar bears and Arctic seals and disrupting ecosystem dynamics (Courtland 2008). Additionally, anthropogenic activities drive shifts in biodiversity hotspots within marine ecosystems, altering species distributions, community composition, and ecosystem functioning, necessitating efforts for conservation and management.

19.5 Marine Life and Physical Characteristics of Oceans

These consequences include elevations in mean global sea-surface temperature, rising at a rate of 0.13°C per decade since 1979 (IPCC 2007a), as well as increases in ocean interior temperature by more than 0.1 °C since 1961 (IPCC 2007a). Moreover, alterations in wind velocity, storm frequency, ocean circulation patterns, vertical structure, and nutrient loads have been observed (IPCC 2007a), alongside a rise in sea level exceeding 15cm over the last century (Rahmstorf 2007), presently accruing at an average rate of about 3.3mm per year. Given the tendency of oceanic and atmospheric gas concentrations to approach equilibrium, heightened $CO_{2[atm]}$ levels prompt greater CO_2 absorption by the oceans, inducing a rise in ocean acidity. This process results from the formation of carbonic acid (H_2CO_3), leading to a reduction in oceanic pH by 0.1 over the past two centuries, equivalent to a 30% increase in hydrogen ion concentration (The Royal Society 2005).

The pace of physical transformations instigated by climate change suggests that corresponding impacts on marine organisms and ecosystems are likely to be unprecedented as well. It has been postulated that surpassing a $CO_{2[atm]}$ threshold of 450 ppm may precipitate irreversible alterations (Hansen et al. 2007), resulting in a global mean temperature increase of 2°C above pre industrial levels. At current rates, this threshold is projected to be exceeded by 2040, although climate-related systems exhibit nonlinear behavior (Lenton et al. 2008). By 2040, sensitive marine ecosystems such as coral reefs and ice-covered polar seas could face irreversible losses (Hester et al. 2008).

Temperature shifts exert effects on biological and physical processes (Brinkhuis et al. 2006). For instance, a mere 1°C increase can lead to a doubling of biochemical reaction rates, and the peculiar nonlinear relationship between temperature and water density underpins phenomena such as the sinking of cold seawater and the floating of ice (Clarke 1993). Despite the disparity in the rate of increase between global mean sea-surface temperatures and land temperatures since 1979 (IPCC 2007a), rising temperatures wield the most pervasive influence on contemporary marine systems (Halpern et al. 2008).

Cyclic variations in solar activity and Earth's orbital parameters, known as Milankovitch cycles, contribute to temperature oscillations occurring over tens to hundreds of thousands of years (Lisiecki et al. 2008). Feedback mechanisms involving greenhouse gases, ocean circulation patterns, and ice extent, along with Milankovitch cyclicity, underlie the Quaternary cycles of glaciation that have persisted over the past 2.5 million years (Crowley and Hyde 2008). Within shorter timeframes, climatic fluctuations known as stadials and interstadials manifest, as observed in events such as Dansgaard-Oeschger and Bond events occurring approximately every 1500years in the North Atlantic region (Bond et al. 1997). These events, including transitions like the onset of the Younger Dryas and the Little Ice Age, are influenced by fluctuating ocean circulation patterns and associated variations in greenhouse gas concentrations (Schmittner and Galbraith 2008).

19.6 Spatial Variability in Temperature Change

The ongoing rise in global mean temperature is uneven across different geographic regions, as observed by scholars such as Hansen et al. (2007). Notably, the Arctic Ocean has experienced warming since the 1950s, with some areas witnessing temperature increases exceeding 4°C. In contrast, the temperature trends around Antarctica vary, with certain locales displaying warming trends while others show cooling trends. For instance, the Weddell Sea has experienced a decrease in sea-surface temperatures by 2 °C, whereas regions like South Georgia have seen a warming of 2 °C (Whitehouse et al. 2008).

Furthermore, observable changes in oceanic currents, like the southward shift of the East Australia Current by approximately 360km over the past six decades, have led to temperature spikes surpassing 2 °C in affected regions.

These regional disparities carry significant implications for local communities. Although oceanic depths typically exhibit relative thermal stability compared to terrestrial habitats, they are not immune to temperature increases. Despite their thermal inertia, it is anticipated that temperatures in these depths will rise for several decades, even if carbon emissions were abruptly halted (Ramanathan and Feng 2008). Observations in the Drake Passage reveal varying temperature alterations at different depths over the last 30 years. For example, while there has been a warming trend of 0.6 °C observed down to 700m below the surface, surface waters have simultaneously experienced a cooling of 2.1 °C (Sprintall 2008). Historical data indicates a disconnect between surface and deep-water temperature changes, as seen in deep-water extinctions during the Paleocene epoch while near-surface planktonic organisms appeared unaffected (Kennett and Stott 1991; Clarke and Johnston 1996).

Intertidal habitats are particularly susceptible to temperature fluctuations, especially in areas where increased daytime temperatures coincide with low spring tides, potentially resulting in die-offs despite the high-stress tolerance of certain intertidal organisms (Helmuth and Hofmann 2000). The decline in biodiversity within rock pools and other ecosystems may disrupt ecosystem functioning (Bracken et al. 2008). Nonetheless, colonization dynamics at the scale of rock pools, combined with variable resistance among species, could improve community stability for certain organisms (Altermatt et al. 2008). Consequently, forecasting winners and losers in the face of such changes remains a daunting task.

Table 19.2 outlines coastal ecosystems and the different threats that they face, including sea level rise, increasing temperatures, elevated atmospheric CO_2 levels, more frequent storms, and shifts in species composition (Duarte et al. 2021). These challenges result in significant consequences such as habitat loss, altered productivity, and changes in growth rates and photosynthetic processes. For example, in Southeast Asia, mangroves have experienced substantial habitat loss due to rising sea levels, leading to coastal erosion and loss of biodiversity (Richards and Friess 2016). Similarly, coral reefs are at risk due to rising temperatures, acidification, and storms, resulting in bleaching events, habitat degradation, and reduced calcification. Recent studies have shown that coral reefs are particularly vulnerable to climate change impacts, with even brief periods of extreme heat causing widespread bleaching and mortality (Hughes et al. 2020). Sandy shores and seagrass meadows also endure threats like increasing storm frequency and rising temperatures, leading to habitat alteration and loss (Dafforn et al. 2021).

The depletion of dissolved oxygen levels in oceanic environments has rendered certain areas unsuitable for multicellular life. This decrease in oxygen solubility in seawater is primarily ascribed to the escalation in temperatures, a phenomenon consistently observed since the 1950s (Garcia et al. 2005). The concentration of dissolved oxygen in seawater demonstrates a direct correlation

Table 19.2 Climate-related threats to major marine ecosystems up to the year 2025.

Ecosystem	Threat	Consequence
Mangroves	1) Sea level rise 2) Rising temperature and $CO_{2[atm]}$ 3) Increasing C3cf. C4 species 4) Increasing storm frequency	1) Landward progression or habitat loss; altered productivity 2) Changing growth rates; photosynthetic change
Coral reefs	1) Increasing storm frequency 2) Rising temperature 3) Increasing acidity 4) Rising sea level 5) Increasing storm frequency 6) Altered circulation and connectivity	1) Upper and nearshore limits affected negatively 2) Bleaching; distribution range contraction; overgrowth by algae 3) Reducing calcification and skeletal structure compromise 4) Drowning reefs; habitat loss; human abandonment off-loading atolls may reduce fishing pressure 5) Physical damage and runoff/silting increase 6) Larval transport disrupted and nutrient availability altered
Sandy shores	1) Increasing storm frequency 2) Rising temperature 3) Rising sea level	1) Changing beach structure, loss of habitat, and perturbed interspecies competition 2) Changing productivity; species range changes 3) Loss of habitat; changings and transport
Seagrass meadows	1) Rising temperature 2) Increasing storm frequency 3) Rising $CO_{2[atm]}$ 4) Sea level rise	1) Metabolic stress; range changes 2) Increased nutrient loading; decreased water clarity; increasing runoff/sedimentation 3) Increased growth rates; increased competitive advantage cf. algae; increased productivity 4) Loss of habitat

Source: Polunin (2008).

with temperature, decreasing by approximately 6% per degree Celsius rise within the temperature range of 0-15 °C. Forecasts indicate that the ongoing warming of the oceans, coupled with escalating atmospheric CO_2 levels, will result in a substantial expansion of low oxygen zones, potentially encompassing more than 50% of their current volume by the end of this century (Diaz and Rosenberg 2008; Oschlies et al. 2008). This expansion is poised to significantly impact crucial fishing areas, potentially leading to both ecological and economic consequences.

Various marine organisms exhibit behavioral adaptations to fluctuating oxygen levels, with notable impacts observed in fish schooling behavior and krill swarms (Johnson et al. 1984; Bertrand et al. 2008). Alterations in group behavior, driven by diminishing oxygen availability, may also influence predators targeting aggregated pelagic prey (Abrahams et al. 2007). Furthermore, coastal eutrophication, resulting from heightened fertilizer runoff and sea level elevation, is anticipated to exacerbate the situation by stimulating microbial activity, thereby further depleting dissolved oxygen levels (Diaz and Rosenberg 2008). While mobile organisms have the capability to evade areas with low oxygen concentrations, sedentary species are compelled to either endure hypoxic conditions or succumb to mortality. Paradoxically, those organisms capable of tolerating such conditions might encounter reduced predation pressure if oxygen-intolerant predators are excluded from the habitat (Altieri 2008).

Sea level fluctuations are linked to global temperature variations, influencing water and ice volumes (Rahmstorf 2007). These changes in sea level have implications for coastal habitats and ecosystems, impacting inundation patterns and habitat availability (Dawson and Hamner 2005; Peters 2008). Historical data suggests that the rate of sea level rise correlates with the degree of warming above pre industrial temperatures (Rahmstorf 2007), with projections indicating further increases of 0.5-1.4m above 1990 levels by 2100. Such changes not only affect habitat space but also drive speciation, alter biodiversity patterns, and modify local nutrient fluxes (Peters 2008). While rising sea levels pose existential threats to some islands, they may offer relief to coral reefs by reducing human-induced pressures in certain atolls. However, elsewhere, rising waters may compel organisms toward artificial sea defenses, with significant implications for intertidal and rocky-shore ecosystems as well as nursery habitats (Fujii and Raffaelli 2008; Vaselli et al. 2008).

19.7 Resulting Biological Disruptions in Marine Ecosystems

The assessment of how marine organisms and ecosystems may be affected by climate change necessitates an examination of impacts at various levels of biological organization. These include genetic expression, physiological processes at the cellular and organismal levels, structural changes in skeletal formations, individual behavior patterns, population dynamics, the structure of communities and ecosystems, as well as interactions within food webs. In accordance with ecological niche theory, the adaptability of species reflects a combination of physiological attributes, environmental conditions, competition with other species, and the ability to disperse. Different levels of biological organization may exhibit varying degrees of sensitivity to climate change, and responses to shifting global stressors may either be independent across these levels or intricately interconnected. Disparities in sensitivities among competing species could disrupt ecological interactions, leading to complex, indirect effects on community dynamics (Poloczanska et al. 2008).

Rare, abrupt events exert distinct evolutionary pressures compared to gradual, long-term changes. Short-lived "pulse" events, such as sporadic warming episodes brought about by transient incursions of warm water into typically cooler regions, may precede more enduring "press" events, such as widespread and sustained warming (Bender et al. 1984; Ridgway 2007). Systems characterized as "resistant" can withstand environmental changes due to broad physiological tolerances or ecological versatility, while "resilient" systems can recover to their pre-disturbance conditions once the stressor has dissipated (Harrison 1979). Biodiversity plays a critical role in determining ecosystem functioning (Danovaro et al. 2008). Ecosystems exhibiting high functional redundancy, where essential roles are fulfilled by multiple species, are anticipated to be more resilient to disturbances. However, given the substantial temperature increase already committed to the Earth (at least 2.4 °C) and the likelihood of its persistence even with immediate cessation of CO_2 emissions, achieving resilience in the face of ongoing climate change poses considerable challenges (Ramanathan and Feng 2008).

Temperature exerts the most significant influence on biological functions (Johnston and Bennett 1996). Within species, gene expression exhibits variability across their geographical distribution in response to temperature fluctuations (Place et al. 2008). However, this variation does not consistently align with latitudinal gradients, posing challenges for predicting species' responses to climate change (Hoffman and Willi 2008). For example, temperature influences muscle development (Johnston 2006), impacting the mobility of organisms. Changes in temperature can alter movement capabilities, potentially affecting predator-prey dynamics (Abrahams et al. 2007). Even slight temperature changes can impair physiological functions, leading to mortality as organisms struggle to perform essential tasks (Peck et al. 2004). Furthermore, growth rates are contingent upon temperature, with initial warming phases causing shifts in community size and biomass (Wiedenmann et al. 2008). However, contrary to expectation, warming does not uniformly enhance growth rates (Pratchett et al. 2008).

Temperature fluctuations also influence reproductive timing, output, and the condition of juveniles and larvae (Munday et al. 2008). Juvenile marine organisms, in particular, are vulnerable to changes in temperature, salinity, and pH, with larvae being more susceptible to elevated temperatures than their adult counterparts (Gagliano et al. 2007). Timely hatching of eggs is crucial for larvae survival, as mismatches with food availability can impact their chances of survival. Changes in plankton bloom timings, influenced by temperature, may disrupt food chains (Platt et al. 2003; Edwards and Richardson 2004), potentially affecting predator-prey dynamics (Richardson and Schoeman 2004; Hipfner 2008). Such disruptions may lead to shifts in prey species abundance, impacting predator populations, including commercially important fish species (Beaugrand et al. 2003). While warming may prompt shifts in species ranges, it does not necessarily lead to declines in community richness; for instance, fish species richness increased in the North Sea as the region warmed (Hiddink and Hofstede 2008). This increase aligns with the expectation that species richness generally decreases with increasing latitude, and local warming effectively simulates a reduction in latitude.

Climate change-induced alterations in demographic characteristics, mortality rates, and allele frequencies can affect the growth, size, resistance, and resilience of populations. Changes in population sizes may disrupt species interactions, potentially leading to regime shifts where ecosystems undergo rapid and potentially irreversible transitions (Litzow and Ciannelli 2007). For example, disturbances such as coral bleaching and storm-induced damage can lead to the proliferation of macroalgae, altering coral reef ecosystems (Mumby et al. 2007). If grazing pressure on algae is insufficient due to factors like disease affecting key grazers like the sea urchin *Diadema antillarum*, ecosystem resilience can be compromised, resulting in a shift to an alternative stable state dominated by algae.

19.8 Pelagic Ecosystems and Climate Dynamics

Open ocean environments play a crucial role in Earth's annual global primary production, accounting for approximately half of the fixed carbon (Field et al. 1998). However, since 1999, there has been a decline in global ocean primary production, with a reduction of 0.19 PgC per year (Behrenfeld et al. 2006). This decline is primarily observed in the permanently-stratified low latitude oceans (approximately 45°N to 45°S), attributed to warming and strengthened stratification resulting in nutrient limitation. A study by Laws et al. (2021) utilized advanced oceanographic techniques to analyze primary production rates across various oceanic regions. Their research further highlights the importance of open ocean ecosystems in sustaining global carbon cycling processes. One of the significant findings of the study was that primary production rates in the ocean were influenced by factors such as organism size, phylogeny (evolutionary relationships among organisms), and oceanographic conditions. The researchers observed variations in primary production rates depending on these factors, highlighting the complex dynamics of oceanic ecosystems; factors of which can be driven by climate change. The research focused on understanding the factors influencing primary production and the variability observed across different oceanic regions. The findings suggested that primary production rates were influenced by a combination of biological and environmental factors.

Phytoplankton and seaweeds, essential components of oceanic ecosystems, rely on nutrients such as phosphate, silicate, nitrate, and trace elements like iron (Boyd et al. 2007). Reduced mixing in the oceanic surface layers leads to decreased nutrient availability, particularly in the illuminated near-surface zone. However, polar regions experience increased wind strength, leading to enhanced mixing and nutrient replenishment, potentially increasing open ocean primary production (Arrigo et al. 2008a). Nevertheless, these increases may not offset the reductions observed in mid-latitude regions. Changes in mixed-layer depth and nutrient availability are also influenced by large-scale ocean circulation patterns (Barker et al. 2009). Modeling studies suggest that surface freshening due to melting of the Greenland ice cap could significantly reduce Atlantic meridional overturning circulation (Schmittner 2005). This reduction would result in a shallowing of the mixed layer globally, imposing nutrient restrictions and causing a 20% reduction in carbon export production.

In the North Atlantic, such changes could lead to a substantial decrease in phytoplankton biomass and subsequent reductions in secondary production, as zooplankton populations are primarily limited by phytoplankton availability (Richardson and Schoeman 2004). Consequently, fisheries production in affected areas is also expected to decline (Cheung et al. 2008). The rise in atmospheric carbon dioxide (CO_2) levels and temperatures is anticipated to have complex effects on phytoplankton dynamics. While these factors may initially stimulate phytoplankton growth rates, they also trigger physiological changes that could increase the vulnerability of cells to ultraviolet (UV) damage (Sobrino et al. 2008).

These ongoing environmental shifts are likely to exert influences on primary production, particularly in coastal regions where nutrient availability is supplemented by river runoff (Gowen et al. 2008). Furthermore, the anticipated rise in sea levels could impact nutrient loading in these coastal areas. Phytoplankton growth plays a crucial role in marine ecosystems by serving as the base of the food chain and by actively sequestering atmospheric CO_2 through the biological pump mechanism, which transports carbon to the ocean depths (Lutz et al. 2007). However, as temperatures increase, nutrient limitation may lead to the production of smaller phytoplankton cells that sink more slowly (Winder et al. 2008). Consequently, this could diminish the efficiency of carbon drawdown from the atmosphere, thereby exacerbating climate change impacts (Sarmiento et al. 1998).

The consequences of these changes extend beyond surface waters. Plankton sinking from the epipelagic zone constitutes a significant source of nutrition for deep-sea benthic organisms. Alterations in the composition of surface plankton communities due to temperature shifts could thus have cascading effects on deep-sea ecosystems (Yasuhara et al. 2008). Shifts in open ocean temperatures have already prompted substantial biogeographic rearrangements in species distributions. For instance, plankton communities in the North Atlantic have migrated by more than 10 degrees of latitude since the 1960s, reflecting the profound impact of warming on marine ecosystems (Beaugrand et al. 2002). Similarly, Arctic warming has facilitated the invasion of Pacific plankton species into the Atlantic, leading to trans-polar species movements (Reid et al. 2007).

These environmental changes also reverberate through marine food webs, affecting even large, wide-ranging species like baleen whales. Despite their size and longevity, these whales are not immune to the rapid responses of their prey, such as zooplankton, to climate-driven alterations (Greene and Pershing 2004; Leaper et al. 2006).

Table 19.3 shows the factors that collectively illustrate the complex interactions between climate change, oceanic processes, and pelagic ecosystems, highlighting the need for comprehensive management and conservation efforts to mitigate their impacts. This analysis underscores the interconnected nature of environmental changes in deep-sea ecosystems and highlights the importance of addressing these factors to mitigate their impacts on marine biodiversity and ecosystem functioning.

These environmental changes shown in Table 19.4 collectively pose significant challenges to deep-sea ecosystems, potentially leading to shifts in species distribution, altered community

Table 19.3 Climate change and environmental shifts on pelagic ecosystems (deep-sea).

Environmental change	Description	Analysis
Rising temperature	Global warming leads to increased sea temperatures, impacting deep-sea ecosystems and altering species distribution and behavior.	Higher temperatures can disrupt thermal gradients and affect the metabolism, reproduction, and distribution of deep-sea organisms.
Changing nutrient availability	Changes in ocean currents and runoff influence nutrient availability, affecting growth and productivity in deep-sea ecosystems.	Nutrient availability is crucial for deep-sea food webs; alterations may disrupt trophic interactions and lead to shifts in community composition.
Changes in carbon flux	Alterations in carbon flux, particularly increased CO_2, influence deep-sea water chemistry, impacting calcium carbonate formation in organisms.	Reduced carbonate availability affects calcifying organisms, potentially leading to structural changes in deep-sea ecosystems.
Impact on community composition	Environmental changes can cause shifts in deep-sea community composition, favoring more resilient species.	Species better adapted to changing conditions may dominate, altering ecosystem structure and function.
Increasing acidity	Elevated CO_2 levels lead to ocean acidification, reducing pH and affecting organisms with calcium carbonate-based structures.	Acidification disrupts marine food webs and threatens organisms like corals and mollusks that rely on calcium carbonate for structure.
Reducing carbonate availability	Ocean acidification reduces carbonate ions, impacting the ability of organisms to build calcium carbonate structures.	Reduced carbonate availability can lead to weakened shells and skeletons, making organisms more vulnerable to predation and environmental stress.
Decreasing oxygen levels	Factors like eutrophication and climate change lead to oxygen-depleted zones, posing challenges for aerobic organisms.	Deoxygenation can cause mass mortality events and disrupt deep-sea habitats, particularly in regions with naturally low oxygen levels.
Expanding anoxic dead zones	Anoxic dead zones expand due to nutrient runoff and circulation changes, creating inhospitable conditions for many marine species.	Anoxic conditions threaten biodiversity and can lead to "dead zones" where marine life cannot survive, impacting ecosystem health and fisheries.

dynamics, and impacts on ecosystem services provided by these environments. Efforts to mitigate these effects often involve understanding and addressing the root causes, such as reducing carbon emissions and improving marine conservation practices.

19.9 Coral Reefs and Ocean Acidification

Ocean acidification is altering ecosystems and poses a peril to marine life. Nonetheless, accurately forecasting the impacts on ecosystems remains highly uncertain, and governmental responses to ocean acidification are still lacking. Adaptive governance offers a potential approach to address this uncertainty.

Coral reefs represent some of the planet's most diverse (Kleypas et al. 2006) and economically significant ecosystems. However, they face a multitude of climate-related challenges, including

Table 19.4 Climate change and environmental shifts on pelagic ecosystems (open ocean).

Environmental impact	Description	Analysis
Rising temperature	Warmer ocean temperatures can lead to changes in species distributions and timing of peak production, disrupting biological processes.	Rising temperatures impact the thermal preferences of pelagic species, potentially leading to shifts in their distribution and altered predator-prey dynamics. Changes in peak production timing can affect the availability of food resources, influencing ecosystem structure.
Changing species distributions	Temperature-induced shifts in species distributions may alter predator-prey dynamics and ecosystem structure, potentially reducing fish production.	Distribution shifts impact community composition and trophic interactions. Reduced fish production may have cascading effects on fisheries and marine ecosystems.
Rising CO_2 levels (atmospheric)	Elevated CO_2 levels contribute to ocean acidification, affecting physiology and behavior of pelagic organisms.	Ocean acidification can disrupt planktic calcification, leading to ecosystem-wide impacts on food webs and nutrient cycling.
Increased primary production	Elevated CO_2 levels can stimulate primary production, potentially altering ecosystem structure and nutrient cycling.	While increased primary production may enhance food availability, its consequences depend on factors such as nutrient limitation and other environmental changes.
Increasing acidity	Ocean acidification affects planktic calcification, disrupting food webs and altering ecosystem functioning.	Changes in acidity can have widespread impacts on pelagic ecosystems, influencing species' ability to form calcium carbonate structures and their overall resilience to environmental stressors.
Altered circulation/ upwelling/ stratification	Changes in ocean circulation patterns affect nutrient availability and productivity, influencing the distribution and abundance of marine organisms.	Alterations in circulation patterns can lead to spatial and temporal variations in nutrient availability, impacting ecosystem productivity and species distributions.
Nutrient limitation and increasing bottom-up control on food chains	Nutrient limitation due to changing conditions increases bottom-up control on food chains, potentially affecting higher trophic levels.	Nutrient limitation can alter energy flow within pelagic ecosystems, leading to shifts in trophic structure and potentially impacting species at higher trophic levels.
Increasing storm frequency	More frequent and intense storms disrupt pelagic ecosystems, affecting nutrient mixing, productivity, and species distribution.	Storms can lead to physical damage to habitats and changes in community structure, influencing the resilience of pelagic ecosystems to environmental stressors.
Increased nutrient availability and production	Changes in nutrient availability influence primary production and ecosystem dynamics, impacting pelagic regions.	Nutrient availability plays a crucial role in regulating primary production and supporting marine food webs. Changes in nutrient inputs can lead to shifts in ecosystem structure and productivity, affecting the overall health of pelagic ecosystems.

Table 19.4 (Continued)

Environmental impact	Description	Analysis
Decreasing oxygen levels ([O_2])	Deoxygenation of ocean waters due to climate change and eutrophication negatively affects pelagic organisms sensitive to low oxygen conditions (Keeling et al. 2010).	Decreased oxygen levels pose significant challenges to pelagic organisms, particularly those with high metabolic demands, potentially leading to disruptions in ecosystem functioning and biodiversity loss (Heller and Zavaleta 2009).
Expanding anoxic dead zones	Anoxic dead zones, with extremely low oxygen levels, expand due to nutrient runoff, posing significant challenges to pelagic organisms and ecosystem health.	Expansion of anoxic zones can lead to habitat loss and declines in biodiversity, impacting the structure and functioning of pelagic ecosystems.

increasing sea temperatures, rising sea levels, and acidification. Crucial for coral formation, calcium carbonate, specifically aragonite, is becoming less available due to acidification. Research indicates a concerning 14.2% decrease in the calcification of corals within the Great Barrier Reef since 1990 (Reid and Croxall 2001). Reefs, in a study by Davis et al. (2021), show a declining calcification of $4.3 \pm 1.9\%$ per year ($\bar{x} = 1.8 \pm 0.7$ mmol/m^2/d/yr) and increasing organic productivity at 3.0 ± 0.8 mmol/m^2/d per year due to climate impacts. If the current trend of increasing atmospheric CO_2 persists, carbonate accretion is projected to decline further, potentially rendering corals exceedingly rare beyond 2050 (Hoegh-Guldberg et al. 2007).

Corals, particularly their symbiotic zooxanthellae algae, exhibit high sensitivity to temperature elevations. When temperatures surpass 31 °C, zooxanthellae are expelled, leading to coral bleaching (Hoegh-Guldberg et al. 2007). The frequency and scale of bleaching incidents have intensified since the 1960s, with notable events in 1998 and 2002 impacting entire reef systems (Berkelmans et al. 2004). The susceptibility of coral reefs to climate change has earned them the designation of being the "canary in the coal mine"for environmental shifts such as warming and acidification. Compounding this vulnerability are various anthropogenic stressors like overfishing (Halpern et al. 2008; Jackson 2010), pollution, sedimentation, and invasive species, which further threaten the health and resilience of coral reef ecosystems.

Even slight increases in temperature, as low as 1 °C, can trigger coral bleaching-a phenomenon where corals expel their symbiotic algae, causing them to lose their color and eventually leading to coral death (Donner, 2009). Bleaching events, ranging from moderate stress to severe mortality, are exacerbated by factors such as reduced growth rates and increased susceptibility to diseases (Muller et al. 2007). While some genetic diversity within coral populations and their symbiotic algae may confer resilience to moderate temperature changes, severe anomalies often result in widespread bleaching and mortality.

As ocean waters become more acidic due to increased carbon dioxide absorption, corals face additional challenges in maintaining their skeletal structures (Salvat and Allemand 2009). Some species may even lose their skeletons entirely, while others experience reduced growth rates or less dense skeletons, rendering them more susceptible to storm damage (Fine and Tchernov 2007). Despite these threats, certain coral species and their symbionts may exhibit varying degrees of resilience to low pH conditions. Ocean acidification negatively impacts various other reef organisms (Halpern et al. 2008; Jackson 2008), including coralline algae, which serve as crucial substrates for coral larval settlement (Doney et al. 2009). Additionally, changes in ocean chemistry can disrupt the sensory abilities of reef fish and have adverse effects on their reproductive processes (Munday et al. 2009). pH fluctuations can also disrupt ion exchange, suppressing metabolic processes and narrowing the range of thermal tolerance (Portner et al. 2005).

19.10 Climate Migration and Adaptation in Marine Organisms

The primary determinant of an organism's ability to withstand environmental variability and change lies in its physiological performance (Baumgartner et al. 1992). As environmental conditions, such as climate, shift over time, organisms respond initially through a combination of physiological and behavioral adaptations shaped by their evolutionary past (Somero 2012). These responses can vary: the new conditions might be physiologically tolerable, allowing for acclimatization such as adjustments within individual organisms or adaptation where tolerant genotypes become more abundant through generations (Jones and Berkelmans 2010). However, conditions may be intolerable, leading to migration of individuals or populations, shifts in the timing of annual events (phenology), or even local extinction if adaptation is not feasible (Parmesan 2006; Carpenter et al. 2008).

Environmental changes can confer benefits to certain species or populations. These advantages may include increased food availability, reduced physiological costs such as lower energy expenditure for respiration, acid–base balance, or calcification, or decreased competition and predation pressure. Consequently, these species may experience enhanced survival, growth, and reproduction, thus emerging as "winners" in the face of environmental change. However, many organisms experience stress when confronted with conditions beyond the normal range of variability. This stress can lead to suboptimal physiological performance, creating "losers" in the context of environmental change. For such individuals, heightened stress levels may result in increased mortality, diminished growth, smaller body size, and reduced reproductive success.

Generally, the metabolic rates of ectothermic organisms exhibit an exponential increase with temperature, impacting various physiological processes, including photosynthesis and respiration, within the organism's tolerance range. This relationship is encapsulated in the concept of Q10, which denotes the rate of increase resulting from a 10°C temperature rise. For example, Eppley (1972) observed an average Q10 of 1.9 for growth rates across approximately 130 species and clones of phytoplankton. This translates to a 37% increase in growth rate for a 2 °C warming. Based on this principle, one might anticipate increased primary production, as well as enhanced growth rates among ectothermic animals and pathogens, in a warmer ocean (Garcia et al. 2018). However, factors such as nutritional status, thermal tolerance, oxygen availability, environmental chemistry, and food availability can impose limitations on growth, production, and other biological processes, notwithstanding the increase in metabolic rate.

Physiological responses to climate change are anticipated to have wide-ranging implications at the population level, encompassing alterations in abundance, phenology (the timing of annual events), and the spatial distribution and dispersion of organisms. Suboptimal environmental conditions and diminished individual performance can lead to reduced population abundance and productivity, as well as diminished resilience to perturbations (Hare et al. 2010).

One of the well-documented biological consequences of climate change is the shifting range and distribution of populations. Firstly, the ability of species to disperse from a regional species pool determines their capacity to colonize or occupy a particular habitat. Second, the range of abiotic environmental conditions that a species can physiologically tolerate defines its fundamental niche, with climate-induced alterations in oceanic conditions potentially leading to changes in niche boundaries. Third, biological interactions such as predation and competition can either restrict or broaden the range of conditions suitable for a species, shaping its realized niche (Bruno et al. 2003). The assemblage of species sharing overlapping realized niches forms a community, and climate change can significantly impact the composition and functioning of marine communities by influencing these processes (Hoegh-Guldberg and Bruno 2010). Both empirical observations and theoretical models indicate that marine fish and invertebrates tend to relocate to higher latitudes and/or deeper depths in response to warming conditions (Mueter and Litzow 2008; Spencer 2008). This trend is corroborated by observational evidence from regions such as the North Sea and the Mid-Atlantic Bight (Perry et al. 2005; Nye et al. 2009; Lucey and Nye 2010). Climate-induced shifts in species ranges may also coincide with alterations in demographic patterns, organism abundance, and size, as indicated by integrated climate-population models (Hare et al. 2010).

Ecosystems serve as complex integrators of organismal physiologies and ecological interactions, responding to alterations in climate patterns and atmospheric CO_2 levels (Díaz et al. 2019). Consequently, the mechanisms linking changes at the level of populations and communities to broader shifts in ecosystem properties, such as trophic structure, food web dynamics, energy flow, and biogeochemical cycles, exhibit considerable diversity (Doney 2010). Disruptions to existing biological interactions can manifest through various pathways, including asynchronous shifts in the seasonal phenologies of interacting predator and prey populations, known as the match–mismatch hypothesis (Durant et al. 2007). Additionally, biogeographic reorganizations can lead to alterations in community composition and biodiversity, while the loss of functionally prominent species can have cascading effects (Bellwood et al. 2004). These processes manifest in both bottom-up and top-down impacts. Bottom-up impacts may include declines in water-column primary production or shifts toward smaller cells in planktonic communities. Conversely, top-down impacts can cascade from the loss of ecologically dominant consumers (Estes et al. 2011). Even if overall productivity and diversity remain relatively constant, alterations in biogeochemical cycling can occur due to the replacement of key functional groups, such as calcifiers (Kwiatkowski and Orr 2018).

The extent to which individual ecosystems experience such changes is influenced by factors like functional redundancy, the rate and sequence of community change, and other aspects of population and community organization (Hooper et al., 2005). Geographic variability also plays a role in modulating ecosystem responses to climate-induced structural changes (Navarrete et al. 2005). Navarrete et al. (2005) observed top-down control in the southern part of the upwelling region along the western coast of Chile, while communities in the north were recruitment-limited, resulting in negligible effects of predators on community composition.

Significant strides have been made in developing methods to identify reliable indicators of marine ecosystem function and resistance to perturbations (Gaichas et al. 2009; Samhouri et al. 2009; Bundy et al. 2010). These indicators, such as the mean length of fish or the proportion of predatory fish in a community, provide valuable insights into ecosystem health and resilience.

19.11 Case Study: Influx of Sargassum across the Caribbean

Intermittently, accumulations of brown algae are observed drifting along the coastlines of numerous Caribbean island nations. This occurrence typically transpires during the winter months when prevailing trade winds are in the area, leading to the significant accumulation of drift algae along the shores (Gavio et al. 2015). These algae primarily consist of two species, namely, *Sargassum fluitans* and *Sargassum natans*. These species are classified as pelagic algae, meaning they grow freely in the open sea without attachment to substrates. Typically, high concentrations of these algae are observed in an area known as the Sargasso Sea (Moreira and Alfonso, 2013). The Sargasso Sea, situated in the North Atlantic Ocean between the latitudes 25° and 35° N and longitudes 70° and 40° W, is a distinctive region characterized by its lack of coastal boundaries. Its perimeter is defined by the North Atlantic Drift to the north, the Gulf Stream to the west, the Canary Current to the east, and the northern equatorial current to the south (Laffoley et al. 2011).

In recent decades, there has been a notable increase in the quantity of floating algae washing up along the coastlines of the Caribbean (Johnson et al. 2013). A surge in Sargassum seaweed, as documented by Smetacek and Zingone (2013) in the Gulf of Mexico during the 1990s, was found to be linked to elevated nutrient discharges into the Mississippi River. Another investigation by Gower and King (2011) revealed an increase of seaweed in the Gulf of Mexico, prompting regulatory authorities to designate 2005 as the "Sargassum Year." By 2011, Sargassum biomass drifts in the Caribbean region had reached a pinnacle, with recorded biomass being 200 times greater than in the preceding eight years. This marked the first instance in recent memory where washed-up Sargassum reached the African coast (Smetacek and Zingone 2013).

Underlying the Sargassum invasion are several hypotheses proposed to be its causes. Some sources suggest that global climate change may have driven the Sargasso Sea further south, thereby bringing the seaweed closer to the Caribbean Sea (Velasco 2015). Sargassum migration, influenced by climate change, refers to the movement patterns of Sargassum populations in response to alterations in environmental conditions driven by climate variability. Rising sea-surface temperatures influence the growth, reproduction, and distribution of Sargassum. Warmer temperatures can accelerate the growth rate of Sargassum, potentially leading to shifts in its distribution range as it seeks optimal temperature conditions. Changes in ocean currents, influenced by climate phenomena such as El Niño and the Atlantic Meridional Overturning Circulation (AMOC), can affect the dispersal patterns of Sargassum. Alterations in current strength and direction may lead to deviations from typical migration routes or the arrival of Sargassum in new regions. Climate change can impact nutrient availability in the ocean, affecting the productivity of Sargassum. Changes in nutrient concentrations, including nitrogen and phosphorus, can influence the abundance and distribution of Sargassum populations, as they rely on these nutrients for growth. Moreover, changes in pH levels may influence the growth and survival of Sargassum populations, potentially impacting their migratory behavior.

Another theory implicates the increased deposition of nutrient-rich Sahara dust (including Fe, K, and phosphate) as a major driver facilitating algae growth in the sea (Moanga 2015). A fourth hypothesis attributes the Sargassum invasion to nutrient overload resulting from deforestation in the Amazon basin in Brazil (Newsweek 2015). Due to deforestation, the Amazon River delivers copious amounts of organic matter and nutrients to the sea, thus promoting excessive algae growth.

Contrary to theoretical expectations of benefits associated with increased availability of native *Sargassum spp.* in pelagic habitats, negative impacts have been extensively documented along coastlines. This suggests a potential dichotomy in the ecological effects of Sargassum, with negative impacts predominating in coastal areas. These impacts include the formation of large masses of Sargassum in bays, leading to decomposition and the creation of hypoxic conditions, as well as detrimental effects on nearshore and neritic communities, such as coral reefs. Additionally, the physical presence of Sargassum can alter coastal dynamics, affecting erosion-accretion processes and restricting the movement and habitat utilization of mobile organisms.

19.12 Management and Government Policies for Marine Ecosystems: The Way Forward

The rapid pace of environmental changes in marine ecosystems is outstripping the adaptive capacity of many organisms. Without significant climate mitigation efforts, these changes are set to intensify in the coming decades. Climate change induces shifts in species' ranges as they adapt to changing environmental conditions, making it imperative to monitor these changes along environmental gradients to understand species' responses to climate-driven pressures (Hillyer and Silman 2010). In addition to affecting temperature and chemistry, these alterations disrupt the physiology, behavior, and productivity of marine life, reshaping population dynamics and trophic interactions from primary producers to top predators. While understanding individual species' responses to specific stressors is vital, it's insufficient for fully assessing ecosystem dynamics. A holistic approach, incorporating multispecies and ecosystem-level analyses, is essential to grasp the impacts of climate change and other stressors on marine ecosystems.

To address these anthropogenic impacts effectively, integrated ecosystem-based management approaches are necessary, complemented by policies and regulations aimed at reducing nutrient pollution, safeguarding critical habitats, and mitigating climate change impacts. An example of an integrated ecosystem-based management system is a marine protected area (MPA). MPAs are designated areas where human activities are regulated to protect marine ecosystems and biodiversity. One example is the establishment of the Palau National Marine Sanctuary, which covers 80% of Palau's exclusive economic zone and prohibits commercial fishing (Victor et al. 2020). The Palau National Marine Sanctuary continues to demonstrate success in marine conservation and sustainable fisheries management. A key achievement is the enforcement of the sanctuary's no-take zone, which has led to the recovery of fish populations and improved coral reef health within protected areas (Victor et al. 2020).

Ecosystem-based fisheries management (EBFM) integrates ecological principles into fisheries management to ensure the sustainability of fish stocks and marine ecosystems. The California Collaborative Fisheries Research Program (CCFRP) is an example of EBFM in action. It brings together scientists, fishermen, and resource managers to collect data, assess fisheries health, and develop management strategies based on ecosystem dynamics (Kaplan et al. 2020).

Integrated Coastal Zone Management encompasses the coordinated planning and management of coastal areas to effectively balance economic development with environmental conservation. The Gulf of Mexico Alliance serves as a partnership between U.S states, federal agencies, and stakeholders in addressing coastal issues through collaborative efforts. GOMA's (Gulf of Maine Alliance) initiatives encompass habitat conservation, water quality improvement, and resilience planning aimed at mitigating the impacts of climate change. Furthermore, recent reports emphasize the importance of incorporating current research findings for efficient decision-making processes within ICZM (Integrated Coastal Zone Management) frameworks (Elgendy et al. 2023).

Numerous programs are working to reduce nutrient pollution from agricultural runoff and other sources in order to protect coastal ecosystems. The Chesapeake Bay Program is a collaborative effort involving local governments, nonprofit organizations, and academic institutions with the goal of restoring the largest estuary in the continental United States–the Chesapeake Bay watershed (Hood et al. 2021). Population growth has led to development pressures and intensified agricultural activities that result in excessive nutrient inputs negatively affecting the bay's ecosystem health. Implementing best management practices on farms has helped to reduce these harmful inputs alongside upgrading sewage treatment facilities.

A challenge for scholars is understanding the development of adaptive governance within highly regulated systems, particularly the role of law in either hindering or facilitating adaptive governance

(Cosens et al. 2017; Cosens et al. 2018). Legal frameworks dictate decision-making processes, including the scale of decisions, authorized participants, and mechanisms for adjusting to change (Craig 2019). The concept of legal adaptive capacity, encompassing both substantive and procedural legal mechanisms, is important in enabling governance systems to adapt (Camacho and Glicksman 2016). Craig (2019) puts forward marine spatial planning (MSP) as a highly adaptable process.

Despite well-documented adverse effects of ocean acidification, global initiatives to combat this phenomenon remain scarce. Addressing ocean acidification within the United Nations Framework Convention on Climate Change (UNFCCC) framework has been proposed, as ocean acidification is regarded as a consequence of alterations in the climate system's state. The UNFCCC, initially aimed at stabilizing greenhouse gas (GHG) emissions to prevent hazardous interference with the climate system, led to the adoption of the Kyoto Protocol in 1997, mandating binding GHG emission reductions for developed nations from 2008 to 2012. Subsequently, the Paris Agreement of 2015 aimed to limit global average temperature increase to well below 2 °C above pre industrial levels, with aspirations toward achieving a 1.5 °C rise.

Another milestone in global governance emerged in 2015 with the adoption of the United Nations Sustainable Development Goals (SDGs), marking significant social and political momentum, particularly within the marine realm, over the past 15years. Notably, SDG 14-Life Below Water-focuses on promoting ocean sustainability, with Target 14.3 specifically aimed at mitigating ocean acidification's impacts through enhanced scientific collaboration across all levels.

19.13 Conclusions and Key Recommendations

Climate change has significantly impacted marine ecosystems, leading to rising sea temperatures, ocean acidification, and loss of sea ice. These changes have profound effects on marine species and their feeding relationships. Rising sea temperatures have caused shifts in the distribution of marine species, affecting their reproduction and migration patterns. Additionally, ocean acidification has impacted the ability of marine organisms such as coral reefs and shellfish to build and maintain their calcium carbonate structures, affecting their reproduction and ultimately altering the feeding relationships within the ecosystem. This disruption in feeding relationships has major effects throughout the food web, impacting the population dynamics of various species. It has been found that changes in ocean currents and sea ice cover due to climate change also influence the migration patterns of marine species.

It is crucial for governments and organizations to prioritize conservation efforts and implement measures to mitigate the impacts of climate change on marine ecosystems. Governments have to take decisive action and implement effective management and policies to safeguard marine environments.

Recent advancements in climate change adaptation strategies emphasize the importance of integrating adaptation measures into broader government policies and functions. This approach illustrates the need for collaborative processes that span different scales of governance (Cosens et al. 2014). Innovative models of coastal and marine governance, such as partnership-led regional marine planning processes and the devolution of seabed management rights, offer promising avenues for promoting locally driven and learning-based approaches to planning and management. By empowering local communities and stakeholders, these initiatives can promote greater resilience and adaptive capacity within marine ecosystems (Cosens et al. 2018).

The expansion of the marine industry presents both social and ecological constraints, compounded by uncertainties surrounding the local effects of ocean acidification. To address these challenges, it is essential to adopt holistic and cooperative management tools, such as MPAs, which can help mitigate the impacts of human activities while promoting sustainable development.

References

Abrahams, M.V., Mangel, M., and Hedges, K. (2007). Predator-prey interactions and changing environments: who benefits? *Philosophical Transactions of the Royal Society, B: Biological Sciences* 362: 2095–2104.

Alongi, D.M. (2008). Mangrove forests: resilience, protection from tsunamis, and responses to global climate change. *Estuarine, Coastal and Shelf Science* 76: 1–13.

Altermatt, F., Pajunen, V.I., and Ebert, D. (2008). Climate change affects colonization dynamics in a metacommunity of three *Daphnia* species. *Global Change Biology* 14: 1209–1220.

Altieri, A.H. (2008). Dead zones enhance key fisheries species by providing predation refuge. *Ecology* 89: 2808–2818.

Arrigo, K.R., van Dijken, G.L., and Bushinsky, S. (2008a). Primary production in the Southern Ocean, 1997–2006. *Journal of Geophysical Research: Oceans* 113: 27.

Barker, S., Diz, P., Vautravers, M.J. et al. (2009). Interhemispheric Atlantic seesaw response during the last deglaciation. *Nature* 457: 1097–1102.

Baumgartner, T.R., Soutar, A., and Ferreira-Bartrina, V. (1992). Reconstructions of the history of Pacific sardine and northern anchovy populations over the past two millennia from sediments of the Santa Barbara Basin, California. *California Cooperative Oceanic Fisheries Investigations Reports* 33: 24–40.

Beaugrand, G., Reid, P.C., Ibañez, F. et al. (2002). Reorganization of North Atlantic marine copepod biodiversity and climate. *Science* 2002 296 (5573): 1692–1694.

Beaugrand, G., Brander, K.M., Lindley, J.A. et al. (2003). Plankton effect on cod recruitment in the North Sea. *Nature* 426: 661–664.

Behrenfeld, M.J., O'Malley, R.T., Siegel, D.A. et al. (2006). Climate-driven trends in contemporary ocean productivity. *Nature* 444: 752–755.

Bellwood, D.R., Hughes, T.P., Folke, C., and Nyström, M. (2004). Confronting the coral reef crisis. *Nature* 429 (6994): 827–833.

Bender, E.A., Case, T.J., and Gilpin, M.E. (1984). Perturbation experiments in community ecology: theory and practice. *Ecology* 65: 1–13.

Berkelmans, R., De'ath, G., Kininmonth, S., and Skirving, W.J. (2004). A comparison of the 1998 and 2002 coral bleaching events on the Great Barrier Reef: spatial correlation, patterns, and predictions. *Coral Reefs* 23: 74–83.

Bertrand, A., Gerlotto, F., Bertrand, S. et al. (2008). Schooling behaviour and environmental forcing in relation to anchoveta distribution: an analysis across multiple spatial scales. *Progress in Oceanography* 79: 264–277.

Bindoff, N.L., Willebrand, J., Artale, V. et al. (2007). Observations: oceanic climate change and sea level. In: *Climate Change 2007: The Physical Science Basis: Contribution of Working Group I to the Fourth Assessment Report of the Intergovernmental Panel on Climate Change* (ed. S. Solomon, D. Qin, M. Manning, et al.), 385–432. Cambridge: Cambridge University Press.

Boden, T.A., Marland, G., and Andres, R.J. (2017). *Global, Regional, and National Fossil-Fuel CO2 Emissions*. Oak Ridge, TN: Carbon Dioxide Information Analysis Center, Oak Ridge National Laboratory, U.S. Department of Energy.

Bond, G., Showers, W., Cheseby, M. et al. (1997). A pervasive millennial-scale cycle in North Atlantic Holocene and glacial climates. *Science* 278 (5341): 1257–1266.

Boyd, P.W., Jickells, T., Law, C.S. et al. (2007). Mesoscale iron enrichment experiments 1993–2005: synthesis and future directions. *Science* 315: 612–617.

Boyer, T.P., Levitus, S., Antonov, J.I. et al. (2005). Linear trends in salinity for the World Ocean, 1955–1998. *Geophysical Research Letters* 32: L01604.

Bracken, M.E.S., Friberg, S.E., Gonzalez-Dorantes, C.A., and Williams, S.L. (2008). Functional consequences of realistic biodiversity changes in a marine ecosystem. *Proceedings of the National Academy of Sciences of the United States of America* 105: 924–928.

Brinkhuis, H., Schouten, S., Collinson, M.E. et al. (2006). Episodic fresh surface waters in the Eocene Arctic Ocean. *Nature* 441 (7093): 606–609.

Bruno, J.F., Stachowicz, J.J., and Bertness, M.D. (2003). Inclusion of facilitation into ecological theory. *Trends in Ecology & Evolution* 18: 119–125.

Bundy, A., Shannon, L.J., Rochet, M. et al. (2010). The good(ish), the bad, and the ugly: a tripartite classification of ecosystem trends. *ICES Journal of Marine Science* 67 (4): 745–768.

Camacho, A.E. and Glicksman, R.L. (2016). Legal adaptive capacity: how program goals and processes shape federal land adaptation to climate change. *University of Colorado Law Review* 87: 211.

Carpenter, K.E., Abrar, M., Aeby, G. et al. (2008). One-third of reef-building corals face elevated extinction risk from climate change and local impacts. *Science* 321: 560–563.

Cazenave, A., Lombard, A., and Llovel, W. (2008). Present-day sea level rise: a synthesis. *Comptes Rendus Geoscience* 340: 761–770.

Cheung, W.W.L., Close, C., Lam, V. et al. (2008). Application of macroecological theory to predict effects of climate change on global fisheries potential. *Marine Ecology Progress Series* 365: 187–197.

Clarke, A. (1993). Temperature and extinction in the sea- a physiologist's view. *Paleobiology* 19 (4): 499–518.

Clarke, A. and Johnston, I.A. (1996). Evolution and adaptive radiation of Antarctic fishes. *Trends in Ecology & Evolution* 11: 212–218.

Conley, D.J., Carstensen, J., Aigars, J. et al. (2009). Hypoxia is increasing in the coastal zone of the Baltic Sea. *Environmental Science & Technology* 45: 6777–6783.

Cosens, B., Gunderson, L., and Chaffin, B. (2014). Adaptive water governance project: assessing law, resilience and governance in regional socio-ecological water systems facing a changing climate. *Idaho Law Review* 51: 1.

Cosens, B.A., Craig, R.K., Hirsch, S.L. et al. (2017). The role of law in adaptive governance. *Ecology and Society* 22 (1).

Cosens, B., Gunderson, L., and Chaffin, B.C. (2018). Introduction to the Special Feature Practicing Panarchy: assessing legal flexibility, ecological resilience, and adaptive governance in regional water systems experiencing rapid environmental change. *Ecology and Society* 23 (1): 1–6.

Courtland, R. (2008). Polar bear numbers set to fall. *Nature* 453: 432–433.

Craig, R.K. (2019). Fostering adaptive marine aquaculture through procedural innovation in marine spatial planning. *Marine Policy* 110: 103555.

Crowley, T.J. and Hyde, W.T. (2008). Transient nature of late Pleistocene climate variability. *Nature* 456 (7220): 226–230.

Dafforn, K.A., Johnston, E.L., Ferguson, A. et al. (2021). Marine urbanization: an ecological framework for designing multifunctional artificial structures. *Frontiers in Ecology and the Environment* 19 (7): 380–388.

Danovaro, R., Gambi, C., Dell'Anno, A. et al. (2008). Exponential decline of deep-sea ecosystem functioning linked to benthic biodiversity loss. *Current Biology* 18: 1–8.

Daskalov, G.M., Grishin, A.N., Rodionov, S., and Mihneva, V. (2007). Trophic cascades triggered by overfishing reveal possible mechanisms of ecosystem regime shifts. *Proceedings of the National Academy of Sciences of the United States of America* 104: 10518–10523.

Davis, K.L., Colefax, A.P., Tucker, J.P. et al. (2021). Global coral reef ecosystems exhibit declining calcification and increasing primary productivity. *Communications Earth & Environment* 2: 105. https://doi.org/10.1038/s43247-021-00168-w.

Dawson, M.N. and Hamner, W.M. (2005). Rapid evolutionary radiation of marine zooplankton in peripheral environments. *Proceedings of the National Academy of Sciences of the United States of America* 102: 9235–9240.

De'ath, G., Lough, J.M., and Fabricius, K.E. (2009). Declining coral calcification on the Great Barrier Reef. *Science* 323: 116–119.

Diaz, R.J. and Rosenberg, R. (2008). Spreading dead zones and consequences for marine ecosystems. *Science* 321: 926–929.

Díaz, S., Settele, J., Brondízio, E.S. et al. (2019). Pervasive human-driven decline of life on Earth points to the need for transformative change. *Science* 366 (6471): eaax3100.

Doney, S.C. (2010). The growing human footprint on coastal and open-ocean biogeochemistry. *Science* 328 (5985): 1512–1516.

Doney, S.C., Fabry, V.J., Feely, R.A., and Kleypas, J.A. (2009). Ocean acidification: the other CO_2 problem. *Annual Review of Marine Science* 1: 169–192.

Doney, S.C., Ruckelshaus, M., Duffy, J.E. et al. (2020). Climate change impacts on marine ecosystems. *Annual Review of Marine Science* 12: 11–37.

Donner, S.D. (2009). Coping with commitment: projected thermal stress on coral reefs under different future scenarios. *PLoS One* 4 (6): e5712.

Duarte, C.M., Losada, I.J., Hendriks, I.E. et al. (2021). The role of coastal plant communities for climate change mitigation and adaptation. *Nature Reviews Earth and Environment* 2 (9): 550–565.

Durack, P.J. and Wijffels, S.E. (2010). Fifty-year trends in global ocean salinities and their relationship to broad-scale warming. *Journal of Climate* 23: 4342–4362.

Durant, J.M., Hjermann, D.Ø., Ottersen, G., and Stenseth, N.C. (2007). Climate and the match or mismatch between predator requirements and resource availability. *Climate Research* 33 (3): 271–283.

Edwards, M. and Richardson, A.J. (2004). Impact of climate change on marine pelagic phenology and trophic mismatch. *Nature* 430: 881–884.

Elgendy, M., Hassini, S., and Coulibaly, P. (2023). Review of climate change adaptation strategies in water management. *Journal of Hydrologic Engineering* 29 (1): https://doi.org/10.1061/JHYEFF.HEENG-6014.

Enghoff, I.B., MacKenzie, B.R., and Nielsen, E.E. (2007). The Danish fish fauna during the warm Atlantic period (ca. 7000–3900 BC): forerunner of future changes? *Fisheries Research* 87: 167–180.

Eppley, R.E. (1972). Temperature and phytoplankton growth in the sea. *Fishery Bulletin* 70: 1063–1085.

Estes, J.A., Terborgh, J., Brashares, J.S. et al. (2011). Trophic downgrading of planet Earth. *Science* 333 (6040): 301–306.

Field, C.B., Behrenfeld, M.J., Randerson, J.T., and Falkowski, P. (1998). Primary production of the biosphere: integrating terrestrial and oceanic components. *Science* 281: 237–240.

Fine, M. and Tchernov, D. (2007). Scleractinian coral species survive and recover from decalcification. *Science* 315: 1811.

Fujii, T. and Raffaelli, D. (2008). Sea-level rise, expected environmental changes, and responses of intertidal benthic macrofauna in the Humber estuary, UK. *Marine Ecology Progress Series* 371: 23–35.

Gagliano, M., McCormick, M.I., and Meekan, M.G. (2007). Temperature-induced shifts in selective pressure at a critical developmental transition. *Oecologia* 152: 219–225.

Gaichas, S., Skaret, G., Falk-Petersen, J. et al. (2009). A comparison of community and trophic structure in five marine ecosystems based on energy budgets and system metrics. *Progress in Oceanography* 81: 47–62.

Garcia, H.E., Boyer, T.P., Levitus, S. et al. (2005). On the variability of dissolved oxygen and apparent oxygen utilization content for the upper world ocean: 1955 to 1998. *Geophysical Research Letters* 32: LO9604.

Garcia, H.E., Bianchi, D., Llort, J. et al. (2018). Enhanced primary production in the Southern Ocean could offset anticipated climate change impacts on global marine ecosystems. *Proceedings of the National Academy of Sciences of the United States of America* 115 (27): 7009–7014.

Gavio, B., Rincon-Diaz, M.N., and Santos-Martinez, A. (2015). Massive quantities of pelagic Sargassum on the shores of San Andres Island, Southwestern Caribbean. *Acta Biológica Colombiana* 20 (1): 239–241.

Gowen, R.J., Tett, P., Kennington, K. et al. (2008). The Irish Sea: is it eutrophic? *Estuarine, Coastal and Shelf Science* 76: 239–254.

Gower, J.F.R. and King, S.A. (2011). Distribution of floating Sargassum in the Gulf of Mexico and the Atlantic Ocean mapped using MERIS. *International Journal of Remote Sensing* 32: 1917–1929.

Greene, C.H. and Pershing, A.J. (2004). Climate and the conservation biology of North Atlantic right whales: the right whale at the wrong time? *Frontiers in Ecology and the Environment* 2: 29–34.

Halpern, B.S., Walbridge, S., Selkoe, K.A. et al. (2008). A global map of human impact on marine ecosystems. *Science* 319: 948–952.

Halpern, B.S., Frazier, M., Potapenko, J. et al. (2019). Recent pace of change in human impact on the world's ocean. *Scientific Reports* 9 (1): 1–8.

Hamilton, L.C., Brown, B.C., and Rasmussen, R.O. (2003). West Greenland's cod-to-shrimp transition: local dimensions of climatic change. *Arctic* 56: 271–282.

Hansen, J., Sato, M., Ruedy, R. et al. (2007). Dangerous human-made interference with climate: a GISS modelE study. *Atmospheric Chemistry and Physics* 7: 2287–2312.

Hare, S.R. and Mantua, N.J. (2000). Empirical evidence for North Pacific regime shifts in 1977 and 1989. *Progress in Oceanography* 47: 103–145.

Hare, J.A., Alexander, M.A., Fogarty, M.J. et al. (2010). Forecasting the dynamics of a coastal fishery species using a coupled climate-population model. *Ecological Applications* 20: 452–464.

Harrison, G.W. (1979). Stability under environmental stress- resistance, resilience, persistence and variability. *The American Naturalist* 113: 659–669.

Heller, N.E. and Zavaleta, E.S. (2009). Biodiversity management in the face of climate change: a review of 22 years of recommendations. *Biological Conservation* 142: 14–32.

Helmuth, B. and Hofmann, G.E. (2000). Defining thermal stress in the rocky intertidal: linking ecology and physiology through biophysics. *American Zoologist* 40: 1051–1054.

Hester, K.C., Peltzer, E.T., Kirkwood, W.J., and Brewer, P.G. (2008). Unanticipated consequences of ocean acidification: a noisier ocean at lower pH. *Geophysical Research Letters* 35: 19601.

Hiddink, J.D. and Hofstede, R.T. (2008). Climate induced increases in species richness of marine fish. *Global Change Biology* 14: 453–460.

Hillyer, R. and Silman, M.R. (2010). Changes in species interactions across a 2.5 km elevation gradient: effects on plant migration in response to climate change. *Global Change Biology* 16: 3205–3214.

Hipfner, J.M. (2008). Matches and mismatches: ocean climate, prey phenology and breeding success in a zooplanktivorous seabird. *Marine Ecology Progress Series* 368: 295–304.

Hoegh-Guldberg, O. and Bruno, J.F. (2010). The impact of climate change on the world's marine ecosystems. *Science* 328: 1523–1528.

Hoegh-Guldberg, H., Mumby, P.J., Hooten, A.J. et al. (2007). Coral Reefs under rapid climate change and ocean acidification. *Science* 318: 1737–1742.

Hoegh-Guldberg, O., Cai, R., Poloczanska, E. S., Brewer, P. G., Sundby, S., Hilmi, K. (2018). The Ocean. In *Global Warming of 1.5°C. An IPCC Special Report on the impacts of global warming of 1.5°C above pre-industrial levels and related global greenhouse gas emission pathways, in the context of strengthening the global response to the threat of climate change, sustainable development, and efforts to eradicate poverty*, 1–93.

Hoffman, A.A. and Willi, W. (2008). Detecting genetic responses to environmental change. *Nature Reviews Genetics* 9: 421–432.

Hood, R.R., Shenk, G.W., Dixon, R.L. et al. (2021). The Chesapeake Bay program modeling system: overview and recommendations for future development. *Ecological Modelling* 456: 109635.

Hooper, D.U., Chapin, F.S., Ewel, J.J. et al. (2005). Effects of biodiversity on ecosystem functioning: a consensus of current knowledge. *Ecological Monographs* 75 (1): 3–35.

Hughes, T., Kerry, J.T., Baird, A.H. et al. (2018). Global warming transforms coral reef assemblages. *Nature* 556: 492–496.

Hughes, T.P., Kerry, J.T., Baird, A.H. et al. (2020). Global warming impairs stock–recruitment dynamics of corals. *Nature* 588 (7836): 366–369.

IPCC (2007a). Summary for policymakers emission scenarios. In: *(2000) Special Report on Emissions Scenarios. A Special Report of Working Group III of the Intergovernmental Panel on Climate Change* (ed. N. Nakicenovic and R. Swart), 570. Cambridge: Cambridge University Press.

IPCC (2007b). *Climate Change 2007: The Physical Science Basis. Contribution of Working Group I to the Fourth Assessment Report of the Intergovernmental Panel on Climate Change* (ed. S. Solomon, D. Qin, M. Manning, et al.), 996. Cambridge: Cambridge University Press.

IPCC. (2021). Summary for policymakers. In *Climate Change 2021: The Physical Science Basis Contribution of Working Group I to the Sixth Assessment Report of the Intergovernmental Panel on Climate Change*, 3–32

Ishii, M., Kimoto, M., Sakamoto, K., and Iwasaki, S.I. (2006). Steric sea level changes estimated from historical ocean subsurface temperature and salinity analyses. *Journal of Oceanography* 62: 155–170.

Jackson, J.B. (2008). Ecological extinction and evolution in the brave new ocean. *Proceedings of the National Academy of Sciences of the United States of America* 105 (Suppl 1): 11458–11465.

Jackson, J.B.C. (2010). The future of the ocean's past. *Philosophical Transactions of the Royal Society, B: Biological Sciences* 365: 3765–3778.

Johnson, M.A., Macaulay, M.C., and Biggs, D.C. (1984). Respiration and excretion within a mass aggregation of *Euphausia superba*: implications for krill distribution. *Journal of Crustacean Biology* 4: 174–184.

Johnson, D., Ko, D.S., Franks, J.S. et al. (2013). The Sargassum invasion of the Eastern Caribbean and dynamics of the Equatorial North Atlantic.

Johnston, I.A. (2006). Environment and plasticity of myogenesis in teleost fish. *Journal of Experimental Biology* 209: 2249–2264.

Johnston, I.A. and Bennett, A.F. (1996). *Animals and Temperature: Phenotypic and Evolutionary Adaptation*. Cambridge University Press.

Jones, A. and Berkelmans, R. (2010). Potential costs of acclimatization to a warmer climate: growth of a reef coral with heat tolerant versus sensitive symbiont types. *PLoS One* 5: e10437.

Joos, F. and Spahni, R. (2008). Rates of change in natural and anthropogenic radiative forcing over the past 20,000 years. *Proceedings of the National Academy of Sciences of the United States of America* 105: 1425–1430.

Kaplan, I.C., Botsford, L.W., Carr, M.H. et al. (2020). From anecdote to action: engaging in-demand seafood stakeholders in fisheries research and management. *Fisheries Research* 220: 105360. https://doi.org/10.1016/j.fishres.2019.105360.

Keeling, R.F., Körtzinger, A., and Gruber, N. (2010). Ocean deoxygenation in a warming world. *Annual Review of Marine Science* 2: 199–229.

Kemp, W.M., Boynton, W.R., Adolf, J. et al. (2005). Eutrophication of chesapeake bay: historical trends and ecological interactions. *Marine Ecology Progress Series* 303: 1–29.

Kennett, J.P. and Stott, L.D. (1991). Abrupt deep-sea warming, paleoceanographic changes and benthic extinctions at the end of the paleocene. *Nature* 353: 225–229.

Kleypas, J.A., Feely, R.A., Fabry, V.J. et al. (2006). Impacts of ocean acidification on coral reefs an Ocean acidification due to increasing atmospheric carbon dioxided other marine calcifiers: a guide for future research. *Report, National Science Foundation, National Oceanic and Atmospheric Administration, and US Geological Survey*, Washington, DC.

Kwiatkowski, L. and Orr, J.C. (2018). Diverging seasonal extremes for ocean acidification during the twenty-first century. *Nature* 8 (2): 141–145.

Laffoley, D.d'A., Roe, H.S.J., Angel, M.V. et al. (2011). *The Protection and Management of the Sargasso Sea: The Golden Floating Rainforest of the Atlantic Ocean*, 633. Marine Science Faculty Publications.

Lawrence, D.M., Slater, A.G., Tomas, R.A. et al. (2008). Accelerated Arctic land warming and permafrost degradation during rapid sea ice loss. *Geophysical Research Letters* 35: L11506.

Laws, E.A., Landry, M.R., and Redalje, D.G. (2021). Size, phylogeny, and oceanography govern rates of primary production in the ocean. *Proceedings of the National Academy of Sciences of the United States of America* 118 (6): e2019034118.

Le Quéré, C., Raupach, M.R., Canadell, J.G. et al. (2009). Trends in the sources and sinks of carbon dioxide. *Nature Geoscience* 2: 831–836.

Leaper, R., Cooke, J., Trathan, P. et al. (2006). Global climate drives southern right whale (*Eubalaena australis*) population dynamics. *Biology Letters* 2: 289–292.

Lenton, T.M., Held, H., Kriegler, E. et al. (2008). Tipping elements in the Earth's climate system. *Proceedings of the National Academy of Sciences of the United States of America* 105: 1786–1793.

Levitus, S., Antonov, J.I., Boyer, T.P. et al. (2009). Global ocean heat content 1955–2008 in light of recently revealed instrumentation problems. *Geophysical Research Letters* 36: L07608.

Lisiecki, L.E., Raymo, M.E., and Curry, W.B. (2008). Atlantic overturning responses to Late Pleistocene climate forcings. *Nature* 456 (7223): 85–88.

Litzow, M.A. and Ciannelli, L. (2007). Oscillating trophic control induces community reorganization in a marine ecosystem. *Ecology Letters* 10: 1124–1134.

Lucey, S.M. and Nye, J.A. (2010). Shifting species assemblages in the Northeast US Continental Shelf Large Marine Ecosystem. *Marine Ecology Progress Series* 415: 23–33.

Lutz, M.J., Caldeira, K., Dunbar, R.B., and Behrenfeld, M.J. (2007). Seasonal rhythms of net primary production and particulate organic carbon flux to depth describe the efficiency of biological pump in the global ocean. *Journal of Geophysical Research, Oceans* 112: C10011.

Mann, M.E., Zhang, Z., Hughes, M.K. et al. (2008). Proxy-based reconstructions of hemispheric and global surface temperature variations over the past two millennia. *Proceedings of the National Academy of Sciences of the United States of America* 105 (36): 13252–13257.

Maslanik, J.A., Fowler, C., Stroeve, J. et al. (2007). A younger, thinner Arctic ice cover: increased potential for rapid, extensive sea-ice loss. *Geophysical Research Letters* 34: L24501.

Meure, C.M., Etheridge, D., Trudinger, C. et al. (2006). Law Dome CO_2, CH_4 and N_2O ice core records extended to 2000 years BP. *Geophysical Research Letters* 33: L14810.

Moanga, D.A. (2015). Karenia brevis hot spots in the west Florida shelf and their associated socioeconomic implications. MS Theses. University of Miami.

Moreira, A. and Alfonso, G. (2013). Inusual arribazón de *Sargassum fluitans* (Børgesen) Børgesen en la costa centro-sur de Cuba. *Revista de Investigaciones Marinas* 33: 17–20.

Mueter, F.J. and Litzow, M.A. (2008). Sea ice retreat alters the biogeography of the Bering Sea continental shelf. *Ecological Applications* 18: 309–320.

Muller, E.M., Rogers, C.S., Spitzack, A.S., and van Woesik, R. (2007). Bleaching increases likelihood of disease on *Acropora palmata* (Lamarck) in Hawksnest Bay, St. John, US Virgin Islands. *Coral Reefs* 27: 191–195.

Mumby, P.J., Hastings, A., and Edwards, H.J. (2007). Thresholds and the resilience of Caribbean coral reefs. *Nature* 450: 98–101.

Munday, P.L., Jones, G.P., Pratchett, M.S., and Williams, A.J. (2008). Climate change and the future for coral reef fishes. *Fish and Fisheries* 9: 261–285.

Munday, P.L., Dixson, D.L., Donelson, J.M. et al. (2009). Ocean acidification impairs olfactory discrimination and homing ability of a marine fish. *Proceedings of the National Academy of Sciences of the United States of America* 106: 1848–1852.

Navarrete, S.A., Wieters, E.A., Broitman, B.R., and Castilla, J.C. (2005). Scales of benthic-pelagic coupling and the intensity of species interactions: from recruitment limitation to top-down control. *Proceedings of the National Academy of Sciences of the United States of America* 102: 18046–18051.

Nye, J.A., Link, J.S., Hare, J.A., and Overholtz, W.J. (2009). Changing spatial distribution of fish stocks in relation to climate and population size on the Northeast United States continental shelf. *Marine Ecology Progress Series* 393: 111–129.

Oschlies, A., Schulz, K.G., Riebesell, U., and Schmittner, A. (2008). Simulated 21st century's increase in oceanic suboxia by CO_2-enhanced biotic carbon export. *Global Biogeochemical Cycles* 22: GB4008.

Parmesan, C. (2006). Ecological and evolutionary responses to recent climate change. *Annual Review of Ecology, Evolution, and Systematics* 37: 637–669.

Peck, L.S., Webb, K.E., and Bailey, D.M. (2004). Extreme sensitivity of biological function to temperature in Antarctic marine species. *Functional Ecology* 18: 625–630.

Perry, A.L., Low, P.J., Ellis, J.R., and Reynolds, J.D. (2005). Climate change and distribution shifts in marine fishes. *Science* 308: 912–915.

Peters, S.E. (2008). Environmental determinants of extinction selectivity in the fossil record. *Nature* 454: 626–U638.

Place, S.P., O'Donnell, M.J., and Hofmann, G.E. (2008). Gene expression in the intertidal mussel *Mytilus californianus*: physiological response to environmental factors on a biogeographic scale. *Marine Ecology Progress Series* 356: 1–14.

Platt, T., Fuentes-Yaco, C., and Frank, K.T. (2003). Spring algal bloom and larval fish survival. *Nature* 423: 398–399.

Poloczanska, E.S., Hawkins, S.J., Southward, A.J., and Burrows, M.T. (2008). Modelling the response of populations of competing species to climate change. *Ecology* 89: 3138–3149.

Polunin, N.V.C. (2008). *Aquatic Ecosystems: Trends and Global Prospects*. Cambridge University Press.

Polyakov, I.V., Timokhov, L.A., Alexeev, V.A. et al. (2010). Arctic ocean warming contributes to reduced polar ice cap. *Journal of Physical Oceanography* 40: 2743–2756.

Portner, H.O., Langenbuch, M., and Michaelidis, B. (2005). Synergistic effects of temperature extremes, hypoxia, and increases in CO_2 on marine animals: from Earth history to global change. *Journal of Geophysical Research: Oceans* 110: 15.

Pratchett, M.S., Munday, P.L., Wilson, S.K. et al. (2008). Effects of climate-induced coral bleaching on coral-reef fish ecological and economic consequences. *Oceanography and Marine Biology* 46: 251–296.

Rabalais, N.N., Diaz, R.J., Levin, L.A. et al. (2010a). Dynamics and distribution of natural and human-caused hypoxia. *Biogeosciences* 7: 585–619.

Rabalais, N.N., Gupta, B.J., Boesch, D.F., and Turner, R.E. (2010b). Forecasting the impact of nutrient reduction on the Gulf of Mexico hypoxic zone. *Hypoxia in the Northern Gulf of Mexico* 30 (5): 753–772.

Rahmstorf, S. (2007). A semi-empirical approach to projecting future sea-level rise. *Science* 315: 368–370.

Ramanathan, V. and Feng, Y. (2008). On avoiding dangerous anthropogenic interference with the climate system: formidable challenges ahead. *Proceedings of the National Academy of Sciences of the United States of America* 105: 14245–14250.

Reid, K. and Croxall, J.P. (2001). Environmental response of upper trophic level predators reveals a system change in an Antarctic marine ecosystem. *Proceedings of the Royal Society of London, Series B: Biological Sciences* 268: 377–384.

Reid, P.C., Johns, D.G., Edwards, M. et al. (2007). A biological consequence of reducing Arctic ice cover: arrival of the Pacific diatom *Neodenticula seminae* in the North Atlantic for the first time in 800,000 years. *Global Change Biology* 13: 1910–1921.

Richards, D.R. and Friess, D.A. (2016). Rates and drivers of mangrove deforestation in Southeast Asia, 2000–2012. *Proceedings of the National Academy of Sciences of the United States of America* 113 (2): 344–349.

Richardson, A.J. and Schoeman, D.S. (2004). Climate impact on plankton ecosystems in the Northeast Atlantic. *Science* 305: 1609–1612.

Ridgway, K.R. (2007). Long-term trend and decadal variability of the southward penetration of the East Australian current. *Geophysical Research Letters* 34: 22921–22936.

Salvat, B. and Allemand, D. (2009). *Acidification and Coral Reefs. Coral Reef Initiative Pacific*, 32. New Caledonia: Coral Reef Initiatives for the Pacific (CRISP).

Samhouri, J.F., Levin, P.S., and Harvey, C.J. (2009). Quantitative evaluation of marine ecosystem indicator performance using food web models. *Ecosystems* 12: 1283–1298.

Sarmiento, J.L., Hughes, T.M.C., Stouffer, R.J., and Manabe, S. (1998). Simulated response of the ocean carbon cycle to anthropogenic climate warming. *Nature* 393: 245–249.

Schmittner, A. (2005). Decline of the marine ecosystem caused by a reduction in the Atlantic overturning circulation. *Nature* 434: 628–633.

Schmittner, A. and Galbraith, E.D. (2008). Glacial greenhouse-gas fluctuations controlled by ocean circulation changes. *Nature* 456 (7220): 373–376.

Shaffer, G., Olsen, S.M., and Pedersen, J.O.P. (2009). Long-term ocean oxygen depletion in response to carbon dioxide emissions from fossil fuels. *Nature Geoscience* 2: 105–109.

Smetacek, V. and Zingone, A. (2013). Green and golden seaweed tides on the rise. *Nature* 504 (7478): 84–88.

Sobrino, C., Ward, M.L., and Neale, P.J. (2008). Acclimation to elevated carbon dioxide and ultraviolet radiation in the diatom *Thalassiosira pseudonana*: effects on growth, photosynthesis, and spectral sensitivity of photoinhibition. *Limnology and Oceanography* 53: 494–505.

Solomon, S., Qin, D., Manning, M. et al. (2007). Climate change 2007: the physical science basis. *Working Group I Contribution to the Fourth Assessment Report of the Intergovernmental Panel on Climate Change.*

Somero, G. (2012). The physiology of global change: linking patterns to mechanisms. *Annual Review of Marine Science* 4: 39–61.

Spencer, R.W. (2008). *The Discovery of Global Warming*, 3. Cambridge: American Institute of Physics, Harvard University Press.

Sprintall, J. (2008). Long-term trends and interannual variability of temperature in Drake Passage. *Progress in Oceanography* 77: 316–330.

Stammerjohn, S.E., Martinson, D.G., Smith, R.C. et al. (2008). Trends in Antarctic annual sea ice retreat and advance and their relation to El Niño–Southern Oscillation and Southern Annular Mode variability. *Journal of Geophysical Research* 108: C03S90.

Stroeve, J., Holland, M.M., Meier, W. et al. (2007). Arctic sea ice decline: faster than forecast. *Geophysical Research Letters* 34: L09501.

Stroeve, J., Kattsov, V., Barrett, A. et al. (2012). Trends in Arctic sea ice extent from CMIP5, CMIP3 and observations. *Geophysical Research Letters* 39 (16): 1–6.

The Royal Society (2005). Ocean acidification due to increasing atmospheric carbon dioxide, 60.

Vaselli, S., Bertocci, I., Maggi, E., and Benedetti-Cecchi, L. (2008). Assessing the consequences of sea level rise: effects of changes in the slope of the substratum on sessile assemblages of rocky seashores. *Marine Ecology Progress Series* 368: 9–22.

Velasco, A. (2015). Sargazo invade el Caribe: ¿Qué es y cómo se combate? El Financiero. Retrieved December 04, 2020, from https://www.elfinanciero.com.mx/nacional/sargazo-invade-el-caribe-quees-y-como-se-combate.

Victor, S., Golbuu, Y., and Wolanski, E. (2020). The Palau National Marine Sanctuary: ecological, social, and economic impacts. *Frontiers in Marine Science* 7: 213. https://doi.org/10.3389/fmars.2020.00213.

Waldbusser, G.G., Hales, B., Langdon, C.J. et al. (2015). Ocean acidification has multiple modes of action on bivalve larvae. *PLoS One* 10: e0128376.

Whitehouse, M.J., Meredith, M.P., Rothery, P. et al. (2008). Rapid warming of the ocean around South Georgia, Southern Ocean, during the 20th century: forcings, characteristics and implications for lower trophic levels. *Deep-Sea Research Part I: Oceanographic Research Papers* 55: 1218–1228.

Wiedenmann, J., Cresswell, K., and Mangel, M. (2008). Temperature-dependent growth of Antarctic krill: predictions for a changing climate from a cohort model. *Marine Ecology Progress Series* 358: 191–202.

Winder, M., Reuter, J.E., and Schladow, S.G. (2008). Lake warming favours small-sized planktonic diatom species. *Proceedings of the Royal Society B: Biological Sciences* 276: 427–435.

Yasuhara, M., Cronin, T.M., deMenocal, P.B. et al. (2008). Abrupt climate change and collapse of deep-sea ecosystems. *Proceedings of the National Academy of Sciences of the United States of America* 105: 1556–1560.

20

X-Press Pearl Disaster

Uswatta Liyanage Harsha Prabath Perera and Amila Sandaruwan Ratnayake

Department of Applied Earth Sciences, Faculty of Applied Sciences, Uva Wellassa University, Badulla, 90000, Sri Lanka

20.1 Introduction

The MV X-Press Pearl disaster in May 2021 produced the worst maritime pollution ever seen in the Indian Ocean. It has had a significant impact on the sensitive coastal environment, local communities, the economy of Sri Lanka, and the Indian Ocean (Figure 20.1). The cargo vessel MV X-Press Pearl was carrying 25 tons of nitric acid, sodium hydroxide, methanol, 9700 tons of toxic epoxy resin and other chemicals, plastic nurdles, metals such as lead and copper, and 300 tons of fuel in its tanks when it caught fire (de Vos et al. 2021; Ratnayake and Perera, 2022; Sewwandi et al., 2022a; Pathmalal et al., 2023). The vessel was anchored about 9 nautical miles (17 km) northwest of the port of Colombo in Sri Lanka on 20May 2021, due to the emission of chemical fumes, and an explosion occurred on 25 May 2021 (Hassan et al., 2021). After that, the vessel burned for almost two weeks. Several explosions occurred before the ship sank completely off the coast of Sri Lanka on 17 June 2021 (Perera et al., 2022). As a result, in a very short time, the western coast of Sri Lanka and the Indian Ocean were covered with a large amount of debris, tons of plastic nurdles, and the remains of burned plastic (Figure 20.1). As boundaries, unburned, white opaque nurdles were reported on beaches as far as Mannar Island on the northwest coast and Trincomalee on the northeast coast of Sri Lanka (Figure 20.1a). In addition, these plastic nurdles were spread over a large area, including along the coasts of Indonesia, Malaysia, and Somalia (Hassan et al., 2021; Pattiaratchi and Wijeratne, 2021; Perera et al., 2022). Accordingly, the MV X-Press Pearl accident was the worst marine pollution in the Indian Ocean, in which over 1750 tons of plastic nurdles (i.e. approximately 84 billion low-density polyethylene nurdles) were released (Figure 20.1).

As a visible consequence, the disaster had significant impacts on marine life (e.g. deaths of turtles ($n = 258$), dolphins ($n = 43$), and whales ($n = 6$)), beach pollution, and the Sri Lankan economy (e.g. fishing and tourism industries) (Ratnayake and Perera, 2022; Pathmalal et al., 2023). Most of these damages are irreversible. To mitigate the damages, the Sri Lankan government declared a state of emergency and initiated massive manual and mechanical clean-up operations (e.g. Sarakkuwa Beach) under the guidance of the Marine Environmental Protection Agency

Coastal and Marine Pollution: Source to Sink, Mitigation and Management, First Edition.
Edited by Meththika Vithanage, Sameera M. Samarasekara, Bryan D. James, and Christopher M. Reddy.

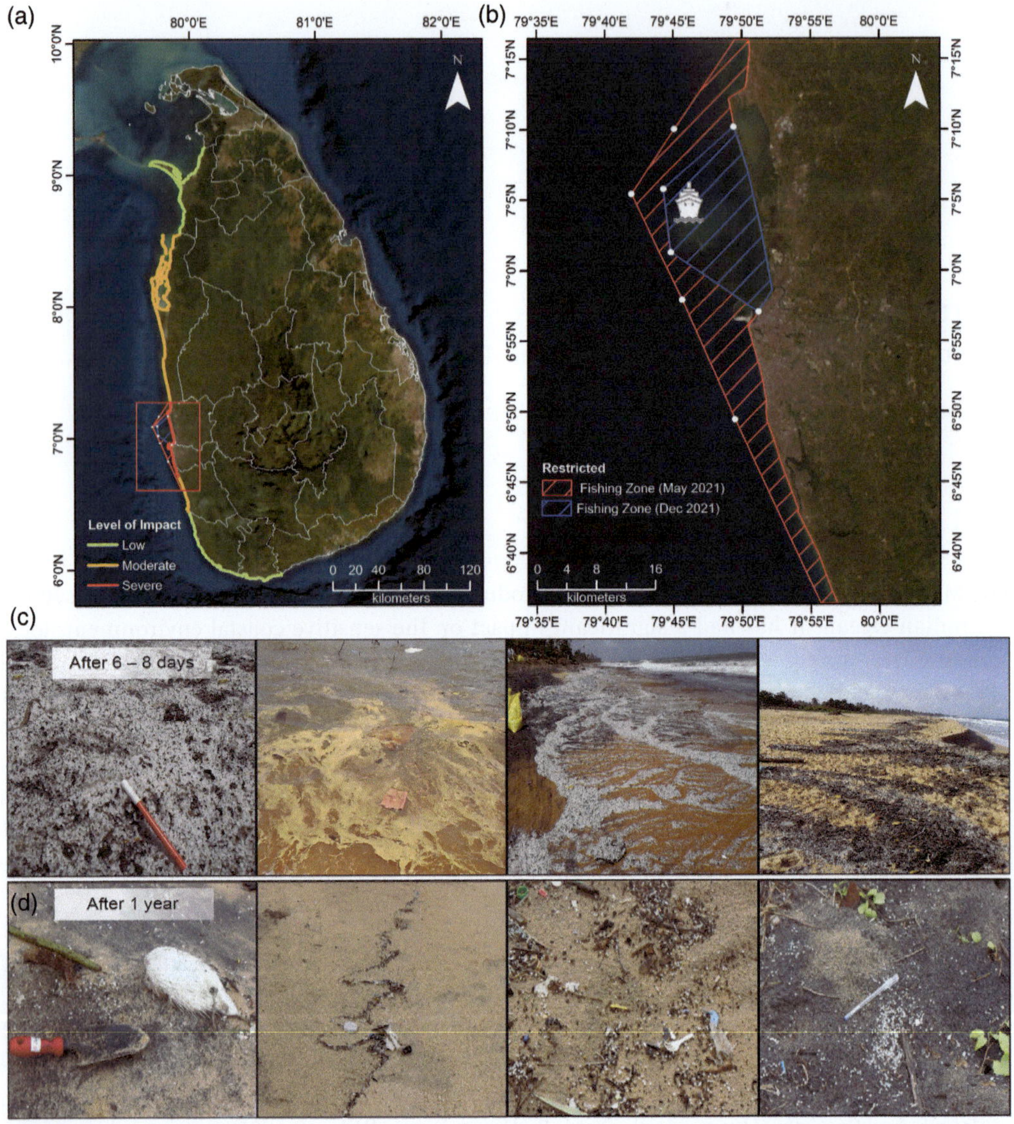

Figure 20.1 (a) Simplified map of the areas along the west coast of Sri Lanka affected by the X-Press Pearl shipwreck and its intensity, (b) restricted fishing zones after 22 May 2021, and revised restricted fishing zones after 1 December 2021, (c) obtained field photographs six to eight days after the explosion, and (d) one year after the disaster.

(MEPA) (Figure 20.2b). As a result, around 1610 tons of debris and polluted sand have been removed from the coastline along Sri Lanka (Hassan et al., 2021). However, the X-Press Pearl disaster released a significant amount of primary microplastics in 1–5 mm diameter into the Indian Ocean, and these buried microplastics can cause a chain of problems for the health of marine animals and humans during the next century (Perera et al., 2022; Sewwandi et al., 2022b; Gunawardhana et al., 2023).

Figure 20.2 (a) Remnants of burned plastics and white-colored unburned nurdles/pellets washed ashore; (b) massive manual and mechanical clean-up operations for plastic pellets removal, and plastic pellets trampling by workers and heavy machinery during clean-up operations; (c) plastic nurdles and remnants of burned plastic mixing with coastal vegetation, reaching inland areas, and damaging fishing nets; and (d) impact to marine life.

20.2 The MV X-Press Pearl Cargo Vessel

The MV X-Press Pearl is a Super Eco 2700 class container ship registered in Singapore and operated by X-Press Feeders Limited. This vessel was built by Zhoushan Changhong International Shipyard Co. Limited in Zhoushan, China, along with its sister ship, X-Press Mekong. The X-Press Pearl was about 186 m long and had a capacity of 37 000 deadweight tons. The ship could carry 2756 twenty-foot (TEU) containers at a time (Rubesinghe et al., 2022; Zhang et al., 2023). The vessel was deployed from Port Klang (Malaysia) to Hamad Port (Qatar) via Singapore and Jebel Ali (Dubai, United Arab Emirates). The return to Malaysia was via Hazira (India) and Colombo (Sri Lanka). The vessel had made three voyages to Colombo on 17 March and 18 April and caught fire shortly after arriving at the port for its third call on 19 May 2021. The vessel was engulfed in flames by 27 May and was declared a total loss (EconomyNext, 2021; Pattiaratchi and Wijeratne, 2021; Zhang et al., 2023). It was still afloat, and as of late 27 May 2021, the fire was considered under control by Sri Lankan and Indian coastguard firefighters. After burning for 12 days, the vessel sank on 2 June while being towed into deep water. The MV X-Press Pearl, which was floating on the Sri Lankan coast off the port of Colombo, was buried forever on the seabed at a depth of 21 m on 17 June 2021 (Rubesinghe et al., 2022). This incident had a great environmental and economic impact on Sri Lanka (Figures 20.1 and 20.2).

20.2.1 Hazardous Chemicals and Goods Contained in the MV X-Press Pearl Vessel

The cargo ship MV X-Press Pearl had 1486 containers, of which at least 81 contained 15 different types of dangerous goods according to the International Maritime Dangerous Goods Code (IMDG Code) at the time of the fire (de Vos et al. 2021; Perera et al., 2022; Sewwandi et al., 2022a; Pathmalal et al., 2023). The containers also contained hazardous and noxious substances (HNS) that could pose a major problem, as per the cargo manifest. HNS included nitric acid (25 tons), caustic soda (1040 tons), methanol (210 tons), and 9700 tons of potentially toxic epoxy resin (Table 20.1). These HNS contain about one-third of the cargo (349 containers) on the vessel. In addition, more than 1750 tons (87 containers) of white opaque plastic nurdles/pellets (<5 mm), high-density polyethylene (HDPE), low-density polyethylene (LDPE), and linear low-density polyethylene (LLDPE) were included in the X-Press Pearl vessel (Table 20.1), which can cause severe environmental damage. It also contained more than 300 tons of fuel in its tanks and metals such as aluminum, copper, iron, and lead in the containers when it caught fire (Table 20.1) (Hassan et al., 2021; Karthik et al., 2022; Rubesinghe et al., 2022; James et al., 2023). The release of these HNS into the environment directly affected marine life, such as dead turtles with "burned" and "bleached" carapace observations and dead fish with plastic pellets stuck in the pharynx and opercle (Cumaranatunga et al., 2023; Ekanayake et al., 2023; Pathmalal et al., 2023).

20.3 The Environmental Impacts of the X-Press Pearl Maritime Debacle

The X-Press Pearl disaster affected the environment at the time of the accident and even many decades after the disaster. The major environmental pollution is caused by the release of plastic pellets into the Indian Ocean (Figures 20.1 and 20.2). In addition, the environment was also affected by oil pollution, chemical pollution, air pollution, shipwrecks, and lost containers.

Table 20.1 Plastics and polymers, chemicals, and metals aboard the MV X-Press Pearl cargo ship.

Plastics and polymers	Amount (MT)	Chemicals	Amount (MT)	Metals	Amount (MT)
Epoxy resins	9700.8	Urea (Dust, Prilled, ...)	1843.3	Aluminum stuffs	2202.2
HDPE	747.8	Inorganic chemicals, Nos	495.3	Copper stuffs	474.6
LDPE	574.0	caustic soda	1126.9	Iron and steel stuffs	74.8
Bare foam pig	1.9	Chemical products, Nos	160.2	Lead	187.0
Polycarbonate	60.9	Nitric acid	28.7	Metal scraps	142.7
Polymer beads	19.3	Perfumery products	8.0		
PVC film	20.4	Assorted perfumes	16.8		
Unsaturated polyester resin	21.0	Quicklime lumps	1196.4		
Vinyl acetate	46.3	Sodium methylate	57.3		
Synthetic resins	177.3	Methanol	235.6		
LLDPE	245.4	Bright yellow Sulfur	562.4		
Packaging of plastics (PS) pellets	31.9	Molybdic oxide	48.0		
Packaging materials	22.8	Env. hazardous substances	10.6		
Plastic pellets	81.3	Pharmaceutical stuffs	7.5		
Polymers of propylene	57.3	Paints	21.2		
ALKYD resin	19.5	Colors for ceramic ware	15.0		
Vinyl copolymer	18.6	Shampoo	16.3		
Polybutadiene	92.7	Cement Conforming	1154.8		
		Modified asphalt	205.2		
		Silicon sealant	12.5		
		Engine coolant and grease	16.9		
		Liquid paraffin	130.2		
		Brake fluid	34.7		
		Lubricating oil	387.8		
		Base oil	47.6		
		Pails of lubricants	145.9		
Total : Plastics	**11939.2**	**Total : Chemicals**	**7985.1**	**Total : Metals**	**3081.3**

20.3.1 Plastic and Microplastic Pollution

The X-Press Pearl disaster was well reflected in the white color of unburned nurdles/pellets (micro < 5 mm), degraded/deformed nurdles, plastic wires, and remnants of burned plastic (macro > 5 mm) washed ashore (Figure 20.2). Table 20.1 shows the types and quantities of plastics and polymers according to the X-Press Pearl ship manifest. The plastics and polymers found on the vessel can be classified as high-density polyethylene (HDPE), low-density polyethylene (LDPE), and linear low-density polyethylene (LLDPE). The washed-up white plastic nurdles on the beach were identified as LDPE, and the remains of burned and/or completely melted nurdles were also identified as LDPE or LLDPE (de Vos et al. 2021; Sewwandi et al., 2022a, 2023; James et al., 2023). Plastic wires are composed of aromatic chains containing polyester or styrene materials (Perera et al., 2022). Although these plastic nurdles and remnants of burned plastic were mainly deposited on sandy beaches, plastic debris was observed below the beach surface (15–30 cm) due to high wave energy and sand movement, as well as trampling by workers and heavy machinery during clean-up operations (Figure 20.2b). In addition, the trucking of plastic nurdles and burned remains allowed them to reach the interior of the country (Figure 20.2c). Preliminary modeling results suggest that only a part of these plastic nurdles have reached the Sri Lanka coastline, and most of the nurdles remain in the Indian Ocean (Hassan et al., 2021). Therefore, plastic nurdles have been transported around Sri Lanka by monsoon and tidal currents and can spread to the coasts of Indonesia, the Maldives, and Somalia (Jayathilaka et al., 2022; Karthik et al., 2022; Pattiaratchi et al., 2022; Sewwandi et al., 2023). Accordingly, their removal from the coastal environment has become more complicated.

20.3.1.1 Microplastic Pollution

Microplastics are defined as small plastics, usually less than 5 mm in diameter (Barboza and Gimenez, 2015; Akdogan and Guven, 2019; Pilapitiya and Ratnayake, 2024). Approximately 84 billion white plastic nurdles (primary microplastics) were released from the X-Press Pearl marine disaster. According to Perera et al. (2022), the alteration of polymer chains in white plastic nurdles begins at temperatures as low as 60 °C. Weathering/degradation of plastics due to UV irradiation, wave action, chemicals, and abrasion against beach sand can also promote the formation of microplastics (Gunawardhana et al., 2023; Pilapitiya and Ratnayake, 2024). According to the weathering/degradation of plastic nurdles about one year after the disaster, the average weight and volume differences of the nurdles were 18.060 mg and 130.119 μm^3, respectively (Perera et al., 2022). Therefore, this diminishing amount of plastic gradually accumulates in the environment as microplastics and nanoplastics. That said, it will add abnormal values for the amount of micro and nanoplastics added to the environment by all the nurdles released from the X-Press Pearl marine disaster. Degraded micro/nanoplastics pose a threat to aquatic and coastal animals through ingestion. The ingested plastics can be bioaccumulated in the gut and stomach of animals (Neves et al., 2015; Guzzetti et al., 2018; Pilapitiya and Ratnayake, 2024). Humans can also ingest microplastics through the consumption of seafood and salt. Consequently, direct or indirect consumption of microplastics causes adverse health effects on the human body, including carcinogens (Rist et al., 2018; Campanale et al., 2020; Pilapitiya and Ratnayake, 2024).

20.3.2 Chemical Pollution

The X-Press Pearl vessel transported various hazardous and harmful substances and burned completely during the accident. It resulted in the release of HNS into the environment. Most of the HNS transported by the vessel X-Press Pearl were flammable and/or soluble (nitric acid (25 tons), caustic soda (1040 tons), methanol (210 tons), sodium methoxide, sodium methylate, and vinyl acetate) in seawater (Rubesinghe et al., 2022; Pathmalal et al., 2023). Therefore, these HNS chemicals have been added to the natural environment by burning in fire and dissolving in water. Both nitric acid and caustic soda products dissolve in water, and their densities are higher than seawater. Accordingly, these chemicals can sink into the seabed. Thus, nitric acid and caustic soda can create a highly corrosive plume on the seafloor. The observations of "burned" and "bleached" carapaces on dead turtles suggest that shells and skins can be disintegrated by contact with moving plumes of nitric acid and/or caustic soda (Hassan et al., 2021; Rubesinghe et al., 2022). Methanol can burn immediately in a fire. A small amount of methanol can mix with seawater and form a plume. Polluted seawater plumes float on water, and evaporation processes can mainly occur. A methanol plume just below the surface water can adversely affect marine life, such as mackerel, tuna, and skipjack (Mannapperuma, 2023; Pathmalal et al., 2023). The amount of epoxy resin added to the marine environment is enormous (9700.8 Mt in Table 20.1). However, the ship manifest does not provide exact information regarding the form of the epoxy resin, that is, liquid, solid, paste, or gel. The liquid epoxy plume can sink as it is denser than seawater. The oceanic currents can carry liquid epoxy with suspended particles. Epoxy resins are toxic to aquatic life and can have long-term effects on marine life, such as skin and eye burns (Liu et al., 2011; Aris et al., 2014; Cantoni et al., 2021). The X-Press Pearl shipwreck can enhance heavy metal concentrations on the western coasts. For example, the concentrations of heavy metals such as Cr, Zn, Cu, Ni, Li, Mo, As, Cd, Pb, Se, Ag, and Fe may increase after the X-Press Pear shipwreck (Perera et al., 2022; Sewwandi et al., 2022a). However, no studies have been carried out to estimate the direct damage caused by these chemicals and heavy metals to marine life and the environment.

20.3.3 Oil Pollution

All the bunker fuel on board the X-Press Pearl has leaked into the sea after the shipwreck. According to satellite images and aerial surveillance, an oil slick measuring 0.51 km^2 and 3.23 km in length was detected near the ship on 8 June, which had extended to 4.3 km by 14 June 2021 (Hassan et al., 2021; Pathmalal et al., 2023). Heavy brown oil slicks were observed within a metallic sheen (0.04–0.30 μm thickness) on the surface of the water, and they were found to be moving toward the northeast coast for several kilometers. Its thickness ranged from 50 to 200 μm, and the oil spill had been continuous for nearly a month, from 8 June to 4 July 2021. The oil slick was derived from bunker fuel rather than diesel or other refined products, based on its physical appearance and properties (James et al., 2023; Mannapperuma, 2023). The X-Press Pearl bunker is composed of intermediate fuel oil 380 (IFO 380). This type of oil is obtained by blending approximately 95% heavy fuel oil (HFO) with 5% gasoline (Hassan et al., 2021; Pallewatta et al., 2023). HFO has a tar-like consistency and is the residue of the oil distillation and cracking processes. This oil waste mainly consists of asphaltenes, resins, solid particles, and low concentrations of heavy metals. HFO is primarily used by the marine industry due to its relatively low cost compared to lighter fuels. However, due to their very high viscosity and high density, their release into the environment poses a greater threat to flora and fauna compared to lighter fuels (Garaniya et al., 2018). Nevertheless, the impact of the oil spill after the X-Press Pearl shipwreck was greatly reduced due to the constant vigilance and safety measures of the MEPA, National Aquatic Resources and Development Agency (NARA), Oil Spill Response Limited (OSRL), Sri Lanka Port Authority (SLPA), and Sri Lanka Armed Forces (Navy, Army, Coast Guard, and Air Force).

20.3.4 Air Pollution

The air pollution caused by the massive fire and explosions was enormous and lasted for about 10 days after the X-Press Pearl shipwreck. The smoke plume, which spread over several kilometers, contained a mixture of pollutants, including nitrogen oxides, sulfur dioxide, carbon monoxide, soot, particulate matter, a range of hydrocarbons, as well as dioxins, furans, and heavy metals (NBRO, 2021; Mannapperuma, 2023). Air pollution was high in the fire zone, based on observations made by the National Building Research Organization (NBRO) on 23 May 2021, at a distance of 50 m from the ship using photoionization detectors (PIDs). In addition, an area of 120 km^2, including a highly populated area, was identified as vulnerable to this air pollution (NBRO, 2021; Chandrasena and Premasiri, 2023). Nitrogen oxides exceeded national ambient air quality levels. Furthermore, ground-level measurements did not report significant pollutant levels. This is because the plume has risen significantly above the ground when it reaches land. The effects of air pollution are not limited to the atmosphere but affect both the sea and the land through the deposition of fire residue (Figure 20.1c). Moreover, some of the air pollutants that reach the upper atmosphere with the smoke can be deposited back into the soil and sea via rain. The air pollutant levels were much lower than on a normal day during the X-Press Pearl accident, due to COVID-19 lockdown restrictions (Hassan et al., 2021).

20.3.5 Impacts on Marine Life and Coastal Environments

The MV X-Press Pearl shipwreck directly affected marine life. Dead marine life was found mainly along the west coast of Sri Lanka, and ingestion of chemicals and/or plastic is believed to be the cause of the sudden increase in the death of marine life (Figures 20.1 and 20.2). In addition to short-term impacts, chemicals like epoxy resin can cause long-term impacts on aquatic animals through genetic modification. For example, partially burned epoxy resins can release bisphenols, which are identified as endocrine-disruptive chemicals. The ingestion of these chemicals (e.g. bisphenol A) is directly responsible for the malfunctioning of the endocrine systems of aquatic animals. Therefore, these chemicals directly affect reproduction, reduce fertility and the quality of sperm, cause secondary sexual feminization, increase mortality, and change the sex ratio (Tsai, 2006; Liu et al., 2011; Aris et al., 2014; Cantoni et al., 2021). White-colored unburned nurdles/pellets, degraded/deformed nurdles, plastic wires, and remnants of burned plastic are other major threats to aquatic life. Ingestion of these micro and/or nanoplastics can bioaccumulate in the gut and stomach of aquatic and coastal animals (Oliveira et al., 2013; Perera et al., 2022). Many marine animals can consume small, decomposed nurdles and microplastics of different colors (Rochman et al., 2019; Sewwandi et al., 2022b). Through the consumption of seafood and salt, microplastics can enter the human body and cause adverse health effects on the human body, including carcinogens (Brandt-Rauf et al., 2012; Barboza et al., 2018; Seth and Shriwastav, 2018; Gruber et al., 2022).

Several coastal ecosystems, such as Negombo Lagoon on the west coast of Sri Lanka, are affected by the MV X-Press Pearl disaster. For example, the Negombo Lagoon and its associated mangrove system have become entangled with large amounts of marine debris and plastics after the MV X-Press Pearl disaster (Hassan et al., 2021; Pathmalal et al., 2023). The west coast of Sri Lanka is an important breeding and nursery site for fish, spiny lobsters, shrimps, crabs, and marine aquarium fish (Wijayaratne and Maldeniya, 2003). Therefore, it is a serious threat to the sensitive flora and fauna of the coastal areas.

20.3.6 Hazard Identification and Risk Analysis

Risk assessment is a process to determine the nature and extent of risk by analyzing hazards and evaluating existing conditions of vulnerability that together could potentially harm exposed people, property, services, livelihoods, and the environment on which they depend (Figure 20.3). Disaster risk assessment involves (i) identification of hazards and a review of the technical characteristics of hazards such as their location, intensity, frequency, and probability; (ii) analysis of exposure and vulnerability, including the physical, social, health, environmental, and economic dimensions; and (iii) a comprehensive assessment of the effectiveness of existing and alternative coping capabilities about potential risk scenarios (UNDP, 2010). Therefore, it is essential to implement a risk analysis system in preparation for future marine disasters such as the MV X-Press Pearl shipwreck (Figure 20.3). The first step in a hazard identification and risk analysis system is to identify potential hazards. After identifying a marine hazard, the next step is to assess the marine risk. Controlling risk and implementing additional risk controls is quite a challenging step. Monitoring and review should be done as a final step (Figure 20.3). Strategies that can be used here include implementing preventive measures to avoid potential risks,

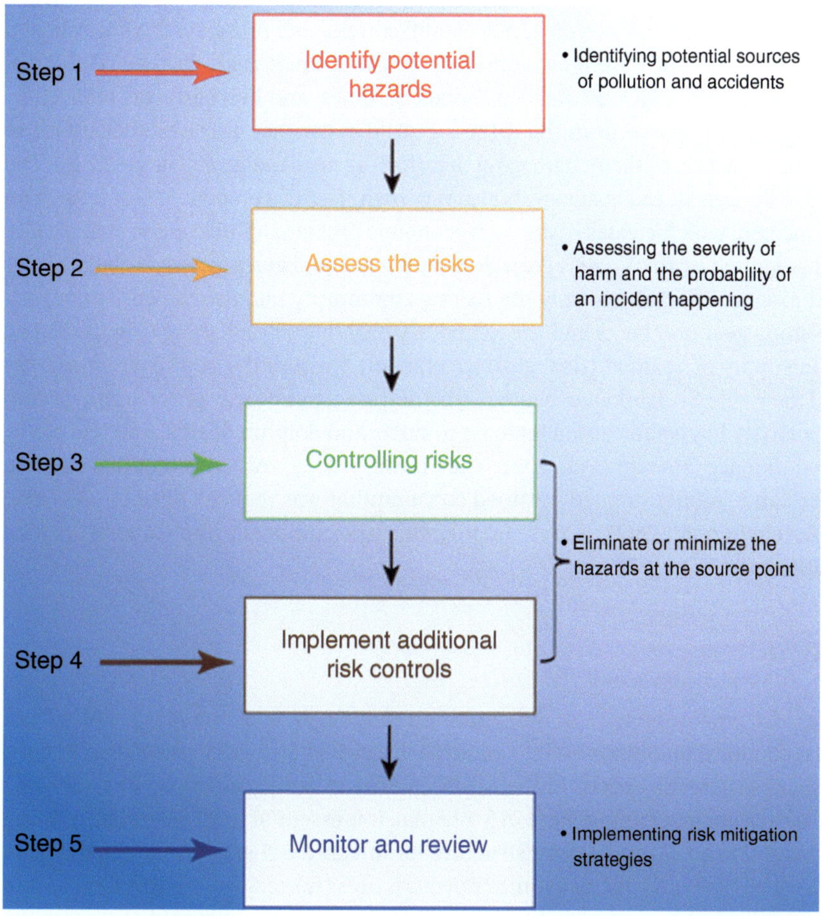

Figure 20.3 Risk analysis system procedure for risk identification and key functions of each step.

establishing emergency response plans in the event of incidents, providing training and resources to responders, improving regulations and enforcement mechanisms, and using alternative technologies and sustainable practices (Erabadupitiya et al., 2023).

20.4 The Socioeconomic Impacts After the X-Press Pearl Disaster

The MV X-Press Pearl disaster created a significant impact on the economy of Sri Lanka (Ratnayake, 2023). For example, the fishing industry is a significant part of the economy of Sri Lanka, contributing 1.3% to the country's gross domestic product (GDP). Approximately 583 000 people are employed in the fishing industry, with a support workforce of 2.7 million (Azmi, 2021; Rubesinghe et al., 2022). The disaster has damaged the livelihood of 20 000 fishing families, and the number of affected fishermen is estimated to be around 16 000 (Gunesekera et al., 2021; The Morning, 2023). The total value of the losses in the fisheries sector per day during the X-Press Pearl disaster period was around LKR 77 877 000 (USD 390 000), due to restrictions on fishing and reduced demand for fish consumption (Gunesekera et al., 2021). Seafood sales and consumption across the country also clearly declined (Arachchi et al., 2023).

Fishing was temporarily prohibited along an 80-km stretch of coastline (Figure 20.1b). Therefore, coastal fishing communities such as fishermen, net manufacturers, fish processors, wholesale and retail merchants, and boat owners were most vulnerable to the X-Press Pearl disaster (Pathmalal et al., 2023). Among them, small fish catching (e.g., sprats, sardines, and herrings) was reduced by 60–70% due to the chemicals released from the X-Press Pearl disaster that dissolved the surface of fishing nets and entangled debris in them, damaging their fishing nets and gear (Figure 20.2c). The cost of replacing nets, fishing gear, and boat parts is also very high. Therefore, about 51% of fishermen employed lost their income, which creates many socioeconomic problems (Rubesinghe et al., 2022). The economic damage caused by the X-Press Pearl disaster has not yet been precisely calculated.

The X-Press Pearl disaster affected not only the fishing community but also the entire country. Due to damage to fishing gear and boats and the temporary prohibition on fishing (Figure 20.1b), the reduction in the supply of seafood (fish, shrimp, crab, etc.) across the country reduced the nutritional needs of the entire Sri Lankan consumer. Furthermore, public fear of consuming "contaminated fish," exacerbated by a substantial increase in turtle and dolphin deaths, had reportedly harmed consumer confidence (Hassan et al., 2021; The Morning, 2023). As a result of this disaster, the "slightly reduced" and "totally stopped" seafood consumption was reduced by about 33% and 10%, respectively (Arachchi et al., 2023). Consequently, this has significant implications for food security and malnutrition.

20.5 Conclusion

This chapter reviews the environmental, social, and economic impacts of the MV X-Press Pearl shipwreck, the worst chemical and plastic-related marine disaster in Sri Lankan maritime history. Such severe maritime disasters may occur in the future in Sri Lankan jurisdiction. To establish a successful coastal disaster management system in Sri Lanka, it is essential to (i) formulate a strong national policy, (ii) establish an administrative structure for integrated coastal zone management, (iii) establish and update new laws for maritime transport, and (iv) recover environmental and social losses while establishing a strong legal framework to recover costs incurred to restore the environment. Finally, such a coastal disaster management system contributes to the development of the coastal environment, human community, marine life, and the economy of Sri Lanka.

References

Akdogan, Z. and Guven, B. (2019). Microplastics in the environment: a critical review of current understanding and identification of future research needs. *Environmental Pollution* 254: 113011. https://doi.org/10.1016/j.envpol.2019.113011.

Arachchi, E.V.F., Jayamali, W.L.L., Sanduni, P.V.I., Jayasooriya, M.C.N., 2023. Impact of sea pollution due to X-Press Pearl ship disaster on seafood consumption patterns of Sri Lankans. 11th International Symposium (IntSym 2023), 21–29. http://ir.lib.seu.ac.lk/handle/123456789/6746

Aris, A.Z., Shamsuddin, A.S., and Praveena, S. (2014). Occurrence of 17α-ethynylestradiol (EE2) in the environment and effect on exposed biota: a review. *Environment International* 69: 104–119. https://doi.org/10.1016/j.envint.2014.04.011.

Azmi, F. (2021). Small scale fishers and transformative potential of fisheries policies in Cambodia, India, and Sri Lanka. *Vidyodaya Journal of Humanities and Social Sciences* 6: 1–18. https://doi.org/10.31357/fhss/vjhss.v06i02.01.

Barboza, L.G.A. and Gimenez, B.C.G. (2015). Microplastics in the marine environment: current trends and future perspectives. *Marine Pollution Bulletin* 97: 5–12. https://doi.org/10.1016/j.marpolbul.2015.06.008.

Barboza, L.G.A., Dick Vethaak, A., Lavorante, B.R.B.O. et al. (2018). Marine microplastic debris: an emerging issue for food security, food safety and human health. *Marine Pollution Bulletin* 133: 336–348. https://doi.org/10.1016/j.marpolbul.2018.05.047.

Brandt-Rauf, P.W., Li, Y., Long, C. et al. (2012). Plastics and carcinogenesis: the example of vinyl chloride. *Journal of Carcinogenesis* 11: 50–60. https://doi.org/10.4103%2F1477-3163.93700.

Campanale, C., Massarelli, C., Savino, I. et al. (2020). A detailed review study on potential effects of microplastics and additives of concern on human health. *International Journal of Environmental Research and Public Health* 17: 1212. https://doi.org/10.3390/ijerph17041212.

Cantoni, B., Cappello Riguzzi, A., Turolla, A., and Antonelli, M. (2021). Bisphenol a leaching from epoxy resins in the drinking water distribution networks as human health risk determinant. *Science of the Total Environment* 783: 146908. https://doi.org/10.1016/j.scitotenv.2021.146908.

Chandrasena, W.K.N. and Premasiri, H.D.S. (2023). Assessment of vapour concentration and estimation of pollutant load for chemical disasters – a case study on worst maritime disaster in Sri Lanka. *12th Annual Research Symposium, National Building Research Organisation* 12: 65–72.

Cumaranatunga, P.R.T., Dalpathadu, K.R., Ayeshya, H.M.U. et al. (2023). Impact of MV X-Press Pearl disaster on coastal and marine fish and fisheries. In: *In Maritime Accidents and Environmental Pollution - The X-Press Pearl Disaster: Causes, Consequences, and Lessons Learned*, vol. 11 (ed. M. Vithanage, A.P. de Alwis, and D. Botheju), 215–272. CRC Press https://doi.org/10.1201/9781003314301-11.

EconomyNext 2021. X-Press Pearl salvors mull boarding as Sri Lanka eyes pollution claim. [online] Available at. https://economynext.com/x-press-pearl-salvors-mull-boarding-as-sri-lanka-eyes-pollution-claim-82567/ (accessed 5 February 2024).

Ekanayake, E.M.L., Liyanage, U.S.P.K., and Amararathna, M.S.O.M. (2023). X-Press Pearl ship impact on sea turtles. In: *Maritime Accidents and Environmental Pollution - The X-Press Pearl Disaster: Causes, Consequences, and Lessons Learned*, vol. 9, 173–189. CRC Press https://doi.org/10.1201/9781003314301-9.

Erabadupitiya, H.R.S.M., Bandara, K.R.D.H., Rathnayaka, K.W.D., and Wijerathne, W.H.M.L.G. (2023). Proposing a hazard identification and risk analysis system to mitigate the impact of shipping on the marine environment. *Journal of Research Technology and Engineering* 4: 25–29.

Garaniya, V., McWilliam, D., Goldsworthy, L., and Ghiji, M. (2018). Extensive chemical characterization of a heavy fuel oil. *Fuel* 227: 67–78. https://doi.org/10.1016/j.fuel.2018.04.094.

Gruber, E.S., Stadlbauer, V., Pichler, V. et al. (2022). To waste or not to waste: questioning potential health risks of micro-and nanoplastics with a focus on their ingestion and potential carcinogenicity. *Exposure and Health* 1–19: https://doi.org/10.1007/s12403-022-00470-8.

Gunawardhana, G.M.S.S., Perera, U.L.H.P., and Ratnayake, A.S. (2023). Sources and fate of plastics into microplastics: degradation and remediation methods. In: *In Maritime Accidents and Environmental Pollution - The X-Press Pearl Disaster: Causes, Consequences, and Lessons Learned*, vol. 8, 155–172. CRC Press https://doi.org/10.1201/9781003314301-8.

Gunesekera, M., Regmi, U., Matheou, A., et al. 2021. Sri Lanka: cargo ship fire – final report, DREF n° MDRLK013. https://reliefweb.int/report/sri-lanka/sri-lanka-cargo-ship-fire-emergency-operation-update-report-n-1-22-september-2021

Guzzetti, E., Sureda, A., Tejada, S., and Faggio, C. (2018). Microplastic in marine organism: environmental and toxicological effects. *Environmental Toxicology and Pharmacology* 64: 164–171. https://doi.org/10.1016/j.etap.2018.10.009.

Hassan, P., Camille, L., Stephane, L.F., Luigi, A. 2021. X-Press Pearl maritime disaster Sri Lanka-report of the UN environmental advisory mission July 2021, UN environmental advisory mission, UN. https://wedocs.unep.org/20.500.11822/36608

James, B.D., Reddy, C.M., Hahn, M.E. et al. (2023). Fire and oil led to complex mixtures of PAHs on burnt and unburnt plastic during the M/V X-Press Pearl disaster. *ACS Environmental Au* 3: 319–335. https://doi.org/10.1021/acsenvironau.3c00011.

Jayathilaka, R.M.R.M., Weerakoon, W.R.W.M.A.P., Indika, K.W. et al. (2022). Spatio-temporal variation of plastic pellets dispersion in the coastline of Sri Lanka: an assessment of pellets originated from the X-Press Pearl incident during the Southwest monsoon in 2021. *Marine Pollution Bulletin* 184: 114145. https://doi.org/10.1016/j.marpolbul.2022.114145.

Karthik, R., Robin, R.S., Purvaja, R. et al. (2022). Microplastic pollution in fragile coastal ecosystems with special reference to the X-Press Pearl maritime disaster, southeast coast of India. *Environmental Pollution* 305: 119297. https://doi.org/10.1016/j.envpol.2022.119297.

Liu, Y., Tam, N.F., Guan, Y. et al. (2011). Acute toxicity of nonylphenols and bisphenol a to the embryonic development of the abalone Haliotis diversicolor supertexta. *Ecotoxicology* 20: 1233–1245. https://doi.org/10.1007/s10646-011-0672-7.

Mannapperuma, A. 2023. Report of the sectoral oversight committee on environment, natural resources and sustainable development on the "X-Press Pearl disaster and obtaining compensation". Sectoral Oversight Committee on Environment, Natural Resources and Sustainable Development, Parliament of Sri Lanka, 1–56. https://www.parliament.lk/uploads/comreports/1690190230023243.pdf

NBRO 2021. Air pollution impact due to the firing of MV X-Press Pearl ship nearby Colombo port. National Building Research Organization, Sri Lanka. [online] Available at. https://www.nbro.gov.lk/index.php?option=com_content&view=article&id=336%3Aair-pollution-impact-due-to-the-firing-of-mv-x-press-pearl-ship-nearby-colombo port&catid=8&Itemid=190&lang=en (accessed 10 February 2024).

Neves, D., Sobral, P., Ferreira, J.L., and Pereira, T. (2015). Ingestion of microplastics by commercial fish off the Portuguese coast. *Marine Pollution Bulletin* 101: 119–126. https://doi.org/10.1016/j.marpolbul.2015.11.008.

Oliveira, M., Ribeiro, A., Hylland, K., and Guilhermino, L. (2013). Single and combined effects of microplastics and pyrene on juveniles (0^+ group) of the common goby Pomatoschistus microps (Teleostei, Gobiidae). *Ecological Indicators* 34: 641–647. https://doi.org/10.1016/j.ecolind.2013.06.019.

Pallewatta, S., Samarasekara, S.M., Rajapaksha, A.U., and Vithanage, M. (2023). Oil spill remediation by biochar derived from bio-energy industries with a pilot-scale approach during the X-Press Pearl maritime disaster. *Marine Pollution Bulletin* 189: 114813. https://doi.org/10.1016/j.marpolbul.2023.114813.

Pathmalal, M.M., Hemantha, R.S.K.W.D., Dilena, P.K. et al. (2023). Impact of the MV X-Press Pearl ship disaster on the coastal environment from Negambo to Benthota in Sri Lanka. *Regional Studies in Marine Science* 58: 102788. https://doi.org/10.1016/j.rsma.2022.102788.

Pattiaratchi, C. and Wijeratne, S. (2021). Oceanographic aspects of the X-Press Pearl maritime disaster. *Loris* 29: 42–45.

Pattiaratchi, C., van der Mheen, M., Schlundt, C. et al. (2022). Plastics in the Indian Ocean–sources, transport, distribution, and impacts. *Ocean Science* 18: 1–28. https://doi.org/10.5194/os-18-1-2022.

Perera, U.L.H.P., Subasinghe, H.C.S., Ratnayake, A.S. et al. (2022). Maritime pollution in the Indian Ocean after the MV X-Press Pearl accident. *Marine Pollution Bulletin* 185: 114301. https://doi.org/10.1016/j.marpolbul.2022.114301.

Pilapitiya, P.N.T. and Ratnayake, A.S. (2024). The world of plastic waste: a review. *Cleaner Materials* 11: 100220. https://doi.org/10.1016/j.clema.2024.100220.

Ratnayake, A.S. (2023). Marine pollution after the MV X-Press Pearl maritime disaster: societal impacts based on research outcomes. *Societal Impacts* 1: 100003. https://doi.org/10.1016/j.socimp.2023.100003.

Ratnayake, A.S. and Perera, U.L.H.P. (2022). Coastal zone management in Sri Lanka: A lesson after recent naval accidents. *Marine Pollution Bulletin* 182: 113994. https://doi.org/10.1016/j.marpolbul.2022.113994.

Rist, S., Carney Almroth, B., Hartmann, N.B., and Karlsson, T.M. (2018). A critical perspective on early communications concerning human health aspects of microplastics. *Science of the Total Environment* 626: 720–726. https://doi.org/10.1016/j.scitotenv.2018.01.092.

Rochman, C.M., Brookson, C., Bikker, J. et al. (2019). Rethinking microplastics as a diverse contaminant suite. *Environmental Toxicology and Chemistry* 38: 703–711. https://doi.org/10.1002/etc.4371.

Rubesinghe, C., Brosché, S., Withanage, H., et al. 2022. X-Press Pearl, a 'new kind of oil spill' consisting of a toxic mix of plastics and invisible chemicals. International Pollutants Elimination Network (IPEN). https://ipen.org/documents/x-press-pearl-new-kind-oil-spill

Seth, C.K. and Shriwastav, A. (2018). Contamination of Indian sea salts with microplastics and a potential prevention strategy. *Environmental Science and Pollution Research* 25: 30122–30131. https://doi.org/10.1007/s11356-018-3028-5.

Sewwandi, M., Amarathunga, A.A.D., Wijesekara, H. et al. (2022a). Contamination and distribution of buried microplastics in Sarakkuwa beach ensuing the MV X-Press Pearl maritime disaster in Sri Lankan sea. *Marine Pollution Bulletin* 184: 114074. https://doi.org/10.1016/j.marpolbul.2022.114074.

Sewwandi, M., Hettithanthri, O., Egodage, S.M. et al. (2022b). Unprecedented marine microplastic contamination from the Xpress Pearl container vessel disaster. *Science of the Total Environment* 828: 154374. https://doi.org/10.1016/j.scitotenv.2022.154374.

Sewwandi, M., Perera, K.I., Reddy, C.M. et al. (2023). Plastics, nurdles, and pyrogenic microplastics in the coastal marine environment: implications of the X-Press Pearl maritime disaster. In: *Maritime Accidents and Environmental Pollution–The X-Press Pearl Disaster* (ed. M. Vithanage, A.P. de Alwis, and D. Botheju), 134–154. CRC Press.

The Morning 2023. MV X-press pearl maritime disaster: has SL done enough multi-sectoral damage control?https://www.themorning.lk/articles/MVANxlsipNKpi3mxz1My (accessed 10 February 2024).

Tsai, W.T. (2006). Human health risk on environmental exposure to bisphenol-A: a review. *Journal of Environmental Science and Health – Part C* 24: 225–255. https://doi.org/10.1080/10590500600936482.

UNDP 2010. Disaster Risk Assessment. UNDP, New York. https://www.undp.org/sites/g/files/zskgke326/files/migration/ly/2Disaster-Risk-Reduction---Risk-Assessment.pdf

de Vos, A., Aluwihare, L., Youngs, S. et al. (2021). The M/V X-Press Pearl nurdle spill: contamination of burnt plastic and unburnt nurdles along Sri Lanka's beaches. *ACS Environmental Au* 2: 128–135. https://doi.org/10.1021/acsenvironau.1c00031.

Wijayaratne, B. and Maldeniya, R. (2003). The role of fisheries sector in the coastal communities of Sri Lanka. *Assessment, Management and Future Directions of Coastal Fisheries in Asian Countries, WorldFish Center Conference Proceedings* 67: 629–656. https://hdl.handle.net/20.500.12348/2148.

Zhang, X., Jiang, M., Zhu, Y. et al. (2023). The X-Press Pearl disaster underscores gross neglect in the environmental management of shipping: review of future data needs. *Marine Pollution Bulletin* 189: 114728. https://doi.org/10.1016/j.marpolbul.2023.114728.

21

Container Overboard in the Port of New Orleans, LA, USA: The Response and Cleanup of the 2020 *Bianca* Pellet Spill

Bryan D. James and Christopher M. Reddy

Department of Marine Chemistry and Geochemistry, Woods Hole Oceanographic Institution, Woods Hole, MA, 02543, USA

21.1 Introduction

One of the most ubiquitous forms of plastic pollution in aquatic ecosystems is preproduction resin pellets, also called nurdles. They are typically 3–5-mm diameter spherules of plastic (15–25 mg) and sold as the "raw material" to manufacture plastic products. Pellets were first reported in the open ocean and on coastal shores in the early 1970s and are now ubiquitously found in natural waters – presumably from chronic leakage during transport and handling. Industry, governmental, and non-governmental organizations (NGOs) have attempted to mitigate these unintended releases. In the 1990s, the United States Environmental Protection Agency (EPA) recognized nurdle pollution and offered practical solutions to address leakage (United States Environmental Protection Agency 1992). The Society for the Plastics Industry, through its Operation Clean Sweep program, an international, industry-sponsored program, has recommended better management practices (Operation Clean Sweep 2024). Despite these efforts, an estimated 230 000 tons of resin pellets leak annually into the environment throughout the plastic lifecycle (Eunomia 2016).

Acute releases (or spills) of pellets are an additional source of plastics to the environment with their own challenges, short- and long-term impacts, cleanup approaches, and necessary solutions for reducing them. Reviews of pellet spills have documented a total of 43 spills on land and water between 2008 and 2023 within Africa, Asia, Europe, and North America (Oracle Environmental Experts 2023b). Typically, those on land were the result of railway and truck accidents, while those on water were from dock and ship accidents (Cedre 2023; Oracle Environmental Experts 2023b). When listed, the pellets were composed of polyethylene (PE), polypropylene (PP), or expanded polystyrene (EPS) – all of which float in seawater (Cedre 2023).

These acute releases can have widespread environmental and economic impacts. In 2012, the containership *Yong Xin Jie* during severe weather outside of Hong Kong lost six containers, carrying a total of 150 metric tons of PP pellets (Oracle Environmental Experts 2023a). Pellets were found on the Hong Kong and Chinese coastlines (~80 km away) within a few days, linked to a crash in fish stocks, and suspected to remain in "mounds" six years later (Oracle Environmental Experts 2023a). Despite these impacts, no peer-reviewed scientific studies were conducted; only a case study was assembled nearly a decade later. Similarly, in 2020, the containership CGM CMA *Bianca* spilled ~20 tons of pellets into the Mississippi River while docked at the Port of New Orleans,

Coastal and Marine Pollution: Source to Sink, Mitigation and Management, First Edition.
Edited by Meththika Vithanage, Sameera M. Samarasekara, Bryan D. James, and Christopher M. Reddy.

Louisiana, USA. Despite news coverage of the event and accessibility to samples, owing to the spill being in a metropolitan area, few studies followed.

In 2021, the largest known spill of an estimated 1680 metric tons occurred when the M/V *X-Press Pearl* caught fire off the coast of Sri Lanka (Oracle Environmental Experts 2023b). Three-quarters of the coastline was polluted with pellets and burned plastic (de Vos et al. 2022). Unlike with other spills, in total over 25 peer-reviewed papers and book chapters have been published on the spill and provide the only robust scientific basis for pellet spills. While these studies have been useful, the production of burned plastics and the much greater mass released make the spill an outlier among the other maritime spills (Cedre 2023; Oracle Environmental Experts 2023b). While each pellet spill (typically less than 500 metric tons) is minuscule compared to the global estimate of chronic pellet pollution (230 000 metric tons), their impact is magnified due to the rapid, unplanned, localized release, often during foul weather. Collectively, when compared to chronic pellet pollution, acute pellet spills are an important and different source of these synthetic materials.

Once in the environment, pellets threaten wildlife and ecosystems in many ways. Because of their size, shape, and color, they are thought to be mistaken as prey by wildlife (de Vos et al. 2022). Once ingested, the pellets can accumulate in an organism's digestive organs, leading to internal scarring and a false sense of satiation that can result in the eventual starvation of the organism (Santos et al. 2021; Charlton-Howard et al. 2023). To date, nurdles have been found in the stomachs of marine mammals, fish, sea turtles, and seabirds (Santos et al. 2021).

Owing to the hydrophobic nature of most plastics, hydrophobic organic contaminants (HOCs) actively partition into them (Mato et al. 2001). These plastic-associated chemicals may leach from the nurdles upon ingestion by wildlife, exposing an organism to contaminants at concentrations greater than those observed in the surrounding environmental compartment (Landrigan et al. 2023). This "Trojan horse" process results in not only physical harm from the pellets but also toxicological harm from any plastic-associated chemicals and degradation products (Landrigan et al. 2023). These impacts, among others, have motivated calls to recategorize resin pellets and plastic pollution, in general, as hazardous (Rochman et al. 2013).

Because plastic has not been listed as "hazardous" in most countries, the cleanup, damage assessment, and remediation of these spills have received much less governmental action than petroleum and other chemical spills (Durrance 2022). Cleanup campaigns do occur but are often delayed, allowing for greater dispersion and more time for the pellets to have an impact. Scientifically, these time-stamped and geo-located spills offer a unique opportunity to understand the fate of plastic pollution in aquatic ecosystems (de Vos et al. 2022). Because of the foresight of concerned citizens and scientists following the 2020 *Bianca* pellet spill, many samples of nurdles were collected. The spill size and location well position the *Bianca* spill as a representative aquatic pellet spill. Taken together, the *Bianca* pellet spill provided an opportunity to understand the challenges and needs of responding to an acute, geo-referenced, and time-stamped plastic release.

21.2 Materials and Methods

21.2.1 Sample Collection

Nurdles were collected from the beach area of Chalmette Battlefield, New Orleans, Louisiana, USA (Figure 21.1, black circle) on 13 August 2020, and 17 October 2020, and stored at ambient laboratory conditions. The analyzed nurdles were graciously provided by Dr. Mark Benfield. The beach area of Chalmette Battlefield is along the Mississippi River approximately 16 km downriver

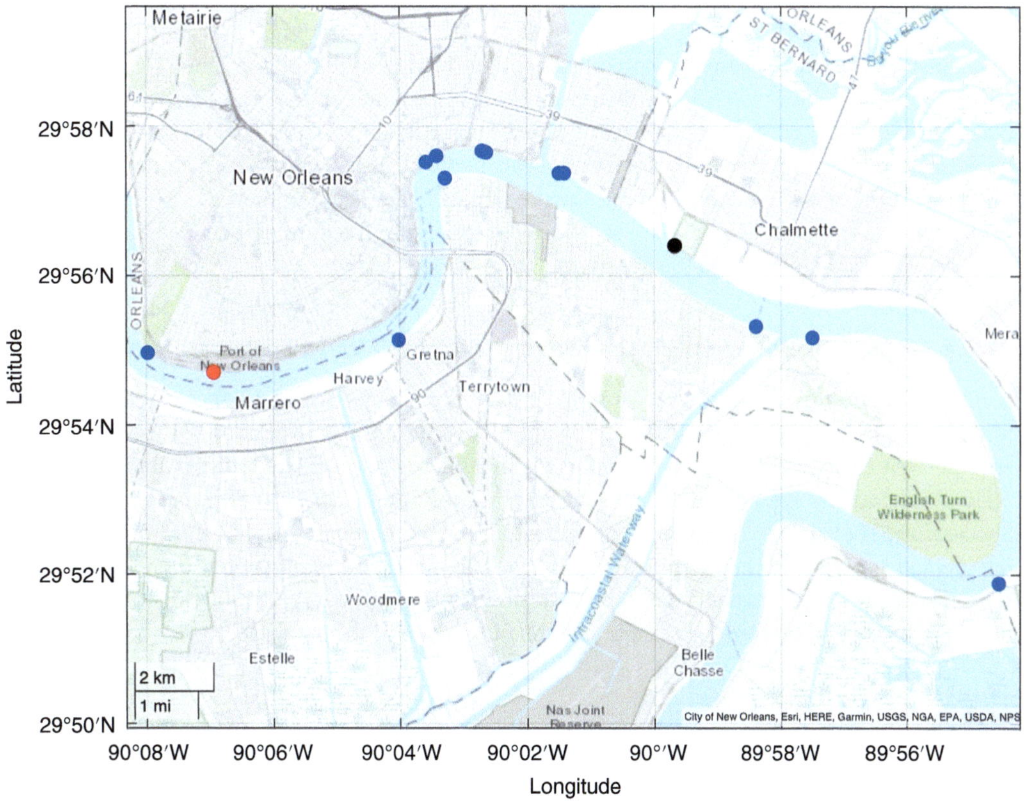

Figure 21.1 Map of the spill. The site of the spill at the Port of New Orleans is indicated in red. The 12 sampling sites for nurdles are indicated in blue. The sampling site of nurdles analyzed herein is indicated in black.

(as measured from the river's centerline) of the spill site in the Port of New Orleans. Based on the time-averaged mean water velocity of the river for the two days leading up, the day of, and the two days after the spill (~0.7 m/s) (United States Geological Survey 2024), a first estimate for the minimum transit time from the spill location to Chalmette Battlefield was calculated to be ~2.5 hours. Thus, the nurdles were presumed to have reached Chalmette Battlefield in relatively little time and spent at most 11 and 76 days at that location since the spill date before their collection.

21.2.2 Morphometric and Colorimetric Analyses

The nurdles were sonicated in Milli-Q water for 50 minutes to remove detritus and other surface adherent material that could bias image-based analyses. This cleaning was also necessary for complementary IR spectroscopic analyses to identify the type of polymer and evaluate the extent of chemical weathering. Imaging and quantitative analyses were performed with a Celestron handheld digital microscope and the United States National Institutes of Health ImageJ software, as previously described for analyzing nurdles (James et al. 2022). Images of the nurdles were taken using the same orientation. Morphometrics of the nurdles included their projected perimeter, projected area, and circularity. Colorimetrics of the nurdles included their average areal hue, saturation, and brightness. Values of hue range from 0 to 360; values of saturation and brightness range from 0 to 100.

21.2.3 Densitometry

The densities of nurdles were measured by adapting the density-titration method (Kolb and Kolb 1991; Bruzan and Baker 1993; Morét-Ferguson et al. 2010; James et al. 2022). Briefly, nurdles were placed in containers with a known mass of water. Then ethanol was titrated with sufficient mixing between additions until the nurdles were neutrally buoyant. The mass of added ethanol was determined, and the mass fraction of ethanol was compared to a standard curve of solution density versus the mass fraction of ethanol in solution. Thus, because the nurdles were neutrally buoyant, their density was equivalent to the density of the solution.

21.2.4 Attenuated Total Reflectance-Fourier Transform Infrared Spectroscopy (ATR-FTIR)

A random set of nurdles from each sampling date was characterized by IR spectroscopy to determine the polymer type and evaluate the extent of weathering (e.g. oxidation) of the nurdles. IR spectra of cleaned nurdles were collected using an Agilent Cary 630 FTIR with a diamond crystal ATR module as the average of 32 scans with $2\,cm^{-1}$ resolution. Spectra were processed in Open Specy (Cowger et al. 2021), applying an eight-order polynomial baseline and third-order smoothing (default settings). In the Open Specy web-based application, sample spectra were compared to the Open Specy IR spectra database, and Pearson's r statistic was automatically calculated to determine the similarity between the sample spectra and spectra in the database.

21.2.5 Statistical Analysis

Quantitative and statistical analyses were performed in Microsoft Excel and GraphPad Prism 10. Metrics are the mean $\pm$ standard deviation (n = sample size). Unpaired t-tests were used to evaluate statistical significance between groups, where appropriate data were transformed to accommodate normality assumptions.

21.2.6 Details of the Spill

The case history of the spill was compiled from primary sources, secondary sources, and personal communications with those involved. Primary sources included published articles by local and national news outlets, federal reports, and presentation materials. Secondary sources included documents prepared by organizations in the spill response community and maritime trade. Personal communications included those with state and federal officials, local researchers, and company representatives.

21.3 Results and Discussion

21.3.1 Timeline

21.3.1.1 The Accident

On 2 August 2020, the container ship M/V CMA CGM *Bianca* was moored and exchanging cargo at the Port of New Orleans (Figure 21.1, red circle). For most of the day, the weather was fair, and the vessel was concurrently loaded and unloaded by two gantry cranes. By early afternoon, it began to rain, rapidly intensifying within minutes, accompanied by high winds that reduced visibility. This extreme weather event, which lasted roughly 20 minutes, was ultimately categorized by the

United States National Transportation Safety Board (NTSB) as a localized downburst. Halfway through the downburst, 10 mooring lines failed and the ship began to drift away from the pier. Footage from security cameras showed the ship drifting from the dock (United States National Transportation Safety Board 2021). At the same time, the cranes carrying containers, buffeted by the wind, began moving along their rails. One of the containers suspended by one of the two cranes collided with containers onboard the ship. The container dropped from the crane, crashed onto the pier, broke open, and fell into the water. This container held an estimated 20 tons of resin pellets stored in 800 plastic sacks of 25 kg each (Gisleson Palmer and Nguyen 2021), which were partially released when the container broke open and fell into the water. A container suspended by the other crane also collided with the ship, damaging the crane and containers onboard. According to the NTSB, the accident resulted in estimated damages of $15 million to the cranes and approximately $60 000 to the ship (United States National Transportation Safety Board 2021). Port officials said the spilled pellets were "irretrievable" (United States National Transportation Safety Board 2021). In total, four containers fell from the ship; three were empty, and one contained a cargo of resin pellets.

21.3.1.2 Notification

In the event of a spill within the United States, the United States Coast Guard National Response Center (NRC) is notified. The NRC serves as the centralized point of contact in the United States federal government for reporting all pollution, oil, chemical, radiological, biological, and other unknown discharges into the environment to federal, state, and local on-scene coordinators (Weaver 2018). The incident was reported to the NRC on 2 August 2020 (Sequence Number 1283508); however, it did not mention the release of any pellets (United States Coast Guard National Response Center 2020). The incident description read,

> "Caller is reporting that [during] a storm two empty containers were blown into the Mississippi River and a third empty container fell onto a barge."

According to reporting by *E&E News*, the United States Coast Guard's Eighth District said they were aware of the spill the day it occurred and had notified Louisiana's Department of Environmental Quality (LDEQ) that same day (Wallace 2020). Within four days, large amounts of nurdles were observed on the banks of the Mississippi River downriver of the Port of New Orleans. Sightings at Chalmette Battlefield (Figure 21.1, black circle) were reported to the NRC on 6 August 2020 (Sequence Number 1283938) (United States Coast Guard National Response Center 2020). A caller reported,

> "the discovery of white industrial plastic pellets on the east bank of the Mississippi River due to an unknown cause. Pellets are spread out over a 1 mile stretch of area."

The United States Coast Guard notified LDEQ again (Wallace 2020).

21.3.1.3 Pellets in the Mississippi River

According to a statement from the Port of New Orleans reported by *The Times-Picayune/The New Orleans Advocate* on 8 August 2020,

> "a container containing plastic resin pellets, a cargo classified as a nonhazardous substance, fell from the vessel into the river... As the damaged container was being retrieved [from the river two days later, August 4, 2020], sacks containing pellets broke out of the container and dispersed in the water." (Schleifstein 2020)

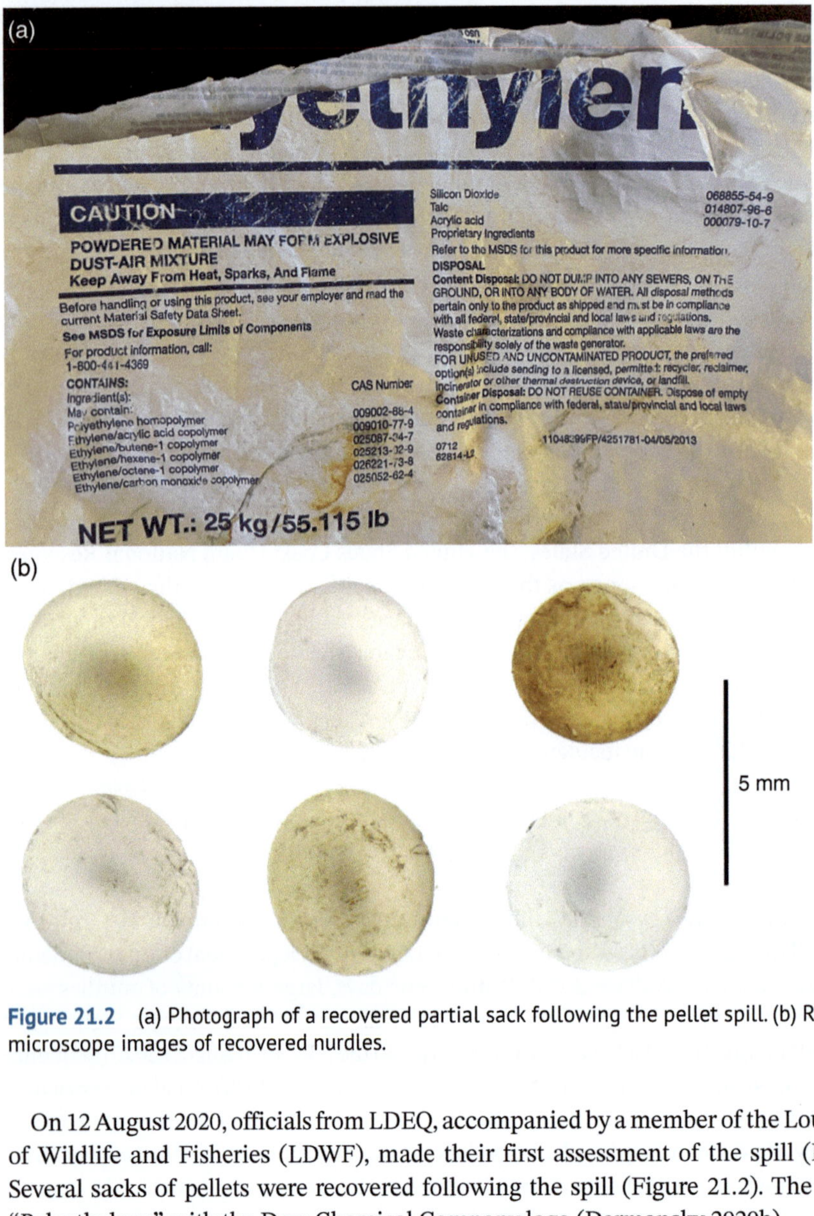

Figure 21.2 (a) Photograph of a recovered partial sack following the pellet spill. (b) Representative microscope images of recovered nurdles.

On 12 August 2020, officials from LDEQ, accompanied by a member of the Louisiana Department of Wildlife and Fisheries (LDWF), made their first assessment of the spill (Dermansky 2020b). Several sacks of pellets were recovered following the spill (Figure 21.2). The sacks were labeled "Polyethylene" with the Dow Chemical Company logo (Dermansky 2020b).

21.3.1.4 Delayed Response

The relatively slow response by all parties to the nurdle spill has been attributed to the lack of defined responsibility for responding to pellet spills by federal, state, and local agencies, as well as the nonhazardous status of resin pellets. As reported by *The Times-Picayune/The New Orleans Advocate* on 18 August 2020,

> "Coast Guard officials determined that they have no jurisdiction over the nurdles and are not responsible for cleanup. That's because the nurdles are not considered a 'hazardous material,' like oil or chemical waste, under the Clean Water Act." (Baurick 2020a)

Similarly, LDEQ spokespeople stated at the time,

> "We're aggressively investigating and assessing the spill and trying to figure out who's responsible for it... This is something different for us,' [they] said. 'It's not like an oil spill or something from a (chemical) plant. We're looking into how to handle it and if we're the correct agency to do it... If it's considered litter, that wouldn't necessarily be us." (Baurick 2020a)

Ultimately, a responding agency was not formally determined (Baurick 2020a; Dermansky 2020a). Discussion between LDEQ, the Coast Guard, and the EPA determined the agencies had no authority to respond under the current regulations (i.e. the Oil Pollution Act, the Comprehensive Environmental Response, Compensation, and Liability Act, and the Federal Water Pollution Control Act) (Todd and Lang 2023). LDWF considered whether there was purview based on the regulations for littering, but because the spill was unintentional, the release was outside their scope (Todd and Lang 2023).

21.3.1.5 Cleanup

As reported by *DeSmog* on 28 August 2020, Dow Chemical learned of the spill on August 18, 2020, and according to the Dow Chemical spokesperson, "[Dow] immediately...began working closely with the State of Louisiana Department of Environmental Quality and the carrier" (Dermansky 2020b). By 22 August 2020, the carrier had voluntarily initiated a formal cleanup of the spilled pellets, though no legally responsible party had been named (Dermansky 2020b). Before that, concerned citizens and NGOs had begun grassroots-level cleanups (Dermansky 2020b; Baurick 2020c; Hawkins 2022). The formal cleanup efforts were conducted by a Texas-based waste management contractor, US Environmental Services (Baurick 2020b). Manual methods were used for the cleanup, consisting of leaf blowers to move the nurdles into the water, where they were collected against a boom before scooping them up with nets. By this point, nurdles had been detected at numerous sites along the banks of the Mississippi River downriver of the port (Figure 21.1, blue circles). Recovery efforts by US Environmental Services continued for more than five months (Baurick 2020b). The LDEQ estimated, at most, that the contents of 80 of the 800 sacks had been recovered by residents and volunteers (Gisleson Palmer and Nguyen 2021). An assessment of fish samples by LDEQ found no pellets within the guts of any sampled fish (Todd and Lang 2023). A similar finding by LDEQ was reported in an evaluation of deceased ducks collected that winter (Todd and Lang 2023). These findings reflected those following the 2020 M/V *Trans Carrier* pellet in the North Sea, a similarly sized spill (Bergstrøm 2023). During the cleanup operation, there was no standard for what was considered "clean" following a pellet spill.

21.3.1.6 Aftermath

Despite the cleanup efforts, nurdles continued appearing on the banks of the Mississippi River months after the spill. On 14 October 2020, a sighting of nurdles was reported again to the NRC (Sequence Number 1289786), having been attributed to the M/V CMA CGA *Bianca* spill two months earlier in the report (United States Coast Guard National Response Center 2020). According to the NRC records, the caller reported 50 kg of polyethylene nurdles on the tide line at the levee on Patterson Drive and Algiers near the tugboat terminal (29.932500° N, 89.997306° W). Approximately six months after the spill, as reported by *The Times-Picayune/The New Orleans Advocate* on 1 February, 2021, nurdles began washing up again on the previously cleaned banks of the Mississippi River on 17 January, 2021 (Baurick 2021). Local groups and scientists hypothesized that the re-nurdling was due to recent high tides remobilizing nurdles lodged in river sediment and

Table 21.1 Timeline of the spill and response.

Date	Days after	Event
2 August 2020	0	Localized downburst led to the loss of four containers from the M/V CMA CGM *Bianca* docked at the Port of New Orleans, Louisiana, USA
2 August 2020	0	Incident reported to NRC
4 August 2020	2	Sacks of pellets first mentioned being lost
6 August 2020	4	Sightings of nurdles on riverbanks downstream of the port reported to NRC
12 August 2020	10	LDEQ and LDWF make their first assessment of the spill
18 August 2020	16	Dow Chemical learns of the spill
22 August 2020	20	Carrier voluntarily initiated a formal cleanup of the spilled pellets
14 October 2020	73	Sighting of nurdles reported again to the NRC
17 January 2021	168	Nurdles wash up again on formerly clean riverbanks

from unrecovered sacks finally tearing open. Sightings and longer-term cleanups for these recurring nurdling events relied on grassroots efforts by local community groups and environmentalists (Dhanesha 2022).

21.3.1.7 Summary of the Spill

The *Bianca* pellet spill resulted from a rare and extreme weather event, causing the damage and loss of a freight container carrying ~20 tons of Dow Chemical resin pellets labeled as "polyethylene." Following the release, state and federal agencies were notified; however, owing to a lack of regulations regarding pellet spills, it was unclear which agency was responsible for overseeing the spill response and whether there was jurisdiction for cleanup. No responsible party was ever designated. Nurdles were observed downriver of the release site several days after, prompting concern by locals. Over several months, grassroots organizations and contractors cleaned the spill. Despite these efforts, nurdles continued to wash up months after the spill but caught the attention of government agencies [e.g. National Oceanic and Atmospheric Administration (NOAA)] to initiate developing plans to respond to these spills. A consolidated timeline of the spill and the response is presented in Table 21.1.

21.4 Understanding the Fate of the Nurdles Following the Spill

Following the spill, no formal monitoring program was established. Dr. Mark Benfield, a professor of oceanography in the Department of Oceanography & Coastal Sciences at Louisiana State University, conducted unofficial monitoring of the spill for several months following the incident (Dhanesha 2022). Samples of nurdles were collected from several locations along the banks of the Mississippi River between 13 August 2020 (11 days after the spill) and 29 October 2020 (88 days after the spill) (Figure 21.1). From this unique and unprecedented collection of samples, we conducted an initial survey of the nurdles, reporting, for the first time, a quantitative description of their shape, size, color, density, and polymer type.

Sacks of pellets were recovered among the nurdles on the riverbanks following the spill, indicating that the spilled pellets were PE produced by the Dow Chemical Company (Figure 21.2a). The exact grade of PE was indeterminate from the sack, as there was no apparent labeling for a brand name, polymer grade, or product number on the sack other than the labeling of "Polyethylene." The sack did state that the contents may include polyethylene homopolymer, ethylene/acrylic acid copolymer, ethylene/butene-1 copolymer, ethylene/hexene-1 copolymer, ethylene/octene-1 copolymer, ethylene/carbon monoxide copolymer, silicon dioxide (CAS# 068855-54-9), talc ($Mg_3Si_4O_{10}(OH)_2$) (CAS# 014807-96-6), acrylic acid (CAS# 000079-10-7), and proprietary ingredients. Ethylene/butene-1 copolymer, ethylene/hexene-1 copolymer, and ethylene/octene-1 copolymer are types of linear low-density PE (LLDPE). Upon calling the product information number listed on the sack (1-800-441-4369), the representatives from Dow Chemical that we corresponded with were unable to identify the product based on the details printed on the sack remnant (Figure 21.2a) (personal communication, 2024).

21.4.1 Shape, Size, and Color of the Nurdles

Nurdles collected 11 days after the spill had a perimeter of 15.3 ± 0.5 mm ($n = 128$), an area of $15.3 \pm 1.0\,mm^2$ ($n = 128$), and a circularity of 0.83 ± 0.03 ($n = 128$) (Figure 21.3a–c). Nurdles collected 65 days later were statistically identical (unpaired t-test, p-value> 0.05) with values of perimeter of 15.3 ± 0.5 mm ($n = 45$), area of $15.5 \pm 0.9\,mm^2$ ($n = 45$), and circularity of 0.83 ± 0.03 ($n = 45$) (Figure 21.3a–c). The color of the nurdles was less uniform; most were white but some were visually orange (Figure 21.2b). Quantitatively, the nurdles collected 11 days after the spill had a hue of

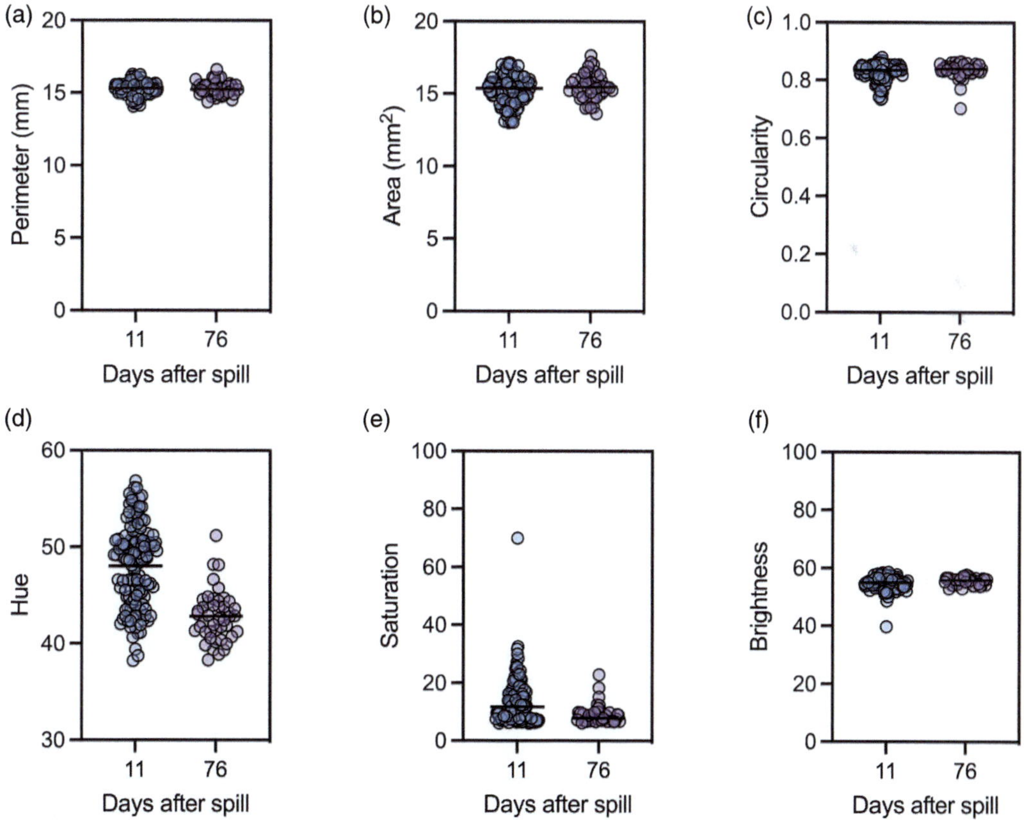

Figure 21.3 Morphometrics and colorimetrics of nurdles collected from Chalmette Battlefield, LA, USA 11 and 76 days after the spill. Morphometrics included perimeter (a), area (b), and circularity (c). Colorimetrics included hue (d), saturation (e), and brightness (f).

48.0 ± 4.3 (n = 128), a saturation of 13.6 ± 8.0 (n = 128), and a brightness of 54.7 ± 2.2 (n = 128) (Figure 21.3d–f). Hue appeared to have a bimodal distribution with modes centered at ~45 and ~50. Saturation followed a lognormal distribution, and brightness followed a normal distribution. Nurdles collected 76 days after the spill had a hue of 42.8 ± 2.6 (n = 45), which followed a normal distribution (Figure 21.3d). The clustering to a lower value of hue at the later timepoint may represent the material homogenizing or another collection of pellets. Their saturation was 9.0 ± 3.1 (n = 45), and their brightness was 55.6 ± 1.2 (n = 45) (Figure 21.3e,f). Both followed the same type of distribution as those of nurdles collected 11 days after the spill. Though minor, the differences in each colorimetric between nurdles collected 11 and 76 days after the spill were statistically significant (unpaired t-test, p-value <0.05). The reasoning for these changes, which trended toward more uniform distributions with greater time in the environment, remains to be determined.

21.4.2 Density of the Nurdles

Density is a valuable property for discriminating between polymers and is a needed parameter for modeling the transport of plastics in the aquatic environment (DiBenedetto et al. 2023; French-McCay 2023; Sutherland et al. 2023). All the tested nurdles floated in Milli-Q water (1.00 g/cm^3). Using solutions of water and ethanol, the densities of the nurdles were determined to range from 0.92 to 0.94 g/cm^3. No deviation from this density range was found for nurdles collected on either sampling date. This density range agrees with the type of plastic spilled based on the sack label, "polyethylene." However, it covers a broad range of PE types (Table 21.2). Owing to the crossover of these density ranges for different types of PE and the density range measured for the pellets, the type of PE was indeterminate from density measurements alone.

21.4.3 IR Assignment of Polymer Type

IR spectroscopy is routinely used to identify the type of polymer in microplastics. The approach relies on spectral matching to known reference materials, spectral interpretation, and ratios of

Table 21.2 Characteristics of polymers mentioned on a recovered sack (Figure 21.2a).

Name	CAS number	Density range (g/cm^3)	Defining IR features[a,b,c] (cm^{-1})
Polyethylene homopolymer (low density)	009002-88-4	0.91–0.94[d]	1377, 718
Polyethylene homopolymer (high density)	009002-88-4	0.94–0.97[d]	730, 720
Ethylene/acrylic acid copolymer	009010-77-9	0.92–0.94[d]	1680–1750
Ethylene/butene-1 copolymer	025087-34-7	0.90–0.94[d]	1378, 729, 719
Ethylene/hexene-1 copolymer	025213-02-9	0.90–0.94[d]	1378, 729, 719
Ethylene/octene-1 copolymer	026221-73-8	0.90–0.94[d]	1378, 729, 719
Ethylene/carbon monoxide copolymer	025052-62-4	0.934–1.105[b]	1680–1750

[a] Based on pure, unweathered materials (Smith 2021).
[b] Brubaker (1952).
[c] Otocka and Kwei (1968).
[d] Peacock (2014).

spectral features. Software packages use Pearson's r statistic to compare a sample spectrum to a database of known materials. Though useful, a shortcoming of this method is that identification is only as good as the coverage of the database. Documented challenges exist for determining the type of PE of environmental plastic samples by this method alone (Brandon et al. 2016; Lynch et al. 2024). Beyond software-based matching, distinct ratios can be used to assign the type of PE. The ratio of 1368 to 1377 cm^{-1} peak areas has been used to assign the type of PE to nurdles following the M/V *X-Press Pearl* spill as low-density PE (LDPE) or high-density PE (HDPE) (James et al. 2022). This ratio correlated well with assignments made using measured densities. IR spectra also have been used to study the weathering of plastic in the environment.

Based on the IR spectra of pellets from unopened sacks recovered following the spill, Benfield et al. assigned the pellets as HDPE (Benfield et al. 2021). The specific library database used and the confidence for this assignment were unreported.

IR spectroscopy confirmed that all the nurdles that were tested were PE based upon the presence of characteristic peaks at 2914 cm^{-1} (CH_2 asymmetric stretching), 2847 cm^{-1} (CH_2 symmetric stretching), 1470 and 1462 cm^{-1} (CH_2 bending deformation), 1375 cm^{-1} (CH_3 bending in LDPE and LLDPE), 1364 cm^{-1} (CH_2 wagging deformation), and 729 and 718 cm^{-1} (CH_2 rocking) (Figure 21.4) (Krimm et al. 1956; Gulmine et al. 2002). Additionally, the absence of well-defined carbonyl peaks discounted that these nurdles were ethylene/acrylic acid copolymer and ethylene/carbon monoxide copolymer as indicated as possibilities from the labeling on the recovered sacks (Figure 21.2a). This is because carbonyl groups are inherent to these copolymers (Table 21.2). Visually, the spectra of the nurdles matched the IR spectra reported by Benfield et al. of pellets from an unopened sack recovered following the spill (Figure 21.4a).

Polymer assignment based on spectral matching was unable (like the density measurements) to confidently assign the type of PE for nurdles collected 11 and 76 days after the spill. Compared to the Open Specy database, our samples yielded determinations of LDPE and HDPE with equal confidence (Pearson's $r \geq 0.98$) for numerous samples. The Open Specy database includes reference spectra of LDPE, LLDPE, and HDPE (Cowger et al. 2021). Using the ratio of the 1368 to 1377 cm^{-1} peak areas yet again gave values for the nurdles consistent with LDPE and HDPE. Spectral interpretation yielded another identification. The presence of a peak at 1375 cm^{-1} and defined peaks at 729 and 718 cm^{-1} suggested the nurdles were LLDPE (Table 21.1). Overall, routine IR methods for polymer identification were inconclusive in confidently discriminating the type of PE in the nurdles. Owing to these challenges, orthogonal characterization methods (e.g. differential scanning calorimetry) should be pursued to resolve this uncertainty (Lynch et al. 2024).

21.4.4 IR Assessment of the Extent of Weathering

Typical peaks indicative of weathering were faint or not present (Gulmine et al. 2003; Brandon et al. 2016; James et al. 2022). The nurdles had either obscure or no peaks around 1650–1750 cm^{-1}, which are often attributed to carbonyl groups from oxidation of the polymer. This was expected as weathering by photodegradation can take much more time for such peaks to develop based on laboratory studies of the weathering behavior of PE (Gulmine et al. 2003; James et al. 2022).

Similarly, the nurdles had either a faint or no broad peak centered at 3300 cm^{-1}. This peak is most often associated with the presence of hydroxyl groups from the oxidation of the polymer (Brandon et al. 2016). The nurdles did tend to have a broad defined peak centered at ~1030 cm^{-1}, spanning from ~900 cm^{-1} to ~1200 cm^{-1} (Figure 21.4). On average, the normalized absorbance at 1030 cm^{-1} was greater for nurdles collected 11 days after the spill than those collected 76 days after the spill

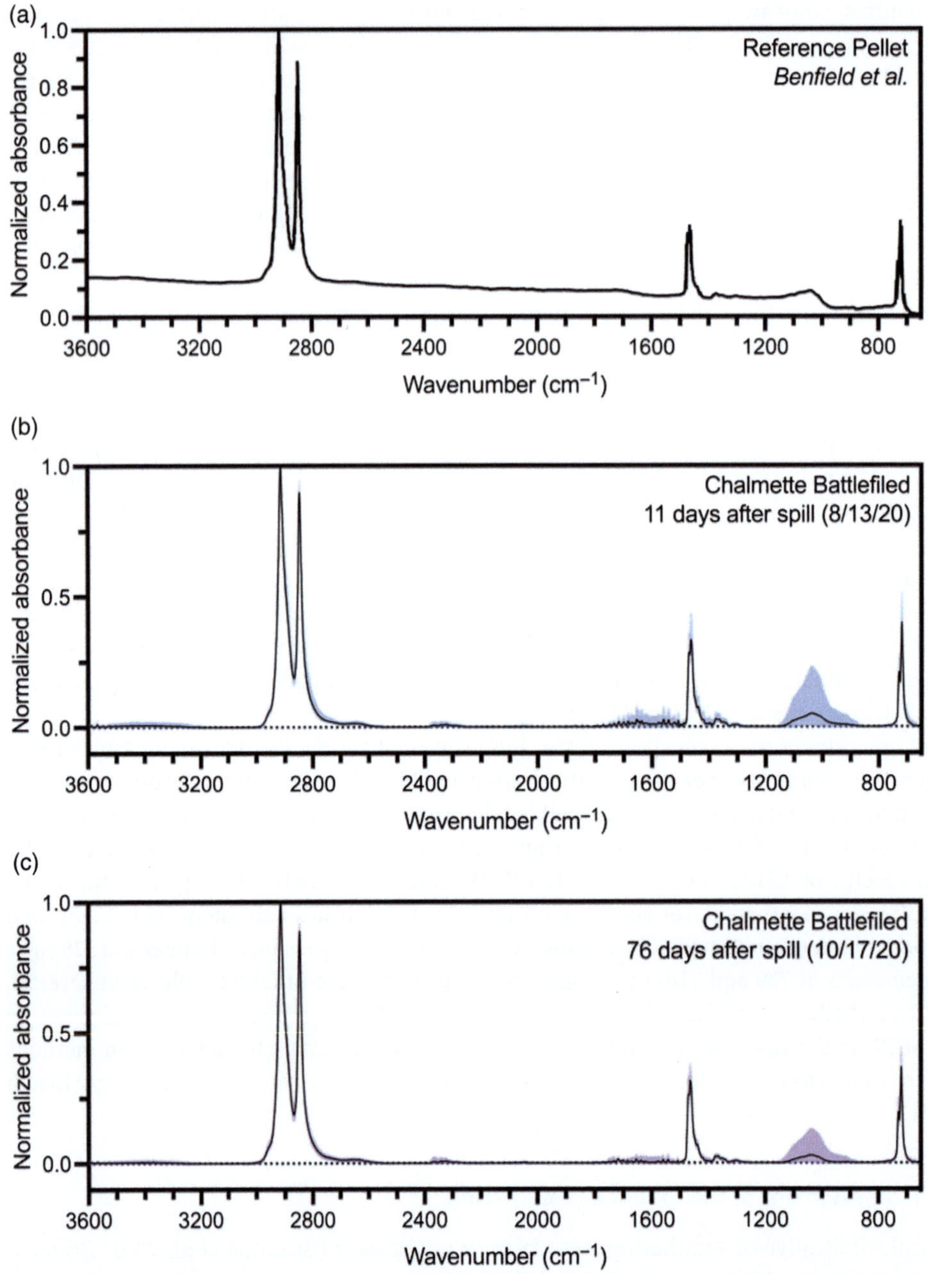

Figure 21.4 IR spectra of the spilled pellets. (a) IR spectrum of pellets from an unopened sack recovered following the spill. *Source:* Adapted from Benfield et al. (2021). The spectrum was collected on a ThermoScientific iS5 FTIR with an ATR accessary. (b) IR spectra of 37 nurdles were collected from Chalmette Battlefield 11 days after the spill (13 August 2020; 8/13/20). (c) IR spectra of 36 nurdles were collected from Chalmette Battlefield 76 days after the spill (17 October 2020; 10/17/20). Shaded regions indicate the range of values at each wavenumber.

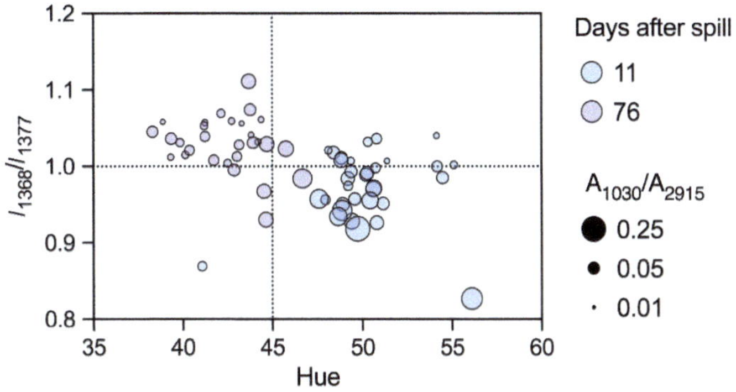

Figure 21.5 Clustering of nurdles based on the type of PE assigned by IR, the hue of the nurdles, the sampling time after the spill, and the normalized absorbance at 1030 cm^{-1} (an indicator of talc and SiO_2). The ratio of peak intensities, 1368 and 1377 cm^{-1}, is often used to identify HDPE (>1) versus LDPE (<1) based on IR spectra of unweathered standard reference materials.

(Figures 21.4, 21.5). This peak is often attributed to carbon-oxygen bonds due to the weathering of PE (Brandon et al. 2016). However, this possibility was less likely because the nurdles lacked a commensurately defined hydroxyl peak.

Many inorganic additives used in plastics absorb light in the IR spectrum. A defined peak at ~1030 cm^{-1} with a large width is characteristic of talc and silicon dioxide (Marzbani et al. 2016; Signoret et al. 2020). Small peaks at ~675 cm^{-1} and ~3680 cm^{-1}, characteristic of the magnesium oxide/magnesium hydroxide component of talc, were present but faint. Silicon dioxide and talc are often added to PE as fillers. The occurrences of these minerals were corroborated by the listing of potential contents on the recovered sacks (Figure 21.2a). Because these minerals have densities about double that of PE, the possible amount of these minerals in the nurdles was at most ~7 wt% (using a rule-of-mixtures); otherwise, the expected density of the nurdles would be greater than that measured. Knowledge of a plastic's inorganic additive content is critical because these components can control a plastic's fate, persistence, and toxicity (Nelson et al. 2021; Walsh et al. 2021, 2022; James et al. 2023).

By the IR ratio method, the nurdles collected 11 days after the spill had values consistent with LDPE, and those collected 76 days after the spill mainly had values consistent with HDPE (Figure 21.5). The difference in hue between the two sampling dates separated according to the type of polyethylene assigned by the IR ratio method (Figure 21.5). These changes could have resulted because of weathering or inhomogeneity of the spilled pellets. Further analyses with larger sample sizes will be necessary to confirm this possibility. Nonetheless, nurdles were still constrained to PE (and not other polymers) based on their density and IR spectra. Even being a geo-located and time-stamped release of pellets with much more provenance than nurdles routinely collected during beach surveys, there remained a similar degree of uncertainty about the spilled material as for any other nurdle collected on shorelines.

21.4.5 Lessons Learned from the Spill

The response to the M/V CMA CGM *Bianca* pellet spill and its cleanup mirrored that of previous spills globally. Common to them has been confusion across government agencies over who is responsible for overseeing the response and handling the cleanup, a reliance on local environmental groups to clean up the nurdles, and a need for defined strategies and approaches for monitoring and cleaning of the spill. A delayed response is nontrivial when dealing with pellet spills. These materials move quickly, can travel far, can blend in with surroundings, persist in the environment, and arguably become more problematic from a toxicological perspective the longer they are in the environment (Cedre 2023). Compared to oil spills, these features and others make pellet spills largely unique (Table 21.3). A key feature of oil spill response is a greater likelihood of decreasing toxicity in the environment with time because of dilution and weathering. Plastics are different. Arguably, pellets are the most benign when they are initially released, but with increasing time in the environment, their potential for injury grows as contaminants associate with the material and the polymer degrades (Karapanagioti and Werner 2019; Yamashita et al. 2019). This fact is of import to the *Bianca* pellet spill due to the substantial loading of HOC in the Mississippi River (Bussan et al. 2017). Because unexposed pellets were recovered and a prolonged sampling campaign was conducted, future studies of this spill should investigate the plastic-associated chemicals of the nurdles and their bioactivity.

In a spill, every second counts. Attempting to remedy the current delay in responding to pellet spills because of jurisdictional confusion among government agencies, United States Senator Richard Durbin introduced bill S.2337, the Plastic Pellet Free Waters Act, to the United States Congress in 2023 (Durbin et al. 2023). If enacted, the new law would require "the Administrator of the Environmental Protection Agency to promulgate certain limitations with respect to preproduction plastic pellet pollution, and for other purposes." For the United States, if enacted, the new law would require,

> "Not later than 60 days after the date of enactment of this Act, the Administrator of the Environmental Protection Agency shall promulgate a final rule to ensure that... the discharge of plastic pellets or other pre-production plastic materials (including discharge into wastewater and other runoff) from a point source (as defined in section 502 of the Federal Water Pollution Control Act (33 U.S.C. 1362)) that makes, uses, packages, or transports those plastic pellets and other pre-production plastic materials is prohibited." (Durbin et al. 2023)

By doing so, the Clean Water Act would regulate the release of pellets into the environment, thus establishing the hierarchy of government agencies for responding to a pellet spill in the United States. At the time of writing, the bill was in committee.

As of 2024, the International Maritime Organization (IMO) initiated efforts to adopt regulations to prevent pellet spills. The proposed directives echo those promoted by Operation Clean Sweep and the US EPA decades ago: (i) packaging pellets in ways that can withstand the standard conditions of transport without spillage, (ii) clearly identifying containers carrying pellets, and (iii) stowing containers with pellets under deck or inboard in sheltered areas of exposed deck (International Maritime Organization 2024). Adopting these simple strategies should help to mitigate ship-based pellet spills.

Nonetheless, in the case of the *Bianca* and *X-Press Pearl* pellet spills, both events would likely still have occurred despite these policies because of the circumstances of the two spills, a rare weather event during container transfer, and a ship fire, respectively. Thus, while positive, these management strategies and policies may fail, so there remains a need to be prepared to respond to pellet spills in a timely and organized manner.

Table 21.3 Comparison of oil and pellet spill regulations and response science in the United States.

Category	Oil spill	Pellet spill
Frequency (2012–2022)[a]	~30 000 spills/4.3 million gallons in waterways	~300 plastic and plastic pellet spills/at least 917 metric tons (37 billion pellets) reported to the US Coast Guard NRC In ~20% of spills, the material entered waterways For ~60% of spills, the amount of spilled material was unknown
Defined as "hazardous" as of writing	Yes	No
Federal regulatory policies	Oil Pollution Act (1990) Federal Water Pollution Control Act (1972) National Oil and Hazardous Substances Pollution Contingency Plan (1968)	None Plastic Pellet Free Waters Act (2023, in Congress)
Response guidelines[b]	Yes	No, but in development (IMO guidelines pending)
Years responding[c]	50+	10+
State of response science[d]	Well developed	Nascent
Complexity of material[e]	Crude oil and refined petroleum products encompass a wide range of bulk and molecular characteristics	Plastics have innumerable formulations (polymer and additives), though, pellets are similarly sized and shaped
Analytical methods[f]	Well-established (e.g. gas chromatography)	Developing (e.g. optical microscopy, IR spectroscopy, Raman spectroscopy, differential scanning calorimetry)
"Fingerprint" characteristics[g]	Well-established (e.g. diagnostic ratios and biomarkers)	Developing (size, shape, color, density, IR spectrum)
Change in mass once released (days to weeks)[h]	>40% of can evaporate in days	Little to none
Change in toxicity once released[i]	Generally, decreases with time	Little and may increase from sorption of HOCs
Long-term fate[j]	In some circumstances, a small fraction of oil can persist for decades – otherwise natural and response actions are effective	Likely persist for years, at least (e.g. Hong Kong spill)

(Continued)

Table 21.3 (Continued)

Category	Oil spill	Pellet spill
Critical property(s) needed for transport modeling[k]	Viscosity and surface tension	Density and shape
Long-range transport[l]	Typically, 10–100 km and possibly 1000 km of travel	1000s km of travel possible

[a] (United States Department of Transportation Bureau of Transportation Statistics 2019; United States Coast Guard 2024)
[b] (ExxonMobil Research and Engineering Company 2014; CEFAS et al. 2023)
[c] (Cedre 2023; Prince 2023)
[d] (Barker et al. 2020; National Academies of Sciences, Engineering, and Medicine 2022; Cedre 2023)
[e] (Marshall and Rodgers 2008; James et al. 2022)
[f] (Wise et al. 2023)
[g] (Aeppli et al. 2014; Stout 2016)
[h] (Lemkau et al. 2010; Ryerson et al. 2012; Gros et al. 2014)
[i] (Mato et al. 2001; Di Toro et al. 2007; Yamashita et al. 2019; Maloney et al. 2021; James et al. 2023)
[j] (Peacock et al. 2005; Yang et al. 2018; Oracle Environmental Experts 2023a)
[k] (Gros et al. 2017; French-McCay 2023; Sutherland et al. 2023)
[l] (Boufadel et al. 2023; Cedre 2023)

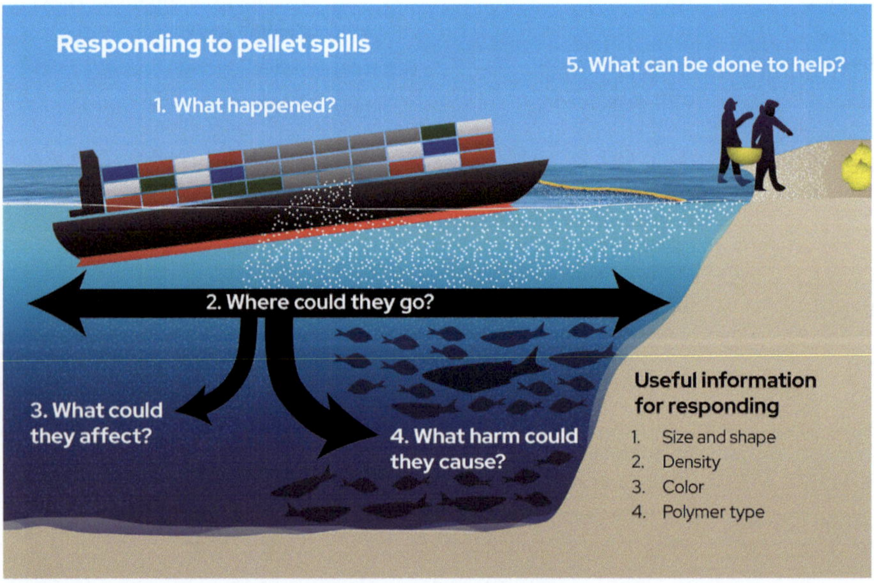

Figure 21.6 Guiding questions for pellet spill response science adapted from NOAA's "Five Questions" for oil spill science recommendations and de Vos et al. (2022).

Taking a lesson from the oil spill response playbook, when confronted with a pellet spill, adapting NOAA's "Five Questions" (Figure 21.6) should help guide decision-making (Barker et al. 2020):

1) **What happened?** How many pellets were spilled? What type of pellets were spilled? What are their density and shape? When and where was the spill? Is there a continued release?
2) **Where could the pellets go?** Once in the water, the pellets will move – responders need to know where the pellets may go to understand the potential impacts and know what response actions can be taken.

3) **What could the pellets affect?** What biota and ecosystems are present in the area where the pellets were released and might go?
4) **What harm could the pellets cause?** Understanding the harm the pellets could cause is critical for prioritizing response actions.
5) **What can be done to help?** Taking action to reduce the harm.

To begin answering these questions requires knowing what spilled. As is done already following oil spills, a "fingerprint" of the pellets released into the environment, like that of spilled oil, needs to be routine. Prompt characterization is necessary for a well-informed response, determination of the spiller, and long-term studies. In the event of an oil spill, immediately the oil is characterized for bulk physical and chemical properties to understand the situation and to confirm who is the spiller, if it is not obvious. At a minimum, the fingerprint for pellets must include characterization of the pellet's shape, size, color, density, polymer type, and IR spectrum. The methods presented herein can be completed easily, and most of them can be conducted in the field, if necessary. By establishing such a record for the spilled material, nurdles recovered during cleaning and monitoring operations can be compared to the fingerprint and assigned a likelihood of whether they originated from the spill. This information is necessary for determining the state of cleanliness of a spill site against background nurdle pollution. Additionally, details of shape, size, and density are critical parameters for modeling the transport of the pellets in the aquatic environment (de Vos et al. 2022; French-McCay 2023). Such modeling efforts can be used in real time to inform response operations and assess risk (French-McCay 2023).

Prompted by the international attention surrounding plastic pollution, maritime and response organizations have begun developing guidelines for responding to pellet spills. Along with the drafted regulations to prevent pellet spills, the IMO has also started to draft guidelines for responding to pellet spills (CEFAS et al. 2023). Similarly, the United States NOAA Office of Response and Restoration convened a workshop to inform the agency on response practices and the state of the art, recognizing a need to develop capacity for responding to pellet spills (United States National Oceanic and Atmospheric Administration 2023). Because of the strides in preparedness being taken by international agencies, it is hopeful that global agencies will be equipped to respond to pellet spills.

References

Aeppli, C., Nelson, R.K., Radović, J.R. et al. (2014). Recalcitrance and Degradation of petroleum biomarkers upon abiotic and biotic natural weathering of deepwater horizon oil. *Environmental Science & Technology* 48 (12): 6726–6734. https://doi.org/10.1021/es500825q.

Barker, C.H., Kourafalou, V.H., Beegle-Krause, C.J. et al. (2020). Progress in operational modeling in support of oil spill response. *Journal of Marine Science and Engineering* 8 (9): 668. https://doi.org/10.3390/jmse8090668.

Baurick, T. (2020a). No cleanup planned as millions of plastic pellets wash up along Mississippi River and flow to the Gulf, *Times-Picayune | New Orleans Advocate*, 18 August. https://www.nola.com/news/environment/no-cleanup-planned-as-millions-of-plastic-pellets-wash-up-along-mississippi-river-and-flow/article_b4fba760-e18d-11ea-9b0b-b3a2123cf48b.html (accessed 21 February 2024).

Baurick, T. (2020b). Nurdle spill cleanup is "too little, too late" as plastic pellets continue to spread, gather in drifts, *Times-Picayune | New Orleans Advocate*, 5 September. https://www.nola.com/news/environment/nurdle-spill-cleanup-is-too-little-too-late-as-plastic-pellets-continue-to-spread-gather/article_9b56e060-ef8c-11ea-bb00-73582b1649de.html (accessed 21 February 2024).

Baurick, T. (2020c). With cleanup stalled for weeks, group of volunteers tackles New Orleans nurdle spill on their own, *Times-Picayune | New Orleans Advocate*, 21 August. https://www.nola.com/news/environment/with-cleanup-stalled-for-weeks-group-of-volunteers-tackles-new-orleans-nurdle-spill-on-their/article_81011dfa-e3d6-11ea-8739-776cd89891a0.html (accessed 21 February 2024).

Baurick, T. (2021). A new crop of nurdles is washing up in New Orleans, renewing calls for cleanup, penalties, *Times-Picayune | New Orleans Advocate*, 1 February. https://www.nola.com/news/environment/a-new-crop-of-nurdles-is-washing-up-in-new-orleans-renewing-calls-for-cleanup/article_820710fc-64cc-11eb-85ce-93b2793068d5.html (accessed 21 February 2024).

Benfield, M., Marchio, E., Boswell, K., Thomas, A., *et al.* (2021). The use of FT-IR to identify putative nurdles from the incident, by Mark C. Benfield from LSU. *Thermo Scientific Spectroscopy & Materials Analysis*, YouTube, 14 December. https://www.youtube.com/watch?v=GBgzc4mTEQw (accessed 21 February 2024).

Bergstrøm, R. (2023). Trans Carrier - plastic pellets pollution in Norway. *Nurdle Spills in Coastal and Marine Environments*, College Park, MD, USA, 7 September. https://crrc.unh.edu/resource/nurdle-spills (accessed 23 February 2024).

Boufadel, M.C., Özgökmen, T., Socolofsky, S.A. et al. (2023). Oil transport following the deepwater horizon blowout. *Annual Review of Marine Science* 15: 67–93. https://doi.org/10.1146/annurev-marine-040821-104411.

Brandon, J., Goldstein, M., and Ohman, M.D. (2016). Long-term aging and degradation of microplastic particles: comparing in situ oceanic and experimental weathering patterns. *Marine Pollution Bulletin* 110 (1): 299–308. https://doi.org/10.1016/j.marpolbul.2016.06.048.

Brubaker, M.M., Coffman, D.D., and Hoehn, H.H. (1952). Synthesis and characterization of ethylene/carbon monoxide copolymers, a new class of polyketones. *Journal of the American Chemical Society* 74 (6): 1509–1515. https://doi.org/10.1021/ja01126a047.

Bruzan, R. and Baker, D. (1993). Plastic density determination by titration. *Journal of Chemical Education* 70 (5): 397. https://doi.org/10.1021/ed070p397.

Bussan, D.D., Ochs, C.A., Jackson, C.R. et al. (2017). Concentrations of select dissolved trace elements and anthropogenic organic compounds in the Mississippi River and major tributaries during the summer of 2012 and 2013. *Environmental Monitoring and Assessment* 189 (2): 73. https://doi.org/10.1007/s10661-017-5785-x.

Cedre (2023). *State of Knowledge on Pollution by Plastic Pellets.* R.23.25.C/3702. https://universitysystemnh-my.sharepoint.com/:b:/g/personal/kathym_usnh_edu/EZmQrD06NFNKvsBwvVlCBRwBPc18lWFJtmjayJBnsMEJ6Q?e=BdeY7E (accessed 27 February 2024).

CEFAS, National Department of Forestry, Fisheries, and the Environment of South Africa, International Group of P&I Clubs, Istituto Superiore per la Protezione e la Ricerca Ambientale, ITOPF, Norwegian Coastal Authority. (2023) Guidelines on clean-up of plastic pellets from ship-source spills. International Maritime Organization. https://universitysystemnh-my.sharepoint.com/:b:/g/personal/kathym_usnh_edu/Ef7mveCY1aRCgzS8mY-tI1UBGtZqpv0QEOztd-pe71Xhyw?e=cduWfJ (accessed 23 February 2024).

Charlton-Howard, H.S., Bond, A.L., Rivers-Auty, J., and Lavers, J.L. (2023). "Plasticosis": characterising macro- and microplastic-associated fibrosis in seabird tissues. *Journal of Hazardous Materials* 450: 131090. https://doi.org/10.1016/j.jhazmat.2023.131090.

Cowger, W., Steinmetz, Z., Gray, A. et al. (2021). Microplastic spectral classification needs an open source community: open specy to the rescue! *Analytical Chemistry* 93 (21): 7543–7548. https://doi.org/10.1021/acs.analchem.1c00123.

Dermansky, J. (2020a). A Plastics Spill on the Mississippi River But No Accountability in Sight, *DeSmog*, 11 August. https://www.desmog.com/2020/08/11/nurdles-plastics-spill-mississippi-river-chalmette-louisiana-formosa/ (accessed 21 February 2024).

Dermansky, J. (2020b) Pollution Scientist Calls Plastic Pellet Spill in the Mississippi River "a Nurdle Apocalypse", *DeSmog*, 28 August. https://www.desmog.com/2020/08/28/new-orleans-louisiana-plastic-spill-mississippi-river-nurdle-apocalypse/ (accessed 21 February 2024).

Dhanesha, N. (2022) What are nurdles? The petrochemical product is infiltrating the environment, *Vox*, 6 May. https://www.vox.com/recode/23056251/nurdles-plastic-pollution-ocean-microplastics (accessed 21 February 2024).

Di Toro, D.M., McGrath, J.A., and Stubblefield, W.A. (2007). Predicting the toxicity of neat and weathered crude oil: toxic potential and the toxicity of saturated mixtures. *Environmental Toxicology and Chemistry* 26 (1): 24–36. https://doi.org/10.1897/06174R.1.

DiBenedetto, M.H., Donohue, J., Tremblay, K.N. et al. (2023). Microplastics segregation by rise velocity at the ocean surface. *Environmental Research Letters* 18 (2): 024036. https://doi.org/10.1088/1748-9326/acb505.

Durbin, R., Merkley, J. and Cardin, B. (2023). S.2337 - Plastic Pellet Free Waters Act. United States Congress. https://www.congress.gov/bill/118th-congress/senate-bill/2337/text (accessed 23 February 2024).

Durrance, S. (2022). Nurdles – one container, a billion problems, *InterSpill 2022*, Amsterdam, Netherlands. https://www.itopf.org/fileadmin/uploads/itopf/data/Documents/Papers/INTERSPILL_2022_Durrance.pdf (accessed 21 February 2024).

Eunomia (2016). *Plastics in the Marine Environment*. http://eunomia.eco/reports/plastics-in-the-marine-environment/ (accessed 23 February 2024).

ExxonMobil Research and Engineering Company (2014). Oil Spill Response Field Manual. https://corporate.exxonmobil.com/-/media/global/files/risk-management-and-safety/oil-spill-response-field-manual_2014.pdf (accessed 23 February 2024).

French-McCay, D. (2023). Behavior Of Nurdles In Marine And Coastal Environments And Approaches For Modeling Nurdle Spills. *Nurdle Spills in Coastal and Marine Environments*, College Park, MD, USA, 6 September. https://crrc.unh.edu/resource/nurdle-spills (accessed 23 February 2024).

Gisleson Palmer, K. and Nguyen, C. (2021). Substitute Resolution NO. R-21-92. New Orleans City Hall. https://cityofno.granicus.com/MetaViewer.php?view_id=42&clip_id=3822&meta_id=532051 (accessed 22 February 2024).

Gros, J., Nabi, D., Würz, B. et al. (2014). First day of an oil spill on the open sea: early mass transfers of hydrocarbons to air and water. *Environmental Science & Technology* 48 (16): 9400–9411. https://doi.org/10.1021/es502437e.

Gros, J., Socolofsky, S.A., Dissanayake, A.L. et al. (2017). Petroleum dynamics in the sea and influence of subsea dispersant injection during deepwater horizon. *Proceedings of the National Academy of Sciences* 114 (38): 10065–10070. https://doi.org/10.1073/pnas.1612518114.

Gulmine, J.V., Janissek, P.R., Heise, H.M., and Akcelrud, L. (2002). Polyethylene characterization by FTIR. *Polymer Testing* 21 (5): 557–563. https://doi.org/10.1016/S0142-9418(01)00124-6.

Gulmine, J.V., Janissek, P.R., Heise, H.M., and Akcelrud, L. (2003). Degradation profile of polyethylene after artificial accelerated weathering. *Polymer Degradation and Stability* 79 (3): 385–397. https://doi.org/10.1016/S0141-3910(02)00338-5.

Hawkins, D. (2022). Tiny Terrors, *ANTIGRAVITY Magazine*, 6 May. https://antigravitymagazine.com/feature/tiny-terrors/ (accessed 21 February 2024).

International Maritime Organization (2024). *IMO agrees new guidance for safe transport of plastic pellets on ships*. https://www.imo.org/en/MediaCentre/Pages/WhatsNew-2043.aspx?utm_source=miragenews&utm_medium=miragenews&utm_campaign=news (accessed 23 February 2024).

James, B.D., de Vos, A., Aluwihare, L.I. et al. (2022). Divergent forms of pyroplastic: lessons learned from the M/V *X-Press pearl* ship fire. *ACS Environmental Au* 2 (5): 467–479. https://doi.org/10.1021/acsenvironau.2c00020.

James, B.D., Reddy, C.M., Hahn, M.E. et al. (2023). Fire and oil led to complex mixtures of PAHs on burnt and unburnt plastic during the M/V X-Press pearl disaster. *ACS Environmental Au* 3 (5): 319–335. https://doi.org/10.1021/acsenvironau.3c00011.

James, B.D., Karchner, S.I., Walsh, A.N. et al. (2023). Formulation controls the potential neuromuscular toxicity of polyethylene photoproducts in developing zebrafish. *Environmental Science & Technology* 57 (21): 7966–7977. https://doi.org/10.1021/acs.est.3c01932.

Karapanagioti, H.K. and Werner, D. (2019). Sorption of hydrophobic organic compounds to plastics in the marine environment: sorption and desorption kinetics. In: *Hazardous Chemicals Associated with Plastics in the Marine Environment* (ed. H. Takada and H.K. Karapanagioti), 205–219. Cham: Springer International Publishing (The Handbook of Environmental Chemistry) https://doi.org/10.1007/698_2018_256.

Kolb, K.E. and Kolb, D.K. (1991). Method for separating or identifying plastics. *Journal of Chemical Education* 68 (4): 348. https://doi.org/10.1021/ed068p348.

Krimm, S., Liang, C.Y., and Sutherland, G.B.B.M. (1956). Infrared spectra of high polymers. II. Polyethylene. *The Journal of Chemical Physics* 25 (3): 549–562. https://doi.org/10.1063/1.1742963.

Landrigan, P.J., Raps, H., Cropper, M. et al. (2023). The Minderoo-Monaco commission on plastics and human health. *Annals of Global Health* 89 (1): https://doi.org/10.5334/aogh.4056.

Lemkau, K.L., Peacock, E.E., Nelson, R.K. et al. (2010). The M/V *Cosco Busan* spill: source identification and short-term fate. *Marine Pollution Bulletin* 60 (11): 2123–2129. https://doi.org/10.1016/j.marpolbul.2010.09.001.

Lynch, J.M., Corniuk, R.N., Brignac, K.C. et al. (2024). Differential scanning calorimetry (DSC): an important tool for polymer identification and characterization of plastic marine debris. *Environmental Pollution* 346: 123607. https://doi.org/10.1016/j.envpol.2024.123607.

Maloney, E.M., Naile, J., and Saunders, D.M.V. (2021). Quantifying the effect of weathering on acute oil toxicity using the PETROTOX model. *Marine Pollution Bulletin* 162: 111849. https://doi.org/10.1016/j.marpolbul.2020.111849.

Marshall, A.G. and Rodgers, R.P. (2008). Petroleomics: chemistry of the underworld. *Proceedings of the National Academy of Sciences* 105 (47): 18090–18095. https://doi.org/10.1073/pnas.0805069105.

Marzbani, P., Resalati, H., Ghassemian, A., and Shakeri, A. (2016). Surface modification of talc particles with phthalimide: study of composite structure and consequences on physical, mechanical, and optical properties of deinked pulp. *Bioresources* 11: 8720–8738. https://doi.org/10.15376/biores.11.4.8720-8738.

Mato, Y., Isobe, T., Takada, H. et al. (2001). Plastic resin pellets as a transport medium for toxic chemicals in the marine environment. *Environmental Science & Technology* 35 (2): 318–324. https://doi.org/10.1021/es0010498.

Morét-Ferguson, S., Law, K.L., Proskurowski, G. et al. (2010). The size, mass, and composition of plastic debris in the western North Atlantic Ocean. *Marine Pollution Bulletin* 60 (10): 1873–1878. https://doi.org/10.1016/j.marpolbul.2010.07.020.

National Academies of Sciences, Engineering, and Medicine (2022). *Oil in the Sea IV: Inputs, Fates, and Effects*. Washington, D.C.: National Academies Press https://doi.org/10.17226/26410.

Nelson, T.F., Reddy, C.M., and Ward, C.P. (2021). Product formulation controls the impact of biofouling on consumer plastic photochemical fate in the ocean. *Environmental Science & Technology* 55 (13): 8898–8907. https://doi.org/10.1021/acs.est.1c02079.

Operation Clean Sweep (2024). *Operation Clean Sweep*. https://operationcleansweep.org/ (accessed 21 February 2024).

Oracle Environmental Experts (2023a). *Acute Plastic Pellet Loss Case Study*. https://hub.nurdlehunt.org/resource/oracle-mapping-the-global-plastic-pellet-supply-chain/ (accessed 27 February 2024).

Oracle Environmental Experts (2023b). *Mapping The Global Plastic Pellet Supply Chain*. https://hub.nurdlehunt.org/resource/oracle-mapping-the-global-plastic-pellet-supply-chain/ (accessed 27 February 2024).

Otocka, E.P. and Kwei, T.K. (1968). Properties of ethylene-acrylic acid copolymers. *Macromolecules* 1 (3): 244–249. https://doi.org/10.1021/ma60003a009.

Peacock, A. (2014). *Handbook of Polyethylene: Structures: Properties, and Applications*. Boca Raton: CRC Press https://doi.org/10.1201/9781482295467.

Peacock, E.E., Nelson, R.K., Solow, A.R. et al. (2005). The west falmouth oil spill: ~100 kg of oil found to persist decades later. *Environmental Forensics* 6 (3): 273–281. https://doi.org/10.1080/15275920500194480.

Prince, R.C. (2023). A half century of oil spill dispersant development, deployment and lingering controversy. *International Biodeterioration & Biodegradation* 176: 105510. https://doi.org/10.1016/j.ibiod.2022.105510.

Rochman, C.M., Browne, M.A., Halpern, B.S. et al. (2013). Classify plastic waste as hazardous. *Nature* 494 (7436): 169–171. https://doi.org/10.1038/494169a.

Ryerson, T.B., Camilli, R., Kessler, J.D. et al. (2012). Chemical data quantify Deepwater Horizon hydrocarbon flow rate and environmental distribution. *Proceedings of the National Academy of Sciences* 109 (50): 20246–20253. https://doi.org/10.1073/pnas.1110564109.

Santos, R.G., Machovsky-Capuska, G.E., and Andrades, R. (2021). Plastic ingestion as an evolutionary trap: toward a holistic understanding. *Science* 373 (6550): 56–60. https://doi.org/10.1126/science.abh0945.

Schleifstein, M. (2020). Plastic pellets litter batture near Chalmette Battlefield after cargo container accident, *Times-Picayune | New Orleans Advocate*, 8 August. https://www.nola.com/news/business/plastic-pellets-litter-batture-near-chalmette-battlefield-after-cargo-container-accident/article_4958b5f8-d8f4-11ea-be69-6b2e1d07d491.html (accessed 21 February 2024).

Signoret, C., Caro-Bretelle, A., Lopez-Cuesta, J. et al. (2020). Alterations of plastics spectra in MIR and the potential impacts on identification towards recycling. *Resources, Conservation and Recycling* 161: 104980. https://doi.org/10.1016/j.resconrec.2020.104980.

Smith, B. (2021). The infrared spectra of polymers II: polyethylene. *Spectroscopy* 36 (9): 24–29. https://doi.org/10.56530/spectroscopy.xp7081p7.

Stout, S.A. (2016). Oil spill fingerprinting method for oily matrices used in the Deepwater Horizon NRDA. *Environmental Forensics* 17 (3): 218–243. https://doi.org/10.1080/15275922.2016.1177759.

Sutherland, B.R., DiBenedetto, M., Kaminski, A. et al. (2023). Fluid dynamics challenges in predicting plastic pollution transport in the ocean: a perspective. *Physical Review Fluids* 8 (7): 070701. https://doi.org/10.1103/PhysRevFluids.8.070701.

Todd, B. and Lang, J. (2023). Mississippi River Nurdle Spill 2020. *Nurdle Spills in Coastal and Marine Environments*, College Park, MD, USA, 6 September. https://crrc.unh.edu/resource/nurdle-spills (accessed 23 February 2024).

United States Coast Guard (2024). *National Response Center*. https://nrc.uscg.mil/ (accessed 8 March 2024).

United States Coast Guard National Response Center (2020). *2020 Reports*. https://nrc.uscg.mil/FOIAFiles/CY22.xlsx (accessed 21 February 2024).

United States Department of Transportation Bureau of Transportation Statistics (2019). Petroleum Oil Spills Impacting Navigable U.S. Waterways. https://doi.org/10.21949/1503663.

United States Environmental Protection Agency (1992). *Plastic Pellets in the Aquatic Environment: Sources and Recommendations: Final Report*, p. 130. https://nepis.epa.gov/Exe/ZyPURL.cgi?Dockey=20004Y95.txt.

United States Geological Survey (2024). USGS 07374525 Mississippi River at Belle Chasse, LA. https://nwis.waterdata.usgs.gov/nwis/uv? (accessed 21 February 2024).

United States National Oceanic and Atmospheric Administration (2023). *Nurdle Spills in Coastal and Marine Environments, Coastal Response Research Center.* https://crrc.unh.edu/resource/nurdle-spills (accessed 27 February 2024).

United States National Transportation Safety Board (2021). *Breakaway of Containership CMA CGM Bianca.* NTSB/MAB-21/18. https://www.ntsb.gov/investigations/AccidentReports/Reports/MAB2118.pdf (accessed 21 February 2024).

de Vos, A., Aluwihare, L.I., Youngs, S. et al. (2022). The M/V X-Press pearl nurdle spill: contamination of burnt plastic and unburnt nurdles along Sri Lanka's beaches. *ACS Environmental Au* 2 (2): 128–135. https://doi.org/10.1021/acsenvironau.1c00031.

Wallace, J. (2020). *When plastic pours into the Mississippi, who's responsible?, E&E News by POLITICO.* https://www.eenews.net/articles/when-plastic-pours-into-the-mississippi-whos-responsible/ (accessed 21 February 2024).

Walsh, A.N., Reddy, C.M., Niles, S.F. et al. (2021). Plastic formulation is an emerging control of its photochemical fate in the ocean. *Environmental Science & Technology* 55 (18): 12383–12392. https://doi.org/10.1021/acs.est.1c02272.

Walsh, A.N., Mazzotta, M.G., Nelson, T.F. et al. (2022). Synergy between sunlight, titanium dioxide, and microbes enhances cellulose diacetate degradation in the ocean. *Environmental Science & Technology* 56 (19): 13810–13819. https://doi.org/10.1021/acs.est.2c04348.

Weaver, J. (2018). *Privacy Impact Assessment Update for the Incident Reporting Information System (IRIS).* DHS/USCG/PIA-023(a). United States Department of Homeland Security. https://www.dhs.gov/sites/default/files/publications/privacy-pia-uscg023-iris-april2018.pdf (accessed 27 February 2024).

Wise, S.A., Rodgers, R.P., Reddy, C.M. et al. (2023). Advances in chemical analysis of oil spills since the deepwater horizon disaster. *Critical Reviews in Analytical Chemistry* 53 (8): 1638–1697. https://doi.org/10.1080/10408347.2022.2039093.

Yamashita, R., Tanaka, K., Yeo, B.G. et al. (2019). Hazardous chemicals in plastics in marine environments: international pellet watch. In: *Hazardous Chemicals Associated with Plastics in the Marine Environment* (ed. H. Takada and H.K. Karapanagioti), 163–183. Cham: Springer International Publishing (The Handbook of Environmental Chemistry) https://doi.org/10.1007/698_2018_299.

Yang, Z., Shah, K., Laforest, S. et al. (2018). A study of the 46-year-old *Arrow* oil spill: Persistence of oil residues and variability in oil contamination along Chedabucto Bay, Nova Scotia, Canada. *Journal of Cleaner Production* 198: 1459–1473. https://doi.org/10.1016/j.jclepro.2018.07.112.

22

Unleashing Potential: Transcending Marine Pollution Forecasts for a Better Future and Critical Thresholds

Sandun Dassanayake[1], *Udari Perera*[2], *and Mahinsasa Rathnayake*[2]

[1] *Department of Decision Sciences, Faculty of Business, University of Moratuwa, Katubedda, 10400, Sri Lanka*
[2] *Department of Chemical and Process Engineering, Faculty of Engineering, University of Moratuwa, Katubedda, 10400, Sri Lanka*

22.1 Introduction

Marine pollution negatively affects the sustainable use and conservation of marine resources. Increasing volumes and veracity of contaminants in marine environments critically alter the harmony in marine environments. Anthropogenic activities elevate the concentration of contaminants, leading to harmful levels of pollution on a global scale (Chapman 2007). In 2017, the United Nations (UN) declared the Ocean Decade in view of countering these harmful effects to restore the global marine environment. The UN Decade of Ocean Science for Sustainable Development, extending through 2030, emphasizes the critical interplay between sustainable development and marine conservation. The decade's goals focus on sustainable development and ecosystem-based management, utilizing innovative methods and data analysis to monitor and ensure ocean sustainability.

The recent developments in remote sensing and biomonitoring techniques have redefined better critical thresholds for advancing the monitoring and assessment of planetary-scale marine pollution and the interoperability of biological observations (Chapman 2007). Additionally, these techniques have been employed to assess the impact of pollution on densely populated, highly urbanized coastal zones that serve as vital transition areas connecting terrestrial and marine environments. The low-elevation coastal zones cover 2% of the world's land area while being home to approximately 10% of the global population (Melet et al. 2020). The projected growth in coastal population is driven by factors like urbanization and coastward migration, with many megacities situated in coastal zones (Mejjad et al. 2023). These areas also support diverse ecosystems, providing essential services like fisheries, shoreline protection, and tourism. The coastal zones face various natural and anthropogenic hazards, including climate change-induced events like sea level rise, marine pollution, and overexploitation of resources, increasing their vulnerability (Melet et al. 2020).

This chapter aims to provide a holistic discussion of marine pollution forecasts and the direction of current research in the ocean and marine forecasting domains. With a scope of presenting the latest developments, the chapter strictly presents the discussion on papers published from 2019 to 2024. To this end, the chapter begins with a section on the global initiative of the Marine Life 2030 program, followed by the introduction of major marine pollution sources. The chapter contains sections on

Coastal and Marine Pollution: Source to Sink, Mitigation and Management, First Edition.
Edited by Meththika Vithanage, Sameera M. Samarasekara, Bryan D. James, and Christopher M. Reddy.

remote sensing-based monitoring techniques, biomonitoring methods, and data-driven assessment approaches. The chapter shall be a guide to multidisciplinary researchers to identify and develop relevant hypotheses to focus on the more pressing research gaps in the pollution monitoring area.

22.2 Critical Thresholds in Marine Life 2030 Program

The Marine Life 2030 program, endorsed by the UN Decade of Ocean Science for Sustainable Development, embodies a transformative vision to enhance the monitoring and understanding of marine environments, crucial for the sustainable use and conservation of ocean resources (Appeltans and Provoost 2021). The initiative seeks to establish a globally coordinated system that can disseminate essential knowledge about ocean life supporting human well-being, sustainable development, and ocean conservation. Through the adoption of a codesign approach, Marine Life 2030 emphasizes the development of interoperable information systems that can transform the observation and forecasting of marine life, benefiting people globally. Holistically, fostering collaboration among diverse stakeholders, such as scientists, resource users, and decision-makers, to create a bottom-up networking structure can counter these local and global marine conservation challenges. The program leverages innovation in "omics, acoustics, imaging, remote sensing, and artificial intelligence to capture biodiversity information more efficiently" (Estes et al. 2021). The emphasis on essential ocean variables helps promote the interoperability of biological observations (Appeltans and Provoost 2021; Muller-Karger et al. 2023).

Research efforts utilize satellite data to assess ocean food availability based on water coloration, which is indicative of food-rich areas. This data is linked to marine biodiversity, particularly the diversity of fish in deeper waters. Collaborations aim to quantify fish diversity and monitor reef ecosystems, addressing challenges like the scarcity of data sets and slow data integration into public databases. The program's codesign process, underscored by the tracking of success through metrics proposed by the Group on Earth Observations (GEO), exemplifies a commitment to not only achieving immediate objectives but also ensuring the enduring impact of capacity-building efforts, service delivery, and the uptake of core metrics. This underscores Marine Life 2030's role in not just advancing scientific understanding and technology but also in fostering a global community dedicated to the sustainable management and conservation of the world's oceans, thereby setting a new benchmark for critical thresholds in marine environment monitoring.

22.3 Major Marine Pollution Sources

Contemporary research publications related to marine pollution cover not only the traditional focus on oil spills and nano and microparticles but also a combination of these. Thus, marine pollution-related studies can be broadly categorized into two areas: pollution monitoring and pollution forecasting. Initially, to identify the major pollution sources of concern in these areas, a scientometric analysis was performed employing contemporary bibliometric data accumulated from 2019 to 2024 in the Scopus database. The search looked for the keyword combination "(ocean **OR** marine **OR** coastal) **AND** pollution **AND** forecasting" in the article title, abstract, and keywords. Figure 22.1 shows the number of published articles that include the search term and the cumulative number of current citations in Scopus database.

Figure 22.2 depicts a word cloud, a visual representation where the prominence of the words reflects their frequency or importance in the related documents. This distribution suggests a

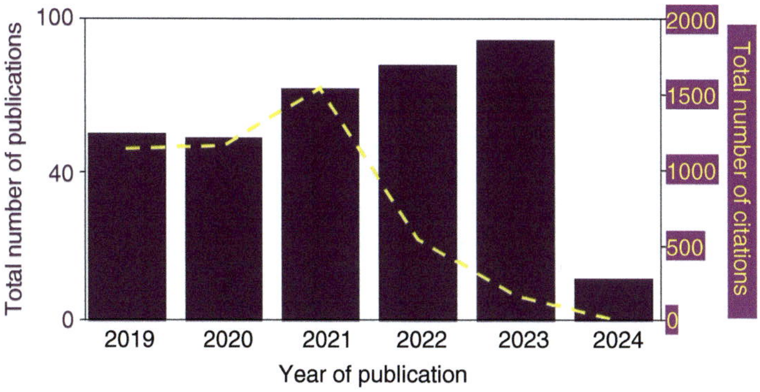

Figure 22.1 Accumulation of published articles and the number of citations in the past five years.

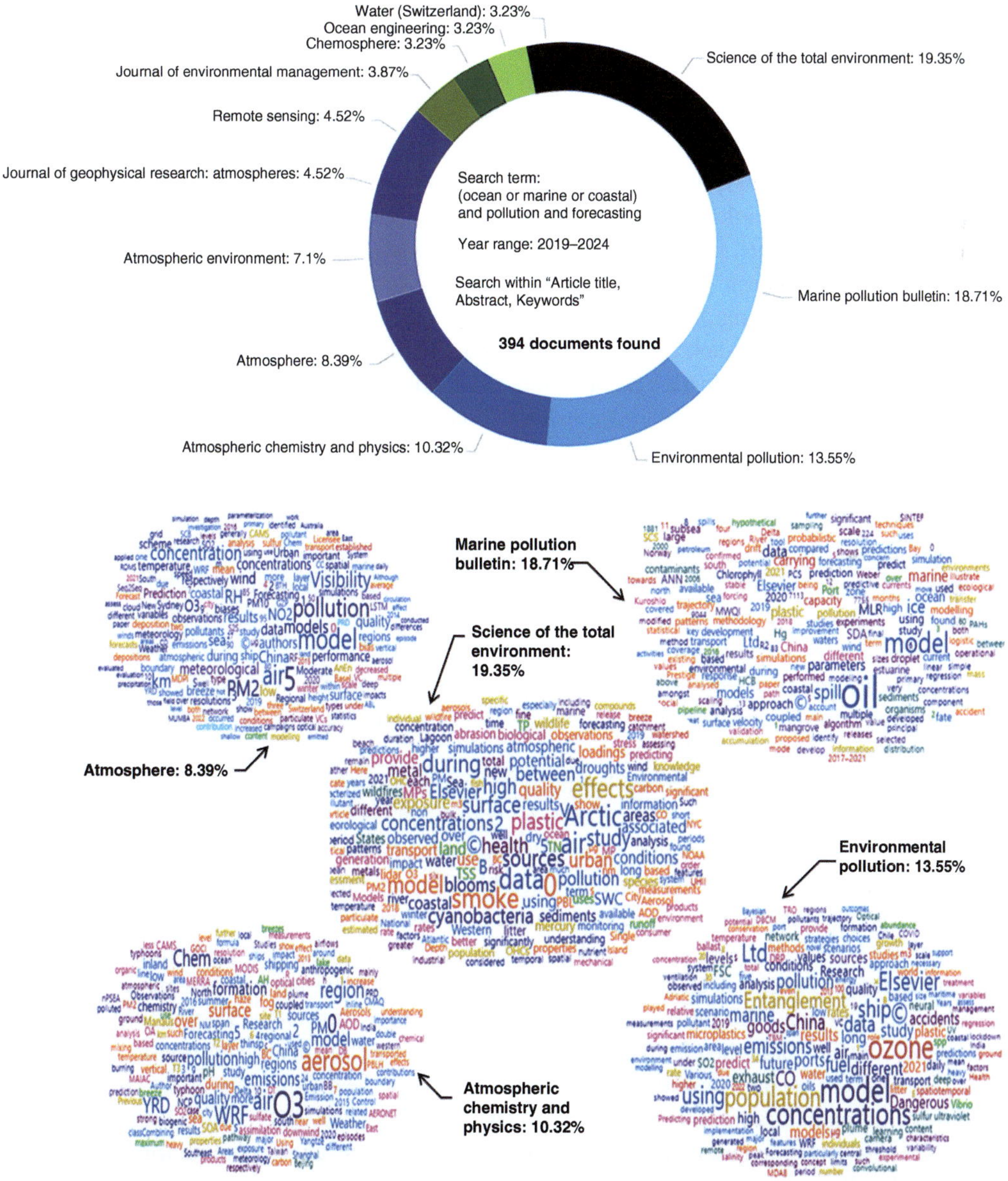

Figure 22.2 Textual data distribution in the articles, abstracts, and keywords of published articles retrieved from a search term related to ocean, marine, coastal pollution, and forecasting from 2019 to 2024.

multidisciplinary approach to marine pollution forecasting, with environmental science, marine pollution, and atmospheric chemistry being the key focus areas. The labels "Science of the Total Environment," "Marine Pollution Bulletin," and "Atmospheric Chemistry and Physics" indicate the most prominent journals or conferences from which the documents were sourced, with "Science of the Total Environment" making up the largest segment. The presence of more specialized terms like "model," "bioaccumulation," "phytoplankton," and "nanoparticles" suggests a strong emphasis on pollution, especially related to marine environments, and a significant reliance on modeling and simulation to study these phenomena. Holistically, monitoring and assessing oil spills and plastic pollution using satellite remote sensing applications have received a growing interest.

22.3.1 Oil Spills

Oil spills resulting from the intentional or accidental release of liquid petroleum hydrocarbons into water can spread rapidly over several hundred kilometers and form a thin crust of oil that can cover beaches, posing threats to human health and having severe environmental and economic impacts on the marine environment, fisheries, wildlife, benthic communities, the human settlement on the beaches, mangrove forests, and other social interests (Sevgili et al. 2022). Oil spills that can occur in varying magnitudes, from minor discharges by ships to catastrophic leaks following offshore drilling accidents, pose significant threats to marine biodiversity and ecosystems. Since the 1970s, there have been 467 major oil spills (over 700 tons each), leading to an estimated 100,000 gallons of oil being discharged into marine ecosystems. For instance, the X-Press Pearl incident resulted in an oil slick covering an area of 0.51 km^2 and stretching 3.23 km in length by 8 June 2021, later expanding to a length of 4.3 km by 14 June 2021 Sewwandi et al. (2023). Moreover, oil spills pose a significant threat to living species. Over 250 000 sea birds were found dead after the Exxon Valdez spill in 1989 and the Deepwater Horizon spill in 2010. The estimated biomass loss was between 25% and 50%, with reef fish populations declining by 40–70% after the Deepwater Horizon spill. Other oil spills like Amoco Cadiz, Prestige, and Rena caused the annihilation of over 20 000 birds. Mangrove ecosystems have suffered significantly, with 5 million tons of oil spilled since the 1950s, destroying 125 000 hectares Sepúlveda et al. Crassostrea (2023). Table 22.1 shows the details of major oil spills that occurred in recent years retrieved from online news sources. Sevgili et al. (2022) developed a forecasting model for underwater oil spilling using nonlinear, time-varying processes and multiple variables. The dynamic nature of oil spills is influenced by factors like chemical transformations and biological oil consumption challenges in accurately locating the leaking point (Wang et al. 2023).

Recent studies have focused on developing advanced machine learning (ML) and deep learning (DL) models to detect, monitor, and manage oil spills using synthetic aperture radar (SAR) data retrieved from sources like the European Space Agency's (ESA) Copernicus Sentinel-1 satellite. Such models include support vector machine (SVM), random forest (RF), and convolutional neural network (CNN) in pinpointing the origin, extent, and trajectory of oil spills using SAR imagery. Moreover, innovative integration of the ML and DL models enhances the capabilities of detecting oil spills through improved pixel-level segmentation and has shown exceptional classification accuracy, particularly when utilizing multiple channels. Mitigating and countering oil spill pollution has significantly benefited from the Decision Support Tools for Oil Spill Response (OSR-DSTs). The proposed General Decision Support System (DSS), leveraging National Oceanic and Atmospheric Administration's (NOAA's) advanced oil spill model (GNOME) and integrating comprehensive met-ocean datasets, exemplifies the strides toward developing comprehensive and effective oil spill management strategies.

Table 22.1 Recent oil spill incidents reported by online news sources.

Year	Spill name	Location	Date	Damage description
2019	MV Solomon Trader fuel oil spill	Rennell Island, Solomon Islands	5-Feb	80 tons of heavy fuel oil
	MV Chrysanthi S bunkering spill	Algoa Bay, Port of Ngqura, South Africa	6-Jul	200–400 l of fuel into the ocean
	Guarello Island diesel spill	Guarello Island, Magallanes Region, Chile	27-Jul	40 000 l of diesel
	North East Brazil oil spill	733 km from Paraiba, Brazil	29-Jul	2 100 km affected
2020	Tanker truck sludge spill	Tauranga, Bay of Plenty, New Zealand	2-Mar	2 000 l of oily sludge
	Pointe D'Esny MV Wakashio oil spill	Ile Aux Aigrettes and Mahebourg, Mauritius	25-Jul	0.23 km^2 observed by 6 August
	Colonial Pipeline oil spill	Huntersville, North Carolina, United States	25-Jul	two million gallons of gasoline
	Oil spill in Golfo Triste	Golfo Triste, Venezuela	2-Aug	Not specified
2021	Chevron Richmond Refinery spill	Richmond, California, United States	9-Feb	600 gallons into the San Francisco Bay
	Mediterranean oil spill	Mediterranean shoreline, Israel, Lebanon	11-Feb	160 km affected
	Orange County oil spill	Orange County, California shoreline, United States	2-Oct	24 696 gallons spilled
	New Orleans oil spill	St. Bernard Parish, outside of New Orleans	27-Dec	300 000 gallons of diesel, faulty pipeline
2022	Callao oil spill	Ventanilla District, Callao, Peru	15-Jan	1 800 490 m^2 of land and 7 139 571 m^2 of sea affected
	Keystone Pipeline oil spill	Washington County, Kansas, United States	7-Dec	14 000 barrels, third major spill in five years
2023	Princess Empress oil spill	Mindoro island	28-Feb	75 miles affected
	Oil spill at Wytch Farm	Wytch Farm, Poole Harbour, Dorset, United Kingdom	26-Mar	200 barrels of reservoir fluid leaked
	Marco Polo oil spill	Baltic Sea, Sweden	22-Oct	Stretches over 5 km

22.3.2 Marine Plastics

Marine plastic pollution refers to the accumulation of plastic debris in marine environments, posing threats to marine health (Issifu and Sumaila 2020). Marine plastic pollution has become a growing nuisance due to the increased use of plastic owing to its lightweight, durability, and low cost (Mejjad et al. 2023). The plastic contaminants in marine environments can be categorized as marine plastic litter (MPL) and microplastics (MMPs) based on the physical size of the contaminant (Sciutteri et al. 2023). MPL refers to all anthropogenic plastic waste present in marine environments, originating from land- and sea-based activities (Sciutteri et al. 2023). MMPs refer to all marine plastic waste particles smaller than 5 mm that can be primary (directly released into oceans) or secondary (resulting from environmental fragmentation of larger debris) (Sciutteri et al. 2023). The main types of plastics that generate MMPs include polyolefins, particularly polyethylene (PE), polypropylene (PP), polystyrene (PS), and polyethylene terephthalate (PET) (Hidalgo-Ruz and

Thiel 2013). Moreover, traces of polyamide (PA) have been identified in marine environmental samples as they originated from fishing nets (Duis and Coors 2016). The nondegradable nature, insolubility in water, and diverse physiochemical properties of MMPs pose threats to the marine environment which are yet to be fully assessed. Currently, enormous interest has been shown in microplastic (MP) research on its prevalence, distribution, fate (how MPs are ingested by organisms and how they contribute to biofouling, bioaccumulation, or biomagnification), and movement (Uzun et al. 2022). Marine organisms and seabirds may digest and carry the contaminants to the food chain or become entangled in floating and suspended debris (Cutroneo et al. 2020; Uzun et al. 2022). Therefore, they endanger marine organisms and may negatively affect human health. Also, they can be a navigational hazard as they tangle ship propellers or clog engine cooling systems (Hong et al. 2017). MMPs can even act as vectors for the spread of foreign species and pathogens and create new habitats for organisms (Zettler et al. 2013). Hence, evaluating the status and progression of plastic pollution with adequate spatial and temporal resolution is critical to assess ecological risk and effectively manage and mitigate marine plastic pollution. A guideline to monitor the micro- and MP in oceans was introduced by the Group of Experts on the Scientific Aspects of Marine Environmental Protection (GESAMP) in 2019 (GESAMP 2019).

Globally, about 8.5 billion metric tons of plastic have been produced since the inception of synthetic plastic production, with marine plastic pollution correspondingly increasing (Mejjad et al. 2023). Middle-income countries are identified as epicenters for plastic leakage into the ocean, with around 12.7 million tons of mismanaged land-based plastic waste estimated to have entered the oceans in 2010 (Issifu and Sumaila 2020). Mejjad et al. (2023) show the multifaceted issue of plastic pollution in Morocco, focusing on the emerging threats of MPs. The significant growth of tourists in urban areas has increased the accumulation of single-use plastic waste. Moreover, the food industry, maritime transport, and fishing activities in urban coastal cities have also contributed to marine plastic pollution (MPP) (Mejjad et al. 2023).

By analyzing the sources, distribution patterns, and impacts of plastic pollution, these methods can help predict areas at high risk and determine the effectiveness of implemented management strategies, thereby guiding efforts to prevent reaching critical ecological and economic thresholds (Mejjad et al. 2023). Sciutteri et al. (2023) examine the critical issue of marine litter (ML) pollution in the Mediterranean Sea, with a focus on the southern Tyrrhenian Sea, highlighting the pressing need to address the widespread problem of anthropogenic waste in marine environments. The Mediterranean is identified as a significant accumulation hotspot for marine litter due to various factors, including its physiography, densely populated coastlines, and busy shipping routes. The study emphasizes the importance of understanding the abundance, composition, and distribution of marine litter on the seafloor, which serves as the ultimate sink for these pollutants. The negative impacts of marine litter, mainly abandoned, lost, or derelict fishing gears (ALDFGs) and MPs, on marine organisms through entanglement, coverage, and ingestion cause significant negative impacts on marine ecosystems (Sousa-Guedes et al. 2024).

Sousa-Guedes et al. (2024) show a recently escalating nuisance of MPL on the marine ecosystems on Santa Luzia Island, Cabo Verde. This sensitive marine protected area has been of significant ecological value, including diverse wildlife and important nesting grounds for loggerhead turtles. However, this environmental integrity has been significantly threatened by the accumulation of MPP (Sousa-Guedes et al. 2024). The physical characteristics of beaches, such as elevation and orientation (northness), influence the accumulation of marine debris (Sousa-Guedes et al. 2024). Sousa-Guedes et al. (2024) applied fine-scale sampling and drone surveys for monitoring and predicting marine pollution levels, facilitating targeted conservation efforts and policymaking. The fine-scale sand sampling and coarse-scale aerial imaging revealed varied accumulation patterns and densities across different beach locations. A study was conducted to analyze the abundance, size, and polymer type of MPs ranging down to 10 μm along a transect from the European Coast to the North Atlantic Subtropical Gyre. The research utilized an underway intake filtration technique and Raman micro-spectrometry, revealing concentrations ranging from 13 to 501 items per cubic meter. The highest concentrations were observed on the European coast, decreasing toward mid-Atlantic waters but remaining elevated in the western North Atlantic Subtropical Gyre (NASG). The study highlighted that small MPs are widely distributed over the ocean surface layer and have a lower residence time compared to larger plastic debris in this compartment (Sousa-Guedes et al. 2024).

Recently, on May 2021, the cargo vessel M/V X-Press Pearl caught fire 18 kilometers off Sri Lanka's west coast, releasing about 1,680 tons of small, white nurdles used in plastic manufacturing that, after melting, fragmenting, charring, and undergoing chemical alterations from the blaze and chemical exposure, resulted in a complex mix of burnt and unburnt pellets varying in color, shape, size, and density, complicating cleanup efforts, disrupting ocean currents, and threatening marine life while the burnt plastic was found to be three times more chemically complex—with traces of combustion-derived polycyclic aromatic hydrocarbons and sometimes petroleum-based compounds—highlighting the urgent need for improved strategies to manage such incidents (De Vos 2021).

Acuña-Ruz et al. (2018) used a spectral library to identify Anthropogenic Marine Debris (AMD) mixtures over beaches on Chiloé Island. Garaba and Dierssen (2018) found that both wet and dry plastic spectra have absorption features around 1215 and 1732 nm, with the reflectance of wet plastic being lower due to water absorption. Topouzelis et al. (2019) conducted an experiment with artificial floating targets consisting of 10×10 m^2 PET water bottles, low density polyethylene (LDPE) plastic bags, and nylon fishing ghost nets. Biermann et al. (2020) used S2 images of various locations where MPL has been reported in scientific publications, mass, or social media. Martinez-Vicente et al. (2020) developed the Normalized Difference Hydrocarbon Index (NDHI) for the detection of plastic on the shoreline using airborne hyperspectral (HS) data. Basu et al. (2021) evaluated the potential of supervised and unsupervised classification algorithms to detect floating marine litter. Kremezi et al. (2021) evaluated several pansharpening methods on PRISMA HS images to increase the ability to detect small-sized plastic targets. Holistically, the researchers in these studies bear the collective opinion that more spatial resolution and more spectral bands, arranged in a dedicated configuration, would significantly improve the detection of marine plastic accumulations from orbit.

22.4 Remote Sensing-Based Monitoring

The selection of an appropriate satellite for oil spill studies hinges on specific research needs, including spatial detail, monitoring frequency, and the temporal scope of the analysis. The ability to discern fine details in satellite imagery is crucial for identifying and tracking oil spills. Satellites such as Quickbird and WorldView-2, which boast high spatial resolutions of 0.61 m and 0.52 m (panchromatic), respectively, are pivotal for detailed observation of oil spill extents and dynamics. Conversely, Sentinel-2 and Landsat-8, with moderate resolutions of 10 m and 15 m, strike a balance between detail and area coverage, proving effective for broader assessments. On the other end of the spectrum, Moderate Resolution Imaging Spectroradiometer (MODIS) and Advanced Very High Resolution Radiometer (AVHRR) satellites offer lower spatial resolutions (up to 1 km for AVHRR and 250 m, 500 m, and 1000 m for MODIS), catering to large-scale environmental monitoring where detail is less critical than coverage and temporal frequency. The frequency at which satellites revisit specific geographic locations directly impacts the timeliness and effectiveness of oil spill monitoring. Satellites like Quickbird, WorldView-2, and RapidEye, with revisit times ranging from 1 to 5.5 days, enable frequent updates on spill developments. Sentinel-2's 5-day revisit time offers a compromise, providing relatively regular monitoring capabilities. In contrast, with a 16-day revisit cycle, the Landsat series is less suited for tracking rapid environmental changes but remains invaluable for longitudinal studies. This classification sheds light on the strategic deployment of satellite resources to enhance our understanding and management of environmental hazards such as oil spills, thereby contributing to preserving and protecting marine and coastal ecosystems.

SAR-equipped satellites represent a critical technological asset. These satellites, distinguished by their operation periods, operators, radar bands, and polarization modes, offer diverse capabilities essential for detecting and tracking oil spills under various conditions, including the ability to operate irrespective of daylight or weather constraints. Early SAR Missions: Earth Resources Satellite-1 (ERS-1) and ERS-2, operated by the ESA from 1991 to 2000 and 1995 to 2011, respectively, utilized C-band radar with single-VV polarization, marking the early adoption of SAR technology for environmental monitoring. The Canadian Space Agency (CSA) advanced SAR applications with RADARSAT-1 (1995–2013) and RADARSAT-2 (since 2007), the latter introducing quad-polarization capabilities in the C-band, significantly enhancing oil spill detection accuracy. ENVISAT ASAR (2002–2012) and TerraSAR-X (since 2007) employed dual and quad-polarization modes, respectively, demonstrating the evolution of SAR technology toward more sophisticated oil spill monitoring methodologies. Satellites like ALOS PALSAR and ALOS-2 (Japan Aerospace Exploration Agency—JAXA), CosmoSkymed-1/2 (Italian Space Agency), and Sentinel-1 (ESA) have expanded the SAR-equipped fleet, offering quad and dual-polarization in L, X, and C bands, respectively, catering to a wide range of applications beyond oil spill detection. The utilization of different radar bands (C, L, X, and S) across these satellites underscores the technological diversification in SAR applications. Each band offers unique advantages in terms of penetration, resolution, and sensitivity to surface features, including oil spills. The progression from single to dual and quad-polarization techniques illustrates the advancements in SAR technology, enabling more detailed and accurate detection of oil spills. These polarization modes enhance the ability to distinguish oil spills from the surrounding seawater under various environmental conditions. The global commitment to environmental monitoring and oil spill detection is evident through the involvement of multiple space agencies, including ESA, CSA, JAXA, the Italian Space Agency, and others. Despite the conclusion of their operations, satellites such as ERS-1, ERS-2, and RADARSAT-1 have left a significant legacy in the field of SAR-based environmental monitoring. The datasets and methodologies developed by these missions continue to support oil spill detection and environmental research. Ongoing missions like RADARSAT-2, TerraSAR-X, Sentinel-1, and others represent the current state-of-the-art SAR technology, offering enhanced capabilities for real-time monitoring and analysis of oil spills. These missions are pivotal in continuously improving oil spill detection strategies, aiding in the timely response and mitigation of environmental impacts.

22.5 Biomonitoring

Recent studies have demonstrated the utility of various marine organisms as biomonitors for environmental pollution and as indicators of ecological health across different ecosystems. Bivalve mollusks, noted for their capacity to filter large quantities of seawater, play a keystone role in nutrient regeneration within benthic environments, with copper isotopes in their tissues used to infer dietary sources and bioavailability changes due to anthropogenic emissions (Araújo et al. 2023). Similarly, *Crassostrea gigas*, through systematic sampling, has been identified as significant for regional aquaculture and shellfish farming, with zinc isotopes in soft tissues providing insights into anthropogenic zinc bioaccumulation (Araújo et al. 2021).

Species such as *Saccostrea palmula* and *Crassostrea corteziensis*, collected from mangrove roots, serve as effective biomonitors of potentially toxic elements (PTEs) in coastal lagoons, with their tissues analyzed for a range of metals, including arsenic, cadmium, copper, iron, lead, and zinc (Sepúlveda et al. 2023). In contrast, Patella sp. (limpets), collected at low tide from intertidal rocky shores, are used as biomonitors for metal pollution through metal analysis in soft tissues, despite being grazers rather than filter feeders (Sánchez-Marín et al. 2022).

The blue mussel (*Mytilus edulis*) and the estuarine acorn barnacle (*Amphibalanus variegatus*), easily collected by scraping, are highlighted for their ability to accumulate trace metals, thus supporting migratory birds and indicating ecological monitoring potential, respectively (Willems et al. 2020). The grouping of *Mytilus* species, including *M. edulis*, *M. trossulus*, and *M. galloprovincialis*, as sessile organisms on mussel beds showcases their high tolerance to environmental stressors and their significant role as indicators of marine pollution and habitat structuring, with their ability to accumulate contaminants and reserve glycogen and fat being of particular interest (Hao 2021).

Moreover, *Cymodocea nodosa*, a primary producer, has been identified as an early detector of trace element (TE) contamination, reflecting sediment pollution and possessing high ecological value in coastal environments. Its tissues, including leaves, rhizomes, and roots, indicate the bioavailability of TEs through bioconcentration and translocation factors (Boutahar et al. 2021). Lastly, Antarctic fish species, *Chionodraco hamatus* and *Trematomus bernacchii*, collected at depths of 80 and 150 m, serve as bioindicators for Antarctic ecosystems, sensitive to contaminants due to their unique eco-physiological adaptations, with studies focusing on metallothionein expression and liver metal content (Marrone et al. 2021). Table 22.2 summarizes common biomarkers and the respective studies in biomonitoring of marine pollution.

22.6 Data-Driven Assessment Techniques

The use of ML, particularly RF classifiers, in oil spill monitoring from satellite images offers a robust method for the early detection and classification of oil spills. Remote sensing-based monitoring and detection of oil spills has been widely used, notably in Europe and the United States, namely, the European Maritime Safety Agency (EMSA) and NOAA, respectively. The predictive capability of the data-driven techniques is crucial for timely intervention and mitigation efforts to prevent environmental damage. Potential real-world applications of these concepts include deploying automated oil spill detection systems for environmental monitoring agencies, enhancing the efficiency of OSR strategies, and contributing to preserving marine ecosystems. By providing tools for the rapid identification of spills, these ML techniques can help minimize oil pollution's ecological and economic impact on marine environments (Conceição et al. 2021). Accurately classifying coral species and conditions helps detect ecosystem changes that may indicate the possibility of other environmental stresses. The high accuracy and robustness of the model can streamline monitoring processes, focusing resources on at-risk or changing areas. ML models can predict marine ecosystem changes by monitoring coral health and diversity, giving an early warning for preventive or mitigation measures. (Ganesan and Santhanam 2022). Figure 22.3 depicts the common ML and DL techniques employed in the marine pollution assessment.

Table 22.2 Key studies in employing different biomarkers for monitoring and assessing marine pollution.

Name	Size	Ease of collection	Filter-feeding behavior	Ecological importance	Biomarker	Source
Bivalve mollusks	N/A	Implied ease due to systematic sampling	Capable of filtering large quantities of seawater	Keystone species in benthic nutrient regeneration	Cu isotopes for Cu bioavailability	Araújo et al. (2023)
Crassostrea gigas	2-to-3-year-old	Systematic sampling, relatively easy	Primary consumers, filters mainly phytoplankton	Significant for regional aquaculture and shellfish farming	Zn isotopes for anthropogenic Zn bioaccumulation	Araújo et al. (2021)
Saccostrea palmula *Crassostrea corteziensis*	72.1 ± 2.7 mm 73.8 ± 4.4 mm	Hand-collected from mangrove roots	Filters PTEs from food, water, sediments	Biomonitor of PTEs in coastal lagoons	As, Cd, Cu, Fe, Pb, and Zn in tissues	Sepúlveda et al. (2023)
Patella sp. (Limpets)	30.4 ± 2.8–49.7 ± 8.4 mm	Hand-collected at low tide	Grazers, not filter feeders	Used as biomonitors for metal pollution	Metals analysis in soft tissues	Sánchez-Marín et al. (2022)
Mytilus edulis (Blue Mussel)	20–110 mm	Easily collected by scraping	Capable of accumulating trace metals	Supports migratory birds, impacts trophic web	Trace metals analysis	Willems et al. (2020)
Amphibalanus variegatus	N/A	Easily collected by scraping	Capable of accumulating trace metals	Indicated ecological monitoring potential	Trace metals analysis	
Mytilus spp. (M. edulis, M. trossulus, M. galloprovincialis)	Up to 6 cm	High, dense populations	Filters phytoplankton, significant filtering capacity	Indicator of marine pollution, habitat structuring	Accumulates contaminants, analysis of reserve glycogen, fat	Hao (2021)
Cymodocea nodosa	N/A	Easy	N/A	Primary producer, early detector of TE contamination	Bioavailability of TEs through bioconcentration	Boutahar et al. (2021)
Antarctic fish (*Chionodraco hamatus, Trematomus bernacchii*)	Varies by year	Collected at 80 and 150 m depths	N/A	Bioindicators for Antarctic ecosystems	Metallothionein expression, liver metal content	Marrone et al. (2021)

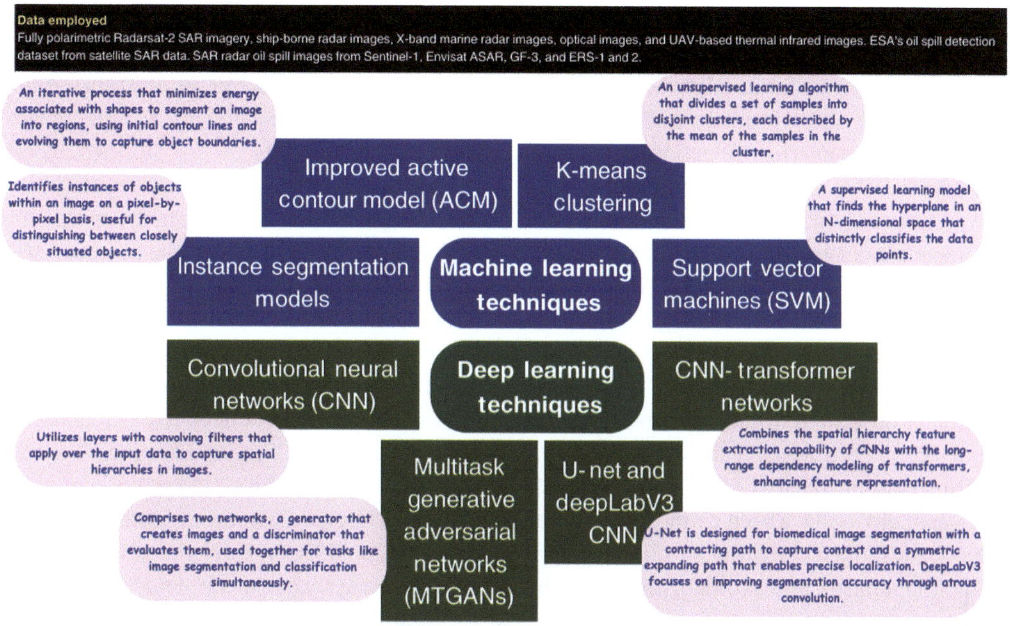

Figure 22.3 Key machine learning and deep learning techniques employed in marine pollution forecasting.

CNNs have been notably utilized for detecting oil spills in the Great Barrier Reef using Envisat ASAR and Sentinel-1 SAR imagery, achieving over 98% accuracy and a low false positive rate, underscoring the method's adaptability and economic efficiency for monitoring regions with significant oil and gas activities (Blondeau-Patissier et al. 2023). Similarly, advancements in segmentation models, SVMs, and artificial neural networks (ANN) have enhanced the accuracy of oil spill detection and discrimination in SAR and optical images, though challenged by the presence of biogenic elements (Shamsudeen 2020).

DL models like U-NET and DeepLabV3 have shown promise in augmenting the dataset of SAR radar oil spill images to improve detection accuracy and model generalization (Ghara et al. 2022). An improved Active Contour Model (ACM) utilizing fully polarimetric RADARSAT-2 SAR imagery has significantly enhanced oil spill identification accuracy through an optimal initial box boundary selection and noise suppression (Song et al. 2018). Moreover, the integration of ANN, decision tree classifier, and adaptive threshold (AT) algorithms with Envisat SAR Imagery has facilitated automatic oil spill detection systems, highlighting the efficiency improvements achievable through shared-memory parallel programming techniques (Mera et al. 2014). Al-Ruzouq et al. (2020) show the key concepts and entities related to ML use in oil spill monitoring and satellite images, including traditional ML techniques, DL models, and the general framework for oil spill detection. This framework encompasses the preprocessing of remotely sensed data, segmentation for dark-spot identification, extraction of discriminative features, and classification to differentiate between oil spills and look-alikes (Al-Ruzouq et al. 2020).

The employment of Genetic Neural Networks on Landsat 8, GF-1, MAMS, and HJ-1 imagery has introduced a novel marine oil spill detection approach, demonstrating high accuracy rates and fast processing times (Li et al. 2017). Multitask Generative Adversarial Networks (MTGANs) have been developed for the classification and segmentation of oil spills, showcasing enhanced accuracy and precision in distinguishing actual oil spills from look-alikes (Fan and Liu 2023). Deep neural network (DNN) techniques based on U-NET and ResNet have leveraged multichannel SAR images to significantly improve oil spill detection accuracy, emphasizing the importance of data augmentation and global normalization in processing complex oil spill shapes (Hasimoto-Beltran et al. 2023).

Hybrid CNN-Transformer Networks, including DeepLabv3 and FC-DenseNet Vision Transformers (FC-ViT), have outperformed standard CNN architectures in detecting oil spills in SAR images, demonstrating superior detection precision through the integration of transformer blocks (Dehghani-Dehcheshmeh et al. 2023). The use of SVM classifiers on unmanned aerial vehicle (UAV) based thermal infrared images and K-means clustering on X-band marine radar images has further advanced the rapid and efficient detection of oil spills, critical for mitigating marine pollution (Chen et al. 2022; Li et al. 2022).

Innovative approaches utilizing HS image classification, such as principal component analysis (PCA) and Deep-CNN, have been explored for mapping oil spills, highlighting the adaptability and noise robustness of DL classifiers (Kang et al. 2022). The AlexNet model's application to SAR images for oil spill information extraction underscores the effectiveness of deep CNNs in enhancing detection accuracy and efficiency (Wang et al. 2021). Furthermore, Feature Merging Networks (FMNet) combined with U-NET and various DL models like U-NET, LinkNet, PSPNet, DeepLabv2, and DeepLabv3+ have been employed on different datasets to improve semantic segmentation and detection accuracy in marine oil spill monitoring (Krestenitis et al. 2019; Fan et al. 2021).

Satellite imagery and ML algorithms continuously monitor vast ocean areas, detecting oil spills early to minimize environmental damage and streamline cleanup operations. However, these studies acknowledge limitations such as high false positive rates, reliance on extensive data augmentation due to limited original image datasets, difficulty distinguishing oil spills from naturally occurring look-alikes, and image noise's negative effects on detection accuracy. Satellite imagery-based oil spill detection methods are more robust, accurate, and applicable when detection algorithms are refined to reduce false positives, training datasets are expanded, environmental condition data is added, and more advanced ML and DL models are explored. This highlights the need for ongoing research and development in this critical area of environmental monitoring. Enhancing image classification precision can help forecast marine pollution and identify critical thresholds. Table 22.3 contains a summary of the contemporary techniques for assessing marine pollution using data-driven techniques.

22.7 Critical Thresholds: Challenges, Opportunities, and Future Directions

Forecasting marine pollution and its ecological impacts requires a comprehensive analysis that covers chemical, biological, ecological, hydrodynamic, toxicological, and socioeconomic factors. Assessing how changing oceans affect living organisms can dramatically improve critical threshold forecasting. Hydrodynamic and physiological processes affect pollutant movement and fate, with uptake and transmission through trophic levels being critical pathways for marine pollution dispersion. Management and conservation strategies are crucial for harnessing the full potential of ecosystems such as Sargassum forests and coral reefs that significantly contribute to counter climate change (Gouvêa et al. 2020). Ocean acidification, caused by increased CO_2 absorption by the world's seas, reduces marine life's survival, growth, calcification, and abundance. Among the many species affected, highly calcified creatures including corals, mollusks, calcified algae, and echinoderm larvae are especially sensitive, suffering significant physiological and developmental damage (Kroeker et al. 2013). Crustaceans, fish, fleshy algae, seagrasses, and diatoms are more resilient or may even profit from shifting conditions (Kroeker et al. 2013).

Table 22.3 Selected studies on marine pollution monitoring and assessment using data-driven techniques.

Dataset	ML/DL algorithms used	Summary	Source
Envisat ASAR (for events between 2002 and 2012), Sentinel-1 SAR (synthetic aperture radar) imagery (for events after April 2014) CSIRO binary dataset of oil-like and non-oil features	Convolutional neural network (CNN)	Semi-Automated Oil Slick Detection System for Great Barrier Reef (GBR) combines CNN and empirical methods for enhanced accuracy. Requires manual verification but achieves over 98% accuracy. Outperforms individual classifier use. Adaptable and economical for oil and gas regions. Suggests future enhancements for comprehensive mapping. Potential for broader application and improvement in marine pollution monitoring and disaster-risk resilience.	Blondeau-Patissier et al. (2023)
A review of the use of SAR and optical images	Instance segmentation models, SVM, ANN	Biogenic elements complicate rapid detection and decision-making. Microwave sensors are used due to their all-weather capabilities. Biogenic elements challenge the accuracy of these sensors, leading to false positives. Machine learning and deep learning models enhance detection accuracy. Need for further research to develop specialized sensors and leverage emerging technologies for accurate oil spill detection and classification.	Shamsudeen (2020)
SAR radar oil spill images from Sentinel-1 and EnviSat radars. A total of 99 original images were collected and augmented to 9801 images using image data augmentation techniques.	U-NET and DeepLabV3 CNN	Deep learning enhances Oil Spill Detection in SAR Images Utilizes deep learning models to improve the model's generalization ability. Data augmentation increases the number of SAR oil spill images that can be used for the analysis. Improves accuracy in oil spill detection through performance metrics analysis. Emphasizes the effectiveness of encoder-decoder networks, particularly DeepLabv3+ with decoder module, for complex SAR imagery segmentation.	Ghara et al. (2022)
Fully Polarimetric RADARSAT-2 SAR Imagery	Improved Active Contour Model (ACM)	Utilizes fully polarimetric synthetic aperture radar (Pol-SAR) for ocean oil spillage identification. Focuses on an active contour model (ACM) with optimal initial box boundary selection and noise suppression process. Uses median level contour as the initial box boundary and Lee filter and Gaussian smooth operator to reduce speckle noise. Experimental tests on RADARSAT-2 Pol-SAR data show improved oil spill identification accuracy.	Song et al. (2018)

(*Continued*)

Table 22.3 (Continued)

Dataset	ML/DL algorithms used	Summary	Source
Envisat (2007-2011) SAR Imagery	ANN, decision tree classifier, and adaptive threshold (AT) algorithm	Automatic Oil Spill Detection System Using SAR Data designed for monitoring international shipping routes off the Galician coast. Incorporates adaptive segmentation and shape-oriented characterization algorithm. Two classifiers were developed for enhanced detection accuracy. Shared-memory parallel programming reduces processing time, improving efficiency by over 25%.	Mera et al. (2014)
Landsat 8, GF-1, MAMS, HJ-1 imagery	Genetic neural network	The Marine Oil Spill Detection Approach combines bottom-up and top-down visual saliency models with high-resolution satellite images. Employs graph-based visual saliency model to identify salient ocean regions. Distinguishes oil spills from other salient features using spectral similarity match model. Classifies regions using a genetic neural network for a high accuracy rate (91.42%). Validates robustness across different satellite data types and complex sea conditions.	Li et al. (2017)
Sentinel-1, GF-3, ERS-1/2 satellite data	Multitask generative adversarial networks	Novel MTGANs Approach for Oil Spill Classification and Segmentation integrates classification and segmentation processes for enhanced accuracy. Uses a discriminator for classification and a fully convolutional symmetric structure generator for segmentation. Distinguishes between actual oil spills and look-alikes, reducing misclassification. Addresses challenges in SAR image analysis, including distinguishing between spills and natural occurrences. Outperforms existing methods on three satellite datasets, achieving higher classification accuracy.	Fan and Liu (2023)

Table 22.3 (Continued)

Dataset	ML/DL algorithms used	Summary	Source
Envisat ASAR images	Deep neural network (M-DNN) based on U-NET and ResNet	Oil Spill Detection Using Multichannel Deep Neural Network utilizes M-DNN and Synthetic Aperture Radar images for accurate oil spill detection. Employs original SAR image, derived variance, and wind field map input channels. Develops a new image dataset with multichannel images and global normalization. Improves detection of complex oil spill shapes and prevents misidentification of high oil seawater levels. Achieves pixel-level classification accuracy of 98.56% and faster training convergence.	Hasimoto-Beltran et al. (2023)
Envisat ASAR, Sentinel-1, and GF-3 SAR images	CNN-Transformer Networks including DeepLabv3 and FC-DenseNet Vision Transformers (FC-ViT)	Hybrid Models for Oil Spill Detection in SAR Images combines convolutional neural networks (CNNs) with transformer networks. Improves accuracy and capability to identify oil spill boundaries in noisy SAR data. Utilizes a comprehensive global dataset of oil spill events. Achieves higher F1-score values, indicating superior detection precision compared to standard CNN architectures.	Dehghani-Dehcheshmeh et al. (2023)
UAV-based thermal infrared images	SVM classifier	Utilizes Histogram of Gradient (HOG) features and SVM for accurate low-altitude oil spill detection. Outperforms traditional methods in recognition accuracy and training efficiency. Achieves 91.3% accuracy from small sample sizes. Shows promise for maritime supervision and environmental protection applications.	Li et al. (2022)
X-band marine radar images	K-means clustering	Novel Approach to Detecting Oil Spills introduces a fully automated oil spill identification method using a K-means clustering model. Integrates texture features and local adaptive threshold segmentation for X-band marine radar images. Outperforms semi-automatic and manual techniques. Enhances efficiency and accuracy of marine radar oil spill detection. Serves as a reference for future research in oil spill extraction and identification. Plans to refine the method with polarization characteristics and additional measured data.	Chen et al. (2022)

(Continued)

Table 22.3 (Continued)

Dataset	ML/DL algorithms used	Summary	Source
Deep Water Horizon Dataset (AVIRIS, 224 spectral bands, 400–2500 nm, images B01 and B02 with size of 350 × 500 pixels) hyperspectral images (258 spectral channels, 400–1000 nm, area of 450 × 800 pixels) Simulated Dataset (Qingdao Scientific Research Base, China, 21 September 2020, Nano-Hyperspec sensor on DJI-M600PRO UAV)	PCA, deep-CNN	Reviews key HSI classification methods. Evaluation of performance across three case studies. Highlights challenges with pixel-level feature extraction methods. Discusses dependence of sparse representation classifiers on dictionary quality. Highlights superior adaptability and noise robustness of deep learning classifiers in varying oceanic conditions. Suggests future research for unsupervised and lightweight models for real-time data applications.	Kang et al. (2022)
SAR images	AlexNet Model	AlexNet for Oil Spill Detection utilizes convolutional neural network model, AlexNet, to extract oil spill information from SAR images. Creates and enhances a dataset from remote sensing oil spills in China. Enhances accuracy and efficiency through preprocessing images and techniques like erosion, dilation, and Gaussian filtering. The modified AlexNet model outperforms others in accuracy and robustness, indicating potential for reliable marine oil spill monitoring.	Wang et al. (2021)
Oil spill detection dataset from the European Space Agency (ESA), using jpg images extracted from satellite SAR data	Feature Merging Network (FMNet) combined with U-Shape Network (U-NET)	FMNet merges global features from threshold segmentation with high-dimensional convolution features. Enhances accuracy of recognizing oil spill areas. Highlights deep learning potential for marine spill detection. Suggests further exploration of network structures and algorithms.	Fan et al. (2021)

Table 22.3 (Continued)

Dataset	ML/DL algorithms used	Summary	Source
Sentinel-1 SAR images	U-NET, LinkNet, PSPNet, DeepLabv2, DeepLabv3+	Utilizes Synthetic Aperture Radar (SAR) images and deep learning models for semantic segmentation. Introduces publicly available SAR image dataset for benchmarking. DeepLabv3+ model shows the best accuracy and inference time performance. Highlights trade-off between accurate oil spill identification and similar-looking features.	Krestenitis et al. (2019)
Ship-borne radar images	Improved Active Contour Model (ACM)	Presents an efficient alternative to existing methods. Enhances Local Binary Fitting (LBF) model with preprocessing techniques. Outperforms other ACMs in calculation speed and accuracy on uniform sea clutter images. Tested on 68 primary radar images from the Dalian "7-16" oil spill. Discusses limitations of ship-borne radar technology for oil spill monitoring. Compares ACM's performance with traditional models.	Xu et al. (2019)
Fully Polarimetric RADARSAT-2 data	Ramp-loss-based Ocean Oil Spillage Classifier (RampOC)	Utilizes Pol-SAR imagery to identify oil spillage. Employs a cost-sensitive risk minimization model and ramp loss function. Addresses imbalance between oil slicks and seawater. Effective even with limited or incorrect labels. Demonstrates superior identification performance across various scenarios. Suggests further exploration into feature selection based on oil slick properties and prior knowledge utilization.	Liang et al. (2019)
Fully Polarimetric RADARSAT-2 data	Optimized Wavelet Neural Network (WNN)	Utilizes fully polarimetric synthetic aperture radar (Pol-SAR) data and optimized wavelet neural network (WNN) classifier. Integrates multiple Pol-SAR features for enhanced classification accuracy. Achieves overall accuracy of 97.67%.	Song et al. (2017)

(Continued)

Table 22.3 (Continued)

Dataset	ML/DL algorithms used	Summary	Source
MODIS (250-m and 500-m resolution), Landsat 7 ETM+ and Landsat 8 (30-m resolution), MERIS (300-m resolution)	Analysis of spectral signatures and visual examination of satellite images	Utilizes MODIS, MERIS, Landsat data, Floating Algae Index (FAI), and hydrodynamic models for oil tracking. Spectral signature varies with viewing geometry, solar angle, oil type, and thickness. Analyzes three oil spill events using a multi-sensor approach and hydrodynamic modeling. Emphasizes the importance of combining satellite observations with hydrodynamic models for early oil pollution response. Showcases synergistic use of different data sources for improved oil spill detection and analysis.	Zhao et al. (2014)
SAR images	Kernel-based anomaly detection using support vector machines	Utilizes spaceborne synthetic aperture radar (SAR) to detect ocean pollution. Leverages oil's damping effect on gravity-capillary waves to reduce backscattering and wave energy. Employs kernel-based anomaly detection within wavelet decomposition of SAR imagery. Accurately identifies oil slicks without signal stationarity or strong backscatter sources. Applied to ENVISAT Advanced SAR images, effectively segmented small slicks with minimal false alarms.	Mercier and Girard-Ardhuin (2006)

Healthy and attractive marine habitats can boost the economy of marine-based sectors such as fisheries, aquaculture, and tourism. Reducing plastic pollution can prevent hazardous compounds from entering the food chain and help improve public health. Promoting plastic waste reuse and recycling can improve sustainable product design and alternative materials, in line with the global circular economy. Understanding the origins, forms, and implications of marine litter can help establish conservation strategies to reduce MP pollution. Raising public awareness about the causes and effects of marine litter and encouraging coastal communities and industries to consume and dispose of waste responsibly can speed up the marine litter pollution assessment. This can help the Marine Life 2030 initiative protect marine biodiversity and sustain marine resources. Pollution hotspots can guide cleanup and conservation efforts, especially in ecologically sensitive places like nesting grounds. Public awareness of distant island plastic pollution can change behavior toward sustainability. Global initiatives such as Marine Life 2030 promote international collaboration, data literacy, and finance to counter marine pollution issues. The key driving force highlighted in these initiatives is the strengthening of data management systems and resources to advance ocean science to preserve marine health and sustainability.

References

Acuña-Ruz, T., Uribe, D., Taylor, R. et al. (2018). Anthropogenic marine debris over beaches: Spectral characterization for remote sensing applications. *Remote Sensing of Environment* 217: 309–322.

Al-Ruzouq, R., Gibril, M.B.A., Shanableh, A. et al. (2020). Sensors, features, and machine learning for oil spill detection and monitoring: a review. *Remote Sens* 12: 3338. https://doi.org/10.3390/rs12203338.

Appeltans, W. and Provoost, P. (2021). Establishing the foundation for the global observing system for marine life. *Frontiers in Marine Science* 8: https://doi.org/10.3389/fmars.2021.737416.

Araújo, D.F., Ponzevera, E., Weiss, D.J. et al. (2021). Application of Zn isotope compositions in oysters to monitor and quantify anthropogenic Zn bioaccumulation in marine environments over four decades: a "Mussel Watch Program" upgrade. *ACS ES&T Water* 1: 1035–1046.

Araújo, D.F., Ponzevera, E., Knoery, J. et al. (2023). Can copper isotope composition in oysters improve marine biomonitoring and seafood traceability? *Journal of Sea Research* 191: 102334.

Basu, B., Sannigrahi, S., Basu, A.S., and Pilla, F. (2021). Development of novel classification algorithms for detection of floating plastic debris in coastal waterbodies using multispectral Sentinel-2 remote sensing imagery. *Remote Sensing* 13: 1598.

Biermann, L., Clewley, D., Martinez-Vicente, V., and Topouzelis, K. (2020). Finding plastic patches in coastal waters using optical satellite data. *Scientific Reports* 10: 5364.

Blondeau-Patissier, D., Schroeder, T., Suresh, G. et al. (2023). Detection of marine oil-like features in Sentinel-1 SAR images by supplementary use of deep learning and empirical methods: performance assessment for the Great Barrier Reef marine park. *Marine Pollution Bulletin* 188: 114598. https://doi.org/10.1016/j.marpolbul.2023.114598.

Boutahar, L., Espinosa, F., Sempere-Valverde, J. et al. (2021). Trace element bioaccumulation in the seagrass Cymodocea nodosa from a polluted coastal lagoon: biomonitoring implications. *Marine Pollution Bulletin* 166: 112209.

Chapman, P.M. (2007). Determining when contamination is pollution—weight of evidence determinations for sediments and effluents. *Environment International* 33: 492–501.

Chen, R., Li, B., Jia, B. et al. (2022). Oil spill identification in X-band marine radar image using K-means and texture feature. *PeerJ Computer Science* 8: e1133. https://doi.org/10.7717/peerj-cs.1133.

Conceição, M.R.A., de LFF, M., CAD, L. et al. (2021). SAR Oil Spill Detection System through Random Forest Classifiers. *Remote Sens* 13: 2044. https://doi.org/10.3390/rs13112044.

Cutroneo, L., Reboa, A., Besio, G. et al. (2020). Microplastics in seawater: sampling strategies, laboratory methodologies, and identification techniques applied to port environment. *Environmental Science and Pollution Research* 27: 8938–8952.

Dehghani-Dehcheshmeh, S., Akhoondzadeh, M., and Homayouni, S. (2023). Oil spills detection from SAR earth observations based on a hybrid CNN transformer networks. *Marine Pollution Bulletin* 190: 114834. https://doi.org/10.1016/j.marpolbul.2023.114834.

De Vos, A., Aluwihare, L., Youngs, S., DiBenedetto, M. H., Ward, C. P., Michel, A. P., ... & James, B. D. (2021). The M/V X-Press Pearl nurdle spill: contamination of burnt plastic and unburnt nurdles along Sri Lanka's beaches. ACS Environmental Au, 2(2), 128–135.

Duis, K. and Coors, A. (2016). Microplastics in the aquatic and terrestrial environment: sources (with a specific focus on personal care products), fate and effects. *Environmental Sciences Europe* 28: 2.

Estes, M., Muller-Karger, F., Forsberg, K. et al. (2021). Integrating biology into ocean observing infrastructure. *Oceanography* 34: 36–43.

Fan, J. and Liu, C. (2023). Multitask GANs for oil spill classification and semantic segmentation based on SAR images. *IEEE Journal of Selected Topics in Applied Earth Observations and Remote Sensing* 16: 2532–2546. https://doi.org/10.1109/jstars.2023.3249680.

Fan, Y., Rui, X., Zhang, G. et al. (2021). Feature merged network for oil spill detection using SAR images. *Remote Sensing* 13: 3174. https://doi.org/10.3390/rs13163174.

Ganesan, A. and Santhanam, S.M. (2022). A novel feature descriptor based coral image classification using extreme learning machine with ameliorated chimp optimization algorithm. *Ecological Informatics* 68: https://doi.org/10.1016/j.ecoinf.2021.101527.

Garaba, S.P. and Dierssen, H.M. (2018). An airborne remote sensing case study of synthetic hydrocarbon detection using short wave infrared absorption features identified from marine-harvested macro-and microplastics. *Remote Sensing of Environment* 205: 224–235.

GESAMP G (2019). Guidelines for the monitoring and assessment of plastic litter in the ocean. *GESAMP Reports Stud* 99: 130.

Ghara, F.M., Shokouhi, S.B., and Akbarizadeh, G. (2022). A new technique for segmentation of the oil spills from synthetic-aperture radar images using convolutional neural network. *IEEE Journal of Selected Topics in Applied Earth Observations and Remote Sensing* 15: 8834–8844.

Gouvêa, L.P., Assis, J., Gurgel, C.F.D. et al. (2020). Golden carbon of Sargassum forests revealed as an opportunity for climate change mitigation. *Science of the Total Environment* 729: 138745. https://doi.org/10.1016/j.scitotenv.2020.138745.

Hao S. (2021). Standard and innovative methods of biomonitoring marine pollution using mussels. Doctor of Philosophy (PhD), RMIT University.

Hasimoto-Beltran, R., Canul-Ku, M., Mendez, G.M.D. et al. (2023). Ocean oil spill detection from SAR images based on multi-channel deep learning semantic segmentation. *Marine Pollution Bulletin* 188: 114651. https://doi.org/10.1016/j.marpolbul.2023.114651.

Hidalgo-Ruz, V. and Thiel, M. (2013). Distribution and abundance of small plastic debris on beaches in the SE Pacific (Chile): a study supported by a citizen science project. *Marine Environmental Research* 87: 12–18.

Hong, S., Lee, J., and Lim, S. (2017). Navigational threats by derelict fishing gear to navy ships in the Korean seas. *Marine Pollution Bulletin* 119: 100–105.

Issifu, I. and Sumaila, U.R. (2020). A review of the production, recycling and management of marine plastic pollution. *Journal of Marine Science Engineering* 8: 945. https://doi.org/10.3390/jmse8110945.

Kang, X., Wang, Z., Duan, P., and Wei, X. (2022). The Potential of Hyperspectral Image Classification for Oil Spill Mapping. *IEEE Transactions on Geoscience and Remote Sensing* 60: 1–15. https://doi.org/10.1109/tgrs.2022.3205966.

Kremezi, M., Kristollari, V., Karathanassi, V. et al. (2021). Pansharpening PRISMA data for marine plastic litter detection using plastic indexes. *IEEE Access* 9: 61955–61971.

Krestenitis, M., Orfanidis, G., Ioannidis, K. et al. (2019). Oil spill identification from satellite images using deep neural networks. *Remote Sensing* 11: 1762. https://doi.org/10.3390/rs11151762.

Kroeker, K.J., Kordas, R.L., Crim, R. et al. (2013). Impacts of ocean acidification on marine organisms: quantifying sensitivities and interaction with warming. *Global Change Biology* 19 (6): 1884–1896.

Li, Y., Cui, C., Liu, Z. et al. (2017). Detection and monitoring of oil spills using moderate/high-resolution remote sensing images. *Archives of Environmental Contamination and Toxicology* 73: 154–169.

Li, K., Yu, H., Xu, Y., and Luo, X. (2022). Detection of marine oil spills based on HOG feature and SVM classifier. *Journal of Sensors* 2022: 1–11. https://doi.org/10.1155/2022/3296495.

Liang, X., Zhen, Z., Song, Y. et al. (2019). Pol-SAR based oil spillage classification with various scenarios of prior knowledge. *IEEE Access* 7: 66895–66909.

Marrone, A., Russa, D.L., Brunelli, E. et al. (2021). Antarctic fish as a global pollution sensor: metals biomonitoring in a twelve-year period. *Frontiers in Molecular Biosciences* 8: 794946.

Martinez-Vicente V., Biermann L., & Mata A. (2020). Optical Methods for Marine Litter Detection (OPTIMAL)-Final Report. Plymouth Marine Laboratory

Mejjad, N., Laissaoui, A., Fekri, A., and Hammoumi, O.E. (2023). Marine plastic pollution in Morocco: state of the knowledge on origin, occurrence, fate, and management. *Environmental Science and Pollution Research* 30: 107371–107389. https://doi.org/10.1007/s11356-023-26973-8.

Melet, A., Teatini, P., Cozannet, G.L. et al. (2020). Earth observations for monitoring marine coastal hazards and their drivers. *Surveys in Geophysics* 41: 1489–1534. https://doi.org/10.1007/s10712-020-09594-5.

Mera, D., Cotos, J.M., Varela-Pet, J. et al. (2014). Automatic decision support system based on SAR data for oil spill detection. *Computers & Geosciences* 72: 184–191. https://doi.org/10.1016/j.cageo.2014.07.015.

Mercier, G. and Girard-Ardhuin, F. (2006). Partially supervised oil-slick detection by SAR imagery using Kernel expansion. *IEEE Transacttions on Geoscience and Remote Sensing* 44: 2839–2846. https://doi.org/10.1109/tgrs.2006.881078.

Muller-Karger, F.E., Canonico, G., Aguilar, C.B. et al. (2023). Marine Life 2030: building global knowledge of marine life for local action in the Ocean Decade. *ICES Journal of Marine Science* 80: 355–357.

Sánchez-Marín, P., Schultze, F., and Besada, V. (2022). Use of limpets as alternative to mussels in metal pollution monitoring; application in the Canary Islands. *Environmental Pollution* 308: 119614.

Sciutteri, V., Pedà, C., Longo, F. et al. (2023). Integrated approach for marine litter pollution assessment in the southern Tyrrhenian Sea: Information from bottom-trawl fishing and plastic ingestion in deep-sea fish. *Marine Pollution Bulletin* 188: 114661. https://doi.org/10.1016/j.marpolbul.2023.114661.

Sepúlveda, C.H., Sotelo-Gonzalez, M.I., Osuna-Martínez, C.C. et al. (2023). Biomonitoring of potentially toxic elements through oysters (*Saccostrea palmula* and *Crassostrea corteziensis*) from coastal lagoons of Southeast Gulf of California, Mexico: health risk assessment. *Environmental Geochemistry and Health* 45: 2329–2348.

Sevgili, C., Fiskin, R., and Cakir, E. (2022). A data-driven Bayesian Network model for oil spill occurrence prediction using tankship accidents. *Journal of Cleaner Production* 370: 133478. https://doi.org/10.1016/j.jclepro.2022.133478.

Sewwandi, M., Perera, K. I., Reddy, C. M., James, B. D., Amarathunga, A. A. D., Wijerathna, I. H. K., & Vithanage, M. (2023). 7 Plastics, Pyrogenic Nurdles, Microplastics and. Maritime Accidents and Environmental Pollution-The X-Press Pearl Disaster: Causes, Consequences, and Lessons Learned, 134.

Shamsudeen, T.Y. (2020). Advances in remote sensing technology, machine learning and deep learning for marine oil spill detection, prediction and vulnerability assessment. *Remote Sensing* 12: 3416.

Song, D., Ding, Y., Li, X. et al. (2017). Ocean oil spill classification with RADARSAT-2 SAR based on an optimized wavelet neural network. *Remote Sensors* 9: 799. https://doi.org/10.3390/rs9080799.

Song, D., Wang, B., Chen, W. et al. (2018). An efficient marine oil spillage identification scheme based on an improved active contour model using fully polarimetric SAR imagery. *IEEE Access* 6: 67959–67981. https://doi.org/10.1109/access.2018.2876173.

Sousa-Guedes, D., Bessa, F., Queiruga, A. et al. (2024). Lost and found: patterns of marine litter accumulation on the remote Island of Santa Luzia, Cabo Verde. *Environmental Pollution* 344: 123338. https://doi.org/10.1016/j.envpol.2024.123338.

Topouzelis, K., Papakonstantinou, A., and Garaba, S.P. (2019). Detection of floating plastics from satellite and unmanned aerial systems (Plastic Litter Project 2018). *International Journal of Applied Earth Observation and Geoinformation* 79: 175–183.

Uzun, P., Farazande, S., and Guven, B. (2022). Mathematical modeling of microplastic abundance, distribution, and transport in water environments: a review. *Chemosphere* 288: 132517.

Wang, X., Liu, J., Zhang, S. et al. (2021). Detection of oil spill using SAR imagery based on AlexNet model. *Computational Intelligence and Neuroscience* 2021 (1): 4812979.

Wang, Y., Chen, X., and Wang, L. (2023). Cyber-physical oil spill monitoring and detection for offshore petroleum risk management service. *Scientific Reports* 13: 4586. https://doi.org/10.1038/s41598-023-30311-w.

Willems, D.J., Reeves, J.M., Morrison, P.D. et al. (2020). Trace metal biomonitoring in the east Gippsland Lakes estuary using the barnacle *Amphibalanus variegatus* and mussel *Mytilus edulis*. *Environmental Science and Pollution Research* 27: 3361–3383.

Xu, J., Wang, H., Cui, C. et al. (2019). Oil spill segmentation in ship-borne radar images with an improved active contour model. *Remote Sensing* 11: 1698.

Zettler, E.R., Mincer, T.J., and Amaral-Zettler, L.A. (2013). Life in the "plastisphere": microbial communities on plastic marine debris. *Environmental Science & Technology* 47: 7137–7146.

Zhao, J., Temimi, M., Ghedira, H., and Hu, C. (2014). Exploring the potential of optical remote sensing for oil spill detection in shallow coastal waters-a case study in the Arabian Gulf. *Optics Express* 22: 13755. https://doi.org/10.1364/oe.22.013755.

23

Impacts of Coastal and Marine Pollution on the Blue Economy: Integrating Blue Finance Perspectives

Gayithri Niluka Kuruppu[1], Nimila Dushyantha[2], Panchala Weerakoon[2], Gayan Amarasooriya[2], Titus Cooray[2], and Harshi Jayasingha[3]

[1] *Department of Industrial Management, Faculty of Business, University of Moratuwa, Moratuwa, 10400, Sri Lanka*
[2] *Department of Applied Earth Sciences, Faculty of Applied Sciences, Uva Wellassa University, Badulla, 90000, Sri Lanka*
[3] *Postgraduate Institute of Agriculture, University of Peradeniya, Peradeniya, 20400, Sri Lanka*

23.1 Introduction

Approximately 40% of the global population resides near coastal regions, with over three billion individuals relying on the oceans for their sustenance. Furthermore, a staggering 80% of worldwide trade is facilitated through maritime routes (United Nations 2022). Therefore, the global ocean economy, including activities such as fishing, aquaculture, tourism, transportation, and extractive industries (e.g. mining, oil, and gas), is essential for global economic development and climate action. However, the ocean confronts significant threats from industrial activities and natural forces, including pollution, overexploitation, and climate change, while land-based natural resource development has become a major concern nowadays (Nham 2023). To address these challenges, the concept of a blue economy has drawn global attention.

The blue economy, defined by the World Bank as the "sustainable use of ocean resources for economic growth while preserving the health of marine ecosystems," has gained a significant focus on the route to economic success (United Nations 2017). It emphasizes environmentally responsible practices to drive economic growth, enhance quality of life, promote social equity, and ensure environmental sustainability that benefits both current and future generations. This marine-based economic concept has led to a surge of interest in pioneering technologies and strategies to expand economic activities in coastal and marine environments (Frohlich et al. 2023; Jouffray et al. 2020). In 2019, the European Union's blue economy employed nearly 4.5 million individuals, contributing €184 billion in gross value added, and recorded a turnover of around €667 billion (European Commission 2023a). Despite the adverse effects of the coronavirus disease 2019 pandemic, the European Union's blue economy managed to yield a gross value of an additional €129 billion in 2020, resulting in a profit of €43.6 billion and an overall turnover of €523 billion (European Commission 2023b).

The blue economy consists of a wide range of components (Figure 23.1), each playing an essential role in exploiting the economic benefits of the coastal and marine environments. Harvesting living resources including fisheries and aquaculture is one of the main pillars of the blue economy,

Coastal and Marine Pollution: Source to Sink, Mitigation and Management, First Edition.
Edited by Meththika Vithanage, Sameera M. Samarasekara, Bryan D. James, and Christopher M. Reddy.

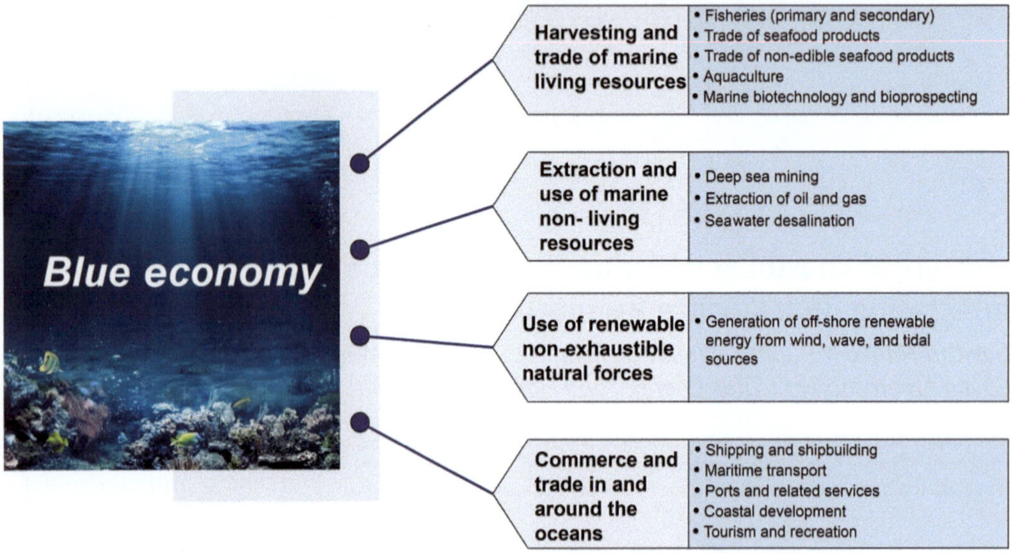

Figure 23.1 Main components of the blue economy. *Source:* Romolo Tavani/Adobe Stock.

supplying necessary proteins for a growing global population (Voyer et al. 2018). Marine biotechnology, involving the study of marine organisms for pharmaceuticals, materials, and other uses, offers eco-friendly novel processes, particularly for the valorization of waste under the zero waste concept (Venugopal 2022). Extraction and generation of energy and nonliving resources is another main pillar of the blue economy. Oil and gas extraction to produce petroleum and natural gas plays a pivotal role in achieving the global energy requirement. Emerging sectors, such as coastal and offshore renewable energies (e.g. wind farms and tidal/wave energy), are key aspects in reducing dependence on fossil fuels and combating climate change. Deep sea mining, while presenting opportunities for accessing critical metals/minerals, demands a careful regulation to prevent marine environmental damage. Seawater desalination to produce pure water also holds promise in meeting growing freshwater demands in coastal regions (Mitra et al. 2021; Voyer et al. 2018). Trade and commerce related to the ocean are also essential in the blue economy. Critical infrastructures, such as ports, shipping, and coastal development, act as the foundation for these, ensuring the smooth flow of goods and services across the maritime domain (Elisha 2019, Hossain 2019).

Marine and coastal tourisms are other significant facets of this marine-based economic concept, attracting millions to coastal destinations for recreation, relaxation, and exploration of marine ecosystems (Kabil et al. 2021). Coastal tourism accounts for approximately 30% of global tourism activities (Ghosh 2012), whereas the United Nations World Tourism Organization (UNWTO) reported that one in every two tourists specifically selects coastal destinations for their travel experiences (UNWTO 2013). This industry generates sufficient economic revenue while fostering conservation awareness. Protection and management of coastal and marine ecosystems, including hazard and habitat protection, surveillance and maritime security, and blue carbon, is also a key component in the blue economy. These pillars, though promising, must adhere to principles of resource efficiency, including reduced waste, energy conservation, quality employment, and innovative financing or technologies. Maintaining a balance between economic gains and environmental preservation is essential, which is guided by global, regional, and national policies. Therefore, the well-being of coastal and marine environments, both socially and environmentally, is essential for the success and sustainability of the blue economy.

Coastal and marine environments experience escalating pollution levels, resulting in ecosystem degradation and biodiversity loss. This also poses risks to human health, well-being, food security, and livelihoods (Tiquio et al. 2017). Prominent examples of coastal and marine pollution include major oil spills, proliferation of hypoxic dead zones due to eutrophication, microplastic pollution even in remote areas such as the Arctic and deep seas, loss of critical ecosystems such as coral reefs and seagrass meadows, climate change, and ocean acidification (Riechers et al. 2021). Despite increased awareness and knowledge, pollution rates continue to increase, requesting for more compelling interventions. Moreover, the complex and interconnected nature of these pollutions poses challenges for effective control and management. Therefore, these pollution scenarios minimize the coastal and marine ecosystem services that lead to economic losses. It indicates that balancing economic growth with the protection of coastal and marine environments is essential for a prosperous and sustainable blue economy.

By examining the wide range of impacts of pollution on key sectors such as fisheries, tourism, shipping, and renewable energy, this chapter seeks to assess the urgency of addressing environmental degradation to ensure the sustainability of the blue economy. Furthermore, this chapter aims to elucidate the role of blue finance in driving transformative change toward a more resilient and sustainable blue economy, highlighting its potential to mitigate the impacts of pollution and unlock the full economic potential of our oceans and coasts.

23.2 Impacts of the Coastal and Marine Pollution on Main Pillars of the Blue Economy

23.2.1 Impacts of the Coastal and Marine Pollution on Fisheries and Aquaculture

Currently, challenges caused by coastal and marine pollution and the lack of law enforcement to minimize them have created severe impacts on fisheries and aquaculture (Sarker et al. 2018). These impacts imposed significant threats to fisheries and aquaculture population, behaviors, reproduction, habitats, and dynamics (Ababouch and Fipi 2015, Daly et al. 2021).

These pollutions cause sea temperature rise, change fishery migration patterns, food habitat, and reproduction activities as well as changes in behaviors and growth (Pauly 2018). Moreover, these pollutions change marine habitats, including coral reefs, and sea grass, which are essential for fish feeding and breeding activities (Carlson et al. 2021; Gregg 2013). As discussed in the study by Kathijotes (2013), certain coastal development models are to shift society from scarcity to abundance through novel ways, such as ecological alterations to coastal and surface waters and contributing to nutrient inputs. That nutrient runoff induces harmful algal blooms, depleting oxygen levels and creating dead zones, adversely affecting fish populations and posing a threat to aquaculture industries (Halpern et al. 2008). These scenarios negatively impact the health and growth of the fish, and pollutants accumulate in aquaculture products, lowering product quality and marketability.

The blue revolution of aquaculture is associated with a wide range of environmental concerns, including habitat destruction, water pollution, eutrophication, biotic depletion, ecological effects, and disease outbreaks by pollutant accumulation (Ahmed and Thompson 2019). Different aquaculture species, such as shellfish (oysters and mussels), are filter feeders susceptible to contamination by induced toxins and pathogens (Impellitteri et al. 2023). Accordingly, coastal and marine pollution impose significant economic costs on the aquaculture industry, including losses due to disease outbreaks, mortalities, and declines in product quality (Hu et al. 2020). Furthermore, pollution-related closures and regulations can disrupt aquaculture operations, leading to revenue losses and increased production costs (Gephart et al. 2020).

23.2.2 Impacts of the Coastal and Marine Pollution on Extraction and Generation of Energy and Nonliving Resources

Coastal and marine pollution poses considerable challenges to the extraction and generation of energy and nonliving resources, creating significant effects on the sustainability of the blue economy. Pollution events such as oil spills in offshore energy extraction activities directly contaminate marine ecosystems while disrupting offshore drilling operations, leading to environmental damage and economic losses (Zhang et al. 2019). The impacts of chemical pollutants, including heavy metals and persistent organic pollutants, are severe as they contaminate the marine water and sediments, potentially affecting the efficiency and safety of energy extraction processes (Atoufi and Lampert 2020).

Furthermore, coastal and marine pollution significantly hinders the development and operation of renewable energy sources, such as wind, tidal, and wave energies. Accumulation of marine debris, including plastic waste and discarded fishing gear, can damage offshore renewable energy infrastructure, leading to operational inefficiencies and costly repairs (Aretoulaki et al. 2021; Bonar et al. 2015). In addition to energy extraction, marine pollution also impacts the extraction of nonliving resources, including minerals and aggregates. Pollution-induced habitat degradation, such as coral reef destruction and sediment contamination, can disrupt mineral mining operations and reduce the availability of valuable mineral resources. Furthermore, pollution-related declines in water quality can impair the health and productivity of marine ecosystems, affecting the abundance and quality of nonliving resources, such as sand, gravel, and minerals (Castelos 2017, Datta 2023).

These impacts are extensive and complex, posing significant challenges to the sustainability and resilience of the blue economy. Effective pollution mitigation and resource management strategies are essential to mitigate these impacts and ensure the long-term viability of coastal and marine resource extraction activities.

23.2.3 Impacts of Coastal and Marine Pollution on Trade and Commerce Related to the Ocean

In 2022, the estimated world seaborne trade volume reached 12 027 million tons, with over one million registered merchant vessels (Equasis Statistics 2022; UNCTAD 2023). Marine transportation includes cargo-carrying commercial shipping (i.e. tankers, container ships, and bulk carriers), noncargo commercial shipping, military, pirate, and fishing vessels (; Halpern et al. 2012). These vessels lead an estimated 33% of global fossil fuel combustion, 3.3% of CO_2 emissions, 10–15% of the anthropogenic SO_x and NO_x emissions, and 1.8 million tons of particulate matter emissions (PMe's) due to noncombustible trace constituents in the fuel (Walker et al. 2019). UNCTAD (2023) indicates that international shipping is responsible for around 3% of global greenhouse gas (GHG) emissions. A study by Wallington and Wiesen (2014) estimated that N_2O emissions are expected to increase by ~20% by 2030 due to marine transportation. From 2012 to 2018, CO_2 emissions increased by 9.6% and the share of shipping emissions increased from 2.76% to 2.89% of global anthropogenic emissions. IMO predicted that this emission may increase from about 90% of 2008 emissions in 2018 to 90–130% of 2008 emissions by 2050 for a range of plausible long-term economic and energy scenarios (Faber et al. 2021).

Due to low cost and the advantage, maritime transport will grow, and it is inevitable to introduce some pollutants into the ocean. The most serious pollutant among all types of marine pollution is the oil pollution caused due to normal navigation (e.g. cabin bilge water, oil tanker ballast water, and washing water) and shipping accidents. These accidents are categorized as structural damage (18%), severe weather conditions (16%), collision (13%), grounding (11%), fire/explosion (9%), loading/unloading (7%), and unknown causes (26%) (Grau 2017). When oil drives into the sea, due to its volatility, a thin oil film forms and makes a barrier between seawater and air; additionally, a vast amount of dissolved oxygen is absorbed by oil, causing seawater hypoxia that leads to marine life death. In the Torrey Canyon Disaster in 1967, 120 000 tons of crude oil spilled and killed thousands of seabirds and other marine organisms (Wells 2017). On 25 July 2020, the oil tanker MV Wakashio spilled more than 1000 metric tons (Mt) of fuel into the Indian Ocean, polluting the nearby coral reef on the southeast coast of the island of Mauritius as well as the surrounding beaches and lagoons (NASA Applied Sciences 2020). In 2009, 100 tons of hydrochloric acid carried a ship sunk in the Yangtze River after colliding with another vessel (Reuters 2009), and on 6 January 2018, the oil tanker SANCHI, an Iranian container ship, collided with the Chinese Hong Kong bulk carrier CF CRYSTAL in the East China Sea and caused 130 000 tons of the condensate oil spill (Ejatlas 2018). On 20 May 2021, X-Press Pearl caught fire and sunk Sri Lankan shores, which was carrying 1486 containers: 81 labeled hazardous containers included 25 Mt of nitric acid and 400 containers of nurdles. The spill of more than 50 billion pellets made this the worst plastic marine pollution event in the world (UNEP 2021).

The International Tanker Owners Pollution Federation Limited (ITOPF) is a not-for-profit organization established on behalf of the world's shipowners to promote an effective response to marine spills of oil, chemicals, and other hazardous substances. ITOPF reported that the number of large and medium tanker spills, and quantities of oil split 7 tons and more from tanker incidents and incident locations where they attended to treat since 1970–2022 (Figure 23.2) (ITOPF 2022).

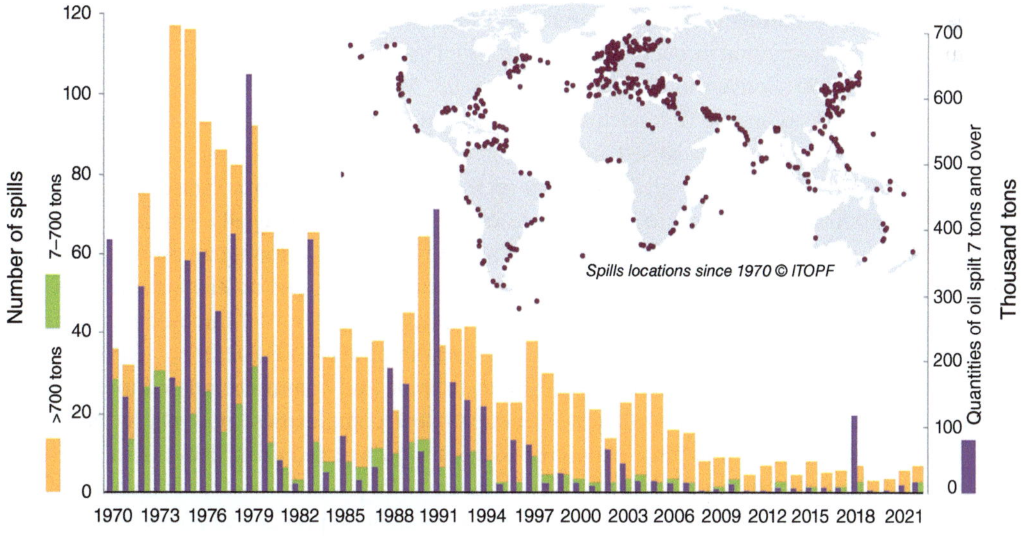

Figure 23.2 The number of large and medium tanker spills, and quantities of oil split 7 tons and more from tanker incidents and incident locations where they attended to treat since 1970–2022. *Source:* Adapted from ITOPF (2022) and ITOPF handbook (2023), London.

In contrast, ship operations, maintenance, and ship-breaking activities also have an impact on the environment, such as oily water discharge, spillage during terminal loading, ballast-water discharge from ships at ports, marine machinery exhaust, discharging harmful chemicals and plastics, industrial waste, cargo residues from bulk carriers, anti-fouling paint, rust, asbestos, sewage, organotins, bacterial contaminants, heavy metals, and possibly more substances into the ocean and at the ports (Al Fartoosi 2013; Byrnes and Dunn 2020; Iduk and Samson 2015). Newark Bay, Newark-Elizabeth Marine Terminal, the largest container shipping facility in Port of New York, and New Jersey estuary are identified as contaminated bays due to significant levels of a wide range of chemicals and water quality stressors, impacted by both industrial and urban expansion for more than 200 years (Bonnevie and Wenning 1995; USEPA 2016), and tidal action has distributed contaminants more than 10,000 acres, leading to higher maintenance costs and disruptions in port operations (NOAA 2023). Shipping and in-port emissions of Thessaloniki, Greece, average PM2.5: 9–13% on both primary and secondary particles (Saraga et al. 2019); moreover, oil spills and untreated wastewater in the port region resulted in environmental damage, necessitating extensive cleanup efforts and affecting the efficiency of port services. Fauziah et al. (2021) reported that marine debris contributing 30–45% of the total waste can be found on the beaches of the Straits of Malacca and the South China Sea, Malaysia, which directly affects the operations of Port Klang and vessels accessed on Straits of Malacca. These examples illustrate the diverse challenges that pollution poses to port operations globally, emphasizing the urgent need for comprehensive mitigation and management strategies.

MARPOL – the International Convention for the Prevention of Pollution from Ships – developed regulatory measures for decarbonizing the shipping industry and reducing GHG emissions from ships (IMO 2018). The short-term decarbonization measures include the Energy Efficiency Existing Ship Index (EEXI) and the Carbon Intensity Indicator (CII) under Annex VI of MARPOL, which was implemented from 2023 onwards and the Energy Efficiency Design Index (EEDI) and the Ship Energy Efficiency Management Plan (SEEMP) were closer to finalization and implemented in future (UNCTAD 2023). Moreover, alternative fuels, dual-fuel vessels, energy-efficient technologies, and slow sailing-like action have been implemented to maintain low-emission shipping, and regulations have been implemented for ports to provide low-emission energy supply infrastructure to reduce the environmental impact (UNCTAD, 2023). The Civil Liability for Bunker Oil Pollution Damage (BUNKER) Convention was adopted on 23 March 2001, and entry into force on 21 November 2008, to ensure that adequate, prompt, and effective compensation is available to persons who endure damage caused by spills of oil, when carried as fuel in ships' bunkers (Griggs 2001). In 2021, IMO adopted a strategy on marine plastic litter from ships, and in March 2022, UNEP adopted a resolution to start work toward an international legally binding instrument to end plastic pollution. COP26, Clydebank Declaration for green shipping zero-emission corridors, aimed to establish maritime routes between 2 (or more) ports and at least 6 green corridors by 2025 (UN-SDGS 2021). Furthermore, the Dhaka–Glasgow Declaration imposed a call for the IMO to work on a mandatory GHG levy on international shipping (CVF 2021).

In future, emerging issues pose additional threats to trade and commerce related to the ocean. Escalating pollution and climate change introduces new complexities, potentially leading to heightened disruptions in shipping, economic losses from damaged coastal infrastructure, and strained global supply chains. Addressing these challenges requires a forward-looking approach, focusing on innovative pollution mitigation technologies and comprehensive policy development to ensure the resilience of ocean-related economic activities. Coastal and marine pollution demands immediate attention for the sustainability of trade and commerce linked to the ocean. The interconnected nature of these challenges requires a holistic approach, integrating research findings, and case studies to inform robust policy development. Ongoing efforts must prioritize mitigating the impact on shipping, safeguarding efficient maritime transport, and addressing the vulnerabilities of port operations. The imperative to strike a balance between economic progress and ecological resilience in coastal developments remains significant. As we navigate the future, proactive measures are crucial to address emerging issues and potential scenarios, ensuring the long-term viability of trade and commerce related to the ocean. Emphasizing the importance of ongoing research and comprehensive policy frameworks becomes paramount in steering ocean-related commerce toward a resilient and environmentally sustainable future.

23.2.4 Impacts of Coastal and Marine Pollution on Tourism

Tourism is one of the main fast-growing global economic activities in coastal areas. Most tourists visit for sun, sand, sea, and beach-related activities, such as swimming, sunbathing, coastal walks, and wildlife watching. Additionally, they engage in sea-related activities, such as whale watching, sailing, scuba diving, surfing, and cruising. The Mediterranean, Caribbean, and Southeast Asia are one of the most popular tourism regions worldwide (UNEP 2009). According to the World Tourism Organization (UNWTO), the majority of tourists visiting coastal regions, estimated at 975 million, traveled internationally between January and September 2023. The revenue generated is expected to reach US$ 152.3 billion by the end of 2023. In the long term, the beach tourism market is projected to reach around US$ 265.1 billion in 2033 (UNWTO 2023; Shah 2023). Due to the rapid development in beach tourism, the volume and the concentration of tourists in coastal areas have dynamic impacts, either directly or indirectly, on coastal regions through land transformation, natural resource utilization, greenhouse gas emissions, sewage, and solid waste generation (Gössling et al. 2018; Zahedi 2008). Moreover, tourism infrastructure development can transform the entire coastal regions into urbanized coastal zones with hotels, restaurants, and ports, leading to an exponentiated population density in these areas due to tourism-associated businesses and employment (Mandić et al. 2018; Gormsen 1997; Petrova et al. 2018).

Due to the high population density, water bodies are rapidly polluted by sewage and solid waste, generating organic matter and nutrients that significantly impact algal growth and eutrophication. For example, the Mexican Caribbean has faced severe changes due to the massive arrival of *sargassum* algae since 2011 (McAdam-Otto 2023). The study by Mohamed (2018) states that five harmful algal blooms were recorded in the Red Sea, where three blooms formed by dinoflagellates (*Noctiluca scintillans*, *Pyrodinium bahamense*, and *Protoperidinium quinquecorne*), one by raphidophytes (*Heterosigma akashiwo*), and one by cyanobacteria (i*Trichodesmium erythraeum*). The study by Thanakorn Jiwarungrueangkul et al. (2024) reported phytoplankton blooms (*Chaetoceros and Eunotogramma*) in urban coastal waters off the coast of the Andaman Sea in Patong Bay, Phuket, Thailand. These harmful algal blooms directly affect some of the most popular tourism regions in the world. Not only algae but also surging populations of species, such as jellyfish, can prevent bathing and swimming activities or negatively affect tourism (Gershwin et al. 2010; Ruiz-Frau 2022). In contrast, most tourists visit tropical coastal environments to observe coral reefs. Due to global climate change, increasing acidification and higher sea temperatures result in coral bleaching and the emergence of new diseases (Riegl et al. 2009). They also point out that 20% of the world's coral reefs have already been lost, and another 50% are at an immediate risk of extinction, causing environmental habitat changes and indirectly affecting the tourism industry economically.

One of the main pillars of marine tourism and travel is the cruise industry, forecasting a 106% increase from 2019 levels in 2023, with 31.5 million passengers sailing, and predicting an increase to 39.5 million by 2027 (CLIA 2023). The studies by Herz (2002) and Butt (2007) show that cruise ships constitute less than 1% of the global merchant fleet. However, their contribution to sewage and waste in the ocean is 25%, owing to higher traveler density. Approximately, 50 tons of sewage are produced per day from an average cruise ship, along with considerable discharges of grey water (from kitchens, laundries, and showers) into the ocean, containing significant amounts of organic matter/nutrients conducive to algal growth and eutrophication (Andersson et al. 2016; Carić 2011). According to MARPOL regulations 73/78 Annex IV, the discharge of sewage into the sea is allowed if the ship is discharging comminuted and disinfected sewage using an approved system under local jurisdiction, at a distance of more than three nautical miles from the nearest land. Ships can discharge sewage that is not comminuted or disinfected at a distance of more than 12 nautical miles from the nearest land.

Another main issue is plastic solid waste dumped into the ocean. These plastics produce microplastics, which directly affect biodiversity and the degradation of natural habitats. Jambeck et al. (2015) calculated, based on worldwide data on solid waste, population density, and economic status, that 275 million Mt of plastic waste were generated in 192 coastal countries in 2010, with 4.8–12.7 million Mt entering the ocean. Law et al. (2020) showed that in 2016, the United States alone generated 42.0 Mt of plastic, and 0.14–0.41 Mt of this waste was illegally dumped into coastal environments, contributing among the highest rates globally. Once plastic reaches the oceans, it forms floating waste, washes up on coastlines, and accumulates on the sea floors. These floating plastics are ingested by fish, dolphins, whales, turtles, and seabirds. Most of the floating waste washes back to the coastlines and pollutes entire coastal areas, affecting fisheries, tourism industries, and the blue economy. To prevent such incidents, some management policies should be implemented. Environmental Impact Assessments, Environmental Management Plans/Systems, and Certification/Award Schemes can be used for command and control by introducing laws and regulations for coastal and marine tourism. For management and economic approaches, taxes, permits, rights, subsidies, grants, deposit-refund schemes, and product and service charges can be applied. Nevertheless, community programs, guidelines, treaties and agreements, research, and education need to be conducted to raise awareness of these issues among the local community and tourists (Craig-Smith et al. 2006; Gössling et al. 2015).

After assessing these impacts, it becomes evident that urgent action is needed to mitigate these challenges. Transitioning from understanding the economic toll of pollution to exploring solutions, such as blue finance, determine the necessity for coordinated efforts to safeguard the oceans' future.

23.3 The Economic Toll of Marine and Coastal Pollution on the Blue Economy

The blue economy, a complex network of economic activities connected to the global oceans and coastlines, has become a critical force of global prosperity. Covering sectors, such as fisheries, shipping, tourism, and renewable energy, and their contribution to employment, trade, and overall economic growth is indispensable. However, the escalating issue of coastal and marine pollution poses a severe threat to the sustainability and economic viability of the blue economy. Plastics, chemicals, and waste are identified as primary causes of the deterioration of marine ecosystems, with profound economic consequences. The economic toll of plastic pollution alone is staggering, estimated at a loss of $330 billion annually (UNEP 2022). Plastics not only act as stressors but also interact synergistically with other threats, causing the damage exponentially (Backhaus and Wagner 2020). Impacts of plastic pollution on tourism, as the mortality of marine life and the degradation of critical habitats, reduced catches, increased fishing costs, and diminished market value (English et al. 2019). It results in direct costs, such as repair expenses and lost productive time, and indirect costs on local governments and volunteer organizations involved in removal efforts (English et al. 2019, IUCN 2023, Just One Ocean 2023).

Fisheries, a linchpin of the blue economy, face a steady decline in stocks due to the adverse effects of marine pollution. Lost and derelict fishing gear contributes to economic losses, both in terms of degraded habitats and financial costs for fishers (US Department of Commerce 2023). Tourism also faces substantial challenges due to marine and coastal pollution. Coastal and marine pollution results in beach closures, deteriorating water quality, and damaged marine ecosystems. These impacts translate into declining tourism revenues, job losses, and reduced income for businesses in the hospitality and tourism sector, adversely affecting local economies and employment (IUCN 2022).

In the shipping industry, the vessel pollution, including oil spills and invasive species discharge, leads to substantial costs in cleanup operations, fines, legal liabilities, and increased insurance premiums. These impacts can potentially raise consumer prices for goods, impacting the profitability of shipping companies and affecting global trade dynamics. The promising shift toward renewable energy sources in the blue economy, such as offshore wind farms and ocean thermal energy conversion, encounters setbacks due to coastal and marine pollution. These pollutions threaten the economic viability of renewable energy initiatives in the blue economy, causing reduced energy production, escalating maintenance costs, and hindering industry growth. Increased insurance premiums, trade challenges, and delays in transition goals further compound the economic consequences, emphasizing the critical need for pollution mitigation strategies (Pires Manso et al. 2023).

Coastal and marine pollution results in indirect economic consequences through public health costs. Contaminated seafood exposes communities to harmful chemicals and toxins, leading to increased medical expenses and loss of productivity (Awuchi and Awuchi 2019). Coastal property values, intrinsically tied to the state of the surrounding environment, suffer from marine pollution, causing financial losses for homeowners and local governments and reducing property tax revenue. These economic impacts of coastal and marine pollution on the blue economy are profound and multifaceted. From fisheries to tourism, shipping, and renewable energy, the detrimental effects resound across sectors, posing a substantial threat to global prosperity. Urgent and coordinated efforts are essential to mitigate pollution and protect the economic pillars of the blue economy.

23.4 Blue Finance for Investing in a Sustainable Maritime Future

23.4.1 Foundations of Blue Finance

Blue finance, providing investments, financial tools, and policies directed toward sustaining marine ecosystems, promoting responsible resource utilization, and ensuring the vitality of coastal communities, is a vital part of the blue economy. Blue finance is defined as the innovative use of financial mechanisms to support ocean-related activities, aiming to reconcile economic development with environmental conservation. Blue finance initiatives encompass a broad variety of activities aimed at promoting ocean conservation, sustainable fisheries, coastal protection, renewable energy, and innovative technology development. The primary objectives revolve around channeling investments to projects that foster marine biodiversity, mitigate pollution, promote sustainable fishing practices, and support coastal resilience against climate change-induced challenges (Figure 23.3).

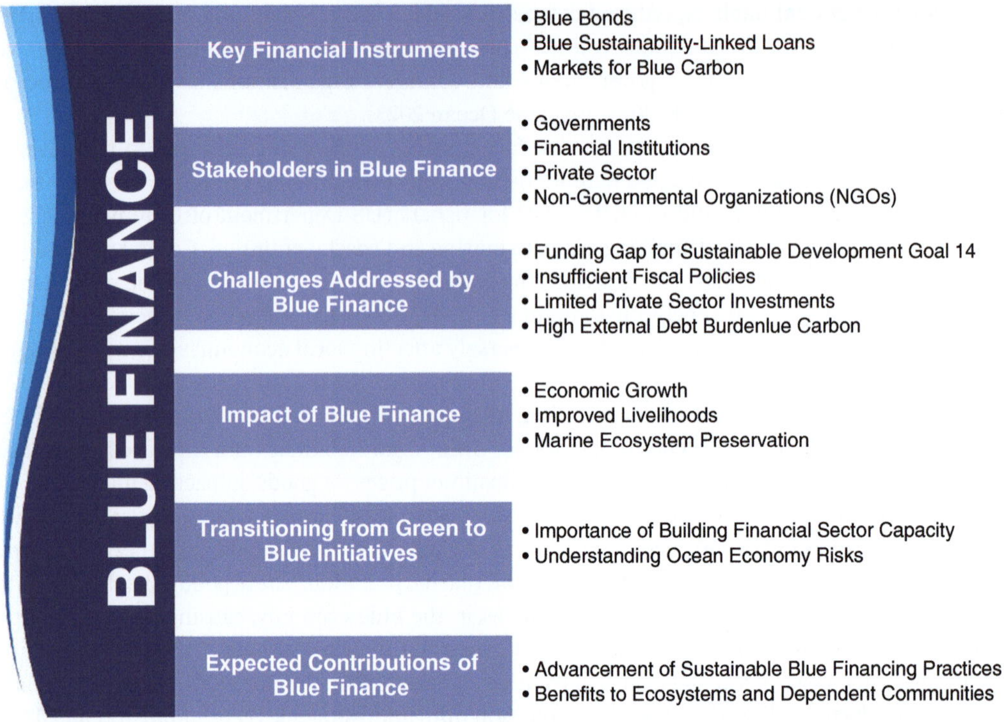

Figure 23.3 The main components of blue finance. *Source:* Kuruppu and Dushyantha (2023)/University of Moratuwa.

Several financial instruments drive the implementation of blue finance. Blue bonds, for instance, operate as debt instruments raised by governments, development banks, or private entities to fund marine- and ocean-based projects. These bonds attract impact investors seeking both financial returns and positive environmental impacts. Additionally, blended finance models combine public and private capital to finance initiatives such as marine conservation, sustainable aquaculture, and coastal infrastructure development (IFC 2022; Khan 2023). The concept of blue finance gained traction in the early twenty-first century as global concerns about the degradation of marine environments escalated. Initially, financial resources allocated for marine and coastal conservation were limited. However, the landmark development of the "Sustainable Blue Economy Finance Principles" in 2018 at the Economist World Ocean Summit signaled a pivotal shift. These principles, formulated by collaborative efforts among the European Committee, the World-Wide Fund for Nature (WWF), the European Investment Bank, and the International Sustainability Unit, provided a foundational framework for sustainable financing initiatives in ocean-related ventures (Shiiba et al. 2022). Since then, there has been significant progress in ocean-oriented financing in recent years, as shown in Tables 23.1 and 23.2.

23.4.2 Addressing the Financing Gap in Blue Finance

Blue finance aligns closely with the United Nations Sustainable Development Goals (SDGs), particularly SDG 14 – Life Below Water. By promoting responsible economic growth, reducing marine pollution, conserving marine ecosystems, and enhancing sustainable resource utilization, blue finance directly contributes to achieving these global goals (Morgan et al. 2022). To meet the 10%

Table 23.1 Summary of recent blue finance practices.

Agent	Name	Year	Scale	Duration
World Bank	PRO BLUE Umbrella 2.0	2018	USD 29 million as of fiscal 2019	
Asian Development Bank (ADB)	Oceans Financing Initiatives	2019	USD 5 billion	5 years
The Nature Conservancy (TNC)	Blue Bonds for Conservation	2019	USD 1.6 billion	5 years
Mirova	Sustainable Ocean Fund	2020	USD 132 million	
Swen Capital Partners, Ifreme, and BpiFrance	Blue Ocean Fund	2021	€ 120 million	
Seventure Partners	Blue Forward Fund	2022	€ 130 million	
Secretary of State for the Sea	Maritime Intervention Fund–FIM	2022	€ 15 million	
CPR Asset Management	CPR Invest Blue Economy	2023		
GO Capital	Impact Ocean Capital Fund	2023	€ 70 million	

Source: Adapted from Berger et al. (2019); Piñeiro-Antelo et al. (2019): Bos et al. (2015); World Bank (2018); Bouillon-Duparc (2023).

Table 23.2 Summary of recent blue bond data.

Issuer	Year	Amount (USD)	Coupon	Redemption	Involved organizations
Seychelles Government	2018	15 million	6.50%	10 years	TNC, World Bank, GEFSeyCCAT and DBS
Nordic Investment Bank (NIB)	2019	200 million	0.37%	5 years	
World Bank	2019	10 million	2019: 2.35% 2020: 2.70% 2021: 3.15%	3 years	Morgan Stanley

Source: Adapted from Berger et al. (2019); Bos et al. (2015); Bouillon-Duparc (2023); Piñeiro-Antelo et al. (2019); World Bank (2018).

protected area target (SDG 14.5), requiring USD$7.7 billion annually for marine protected areas (MPAs), the current investment falls significantly short, being 20–30 times lower. Looking ahead to the global biodiversity framework's goal of 30% protected areas by 2030 suggests an even larger financing gap of approximately USD$27.7 billion annually (Narula et al. 2023). This highlights the insufficiency of government fiscal policies, encompassing taxes, subsidies, and budget allocations, to fully address the financing requirements. Therefore, attracting private capital becomes necessary to bridge this financial gap and facilitate the transition toward blue economies. There is a notable opportunity to exploit and increase the private sector's interest in marine conservation and the blue economy (Morgan et al. 2022).

The impact of blue finance initiatives manifests in various dimensions. Investments in marine conservation and sustainable fisheries support biodiversity, protect endangered species, and restore degraded habitats. Moreover, supporting coastal communities through sustainable tourism and infrastructure development fosters economic growth while preserving natural resources. However, challenges still persist. Limited awareness and understanding of blue finance concepts among stakeholders hinder its widespread adoption. Furthermore, establishing effective governance structures, ensuring transparency in fund allocation, and accurately measuring the impact of investments pose ongoing challenges. In addition, these are some of the challenges faced when transitioning to a blue economy (Narula et al. 2023).

23.4.3 Challenges to Implement Blue Finance Initiatives to Achieve Blue Economy

The absence of comprehensive frameworks, global standards, and consistent methodologies in guiding investors poses a significant challenge to successfully launch blue finance initiatives. Despite several proposed frameworks, such as the blue economy, sustainability framework, and various guidelines by institutions, such as the World Bank and International Finance Corporation, challenges persist due to divergent mechanisms, inadequate information, and poor awareness. The lack of consensus on blue financing definitions, regulatory mechanisms, and metrics for measuring investment impact results in inconsistencies, transparency issues, and credibility concerns. Harmful subsidies in maritime sectors contribute to significant overfishing, illegal fishing, and negative impacts on ocean health. Estimated global annual fishing subsidies of US$35 billion in 2018 and substantial fossil fuel subsidies exacerbate these issues, affecting marine pollution. Capacity-enhancing fishery subsidies lead to risks for fish stocks, species extinction, and economic inequality, and threaten livelihoods and food security for fishing communities. These subsidies counteract conservation efforts and limit incentives for blue economy investments.

Despite increasing finance flows, there is a scarcity of investment-grade projects aligned with blue economy principles and guidelines. Maritime projects often generate poor financial returns, lack appropriate size, possess high risk-return ratios, and struggle to attract adequate capital. Limited stakeholder awareness and understanding of the blue economy, coupled with challenges in project bankability due to poor financial viability and governance issues, contribute to this problem. Commercial lenders experience challenges, such as inadequate profits, limited investment-ready projects, higher risks due to information gaps, and insufficient financial resources, reducing their propensity to fund blue economy projects. Most voluntary blue economy investment commitments stem from NGOs and governments, with minimal contributions from private institutions or public-private partnerships. A significant percentage of investors have not evaluated their portfolios for ocean sustainability impacts, indicating a lack of assessment and awareness regarding ocean-related investment risks.

Insufficient data availability, technical constraints, and capacity limitations hinder the transition to the blue economy. Issues, such as inaccurate marine surveys, limited technical training, and poor political support, in prioritizing investments impede progress. Uncertainty around coastal/marine responsibilities, jurisdiction, and property rights, coupled with inadequate expertise in the latest financing mechanisms and entrepreneurship, also act as barriers to financing initiatives for the blue economy. The trajectory of blue finance suggests an immense potential for growth and evolution. Strengthening partnerships between public and private sectors, raising awareness about the importance of ocean conservation, and integrating innovative financial tools into policy frameworks are critical steps toward realizing the full potential of blue finance.

23.5 Conclusion

As studied in this chapter, the severe impacts of pollution on key sectors, such as fisheries, tourism, and maritime transportation, are profound and far-reaching. Therefore, these pollutions negatively impact the health of marine ecosystems, while causing a significant economic cost, undermining the sustainable development goals of both coastal communities and nations. However, among these challenges, blue finance shows a promising solution to overcome these enduring challenges. By leveraging innovative financial mechanisms and investment strategies, blue finance holds the potential to drive transformative change toward a more sustainable and resilient blue economy. Through the strategic allocation of resources, the collaboration between stakeholders, and the promotion of blue technologies and practices, we can mitigate the impacts of pollution while unlocking the full economic potential of our oceans and coasts.

In the context of the blue economy in the twenty-first century, we must recognize the urgent need for concerted action. By addressing the root causes of coastal and marine pollution, fostering responsible stewardship of our marine resources, and embracing the principles of sustainability and inclusivity, we can navigate toward a brighter, more prosperous future for generations to come.

Acknowledgement

This research was financially supported by the Senate Research Committee (SRC) of the University of Moratuwa, Sri Lanka under grant no SRC/LT/2024/08.

References

Ababouch, L. and Fipi, F. (2015). Fisheries and aquaculture in the context of blue economy. *Feeding Africa* 2 (21–23): 13.

Ahmed, N. and Thompson, S. (2019). The blue dimensions of aquaculture: a global synthesis. *Science of the Total Environment* 652: 851–861.

Al Fartoosi, F.M. (2013). The impact of maritime oil pollution in the marine environment: a case study of maritime oil pollution in the navigational channel of Shatt Al-Arab. World Maritime University Dissertations, 318.

Andersson, K., Brynolf, S., Lindgren, J., and Wilewska-Bien, M. (ed.) (2016). *Shipping and the environment: improving environmental performance in marine transportation*. Berlin Heidelberg: Springer.

Aretoulaki, E., Ponis, S., Plakas, G., and Agalianos, K. (2021). Marine plastic littering: a review of socio-economic impacts. *J. Sustain. Sci. Manag* 16 (3): 277–301.

Atoufi, H.D. and Lampert, D.J. (2020). Impacts of oil and gas production on contaminant levels in sediments. *Current Pollution Reports* 6: 43–53.

Awuchi, C.G. and Awuchi, C.G. (2019). Impacts of plastic pollution on the sustainability of seafood value chain and human health. *International Journal of Advanced Academic Research* 5 (11): 46–138.

Backhaus, T. and Wagner, M. (2020). Microplastics in the environment: much ado about nothing? A debate. *Global Challenges* 4 (6): 1900022.

Berger, M.F., Caruso, V., and Peterson, E. (2019). An updated orientation to marine conservation funding flows. *Marine Policy* 107: 103497.

Bonar, P.A.J., Bryden, I.G., and Borthwick, A.G.L. (2015). Social and ecological impacts of marine energy development. *Renewable and Sustainable Energy Reviews* 47: 486–495.

Bonnevie, N.L. and Wenning, R.J. (1995). Sources of pollution and sediment contamination in Newark Bay, New Jersey. *Ecotoxicology and Environmental Safety* 30 (1): 85–100.

Bos, M., Pressey, R.L., and Stoeckl, N. (2015). Marine conservation finance: the need for and scope of an emerging field. *Ocean & Coastal Management* 114: 116–128.

Bouillon-Duparc, H. (2023). Blue finance is thriving, Available at: https://learnandconnect.pollutec.com/en/blue-finance-is-thriving/ (accessed 25 December 2023).

Butt, N. (2007). The impact of cruise ship generated waste on home ports and ports of call: a study of Southampton. *Marine Policy* 31 (5): 591–598.

Byrnes, T.A. and Dunn, R.J.K. (2020). Boating- and shipping-related environmental impacts and example management measures: a review. *Journal of Marine Science and Engineering* 8 (11): 908. https://doi.org/10.3390/jmse8110908.

Carić, H. (2011). Cruising tourism environmental impacts: case study of Dubrovnik, Croatia. *Journal of Coast Research* 61: 104–113.

Carlson, R.R., Evans, L.J., Foo, S.A. et al. (2021). Synergistic benefits of conserving land-sea ecosystems. *Global Ecology and Conservation* 28: e01684.

Castelos, M.A. (2017). Non-living marine resources. *Spanish Yearbook of International Law* 21: 345–361.

CLIA (2023). State of the Cruise Industry 2023. Available at: https://cruising.org/-/media/clia-media/research/2023/2023-clia-state-of-the-cruise-industry-report_low-res.ashx (accessed 4 February 2024).

Craig-Smith, S.J., Tapper, R., and Font, X. (2006). The coastal and marine environment. In: *Tourism and Global Environmental Change: Ecological, Economic, Social and Political Interrelationships* (ed. S. Gössling and C.M. Hall), 107–127. London: Routledge.

CVF (2021). *Dhaka-glasgow Declaration of the Climate Vulnerable Forum*. Climate Vulnerable Forum org.

Daly, J., Knott, C., Keogh, P., and Singh, G.G. (2021). Changing climates in a blue economy: assessing the climate-responsiveness of Canadian fisheries and oceans policy. *Marine Policy* 131: 104623.

Datta, M.G. (2023). Pollution in marine ecosystem: impact and prevention. In: *Current Status of Marine Water Microbiology* (ed. R. Soni, D.C. Suyal, L. Morales-Oyervides, and M. Fouillaud), 169–191. Springer.

Ejatlas (2018). Iranian tanker disaster and oil spill in East China Sea, China. Ejatlas, 14 June 2023. Available at: https://ejatlas.org/conflict/oil-spill-of-east-china-sea-in-2018china (accessed 5 January 2024).

Elisha, O.D. (2019). The Nigeria blue economy: prospects for economic growth and challenges. *International Journal of Scientific Research in Education* 12 (5): 680–699.

English, E., Wagner, C., and Holmes, J. (2019). The effects of marine debris on beach recreation and regional economies in four coastal communities: A regional pilot study. Final Report.

Equasis Statistics (2022). The world merchant fleet Report - statistics from Equasis. European Maritime Safety Agency. Available at: https://emsa.europa.eu/csn-menu/download/7697/472/23.html (accessed 5 January 2024).

European Commission (2023a). BlueInvest: new investor report features an ocean of investment opportunities in the sustainable blue economy [online]. Directorate-General for Maritime Affairs and Fisheries. Available from: https://oceans-and-fisheries.ec.europa.eu/news/blueinvest-new-investor-report-features-ocean-investment-opportunities-sustainable-blue-economy-2023-03-09_en (accessed 10 October 2023).

European Commission (2023b). European blue economy report 2023: economic crisis takes a toll but doesn't stop the growth [online]. Directorate-General for Maritime Affairs and Fisheries. Available at: https://oceans-and-fisheries.ec.europa.eu/news/european-blue-economy-report-2023-economic-crisis-takes-toll-doesnt-stop-growth-2023-05-24_en (accessed 9 October 2023).

Faber, J. et al. (2021). Fourth IMO GHG Study 2020. International Maritime Organization (IMO).

Fauziah, S.H., Rizman-Idid, M., Cheah, W. et al. (2021). Marine debris in Malaysia: A review on the pollution intensity and mitigating measures. *Marine Pollution Bulletin* 167: 112258. https://doi.org/10.1016/j.marpolbul.2021.112258. Epub 2021 Apr 8. PMID: 33839567.

Frohlich, M., Fidelman, P., Dutton, I. et al. (2023). A network approach to analyse Australias blue economy policy and legislative arrangements. *Marine Policy* 151: 105588.

Gephart, J.A., Golden, C.D., Asche, F. et al. (2020). Scenarios for global aquaculture and its role in human nutrition. *Reviews in Fisheries Science & Aquaculture* 29 (1): 122–138.

Gershwin, L.A., Nardi, M.D., Winkel, K.D., and Fenner, P.J. (2010). Marine stingers: review of an under-recognized global coastal management issue. *Coastal Management* 38 (1): 22–41. https://doi.org/10.1080/08920750903345031.

Ghosh, T. (2012). Sustainable coastal tourism: problems and management options. *Journal of Geography and Geology* 4 (1): 163–169.

Gormsen, E. (1997). The impact of tourism on coastal areas. *GeoJournal* 42: 39–54.

Gössling, S., Hall, C.M., and Scott, D. (2015). *Tourism and Water.* Clevedon: Channel View Publications.

Gössling, S., Hall, C.M., and Scott, D. (2018). Coastal and ocean tourism. In: *Handbook on Marine Environment Protection: Science, Impacts, and Sustainable Management* (ed. M. Salomon and T. Markus), 773–790. Cham, Switzerland: Springer.

Grau, T.F.D. (2017). Analysis of global marine environmental pollution and prevention and control of marine pollution. Master Thesis. Universitat Politècnica de Catalunya.

Gregg, K.L. (2013). 'Literature review and synthesis of land-based sources of pollution affecting essential fish habitats in southeast Florida. '*NOAA Fisheries Southeast Regional Office*, West Palm Beach FL, 1–55.

Griggs, P. (2001). International Convention on Civil Liability for Bunker Oil Pollution Damage. Available at: https://www.bmla.org.uk/documents/imo-bunker-convention.doc (accessed 15 January 2024).

Halpern, B.S., Hengl, T. and Groll, D., (2012). 'Shipping density (commercial).' A Global Map of Human Impacts to Marine Ecosystems, showing relative density (in color) against a black background. Scale, 1.

Halpern, B.S., Walbridge, S., Selkoe, K.A. et al. (2008). A global map of human impact on marine ecosystems. *Science* 319 (5865): 948–952.

Herz, M. (2002). *Cruise Control. A Report on How Cruise Ships Affect the Marine Environment.* Washington: The Ocean Conservancy.

Hossain, M.M. (2019). Future importance of healthy oceans: ecosystem functions and biodiversity, marine pollution, carbon sequestration, ecosystem goods and services. *Journal of Ocean and Coastal Economics* 6 (2): 4.

Hu, Z., Li, R., Xia, X. et al. (2020). A method overview in smart aquaculture. *Environmental Monitoring and Assessment* 192: 1–25.

Iduk, U. and Samson, N. (2015). Effects and solutions of marine pollution from ships in Nigerian Waterways. *International Journal of Science & Engineering Research* 6 (9): 81–90.

IFC (2022). Blue finance safeguarding marine ecosystems, the climate, and livelihoods. Available at: https://www.ifc.org/en/stories/2022/blue-finance-marine-ecosystems-climate-livelihoods. (accessed 10 January 2024).

IMO (2018). International Convention for the Prevention of Pollution from Ships (MARPOL). 7 Feb 2018, International Maritime Organization. Available at: http://www.imo.org/en/about/

conventions/listofconventions/pages/international-32-convention-for-the-prevention-of-pollution-from-ships-(marpol).aspx (accessed 10 January 2024).

Impellitteri, F., Multisanti, C.R., Rusanova, P. et al. (2023). Exploring the impact of contaminants of emerging concern on fish and invertebrates physiology in the mediterranean Sea. *Biology* 12 (6): 767.

ITOPF 2022. ITOPF Handbook 2022/23, London, p. 60. https://www.itopf.org/fileadmin/uploads/itopf/data/Documents/Company_Lit/ITOPF_Handbook22_web.pdf

IUCN (2022). Summary on the economic impact of marine plastic pollution in antigua and barbuda, grenada & saint lucia [online]. International Union for Conservation of Nature (IUCN). Available at: https://www.iucn.org/sites/default/files/2023-03/summary-on-the-economic-impact-of-marine-plastic-pollution-in-antigua-and-barbuda-grenada-and-saint-lucia.pdf (accessed 23 December 2023).

IUCN (2023). The economic impact of plastic pollution in grenada [online]. International Union for Conservation of Nature (IUCN). Available from: https://www.iucn.org/sites/default/files/2023-10/grenada-summary-economic-impact-marine-plastics.pdf (accessed 12 December 2023).

Jambeck, J.R., Geyer, R., Wilcox, C. et al. (2015). Plastic waste inputs from land into the ocean. *Science* 347: 768–771. https://doi.org/10.1126/science.1260352.

Jiwarungrueangkul, T., Kongpuen, O., Sangmanee, C. et al. (2024). Weekly variations of nutrients and their associations with phytoplankton blooms in the urban coastal waters of Andaman Sea coast: a case study in Patong Bay, Phuket, Thailand. *Marine Pollution Bulletin* 198: 115824. https://doi.org/10.1016/j.marpolbul.2023.115824.

Jouffray, J.-B., Blasiak, R., Norström, A.V. et al. (2020). The blue acceleration: the trajectory of human expansion into the ocean. *One Earth* 2 (1): 43–54.

Just One Ocean 2023. Plastic pollution facts [online]. Available from: https://justoneocean.org/ocean-plastic-facts (accessed 11 November 2023).

Kabil, M., Priatmoko, S., Magda, R., and Dávid, L.D. (2021). Blue economy and coastal tourism: a comprehensive visualization bibliometric analysis. *Sustainability*.

Kathijotes, N. (2013). Keynote: blue economy-environmental and behavioural aspects towards sustainable coastal development. *Procedia-Social and Behavioral Sciences* 101: 7–13.

Khan (2023). Maritime Law, Blue Economy, and Blue Finance elements of sustainable marine development. https://www.linkedin.com/pulse/maritime-law-blue-economy-finance-elements-sustainable-khan (accessed 11 November 2023).

Kuruppu, G.N. and Dushyantha, N. (2023). Unlocking the blue economy: financing a sustainable ocean future. *Bolgoda Plains Research Magazine* 3 (2): Published by University of Moratuwa, Sri Lanka.

Law, K.L., Starr, N., Siegler, T.R. et al. (2020). The United States contribution of plastic waste to land and ocean. *Science Advance* 6 (44): eabd0288. https://doi.org/10.1126/sciadv.abd0288.

Mandić, A., Mrnjavac, Ž., and Kordić, L. (2018). Tourism infrastructure, recreational facilities and tourism development. *Tourism and Hospitality Management* 24 (1): 41–62.

McAdam-Otto, L.K. (2023). "It's all about the beaches": Sargassum algae, tourism, and coastal transformations along the Mexican Caribbean. *Coastal Studies & Society* 2 (4): 356–377.

Mitra, A., Gobato, R., Saha, A. et al. (2021). Non-living resources of the sea as the pillars of Blue Economy. *Parana Journal of Science and Education* 7 (3): 28–34.

Mohamed, Z.A. (2018). Potentially harmful microalgae and algal blooms in the Red Sea: current knowledge and research needs. *Marine Environmental Research* 40: 234–242. https://doi.org/10.1016/j.marenvres.2018.06.019.

Morgan, P.J., Huang, M.C., Voyer, M. et al. (2022). *Blue Economy and Blue Finance: Towards Sustainable Development and Ocean Governance*. Asian Development Bank Institute.

Narula, K., Spaiding, M. J., Thiele, T., Dyer, J., Pouponneau, A., & Chauhan, P. (2023). 'Generating finance for blue economy transition.' T20 policy brief.

NASA Applied Sciences (2020). 'OIL SPILLS: Mauritius Oil Spill 2020.'

Nham, N.T.H. (2023). The role of financial development in improving marine living resources towards sustainable blue economy. *Journal of Sea Research* 195: 102417.

NOAA (2023). Lower Passaic River and Greater Newark Bay: Hazardous Waste Site (New Jersey) 1940s to present What Happened? October 17, 2023. Available at: https://darrp.noaa.gov/hazardous-waste/lower-passaic-river-and-greater-newark-bay. (accessed 5 January 2024).

Pauly, D. (2018). A vision for marine fisheries in a global blue economy. *Marine Policy* 87: 371–374.

Petrova, M., Dekhtyar, N., Klok, O., and Loseva, O. (2018). Regional tourism infrastructure development in the state strategies. *Problems and Perspectives in Management* 16 (4): 259–274.

Piñeiro-Antelo, M., Villa, F.R.D., and Santos, X.M. (2019). ODA in Galicia (Spain). The importance of the fisheries sector and the cultural priority. *Marine Policy* 107: 103455.

Pires Manso, J.R., Martínez Vázquez, R.M., Milán García, J., and de Pablo Valenciano, J. (2023). Renewable energies and blue economy: new trends in global research. *Energies* 16 (10): 4210.

Reuters (2009). China cleans oil, chemical spills after ship accidents. Reuters, 2 November 2009. Available at: https://www.reuters.com/article/idUKPEK125639/ (accessed 5 January 2024).

Riechers, M., Brunner, B.P., Dajka, J.-C. et al. (2021). Leverage points for addressing marine and coastal pollution: a review. *Marine Pollution Bulletin* 167: 112263.

Riegl, B., Bruckner, A., Coles, S.L. et al. (2009). Coral reefs: threats and conservation in an era of global change. *Annals of the New York Academy of Sciences* 1162 (1): 136–186.

Ruiz-Frau, A. (2022). Impacts of jellyfish presence on tourists' holiday destination choices and their willingness to pay for mitigation measures. *Journal of Environmental Planning and Management* 66 (10): 2107–2125. https://doi.org/10.1080/09640568.2022.2061926.

Saraga, D.E., Tolis, E.I., Maggos, T. et al. (2019). PM2.5 source apportionment for the port city of Thessaloniki, Greece. *Science of the Total Environment* 650 (2): 2337–2354. https://doi.org/10.1016/j.scitotenv.2018.09.250.

Sarker, S., Bhuyan, M.A.H., Rahman, M.M. et al. (2018). From science to action: exploring the potentials of blue economy for enhancing economic sustainability in Bangladesh. *Ocean & Coastal Management* 157: 180–192.

Shah, R. (2023). Beach Tourism Market Outlook (2023 to 2033), Future Market Insights, May 2023. Available at: https://www.futuremarketinsights.com/reports/beach-tourism-market (accessed 4 February 2024).

Shiiba, N., Wu, H.H., Huang, M.C., and Tanaka, H. (2022). How blue financing can sustain ocean conservation and development: a proposed conceptual framework for blue financing mechanism. *Marine Policy* 139: 104575.

Tiquio, M.G.J.P., Marmier, N., and Francour, P. (2017). Management frameworks for coastal and marine pollution in the European and South East Asian regions. *Ocean & Coastal Management* 135: 65–78.

UNCTAD (2023). Review of Maritime Transport 2023: Towards a Green and Just Transition. United Nations publication. Sales No. E.23.II.D.23. Geneva.

UNEP (2009). Coastal tourism in the Mediterranean: Adapting to climate change. Cagliari (Sardinia), Italy, 8–10 June 2009'. United Nations Environment Programme (UNEP), République Française – Ministère de l'Écologie, du Développement et de l'Aménagement durables, United Nations World Tourism Organization (UNWTO), Priority Actions Programme, Paris.

UNEP (2021) X-Press Pearl Maritime Disaster Sri Lanka - Report of the UN Environmental Advisory Mission (July 2021). 06 August 2021. [online] Available at: https://www.unep.org/resources/

report/x-press-pearl-maritime-disaster-sri-lanka-report-un-environmental-advisory-mission (accessed 5 January 2024).

UNEP (2022) Diving Deep: Finance, Ocean Pollution and Coastal Resilience [online]. United Nations Environment Programme. Available at: https://www.unepfi.org/publications/diving-deep/#:~:text=The%20Diving%20Deep%20Guidance%20helps,solutions%20including%20mangroves%20and%20reefs (accessed 25 November 2023).

United Nations (2017). The Potential of the Blue Economy: Increasing Long-term Benefits of the Sustainable Use of Marine Resources for Small Island Developing States and Coastal Least Developed Countries [online]. World Bank. Available at: https://www.sciencedirect.com/science/article/pii/S1385110123000862 (accessed 6 October 2023).

United Nations (2022). Blue Economy: Oceans as the Next Great Economic Frontier [online]. Regional Information Centre for Western Europe. Available at: https://unric.org/en/blue-economy-oceans-as-the-next-great-economic-frontier/ (accessed 10 October 2023).

UN-SDGS (2021). Canada joined the Clydebank Declaration at the 26th Conference of Parties to the UN Framework Convention on Climate Change. United Kingdom Government.

UNWTO (2013). Sustainable Tourism Governance and Management in Coastal Areas of Africa. Sustainable Tourism Governance and Management in Coastal Areas of Africa.

UNWTO (2023). International Tourism to End 2023 Close to 90% of Pre-Pandemic Levels, NEWS UNWTO, 30 Nov 2023. [online] Available at: https://www.unwto.org/news/international-tourism-to-end-2023-close-to-90-of-pre-pandemic-levels?utm_source=untn&utm_medium=crm (accessed 04 February 2024).

US Department of Commerce (2023). Economic Loss [online]. Marine Debris Program. Available at: https://marinedebris.noaa.gov/why-marine-debris-problem/economic-loss (accessed 25 December 2023).

USEPA (2016) Lower 8.3 Miles of the Lower Passaic River Part of the Diamond Alkali Superfund Site Essex and Hudson Counties, New Jersey. 3 March 2016, United States Environmental Protection Agency, p1000. Available at: https://www.diver.orr.noaa.gov/web/guest/diver-admin-record?diverWorkspaceSiteId=6233 (accessed 5 January 2024).

Venugopal, V. (2022). Green processing of seafood waste biomass towards blue economy. *Current Research in Environmental Sustainability* 4: 100164.

Voyer, M., Quirk, G., McIlgorm, A., and Azmi, K. (2018). Shades of blue: what do competing interpretations of the Blue Economy mean for oceans governance? *Journal of Environmental Policy & Planning* 20 (5): 595–616.

Walker, T.R., Adebambo, O., Del Aguila Feijoo, M.C. et al. (2019). Environmental effects of marine transportation. In: *World Seas: An Environmental Evaluation* (ed. C. Sheppard), 505–530. Elsevier Ltd.

Wallington, T.J. and Wiesen, P. (2014). N_2O emissions from global transportation. *Atmospheric Environment* 94: 258–263.

Wells, P.G. (2017). The iconic Torrey Canyon oil spill of 1967-Marking its legacy. *Marine Pollution Bulletin* 115 (1-2): 1–2.

World Bank (2018). Seychelles Launches World's First Sovereign Blue Bond [Press release]. Retrieved from https://www.worldbank.org/en/news/press-release/2018/10/29/seychelles-launches-worlds-first-sovereign-blue-bond (accessed 20 December 2023).

Zahedi, S. (2008). Tourism impact on coastal environment. *WIT Transactions on the Built Environment* 99: 45–57.

Zhang, B., Matchinski, E.J., Chen, B. et al. (2019). Marine oil spills—oil pollution, sources and effects. In: *World Seas: An Environmental Evaluation* (ed. C. Sheppard), 391–406. Elsevier.

24

Macro Issues of Microplastics: Present Status and Future Challenges

Wijetunge, D.S., Perera, S.P.K.K.A., Karunathilake, N.M.O.A., and Ranatunga, R.R.M.K.P.

Department of Zoology and Aquatic Sciences, Centre for Marine Science and Technology, University of Sri Jayewardenepura, Nugegoda, 10250, Sri Lanka

24.1 Microplastics (MPs): The Global Pollutant

24.1.1 Current Understanding of MPs

Plastic pollution has become one of the most widespread global menaces. The presence of mismanaged plastics poses enduring threats to the marine ecosystems that cover more than 70% of the Earth's surface, supporting a rich biodiversity. Plastics have proven their importance in improving human lives because of their durability, excellent mechanical properties, resistance to weathering, and long service life. Plastic production has been steadily increasing since the 1950s and continues to grow on a massive scale.

Plastics can be categorized as macro-plastics (size > 25 mm), meso-plastics (size 25–5 mm), and MPs (size < 5 mm) depending on size (Andrady, 2011). The presence of MPs in the marine environment has become a global concern because of their omnipresence, minuscule sizes, and potential impacts on marine organisms and human health (Yuan et al., 2022). According to origin, there are two major groups of MPs: primary and secondary. Primary MPs are those purposely manufactured at the microscopic size and are widely used in personal care products such as cosmetics, toothpaste, facial cleansers, deodorant, baby products, peelings, sunscreen, body washes, resin pellets, hair colors, nail polish, bath gels, eye shadows, insects' repellents, air-blasting facilities, synthetic clothing, gas–oil drilling fluids, plastic powders in molding, and countless daily tasks (Alomar et al., 2016). Secondary MPs are broken particles of macro-plastics.

MPs can enter the ocean from various sources, including land-based activities (i.e. wastewater discharge, urban runoff, and agricultural runoff), maritime activities (i.e. fishing, shipping, and aquaculture), and natural phenomena (i.e. weathering and fragmentation) (Meijer et al., 2021). Once in the ocean, the movement of MPs between seawater and marine sediments is affected by tides. Tides transport plastic waste from coastal zones into seas. Moreover, MPs in seas can be brought back to the coast under tidal action. The MP migration on the surface of water bodies is mainly influenced by wind and ocean currents (Wessel et al., 2016; Li et al., 2018).

MPs are hydrophobic (Ivar et al., 2014) and are more liable to adsorb organic matter (even crude oil) and heavy metals with an increase in their degradation degree (Hirai et al., 2011; Zhou et al., 2018; Zhang et al., 2018). As a result, MPs become denser and are deposited on the seafloor.

Coastal and Marine Pollution: Source to Sink, Mitigation and Management, First Edition.
Edited by Meththika Vithanage, Sameera M. Samarasekara, Bryan D. James, and Christopher M. Reddy.

Therefore, more than 70% of MPs are deposited in seafloor sediments, and large quantities are recorded in polar sea ice (Thompson 2015). According to statistics from 2014, up to 93 000–236 000 tons are suspended in global oceans (Law and Thompson 2014).

24.1.2 Eco-toxicological Effects of MPs and the Importance of Long-Term Monitoring

Due to the higher abundance and smaller size of MPs in the ocean, there is a greater probability of being swallowed by organisms like finfish and accumulating along food chains, posing a significant impact on organisms like seabirds. MPs have been detected in biological samples such as the stomach, digestive tract, muscles, other tissues, organs, and even lymphatic systems, which can be transported and enriched in marine organisms (Lusher 2015). After ingestion, these can affect the growth and development of organisms, causing a wide range of negative effects, including cancer, poor immunological function, malformations, and impaired reproductive function in both people and animals (Belioka et al. 2023). Furthermore, these might impact gene expression and heredity, which can result in biological death (Moore 2008).

The toxicological effects of MPs not only impact marine animals but also directly impact seaweed and algae. MP particles decrease the chlorophyll content of algae and thus affect their photosynthesis process (Besseling et al. 2013). MPs can accumulate on the surface of crescent algae through electrostatic interaction, which hinders the absorption and utilization of sunlight and CO_2 by algal cells and impacts the photosynthesis process of algae, thus inhibiting the growth (Bhattacharya et al. 2010; Nolte et al. 2017).

MPs deposited in seafloor sediments negatively impact benthic dwellers. They disrupt the water and gas exchange process at the sediment interface and impact the biochemical reaction process on the sediment surface. Furthermore, soil enzymes are vital components of soil. MPs significantly impact the activity of soil enzymes. Polystyrene (PS) particles significantly reduce soil microbes' biomass and enzyme activity and have an inhibitory effect on the soil microbial community (Awet et al. 2018).

MP particles form porous surfaces due to weathering. These bear persistent organic pollutants (POPs) and heavy metals (such as lead, zinc, copper, and chromium) from complex polluted situations and carry them to areas where they would not naturally occur. This can significantly affect the health of marine organisms and the wider ecosystem (Figure 24.1) (Yuan et al. 2022). Chemical pollutants tend to be more insidious and can bioaccumulate in organisms and biomagnify up the food chain, leading to toxic effects in top predators and potentially impacting the stability of entire ecosystems (Miller et al. 2020). These chemical pollutants can also affect the subtle changes in the reproduction and behavior of marine organisms. Such alterations may disrupt the balance of marine ecosystems and result in declines in biodiversity of the marine ecosystem (Corinaldesi et al. 2021).

MP pollution of the marine environment represents a severe hazard to the survival of marine life and poses a risk to both human health and the economy. Due to their small particle size and difficulty in visualization, hand removal of these particles is exceedingly difficult, thus posing a problem for prevention and control methods. MPs will become more persistent with time. According to the literature, more MPs will be in our oceans than fish by 2050 (Auta et al. 2017). Therefore, long-term monitoring of MP pollution is a timely necessity.

24.1.3 Plastic Biogeochemical Cycle

All environmental compartments, such as soil, vegetation, water bodies (freshwater and sea), and the atmosphere, contain biogeochemical cycles that are relatively well understood regarding

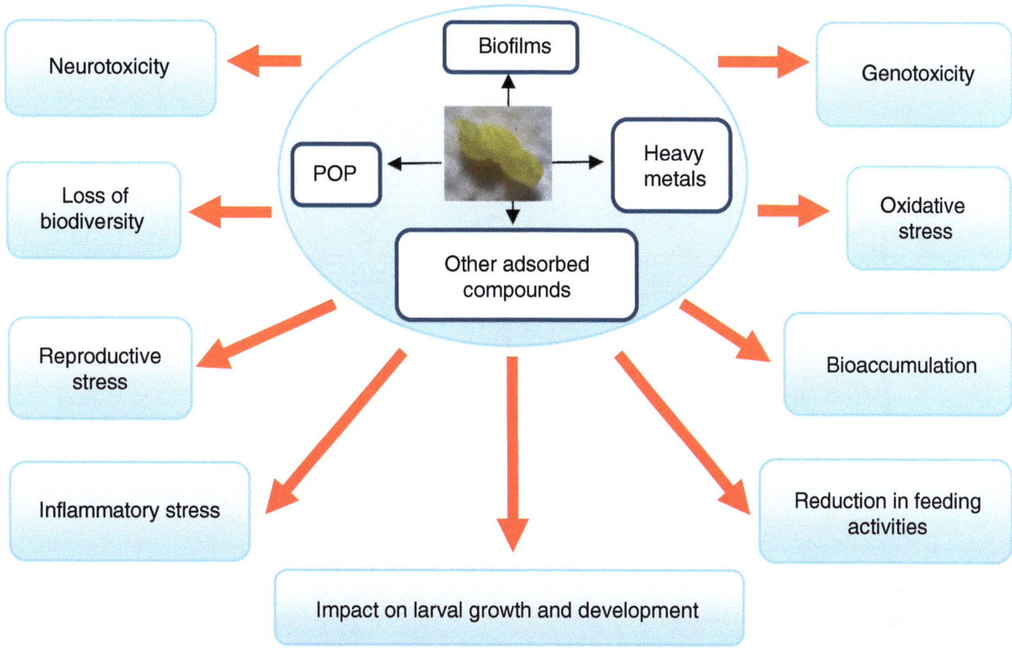

Figure 24.1 Eco-toxicological effects of MPs on the marine ecosystem: Adsorbed POPs, microorganisms, heavy metals, and other compounds increase toxicity in MPs and cause numerous impacts on organisms. *Source:* Center for Marine Science and Technology (CMST).

the major principles governing their destiny and transport within and across ecosystem compartments and environmental reservoirs. Similarly, plastics have their own life cycle as well. Therefore, Horton and Dixon (2017) and Bank and Hansson (2019) introduced the important concept of the plastic cycle to illustrate the biogeochemical cycle of plastics between environmental compartments and stress the global nature of this ubiquitous problem. The plastic cycle is the continuous and complex movement of plastic materials between different abiotic and biotic ecosystem compartments, including humans (Figure 24.2) (Bank and Hansson 2019). However, regarding the movement, transport, and destiny of particulate (and chemical) matter, these environmental compartments are very closely interlinked, with unclear, permeable boundaries. Interactions between compartments can differ depending on weather and environmental conditions. This means the abundance and fate of MPs in any given environment will depend on the degree of connectivity with nearby environments, which can be highly dependent on space and time.

24.1.3.1 Plastic Cycle in the Terrestrial Environment

All types of plastics are manufactured on land, other than plastics that are used for maritime or fishing uses in the ocean. Direct littering and inefficient waste management release the waste of consumer products onto land. All these items break down to form secondary MPs within the environment (Nizzetto et al. 2016). MPs may also be released directly to land along with sewage sludge applied to agricultural land as a fertilizer and wastewater treatment plants in terrestrial environments (Carr et al. 2016; Murphy et al. 2016; Talvitie et al. 2017). The MPs released to the terrestrial environment can also undergo infiltration and be suspended in soil.

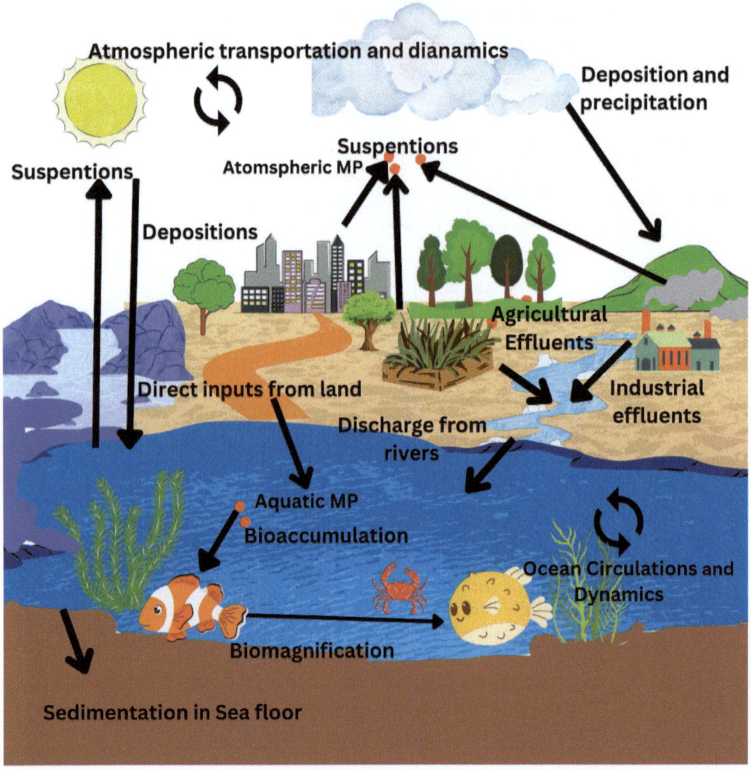

Figure 24.2 Graphical representation of plastic biogeochemical cycle presenting MPs' complex and continuous movements between abiotic and biotic ecosystem compartments.

24.1.3.2 Plastic Cycle in the Freshwater Environment

Freshwater represents the most complex system regarding MP transport and retention as it gets MPs from the terrestrial environment and functions as a pathway to the marine environment. Larger plastic items can be released into the freshwater environment through improper waste disposal, either through littering or through loss from landfills, and transported from land through wind or surface runoff. In addition to macro-plastics, significant direct MP inputs to freshwater systems include agricultural drainage and runoff from farmland, storm drainage and urban run-off, and industrial effluents. In some cases, during very high flow conditions, combined sewer overflows (CSOs) are designed to release untreated sewage into nearby rivers to reduce the pressure on drainage systems, releasing both micro- and macro-plastic wastes (Murphy et al. 2016; Horton et al. 2017). A small fraction of such MPs undergo sedimentation in rivers or lakes, and most discharge to the ocean.

24.1.3.3 Plastic Cycle in the Marine Environment

Sources of MPs to the marine environment are widespread as oceans are generally considered the ultimate sink for all types of plastics in the environment (Browne et al. 2011; Law and Thompson 2014). In addition to the inputs from rivers, plastics will also discharge to the oceans directly via mismanaged maritime or fishing waste, including abandoned fishing gear, accidental cargo loss, and illegal dumping. Also, the macro-plastics released into the ocean may undergo degradation and produce the MPs. Some marine organisms ingest such produced MPs. These MPs undergo bioaccumulation in their body and go through the marine food chain via biomagnification.

24.1.3.4 Plastic Cycle in the Atmosphere

Recently, it has been discovered that many MP particles will become suspended and transported within the air as "urban dust" due to their lightweight nature. These are typically derived from road dust (e.g. tire and paint particles) and fibers from synthetic textiles, especially from soft furnishings. These particles are transported far from the source by air currents and wind. Some processes can lead to the deposition of MPs to land or aquatic environments, and weather events like heavy rainfall will facilitate the deposition of particles back onto land (Zylstra 2013; Dris et al. 2016; Horton et al. 2017; Dris et al. 2017; Geyer et al. 2017).

24.1.4 MPs and Climate Change

Plastic pollution and climate change are two major global threats that impact the pedosphere, hydrosphere, and atmosphere. With the influence of climate change, elevated temperatures may lead to extreme weather events such as glacier thawing and raised ocean temperatures, causing sea-level rise and ocean acidification. Climate change enhances the impact of MPs in the environment by influencing their degradation, distribution, and interaction with ecosystems.

Rising temperatures can expedite the degradation of plastics, leading to enhanced fragmentation and the release of smaller MP particles. Additionally, increased temperatures facilitate the physical breakdown of plastics via thermal degradation, making them more susceptible to other degradation processes (Kamweru et al. 2011).

In the case of glacier thawing, MPs retained in the ice would be released into the sea, and MPs with a density lower than that of the surrounding water would be carried away by ocean currents and redistributed over long distances, thus transporting the MPs released from the melted snow into the aquatic habitat. Furthermore, climate change can also force more frequent and intense extreme weather events like storms, hurricanes, floods, and changes in ocean currents and circulation patterns, which can transport and disperse MPs across larger areas (Cooper and Corcoran 2010). In addition to the horizontal surface water movement, strong winds result in vertical water mixing, which resuspends the settled MPs in the seabed onto the water surface. Floods, storms, and extreme weather events can force the spreading of MPs to areas without plastic pollution, exposing pollution-free zones to MPs (Kukulka et al. 2012; Reisser et al. 2014).

The rising temperatures and alterations in ocean chemistry, such as ocean acidification, can cause the exposure of marine species to MPs, leading to the accumulation of MPs along food chains (Wright et al. 2013; Egbeocha et al. 2018; Guzzetti et al. 2018) and changes in reproductive cycles and behaviors of marine species (Beckwith & Fuentes 2018). Global warming raises the soil temperature of permafrost, forcing the release of MPs into the soil and water bodies due to melting processes (Teufel and Sushama 2019; Chen et al. 2021).

Changes in rainfall directly affect the MP distribution and transportation. An increased rainfall results in enhanced surface water runoffs that transfer MPs from the terrestrial environment to the aquatic environment and also carry atmospheric MPs by trapping them in the raindrops (Napper et al. 2023; Abbasi 2021). Rainfall may impact the water flow hydrodynamics and resuspend the MPs in the sediments, resulting in an inverse relationship between sediments and suspended MPs (Baldwin et al. 2016).

24.2 Importance of Standardized MP Assessments and Accuracy in Data Acquisition

Most publications address the identification, characterization, and quantification of MPs in the marine environment since the destination of most of the inland debris is at sea. These data are important for global environmental concerns with potential widespread biological, ecological, chemical, socioeconomic, and health implications. Data acquisition and analysis in marine biota are crucial and risky for monitoring (Hermsen et al. 2018). Systematic protocols for MP analysis need to be developed and established. Current frameworks and studies in this area show a wide range of methodologies and reporting styles, making comparability among studies difficult. Figure 24.3 highlights the important aspects of standardized MP protocols.

24.2.1 Consistent Framework of Studies

Most MP studies contain protocols for sampling, sample processing and storage, laboratory preparation, pretreatments, chemical treatments, separation, identification, quantification, characterization, and polymer-type identification. Consistency is essential when comparing data from different studies, laboratories, geographical regions, and country to country. Industrial, agricultural, shipping, fisheries, port, and other anthropogenic development activities highly affect the environmental MP contamination. Therefore, consistency in protocols may provide validity for the studies locally and internationally.

24.2.2 Quality Assurance

The methodology must generate results with minimal background contamination for real-world situations. Adhering to quality and established methods minimizes variability and errors for global researchers in sample processing and analysis. The global quality assessment system presented by Klimisch and criteria for reporting and evaluating ecotoxicity data (CRED) includes 10 criteria to assess data reliability (Figure 24.4) (Hermsen et al. 2018). Factors such as sampling method, sample pretreatment, sample processing, and the nature of positive and negative controls can be considered in the quality assessment of standard protocols.

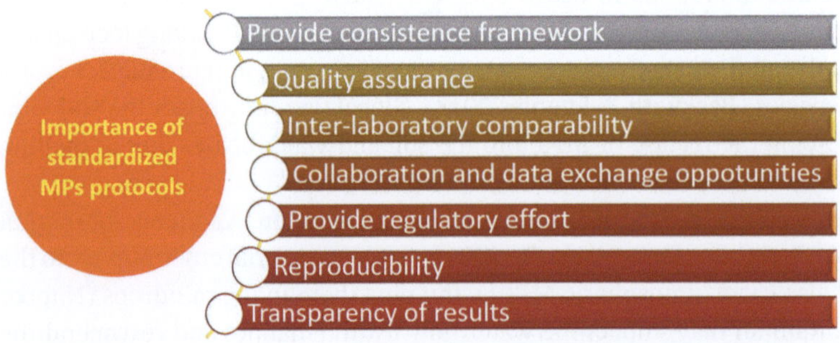

Figure 24.3 The importance of standardization of MP research. Standard MP protocols produce consistent, high-quality results that are conclusive, comparable, and reproducible.

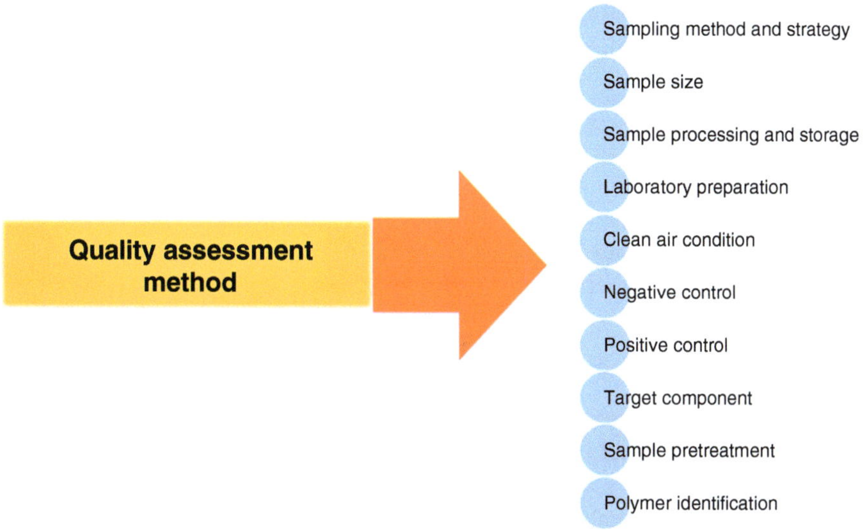

Figure 24.4 Klimisch and CRED 10 criteria for a quality assessment.

24.2.3 Interlaboratory Comparability

Comparability is essential to synchronize findings from multiple studies and evaluate the global distribution, abundance, and contamination of MPs. Standardized protocols enable data comparability among different analyses in laboratories and research groups.

24.2.4 Facilitating Collaboration

Collaborations are effective in terms of knowledge and technology-sharing aspects. Standardized methods increase transparency, streamline data exchange, and enable researchers from different institutions to collaborate more effectively. For instance, the Global Plastics Leakage Project initiated by the Commonwealth Scientific and Industrial Research Organization (CSIRO), Australia, collaborated with numerous countries in the Asia–Pacific region, including Sri Lanka, as well as South America and Africa, in developing the first global, empirical baseline estimate of mismanaged waste entering the coastal and marine environments near major urban centers in countries around the world (CSIRO Marine Debris Team & Center for Marine Science and Technology 2021).

24.2.5 Regulatory Efforts

Standardized methods can inform evidence-based decision-making and regulatory actions aimed at mitigating MP contamination. When addressing policymakers and government regulatory bodies, we need accurate and acceptable documents to make regulatory guidelines on MP pollution. Standardized protocols provide a basis for developing regulatory guidelines and policies related to MP pollution.

24.2.6 Reproducibility

Reproducibility is the ability of an experiment to provide consistent results when carried out by other researchers. It is an essential component of research and is necessary to foster confidence in the reliability and validity of scientific findings. Standardized protocols enable the advancement of scientific knowledge by promoting methodological rigor and reproducibility. Consistent methods allow researchers to build upon existing findings, refine analytical techniques, and address key research questions more effectively.

24.2.7 Transparency of Results

Standardized protocols enhance public trust in scientific research by ensuring transparency, reliability, and accountability. They increase confidence in the accuracy and validity of research findings related to MP pollution (Provencher et al. 2020).

24.2.8 Evolution of MP Analysis Protocols

Researchers started developing standard methods for collecting and analyzing MPs in the early 2000s. Early-stage MP analysis procedures consisted of three common stages: extraction, identification, and quantification. Researchers would manually count and classify particles based on size, shape, and color. This method was highly labor-intensive and led to human errors. However, current methodologies have many challenges, such as density separation and sample treatment, including digestion of organic content and polymer-type identification. The key point in methodological improvements is the need for cost-effective and harmonized protocols.

The International Organization for Standardization (ISO) and the National Oceanic and Atmospheric Administration (NOAA) spearheaded some standardization initiatives (Masura et al. 2015). According to the literature, there is no universally standardized protocol for MP analysis. However, several organizations and research groups have proposed validated protocols that are widely accepted and commonly used.

The common procedure for the MP analysis has evolved as follows.

24.2.8.1 Sample Collection and Preservation

MPs can be found at several depths in the water column and can sink to the bottom of the water body due to density variations. Repetitive samples should be collected from various locations within the study area, ensuring representative coverage. Samples should be carried out with great care to eliminate contamination. Adding preservatives such as ethanol helps to prevent microbial growth and degradation of organic material (Hermsen et al. 2018) (Table 24.1).

24.2.8.2 Sample Processing

Most studies follow modified versions of the standard MP analysis protocol suggested by Masura et al. (2015), which includes wet peroxide oxidation (WPO), density separation, vacuum filtration, and microscopic identification. Cross-contamination via airborne MPs should be avoided throughout the procedure.

Multitier sieving has been successfully employed for initial screening using a series of sieve cascades with a decreasing mesh size to pass the sample. After sieving, particles of different sizes are retained on different sieves. The remaining samples under the desired size ranges undergo WPO to remove any organic elements attached to the surface of plastic particles (Bessa et al. 2019). WPO generally requires oxidizers, acids, or alkaline materials. Earlier stages of MP analysis did not contain the digestion of organic disturbances. Table 24.2 contains information regarding the selection of the most appropriate strategy for the digestion of organic materials.

MPs will be separated from the sample via density separation. The process uses the buoyant force of a liquid with an intermediate density to separate the lighter materials from the denser materials

Table 24.1 Types of MP sampling.

Type of sampling	Methodology
Surface water sampling	Surface skimming devices, i.e. manta trawls or Neuston nets/plankton nets, are used by filtering water through a fine mesh directly to clean glass bottles (Figure 24.5a)
Flowing water sampling	Specifically designed devices can be installed to capture contaminated surface runoff from industrial, agricultural, or domestic settlements
Sediment sampling	A Van Veen bottom grab sampler can be used (Figure 24.5b)
Core sampling	Coring devices can be used to assess MP distribution over time
Biological sampling	Wild capture using fishing rods, nets, and handpicking or purchasing from the market. The organism will be dissected to obtain the testing sample/organ (i.e. skin, gut content, gastrointestinal tract, muscles, and gills), and various digestion techniques can be performed to dissolve organic matter and recover MPs for analysis
Air sampling	High-volume air samplers can be used with air filters (Perera et al. 2022)
Ingestion analysis	Selected organisms are exposed to a known concentration of MPs under controlled laboratory conditions to study ingestion rates and effects

(a) (b)

Figure 24.5 Surface sampling using (a) plankton net (b) Van-Veen Bottom Grab Sampler. *Source:* Center for Marine Science and Technology (CMST).

after thoroughly shaking and settling the mixture of materials in the liquid (Hidalgo-Ruz et al. 2012). Density separation can be performed using sodium chloride (NaCl) (d = 1.0–1.2 g/cm^3), zinc chloride ($ZnCl_2$) (d = 1.6–1.8 g/cm^3), sodium iodide (NaI) (d = 1.8 g/cm^3), and sodium bromide (NaBr) (d = 1.37–1.40 g/cm^3). The solution should be chosen according to the objectives of the experiment (Table 24.3). The supernatant of the density-separated solution will be passed through glass fiber filter (GFF) papers, and MPs will be concentrated (Vermeiren et al. 2020).

Table 24.2 Different digestion strategies of MPs and their pros and cons.

Extraction strategies	Remarks
Without adding oxidizing agents	Visual observation only; prone to human errors; lead to misunderstanding
H_2O_2 15% (peroxidation)	Appropriate for all the samples; time consuming
H_2O_2 30% (peroxidation)	Partial degradation to particles; size and shape of MPs can be changed
KOH 10% (alkaline digestion)	Maximum temperature of 40 °C; fast approach; not appropriate for fat tissues
Digestion with Fenton's reagent[a]	Suitable for digestion of organic material in sediments and soil; highly reactive; explosive reactions can happen
Acid digestion	Degradation, digestion, and deformation of several polymers
Enzymatic digestion	Appropriate for small samples but not for larger tissues; costly

[a] 30% H_2O_2 mixed with Fe(II) solution.
Source: Adapted from Bessa et al. (2019).

Table 24.3 Buoyancy of common polymers compared to seawater.

Polymers	Density (g/cm^3)	Buoyancy concerning seawater
Polyamide (PA)	1.12–1.15	Negative ↓
Nylon 6,6 (PA 6,6)	1.13–1.15	
Polyethylene terephthalate (PET)	1.38–1.41	
Polyvinyl chloride (PVC)	1.38–1.41	
Polytetrafluoroethylene (PTFE)	2.10–2.30	
Seawater	1.025	
Polystyrene (PS)	0.01–1.06	Positive ↑
Polypropylene (PP)	0.85–0.92	
Low-density polyethylene (LDPE)	0.89–0.93	
High-density polyethylene (HDPE)	0.94–0.98	

24.2.8.3 Physical and Chemical Identification of MPs

Visual assessment of potential plastics is the most widely used method for physically identifying MPs. Potential plastics can be confirmed via the hot needle test. Chemical investigation of the polymeric composition typically combines optical, spectroscopic, or thermo-analytical techniques. Spectroscopic techniques such as Fourier transform infrared (FTIR) spectroscopy (for 5–20-μm-sized MPs) (Harrison et al. 2012) and Raman spectroscopy (for 5–10-μm-sized MPs) (Käppler et al. 2015) have become popular for polymer-type identification. These technologies are costly, time consuming, and limited across multiple laboratories. Recent advancements in μ-FTIR imaging have enabled the automatic identification of MPs concentrated on the filter membrane without pre-sorting. Focal plane array (FPA)-based μ-FTIR imaging provides a considerable time-effective analysis and increases the accuracy of the findings from an unbiased measurement of the entire filter surface.

The rapid, automated analysis of MPs via laser-directed infrared spectroscopy (LDIR) has been used lately to identify particles as small as 6–3 μm in size. LDIR is highly efficient, operable, and considered the most standardized MP identification method (Zhang et al. 2022). Thermal analysis techniques such as thermogravimetric coupled with differential scanning colorimetry (TGA-DSC) and pyrolysis–gas chromatography/mass spectrometry (Pyr-GC/MS) have been used to quantify MP in environmental samples chemically. However, the error caused by visual sorting identification might reach 70%, increasing with decreasing particle size (Chen et al. 2020).

Recently, advanced microscopy techniques, such as scanning electron microscopy (SEM) and transmission electron microscopy (TEM), have provided high-resolution images of MPs, providing for detailed morphological analysis; these techniques also help to identify surface features of MPs and potential sources of MPs (Fang et al. 2023). With the evolution of advanced MP protocols, automated image analysis software has been developed recently to streamline the quantification and characterization of MPs from microscopy images (Primpke et al. 2017).

24.2.9 Advanced Integrated Studies on Cryospheric, Aquatic, Terrestrial, Biospheric, and Atmospheric Settings

Plastic waste management has emerged as a significant global issue due to its widespread use, environmental persistence, and adverse impacts on ecosystems and human health. One of the biggest challenges with plastic waste is its persistence in the environment for hundreds to thousands of years, accumulating in landfills, waterways, oceans, and even remote ecosystems such as the polar region and deep-sea trenches. Such accumulations would directly affect the biodiversity and the economics of the world. Due to the universal impact of MPs on the world's systems, advanced integrated studies of MP pollution in cryospheric, aquatic, terrestrial, biospheric, and atmospheric settings are of higher importance.

The cryosphere is the part of the Earth's surface characterized by the presence of frozen water encompassing mountain glaciers, continental ice sheets, seasonal snow, ice cover on land, and sea ice, which are primarily influenced by anthropogenic pollutants through atmospheric deposition (Zhang et al. 2022). Typical cryospheric regions include the Arctic, Antarctic, Alps, Tibetan Plateau, and Andes. Global warming causes rapid shrinking in the cryosphere and thermal expansion of the ocean, resulting in rising sea levels and releasing pollutants from a frozen state to the environment. Though MP sampling strategies may differ due to higher latitudes and altitudes, sample processing and analyzing procedures are the same as in other environments. Since cryospheric regions are located distantly compared to MPs generating sources in other environmental matrices, they are considered temporary sinks of MPs, alarming the risk of contamination in the case of glacier melting. Different colored MPs in snow caps absorb radiation and accelerate glacier melting, thus impacting climate change (Zhang et al. 2023). Ocean currents from the Atlantic Ocean can convey MPs to the Arctic Ocean, and MPs trapped within sea ice can be transported across and within the Arctic Ocean.

The most common and integrated assessments worldwide are aquatic, terrestrial, and biospheric MP studies. MP assessments in terrestrial environments are lesser than those in marine, freshwater, and biospheric studies due to the major focus on mismanaged plastics and marine pollution. Terrestrial environments are more vulnerable to MP contamination due to the direct involvement of anthropogenic activities such as plastic packaging, single-use recreational items, and sewage-treated sludge. Biospheric MP studies use biotic samples extracted from organisms, including humans. These are of high importance since the results indicate the potential impact on the body tissues of organisms and trends of bioaccumulation and biomagnification.

MPs are ubiquitous in the atmosphere and are emitted from various sources, such as city dust, house furniture, industrial emissions, and synthetic clothing (Patil et al. 2021). Atmospheric circulation is the primary cause of MP abundance in pristine environmental compartments like polar regions. Therefore, integrated studies from different environmental compartments help visualize the flow and impact of MPs as a universal pollutant.

24.2.10 Empirical Data Versus Global Metadata

A plethora of scientific, non-scientific, quantitative, and qualitative studies have been conducted worldwide on plastic pollution. Some generated data that might be freely available, while others have restricted access.

Empirical studies are primary assessments presenting original results specifically generated throughout the respective studies. Empirical data are study-specific, limited, and highly objective. Here, the capture, curation, and analysis of large volumes of data are specific and focused on the scientific question (Cheruvelil and Soranno 2018). Globally available metadata layers include secondary data derived from single, few, or more primary assessments.

Conducting empirical studies is not easy. Especially during the past decade, researchers faced many logistical and technical challenges when initiating, processing, analyzing, visualizing, and integrating heterogeneous data and knowledge from diverse and isolated resources (Cheruvelil and Soranno 2018). Empirical studies can be individual or team-based, and they include a diverse set of contributors, such as scientists and collaborators, with the knowledge and perspectives needed to compile data across scales or disciplines. Empirical studies enable the sharing of methods, codes, and data, facilitating the documentation of the data and methods. Table 24.4 summarizes the development of empirical data in Sri Lanka.

Global metadata is widely available worldwide, especially in medical and environmental fields. For instance, scientists add debris counts of beach clean-ups or marine debris surveys into global databases, and medical researchers add medical histories of diagnosed patients suffering from a known sickness over the years. These databases continuously update and contain large volumes of data collected over the years, providing a snapshot of the current scenario. These large volumes of data are highly variable across disciplines (Cheruvelil and Soranno 2018).

Metadata analysis refers to any secondary analysis of the findings of two or more primary research studies. It interrogates information from other scientific research to gain a more integrative understanding of what has been discovered about some topics. Metadata users such as researchers, data scientists, or enthusiasts need to have profound knowledge of the accessibility, availability, and accuracy of these resources on the web. Extensive computational skills and expertise are required to understand and interpret secondary data relevant to the research objectives. To integrate these into the analysis, scientists must spend significant time understanding the background, structures, data patterns, and formats. Global meta-analysis is often used to make generalizations across all available evidence at the global scale (Shackelford et al. 2021). It is important to extrapolate findings from one location to another, to scale up knowledge and processes from local to regional and global extents, and to forecast knowledge from past to current and future states (Cheruvelil and Soranno 2018).

Global meta-analysis can be either team-based or individual because of minimum groundwork. Open science uses a set of transparent, inclusive, and reproducible practices that result in publicly available research, data, and dissemination products. It also expects all data, codes, and other research products to be findable, accessible, interoperable, and reusable. It stands to advance discovery, foster reproducibility, leverage investments in research, democratize science, foster a more inclusive science, and improve communication between scientists, the public, and decision-makers (Cheruvelil and Soranno 2018). Globally available metadata is not recommended for decision-making in specific

Table 24.4 MP assessments conducted in Sri Lanka to date.

Location	Reference(s)	Type of sampling
Marine environment	Ranatunga and Karunarathna (2018); Ranatunga et al. (2021)	Beach sand, finfish (*Sardinella gibbosa*)
	Koongolla et al. (2018)	Surface water, beach sand, sediment
	Weerakoon et al. (2018)	Beach sand
	Dharmadasa et al. (2021)	Coastal and lagoon sediment, surface water
	Viraj et al. (2019)	Beach sand
	Weerakoon et al. (2019)	Surface water
	Wijethunga et al. (2019)	Sessile invertebrates (*Saccostrea* sp., *Littorina* sp., *Patella* sp.)
	Athapaththu et al. (2020)	Surface water
	Athawuda et al. (2020)	Surface water
	Meegahakotuwa et al. (2020)	Surface water
	Weerakoon and Wimalasiri (2020)	Surface water
	Dissanayake and Rathnasuriya (2022)	Brown mussel (*Perna perna*)
	de Vos et al. (2022)	Plastic samples aftermath MV *X-Press Pearl* incident
	Sewwandi et al. (2022b)	Plastic nurdles aftermath MV *X-Press Pearl* incident
	Sewwandi et al. (2022a)	Beach sand
	Weerasekara et al. (2022)	Beach sand, surface water
	Gamage et al. (2023)	Finfish (bluestripe herring)
	Hansani et al. (2023)	Surface water, sediment, live scleractinian corals, soft corals
	Jayarathna et al. (2023)	Pelagic fish species (*Thunnus albacares, Katsuwonus pelamis, Elagatis bipinnulata*)
	Kandeyaya et al. (2023)	Finfish, shellfish (*Sardinella gibbosa, Stolephorus commersonnii, Hemiramphus archipelagicus, Katsuwonus pelamis, Penaeus monodon, Scylla serrata, Sepia* sp., *Perna* sp.)
	Rathnayake et al. (2023)	Seagrass species (*Halodule pinifolia*), microalgae (*Chaetomorpha* sp.)
Freshwater/brackish water environments	Thakshila and Ranatunga (2019)	Surface water, sediment
	Kapukotuwa et al. (2020)	Surface water, sediment
	Karunarathna and Ranatunga (2021)	Surface water, sediment

(*Continued*)

Table 24.4 (Continued)

Location	Reference(s)	Type of sampling
	Kapukotuwa et al. (2022b)	Raw salt from salterns and salt in commercial products
	Athukorala et al. (2023a)	Finfish (*Siganus javus, Leiognathus splendens, Leiognathus blochii, Mugil cephalus, Strongylura leiura, Stolephorus indicus, Caranx heberi*)
	Athukorala et al. (2023b)	Surface water, sediment
	Bandara et al. (2023)	Surface water, sediment, finfish (*Caranx sexfasciatus, Heteropneustes fossilis*)
	Dias et al. (2023)	Mud crab (*Scylla serrata*)
	Lahiru et al. (2023)	Surface water, sediment
	Nawarathna et al. (2023)	Indian backwater oyster (*Magallana bilineata*)
	Weerasuriya et al. (2023)	Common food fish (*Oreochromis niloticus, Heteropneustes fossilis*)
	Lawan et al. (2024)	Shrimps (*Penaeus monodon, Penaeus indicus*)
Other	Gamage et al. (2022)	Personal care and cosmetic products
	Kapukotuwa et al. (2022a)	Bottled water
	Perera et al. (2022)	Outdoor air
	Peiris et al. (2023)	Lichens

contexts unless the situations are known to be generalizable. If the local situation is different from the global situation, comparing data makes no sense and generates false interpretations. For instance, Sri Lanka was ranked fifth in the top 20 mismanaged plastic offenders in the world's oceans by Jambeck et al. (2015). This ranking was primarily based on estimated solid waste data, coastal population density, and economic status of selected countries. In the Sri Lankan case, the source of waste data is unclear. Ranatunga et al. (2023) emphasized the importance of using empirical data versus data in global metadata layers. Arguments against metadata sharing include a reluctance to release control of the data, fear of being "scooped," fear that others will misinterpret data, fear that sharing will restrict the types of data that scientists can use, and a lack of resources or the ability to fully document the data (Table 24.5).

Empirical studies backed by globally available metadata can be effectively used to forecast future trends of MP abundances in different environmental compartments of the world.

24.3 Economics of Waste Management

One of the major issues metropolitan regions worldwide face is the rising amount of municipal solid waste (MSW). Growing per capita income and rapid population expansion have led to the production of massive amounts of solid waste, a major hazard to human health and the environment. It is evident in most developing countries where vast amounts of solid wastes are carelessly

Table 24.5 Empirical studies versus metadata layers

Empirical studies	Metadata layers
Require a team of collaborators	Can be either individual or team-based
Present primary data	Present secondary data
Logistics and technical support needed to conduct the study	Logistics and technical constraints are not relevant
Original data are study-specific	A collection of data collected throughout the years
Can be open access or restricted access	Mostly open access
Can be catalyzed by open or metadata inputs	Contain data derived from one or few primary studies
Clear, study-specific results are presented	Require expertise in extrapolating and interpreting data
Each study presents original results with background information	A significant amount of time and computational skills needed for data sorting
Higher accuracy	Accuracy depends on the ability to generalize data and situations
Study-specific outcomes can be used in decision-making	Difficult to use in decision-making
Applicability[a] and transferability[b] are higher in similar studies	Applicability and transferability depend on similarities of situations
Higher control of data	Less control of data

[a] Applicability is the extent to which an intervention would be feasible in a different context (e.g. MP assessments might not be feasible in an area where sampling is not allowed).
[b] Transferability is the extent to which an intervention would have the same effect in a different context (e.g. MP assessments conducted on organisms in the marine environment might have a different effect in the freshwater environment).

dumped, placing strain on limited water and land resources, and contaminated biota negatively impacts people's health, especially the poor, who are exposed widely.

Waste management economics focuses on the efficient allocation of resources to address the generation, collection, transportation, treatment, and disposal of waste to maximize societal welfare while minimizing environmental impact and economic costs. It encompasses various economic principles, including cost–benefit analysis, pricing mechanisms, market incentives, and regulatory frameworks. The waste management sector represents a significant portion of the global economy, with the market size continually growing due to increasing population, urbanization, and industrialization. The demand for waste management services and technologies is also rising as societies produce more waste.

The waste management economy presents various economic opportunities, including job creation, entrepreneurship, and innovation. Waste management companies employ a diverse workforce in collection, sorting, processing, and management activities. Additionally, there is growing interest in innovative technologies and business models aimed at improving waste management efficiency and sustainability. Government regulations and policies play a significant role in shaping the waste management economy. Regulations govern waste disposal practices, recycling targets, landfill taxes, and extended producer responsibility (EPR) schemes, among other measures. These policies influence investment decisions, market dynamics, and the overall direction of the waste management sector.

Plastic waste management poses a larger impact on the global economy. Due to the widespread nature of plastics, the cost of cleaning up plastic waste from the environment, including oceans, rivers, beaches, and urban areas, is substantial. Governments, communities, and organizations spend billions of dollars annually on clean-up efforts, including waste collection, beach clean-ups, and marine debris removal, diverting resources that could be allocated to other priorities.

Plastic pollution affects the aesthetic appeal and recreational value of natural environments, such as beaches, parks, and waterways, causing negative consequences for tourism, recreation, and outdoor industries, leading to lost revenue, decreased visitor numbers, and damage to local economies that rely on tourism. During a marine debris survey conducted by the Centre for Marine Science and Technology, University of Sri Jayewardenepura, Sri Lanka, in collaboration with the CSIRO, Australia, it was found that lots of plastic debris over the beaches of eastern Sri Lanka originated from neighboring countries like India, Bangladesh, Thailand, and Myanmar (Figure 24.6).

Plastic pollution in marine environments harms fisheries and aquaculture industries through fishing gear entanglement, ingestion by marine species, and habitat degradation, which may disrupt fish populations, reduce catch yields, and undermine the economic viability of fishing communities and seafood industries (Awuchi & Awuchi 2019). Plastic waste in oceans and waterways can threaten shipping and transportation industries, including vessel damage, navigation hazards, and increased maintenance costs.

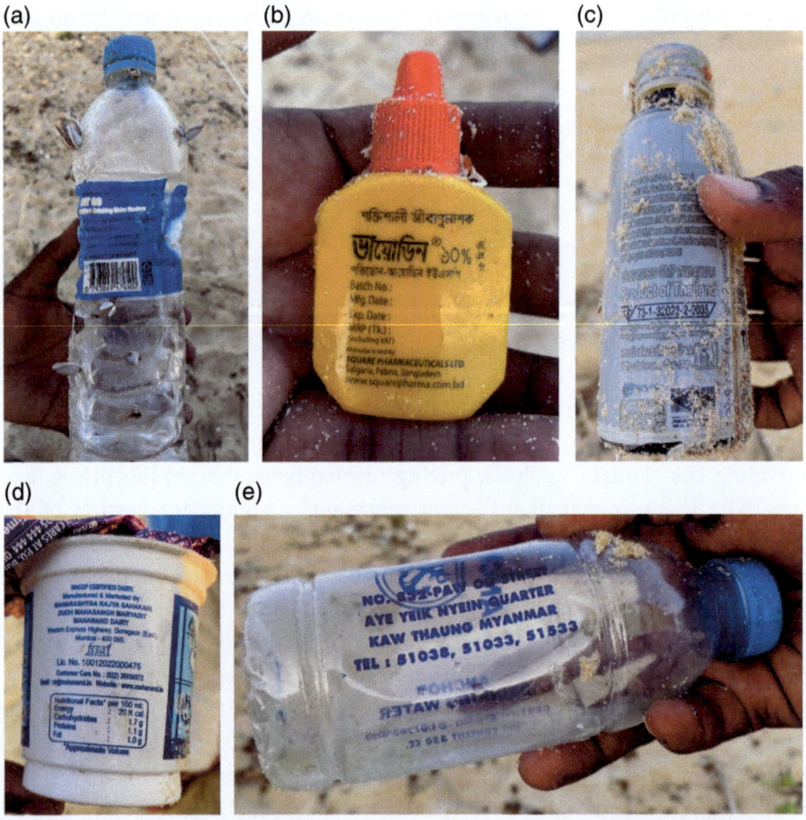

Figure 24.6 Floating plastics with international origin found on Sri Lankan beaches. (a) Myanmar (b) Bangladesh (c) Thailand (d) India (e) Myanmar. *Source:* Center for Marine Science and Technology (CMST).

Plastic debris can foul propellers, clog intake valves, and cause operational disruptions, leading to economic losses for maritime transportation companies (Obebe and Adamu 2020). Plastic waste in the environment can indirectly impact human populations, resulting in increased healthcare costs associated with treating injuries, infections, and diseases related to plastic pollution. For example, injuries from marine debris, contaminated seafood consumption, and exposure to toxic chemicals leached from plastics can all contribute to healthcare expenditures (Vital et al. 2021). Plastic waste poses significant economic challenges to society, with direct costs for cleanups and waste management and indirect costs associated with impacts on industries, public health, and natural resources. Addressing the economic dimensions of plastic pollution requires comprehensive strategies that consider the entire life cycle of plastic products and prioritize solutions that promote sustainability, resource efficiency, and environmental protection.

Single-stream recycling can be applied to waste management. This approach allows businesses and households to commingle different recyclables (plastics, papers, glasses, and metals) in a single bin, simplifying collection but requiring more advanced sorting facilities. These systems involve consumers paying a deposit on beverage containers, which is refunded upon return, encouraging recycling.

When conducting a commercial review of waste management options, businesses should consider factors such as the type and volume of waste generated, local regulations, cost implications, and the environmental impact of each option. Additionally, evaluating the entire life cycle of materials, from production to disposal, is essential for making informed and sustainable decisions. Policymakers and stakeholders may create plans that balance environmental sustainability with profitability by applying economic concepts to waste management, which would ultimately encourage more efficient resource use and reduce the negative impact of waste on society and the environment.

24.4 Mitigating MP Pollution: Sustainable Approaches, Circular Economy, and Policymaking

24.4.1 Sustainable Approaches and Circular Economy

Over 300 million tons of annual plastic waste is produced globally, yet only 9% is systematically recycled. Extreme consumer demands, continuous sheer-scale production, poor consumption patterns, and long lifespans make plastics and MPs a major contributor to greenhouse gas emissions (and eventually global warming), freshwater and marine pollution, loss of biodiversity, and chemical contamination due to degradation. Sustainable approaches for using plastics are a timely need to save the world from a catastrophic future due to the current resource-inefficient linear economic model of plastic waste management. Numerous short-term and long-term solutions are introduced and implemented for better plastic waste management worldwide (Samarasinghe et al. 2021; Tyagi et al. 2021)

Sustainable use of plastic waste can be achieved through initiatives such as using plastic waste as a source of energy, that is, using plastic wastes as a fuel substitute in cement plants, improving mechanical recycling facilities, improving sanitary landfilling methods, and implementing stern laws, regulations, and policies. The 4R strategy (i.e. Refuse ↔ Reduce ↔ Reuse ↔ Recycle) is commonly used among individuals and communities in terms of plastic waste management. Refusing and reusing plastic products directly minimize waste generation at the source by avoiding unnecessary synthetic packaging. Reusing involves repairing, refurbishing, or repurposing items to extend their lifespan and maximize utility. Conscious choices about reducing plastic consumption lead to less purchasing and adapting to a less-plastic lifestyle. Recycling avoids the haphazard dumping of plastic wastes and instead converts them into new products or materials. Recycling helps to conserve natural resources, reduce energy consumption, and decrease the environmental impact of waste disposal. Recycling industries play a crucial role in recovering valuable materials from waste streams, such as metals, plastics, papers, and glasses. These materials are often processed and reintroduced into the manufacturing supply chain, contributing to resource conservation and reducing the need for virgin materials. The 4R strategy impacts minimizing the number of plastics thrown out as mismanaged plastic wastes. More "Rs" have been introduced with variations in number. The "9Rs" strategy denotes R0 – Refuse; R1 – Rethink; R2 – Reduce; R3 – Reuse; R4 – Repair; R5 – Refurbish; R6 – Remanufacture; R7 – Repurpose; R8 – Recycle; and R9 – Recover (Tyagi et al. 2021).

The world is moving toward a zero-plastic waste economy by driving innovation and enabling new technologies and business models (Tyagi et al. 2021). Waste-to-energy (WTE) technologies (Qazi et al. 2018) convert waste into heat, electricity, or fuel through incineration, gasification, and anaerobic digestion. WTE facilities contribute to the energy sector by generating electricity or heat from waste materials, providing an alternative to fossil fuels, and reducing greenhouse gas emissions. WTE technologies provide benefits to global energy production. The waste management economy is increasingly aligning with principles of the circular economy, which emphasizes the reuse, reduction, and recycling of materials to minimize waste and resource depletion.

Circular economy initiatives promote closed-loop systems where materials are continuously circulated within the economy, reducing reliance on finite resources and minimizing environmental impacts (Qazi et al. 2018; Samarasinghe et al. 2021). The current pattern of plastic usage showcases a cradle-to-grave approach, including production, consumption, and disposal of plastic wastes in a linear pathway. Establishing a circular economy for plastics includes several criteria, that is, identification of plastic-related initiatives toward a circular economy, characterization of initiatives and their potential contribution to a circular economy, developing of an evaluation framework to assess the effectiveness of selected initiatives, and identification of knowledge gaps on circularity issues.

24.4.2 Identification of Plastic-related Initiatives Towards a Circular Economy

Plastic-related initiatives should be identified at community, industrial, and regional scales. Initiatives might facilitate the end of a plastic cycle or restrict the beginning of the plastic cycle. Initiatives improving plastic waste collection, recycling, and recovery facilitate ending the plastic cycle. Initiatives such as refusing the use of single-use plastics, extending the lifespan of used plastics by reusing, repairing, and repurposing waste, and promoting eco-friendly alternatives to plastics restrict the beginning of the plastic cycle. Up-cycled products out of plastic waste, such as plastic baskets or eco-bricks filled with mud or plastic used for construction, eco-designing products to reduce plastic content in the packaging, substituting virgin plastics with recycled materials, promoting "no-plastic packaging," increasing "recycled content" of packaging, and eliminating the use of single-use plastics in health and recreational sectors are among positive trends toward the circular plastic economy.

Community-based initiatives include raising awareness on systematic waste collection, segregation, alternatives for single-use plastics, eco-friendly substitute materials such as cloth, and recycling via education and training workshops. India shares success stories on incentivizing recycling by enabling an exchange of recyclable plastic waste instead of school fees or money for a cup of tea, food, or grocery items, or a facility for charging a phone or free meals for waste pickers. This initiative is widely used and can potentially be replicated and scaled up. For instance, Akshar School in Assam, India, collects school fees in the form of plastic waste. The establishment of "waste banks" has been successful in Indonesia (Tyagi et al. 2021).

Industry-based initiatives include policy advocacy, supporting scientific studies and providing a platform for research and collaboration, capacity building and raising awareness, facilitating the implementation of a circular economy in the plastic sector, and financing beginners in the waste management sector. For instance, Unilever is aiming to make 100% of their plastic packaging reusable, recyclable, or compostable by 2025, and "Danone-Aqua" is addressing marine plastic pollution, aiming to produce 50% of its bottles out of recycled plastics by 2025 (Tyagi et al. 2021). These can be considered positive initiatives in the industrial sector.

International organizations facilitate initiatives like establishing self-managed waste processing units where facilities are provided for waste pickers to use low-cost technologies to convert plastic wastes into plastic flakes. Plastic flakes are then sold to end users at a price in a way they receive a fair wage (i.e. SWITCH-Asia Grants Programme). Enhancing municipal capacities, designing and implementing new infrastructure, supporting local enterprises and start-ups involved in waste management, and running educational programs for the local population and visitors to reduce waste (i.e. Project Aviral – Reducing Plastic Waste in the Ganga implemented by German Development Agency GIZ) are among positive measures toward a circular economy (Tyagi et al. 2021).

24.4.3 Characterization of Initiatives and Their Potential Contribution to a Circular Economy

Identified initiatives should be characterized by the issues they encounter and their potential contribution to a circular economy. Sustainable initiatives at the grassroots and industrial level make an equally positive impact on the circular economy of plastics.

24.4.4 Developing an Evaluation Framework to Assess the Effectiveness of Selected Initiatives

Developing evaluation frameworks and research plans to assess the effectiveness of selected initiatives is important in terms of their potential, sustainability, and circularity. Rewarding sustainable initiatives can be further monitored as potential future case studies.

24.4.5 Identification of Knowledge Gaps on Issues of Circularity

Knowledge gaps on circularity issues should be addressed in the long-term implementation of selected initiatives. Less information about the actual outcomes of a circular economy leads to overestimating or underestimating the impact of selected initiatives. Success stories and barriers to implementation should be available for scaling up existing initiatives. Studies on alternative generation, identifying circular strategies not yet being pursued, and reviewing the sustainability of existing initiatives must be promoted and facilitated (Tyagi et al. 2021).

With the rapid urbanization and population growth, Sri Lanka records 265 000 megagrams of plastic consumption, with an annual increase of 16%. Seven thousand megagrams of MSW are generated daily, and 6–10% is from plastics. Samarasinghe et al. (2021) discussed the circular economy potential of plastic waste in Sri Lanka and provided three recommendations for circular economy opportunities for Sri Lanka, including source segregation, landfill or dumpsite mining, and public participation in microlevel plastic waste-related businesses.

24.4.5.1 Source Segregation

Among the challenges Sri Lanka faces in MSW management are poor segregation at source, inefficient waste collection, and improper disposal of generated MSW. Source segregation helps easy resource recovery, waste collection, and disposal of plastic waste. Provision of specific bins and raising awareness facilitate scientific disposal of plastic waste.

24.4.5.2 Landfill or Dumpsite Mining

This refers to the process of reclaiming land from waste dumping and the recovery of recyclable and combustible components. Non-recyclable plastic waste has a high net calorific value and can be recovered to generate WTE options. For instance, nonrecyclable waste can be used for co-fueling cement and pyrolysis factories.

24.4.5.3 Public Participation in Microlevel Plastic Waste-Related Businesses

Public participation in plastic waste management must be encouraged. In addition to providing extra support for recycling facilities, integrating the informal sector into the formal sector for microlevel plastic waste-related businesses will reduce overall MSW management costs and create job opportunities.

Lack of financial commitments, lack of proper education and awareness on impacts of plastic waste accumulation, absence of a strong controlling body, lack of logistics and infrastructure, and poor coordination among responsible authorities can be identified as drawbacks in regulating circular economy strategies in Sri Lanka.

Figure 24.7 represents the circular plastic flow generally practiced in Sri Lanka, according to Samarasinghe et al. (2021). Accordingly, the total plastic circulated within the country is locally produced using recycled plastics and imported pellets or imported plastics. Once discarded, 33% of the plastic waste generated within the country is properly collected, and the rest accumulates as mismanaged plastic waste. Properly handled plastic waste is utilized mainly for pyrolysis, sanitary

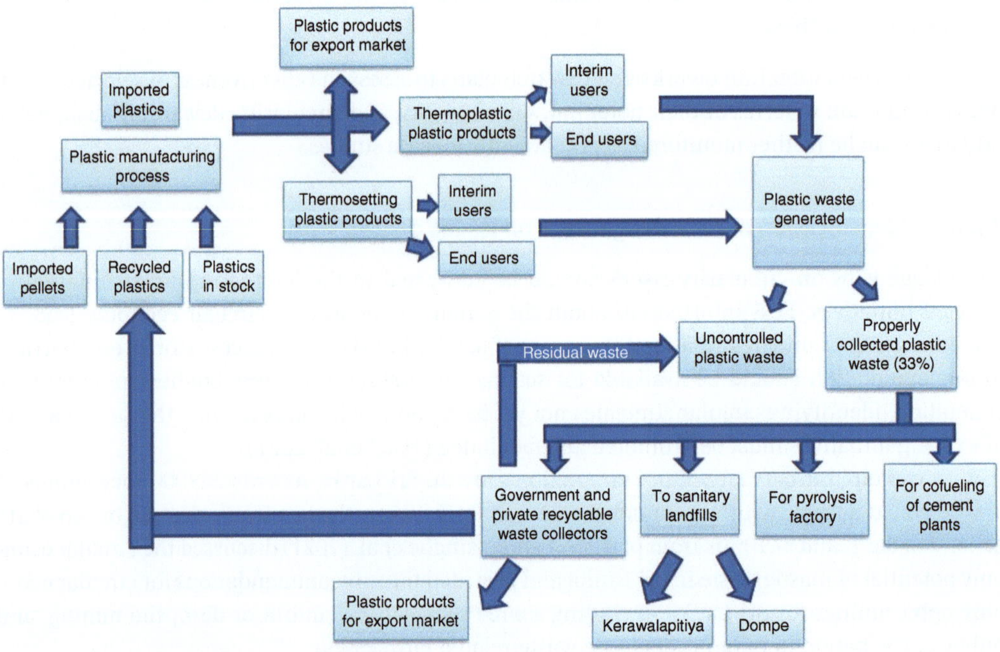

Figure 24.7 Conceptual diagram of plastic flow in Sri Lanka. *Source:* Adapted from Samarasinghe et al. (2021).

landfills, co-fueling, and recycling by government and private recycling companies that manufacture products for export and local markets.

24.4.6 Policymaking

During 2013 and 2020, the policies and initiatives addressing plastic contamination or even banning plastics boomed worldwide. Although there are few policies aimed at reducing primary MPs, interventions have been increasing to reduce single-use plastic bags in many countries, including Sri Lanka, to ensure they do not end up with coastal waters (Athapaththu et al. 2020). Such policies include restrictions on plastic bag sales, charges, and taxes from plastic bag sellers (Walker and Xanthos 2018). Countries like Australia, North America, and the United Kingdom have implemented various local jurisdictions in bans, partial bans, and fees for plastic bags. Some European countries have widespread interventions, such as introducing a fee per bag. Bangladesh, India, and South Africa have progressively introduced restrictions on plastic bag consumption (Dikgang et al. 2012; Xanthos and Walker 2017).

Many countries have taken interventions through taxes, bans, and policies to reduce or manage plastic bags, but there are few interventions for microbeads (Xanthos and Walker 2017; Schnurr et al. 2018). The increase in such microbead-containing cosmetics in the twentieth century gave rise to national bans on the sale and use of microbeads. Only a few countries have taken this step. For instance, Canada has restricted a ban on single-use toiletries and cosmetics containing microbeads from stores (Rochman et al. 2015). The province of Ontario passed legislation banning the production of microbeads in 2015 (Walker and Xanthos 2018). The Netherlands was one of the first countries to declare its intention to exclude microbeads in cosmetics at the end of 2016 and to legislate to prohibit the import, manufacture, and sale of microbeads in washable cosmetics (Schnurr et al. 2018).

While the positive impacts of single-use plastic bag reduction policies are yet to be used in some countries, other countries are currently developing strict policies and regulations to minimize other types of single-use plastics, which will indirectly decrease degradation into secondary MPs. The Canadian federal government banned six "harmful" single-use plastic items such as plastic checkout bags, straws, stir sticks, six-pack rings, cutlery, and food containers made from hard-to-recycle plastics (Walker 2021), which are commonly found in plastics in the marine environment during beach clean-ups or citizen science monitoring surveys (Walker et al. 2006; Ambrose et al. 2019). However, only personal care products were tackled, and microbeads and other MPs in abrasive materials, such as plastic blasting and automotive molding, were unregulated (Pettipas et al. 2016). Furthermore, the Helsinki Commission (HELCOM) suggested a regional action plan to tackle MPs, including recommendations on legal measures to act upon it, encouraging MP-free formulas, and replacing MPs in personal care products (HELCOM 2015).

Plastic bags and microbead interventions significantly implement policies and legislation (Schnurr et al. 2018; Anagnosti et al. 2021). The absence of mandatory laws, weak and poor enforcement, and inadequate cooperation in most developing countries contribute to the continuous discharge of plastics and MPs into the oceans. As there are existing environmental regulations to guide different environmental pollutions, environmental guidelines and standards should guide plastic discharge and address noncompliance with policies. Present laws on environmental pollution and the development of strict plastic discharge regulations need to be strengthened. Plastic industries and firms responsible for production should be monitored or sanctioned if necessary (Onyena et al. 2022).

In terms of policymaking, the Government of Sri Lanka passed a bill to ban the use of PE and other plastics. Accordingly, the gazette notification Nos. 2034/33, 2034/34, 2034/35, 2034/36, 2034/37, and 2034/ 38 issued orders and regulations to prohibit the manufacture of polythene or any polythene product with 20 μm and below, food wrappers (lunch sheets), any bag of high density (grocery bags), burning of refuse and other combustible matters inclusive of plastic, the use of polyethylene products as decorations, and the manufacture of food containers, plates, cups, and spoons from expanded PS under the National Environment Act No. 47 of 1980 and Act No. 53 of 2000 in 2017 (CEA 2017). Gazette notifications in 1994, 2016, and 2017 also ban the production and use of polyethylene sheets, PS-based packaging, and cutlery. However, the Sri Lankan government struggled to implement policies restricting the use of plastic, succumbing to pressure from plastic manufacturers and retailers (Athapaththu et al. 2020).

24.5 Conclusion

Mismanaged plastic waste poses enduring threats to the world's marine ecosystems. Besides macro-plastics, MPs in the ocean significantly impact resident organisms due to ingestion and, eventually, bioaccumulation along food chains and biomagnification. Potential threat toward humans as apex predators alarms the timely necessity of long-term monitoring of MP pollution.

Scientific and nonscientific MP assessments provide invaluable information regarding current situations of MP contamination. However, systematic MP protocols have yet to be developed and established. Climate change impacts plastic degradation and the distribution of MPs over the world's systems. Therefore, advanced integrated studies of MP pollution in cryospheric, biospheric, aquatic, terrestrial, and atmospheric settings are highly important. Empirical studies supported by global metadata are recommended in forecasting the future of MP contamination across different compartments of the world.

Plastic waste management demands a significant portion of the global economy due to rapid population growth, urbanization, and industrialization. Therefore, introducing sustainable approaches like a carbon-neutral circular economy through innovation, enabling new technologies, and enforcing policies to mitigate plastic pollution are timely needs to save the world from a devastating future.

References

Abbasi, S. (2021). Prevalence and physicochemical characteristics of microplastics in the sediment and water of Hashilan Wetland, a national heritage in NW Iran. *Environmental Technology and Innovation* 23: 101782. https://doi.org/10.1016/j.eti.2021.101782.

Alomar, C., Estarellas, F., and Deudero, S. (2016). Microplastics in the Mediterranean sea: deposition in coastal shallow sediments, spatial variation, and preferential grain size. *Marine Environmental Research* 115: 1–10. https://doi.org/10.1016/j.marenvres.2016.01.005.

Ambrose, K.K., Box, C., Boxall, J. et al. (2019). Spatial trends and drivers of marine debris accumulation on shorelines in South Eleuthera, The Bahamas using citizen science. *Marine Pollution Bulletin* 142: 145–154. https://doi.org/10.1016/j.marpolbul.2019.03.036.

Anagnosti, L., Varvaresou, A., Pavlou, P. et al. (2021). Worldwide actions against plastic pollution from microbeads and microplastics in cosmetics focusing on European policies. Has the issue been handled effectively? *Marine Pollution Bulletin* 162: 111883. https://doi.org/10.1016/j.marpolbul.2020.111883.

Andrady, A.L. (2011). Microplastics in the marine environment. *Marine Pollution Bulletin* 62 (8): 1596–1605.

Athapaththu, A.M.A.I.K., Thushari, G.G.N., Dias, P.C.B. et al. (2020). Plastics in surface water of southern coastal belt of Sri Lanka (Northern Indian Ocean): distribution and characterization by FTIR. *Marine Pollution Bulletin* 161 (Pt A): 111750.

Athawuda, A.M.G.A.D., Jayasiri, H.B., Thushari, G.G.N., and Guruge, K.P.G.K.P. (2020). Quantification and morphological characterization of plastic litter (0.30-100 mm) in surface waters of off Colombo, west coast of Sri Lanka. *Environmental Monitoring Assessment* 192 (8): 509.

Athukorala, A., Amarathunga, A.A.D., De Silva, D.S.M. et al. (2023a). Microplastic contamination in selected commercial fish species in Negombo Lagoon, Sri Lanka. International Conference on Applied and Pure Sciences, 2023, Faculty of Science, University of Kelaniya, Sri Lanka.

Athukorala, A., Amarathunga, A.A.D., De Silva, D.S.M. et al. (2023b). Occurrence and abundance of microplastics in surface waters and sediments from the Negombo Lagoon, Sri Lanka. 16th International Conference on the Environmental Effects of Nanoparticles and Nanomaterials. Plymouth: 32.

Auta, H.S., Emenike, C.U., and Fauziah, S.H. (2017). Distribution and importance of microplastics in the marine environment: a review of the sources, fate, effects, and potential solutions. *Environment International* 102: 165–176. https://doi.org/10.1016/j.envint.2017.02.

Awet, T.T., Kohl, Y., Meier, F. et al. (2018). Effects of polystyrene nanoparticles on the microbiota and functional diversity of enzymes in soil. *Environmental Sciences Europe* 30 (1): 1–10.

Awuchi, C.G. and Awuchi, C.G. (2019). Impacts of plastic pollution on the sustainability of seafood value chain and human health. *International Journal of Advance Academic Research* 5 (11): 46–138.

Baldwin, A.K., Corsi, S.R., and Mason, S.A. (2016). Plastic debris in 29 Great Lakes Tributaries: relations to watershed attributes and hydrology. *Environmental Science & Technology* 50 (19): 10377–10385. https://doi.org/10.1021/acs.est.6b02917.

Bandara, H.M.N.N., De Silva, H.S.D.R., Wijayathilaka, H.D.et al. (2023). Microplastic pollution in Dondra lagoon, Southern coast of Sri Lanka: Influence on habitat and feeding modes of selected fish species. PGIS research congress 2023, Postgraduate Institute of Science, University of Peradeniya, Sri Lanka.

Bank, M.S. and Hansson, S.V. (2019). The plastic cycle: a novel and holistic paradigm for the Anthropocene. *Environmental Science & Technology* 53 (13): 7177–7179. https://doi.org/10.1021/acs.est.9b02942.

Beckwith, V.K. and Fuentes, M.M.P.B. (2018). Microplastic at nesting grounds used by the Northern Gulf of Mexico loggerhead recovery unit. *Marine Pollution Bulletin* 131: 32–37. https://doi.org/10.1016/j.marpolbul.2018.04.001.

Belioka, M., Markozanne, G., Chrissopoulou, K., and Achilias, D.S. (2023). Advanced plastic waste recycling—the effect of clay on the morphological and thermal behavior of recycled PET/PLA sustainable blends. *Polymers* 15 (14): 3145. https://doi.org/10.3390/polym15143145.

Bessa, F., Frias, J., Kögel, T. et al. (2019). Harmonized protocol for monitoring microplastics in biota. JPI-Oceans BASEMAN Project, 30. https://doi.org/10.13140/RG.2.2.28588.72321/1

Besseling, E., Wegner, A., Foekema, E.M. et al. (2013). Effects of Microplastic on Fitness and PCB Bioaccumulation by the Lugworm Arenicola marina (L). *Environmental Science & Technology* 47: 593–600. https://doi.org/10.1021/es302763x.

Bhattacharya, P., Lin, S., Turner, J.P., and Ke, P.C. (2010). Physical adsorption of charged plastic nanoparticles affects algal photosynthesis. *The Journal of Physical Chemistry C* 114 (39): 16556–16561. https://doi.org/10.1021/jp1054759.

Browne, M.A., Crump, P., Niven, S.J. et al. (2011). Accumulation of microplastic on shorelines worldwide: sources and sinks. *Environmental Science & Technology* 45: 9175–9179. https://doi.org/10.1021/es201811s.

Carr, S., Jin, L., and Tesoro, A.G. (2016). Transport and fate of microplastic particles in wastewater treatment plants. *Water Research* 91: 174–182. https://doi.org/10.1016/j.watres.2016.01.002.

CEA (2017). Regulations on Polythene & Plastic Management 2017. Retrieved 20.02.2024, from https://www.cea.lk/web/?option=com_content&view=article&layout=edit&id=1080 (accessed 20 February 2024).

Chen, Y., Wen, D., Pei, J. et al. (2020). Identification and quantification of microplastics using Fourier-transform infrared spectroscopy: current status and future prospects. *Current Opinion in Environmental Science and Health* 18: 14–19.

Chen, X., Huang, G., Gao, S., and Wu, Y. (2021). Effects of permafrost degradation on global microplastic cycling under climate change. *Journal of Environmental Chemical Engineering* 9 (5): 106000. https://doi.org/10.1016/j.jece.2021.106000.

Cheruvelil, K.S. and Soranno, P.A. (2018). Data-intensive ecological research is catalyzed by open science and team science. *BioScience* 68 (10): 813–822.

Cooper, D. and Corcoran, P.L. (2010). Effects of mechanical and chemical processes on the degradation of plastic beach debris on the island of Kauai, Hawaii. *Marine Pollution Bulletin* 60 (5): 650–654. https://doi.org/10.1016/j.marpolbul.2009.12.026.

Corinaldesi, C., Canensi, S., Dell'Anno, A. et al. (2021). Multiple impacts of microplastics can threaten marine habitat-forming species. *Communications Biology* 4 (1): https://doi.org/10.1038/s42003-021-01961-1.

CSIRO Marine Debris Team & Center for Marine Science and Technology, University of Sri Jayewardenepura, Nugegoda, Sri Lanka team (2021). Global plastics leakage baseline data summary report, Sri Lanka.

Dharmadasa, W.L.S.S., Andrady, A.L., Kumara, P.B.T.P. et al. (2021). Microplastic pollution in marine protected areas of southern Sri Lanka. *Marine Pollution Bulletin* 168: 112462.

Dias, B.C.G., Amarathunga, A.A.D., De Silva, D.S.M. et al. (2023). Abundance and microplastic characterization found in the mud crab *Scylla serrata* inhabiting Negombo Lagoon, Sri Lanka. International Conference on Applied and Pure Sciences, 2023, Faculty of Science, University of Kelaniya, Sri Lanka.

Dikgang, J., Leiman, A., and Visser, M. (2012). Analysis of the plastic-bag levy in South Africa. *Resources, Conservation and Recycling* 66: 59–65. https://doi.org/10.1016/j.resconrec.2012.06.009.

Dissanayake, D.M.T.N. & Rathnasuriya, M.I.G. (2022). Microplastic contamination in Brown mussels (Perna perna) along the Southern coastal waters of Sri Lanka. Fourth Research Symposium of Ocean University of Sri Lanka-2022.

Dris, R., Gaspéri, J., Saad, M. et al. (2016). Synthetic fibers in atmospheric fallout: a source of microplastics in the environment? *Marine Pollution Bulletin* 104 (1–2): 290–293. https://doi.org/10.1016/j.marpolbul.2016.01.006.

Dris, R., Gaspéri, J., Mirande, C. et al. (2017). A first overview of textile fibers, including microplastics, in indoor and outdoor environments. *Environmental Pollution* 221: 453–458. https://doi.org/10.1016/j.envpol.2016.12.013.

Egbeocha, C.O., Malek, S., Emenike, C.U., and Milow, P. (2018). Feasting on microplastics: ingestion by and effects on marine organisms. *Aquatic Biology* 27: 93–106. https://doi.org/10.3354/ab00701.

Fang, C., Luo, Y., and Naidu, R. (2023). Microplastics and nanoplastics analysis: options, imaging, advancements and challenges. *TrAC Trends in Analytical Chemistry* 166: 117158.

Gamage, S.G., Mahagamage, M.G.Y.L. & Prasadika, H.K.A.E. (2022). Microplastics in cosmetics and personal care products: First study from Sri Lanka. *6th International Research Conference of Uva Wellassa University*, IRCUWU 2022, Uva Wellassa University of Sri Lanka, Badulla, Sri Lanka.

Gamage, T.R.I., Rathnasuriya, M.I.G. & Kumara, W.A.A.U. (2023). Microplastics occurrence in the gut content of Bluestripe herrings *Herklotsichthys quadrimaculatus* (Rüppell, 1837) along the Southern coastal waters of Sri Lanka. Proceedings of the *79th Annual Scientific Sessions*, 2023 – *Part I*, Sri Lanka Association for the Advancement of Science, Colombo, Sri Lanka.

Geyer, R., Jambeck, J., and Law, K.L. (2017). Production, use, and fate of all plastics ever made. *Science Advances* 3 (7): https://doi.org/10.1126/sciadv.1700782.

Guzzetti, E., Sureda, A., Tejada, S., and Faggio, C. (2018). Microplastic in marine organism: environmental and toxicological effects. *Environmental Toxicology and Pharmacology* 64: 164–171. https://doi.org/10.1016/j.etap.2018.10.009.

Hansani, K., Thilakarathne, E.P.D.N., Koongolla, B.J. et al. (2023). Contamination of microplastics in tropical coral reef ecosystems of Sri Lanka. *Marine Pollution Bulletin* 194 (Pt A): 115299.

Harrison, J.P., Ojeda, J.J., and Romero-Gonzalez, M.E. (2012). The applicability of reflectance micro-Fourier-transform infrared spectroscopy for the detection of synthetic microplastics in marine sediments. *Science of the Total Environment* 416: 455–463.

HELCOM (2015). Regional Action Plan for Marine Litter in the Baltic Sea, 20. Available online at https://cn.bing.com/search?q=Regional+Action+Plan+for+Marine+Litter+in+the+Baltic+Sea%3B+Baltic+Marine+Environment+Protection+Commission&qs=n&form=QBRE&sp=-1&pq=&sc=0-0&sk=&cvid=ACF8E392A1454D8597993C2698AF1C66&ghsh=0&ghacc=0&ghpl= (accessed 19 January 2024)

Hermsen, E., Mintenig, S.M., Besseling, E., and Koelmans, A.A. (2018). Quality criteria for the analysis of microplastic in biota samples: a critical review. *Environmental Science and Technology* 52 (18): 10230–10240.

Hidalgo-Ruz, V., Gutow, L., Thompson, R.C., and Thiel, M. (2012). Microplastics in the marine environment: a review of the methods used for identification and quantification. *Environmental Science and Technology* 46 (6): 3060–3075.

Hirai, H., Takada, H., Ogata, Y. et al. (2011). Organic micropollutants in marine plastics debris from the open ocean and remote and urban beaches. *Marine Pollution Bulletin* 62 (8): 1683–1692. https://doi.org/10.1016/j.marpolbul.2011.06.004.

Horton, A.A. and Dixon, S. (2017). Microplastics: an introduction to environmental transport processes. *WIREs Water* 5 (2): https://doi.org/10.1002/wat2.1268.

Horton, A.A., Svendsen, C., Williams, R.J. et al. (2017). Large microplastic particles in sediments of tributaries of the River Thames, UK – Abundance, sources and methods for effective quantification. *Marine Pollution Bulletin* 114 (1): 218–226. https://doi.org/10.1016/j.marpolbul.2016.09.004.

Ivar do Sul, J.A., Costa, M.F., and Fillmann, G. (2014). Microplastics in the pelagic environment around oceanic islands of the Western Tropical Atlantic Ocean. *Water, Air, & Soil Pollution* 225 (7): https://doi.org/10.1007/s11270-014-2004-z.

Jambeck, J.R., Geyer, R., Wilcox, C. et al. (2015). Plastic waste inputs from land into the ocean. *Science* 347 (6223): 768–771.

Jayarathna, N.K.D.M.P., Walpita, C.N., Amarathunga, A.A.D. et al. (2023). Microplastics in selected offshore pelagic fish in Indian Ocean. *International Conference on Applied and Pure Sciences*, 2023, Faculty of Science, University of Kelaniya, Sri Lanka.

Kamweru, P.K., Ndiritu, F.G., Kinyanjui, T. et al. (2011). Study of temperature and UV wavelength range effects on degradation of photo-irradiated polyethylene films using DMA. *Journal of Macromolecular Science, Part B* 50 (7): 1338–1349. https://doi.org/10.1080/00222348.2010.516172.

Kandeyaya, K.B.K.D.K., Ranatunga, S., and Ranatunga, R.R.M.K.P. (2023). Occurrence of microplastics in some commercially important seafood varieties from Negombo, Sri Lanka. *Regional Studies in Marine Science* 62: 102958.

Käppler, A., Windrich, F., Loder, M.G.J. et al. (2015). Identification of microplastics by FTIR and Raman microscopy: a novel silicon filter substrate opens the important spectral range below 1300 cm^{-1} for FTIR transmission measurements. *Analytical and Bioanalytical Chemistry* 407: 6791–6801.

Kapukotuwa, R.W.M.G.K., Jayasena N., Abayasekara C.L. et al. (2020). Microplastics in stream water and sediments of Rawan-oya tributary of Mahaweli river in Kandy district. *Proceedings of the Postgraduate Institute of Science Research Congress*, Sri Lanka, Postgraduate Institute of Science, University of Peradeniya, Sri Lanka, 7.

Kapukotuwa, G.K., Premathilake, B.A.P.C., Jayasena, N. & Rajakaruna, R.S. (2022a). Microplastics in commercially bottled water in Sri Lanka. *Proceedings of the Postgraduate Institute of Science Research Congress*, Sri Lanka, University of Peradeniya, Sri Lanka.

Kapukotuwa, R.W.M.G.K., Jayasena, N., Weerakoon, K.C. et al. (2022b). High levels of microplastics in commercial salt and industrial salterns in Sri Lanka. *Marine Pollution Bulletin* 174: 113239.

Karunarathna, Y.D.T.I. & Ranatunga, R.R.M.K.P. (2021). Microplastic pollution along the Attanagalla river, Sri Lanka. *8th International Conference of Multidisciplinary Approaches (iCMA)*, 2021, Faculty of Graduate Studies, University of Sri Jayewardenepura, Sri Lanka.

Koongolla, J.B., Andrady, A.L., Kumara, T.P., and P.B. & Gangabadage C.S. (2018). Evidence of microplastics pollution in coastal beaches and waters in southern Sri Lanka. *Marine Pollution Bulletin* 137: 277–284.

Kukulka, T., Proskurowski, G., Morét-Ferguson, S. et al. (2012). The effect of wind mixing on the vertical distribution of buoyant plastic debris. *Geophysical Research Letters* 39 (7): https://doi.org/10.1029/2012gl051116.

Lahiru, A.A.S., Amarathunga, A.A.D., Malavipathirana, S. et al. (2023). Assessment of litter and microplastic pollution in the water and sediment of Hirikatuoya stream, within the Walawe River basin, Sri Lanka. *Proceedings of the International Conference on Applied and Pure Sciences*, 2023, Faculty of Science, University of Kelaniya, Sri Lanka, 23.

Law, K.L. and Thompson, R.C. (2014). Microplastics in the seas. *Science* 345 (6193): 144–145. https://doi.org/10.1126/science.1254065.

Lawan, P.L.M.J.H., De Silva, D.S.M., Amarathunga, A.A.D. et al. (2024). Microplastic contamination in shrimps from the Negombo Lagoon – Sri Lanka. *Water* 16 (3): 447.

Li, J., Liu, H., and Chen, J. (2018). Microplastics in freshwater systems: a review on occurrence, environmental effects, and methods for microplastics detection. *Water Research* 137: 362–374. https://doi.org/10.1016/j.watres.2017.12.056.

Lusher, A. (2015). Microplastics in the marine environment: distribution, interactions, and effects. In: *Marine Anthropogenic Litter* (ed. M. Bergmann, L. Gutow, and M. Klages), 245–307. Cham: Springer https://doi.org/10.1007/978-3-319-16510-3_10.

Masura, J., Baker, J., Foster, G., & Arthur, C. (2015). Laboratory methods for the analysis of microplastics in the marine environment: recommendations for quantifying synthetic particles in waters and sediments. NOAA Technical Memorandum NOS-OR&R-48.

Meegahakotuwa, B.M.B., Ranatunga, R.R.M.K.P., and Gunasekara, A.J.M. (2020). Microplastic contamination in coastal waters off Maha Oya estuary in Negombo. *Uni-In-Alliance* 2020: 49.

Meijer, L.J.J., Van Emmerik, T., Van Der Ent, R.J. et al. (2021). More than 1000 rivers account for 80% of global riverine plastic emissions into the ocean. *Science Advances* 7 (18): https://doi.org/10.1126/sciadv.aaz5803.

Miller, M.E., Hamann, M., and Kroon, F.J. (2020). Bioaccumulation and biomagnification of microplastics in marine organisms: a review and meta-analysis of current data. *PLOS ONE* 15 (10): e0240792. https://doi.org/10.1371/journal.pone.0240792.

Moore, C.J. (2008). Synthetic polymers in the marine environment: a rapidly increasing, long-term threat. *Environmental Research* 108: 131–139.

Murphy, F., Ewins, C., Carbonnier, F., and Quinn, B. (2016). Wastewater treatment works (WWTW) as a source of microplastics in the aquatic environment. *Environmental Science & Technology* 50 (11): 5800–5808. https://doi.org/10.1021/acs.est.5b05416.

Napper, I.E., Baroth, A., Barrett, A.C. et al. (2023). The distribution and characterisation of microplastics in air, surface water, and sediment within a major river system. *Science of the Total Environment* 901: 166640. https://doi.org/10.1016/j.scitotenv.2023.166640.

Nawarathna, M.K.R.A.W.S.N., Rathnasuriya, M.I.G., Kapukotuwa, G.K. & Rajakaruna, R.S. (2023). Microplastics contamination in Indian backwater oysters (*Magallana bilineata*) in the Panadura estuary, Sri Lanka. *Proceedings of the Postgraduate Institute of Science Research Congress*, Sri Lanka, 2023, Postgraduate Institute of Science, University of Peradeniya, Sri Lanka, 1.

Nizzetto, L., Futter, M.N., and Langaas, S. (2016). Are agricultural soils dumps for microplastics of urban origin? *Environmental Science & Technology* 50 (20): 10777–10779. https://doi.org/10.1021/acs.est.6b04140.

Nolte, T.M., Hartmann, N.B., Kleijn, J.M. et al. (2017). The toxicity of plastic nanoparticles to green algae as influenced by surface modification, medium hardness, and cellular adsorption. *Aquatic Toxicology* 183: 11–20. https://doi.org/10.1016/j.aquatox.2016.12.005.

Obebe, S.B. and Adamu, A.A. (2020). Plastic pollution: causes, effects, and preventions. *International Journal of Engineering Applied Sciences and Technology* 4 (12): 85–95.

Onyena, A.P., Aniche, D.C., Ogbolu, B.O. et al. (2022). Governance strategies for mitigating microplastic pollution in the marine environment: a review. *Microplastics* 1 (1): 15–46. https://doi.org/10.3390/microplastics1010003.

Patil, S., Bafana, A., Naoghare, P.K. Krishnamurthi, K., & Sivanesan, S. (2021). Environmental prevalence, fate, impacts, and mitigation of microplastics—a critical review on present understanding and future research scope. *Environmental Science and Pollution Research*, 28, 4951–4974. https://doi.org/10.1007/s11356-020-11700-4

Peiris, B.H.P.H., Jayalal, R.G.U., Wijesekara, S.S.R.M.D.H.R. (2023). An Assessment of Airborne Microplastic Distribution in Sri Lanka by Using Lichens. Applied Sciences Undergraduate Research Symposium. Sabaragamuwa University of Sri Lanka, 42.

Perera, K., Ziajahromi, S., Nash, S.B. et al. (2022). Airborne microplastics in indoor and outdoor environments of a developing country in South Asia: abundance, distribution, morphology, and possible sources. *Environmental Science and Technology* 56 (23): 16676–16685.

Pettipas, S., Bernier, M., and Walker, T.R. (2016). A Canadian policy framework to mitigate plastic marine pollution. *Marine Policy* 68: 117–122. https://doi.org/10.1016/j.marpol.2016.02.025.

Primpke, S., Lorenz, C., Rascher-Friesenhausen, R., and Gredts, G. (2017). An automated approach for microplastics analysis using focal plane array (FPA) FTIR microscopy and image analysis. *Analytical Methods* 9 (9): 1499–1511.

Provencher, J.F., Covernton, G.A., Moore, R.C. et al. (2020). Proceed with caution: the need to raise the publication bar for microplastics research. *Science of the Total Environment* 748: 141426. https://doi.org/10.1016/j.scitotenv.2020.141426.

Qazi, W. A., Abushammala, M.F.M., Mohammed-HashamAzam, & Younes, M.K. (2018). Waste-to-energy technologies: a literature review. *The Journal of Solid Waste Technology and Management*, 44(4), 387–409. https://doi.org/10.5276/JSWTM.2018.387

Ranatunga, R.R.M.K.P. & Karunarathna, K.P.R. (2018). First evidence of microplastics in beach sand from Negombo, Sri Lanka, and the potential accumulation in marine fish. *International Conference on Multidisciplinary Approaches (iCMA)*, University of Sri Jayewardenepura, Sri Lanka.

Ranatunga, R.R.M.K.P., Wijetunge, D.S., and Karunarathna, K.P.R. (2021). Microplastics in beach sand and potential contamination of planktivorous fish *Sardinella gibbosa* inhabiting in coastal waters of Negombo, Sri Lanka. *Sri Lanka Journal of Aquatic Sciences* 26 (1): 37–54.

Ranatunga, R.R.M.K.P., Wijetunge, D., Ranaweera, W.V.P.H. et al. (2023). Ranking Sri Lanka among the world's top mismanaged waste polluters: does model data change the story? *Sustainability* 15: 2687.

Rathnayake, R.M.L.I., Amarathunga, A.A.D., De Silva, D.S.M. et al. (2023). Accumulation of Microplastics in Macrophytes of Negombo lagoon, Sri Lanka. 16*th International Conference on the Environmental Effects of Nanoparticles and Nanomaterials*, pp. 48–49. Plymouth.

Reisser, J., Slat, B., Noble, K. et al. (2014). The vertical distribution of buoyant plastics at sea: an observational study in the North Atlantic Gyre. *Biogeosciences* 12: 1249–2015.

Rochman, C.M., Tahir, A., Williams, S.L. et al. (2015). Anthropogenic debris in seafood: plastic debris and fibers from textiles in fish and bivalves sold for human consumption. *Scientific Reports* 5 (1): https://doi.org/10.1038/srep14340.

Samarasinghe, K., Kumar, S.P., and Visvanathan, C. (2021). Evaluation of circular economy potential of plastic waste in Sri Lanka. *Environmental Quality Management* 31 (1): 99–107.

Schnurr, R.E., Alboiu, V., Chaudhary, M. et al. (2018). Reducing marine pollution from single-use plastics (SUPs): a review. *Marine Pollution Bulletin* 137: 157–171. https://doi.org/10.1016/j.marpolbul.2018.10.001.

Sewwandi, M., Amarathunga, A.A.D., Wijesekara, H. et al. (2022a). Contamination and distribution of buried microplastics in Sarakkuwa beach ensuing the MV X-Press Pearl maritime disaster in Sri Lankan sea. *Marine Pollution Bulletin* 184: 114074.

Sewwandi, M., Hettithanthri, O., Egodage, S.M. et al. (2022b). Unprecedented marine microplastic contamination from the X-Press Pearl container vessel disaster. *Science of the Total Environment* 828: 154374.

Shackelford, G.E., Martin, P.A., Hood, A.S.C. et al. (2021). Dynamic meta-analysis: a method of using global evidence for local decision making. *BMC Biology* 19 (1): 33.

Talvitie, J., Mikola, A., Setälä, O. et al. (2017). How well is microlitter purified from wastewater? A detailed study on the stepwise removal of microlitter in a tertiary level wastewater treatment plant. *Water Research* 109: 164–172. https://doi.org/10.1016/j.watres.2016.11.046.

Teufel, B. and Sushama, L. (2019). Abrupt changes across the Arctic permafrost region endanger northern development. *Nature Climate Change* 9 (11): 858–862.

Thakshila, V.G.N. & Ranatunga, R.R.M.K.P. (2019). Evidence of Microplastic Contamination in Surface Waters and Sediment of Kelani River Estuary. 24*th International Forestry and Environment Symposium*-2019, Department of Forestry and Environmental Science, University of Sri Jayewardenepura, Sri Lanka.

Thompson, R.C. (2015). Microplastics in the marine environment: sources, consequences and solutions. In: *Marine Anthropogenic Litter* (ed. M. Bergmann, L. Gutow, and M. Klages), 185–200. Cham: Springer https://doi.org/10.1007/978-3-319-16510-3_7.

Tyagi, A., Ruoso, L.E., Retamal, M. et al. (2021). Towards a circular economy for plastics in India: A review of community, industry, and public sector initiatives. Report Number 2021-2. Development Alternatives, UTS and TERI, Australia and India.

Vermeiren, P., Munoz, C., and Ikejima, K. (2020). Microplastic identification and quantification from organic rich sediments: a validated laboratory protocol. *Environmental Pollution* 262: 114298.

Viraj, R.K.L., Jayasiri, H.B., Devmali, N.L.D.H. et al. (2019). Plastic contamination in selected beaches of Sri Lanka with special reference to microplastics. National Aquatic Resources Research and Development Agency (NARA), Scientific sessions 2019.

Vital, S.A., Cardoso, C., Avio, C. et al. (2021). Do microplastic contaminated seafood consumption pose a potential risk to human health? *Marine Pollution Bulletin* 171: 112769.

de Vos, A., Aluwihare, L., Youngs, S. et al. (2022). The M/V X-Press pearl nurdle spill: contamination of burnt plastic and unburnt nurdles along Sri Lanka's beaches. *ACS Environmental Au* 2 (2): 128–135.

Walker, T.R. (2021). Plastic industry plan to sue the Canadian federal government for listing plastic as toxic may increase plastic marine pollution. *Marine Pollution Bulletin* 169: 112583. https://doi.org/10.1016/j.marpolbul.2021.112583.

Walker, T.R. and Xanthos, D. (2018). A call for Canada to move toward zero plastic waste by reducing and recycling single-use plastics. *Resources, Conservation and Recycling* 133: 99–100. https://doi.org/10.1016/j.resconrec.2018.02.014.

Walker, T.R., Grant, J.E., and Archambault, M. (2006). Accumulation of marine debris on an intertidal beach in an urban park (Halifax Harbour, Nova Scotia). *Water Quality Research Journal of Canada* 41 (3): 256–262. https://doi.org/10.2166/wqrj.2006.029.

Weerakoon, W.R.W.M.A.P. & Wimalasiri, H.B.U.G.M. (2020). Aggregation of floating microplastics in coral-dominated ecosystems in Sri Lanka. NARA Scientific Sessions - 2020, National Aquatic Resources Research and Development Agency (NARA), Crow Island, Colombo 15, Sri Lanka.

Weerakoon, W.R.W.M.A.P., Samaranayaka, T.B.D.T., Jayasiri, H.B. & Arulanathan, K. (2018). Quantitative analysis of micro-plastic contamination in beach sand at the Western and Southwestern coastal stretches in Sri Lanka. National Aquatic Resources Research and Development Agency (NARA) International Scientific Sessions - 2018.

Weerakoon, W.R.W.M.A.P., Grosvik, B.E., Dalpadado, P. et al. (2019). Enumeration of microplastics in Sri Lankan waters: Preliminary findings from the RV Dr. Fridtj Nansen Ecosystem Survey, 2018. National Aquatic Resources Research and Development Agency (NARA), Scientific sessions 2019.

Weerasekara, P.G.Y.W., De Silva, D.S.M., De Silva, R.C.L., & Amarathunga, A.A.D. (2022). Sampling and analysis of microplastics in the coastal area from the Estuary of Kelani River to the Estuary of Mahaoya. International Conference on Applied and Pure Sciences, 2022, Faculty of Science, University of Kelaniya, Sri Lanka.

Weerasuriya, W.A.M.B., Amarathunga, A.A.D., Walpita, C.N. & Sivyer, D. (2023). Microplastic contamination of food fish in Bellanwila-Attidiya Wetland, Sri Lanka. *9th International Conference of Sabaragamuwa University of Sri Lanka*, Sabaragamuwa University, Sri Lanka.

Wessel, C., Lockridge, G., Battiste, D.R., and Cebrián, J. (2016). Abundance and characteristics of microplastics in beach sediments: insights into microplastic accumulation in northern Gulf of Mexico estuaries. *Marine Pollution Bulletin* 109 (1): 178–183. https://doi.org/10.1016/j.marpolbul.2016.06.002.

Wijethunga, H.N.S., Athawuda, A.M.G.A.D., Dias, P.C.B. et al. (2019). Screening the effects of microplastics on selected invertebrates along Southern coastal belt in Sri Lanka: A preliminary approach to coastal pollution control. Uva Wellassa University of Sri Lanka, International Research Conference 2019.

Wright, S., Rowe, D., Thompson, R.C., and Galloway, T.S. (2013). Microplastic ingestion decreases energy reserves in marine worms. *Current Biology* 23 (23): R1031–R1033. https://doi.org/10.1016/j.cub.2013.10.068.

Xanthos, D. and Walker, T.R. (2017). International policies to reduce plastic marine pollution from single-use plastics (plastic bags and microbeads): a review. *Marine Pollution Bulletin* 118 (1–2): 17. https://doi.org/10.1016/j.marpolbul.2017.02.048.

Yuan, Z., Nag, R., and Cummins, E. (2022). Human health concerns regarding microplastics in the aquatic environment—From marine to food systems. *Science of the Total Environment* 823: 153730. https://doi.org/10.1016/j.scitotenv.2022.153730.

Zhang, J., Yuan, D., Zhao, Q. et al. (2018). Tunable particle separation in a hybrid dielectrophoresis (DEP)- inertial microfluidic device. *Sensors and Actuators B: Chemical* 267: 14–25. https://doi.org/10.1016/j.snb.2018.04.020.

Zhang, Y., Gao, T., Kang, S. et al. (2022). Current status and future perspectives of microplastic pollution in typical cryospheric regions. *Earth-Science Reviews* 226: 103924.

Zhang, Y., Gao, T., Kang, S. et al. (2023). Cryosphere as a temporal sink and source of microplastics in the Arctic region. *Geoscience Frontiers* 14 (4): 101566. https://doi.org/10.1016/j.gsf.2023.101566.

Zhou, Q., Zhang, H., Fu, C. et al. (2018). The distribution and morphology of microplastics in coastal soils adjacent to the Bohai Sea and the Yellow Sea. *Geoderma* 322: 201–208. https://doi.org/10.1016/j.geoderma.2018.02.015.

Zylstra, E.R. (2013). Accumulation of wind-dispersed trash in desert environments. *Journal of Arid Environments* 89: 13–15: https://doi.org/10.1016/j.jaridenv.2012.10.004.

Index

Page numbers followed by 'f' refer to figures. Page numbers followed by 't' refer to tables.

Coastal and Marine Pollution: Source to Sink, Mitigation and Management, First Edition.
Edited by Meththika Vithanage, Sameera M. Samarasekara, Bryan D. James, and Christopher M. Reddy.

d

h

i

m

o

p

q

r

t

u

x

y

z